NASA SP-4006

ASTRONAUTICS AND AERONAUTICS, 1965

Chronology on Science, Technology, and Policy

NASA Historical Staff,
Office of Policy Analysis

Scientific and Technical Information Division   1966
NATIONAL AERONAUTICS AND SPACE ADMINISTRATION
Washington, D.C.

Foreword

The year 1965 recounted by this volume was an outstanding one in the U.S. space program. In his space report to Congress, President Johnson called it "the most successful year in our history." It was one filled with noteworthy milestones deriving from less noticed decisions, actions, and labors of previous years. In the same way, milestones of the future are to be seen in their formative stages in this chronology for 1965.

Man received his first close-up view of our neighboring planet Mars when on July 14, MARINER IV relayed to earth its photographs of lunar-like craters on the Martian surface. The conclusion of the Ranger program was witnessed by millions of Americans who watched on live television as the cameras of RANGER IX approached the moon on March 24. ALOUETTE II was orbited by NASA for Canada in November and FR-1 for France in December.

The orbiting of ten Gemini astronauts in a series of five spectacular flights during the year ended a 22-month gap since the last Mercury flight, FAITH 7. The man-rated version of the Air Force Titan II reliably launched all Gemini flights, Astronauts Grissom and Young in March, as well as McDivitt and White in June, the latter marking his "space walk" outside the GEMINI IV spacecraft. The eight-day mission of Astronauts Cooper and Conrad in GEMINI V demonstrated that trained space pilots were physically capable of a lunar mission. Orbital rendezvous techniques were thoroughly demonstrated by the flights of GEMINI VII and VI. Astronauts Borman and Lovell in GEMINI VII took another long step in astronautics with their fourteen-day mission in December of 206 revolutions. Throughout the Gemini operation, the team effort involved closest cooperation of all of NASA and the military services, contractors, and the scientific community.

Milestones in the lunar-landing Apollo program were not as well publicized as Gemini but marked significant progress. The Project Fire success in atmospheric entry of a test vehicle at speeds simulating a return from the moon provided a geometric jump in reentry physics. The first full-duration test of the gigantic Saturn V first stage of 7.5 million pounds of thrust on August 5 was a significant milestone in an engine program begun in 1958. As the Saturn I concluded its operational life with ten straight successes with the orbiting of Apollo boilerplate capsules and three Pegasus micrometeoroid satellites, the Saturn IB was being erected on the launch pad to begin its flight tests in 1966.

Spectacular scientific and manned spaceflight events of 1965 could not overshadow the practical utility to man on earth of communications and meteorological satellites. TIROS X, placed in orbit for the Weather Bureau, maintained service to worldwide needs for weather data, while TIROS IX provided the first complete picture of the cloud-cover over the entire earth on February 13. NASA launched EARLY BIRD I for the Communications Satellite Corporation in April. During the same month, NASA turned operational control of SYNCOM III over to the Department of Defense for service in important Far East communications.

A chronology is not an adequate substitute for a documented narrative history. But in this chronology, spliced alongside the U.S. space events of 1965, one can note the less publicized decisions, actions, and discussions concerned with the shaping of the future. About 90 percent of NASA's $5.175 billion went to contractors for work done by almost 400,000 people in the factories and laboratories of some 20,000 prime contractors and subcontractors. In the university program about 10,000 scholars at 100 universities in all fifty states were working on space-related topics.

In addition to NASA-related events the chronology gives some of the impact on the American scene of the space effort, including critical comment testing in democratic fashion the pace and scale of space efforts. Actions, deliberations, and comment as part of international cooperation and competition are likewise represented in these pages. Hopefully this volume will serve the serious student of today as he seeks knowledge of past events so as to better understand the future.

The late Hugh L. Dryden once wrote:

> Free peoples everywhere must retain a reliable perspective from which to discern better the future scientific, social, economic, political, and strategic consequences of dynamic advances now underway. The manner of the impact of technology upon society in the future will partly result from the broadest possible appreciation of its full significance.

His passing on December 2 of this eighth year of the Space Age was noted throughout the world. He leaves lasting contributions to the development of space technology and of a sound philosophy of astronautics. This volume helps to document 12 months marking what Dr. Dryden called "the opening of a brilliant new stage in man's evolution." It should assist its readers in gaining helpful perspective upon man's challenging venture into space.

JAMES E. WEBB
Administrator
National Aeronautics and
Space Administration

Contents

	PAGE
FOREWORD	III
Administrator James E. Webb	
PREFACE	VII
JANUARY	1
FEBRUARY	48
MARCH	100
APRIL	162
MAY	213
JUNE	264
JULY	307
AUGUST	363
SEPTEMBER	410
OCTOBER	458
NOVEMBER	500
DECEMBER	533
APPENDIX A: satellites, space probes, and manned space flights, 1965	575
APPENDIX B: major nasa launchings, 1965	605
APPENDIX C: summary chronology of manned space flights, 1961–1965	609
APPENDIX D: abbreviations of references	619
INDEX	623

Preface

This chronology is designed to collect in preliminary form pertinent information on aeronautical and space affairs. Future historical research and narratives will of course deepen the process of documentation and enrich perspective on the high velocity of contemporary science and technology, as well as their impact and implications. The volume was prepared from open public sources to provide a reference for future historians and other analysts, scholars, students, and writers. Its detailed index was intended to provide ready access to most specialized needs.

The entire NASA Historical Staff in Headquarters participated in source collection, review, and publication. The Science and Technology Division of the Library of Congress was responsible for drafting of the text proper, in the persons of Miss Lynn Catoe, Mrs. Anne Horton, and Miss Shirley Medley. The index was prepared by Arthur G. Renstrom, also of the Library of Congress. General editor of the entire *Astronautics and Aeronautics, 1965* project was Dr. Frank W. Anderson, Jr., Deputy NASA Historian; Mrs. Helen T. Wells was technical editor. Lloyd Robbins and Creston Whiting (ATSS-T) provided timely translations of Russian materials. Historians and historical monitors throughout NASA contributed useful inputs: validation was the constant concern of many busy persons throughout NASA.

Appendix A, "Satellites, Space Probes, and Manned Space Flights-1965," and Appendix B, "Major NASA Launchings, 1965," were prepared by Dr. Anderson. Appendix C, "Summary Chronology of Manned Space Flights, 1961–1965," was prepared by William D. Putnam, Assistant NASA Historian for Manned Space Flight. Mrs. Wells prepared Appendix D, "Abbreviations of References."

This preliminary chronicle is but a first step in the historical process of documenting the dynamic and complex events of space exploration and exploitation. Comments, additions, and criticism are welcomed at any time.

EUGENE M. EMME
NASA Historian (EPH)
Office of Policy Analysis

January 1965

January 1: Operation of SYNCOM II and SYNCOM III communications satellites was transferred to DOD by NASA, which had completed its R&D experiments. Telemetry and command stations and range and range-rate equipment operated by NASA for the Syncom program would be transferred to DOD along with the satellites. DOD had furnished the communications ground stations used to relay transmissions via the two Syncoms for the past two years and would provide NASA with certain telemetry and ranging data of continuing scientific and engineering interest. SYNCOM III was to prove useful in DOD's Vietnam communications. (NASA Release 65-5)

- About 500 employees of the Manned Spacecraft Center's Florida Operations were transferred to the Kennedy Space Center, effective today, under a realignment announced Dec. 24, 1964, by NASA Hq. Elements of the manned space flight organization were regrouped to meet the requirements imposed by concurrent Gemini and Apollo launch schedules. (MSC *Roundup*, 1/6/65, 1)
- Two hrs. and 20 min. of radio signals from Jupiter were received around midnight New Year's Eve as predicted by George A. Dulk of the Univ. of Colorado. The signals were received at the Altitude Observatory of the National Center for Atmospheric Research at Boulder, which had kept its radiotelescope operating for the event. (Osmundsen, *NYT*, 1/2/65, 1)

January 2: NASA had compromised the scientific value of the interplanetary research program by spending too little on the Deep Space Net communications system, according to Frank Drake, prof. at Cornell Univ., in *Saturday Review* article. Drake noted that MARINER IV would only be able to relay 22 photos of Mars back to earth and that these would be of lesser quality—all because of communications limitations: ". . . one concludes that the space program could well use an array containing a hundred or more 85-ft. antennas. One array might cost $40,000,000, still only a few per cent of what will almost certainly be spent on planetary exploration in the next ten years." (*SR*, 1/2/65)

- Soviet cosmonaut Col. Vladimir Komarov, who commanded the three-man spacecraft VOSKHOD I on its orbital flight, told a Havana newspaper: "I believe I will take part in a similar trip—if not to the moon, then to another place." Komarov was a member of the Soviet delegation in Havana for celebration of the sixth anniversary of Fidel Castro's revolution. (*New Orleans Times-Picayune*, 1/3/65; AP, *Hartford Courant*, 1/3/65)

January 2: Bell Telephone Laboratories, Inc., received the first patent for a satellite communication system with its own orbit pattern. The satellite would linger for a considerable period over each of two widely separated areas; while hovering virtually stationary, it could relay television and radio programs within its range, and also store programs from one area to play later on the other side of the globe. (Jones, *NYT,* 1/2/65; *Chic. Trib.,* 1/3/65)

- U.K. was said to be considering the possibility of a licensing agreement with the U.S. that would enable British manufacturers to make parts of late-model aircraft produced in the U.S. American planes under consideration were: McDonnell Aircraft Corp's F4c (Phantom II) carrier-based attack aircraft; F-111 low-level strike aircraft made by General Dynamics Corp. and Grumman; and c-141 and Orion, both made by Lockheed Aircraft Corp. (Farnsworth, *NYT,* 1/3/65, 13)

January 3: MARINER IV changed the rate of sending scientific data from $33\frac{1}{3}$ to $8\frac{1}{3}$ bits of information per second by an automatic switching operation. This was the first command initiated by the spacecraft itself since it performed its mid-course maneuver Dec. 5. MARINER IV had traveled nearly 63 million miles in its 325-million-mile flight to Mars; the straight-line distance between earth and the spacecraft was 6,156,-704 miles. Systems were operating normally after 36 days in space. (NASA Release 65–4)

- More than 50 million Europeans—including viewers behind the Iron Curtain—had received same-day transmission of the Tokyo Olympic Games via U.S. satellites SYNCOM III and RELAY I last October, NASA announced. (NASA Release 65–2)

- Japan's Ministry of Telecommunications said signals from what they had thought a new Soviet satellite turned out to be Italian-U.S. SAN MARCO I, launched Dec. 15, 1964. (AP, Wash. *Eve. Star,* 1/2/65; AP, Wash. *Sun. Star,* 1/3/65)

- Dr. Albert J. Kelley, Deputy Director of NASA Electronics Research Center, said in an article in Boston *Sunday Globe:* "The need for increased electronics research to develop devices which will meet the demands and rigors of long space flights will affect our industrial outlook and economy in many ways. By requiring a 'new look' at electronics, NASA, led by ERC, will provide a research emphasis such as we have not had since World War II when the golden age of electronics started.

    "We have been in the 'rocket phase' and are now entering the 'electronic phase' of space flight development, a phase which will affect us dramatically over many years." (Boston *Sun. Globe,* 1/3/65)

- British designers had perfected a miniature rocket costing only $2,240 per copy, it was reported. Nine ft. in length with a $7\frac{1}{2}$-in. diameter, the rocket would use solid fuel and reach a speed of 3,500 mph, sending the casing containing scientific instruments to maximum altitude of 80 mi. plus. (AP, *Kansas City Times,* 1/4/65)

- Scientists concluded that explosions and resultant earth-craters created by giant meteorites bore a striking similarity to the effect produced by the larger nuclear weapons; hence a meteorite fall might be mistaken for a nuclear explosion. Opinions varied as to the size of the body that could gouge a crater as large as the Meteor Crater of Arizona—

anywhere from 30,000 tons to 2.6 million tons, with an explosive force of 20 million tons of TNT. Both the size of the meteorite and its velocity on impact would be factors in producing a crater. (Sullivan, *NYT*, 1/3/65, 6E)

January 3: Semyon A. Kosberg, 61, one of the Soviet Union's leading designers of airplane engines, was killed in an automobile accident. He had been given the title "Hero of Socialist Labor" and had won a Lenin prize for his designs. (*NYT*, 1/5/65, 12)

- Writing in *Pravda*, I. Akulinichev, Dr. of Medical Sciences, said: ". . . Of course, the question of lunar laboratories is now only at the level of scientific planning. . . . To bring this possibility closer to our times, it is necessary to accomplish manned flights to the region of the Moon. Further, we need to solve reliably the question of methods to use for a successful lunar landing of a spacecraft and the return of the cosmonauts to Earth. In my view, the first lunar laboratories will initially study the possibilities of the prolonged sojourn of man on the Moon. Scientists will investigate ways of using the lunar conditions for assisting the normal life activity of people. . . . Finally, the scientists will study the conditions of orientation on the Moon and the possibilities of the navigation of interplanetary spacecraft."

  In the same issue of *Pravda*, Soviet Academician B. Konstantinov wrote: "In this New Year's article, I wish to dwell on the possibility of international cooperation in the use of solar energy. . . . What appears most attractive is the conversion of solar energy into electricity. In the foreseeable future, man may solve this problem; along with this, it is conceivable that the problems of controlling the weather and climate will also be solved." (*Pravda*, 1/3/65, 4, ATSS-T Trans.)

January 4: Gemini GT-3 spacecraft arrived at NASA Kennedy Space Center for final flight preparations before the nation's first two-man flight this spring. (KSC Release 3-65)

- According to Dr. Harold B. Finger, Manager of AEC-NASA Space Nuclear Propulsion Office (SNPO), NASA would not spend any further funds on Project Orion (nuclear-pulse propulsion project). The decision was based on the fact that such a system could not be used while the nuclear test ban treaty was in effect. In addition, NASA felt there were more urgent projects on which to spend the money. (*M&R*, 1/4/65, 9)

- Dr. Barry Commoner, professor of plant physiology at Washington Univ. in St. Louis and chairman of the AAAS Committee on Science in the Promotion of Human Welfare, told *Aviation Week and Space Technology* that the question of the probability of finding life on Mars had not been "fully and fairly aired," and that an "overbalance of the positive viewpoint has been presented to Congress and the public by NASA officials." Dr. Commoner said that if asked his views on Voyager as a tax-paying citizen, his feeling would be that "the value of pursuing a program to find life on Mars at this time is not worth the $1.25 billion to be invested because the problem of finding life there has not been adequately explored." He had made similar charges in a speech at the AAAS meeting last December in Montreal.

  Dr. Homer E. Newell, NASA Associate Administrator for Space Science and Applications, was reported by *Aviation Week and Space Technology* as listing six major points in defending NASA's position regarding Mars exploration: (1) Numerous competent scientists had said there was little liquid water on Mars and that the planet had a dry, dusty surface with high ultraviolet radiation. Changing patterns on the planet indicated some form of seasonal change, however. (2) With the evidence at hand, it was not possible to say there was life on Mars, only that life might be there. (3) If there were life on Mars it might be similar to basic life forms on earth. (4) The only reasonable approach we could take to the exploration of Mars would be to make sure we looked for life before the planet was contaminated from earth. If life was not found on Mars, it still would be valuable to determine how far the planet's chemical processes had progressed toward life formation. (5) The Voyager program had not been sold to Congress on the basis that there was life on Mars. It has been pointed out during budget hearings that there might not be life on the planet but nobody could responsibly take the position that there wasn't. Therefore, the early emphasis of Project Voyager was on bioscience. (6) The Mars exploration was part of an overall program to explore the solar system, including the moon, comets, and other planets. Mars happened to be the planet NASA was focusing its attention on because it would be in the optimum launch position through the mid-1970's. (*Av. Wk.*, 1/4/65, 18)

January 4: Dr. Gerard P. Kuiper, director of the Lunar and Planetary Laboratory of the Univ. of Arizona and principal scientific investigator on the Ranger project, replied to Robert C. Cowen's article, "Was the Ranger Worth the Cost?", which appeared in the *Christian Science Monitor* Nov. 18. Mr. Cowen had raised four principal questions: (1) Was the recent RANGER VII mission scientifically justifiable? (2) Was it well planned and executed? (3) Were the results up to expectations? (4) Where do we go from here? Dr. Kuiper said in letter to *CSM*: "Ranger was the U.S. pioneering program of deep-space research and accomplished much more than getting the 4,300 lunar photographs. It established the worth and feasibility of the 'parking orbit' and other concepts of space ballistics, power supply, and communication, as well as preparation for Mars and Venus probes.... The cost of the 4,300 lunar records is therefore not the full $270 million (which moreover includes Rangers VIII and IX, not yet flown) but, say, $50–$100 million. No ground-based effort, even with the 300–400–inch telescope costing over $100 million, would, even in the absence of our disturbing atmosphere, have yielded 100th of the magnification (resolution) obtained in Ranger VII. I definitely know of no better and cheaper way to get high-resolution photographs...."

In a reply to Dr. Kuiper, Mr. Cowen quoted from a letter by Dr. Andrew T. Young of Harvard College Observatory and published in *Science*: "... It is clear that there are some things that can only be learned above the atmosphere, and it is important that we have a program directed at learning them.... [But] many things that can be learned from above the atmosphere can also be learned, much more cheaply, by ground-based techniques. For example, some of the most convincing evidence for life on Mars is based on a few hours of twilight observations with the 200-inch telescope. . . . But the 200-inch telescope has been available for planetary research only a few times, generally during daylight or twilight. . . . Rocket-borne research involves many costly failures, but a duplicate 200-inch telescope could easily be built and staffed for the $28 million that Ranger 7 alone cost. . . ." (CSM, 1/4/65)

January 4: Gen. Bernard A. Schriever (USAF) announced the activation of the Contract Management Div., Air Force Systems Command (AFSC), under the command of Col. Fred L. Rennels, Jr. (USAF). Located at Los Angeles Air Force Station, the new division would be responsible for DOD contract management activities in those plants assigned to the Air Force under the DOD National Plant Cognizance program. (AFSC Release 61.64)

- USAF announced that Electro-Optical Systems, Inc., was receiving a $1,056,700 final increment to an existing contract for production of ion thrustor systems for orbital flight. (DOD Release 917–65)
- Col. John H. Glenn, Jr., former NASA astronaut and first American to orbit the earth, retired from the Marine Corps after 22 yrs. in the service. Glenn said he would spend much of his time as a consultant to NASA. He would also be a director of Royal Crown Cola Co. (DOD Release 912–64; Wash. Eve. Star, 1/4/65; Wash. Post, 1/5/65; Balt. Sun, 1/5/65; Chic. Trib., 1/5/65)

January 5: NASA announced plans to negotiate with Lockheed Missile and Space Co. to modify five Agena D second-stage launch vehicles for use in Lunar Orbiter missions. Modifications under the incentive contract would include vehicle engineering support; systems testing; overall system integration functions; shroud, adapter and interface coordination; and design fabrication of ground equipment. The Lunar Orbiter program would secure topography data of the moon's surface to extend scientific knowledge and to help select and confirm landing sites for the Apollo manned moon landings. (NASA Release 65–6)

- NASA Manned Spacecraft Center had received an estimated 1,351 applications or letters of interest relating to the scientist-astronaut program. The deadline for filing applications had been Dec. 31, 1964. (*Houston Post*, 1/5/65)
- J. Stalony-Dobrazanski of the Northrop Corp. reported at AIAA meeting in New York that spaceships could be kept cool automatically during reentry by a new guidance system. Network of supersensitive thermometers imbedded in the outer skin of the spacecraft would monitor the temperature, then computer would order correction in vehicle's trajectory or orientation if friction of the atmosphere raised skin temperature above a certain point. (Wash. *Daily News*, 1/26/65)
- Western Electric Company had received a $90,644,200 modification to an existing cost-plus-incentive-fee contract for research and development of Nike-X missile system, DOD announced. (DOD Release 3–65)
- Federal Aviation Agency (FAA) announced completion of the new Federal Aviation Regulations (FARs)—a simplification of rules governing the Nation's pilots, airlines, and airplane manufacturers. Number of regulations was reduced from 125 to 55. (FAA Release 65–2)

January 5: In a television interview, Israeli Premier Levi Eshkol urged West Germans to end the activity of German rocket experts in the United Arab Republic, said that these experts were helping the Arabs to prepare a war against Israel. The West German government had officially deplored the participation of German scientists and military experts in Arab rocket projects, but had not interfered on the grounds that the group was composed of private citizens who, according to the German Constitution, could work where they pleased. German rocket expert Prof. Wolfgang Pilz, leader of Germans working for the U.A.R., spoke in an interview of the pressure brought to bear by the Israeli Government, particularly the terrorist tactics of Israeli secret agents which made it necessary for Germans to be accompanied by body guards at all times. (*NYT*, 1/7/65, 5; Buchalla, *NYT*, 1/8/65, 1)

January 6: NASA Nike-Apache sounding rocket reached a peak altitude of 91.1 mi. from Wallops Island, Va. Purpose was to simultaneously measure the altitude of sodium airglow with sodium vapor and interference filters and determine atmospheric density with a 26-in., metallized, inflated mylar sphere. (NASA Rpt. SRL)

- F-111A was flown successfully for the second time from Carswell AFB, Tex. Flight data: maximum altitude, 27,000 ft.; maximum speed, 400 knots (460 mph); flight time, 1 hr. and 2 min. General Dynamics test pilots Richard L. Johnson and Val E. Prahl conducted stability and control tests at 10,000 and 20,000 ft., operating the wing sweep mechanism from 16° takeoff position to 26° position, then 43°, back to 40° to make sure the system worked, and finally to full-swept 72.5° position. This was the first time that wing position was varied in the flight of a military aircraft. The major test objective of the flight was accomplished—10 min. of flight with wings fully aft. Flight plans calling for an evaluation of stability at 30,000 ft. were called off because fuel flow and temperature in one of the two jet engines appeared to be outside normal limits, but this involved no reduction in flight time. General Dynamics reportedly would receive a bonus amounting to more than $800,000 for completing this milestone flight 24 days ahead of schedule. (Thomis, *Chic. Trib.*, 1/7/65; Witkin, *NYT*, 1/7/65, 1; *Av. Wk.*, 1/11/65, 19)

- Air Force Secretary Eugene M. Zuckert placed further restrictions on simulated bombing missions of B-58 Hustlers over Chicago: the supersonic bombers would fly at higher altitudes (48,000–49,000 ft. instead of 41,000–44,000-ft. range originally programed) to reduce impacts of sonic booms; flights would be canceled during bad weather. It had been announced earlier that the number of training missions per day would be reduced from a maximum of four to two. (*Chic. Trib.*, 1/7/65)

- Federal Aviation Agency Administrator Najeeb E. Halaby proposed that a 10-day international aerospace and science exposition be held at the Dulles International Airport in June 1966. Purpose of the exposition would be to stimulate aerospace exports. (*NYT*, 1/8/65, 10)

- Indonesian Air Vice Admiral Budiardjo, deputy air force chief for logistics, claimed that Indonesia had begun surveys for space flights and would be able to launch its first astronaut by 1968. (AP, *Wash. Post*, 1/7/65, A13)

January 7: MARINER IV was 70 million mi. on its 325-million-mi. flight to Mars after 40 days in space. All systems were operating normally (AP, Phil. *Eve. Bull.*, 1/7/65)

- Dr. William A. Lee of NASA Manned Spacecraft Center announced new launch schedule for Saturn IB and Saturn V: 1966, three unmanned and one manned launches of Saturn IB; 1967, two unmanned Saturn V launches, one manned Saturn IB, Lem test with Saturn IB, one manned flight with complete Apollo spacecraft, using Saturn IB, and one manned flight using either Saturn IB or Saturn V, whichever was farthest along in development; 1968, a dress rehearsal for the lunar mission in earth orbit for one week with astronauts participating. "Then the moon," said Dr. Lee. "We have a fighting chance to make it by 1970 and also stay within the $20 billion price tag set for the mission by former President Kennedy." (AP, Wash. *Eve. Star*, 1/7/65)
- NASA Administrator James E. Webb swore in R. Walter Riehlman, former Republican member of the House of Representatives from New York's 34th District (Syracuse) as a consultant on policy matters. (NASA Release 65-9)
- ComSatCorp asked nine foreign companies to propose studies of launch vehicles for medium altitude communications satellites in addition to the 16 American companies approached a month ago. The deadline for submitting proposals was extended from Jan. 11 to Feb. 1. (ComSatCorp)
- AEC report said that nuclear fuel aboard a spacecraft which failed to go into orbit last April 21 had burned up harmlessly at high altitude. This was a reply to Russian and other critics who had accused the U.S. of causing radiation hazards by putting atomic generators aboard spacecraft. The generator involved was a Snap-9A aboard a Navy navigation satellite launched from Vandenberg AFB, Calif. (UPI, Phil. *Eve. Bull.*, 1/8/65)
- Sen. Leverett Saltonstall (R.—Mass.) introduced in the Senate a bill designed to set aside March 16 of every year in honor of Dr. Robert H. Goddard, "the father of modern rocketry." The bill was referred to the Committee on the Judiciary. (*CR*, 1/7/65, 283)
- Vice Adm. H. G. Rickover (USN) spoke before the Publishers' Lunch Club of New York. In his speech Admiral Rickover said: "How to resolve the antithesis between technology and individual liberty; how to insure that technology will be beneficial, not harmful, to man, to society, and to our democratic institutions—this, I would say, is a public question. I raise it here because I believe the members of this audience are particularly well qualified to explore this problem. In your business the conflict between technology and liberty —so prevalent everywhere else in our society—is muted, if not absent altogether.

    "Improvements in the mechanics of producing and selling books have not diminished the importance of the author. Your success still depends on him. He cannot be rendered obsolete by automation. The human factor therefore continues to outweigh the technical. As in the past, your main function is to discover talent and help bring it to fruition. You know that liberty enhances creativity, that men with a special competence must be allowed to follow their own judgment. . . .

"How to make technology most useful to ourselves and our society, yet prevent it from controlling our lives—that is the problem. The problem is aggravated by the bureau-cratization of American life, itself largely a result of technology." . (Text, *CR*, 1/29/65, 1522–24)

January 7: USAF announced that AFSC Space Systems Div. had awarded a $1,783,500 increment to an existing contract for procurement of standard launch vehicle boosters to Douglas Aircraft Co., Inc. (DOD Release 8–65)

- Britain would go ahead with the $880-million U.K.-France project to build the Concorde supersonic airliner, according to the London *Daily Express*. Two Concorde prototypes, and possibly as many as six, would be built with work shared by the British Aircraft Corp. and France's Sud Aviation. There had been no official French response to British Labor government's proposal that the Concorde project be cut back, but French government as well as British union leaders were said to be hostile to the proposed "review." (AP, Wash. *Eve. Star*, 1/7/65; *Av. Wk.*, 1/11/65, 32)

- Julius E. Kuczma, executive secretary of the U.S. Labor-Management Government Commission, said his group had decided to hold a hearing and take any steps necessary to resolve the labor dispute that had halted construction work at Cape Kennedy last month. (UPI, *Orl. Sen.*, 1/8/65)

January 8: FAA announced that contracts for industry study in the supersonic transport program had been extended an additional two months. (AP, *Balt. Sun*, 1/9/65)

- Dr. Richard Shorthill of Boeing Scientific Research Laboratories reported that from 400 to 800 "hot spots" were observed on the moon during the eclipse of December 18, 1964. The lunar face had been scanned at infrared wavelengths from the Helwan Observatory near Cairo, Egypt. Recent impacts from meteors, which would create rocky craters slower to cool after the sunlight was obscured, might account for the "hot spots." It was already known that prominent craters from which rays radiated in all directions, such as Tycho, were slow to cool, compared to the normal surface, which was thought to be carpeted with dust. While the total number of slow-cooling locations would remain uncertain until the tape-recorded results had been plotted by computer, Dr. Shorthill felt that if the technique produced an inventory of young craters, it would help in spotting new ones when they occurred and in estimating the rate at which the moon and the earth were bombarded by debris from space. (Sullivan, *NYT*, 1/9/65)

- Application for patents on a recoverable single-stage spacecraft booster was filed with the U.S. Patent Office by NASA Marshall Space Flight Center. Invented by Philip Bono, a space engineer at Douglas Missile and Space Systems Div., the booster was called Rombus (Reusable Orbital Module—Booster and Utility Shuttle) and would have the capability of placing approximately 1 million lb. in circular orbit 175 mi. high and could be reused 20 times. Rombus would have its own propulsion for orbiting, deorbiting, and landing retrothrust, would employ eight strap-on, jettisonable liquid hydrogen fuel tanks. The vehicle resulted from a NASA-funded study but was not presently being developed. (*Marshall Star*, 1/13/65, 1–2; Seattle *Post-Intelligence*, 1/8/65)

January 9: At Vatican City, Pope Paul VI saw a movie made up of photos taken by RANGER VII as it neared the moon. NASA Associate Administrator Robert Seamans, Jr., in Europe on other business, and NASA European representative, Gilbert W. Ousley, were received by the Pope, showed him the movie, and answered his questions. (*N.Y. Herald Trib.*, 1/11/65; AP, Balt. *Sun.*, 1/11/65)

- Dr. Eric Ogden, Chief of the Environmental Biology Division at NASA Ames Research Center, was recipient of a Research Committee Citation presented by the American Heart Association in New York. His work for the Heart Association had been primarily in planning and evaluating heart research projects. (ARC Release 65–1)
- Tass announced that the Soviet Union would launch new types of space rockets into the Pacific Ocean from Jan. 11 until Mar. 1 to gather experimental data, and had asked other governments using sea or air routes in the Pacific to make arrangements for ships and aircraft not to enter the impact area between noon and midnight during the launching period. The carrier rockets would be fired to a point within a radius of 74 mi. from a center with coordinates of 1.58° north latitude and 164.17° west longitude. (Reuters, *NYT*, 1/10/65; Tass, *Izvestia*, 1/12/65, 4, ATSS–T Trans.)
- Working on the assumption that a leveling off of defense expenditures in the Federal budget would be accompanied by diversion of some defense funds for other public needs, California was taking steps to find new customers for its aerospace industries. 37 per cent of California's manufacturing industry was concentrated in ordnance, aircraft, electrical, and instrument production, all of which, according to Gov. Edmund G. Brown, would be vulnerable to cutbacks and phaseouts in the Government's space and defense programs. The state was prepared to finance study contracts in four major problem areas: waste management, data collection, care of the mentally and criminally ill, and transportation systems. Aerojet-General Corp. had already signed a six-month, $100,000 contract to develop long-range state plans to manage all kinds of waste, including air and water pollution. (Davies, *NYT*, 1/10/65, 12)
- Univ. of Louisville would be the first engineering school in the U.S. to have installed an electric system linking its computers with all laboratories and classrooms in its Speed Scientific School. Students working on experiments would signal measurements directly to a computer for immediate calculation and correlation. Experiments could be shown on closed circuit TV. Eventually the computers would be programed to direct experiments by automatically changing temperatures, mixtures, pressure rates, or liquid flows. (*NYT*, 1/10/65, 44)

January 10: NASA signed a one-year $70,000 contract with Flight Safety Foundation to report and evaluate research and development projects and events related to rough air in the atmosphere. The study would be conducted from FSF Offices in New York City, Phoenix, Ariz., and Los Angeles. (NASA Release 65–10)

January 10: In an article entitled "The Pentagon, the 'Madmen,' and the Moon," Maj. Gen. of the Soviet Air Force B. Teplinskiy said: "Sober voices in the United States call for collaboration with the U.S.S.R. in space research. The *Saturday Evening Post* said: 'When we reach the moon and the stars, we shall find the solutions to the most profound secrets of the universe. How much more easily accessible all this would be if we would fly there together.'

"It is known throughout the entire world that the lag in this respect does not depend on the Soviet Union. It is the spiteful policy of those U.S. circles, which do not hide their military space plans, which constitute the obstacle. These plans are widely trumpeted by the press, television, and radio. Such a position is not accidental. On the one hand it allegedly pursues the aim of enhancing U.S. prestige while it actually is aimed at blowing up the psychosis around the space armaments race and at trying to provoke the Soviet Union into retalatory measures or to intimidate it by the alleged U.S. possibilities. A naive scheme." (*Krasnaya Zvezda*, 1/10/65, 3)

- Data from SOLRAD, the Naval Research Laboratory's satellite monitoring the sun's x-ray behavior during the 1964–65 International Years of the Quiet Sun (IQSY), indicated that the sun was at its quietest during May, June, and July, 1964. Information from SOLRAD also suggested that the x-ray region of the corona, instead of being a homogeneous region of a million miles or so, was a series of small cells that flared up to emit hard x-rays and then decayed rapidly. What was seen on earth was the net effect of many knots of very hot, flashing gas giving the appearance of a homogeneous region. (Simons, *Wash. Post*, 1/11/65; Hines, Wash. *Eve. Star*, 1/11/65; *M&R*, 1/18/65)

- Eight NASA astronauts began geology field training in Hawaii, where they visited lava fields of Mauna Loa and Kilauea, active volcanoes, as well as upper elevations of dormant Mauna Kea. Geologists believed that these shield volcanoes contained features similar to those of the lunar surface.

  Study emphasis was on mechanics of lava flow, fissure eruption, deep lava lakes; examples of hot and cold basaltic flows; physical composition of lava rock; and topographic forms of shield volcanoes. Underfoot textures theorized as being typical of lunar terrain ranged from the glassy form of "pahoehoe" lava, through the crusty snow effect of "aa" lava, to the sinking feeling of loose cinders and pumice.

  The study was conducted by Dr. Ted Foss, head of the Geology and Geochemistry Section at NASA Manned Spacecraft Center. Astronauts were Charles Conrad, Jr., Clifton C. Williams, David R. Scott, Edwin E. Aldrin, Jr., Alan L. Bean, Donn F. Eisele, Roger Chaffee, and Richard Gordon. (UPI, *Houston Chron.*, 1/11/65; Bryan, *Houston Post*, 1/14/65)

- U.S. Chamber of Commerce released a report entitled "Criteria for Federal Support of Research and Development," which proposed the establishment of a forum for debating scientific and technical issues (such as space exploration and desalting of the oceans) before they became national policy. The council, to be composed of representatives of industry, labor, the Government, and the academic industry, would investigate the inherent worth of proposed programs and their value to society to increase public understanding of issues that were usually decided by the Government alone and debated afterward. (Clark, *NYT*, 1/11/65, 46)

January 11: Dr. Edward C. Welsh, Executive Secretary of the National Aeronautics and Space Council, said before the New York Academy of Sciences that "scientists should not set themselves up to judge the overall value" of the national space program. Past advice from scientists had not always been sound advice, he noted. ". . . Organized science has not always been outstanding for its courage, its vision, or its optimism regarding goals for human efforts. Elements of conservatism, parochialism, and even reactionary thinking do appear among scientists just as they do among many other groups in our society."

Dr. Welsh was also critical of the practice of criticizing the space program "by narrowly comparing" the dollars spent for space with what those same dollars might accomplish "if devoted to other endeavors, scientific or otherwise." He said that often such dollars were not transferable; that space dollars might change the general climate to one favoring broader aid to the whole spectrum of science; and that since space expenditures sought broader goals than those of science, "the comparison may well be invalid on the face of it."

He continued: "The visionaries, whether primarily scientists or policy makers, must be given the opportunity to point out the many benefits which can flow from the manned and unmanned uses of aerospace. But, given such opportunity, they should use it effectively and affirmatively. Regardless of their motivations, the pessimists who cry out against aerospace research and technological endeavors have clearly set themselves against progress. The United States can no longer relax and rest on its past industrial laurels. The race for survival, literally and philosophically, is on. Of course, we would all like to believe in the solely non-aggressive uses of aerospace by all countries which have the needed technology. However, the realities of life dictate adequate preparation to preserve our national and Free World security. We should follow the axiom that a pound of prevention is worth mega-tons of cure."

Howard Simons commented in the *Washington Post* that these remarks were probably precipitated by a report from a committee of the American Association for the Advancement of Science which had charged that social, economic, military, and political pressures were distorting the traditional values and effectiveness of science. The report was highly critical of Project Apollo: "The Apollo program, in its present form, does not appear to be based on the orderly, systematic extension of basic scientific investigation." (Text, *CR*, 1/28/65, A364–65; Simons, *Wash. Post*, 1/12/65)

- NASA announced that Launch Complex 16 at Cape Kennedy would be modified to convert the former Titan missile facility into static test stands for the Apollo manned lunar spacecraft. Construction bids were expected to be opened by Army Corps of Engineers, late this month. The modified test facility would replace an Apollo static test stand originally planned for the NASA Kennedy Space Center's Merritt Island facility. Officials estimated that the modification of Complex

16 would represent a cost reduction of about 72 per cent under the original $7 million construction estimate for test stands on Merritt Island. (KSC Release 7–65)

January 11: NASA Langley Research Center scientist Windsor L. Sherman proposed conversion of Project Mercury spacecraft into unmanned, recoverable orbiting telescope platforms. Equipment would include a 76-cm. Cassegrainian telescope, a camera recording system, and an attitude control system. The system would weigh approximately 4,700 lb. and would be aimed for a 300-mi. orbit. The observatory would remain in orbit 100–200 days, exposing four frames of film on each orbit for a total of 6,000 frames. After all film was exposed, the system would be braked out of orbit and would descend into the Bermuda recovery area of the Eastern Test Range, using the same recovery techniques developed for the manned Mercury landings. In addition to its capacity to perform a variety of such astronomical observations as high resolution photography, photometry, and spectroscopy, Sherman said, the recoverable observatory would permit reuse of capsule, optical, and control systems. It would allow study of space effects on equipment, and the system could serve as a test bed for advanced orbiting telescopes. (*Av. Wk.*, 1/11/65, 23)

• Dr. John J. Brennan, Jr., Chairman of the Committee for the Preservation of Cambridge Industries, said he would take to Washington the committee's fight to keep the NASA Electronic Research Center out of Cambridge. Dr. Brennan said the City of Cambridge's claim that the renewal project would cost the Federal government $15 million was way off. He said costs would be between $40 million and $50 million. In a letter to the House and Senate Appropriations Committees and the House and Senate space committees, Brennan stated: "We are taking every proper course of action, legal and otherwise, to stop this senseless destruction. . . .We do not believe that the overall destruction will bear judicial scrutiny."

Paul Frank, director of the Cambridge Urban Redevelopment Authority, said Brennan's figures were inaccurate and that the $40–$50 million figure was wrong. He claimed the overall cost would bring it down to $14,500,000. Of this figure, the Federal government cost would be $9,600,000 with the remaining $4,900,000 paid by the City of Cambridge, he asserted. (*Boston Globe,* 1/11/65)

During the week of January 11: Titan III program director Brig. Gen. Joseph S. Bleymaier (USAF) said at a meeting of the New York Academy of Sciences that the launch of the Titan III-A, on Dec. 10, 1964, may have gained the most accurate orbit ever achieved in the U.S. space program. The vehicle achieved an orbit with 102-n. mi. apogee and a 99-n. mi. perigee against a planned 100-n. mi. nominal orbital altitude. Deviation from a true circle was 0.00075 against a predicted value of 0.00050. Time for a single orbit was 88.2 min., within 0.04 min. of the time predicted. (*M&R,* 1/18/65, 10)

January 11: U.S.S.R. orbited COSMOS LII earth satellite. Orbital data: apogee, 304 km. (188.9 mi.); perigee, 205 km. (127.4 mi.); period 89.5 min.; inclination to the equator, 65°. The satellite carried scientific equipment "for the further investigation of outer space in accordance with the program announced by Tass on the 16th of

March, 1962." (Tass, *Komsomolskaya Pravda*, 1/12/65, 1, ATSS–T Trans.)

January 11: North American Air Defense Command (NORAD) tracked a new Russian satellite (COSMOS LII) for several hours before Moscow announced the launching. As of this date, NORAD's space detection and tracking system was observing 488 man-made objects in space, of which 29 were actual payload satellites and the rest debris from previous launchings. (AP, Balt. *Sun*, 1/12/65)

- Arthur D. Little, Inc., released a 54-page study entitled "Strategies for Survival in the Aerospace Industry," which predicted that in the next five years the production portion of the defense budget would decline about 30 per cent and research and development would decline about 15 per cent. The report recommended that "in view of a declining market and fewer opportunities within the market, the aerospace industry's principal objective within the next few years should be to achieve stability, rather than to search for growth opportunities." (Duggan, N.Y. *Her. Trib.*, 1/12/65)

- In January, Dr. Donald F. Hornig began his second year as science adviser to President Johnson and director of the White House Office of Science and Technology. In interview he mentioned that his job was created to prevent a recurrence of the kind of official surprise that greeted Russia's launching of the SPUTNIK I on Oct. 4, 1957. Hornig said the policy questions that he encountered were not ones of "right or wrong, but wise or less wise." (*Av. Wk.*, 1/11/65, 16)

- U.S. Junior Chamber of Commerce named Capt. Joseph H. Engle (USAF) one of the ten outstanding young men of 1964. Captain Engle, the youngest of the X–15 pilots, had logged nine flights in the X–15. Awardees would be honored at an awards congress Jan. 15–16 in Santa Monica, Calif. (AP, *Des Moines Register*, 1/12/65)

January 12: Kiwi-TNT (Transient Nuclear Test) was successfully completed at Jackass Flats, Nev. This was a safety test to verify predictions of behavior of graphite nuclear reactor during a maximum power excursion. Using data from the test scientists would establish safety standards, particularly for launching nuclear-powered rockets. Nuclear energy released in the test was well within the designated maximum of nuclear test ban treaty of 1963. Preliminary test results indicated: (1) from ½-mi. to 50-mi. downwind from the test site, radiation did not approach accepted danger levels; (2) lethal radiation was confined to 200-to-300-ft. radius of the site, and beyond 500-to-600-ft. radius "a person would probably have survived unhurt unless struck by a piece of debris"; (3) pre-test predictions of the reactor's behavior were accurate; and (4) cleaning up radioactivity at the site was easier than expected. Kiwi ground-test version of a nuclear-reactor rocket engine was a NASA-AEC project. (UPI, *Wash. Post*, 1/13/65; *NYT*, 1/13/65; AP, Balt. *Sun*, 1/13/65; *JAMA*, 2/8/65, 27–29; *Rover Chron.*, n.d.)

- USN announced the Transit navigational satellite system was operational and had been in use since July 1964. The three gravity-gradient-stabilized satellites, weighing between 110 and 160 lbs. each, were launched on Thor-Able-Star boosters into near-circular 600-mi. polar orbits from Pt. Mugu, Calif. Operational lifetime of the satellites was expected to be about two years. The satellites emitted radio signals which ships used to determine their positions, and could provide ships with navigational fixes—accurate to 0.1 mi.—about every 90 min. The shipboard computer operated automatically, beginning when the satellite approached, receiving the data, computing the ship's position, and typing the results for the navigator. A number of fleet units were reported to be using the system. Capt. F. H. Price, Jr. (USN), who tested the system from the nuclear-powered cruiser U.S.S. *Long Beach*, called the system "the most reliable means of providing navigational information" and said it met the requirement of an "accurate, dependable, worldwide, all-weather, 24-hour-a-day capability." This was the first continuous use of space technology in direct support of the fleet. It was predicted, but not officially confirmed, that the Polaris missile-firing submarines would adopt the navigational satellite system. NASA was studying commercial applications of a navigational satellite system and considering the possibility of developing its own system if it proved economically feasible. (DOD Release 16–65; AP, *Chic. Trib.*, 1/13/65; Watson, Balt. *Sun*, 1/13/65; *M&R*, 1/18/65, 14)

January 12: S. Walter Hixon, Jr., Supervisory Employee Development Officer at the NASA Langley Research Center, was selected for his educational activities as the Federal Civil Service Employee of the Year in the Hampton Roads area. Hixon had conducted four major programs at Langley including graduate study, advanced in-house training, a cooperative college education plan, and an apprenticeship training system. (LaRC Release)

- France's newest satellite tracking station, located outside Pretoria, South Africa, was nearing completion and would probably be operational by July 1965. The $840,000 station would be used to track France's first satellite, scheduled to be orbited around the earth in 1965. (AP, Balt. *Sun*, 1/13/65)

- The first 95-passenger DC-9 jet liner rolled off the Douglas Aircraft Co. assembly line. A short-haul, twin-engine jet, the DC-9 would be able to land on most conventional airstrips and would, therefore, serve 98 per cent of the Nation's civil airports. 58 planes had been ordered and options were taken on 60 more, but development costs would not be met until the 200 mark was reached. Flight tests would begin in March 1965 or sooner, and airlines operating the new jet expected to start passenger service early in 1966. (UPI, *NYT*, 1/12/65, 72)

- DOD announced Peter Kiewit Sons Company had received a $9,495,000 contract for modification of Titan II launch facilities in the vicinity of Davis-Monthan AFB, Ariz.; Little Rock AFB, Ark.; McConnell AFB, Kan.; and at Vandenberg AFB, Calif. The Army Corps of Engineers awarded the contract. (DOD Release 18–65)

- A Canadian company, Jarry Hydraulics, Ltd., designed and built the variable-wing sweep device for the USAF's F-111 fighter bomber. The actuator, consisting of a unit in the fuselage which controlled two booms, could withstand more than 500,000 lbs. tension and could set the wings within .015 of an inch of the position selected by the pilot, at a rate of 200° per minute. (Toronto *Globe and Mail*, 1/12/65)

- DOD would be using 1,274 computers by the end of FY 1965, compared to the 815 computers which were in use when Robert S. McNamara first became Secretary of Defense. NASA would be using 224 computers in various branches of its operations. (Fay, Wash. *Eve. Star*, 1/12/65)

January 13: X-15 No. 3 flown by NASA pilot Milton O. Thompson to maximum altitude of 99,400 ft. and maximum speed of 3,712 mph (mach 5.48). Purpose of the flight was to collect air flow data and record measurements of skin friction on the aircraft's surface. (NASA X-15 Proj. Off.; FRC Release; *X-15 Flight Log*)

- NASA launched a two part 99-lb. sounding rocket payload from NASA Wallops Station which reached an altitude of 614 mi. but did not separate in flight as planned. Launched on a four-stage Javelin (Argo D-4) and designed as "mother-daughter" experiment, the payload was to separate into two sections at about 170-mi. altitude with radio signals to be sent from daughter to mother as they continued to rise separately. The technique was devised to provide more accurate profiles of electron density in the upper atmosphere. Telemetry data would be analyzed to determine why the sections did not separate. (Wallops Release 65-3; NASA Rpt. SRL)

- NASA successfully launched an Aerobee 150A sounding rocket to peak altitude of 110 mi. from Wallops Island, Va., with instrumented payload to measure the ultraviolet and visible light emitted from the earth's atmosphere between 37 mi. and 125 mi. An Attitude Control System (ACS) was also flown. Good spectral data were collected. (NASA Rpt. SRL)

- Reported that NASA Administrator James E. Webb had ruled against a protest by a group of NASA astronauts of the NASA decision to limit the first manned Gemini flight to three orbits. The astronauts had requested that the GT-3 flight should be "open-end," leaving it to the astronauts as to whether they should go for three or even 30 orbits. (Macomber, Copley News Service, *San Diego Union*, 1/13/65)

- XC-142A V/Stol, flown by Ling-Temco-Vought test pilots John Konrad and Stuart Madison, made a flawless first transition flight. The transport aircraft took off like a helicopter, adjusted its wings for conventional flight, and then circled the field, reversed the process, and made a vertical landing. The XC-142A's first transition flight came only six flights after its initial hover flight on Dec. 29, 1964. It was the Nation's first V/Stol built for operational evaluation rather than research. (AP, *CSM*, 1/13/65)

- NASA Langley Research Center scientists Harry W. Carlson and Francis E. McLean said that for the first time there was hope for a significant reduction in the sonic booms expected from proposed supersonic airliners. A plane flying faster than the speed of sound compresses the air around it into shock waves trailing from the nose, wings, engine inlets, tail, and any other protuberances. Near the plane there would be separate waves, producing "near field effects." Traced on a graph to show changes in pressure, the waves would make a jagged line resembling the letter "N." As the waves traveled toward the ground, they would coalesce into two powerful waves—one appearing to trail from the nose and one from the tail—producing "far-field effects" also shaped as a letter "N" in terms of pressure patterns. The sharp peaks of this N-shaped wave were suspected of causing most of the annoyance and structural damage possible from sonic booms. Carlson and McLean discovered that planes the length and shape of supersonic airliner designs would not fly far enough away from the earth for their far–field effects to be felt on the ground, leaving only the less bothersome near–field effect to be taken into account.

It was hoped that this new finding would mean that designs currently submitted to the Government in the design competition for supersonic transport, or minor refinements of them, would fit within Government-imposed sonic boom limitations and that still further improvement through design changes would bring further decreases in the boom.

Dr. Floyd L. Thompson, LaRC Director, called what had been learned "significant new knowledge" and said it could, under the best of circumstances, "have great significance." He pointed out that the best of circumstances were seldom found in designing an airplane—particularly the supersonic transport, which he said was "at least as sophisticated technically as the Apollo." (Clark, *NYT*, 1/14/65, 1, 12)

January 13: DOD announced that during the next six months 150 ICBMs scheduled for deactivation (27 Atlas E, 69 Atlas F, and 54 Titan I missiles) would be put into storage at Norton AFB, Calif. Some of these missiles would be used eventually as spacecraft boosters, others would be employed in the Nike-X program. They would be replaced by the more advanced Minuteman ICBMs, of which a total of 1,000 were authorized by Congress. It had cost almost $1 million a year to keep each of the older ICBMS combat-ready, as compared to $100,000 a year for each Minuteman. (Sehlstedt, Balt. *Sun*, 1/14/65; *A&A*, 1/65, 92)

• Dr. John C. Evvard, Deputy Associate Director for Research at NASA Lewis Research Center, discussed possible propulsion systems for future space-flight beyond the moon before the Conference on Civilian and Military Uses of Aerospace sponsored by the New York Academy of Sciences. He cited a manned Mars project as a prime example of a mission that could be performed by a number of different propulsion concepts. For example, manned trips by chemical rockets would be weight-restricted, but chemical rocket systems would have the advantage of having been extensively flight-tested on many other missions. Although the reactor for planned nuclear propulsion systems had only been ground tested, evaluations of complete nuclear rocket engine systems were expected within the next few years. Electric propulsion systems for manned spaceflight were even further in the future and might not be ready by 1980; but by then the mission capability of the nuclear rocket would have been so thoroughly demonstrated that it would be more attractive than chemical engines for those missions requiring increased propulsion capability. Even further into the future were nuclear systems such as the gaseous-core-cavity reactor which would yield higher performance. (LRC Release 65–5)

January 14: MARINER IV had functioned in space for more than 1,100 hr. on its 6,000-hr. flight to Mars, and had flown 81.3 million mi., leaving some 245 million mi. to be travelled before the spacecraft would encounter Mars next July. The earth–MARINER distance was 8,342,946 mi. at 9 a.m. EST with the spacecraft travelling 9,276 mph relative to the earth and 69,462 mph relative to the sun. (NASA Release 65–12, 1/14/65)

- Vincent R. Lalli of NASA Lewis Research Center described to the 11th National Symposium on Reliability and Quality Control in Miami Beach the R&QA procedure applied at Lewis to engine subsystems of the Sert-I (Space Electric Rocket Test) spacecraft to establish reliability standards for equipment never flown in space before. He said an experimental assembly of components, or "electrical breadboard," was built for electrical stress measurements; once the analysis of stresses during operation was complete, safety factor could be defined. "Stress" did not refer to mechanical stress but to all physical factors—fatigue, corrosion, current, temperature, etc.—that could degrade or destroy equipment.

  Lalli pointed out: "The real uniqueness of this process is revealed in the stress analysis area where the role of the reliability engineer is extended beyond the analytical approach into obtaining transient experimental stress data." (LRC Release 65–4)

- *Houston Chronicle* reported that preliminary funds for the unmanned exploration of Mars would be included in the NASA FY 1966 budget. On Oct. 30, 1964, the Space Science Board of NAS had recommended to NASA that Mars be the next goal because it was the likeliest of the planets to be inhabited by living things and would therefore be of greater scientific importance than the moon or proposed manning orbiting laboratories. (Mackaye, *Houston Chron.*, 1/14/65)

- The Enrico Fermi Medal was conferred on Vice Adm. Hyman G. Rickover (USN) by President Johnson. Adm. Rickover, the first nonscientist to receive the award, was cited for "engineering and administrative leadership in the development of safe and reliable nuclear power and its successful application to our national security and economic needs." He was also credited with almost single-handedly convincing Congress and DOD to start the nuclear submarine program. (UPI, *NYT*, 1/14/65, 14)

- In London, 10,000 British aircraft workers marched to protest the rumored intention of the Labor Government to curtail production of British military planes. Defense Minister Denis Healey reportedly recommended that development and production of the TSR–2 (tactical-strike-reconnaissance) aircraft be canceled and that Britain buy F–111's from U.S., thus cutting defense costs. Two other projects subject to cancellation were the P–1154 vertical-takeoff fighter and a short-takeoff fighter, both at a less advanced stage of development than the TSR–2. Leaders of the British aircraft industry, which employed slightly more than one per cent of the nation's work force, said such a cutback would cause widespread unemployment in the industry. (Lewis, *NYT*, 1/13/65, 9; Lewis, *NYT*, 1/15/65; Farnsworth, *NYT*, 1/16/65)

January 15: USAF launched a Thor-Agena D booster with an unidentified satellite toward polar orbit from Vandenberg AFB. (UPI, *Denver Post,* 1/17/65)

- USAF successfully launched a four-stage Athena reentry research vehicle from Green River, Utah. Impact occurred within a predetermined target area in the White Sands Missile Range, N. Mex. (*M&R,* 1/25/65, 8)
- The U.S.S.R. filed a brief report with the International Aviation Federation on the flight of VOSKHOD I (Oct. 12–13, 1964) for confirmation of the flight achievements as absolute world records, and of world records in the orbital flight class in multiseat spacecraft: duration of flight, 24 hrs., 17 min., 0.3 sec.; flight distance, 416,195,878 mi. (669,784,027 km.); flight height, 254 mi. (408 km.); and maximum weight raised to the flight height, 11,729 lbs. (5,320 kg.). (*Pravda,* 1/15/65, 6; *Krasnaya Zvezda,* 1/15/65, 4, ATSS–T Trans.)
- Top fuel experts of the Coordinating Research Council of New York reported that adoption of a single type of jet fuel by the entire airline industry "would not significantly improve the over-all excellent safety record of commercial aviation." The study on fuel safety was requested by Federal Aviation Agency Administrator Najeeb E. Halaby following the fatal in-flight explosion that occurred in a jet airliner December 8, 1963, near Elkton, Md. The aircraft was carrying a mixture of JP–4 and kerosene when it exploded in a lightning storm, giving rise to the question of the relative safety of the two fuels including the effects of mixing the two. Consensus of the group was that the airlines should continue their policy of being individually responsible for selecting fuels and for safety practices associated with handling such fuels. Another conclusion was that aircraft safety depended less upon the particular type of fuel used than upon equipment design and proper fueling techniques. (FAA Release 65–9)
- AEC entered into 33 mo. contracts with Combustion Engineering, Windsor, Conn., and Atomics International, Canoga Park, Calif., for joint research and development work on the heavy water-moderated, organic-cooled reactor concept. This concept could lead to construction of large central station power plants and applications to large-scale water desalting operations. (AEC Release H–12)
- U.S. recorded seismic signals from an underground event in the Soviet nuclear testing area in the Semipalatinsk region. The event was reportedly 75 times stronger than previous explosions registered from the same area. (AEC Release H–13; AP, *New Orleans Times-Picayune,* 1/17/65)

January 16: NASA announced it would request preliminary design proposals from private industry for the unmanned Voyager spacecraft that would land scientific instruments on Mars in 1971. From these proposals, several contractors would be chosen to perform a 3-mo. program design definition. Previous NASA studies had indicated the system might consist of a spacecraft "bus" or main body, a propulsion and braking system, and a landing capsule. (NASA Release 65–15)

- Addressing the Houston Junior Chamber of Commerce, Gen. Bernard A. Schriever (USAF) emphasized the importance of technology in maintaining national security: "Recent events show a number of applications of technology designed to increase our national security. These include the first flights of the supersonic XB-70 aircraft, the YF-12A long-range interceptor, the F-111 supersonic fighter, the Titan IIIA space booster, and the Minuteman II missile. . . .

"Research not only supports today's weapon systems but also provides the advanced technology from which new systems will emerge. . . .

"To name some specifics, a new high-strength, lightweight material —formed from boron fibers and a plastic binder—would make possible great weight savings in aircraft and space vehicle structures with no sacrifice of either strength or stiffness. We have already produced laboratory samples of this boron composite. It is potentially as strong as the high-strength steels, structurally rigid, and as light as magnesium. It may have higher temperature capabilities than aluminum and magnesium, should be easy to fabricate, and should have a high resistance to corrosion.

"Another advance in the materials area is the use of oxide-dispersed metals in aircraft engines to provide strength at high temperatures. This development will make possible a substantial increase in the operating temperature of turbojet engines, which in turn will make for greater operating efficiency and improved thrust-to-weight ratios." (Text, AFSC Release)

January 17: Robert L. Sohn, scientist at Space Technology Laboratories, proposed to use the gravity field of Venus as a brake for manned spacecraft returning from Mars.

"We don't expect to have boosters powerful enough to launch spacecraft of the 1970s that can carry extra propulsion to brake reentry speeds. . . . The landing corridor will be so narrow that a small fractional error in navigation would send the spacecraft into an eternal orbit around the sun." He said traveling near Venus on the return journey from Mars would slow a spacecraft as it passed through the Venutian gravity field. Then, with some midcourse maneuvering and navigation, the astronaut could return to earth and reenter earth's atmosphere with greater margin of error. (Macomber, *San Diego Union*, 1/17/65)

• Dr. I. M. Levitt, Director of the Fels Planetarium, said in the *Philadelphia Inquirer:* "As of this moment, the Soviets have tentatively determined that the maximum 'safe' period of weightlessness is 24 hr. They hold that after this period, 'irreversible physiological changes begin to occur in the human system which, if not corrected, will eventually lead to death'. . . .

"The Soviets have also discovered a correlation between high accelerations and weightlessness. They believe that when an astronaut is subjected to high accelerations on launch he tends to overestimate or to overcompensate for his movements. Once the astronaut is weightless, then a radical reversal takes place in which the astronaut undercompensates and may suffer disorientation. . . .

"The Soviets appear to have concluded that flight crews of the future will be selected as medical teams, and they will further be selected on the basis of biological and bacteriological compatibility. The crew will be concerned with developing means for forecasting their own health during the entire trip so as to preserve it." (*Phil. Inq.*, 1/17/65)

January 17: Tass reported that a Soviet archeologist had discovered a Neolithic drawing in a cliff gallery in Soviet Central Asia resembling a cosmonaut. The figure carried "something resembling an airtight helmet with antennae on its head" and "some sort of contraption for flight" on its back. (Reuters, *Wash. Post*, 1/18/65; *NYT*, 1/23/65)

January 18: USAF launched an unidentified satellite on a Thor-Altair booster from Vandenberg AFB, Calif. Altair was normally the solid-fuel fourth stage of the Scout booster. (AP, Wash. *Eve. Star*, 1/19/65)

- In an editorial in *Aviation Week and Space Technology*, Editor Robert Hotz said: "This is a year in which we will hear much about the growing pains of Apollo. It would be most amazing if we didn't. For Apollo is now in the midst of that difficult period when the problems of creating this incredibly intricate and complex technical system are being hammered the hardest toward solutions. It is also the period when the effectiveness of the management structure in welding all of the complex subsystems into a successfully functioning overall system within the time and money boundaries already established becomes most vital." (*Av. Wk.*, 1/18/65, 17)

- In his defense message to Congress, President Johnson cited major new developments in strategic weapon systems slated to begin this year:

"A new missile system, the Poseidon [new name for Polaris B-3], to increase the striking power of our missile-carrying nuclear submarines. The Poseidon missile will have double the payload of the highly successful Polaris A-3. The increased accuracy and flexibility of the Poseidon will permit its use effectively against a broader range of penetration of enemy defenses.

"A new Short Range Attack Missile (SRAM) that can, if needed, be deployed operationally with the B-52 or other bombers. This aerodynamic missile—a vast improvement over existing systems—would permit the bomber to attack a far larger number of targets and to do so from beyond the range of their local defenses.

"A series of remarkable new payloads for strategic missiles. These include: penetration aids, to assure that the missile reaches its target through any defense; guidance and re-entry vehicle designs, to increase many-fold the effectiveness of our missiles against various kinds of targets; and methods of reporting the arrival of our missiles on target, up to and even including the time of explosion."

In addition, he said that development of the C-5A (formerly the CX) cargo transport and procurement of the Air Force F-111 fighter-bomber and new A-7 Navy attack aircraft would begin.

Finally, regarding the role of science and technology in the Nation's security, the President said:

"We are currently investing more than $6 billion per year for military research and development. . . . About $2 billion a year of this program is invested in innovations in technology and in experimental programs. Thus, we provide full play for the ingenuity and inventiveness of the best scientific and technical talent in our Nation and the Free World.

"American science, industry, and technology are foremost in the world. Their resources represent a prime asset to our national security." (Text, *Wash. Post*, 1/19/65; AP, *NYT*, 1/19/65, 16; Norris, *Wash. Post*, 1/22/65)

January 18: The new Sram (short-range attack missile), cited by President Johnson in his defense message to Congress, would be expected to travel 150 mi. from the launching plane to its target. The Sram would be designed for launching initially from a B–52, but later from smaller aircraft such as the F–4C or the F–111. It would be launched toward the rear after the aircraft had passed its target, would climb to 100,000-ft. altitude, powered by its own solid-propellant motor, then plunge vertically toward its target having allowed the launch plane time to escape its nuclear warhead detonation. (Watson, *Balt. Sun*, 1/19/65; Miles, *Wash. Post*, 1/20/65)

- Alfred Gessow, Chief of Fluid Physics Research, NASA, discussed before the Compressed Gas Association in New York City the problems of spacecraft deceleration and heating involved in return through the earth's atmosphere. He explained why the blunt shape solved deceleration and much of the heat problem in returning Mercury spacecraft from orbit through the atmosphere to earth. Looking beyond the satellite return speed (Mercury and Gemini) and lunar return speed (Apollo), return from interplanetary flight poses the problem of much higher spacecraft speed (and thus heating). Research indicates "that the more pointed shape, although it doesn't show up too well at the lower re-entry speeds, is better than the blunt nose at the higher speeds because the bow shock is weaker, thus producing lower radiant heating losses. Thus, in a very short time scale, but taking a big leap forward in the velocity-temperature scale, we find ourselves going into another phase of the blunt vs. pointed nose cycle.

  ". . . The switching between slender and blunt shapes is not new in the race for higher speeds at all times of history. Going through history, compact rocks were replaced by slender arrows; the concept of powder guns created round cannonballs; the rocket age produced slender forms again, which ironically, finally got blunt noses. It is interesting to see how long it took to make such changes empirically and how rapidly these variations have been made by following scientific principles. . . ." (Text)

- Japan expected to orbit a satellite within the next three years, *New York Times* reported. Although Japan's progress in the missile field had been slowed by the limited annual budget allocations of the Defense Forces, scientific advances, particularly in the field of electronics, plus stimulus to Japanese industry provided by the Korean War, had brought marked advances in rocketry and missiles. (*NYT*, 1/18/65)

- The Communist New China News Agency (NCNA) said in a broadcast that Indonesia had successfully launched a two-stage scientific rocket Jan. 5 from somewhere in West Java. The rocket was reportedly made by the Indonesian air force. There were no other details. (UPI, *Miami Her.*, 1/18/65)

January 19: An unmanned instrument-packed Gemini spacecraft (GT–2) was launched from Cape Kennedy on Titan II launch vehicle in suborbital shot preliminary to U.S.'s first two-man venture. Aboard was an automatic sequencer which issued orders at precise times en route to fire the rocket's second stage, to separate the spacecraft from the rocket, to jettison the spacecraft's storage section, to cartwheel the spacecraft into a reentry attitude, and to open the spacecraft's parachutes.

The rocket reached a maximum altitude of 98.9 mi. and a speed of 16,708.9 mph before impacting 2,127.1 mi. downrange. The Gemini spacecraft descended by parachute into the Atlantic 16 mi. short of the planned impact point and 52 mi. from the carrier U.S.S. *Lake Champlain* which recovered the capsule an hour and 45 min. after launch. The capsule was reported in excellent condition.

Major experiments for which the test was intended were apparently complete successes: a test of the heat shield; a test of the retrorocket system; and a test of the sequencing system.

Despite its successes, the test had some difficulties: a fuel cell that would be the primary electrical system in the spacecraft during long-duration manned flights failed to operate before launching because of a stuck valve; the temperature was found to be too high in the cooling system of the spacecraft. (NASA Release 64-296; MSC *Roundup*, 1/3/65, 1; Wash. *Eve. Star*, 1/19/65; *Houston Chron.*, 1/19/65; UPI, Rossiter, *Wash. Post*, 1/20/65; AP, Balt. *Sun*, 1/20/65)

January 19: Dr. Burton I. Edelson, staff member of the National Aeronautics and Space Council, spoke on communications satellites at the AIAA meeting in Las Cruces, N.Mex. He said: "There is a general growing interdependence of politics, economics, and technology, and in no area do these forces interact more noticeably, than in international communications. When we try to predict the course that communications satellites systems will follow in the years to come we must consider not only decibels and megacycles, rocket thrusts and orbital elements, but the competitive economic pressure of transoceanic cables and the political aspirations of developing nations. . . .

"Finally, I believe the words of Arthur Clarke, the visionary who first conceived of the communications satellite, will be fulfilled: 'Comsats will end ages of isolation making us all members of a single family, teaching us to read and speak, however imperfectly, a single language. Thanks to some electronic gear twenty thousand miles above the equator, ours will be the last century of the savage.'" (Text)

January 20: President Lyndon B. Johnson was inaugurated. In his Inaugural Address, he said:

"For every generation, there is a destiny. For some, history decides. For this generation, the choice must be our own.

"Even now, a rocket moves toward Mars. It reminds us that the world will not be the same for our children, or even for ourselves in a short span of years. The next man to stand here will look out on a scene different from our own.

"Ours is a time of change—rapid and fantastic change—baring the secrets of nature—multiplying the nations—placing in uncertain hands new weapons for mastery and destruction—shaking old values and uprooting old ways. . . .

"Change has brought new meaning to that old mission. We can never again stand aside, prideful in isolation. Dangers and troubles we once called 'foreign' now live among us. If American lives must end, and American treasure be spilled, in countries we barely know, that is the price that change has demanded of conviction.

"Think of our world as it looks from that rocket heading toward Mars.

"It is like a child's globe, hanging in space, the continents stuck to its side like colored maps. We are all fellow passengers on a dot of earth. And each of us, in the span of time, has only a moment among his companions.

"How incredible it is that in this fragile existence we should hate and destroy one another. There are possibilities enough for all who will abandon mastery over others to pursue mastery over nature. There is world enough for all to seek their happiness in their own way.

"Our own course is clear. We aspire to nothing that belongs to others. We seek no dominion over our fellow man, but man's dominion over tyranny and misery. . . ." (Text)

January 20: Dr. Robert Jastrow, Director of NASA's Goddard Institute for Space Studies, said at the annual meeting of the Franklin Institute in Philadelphia: "Beyond military and political advantages of getting to the moon are possibilities we cannot conceive." The moon, he said could prove to be "the Rosetta stone of the universe. Its lifeless surface could give us the clue to the process of life." (*Phil. Eve. Bull.,* 1/21/65)

- Lockheed Missiles and Space Co. conducted successful static firings of the Agena target vehicle for Project Gemini. The firing tests, which included simulated maneuvers to be made by Agena during rendezvous with the Gemini spacecraft, included five separate firings of the main engine and of the secondary propulsion system. The tests lasted some 12 hrs. and were termed by Lockheed "complete captive flight." All systems of the actual flight Agena were tested, including command from earth transmitters, programmed commands within the Agena, telemetry, and docking simulation. Previous Gemini Agena firings had tested the vehicle's engines only. (*Huntsville Times,* 1/22/65)

- Dr. M. P. Lansberg of the National Aeromedical Center, Soesterberg, The Netherlands, told scientists attending the symposium on the inner ear at the Naval School of Aviation Medicine at Pensacola Air Station that one role of space flight would be the exploration of the functioning of the vestibular organ. "This might well be the most important and fascinating side of space flight," said Dr. Lansberg. "Not what it will reveal to us of distant worlds, but what it will unveil to us about ourselves."

  Dr. Lansberg also warned against expecting too much from experiments conducted here on earth in trying to determine how much gravity-producing spinning man could stand. In recommending rates of speed to space engineers, he said "we should be conservative." (Harris, *Pensacola Journal,* 1/21/65)

- In an article in *The Huntsville Times,* Richard Lewis said: "If Project Apollo continues at its present pace, the United States will be able to attempt the landing of astronauts on the moon in 1968. . . .

  "This impression of the status of Apollo . . . was gained by this reporter in tours of both industrial and test centers for the mammoth project. . . .

  "The story at these centers is this: no new breakthroughs in electronics, mechanics, metallurgy, propulsion or guidance and navigation are required for the program. All major problems are settled. They have been solved or 'worked around.' . . .

"So well does Apollo appear to be running that there is a strong probability it will overtake the later flights of Project Gemini, the two-man spacecraft program." (Lewis, Chicago *Sun-Times, Huntsville Times*, 1/20/65)

January 20: It was reported that Lockheed Propulsion Co. had successfully test-fired a new solid-propellant rocket motor at the proving ground in Redlands, Calif. The lightweight "pulse motor" measured 10 ft. in length, 2 ft. in dia., and contained 40 solid-propellant wafers, each of which could develop more than 1,000 lbs. of thrust. This was possibly the rocket motor that would power the Sram (short-range attack missile) mentioned by President Johnson in his defense message to Congress [See Jan. 18, 1965]. (Miles, *Wash. Post*, 1/20/65; *SBD*, 1/18/65, 74)

- USAF successfully launched its first Minuteman ICBM of 1965 from Vandenberg AFB, Calif. The missile was sent on a 5,000-mi. course toward a target in the Pacific. (UPI, L.A. *Herald Examiner*, 1/21/65)

January 21: MARINER IV completed nearly one-quarter of its 7½-mo. journey to Mars and was more than 10 million mi. from earth. The craft was traveling 10,680 mph relative to the earth; velocity relative to the sun was 68,255 mph; total distance traveled was over 93 million mi. After 54 days in space, all systems were functioning normally except the solar plasma probe which ceased returning intelligible data one week after launch. (NASA Release 65-17)

- Laser beam was bounced off NASA's EXPLORER XXII ionosphere satellite and photographed by Air Force Cambridge Research Laboratories scientists Robert Iliff and Theodore Wittanen. This was first such photo and was important verification of feasibility of use of laser for both satellite tracking and geodetic purposes. When such laser reflections off satellites were photographed against a star background from two ground stations of known locations and other ground stations in the field, triangulation of the simultaneous photos would locate the position of field stations with an accuracy hitherto not possible by other means. This success with Largos (Laser Activated Reflecting Geodetic Optical Satellite) also set a distance record for photo or photoelectric detection of reflected laser signals; slant range to satellite was 950 mi. (AFCRL Release 2-65-2)

- USAF launched a 100-lb. ARV (Aerospace Research Vehicle) satellite pickaback aboard an Atlas ICBM from Vandenberg AFB, Calif. The satellite, carrying instrumentation to sample radiation and micrometeoroids, was the first to be sent toward westward orbit around the earth. *Satellite Situation Report* for January 31, 1965, did not indicate that the satellite had achieved orbit. (AP, *Wash. Post*, 1/22/65; *M&R*, 2/1/65, 9; *SSR*, 1/31/65, 13)

- Sen. Margaret Chase Smith (R–Me.), ranking member of the Senate Aeronautical and Space Sciences Committee, told NANA in an interview that the United States was giving more to the Soviets than it got in a lopsided exchange of space data. She said that for several months weather information derived from "conventional" sources in the Soviet Union had been sent through a communications link between Moscow and Suitland, Md. "The weather information is not that derived from a satellite as provided for by the agreement," she asserted.

"Up to the present time, based on the information I have available, the Soviets are realizing more from the 1962 Geneva Agreement than we are."

Senator Smith added that measuring the results of the Geneva Agreement strictly on scientific knowledge gained "is not a broad enough yardstick. Any real plusses, it seems to me, must be measured in the light of what we seek to accomplish, namely, the mastering of space for the benefit of all mankind. The fact that the Geneva Agreement ever came into existence shows an awareness of the magnitude of the task confronting man if he expects to operate successfully in space." (Glaser, NANA, *Indianapolis Star*, 1/21/65)

January 21: NASA Administrator James E. Webb received an honorary doctorate from Wayne State Univ., Detroit, Mich. During a speech there he said:

"Our goal is 100 per cent assurance of [space booster] success. This is difficult to achieve, but until we are certain in our own minds that we can count on success we do not go [on] with a manned shot. My directive on this is very clear. It came first from President Kennedy and has been restated by President Johnson. It is 'Go when ready and don't go until ready.'" (Text)

- As part of the ceremonies dedicating the Capt. Theodore C. Freeman Memorial Library of Astronautics at the Houston Baptist College, Faith L. Freeman, 10-yr.-old daughter of the late astronaut, was awarded a scholarship to the college. (*Houston Post*, 1/22/65; MSC *Roundup*, 2/3/65, 8)

- Federal Aviation Agency announced that Alitalia had reserved three additional delivery positions for the U.S. supersonic transport plane, bringing the Italian carrier's total to six. The new total of reserved positions for the SST was 96; the number of airlines holding positions was 21. (FAA Release 65–12)

- The newspaper *La Mañana* said "flying saucers" had appeared in Uruguay. Several readers had reported saucers zigging and zagging at great speed, and said they "could only be manned space ships." (UPI, Wash. *Daily News*, 1/22/65)

January 22: NASA's TIROS IX successfully injected into a polar orbit by a three-stage Delta rocket launched from Cape Kennedy. The spacecraft was to have gone into a circular orbit about 460 mi. above the earth but the second stage of Delta burned 11 sec. too long and pushed TIROS IX into an elliptical orbit with apogee 1,602 mi., perigee 426 mi., inclination 81.6°, and period 119 min. First NASA attempt to place a satellite in near-polar sun-synchronous orbit from Cape Kennedy involved three dog-leg maneuvers. In a sun-synchronous orbit the precession (westward drift) of the satellite would be about 1° daily, the same rate and direction as the earth moves around the sun.

A hat-box shaped structure, TIROS IX was an 18-sided polygon, 22-in. high, 42-in. in dia., weighing 305 lbs., with one of its flat sides facing earth when initially injected into orbit. Ground signals to the control system tipped the craft up 90° so that it assumed the appearance of a fat wheel rolling on a track around the earth. Two cameras were placed on the perimeter opposite each other so that as the wheel rolled

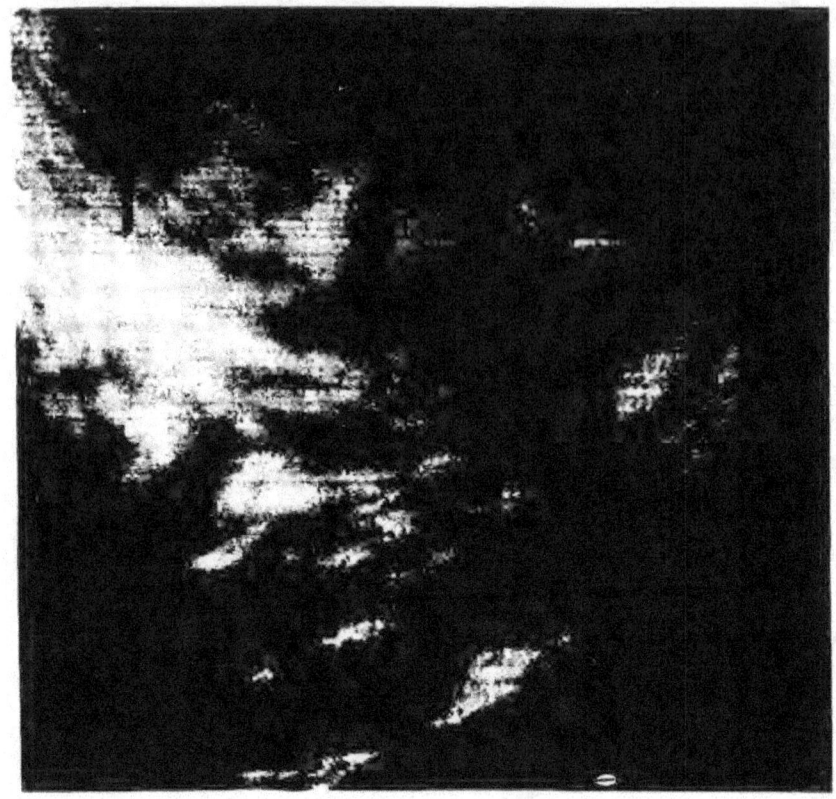

January 22: TIROS IX photograph of ice-covered U.S. Great Lakes area.

at 10 rpm, each camera, in turn, would roll into position and snap a picture, triggered by an infrared horizon sensor.

The combination of TIROS IX's polar orbit (83.4°) and rolling wheel was expected to provide 100% photographic coverage of the earth's cloud cover during daylight hours.

Primary purpose of the TIROS IX launching was to test the new cartwheel concept as a forerunner of a joint NASA-Weather Bureau Tiros Operational System (TOS) of weather satellites. (NASA Release 65–7; *Goddard News*, 1/25/65; AP, Wash. *Eve. Star*, 1/22/65; UPI, Wash. *Daily News*, 1/22/65; AP, Balt. *Sun*, 1/23/65; Appel, *NYT*, 1/23/65, 9; Hixson, N.Y. *Her. Trib.*, 1/23/65)

January 22: A $5,178,000 contract was awarded to a joint venture of Blount Brothers Corp., Montgomery, Ala., and Chicago Bridge and Iron Co., Oak Park, Ill., for a large space chamber to be built at NASA Lewis Research Center's Plum Brook Station. Facility would be used for evaluation and developmental testing of complete spacecraft, as well as nuclear electric power generation and propulsion systems. It would be one of the world's largest space environment chambers (cylindrical chamber 100 ft. in diameter and 122 ft. to the top of its hemispherical dome). (*Lewis News*, 1/22/65, 1)

January 22: On the Les Crane Show (ABC–TV), Dr. Charles S. Sheldon of the National Aeronautics and Space Council staff said in his opening debate statement: ". . . what is the space program?

"It is a program in general science which seeks answers to the most fundamental processes of nature, and will support a great jump forward in our mastery of these forces for human betterment.

"The space program is one of practical applications. . . .

"The space program is one of exploration, opening the whole solar system to the coming generation. . . .

"Space science is neither good nor evil. It is what men choose to do with such knowledge. This country's intent is to develop space for the benefit of all mankind, and space offers us new opportunities for international cooperation. Our hope is space can become a substitute for war by diverting man's restless energies into a supreme challenge of a constructive nature." (Text)

- Maj. Gen. Samuel C. Phillips, Apollo program director in NASA's Office of Manned Space Flight, said that 1965 would be a year of "heavy ground testing" in NASA's lunar program. Among the major events he anticipated were completion of testing of the Apollo spacecraft for the first manned flight; qualification of all elements of the Saturn IB launch vehicle and delivery of first flight stages to Cape Kennedy; and initial testing of Saturn V elements. (NAA *S&ID Skywriter*, 1/22/65, 4)

January 23: Atlas-Agena D launch vehicle with unidentified satellite payload was launched by USAF from Western Test Range. (*U.S. Aeron. & Space Act., 1965,* 132)

- Secretary of Defense Robert S. McNamara announced proposals were being requested from industry for design studies to assist in developing cost and technical information required to proceed with development of the manned orbiting laboratory (Mol). Three contractors would be selected. Decision whether to proceed with full-scale development of Mol would be made upon completion of the design studies. (DOD Release 42–65)

January 24: French scientists bounced laser beams off NASA satellite EXPLORER XXII three times, according to French Ministry of Scientific Research on Feb. 3. Laser beams were reflected from glass prisms on the satellite. (AP, *NYT*, 2/4/65, 3)

- Eldridge H. Derring, Executive Assistant to the Associate Director and head of the Research Staff Office, LaRC, died after an illness of several months. (*Langley Researcher*, 1/29/65, 8)

January 25: President Johnson sent FY 1966 Budget Request to Congress, recommending a total space budget of $7.114 billion. Of this sum, NASA would receive $5.26 billion, DOD $1.6 billion, AEC $236 million, Weather Bureau $33 million, and National Science Foundation $3 million.

The NASA request provided for initiation of a major new project—Project Voyager, budgeted at $43 million—and intensive study of Apollo-X, with funding of $50 million. Hardware development funds were requested for the Advanced Orbiting Solar Observatory ($25.1 million), and the Radio Astronomy Explorer Satellite. Advanced research was reduced by cancellation of development of the 260-in.-dia. solid-fuel rocket motor, the M–1 liquid-hydrogen engine (1.2-million-lb.-thrust), and Snap-8 nuclear electric power unit.

NASA Associate Administrator Dr. Robert C. Seamans, Jr., labeled the budget an austere one, but said the chances of landing a man on the moon by 1970 were still good. In discussing the new programs, Dr. Seamans said the requested $43 million for Voyager would be spent on project definition of the spacecraft bus and landing capsule to explore Mars in the next decade. This funding would also enable NASA to make a Martian fly-by in 1969 to test the spacecraft and launch vehicle prior to the 1971 and 1973 missions.

Major portions of the DOD space budget were allotted for the following: (1) pre-program definition phase of the Manned Orbiting Laboratory (MOL); (2) accelerated research on reentry and recovery of spacecraft; (3) continued development of the Titan III space booster; (4) development of the Defense Communications Satellite System.

Two thirds of AEC's budget request was earmarked for development of nuclear rocket propulsion and nuclear power sources for space applications. The nuclear propulsion program, Project Rover, was alloted $84.1 million; the nuclear power source program, Snap, $70.5 million; and advanced projects applicable to space, $12 million. The Pluto reactor program was not included in the budget request.

The Weather Bureau would start its investment in an advanced weather satellite system in FY 66 with a $500,000 request for sensors and subsystem studies in conjunction with NASA studies. Funds for three Tiros Operational System (TOS) satellites and four Delta launch vehicles to be delivered in two years, $21.6 million, were included in the budget request. Most of the rest was requested for the National Weather Satellite Center (NWSC) and would be spent to convert the present Tiros command and data acquisition facilities to full-time, operational centers run solely by the Weather Bureau.

President Johnson asked Congress for $650 million as a White House contingency fund to meet the possible need to accelerate supersonic transport development. (Text, *M&R*, 2/1/65, 10–17; Text, *NYT*, 1/26/65, 26–28; *Av. Wk.*, 2/1/65, 16–17; NASA Budget Briefing FY 1966)

January 25: President's message sending budget for Fiscal Year 1966 included the following remarks: "Space research and technology: This Nation has embarked on a bold program of space exploration and research which holds promise of rich rewards in many fields of American life. Our boldness is clearly indicated by the broad scope of our program and by our intent to send men to the moon within this decade.

"The costs are high—as we knew they would be when we launched this effort. We have seen a rise in annual expenditures for the space program from less than one-half billion dollars in 1960 to over $4 billion in 1964.

"Expenditures are continuing to increase. However, we have built up momentum and are concentrating on our highest priority goals. Therefore, we will no longer need to increase space out-lays by huge sums each year in order to meet our present objectives.

"This budget proposes that expenditures increase by $22 million in 1966 over 1965. This is the smallest annual increase since 1959. The new obligational authority requested is about the same as enacted for 1965." (NASA LAR IV/16)

January 25: NASA Administrator James E. Webb and DOD Secretary Robert S. McNamara announced NASA–DOD agreement on the Manned Orbiting Laboratory (Mol), released in conjunction with FY 1966 budget:

". . . Planning for the Defense manned orbiting laboratory program will also consider, in cooperation with NASA, broader objectives of scientific and general technological significance.

"To determine the essential characteristics of the vehicle that will be required, the DOD will continue intensive studies and design of experiments and systems aimed at the primary military objectives.

"Cooperative studies, by NASA and Defense, will identify and define scientific and general technological experiments which might be carried out, with NASA participation, in conjunction with the military program.

"DOD, with assistance from NASA, will compare configurations of Apollo which may be suitable for military experiments with the Gemini B–MOL configuration to determine the complete system that can meet the primary military objectives in a more efficient, less costly, or more timely fashion.

"On the basis of these studies, a decision will be made whether to proceed with full-scale development by Defense of a manned orbiting laboratory system and what the specific developments and vehicle configurations are to be. The Defense budget includes $150 million in FY 1966 for the program. . . .

"Depending upon the manned orbiting laboratory decision, upon the progress in the Gemini and Apollo programs, and upon the results of NASA studies, a decision will be made whether to proceed with modifications to the Apollo system and the nature and timing of necessary specific developments. The NASA 1966 budget includes about $50 million for proceeding with design and pacing developments. . . ."
(NASA Budget Briefing FY 1966)

• NASA Associate Administrator Dr. Robert Seamans said during FY 1966 budget briefing: ". . . it is conceivable . . . that the lunar landing would occur in early 1970 . . . we feel actually greatly encouraged at the progress that has been made freezing the design, and we feel very reassured at the test results we are achieving on our propulsion systems and with our stages. So that we really feel that there is more chance that we can get off the flight on an earlier mission than I would have said a year ago."

Dr. Seamans said Apollo gave the nation a capability for a wide variety of scientific and technological flights in earth orbit, in orbit around the moon, and also for an extended lunar stay time. He commented that the objectives of the current extended Apollo (Apollo-X) design and feasibility studies were to extend the time of the lunar mission out to the order of two weeks. He also said that Apollo-X circumlunar flights, in polar orbit about the moon and taking photographs of the entire lunar surface, on missions that could involve staytimes on the moon of up to one or two weeks, all would have great possibility and would offer great interest scientifically. In commenting on an earlier agreement (1963) with the Pentagon for developing of a manned orbiting laboratory, Dr. Seamans said: "At the time of that agreement, we were really thinking of something that we now realize is further out in time, namely, a more permanent space station that could stay in orbit for a year's time and could be resupplied, and would permit the crew to be ferried into orbit and bring them back. The study really related to that kind of possibility which we now realize is much further out in time . . . we may end up with what is called the MOL, and we may also find that there are important uses for the Apollo system beyond the present manned lunar landing program."

He said NASA studies of improving both the Saturn IB and the Saturn V launch vehicles indicated that "these two launch vehicles can take care of our needs for an extended period of time." (NASA Budget Briefing FY 1966)

January 25: NASA announced two Radio Astronomy Explorer satellites (RAE–A and RAE–B) would be designed to investigate low-frequency (long wavelength) emissions from our galaxy, its planets, and the stars. These emissions are mostly intercepted by the ionosphere so that little can be learned about them from ground-based receivers. This would be the first attempt to map the galaxy for low-frequency emissions. The 280-lb. spacecraft would be launched by Thrust-Augmented Delta into circular orbits at altitudes of about 3,700 mi. and would measure the intensity of the signals, their frequency, times of emission and, within limitations, define the regions of space in which they originated. Proposed designs called for the development of two 750-ft., V-shaped antennas that would be mounted opposite each other, forming a giant X. They would be anchored to the basic spacecraft, a cylinder of about 40-by-40 in., capped by two truncated cones. NASA Goddard Space Flight Center would design, integrate, and test the two spacecraft. First launch was not expected before 1967. (NASA Release 65–20)

- AEC announced that the Snap-10A nuclear generator designed for spacecraft had produced electricity for the first time in a ground test at Canoga Park, Calif., by its builder, Atomics International. The system would ultimately provide power for spaceship propulsion systems such as the ion engine. (AEC Release H–18; *Wash. Post,* 1/26/65)

- Univ. of Miami, Coral Gables, Fla., bestowed an honorary dectorate upon NASA Administrator James E. Webb. Mr. Webb said in a speech there:

    ". . . space science and technology are not remote and esoteric pursuits but rather are deeply woven into the fabric of our society. The space scientist does not practice a new art. He is an astronomer, a physicist, a chemist, a geologist, rooted in our university system of vigorous effort to expand our knowledge of the universe in which we live. The space technologist is an engineer of materials, structures, fuels, power sources, electronics, rooted in our industrial and government laboratory systems. Both, however, are directing their interests and talents to the newest and most exciting frontiers—where the most rapid progress is made and the breakthroughs scored. The knowledge they gain feeds back into our scientific and technical communities and into our industrial laboratories. . . . Thus, the talents, the skills, and the funds for space exploration are all drawn broadly from our society and continue to feed back into it, in the forefront of scientific and technical progress—the unique hallmark of the American way." (Text)

January 25: ComSatCorp filed with the FCC its intent to contract for 24 satellites that could be used by ComSatCorp to provide a global communications service for DOD. The satellites would be made available for three launchings which DOD had slated for the early part of 1966 on either a Titan IIIC or Atlas-Agena D launch vehicle. DOD would pay only for service rendered following successful launch, with ComSatCorp assuming the risk if the satellites did not work satisfactorily in orbit. This proposal was separate from the program managed by ComSatCorp to develop an international commercial communications satellite system. (ComSatCorp Release)

- USAF was reported to be considering the use of surplus Wing 1 Minuteman ICBMS as Guidance Error Analysis Vehicles (Geav). According to Air Force Central Inertial Guidance Test Facility (CIGTF), surplus Minuteman boosters could be the cheapest means to evaluate future inertial guidance systems in a true missile environment. ETR was selected for Geav because no other range could measure missile velocity in three axes to the required accuracy. The Minuteman guidance system itself would be reprogramed and located in a recoverable payload for reasons of economy. (*M&R*, 1/25/65, 34)
- Dr. A. J. Drummond of Eppley Laboratory, Newport, R.I., told *Missiles and Rockets* that a number of Russian cosmonauts were said to have died in booster failures at launch. Dr. Drummond got his information through unofficial sources while attending a technical meeting in Leningrad last year. He also said there were no large solar-simulation testing facilities in the Soviet Union and that Soviet spacecraft used crude bulk insulation for thermal control instead of emission-absorption coatings. (*M&R*, 1/25/65, 7)

January 26: USN fired a Hydra-Iris sea-launched sounding rocket to 184-mi. altitude carrying a 100-lb. payload. The rocket was launched from a point about 1,400 mi. east of Montevideo, Uruguay. Mission was to measure radiation intensity within the inner Van Allen radiation belt. (*M&R*, 2/8/65, 8)

- The first J-2 liquid-hydrogen rocket engine built to flight configuration was delivered to Douglas Aircraft Co., Sacramento, for installation and testing in the Saturn S-IVB battleship stage. The 200,000-lb.-thrust engine had been recently accepted by NASA from Rocketdyne Div., North American Aviation, Inc. (*Marshall Star*, 1/27/65, 1, 6)
- Dr. John D. Nicolaides, Chairman of Notre Dame's Aerospace Engineering Dept., formerly Special Assistant to the NASA Associate Administrator for Space Science and Applications, told National Space Club at a Washington, D.C., luncheon that we must realize we were "not yet first in the race for space supremacy. . . . The [Soviet] lead in both numbers and weights of unmanned launchings continues to increase. They are publishing just as many scientific papers as we are and they are just as good." Nicolaides added that he was not including their work in life sciences "which is well ahead of ours by virtue of the simple fact that they have been experimenting in space."

Dr. Nicolaides said he was alarmed by the U.S.S.R.'s "extensive planetary program." They started early and continued a truly massive effort compared to ours, he said. "They are launching their heavy spacecraft at each opportunity to both Mars and Venus, while we have abandoned Venus completely and are only studying scientific measurement on Mars in 1971. . . ." (NSC *Newsletter*, 2/65)

January 26: USAF selected Thiokol Chemical Corp. and Lockheed Propulsion Co. to develop and test new 156-in.-dia. solid-propellant motors during 1965. Lockheed would develop two of the three motors. The first would be a flight-weight motor with thrust in excess of three million pounds. The second motor would be in the one-million-pound-thrust class and would incorporate a submerged nozzle. Both motors would use advanced liquid injection thrust vector control to explore methods of guiding huge motors of this size. Thiokol's Wasatch Div. of Brigham City, Utah, would develop the third motor. This flight-weight motor would have a thrust of over 320,000 lb., and incorporate a deeply submerged nozzle permitting the total motor length to be under 21 ft. (DOD Release 52-65)

- At the AIAA Aerospace Sciences Meeting and Honors Convocation in New York, awards were made to men who had made valuable contributions to development of the aerospace industry:

  Dr. Eugene N. Parker, associate professor at the Enrico Fermi Institute of Nuclear Studies, Univ. of Chicago, received the Space Science Award "for distinguished individual research on the causes and properties of the solar wind."

  Arthur E. Raymond, responsible for the design of the Douglas DC series of commercial transports received the Sylvanus Albert Reed Award. He was honored for "numerous and distinguished contributions to the aeronautical sciences and the development of aircraft during the last 30 years."

  Igor I. Sikorsky and Michael Gluhareff were given the 1964 Elmer A. Sperry Award. Mr. Sikorsky was cited as a helicopter pioneer for "the concept and development of a new form of aerial transportation capable of carrying and placing large external loads over any terrain." Mr. Gluhareff was honored for his engineering contributions in the development of the multipurpose helicopter.

  Dr. Wallace D. Hayes, professor of aerospace engineering at Princeton University, received AIAA's fourth annual Research Award for his leading role in the development of supersonic and hypersonic flow theory.

  Sir Frank Whittle, British engineer, was named first recipient of the Goddard Award for his "imagination, skill, persistence, and courage in pioneering the gas turbine as a jet propulsion aircraft engine, thus revolutionizing military and commercial aviation for all time."

  Harry F. Guggenheim, who had supported aerospace endeavors, received a special commendation for his "contributions, encouragement, and personal participation in the development of aviation and rocketry." (*NYT*, 1/21/65, 53M; *NYT*, 1/27/65, 58; *NYT*, 1/9/65, 50; *Av. Wk.*, 1/25/65; *Av. Wk.*, 1/11/65, 13; *Langley Researcher* 1/29/65)

- An article by Omer Anderson on U.A.R. rocket program was inserted in the *Congressional Record* by Rep. Silvio Conte (R-Mass.). Based on interviews with German scientists just back from Egypt and with

West German defense ministry officials who debriefed them after their return, the article said: "Egypt's missile program is considerably further advanced than is generally realized in the West.

"Some of these scientists who have returned to West Germany say that Nasser will have the missiles to devastate wide areas of Israel by late 1967 and that he will have rockets with a 1-ton payload by the end of 1965.

"West German defense ministry experts who have questioned the returning rocket scientists regard their assessment of Nasser's rocket potential as entirely realistic and possibly too conservative.

"The scientists say Nasser has accelerated greatly his rocket program since the first test firing of four missiles on July 22, 1962." (CR, 1/26/65, 1160)

January 26: Thompson-Ramo-Wooldridge Space Technology Laboratories chosen as the winner in a two-year design competition to produce the rocket engine for Apollo Lunar Excursion Module (Lem). The liquid-propellant engine was designed to vary its power output between a low of 1,000 lb. thrust and a high of 10,000 lb. (*NYT,* 1/29/65)

• House Committee on Science and Astronautics began a two-day seminar with a panel discussion on science and technology, with specific reference to aeronautics. Speaker of the House John W. McCormack opened the seminar.

Senate Committee on Aeronautical and Space Sciences began executive hearings on the subject of launch vehicles. (NASA LAR IV/17)

• In U.S. launch vehicles hearings before Senate Committee on Aeronautical and Space Sciences, NASA Administrator James E. Webb discussed recent study by the Aeronautics and Astronautics Coordinating Board's Launch Vehicles Panel:

"In considering the merits of canceling certain vehicles in order to provide quantity production of the remaining vehicles, the Launch Vehicle Panel of the AACB evaluated several alternatives against a forecast of DOD's and NASA's needs over the next 10 years. This space-mission forecast served as a basis for determining the number of launch vehicles required and the cost of producing the various combinations of these launch vehicles.

"The result of the study is particularly interesting in that it shows a cost difference of less than 1 per cent among the alternative options. This difference is less than the accuracy of the data used in the analysis. The results indicate that any economies that might be realized by increased quantity production of boosters would be lost through cost of adapting specific mission spacecraft to a new vehicle where the costs of such work have already been incurred. . . .

"The major advantages of the recent comprehensive study . . . , as distinct from previous reviews, were the development of much improved methods for estimating the costs of launch vehicles considering the effects of quantity production, variety of vehicles, and inplant workload; the use of an inclusive or overall forecast as a basis for determination of both DOD and NASA space missions against which total launch vehicles costs could be calculated; and the value of the results of the study to NASA to confirm our judgment on the use of the SATURN I–B for the APOLLO and VOYAGER missions.

". . . we are making extensive use of DOD-developed launch vehicles and will continue to do so for some time to come. However, a wider variety of first-stage boosters and upper stages is required by NASA space missions than by those of the DOD. We have requirements for a wider range of variety of size, payload, and velocity for our missions. We have been carefully investigating our future vehicle needs; optimum vehicle configurations; and the most promising advanced propulsion methods to be sure that our program will provide the options that the country will need in making decisions to undertake future missions.

". . . we are utilizing the channels and procedures established by the DOD–NASA launch vehicle agreement and by the AACB to coordinate the needs and activities of NASA and the DOD to assure the most effective national launch vehicle program. However, we are presenting to the Congress, in our budgets each year, the specific booster needs we have over and above those which can be met by DOD-developed systems. . . ." (*Hearings . . . National Space Launch Vehicles*, 6–19)

January 26: USN began tests of two new air-cushion vehicles variously called hydro-skimmers, hovercraft, or ground effects machines. The craft were lifted a few feet above the surface by cushions of air trapped beneath their hulls and were driven at speeds up to 50 knots by aircraft propellers. The vehicles would be tested during the next three to six months to determine their potential usefulness and operational suitability for naval operations. (Baldwin, *NYT*, 1/31/65, 88)

- Federal Aviation Agency (FAA) Administrator Najeeb E. Halaby told the House Science and Astronautics Committee that designs of U.S. manufacturers for the proposed supersonic airliner "demonstrated clearly the feasibility" of building a plane that would prove as profitable, if not more so, over transcontinental or greater ranges as current jet airliners. Presidential Committee to evaluate Sst program would begin its extensive critical review late next month. (Clark, *NYT*, 1/27/65, 19)

- Sen. A. S. Monroney (D–Okla.) suggested in a speech before the Aero Club of Washington that the experimental RS–70 bomber be used as a test plane for U.S. supersonic transport program. Monroney was interested in more extensive use of the RS–70 for civil airliner studies than had been made by NASA. He said use of the RS–70 could produce savings in both development and construction costs of the proposed airliner. (Sehlstedt, Balt. *Sun*, 1/27/65)

- William C. Foster, Director of U.S. Arms Control and Disarmament Agency, said in testimony before House Foreign Affairs Committee that radioactive leakage from Soviet underground nuclear test Jan. 15 was apparently accidental. The radioactive fallout apparently did not violate the intent of the 1963 nuclear test-ban treaty. (*FonF*, 1965, 61)

- The British Defense Ministry announced that its fleet of Valiant bombers would be scrapped because of weakened structure caused by metal fatigue. Valiant was the first of the three "V" types of jet bombers built by U.K. following World War II. They had been in service nine years and only about half the original force remained in service as reconnaissance or tankers. This would not affect Britain's contribution to NATO or its proposal for an Atlantic nuclear force. (*NYT*, 1/27/65)

January 26: Richard E. Horner, former Associate Administrator of NASA (1959–60), was installed as 1965 president of the American Institute of Aeronautics and Astronautics. (*Av. Wk.*, 2/8/65, 13)

January 27: President Johnson sent to the House and Senate his message transmitting annual report on the U.S. space activities. In his letter, President Johnson said: "The advances of 1964 were gratifying and heartening omens of the gains and good to come from our determined national undertaking in exploring the frontiers of space. While this great enterprise is still young, we began during the year past to realize its potential in our life on earth. As this report notes, practical uses of the benefits of space technology were almost commonplace around the globe—warning us of gathering storms, guiding our ships at sea, assisting our mapmakers and serving, most valuably of all, to bring the peoples of many nations closer together in joint peaceful endeavors.

"Substantial strides have been made in a very brief span of time—and more are to come. We expect to explore the moon, not just visit it or photograph it. We plan to explore and chart planets as well. We shall expand our earth laboratories into space laboratories and extend our national strength into the space dimension."

A hypersonic aircraft—one that could fly the Atlantic in less than an hour—had reached the stage where models were being constructed for wind tunnel tests. President Johnson's report disclosed: "Two structural models embodying design concepts applicable to the fuselage of a hydrogen-fueled hypersonic aircraft were being constructed for testing at 1,500–2,500° F—temperatures likely to be encountered in hypersonic flight. Equipment was developed for inducing angular oscillations in the test section flow of a large transonic wind tunnel and will be used to obtain the dynamic response of wind tunnel models." (*CR*, 1/27/65, 1366; *U.S. News*, 2/2/65)

- NASA launched a Nike-Cajun with acoustic grenade experiment at Point Barrow, Alaska, to obtain upper atmospheric meterological data within the Arctic Circle. 12 grenades were ejected and detonated at intervals from about 25 to 56 mi. altitude as the rocket ascended. By recording the sounds on five sensitive microphones on the ground, scientists could obtain wind direction and velocity, atmospheric temperature, density, and pressure data. This was the first of a series of such experiments to gather upper atmospheric data within the Arctic Circle. Point Barrow was 1,100 mi. from the North Pole and 300 mi. within the Arctic Circle, at 71° north latitude. (Wallops Release 65–4; AP, *NYT*, 1/29/65)

- NASA Langley Research Center requested G. T. Schjeldahl Co. to submit a bid for construction of six inflatable 100-ft., 130-lb. spherical satellites to be used in the national geodetic satellite program. They would be nearly identical to ECHO I and would be named Pageos (Passive Geodetic Satellite).

  Pageos would be launched in 1966 into a near-polar orbit at an altitude of about 2,300 mi. Ground camera stations would simultaneously photograph it against a star background to gather precise data for locating any point on Earth.

The other two types of spacecraft to be used in the geodetic satellite program would be the 350-lb. Geos and the 120-lb. Beacon Explorer-B. (NASA Release 65–22; Beacon Explorer-B Press Kit)

January 27: NASA Ames Research Center discussed for the press the major significant advances in aeronautics and space-oriented research accomplished by the Center during 1964. Accomplishments cited were: (1) establishing feasibility of manned control of large boosters; (2) problem definition for hypersonic transport; (3) design of new take-off and landing aid to precisely locate aircraft position on the runway; (4) design of probe vehicle to define Mars atmosphere; (5) improvement of M–2 maneuverable atmosphere entry craft; (6) discovery that Mars contamination problem is probably not severe; (7) development of system for measuring stress in humans; (8) demonstration of need for special training for jet transport pilots to combat severe air turbulence; (9) discovery in meteorite of an extraterrestrial mineral unknown on earth; (10) feasibility demonstration of moon and planet mission navigation by hand-held sextant; (11) formulation of certification requirements for supersonic transport take-off; (12) design of ducted-fan to provide efficient airflow for flight from hover to high subsonic speeds; (13) formation of organic material under Martian conditions; (14) development of a new magnetic field chamber; (15) derivation of formula for simple calculation of convective (friction) heating of spacecraft in planet atmospheres; (16) tests of radiative heating in simulated planet atmospheres; (17) improvement of techniques for prediction of heat shield performance; (18) development of a low-power, high-performance magnetometer; (19) measurements of solar wind on IMP–B and OGO II; (20) feasibility demonstration of new pod for vertical-lift engines for Vtol aircraft at flight speeds up to 170 mph. (ARC Release)

- Experiments at NASA Ames Research Center by Dr. John Young and Dr. Cyril Ponnamperuma indicated that Mars may lie under a steady rain of edible sugars produced photochemically in the Martian atmosphere. It was speculated that the sugars and other compounds might drift to the Martian surface, seep into the soil, and form underground reservoirs of nutrients.

  Results of tests for survival of 50 strains of earth bacteria in simulated Martian atmosphere indicated that the strains of bacteria which form hard spores and are thus most likely to survive space flight are most sensitive to the freeze-thaw extremes of temperature that prevail on Mars. Thus, while the bacteria might survive on Mars in spore form, they would not grow there and would not contaminate the planet. Other bacteria would die en route. (*S.F. Chron.*, 1/28/65)

- In U.S. launch vehicles hearings before Senate Committee on Aeronautical and Space Sciences, Dr. Alexander Flax, Assistant Secretary of the Air Force (R&D), said:

  "In general, the joint DOD–NASA study [by the AACB] has shown that no drastic revisions to the national launch vehicle family are required to meet the mission demands that we can project for the immediate future and that further no drastic revisions can be justified on purely economic grounds. In addition, it is clear that the extensive effort on the part of both the DOD and the NASA in improving, launch vehicle system reliability has been paying off, and that we can expect a continuing trend in this regard. However, it is important to also recognize some of the natural limitations inherent in any long-range projection of requirements such as we have had to make for this study period. There are a number of areas in which unforeseen increased mission capability demands could react on our launch vehicle performance requirements. We must, therefore, continually maintain effective exploratory and advanced development programs which will provide us with the technology to meet such demands in the future. . . ." (*Hearings . . . National Space Launch Vehicles*, 87)

January 27: William M. Allen, president at the Boeing Co., addressed National Defense Transportation Association in Washington, D.C.: "Our first Boeing study of a supersonic transport was made in 1952. Preliminary design effort was started more seriously in 1956 and 1957. Then in 1958 the SST became a major engineering project and it has continued in that status ever since, involving many of our top engineers.

"From the start of our effort to the present, design determinations have come in an orderly and unhurried progression, as a result of the integration of mountains of test data, much of it worked out in close conjunction with NASA laboratories which, incidentally, deserve the sincere thanks of the American people for their pioneering work in this field.

"In the process we explored 290 configurations, and completed wind tunnel testing on 56 different high-speed wings. . . ." (*CR*, 1/28/65, 1454–56)

- Gen. Bernard A. Schriever (USAF), Commander of the Air Force Systems Command, described to members of the Charlotte, N.C., Chamber of Commerce the development of the U.S. ballistic missile program: "In the ballistic missile program, of course, we were not concerned with the missile alone—complicatd as it was—but also had the problem of constructing the facilities to test the missiles; building the ground support equipment; and training crews to install, service, and launch the missiles. This was a $17 billion program, and was larger in scope than the Manhattan Project which produced the atomic bomb during World War II.

  "To give you some idea of the size of the task, imagine that Henry Ford in the early days had not only had the problem of designing and building his automobiles, but at the same time had to construct all the highways and bridges, build all the service stations and garages, and plan and conduct driver training programs. . . ." (Text)

- R. E. Clarson, Inc., St. Petersburg, Florida, was awarded a $2,179,000 NASA contract for miscellaneous additions and changes at Launch Complex 34 for the Saturn IB. Work would be done at Cape Kennedy, Florida. The contract was awarded by the Army Corps of Engineers. (DOD Release 53–65)

- National Science Foundation announced that an ocean area 100 mi. NNE of Maui Island of the Hawaiian Islands had been tentatively selected as the site for the attempt to drill a six-mi.-deep hole into the ocean bottom to penetrate beyond the earth's crust. The operation would be called Project Mohole and would be an attempt to gain knowledge of the earth's origin and structure, the formation of minerals, and the causes of earthquakes. Drilling was expected to begin in 1968. (Clark, *NYT*, 1/28/65, 50; AP, *Wash. Post*, 1/28/65)

January 27: Stellar objects dubbed "interlopers" had been discovered by the Mt. Wilson-Palomar Observatories in California. Dr. Allan R. Sandage of Mt. Wilson said an effort would be made to determine whether the new objects were a form of quasar. He said an alternative possibility was that the objects were a rare form of star system in which two stars lay so close to each other that the presence of one caused explosions on the surface of the other. The resulting strongly ultraviolet light would superficially resemble that of quasars. Dr. Sandage reported that so far about 45 quasars had been identified. The "interlopers," so called because of their close resemblance to quasars, had been found at the rate of two to a square degree of sky in the limited region studied. So far, they totaled four. (Sullivan, *NYT*, 1/27/65, 31)

- J. Gordon Vaeth of the U.S. Weather Bureau's National Weather Satellite Center told the American Meteorological Society that the Weather Bureau was developing a system in which buoys moored in the ocean would broadcast weather data to communications satellites that would rebroadcast it almost instantaneously to almost any point on earth.

    Mr. Vaeth said the initial optimum number of buoys would be 300, spaced about 600 mi. apart in major ocean regions. They would be moored at known, fixed points and would send data on sea and air temperature, wind direction and velocity, and barometric pressure. Relays from the satellites would be by very-high-frequency radio and would be picked up on the ground by inexpensive receiving stations, aircraft, or ships at sea.

    Mr. Vaeth saw the buoy network as an ideal vehicle for international cooperation in meteorology. (Schmeck, *NYT*, 1/28/65, 50)

- France announced it would build a launching site for spacecraft in French Guiana, on the northeast coast of South America, to be ready Jan. 1, 1968. (Reuters, *Wash. Post*, 1/28/65)

- USAF said in its *Project Blue Book* that no Ufo "has ever given any indication of a threat to our national security" or displayed "technological developments or principals beyond the range of present day scientific knowledge." Report covered 8,908 sightings during past 18 yrs, including 532 during 1964. (Noyes, Wash. *Eve. Star*, 1/27/65, 28)

- A new theory for the behavior of matter, called SU–6, was presented in New York at the annual meeting of the American Physical Society by Dr. Abraham Pais of the Rockefeller Institute. The concept, based on a branch of mathematics known as symmetry group theory, supported views that all matter might be composed of basic building blocks, or "quarks," that could be either real fragments or mathematical entities smaller than the electron. It grouped the 100+ known fragments of matter into groups and then predicted behavior. A modification makes the theory also compatible with Einstein's relativity theory. (Sullivan, *NYT*, 1/28/65, 1, 10)

January 28: MARINER IV, launched two months ago, was 11,873,789 mi. from earth and moving toward Mars at a speed of 12,291 mph relative to the earth at 9 a.m. EST. Velocity relative to the sun was 67,086 mph. (NASA Release 65–21)

- The first major Saturn V flight component, a 33-ft.-dia., 60,000-lb. corrugated tail section which would support the booster's five 1.5-million-lb.-thrust engines, arrived at NASA Marshall Space Flight Center from NASA Michoud Operations, near New Orleans. The section was one of five major structural units comprising Saturn V's first stage. (*Marshall Star*, 1/27/65, 1)
- USAF announced a four-stage Blue Scout Jr. rocket combination with a scientific payload had failed after launch from Cape Kennedy. The second stage developed trouble about 100 sec. after launching, causing the range safety officer to send destruct signal. The stage broke apart on its own. The third stage, meanwhile, separated from the second stage, ignited, and followed approximately its preplanned path. The fourth stage failed to ignite; it and the payload plummeted harmlessly into the Atlantic Ocean southwest of Ascension Island. The probe was to have sent its instrumented payload 24,500 mi. into space to study earth's magnetic field. (*NYT*, 1/29/65; *U.S. Aeron. & Space Act., 1965*, 132)
- Construction work at Cape Kennedy halted as 3,700 building trade workers stayed off the job in a two-year-old contract dispute with NASA. The present dispute was between building trades unions and the Marion Power Shovel Co., a NASA contractor, over pay scales. Work on 44 projects involving contracts totaling $192 million had been brought to a standstill. The biggest project affected was the 52-story Saturn V moon rocket assembly building that was to be ready for the first of these rockets within two years. (UPI, *NYT*, 1/29/65, 6; AP, *Houston Post*, 1/29/65)
- President Johnson, on the advice of Defense Secretary McNamara, and contrary to the opinion of the Joint Chiefs of Staff, had decided to postpone the production order for the Nike-X missile defense system, Neal Stanford of the *Christian Science Monitor* asserted. $2 billion had already been spent on the R&D phase of the Nike-X and an additional $20 billion would be required to produce and deploy it. The FY 1966 budget provided approximately $400 million for continued research and development on the Nike-X system pending the decision on whether to put Nike-X into production. (Stanford, *CSM*, 1/28/65)
- Army XV–9A experimental pressure jet helicopter, which was first flown on November 5, 1964, gave its first public demonstration in Culver City, Calif. It was designed and developed under a U.S. Army Transportation Research Command contract with the Hughes Tool Company to evaluate the hot-cycle pressure jet system which would eliminate the requirement for heavy gear boxes, complex mechanical drive components, and an antitorque tail rotor. Aircraft based on this concept could carry payloads greater than the empty weight of the aircraft itself. (DOD Release 55–65)

January 29: AEC said in its Annual Report to Congress that the United States now had four Vela satellites in distant orbits to detect nuclear explosions in space. Two more would be launched this year. U.S. facilities in the Pacific had been brought to a state of instant readiness to resume atmospheric testing should the Soviet Union violate the limited nuclear test ban treaty. (AEC *Annual Report*, 76–77)

January 29: Speaking on the Senate floor, Sen. Richard B. Russell (D-Ga.) said: "I am greatly disturbed that funds for the continuation of the large solid rocket engine program have been eliminated from the 1966 budget for the National Aeronautics and Space Administration.

". . . I am concerned about the effect that the proposed termination of this program will have over our long-range space effort and upon the security of the country. For it will cut off, at a particularly inappropriate time, a crucial research and development program that already has shown significant potential for fulfilling future space booster needs—for both defense and nondefense purposes. This is particularly true for launching large payloads and missions into deep space that are contemplated in the not-so-distant future.

"The booster technology and capability that we are developing under the large solid rocket engine program could become a vital factor in preventing the Russians from achieving a position of dominance in space. . . .

"Indeed, the decision to terminate this program appears to be a direct contradiction of Mr. Webb's own views, as expressed as recently as Tuesday of this week to the Committee on Aeronautical and Space Sciences. He said unequivocally that space missions contemplated for the next decade and beyond will require 'new launch vehicles and new space vehicle developments.' He said our experience with the Apollo moon program has shown that 'a policy of support for the development of carefully selected advanced propulsion systems must be followed if we are to assure they will be available when needed.'

"It is highly inconsistent—to say the least—to speak boldly of exploring the moon, reaching and charting the planets, establishing manned stations in space, and extending our national strength into the space dimensions, while at the same time killing off one of the most promising programs for the achievement of these very goals. . . ." (*CR*, 1/29/65, 1535)

- NASA approved a contract with the McDonnell Aircraft Corp. converting the $712-million Gemini spacecraft contract from a cost-plus-fixed-fee to a cost-plus-incentive-fee. This was the largest incentive contract that NASA had negotiated; it provided profit incentives for outstanding performance, control of costs, and timely delivery as well as potential profit reductions when performance, cost, and schedule requirements were not met. (NASA Release 65–26)

- The National Commission on Technology, Automation and Economic Progress, established by law in 1964 to find out what technological change was doing to the economic and social fabric of the country and how to obtain its maximum benefits with the least possible harm, met for the first time with Vice President Hubert Humphrey. The Commission would meet again Feb. 18–19 to determine the areas to be explored intensively and possibly to select outside personnel to help with basic research in these studies. (*NYT*, 1/30/65, 6)

January 29: C-141A was certified as a commercial cargo jet. FAA Administrator, Najeeb E. Halaby, said the Lockheed fanjet could "help give civil freight transportation the kind of mobility that brings success to the armed forces." The plane could operate at 550 mph, and needed only a 6,000-ft. runway. The certification climaxed an unusual program in which the FAA, USAF, and industry had jointly developed, produced, and tested the new craft. (AP, *NYT*, 1/31/65, 27)

- U.S. Army formally accepted the first two XV-5A V/Stol (Vertical/Short Take-Off and Landing) lift-fan research aircraft at Edwards AFB where they were being readied for a six-month Army flight evaluation. Test pilots from NASA, USAF, USN, and the FAA would assist in the evaluation. (DOD Release 59-65)

January 30: COSMOS LIII, an unmanned satellite containing scientific equipment for outer space research, was orbited by the Soviet Union. Preliminary orbital data: period, 98.7 min.; apogee, 741 mi. (1,192 km.); perigee, 141 mi. (227 km.); inclination, 48.8°. Equipment on board was operating normally. (Tass, *Pravda*, 1/31/65, 4, ATSS-T Trans.)

- Funeral services for Sir Winston Churchill were televised live and by delayed transmission from London via TELSTAR II communications satellite. Churchill died on Jan. 24. Earlier in the week, pictures of Sir Winston's body lying in state in Westminster Hall had also been transmitted live via TELSTAR II. (NBC; CBS; *Wash. Post*, 1/27/65)

- NASA Ames Research Center was conducting tests on a Douglas F5D aircraft with a specially designed planform wing that might minimize landing speeds for the proposed supersonic transport. A tornado-like flow, called "vortex airflow," and resulting from the sharp difference between the low pressure on the top of the wing and the high pressure on the underside, was generated along the leading edges of the "S"-shaped wing. Engineers said use of the sharply angled wings with tornado effect on top had these advantages: (1) the tornadoes affected air flow over the entire aircraft and eliminated turbulence that would make other aircraft directionally unstable when coming in nose high for a landing; (2) the tornadoes made it almost impossible for the wings to lose their lift completely. Also, it was felt this wing shape took maximum advantage of the cushioning effect produced in compressing air between the underside of the wings and the ground which would make it necessary to level off sharply at the last moment before touching down.

    Existence of this tornado-like flow along the leading edges of the wing encouraged the belief that a supersonic airliner might be built without resorting to variable-sweep wings.

    In current design competition for supersonic transport under Government auspices, the Lockheed Aircraft Corp. had taken the first approach. The Boeing Co. had a design with movable wings. (ARC Release 65-3; Witkin, *NYT*, 1/30/65)

- An article published in *The New Scientist* reported that experts at the Royal Radar Establishment at Malvern, England, believed that the U.S. communications satellite ECHO II—launched Jan. 25, 1964, and still in orbit—had been pierced by its own launching canister shortly after injection into orbit.

According to *The New Scientist*, the shape of ECHO II after launching was flabby and elongated rather than the perfect sphere wanted for some of its communication experiments.

Analysis of Malvern's radar tracks on ECHO II revealed writhing echoes that, according to their theory, arose when the very short radar pulses entered a hole and rebounded from the aluminum-coated interior of the balloon. The Malvern team thought the balloon had a puncture about 18 in. long and 27 in. wide in one side.

NASA spokesmen said they did not believe ECHO II had been punctured by either its launching canister or its launching vehicle and that sightings from more than a dozen radar stations had contradicted the Malvern theory. They added that ECHO II's ability to reflect radio signals had not been seriously impaired and many messages had been bounced off in the last year. (Hillaby, *NYT*, 1/31/65, 29)

January 30: Soviet Union launched a "new type" space booster that spanned more than 8,000 mi. of the Pacific, according to Tass. The firing was said to have been so successful that a second planned shot was canceled. (*M&R*, 1/8/65, 8)

- Dr. Joseph Charyk, president of Communications Satellite Corporation, speaking in Kaanapali, Hawaii, said the geographic location of Hawaii ensured that the impact of Early Bird comsat would be "more profound there than in any of the other states of the union." Hawaii, he noted, would not have to wait, as it does now, to see mainland television programs. Dr. Charyk envisioned a full global communications system by 1967. He predicted Hawaii would become a center for communications traffic of all types. (*NYT*, 1/31/65, 13)

January 30: Columnist James J. Haggerty, Jr., said: "It is all but incredible that after seven years of space research no manned military project has reached the hardware stage. . . ." (Haggerty, *J/Armed Forces*, 1/30/65, 9)

January 31: Seventh anniversary of the first U.S. satellite, EXPLORER I. In defiance of the original predicted lifespan that should have ended some two years ago, the satellite continued to pass overhead every 104 min., with perigee of 214 mi. and apogee of 983 mi. Trajectory plotters at NASA Marshall Space Flight Center believed EXPLORER I would plunge into the atmosphere and burn in 1968. It had slowed down since launch but had logged 904 million mi. around the earth. (*Marshall Star*, 1/27/65, 1, 6)

- Japan launched Lambda III-2, the largest rocket that country had yet developed. The 62-ft., three-stage rocket attained an altitude of 620 mi. and impacted northwest of the Mariana Islands, some 1,130 mi. from the launch site at Tokyo University's space center on Kyushu. (*M&R*, 2/8/65, 8)

- In an interview on the eve of his retirement as Air Force Chief of Staff, Gen. Curtis E. LeMay discussed the role of the military in space: "Developing military capabilities in space is a task that I think we ought to accept as an unavoidable requirement. It is the only way that we can establish control over corridors of access to our country that would otherwise be open to exploitation by aggressor forces. . . .

"I am confident that man will prove useful in this medium. Just as he has adapted aircraft to tasks no one could foresee in 1903, he will undoubtedly discover uses for space systems over the years ahead that go far beyond the observation and inspection functions we envision at this time." (AP, Haugland, Balt. *Sun*, 2/1/65)

January 31: Tass had reported that Soviet astronomers believed the upper layer of the moon's surface was saturated with meteoric matter distinguished chemically and in mineral content from deeper layers.

"Highly accurate and reliable" observation had been made by a Gorky University team headed by Vsevolod S. Troitsky, the Soviet Union's leading authority on radio emanations of the moon. (Shabad, Louisville *Courier-Journal*, 1/31/65)

- Two U.S. physicists, Prof. Robert V. Pound of Harvard and Assistant Professor Glen A. Rebka, Jr., of Yale, were awarded the Eddington Medal of the Royal Astronomical Society of London for gravitational red shift experiments reported in 1960 that confirmed Einstein's principle of equivalence, one of the basic assumptions of the general relativity theory. (*NYT*, 2/1/65, 12)
- Smithsonian Institution disclosed architectural plans for a national air and space museum to be built in Washington, D. C., opposite the National Gallery of Art. Designed by Gyo Obata, a St. Louis architect, the building would be modern in concept with an internal design that would provide a sweeping vista of exhibit areas. Smithsonian officials hoped to receive Congressional authorization to build the museum at a cost of $42 million. (*NYT*, 1/31/65)

During January: Reviewing Apollo program progress, Dr. Joseph Shea, Manager of the Apollo Spacecraft Program Office at NASA Manned Spacecraft Center, said that NASA had characterized the program as a series of phases. He explained that 1963 and 1964 were years of detailed designs and initial developmental testing; 1964 and 1965 were years of extensive ground tests and qualifications for flight; from 1966 on, ground tests would be supplemental to extensive flight tests, initially on the Saturn IB and later on the Saturn V. From his visits to almost all of the major Apollo hardware contractors, Dr. Shea said he could report that all of the subsystems associated with the command and service modules "are well along in their ground test programs.

"Almost all elements are on schedule and the test results indicate that the designs will meet our program objectives. By early this year, all subsystem hardware will be undergoing the rigorous qualification tests which we require before certifying such hardware ready for flight. . . . By the end of 1965, there will be three Apollo spacecraft in continuous ground testing. 1964 was, in retrospect, a year where milestone by milestone, we have achieved Apollo objectives." (NAA *S&ID Skywriter*, 1/15/65, 1, 4; Witkin, *NYT*, 1/24/65, 60)

- Dr. William H. Pickering, Director of the Jet Propulsion Laboratory, said in an article in *Astronautics and Aeronautics* for January:

". . . With Ranger 7, the prime factor was the expectation that the Apollo mission would choose a landing area on one of the smooth 'maria.' So it was of great value to this program to find out as much as possible about the mare topography. In particular, it was necessary to know if these areas were really lava flows and, if so, how much was exposed lava and how densely the small craters were scattered over the surface.

"Ranger gave some of the answers. In some areas, at least, small craters were indeed strewn very thickly. Probably such areas lie along the rays which radiate from some of the more recent large craters. Between these ray regions the surface of the mare appears to be quite smooth. No large rocks or fissures are apparent, although the general roughness may be comparable to some terrestrial lava fields where the lava is of the 'pahoehoe,' or fluid variety. However, the absence of any significant number of features showing edges with a small radius of curvature, and the presence of small craters which have been filled with debris, point to erosion as a significant modifier of the primeval lunar surface. This erosion could arise from meteoric bombardment and the effects of solar radiations. Estimates of the depth of surface which has been eroded away range from 5 to 50 ft. . . ." (*A&A*, 1/65, 18–20)

During January: GAO charged that mismanagement of the Nimbus meteorological satellite project resulted in unnecessary costs of $1.2 million. The report claimed that Nimbus' project manager at NASA Goddard Space Flight Center "did not effectively carry out his responsibility" for flight planning when it became evident that the spacecraft had become overweight and that he allowed the contractor to continue working toward the original design goal "even though it was clear [the effort] would be futile" because of booster limitations.

Dr. Homer E. Newell, NASA Associate Administrator for Space Science and Applications, rebutted the GAO allegations: "The costs which were incurred on the Nimbus project during the 5½ months between May 1961 and November 20, 1961, were for the development of the fully redundant Nimbus system to satisfy the requirements of the Plan for a National Operational Meteorological Satellite System. . . . Our effort to achieve the redundant system in the first Nimbus flight was continued as long as possible. . . . we did not want to take the step of dropping the redundant system, even for the first flight, until we were sure we had to." (GAO Nimbus Rpt., 1/65; *M&R*, 2/8/65, 9)

• Writing in the January 1965 issue of *Astronautics and Aeronautics*, Dr. Harold B. Finger, Manager of AEC–NASA Space Nuclear Propulsion Office (SNPO), summed up the various components of the advanced nuclear propulsion program and emphasized the importance of the solid-core nuclear rocket within the field: "Solid-core nuclear rockets are the best understood and most nearly developed of the many advanced nuclear-propulsion concepts being investigated in this country. They offer the most assured and earliest possible means for very substantial improvements and advances in space-flight propulsion capability. Furthermore, because solid-core nuclear rockets rely heavily on technology and techniques of chemical rocket engines and although much extension of these techniques is required, no fundamentally new engineering approaches are required to develop this new breed of substantially improved rocketry for actual flight use. Solid-core nuclear rockets can be relied on for our future space missions.

"Progress has been made in electric propulsion, particularly in the thruster area, and important research data and technology are also beginning to be provided in the difficult area of nuclear-reactor electric generating systems required for prime electric propulsion.

"Beyond these systems, other advanced nuclear propulsion concepts are not yet well-enough understood to justify undertaking significant development efforts." (A&A, 1/65, 30–35)

During January: Nine areas of scientific experiments for the first manned Apollo lunar landing mission had been summarized and experimenters were defining them for NASA. Space sciences project group expected to publish the complete report by Mar. 1, to be followed by requests for proposals from industry on designing and producing instrument packages. A major effort was under way by a NASA task force making a time-motion study of how best to use the limited lunar stay-time of 2 hr. minimum for the first flight. (*Av. Wk.*, 2/1/65, 13)

- NASA Goddard Space Flight Center announced it would negotiate with Radiation, Inc., of Melbourne, Florida, for a contract to develop a new weather measuring system to be tested aboard the Nimbus B meteorological satellite. The new equipment, Interrogation Recording and Location System (IRLS), would tie together weather observations made on the ground and in space as well as oceanographic measurements. (GSFC Release G–1–65)

- In an article in the *Indianapolis Star* discussing Soviet medical practices observed during his visit to Russia at the invitation of the Soviet Academy of Science, Dr. John M. Keshishian, associate in surgery on the George Washington University School of Medicine faculty, said: "It is not generally known that just before Voskhod was ordered into reentry, the pulse rate of one cosmonaut dropped to 40.

  "When your pulse rate drops below 40 heartbeats a minute, you're in trouble.

  "The Russians haven't said anything about this . . . but it could be another one of the problems their space medicine is encountering for which there seems to be no ready solutions.

  "For example, some Russian cosmonauts have suffered severe. hallucinations, both in flight and afterwards. Others have suffered equally severe and, thus far, inexplicable vertigo during which they can't be certain whether the floor's coming up to meet them or vice versa, or whether they're spinning, or the room is. And Russian physicians have found that . . . space flight environment—possibly weightlessness—draws calcium from the blood and expels it in the urine." (World Book Encyclopedia Science Service, Inc., Keshishian, *Indianapolis Star*)

- In an article in *Foreign Affairs* entitled "Slowdown in the Pentagon," Hanson W. Baldwin said: "The sprawling bureaucracy of big government; the control of major military or paramilitary projects by agencies over which the Defense Department has no direct authority, including the Atomic Energy Commission, the National Aeronautics and Space Administration, the Central Intelligence Agency, the Bureau of the Budget; congressional legislation and executive regulation—social, political and economic; the tremendous size and complexity of the Armed Forces; overcentralization and overregulation in the Pentagon; too much service rivalry and not enough service competition—all these and other factors have become builtin roadblocks in defense development and contracting.

"The creation of the National Aeronautics and Space Administration has provided another type of problem. NASA stemmed from the same kind of political philosophy that nurtured the AEC. Atom bombs were too powerful to allow the generals to play with them; ergo, a civilian agency must control nuclear power—and it must be channeled away from nasty military purposes. The same scientific-political pressure groups that advocated this concept helped (with President Eisenhower's approval) to establish NASA, again on the theory that space efforts must be controlled by civilians and that space must not be used for military purposes. . . .

"But in the case of NASA, the problem has been compounded. For while the AEC is essentially a research and production agency, NASA is an operating agency as well. From a small highly efficient aeronautical research agency, it has now expanded into a gargantuan multibillion-dollar empire, with tentacles all over the country, managing the biggest program on which the United States has ever embarked—to place a man on the moon.

"In its early years, NASA was sluggishly if at all responsive to military needs, and the Pentagon itself was inhibited from any effective space developments (though, curiously, the only effective space boosters available were military ballistic missiles). Gradually the liaison, due to Dr. [Edward C.] Welsh and others, has been greatly improved. Numerous military officers, active and retired, now hold some of the most important positions in NASA, and in addition the Armed Forces have furnished most of the astronauts and by far the most important part of the facilities and services used by the agency. The two-headed control still offers difficulties, but today the main stumbling blocks to the rapid development of military space projects are Secretary McNamara and his Director of Defense Research and Engineering, Dr. Harold Brown, who in his new political role in the Pentagon has become a remarkably unadventurous scientist.

"Often the President's Scientific Adviser, whose contacts with Pentagon and other Government scientists cut squarely across organizational lines, has also acted as roadblock to new developments. He exercises tremendous power without either specific responsibility or specific authority; therefore, his intervention often not only delays but confuses. The Adviser's great power stems largely from his White House status; unfortunately around him has grown up a small but important office manned by men more impressive as bureaucrats than as scientists, who represent, in effect, another echelon of delay. . . ." (*Foreign Affairs*, 1/65; *CR*, 2/4/65, 2007)

During January: Committee assignments for both parties were made in both Houses of Congress. New members on the Senate Committee on Aeronautical and Space Sciences: Walter F. Mondale (D–Minn.), Joseph Tydings (D–Md.), Len B. Jordan (R–Ida.), and George D. Aiken (R–Vt.).

New members of the House Committee on Science and Astronautics: Roy A. Taylor (D–N.C.), George E. Brown, Jr. (D–Calif.), Walter H. Moeller (D–Ohio), William R. Anderson (D–Tenn.), Brock Adams (D–Wash.), Lester L. Wolff (D–N.Y.), Weston E. Vivian (D–Mich.), Gale Schisler (D–Ill.), and Barber B. Conable, Jr. (R–N.Y.). (Comm. Off.)

- *During January:* Marvin L. White. AFCRL's Space Physics Laboratory, predicted the sun was encircled by "rings" of electric current totaling nearly 200 billion amps. Although total current was high, White postulated that the current density was low because the current was spread over a large area; he predicted current density to be about three trillionths of an ampere per square centimeter, the same order of magnitude as in the earth's atmosphere. White's calculations were based on particle flux data from MARINER II. (OAR *Research Review*, 1/65, 1-2)
- With launch of two balloons to 87,000-ft. altitudes from Chico, Calif., AFCRL began one-year series of high-altitude balloon flights to measure moisture in the stratosphere. Series would consist of vertical soundings—25 in all, at the rate of two per month—in which all data would be obtained in recoverable instrumented payload parachuted to earth when balloon descended to 30,000 ft., and horizontal soundings—five 11-day flights at float altitudes averaging 75,000 ft.—in which data gathered over thousands of miles would be telemetered every two hours to ground stations. (OAR *Research Review*, 5/65, 15-16)
- In an article on detecting extraterrestrial life, William R. Corliss in *International Science and Technology* described some of the plans for collecting data on possible life-forms elsewhere and some of the factors making the search for extraterrestrial life so challenging. He noted the complications for Martian life-detection if retrorockets were necessary to brake the landing of a scientific package: "First, of course, they add weight to the landing package, right where it hurts the most. Also, their control adds complexity and increases the chance of failure. Finally and perhaps most importantly, they would make the problem of life-detection more difficult and any results more ambiguous; the rocket exhaust would tend both to fuse the surface of the landing area (maybe even killing any existing organisms), and to add combustion contaminants of its own in the most crucial area—around the lander." He listed the variety of experimental instruments proposed for detecting extraterrestrial life (or clues of life) and explained why the dependability of these instruments—based on different physical and chemical principles—varied widely. (*Int. Sci. & Tech.*, 1/65, 28-34)

February 1965

February 1: The second meeting of the French-Anglo-United States Supersonic Transport (FAUSST) group was held in Washington to discuss airworthiness objectives in connection with commercial supersonic transports (SST). Agenda items included a discussion of atmospheric problems, structures, and sonic boom as related to SST flight. (FAA Release T-65-4)

- Gen. Curtis E. LeMay, retiring Air Force Chief of Staff, received a fourth Distinguished Service Medal from President Johnson at the White House. Later, during formal retirement ceremonies at Andrews AFB, a letter from the President was read: "All the world can be grateful to you for your courage, tenacity and exacting standards of professionalism." Gen. LeMay was succeeded by Gen. John Paul McConnell. (Loftus, *NYT*, 2/2/65, 13; *NYT*, 2/2/65, 13)

- Sealed brushless DC motor, originally developed to power instrumentation on unmanned spacecraft, was selected for use in the Apollo two-man Lunar Excursion Module (Lem) and the Gemini two-man spacecraft. The new motor utilized photo-optical detectors and transistorized switching elements which duplicated the functions of conventional brushes and commutator without physical contact of the rotating parts. Environmental tests had shown the brushless motor had a predicted operational life of one year. A barrier to DC motors had been the short life of conventional brushes in the space vacuum because of the lack of lubricating moisture necessary to prevent excessive friction. Motor was developed by NASA Goddard Space Flight Center under contract with Sperry Farragut. (GSFC Release G-2-65)

- NASA Flight Research Center issued requests for proposals for two preliminary feasibility studies of a manned lifting reentry vehicle to 16 industrial firms. Primary objective of the proposed studies would be to determine problem areas and their influence on design and to provide accurate estimates of the weight, cost, and developmental schedule involved with such a research craft. (FRC Release 5-65)

- NASA awarded $8.3 million contract to Pacific Crane and Rigging Co. for installation of ground support equipment at Kennedy Space Center's Apollo-Saturn V Launch Complex 39 on Merritt Island. The contract called for purchase, fabrication, assembly, installation, cleaning, and testing of electrical, mechanical, pneumatic, and hydraulic systems, valves and control modules, pipe assemblies, and support hardware. (KSC Release 17-65)

- Transfer of USN's Pacific Missile Range and instrumentation facilities at Point Arguello, Calif., to USAF operational control became effective. The Navy also turned over its Point Pillar tracking stations in California and mid-Pacific stations at Canton Island, Eniwetok, and at

South Point and Kokee Park, Hawaii. Missile impact location stations (MILS) at Wake and Midway were still under Navy control. For operation of its Pacific Missile Range, Navy retained tracking stations at Barking Sands missile tracking facility, Kauai, Hawaii, and was a tenant at Johnston Island. Other PMR stations included those at St. Nicolas and San Clemente Islands on the Sea Test Range. (Zylstra, *M&R*, 3/8/65, 33–34)

February 1: Astronaut Walter M. Schirra, Jr. (USN), said he probably would be assigned as command pilot to the sixth Gemini flight, which would be the first U.S. attempt to meet and join two vehicles in space. (AP, Balt. *Sun*, 2/2/65)

- USAF successfully launched an Athena test missile from Green River, Utah, to White Sands Missile Range, N.Mex. (AP, *Wash. Post*, 2/3/65)
- Construction unions' strike, that had shut down all NASA construction at Merritt Island and Cape Kennedy since Jan. 28, ended when the President's Missile Sites Labor Commission set a date for hearing the grievances of the unions involved. It had been the fifth walkout within a year. (UPI, *Chic. Trib.*, 2/2/65; *Wash. Post*, 2/2/65)
- FAA predicted continued aviation growth over the next five years: U.S. airline revenue passenger miles would increase 30 billion over the 54 billion flown in FY 1964; general aviation, measured in estimated hours of flying, would increase by four million hours over the estimated 15.5 million flown in FY 1964; general aviation fleet would number 105,000 aircraft by 1970, compared to 85,088 aircraft as of Jan. 1, 1964. (FAA Release T–65–3)

February 2: Capt. Joseph H. Engle (USAF) flew X–15 No. 3 to 98,200 ft. altitude and a maximum speed of 3,886 mph (mach 5.7) in a test to determine how ablative material reacted to intense heat. (NASA X–15 Proj. Off.; UPI, *Wash. Post*, 2/3/65; *X–15 Flight Log*)

- NASA announced it would negotiate a two-phase contract with Aerojet General Corp. for design and development of a liquid-hydrogen, regeneratively cooled exhaust nozzle for the Phoebus nuclear rocket reactor test program. First phase of the contract would include a four-month preliminary design study of nozzle concepts and an evaluation of fabrication and testing methods. This phase would be negotiated on a cost-plus-fixed-fee basis at an estimated value of $400,000. Using results of Phase I, the contractor would design, develop, test, and deliver three nozzles to the Nuclear Rocket Development Station at Jackass Flats, Nev. Phase II would be awarded as an incentive contract with an estimated value of $10 million.

    Phoebus, a 5,000-megawatt reactor, would be tested as part of the program to develop nuclear propulsion devices for space missions. (NASA Release 65–28)

- Discussing the missions and plans of NASA's new Electronics Research Center, Dr. Winston E. Kock, ERC Director, told the Harvard Engineers Club in Cambridge: "I believe that the recent strengthening of research in NASA can act to overcome any such braking of scientific enthusiasm which the recent changes in our defense program . . . may have instigated. I have seen at first hand true research enthusiasm at two NASA Research Centers, Lewis in Cleveland, and Ames in California, and, at Cambridge's new NASA Research Center, the response we have had from inventive, research-minded individuals, expressing an interest in association with the Center has been phenomenal. I have always believed in the saying 'necessity is the mother of invention,' and I feel certain that it was the necessities of World War II that brought into bloom radar, the jet aircraft, nuclear power, the V-2 rocket (which led to our present missile and space rockets), and many other developments which have proved to be of vital importance to our way of life. So, I feel that a counterbalance to today's reduced necessities of the broad, new opening field of *space* research . . . will help to keep our nation's research talent active and enthusiastic, and maintain it in the strong virility it has exhibited since the start of World War II." (Text)

- *February 2:* Charles W. Harper, Director of NASA Hq. Aeronautics Div., discussed aeronautical research at a luncheon of the Aviation/Space Writers Association. He said: " . . . aviation has a tremendous potential in the short-haul 'aerial bus.' Both VTOL (vertical take-off and landing) and STOL (short take-off and landing) are being considered for this job. . . . On the basis of our current knowledge I would conclude the VTOL commercial transport offers tremendous potential but requires additional research . . . before it is ready for detailed feasibility study as a commercial transport. On the other hand the STOL machine is ready for a careful examination since the major problems seem to be in hand.

  "A 20 to 50 passenger STOL machine should, or could, have a top speed of 300 to 400 knots, a steep approach with a touch down at 45 knots and an operational field length of some 1200 to 1500 feet. All-weather operation is required and, with the aid of space technology advances, this appears quite possible. We think we can display electronically to the pilot the important features of the airport so that he can approach it and land using the same information that he does in clear weather.

  "We see two large markets for vehicles of this type. In a smaller simple version, perhaps bearing a little sacrifice in performance, an air transport well suited for use in underdeveloped areas. Rugged, easy to fly and simple to maintain, it could enable these countries to jump from jungle or desert trails to modern transport system without building enormously expensive railways and highways. This would be a good market for U.S. industry. In a larger sophisticated version it could be the vehicle to make the present short haul feeder lines self sufficient, not depending on connecting traffic from the trunk lines. This too would be a desirable situation for American industry. NASA plans to pursue both of these potentials actively until the air industry has enough confidence in success to proceed on its own. . . ." (Text)

- Alfred J. Eggers, Jr., NASA Deputy Associate Administrator for Advanced Research and Technology, addressed the Science Teachers' Association of Santa Clara County, Calif., at NASA Ames Research Center. He said: "The question then is, what has man done in space to date? According to the eminent archaeologist, V. Gordon Childe, whatever man has done in the relatively short evolutionary history documented by his fossil remains, he has done without significantly improving his inherited equipment by bodily changes detectable in his skeleton. Moreover, this equipment is inadequately adapted for survival in any particular environment, and indeed it is inferior to that of most animals for coping with any special set of conditions. Yet in spite of his physical inferiority, man has been able to adjust himself to a greater range of environment than any other creature, to multiply much faster than any near relative amongst the higher mammals, and indeed to beat them all at their special tricks. Thus he learned to control fire, and he developed the skills to make clothes and houses, with the result that he lives and thrives from pole to pole on earth, and already he is concerned with a population explosion. He has developed trains and cars that can outstrip the fleetest cheetah, and he has developed the airplane so that he can mount higher than the eagle. Moreover, he developed telescopes to see further than the hawk, and firearms to lay low the elephant or any other animal, including himself. But whatever their use, the important point is that fire, clothes, houses, trains, airplanes, telescopes, and guns are not part of man's body. He can set them aside at will. They are not inherited in the biological sense, but rather the skill needed for their conception, production, and use is part of our intellectual heritage, the result of a tradition built up over many generations and transmitted not in the blood but through speech and writing.

"The true stepping stones to the moon are ourselves and our forefathers. The stepping stones beyond are our children, and much of what they will be and where they will lead the human race, is up to you and your kind. If you succeed in your work, you will have made an invaluable contribution to the betterment of man's ability to *make himself*, to *master himself*, and finally to *understand himself* in his environment. Indeed, if you are especially successful, you may, in the words of V. R. Potter, 'develop a new breed of scholars, men who combine a knowledge of new science and old wisdom, men who have the courage of the men of the Renaissance who thought truth was absolute and attainable,' and who may yet be right. I submit we can do no less than find out." (Text)

February 2: Philco Corp., opposing the bid by the Communications Satellite Corporation to supply DOD with communications satellite service, asked the FCC to prevent ComSatCorp from signing a "sole source" contract with Hughes Aircraft Co. Philco, which was already preparing a satellite system for DOD under a contract awarded in July 1963, said ComSatCorp's proposed contract "is in violation of the letter and spirit of the FCC rules and regulations which require competition in ComSatCorp procurement." Since ComSatCorp apparently had been negotiating the matter for some time, "its present statement that stringent time requirements impel waiver of the FCC's rules and regulations is insupportable," Philco said.

Under ComSatCorp's plan, DOD would be supplied 24 satellites built by Hughes Aircraft Co. and would pay for service only if the satellites worked. ComSatCorp would absorb the costs if they did not. DOD had made no decision for or against the offer. (Wash. *Eve. Star,* 2/2/65; UPI, *NYT,* 2/3/65, 54)

- Sen. Warren G. Magnuson (D–Wash.) introduced in the Senate a bill to provide for a national oceanographic program and the establishment of a National Oceanographic Council. Senator Magnuson said the National Oceanographic Council would have "certain key responsibilities and functions . . . in the oceanographic field [which] would be similar to those of the National Aeronautics and Space Council in the space program. . . ."

He noted that "a number of departments and agencies have separate missions in the aeronautics and space program," and that the National Aeronautics and Space Council "takes precedence over" the operating agencies to coordinate the national aeronautics and space program. Similarly, 6 departments and 22 agencies "are engaged or have a direct interest in the seas. . . ." (CR, 2/2/65, 1754–57)

February 2: R. E. Clarson, Inc., of St. Petersburg, Fla., was awarded a $2,179,000 contract for alterations to Launch Complex 34, Cape Kennedy, to accommodate the Saturn IB rocket. Army Corps of Engineers made the award. (AP, *Miami Her.*, 2/3/65)

- Editorializing in the Washington *Evening Star* about "lean years" beginning for the aerospace industry, William Hines said: ". . . Since the '50s, aerospace companies have become accustomed to a diet of caviar, filet and champagne. The government has poured something like 100 billion into rockets, missiles and spacecraft since the Soviet Union's Sputnik went up in October, 1957. The torrent of funds is now being reduced, if not precisely to a trickle, certainly to a more moderate flow. . . .

  "The aerospace crisis is serious enough that the management-consultant firm of Arthur D. Little, Inc., has just published a study, 'Strategies for Survival in the Aerospace Industry.' It makes the following revealing point:

  "'The period 1954–1963 was one of remarkably steady growth in the funding of military and space systems. In fact, it was so steady that many participants perhaps forgot that there were concrete, finite objectives to be achieved with these funds.'. . ." (Hines, Wash. *Eve. Star*, 2/2/65)

- Prime Minister Harold Wilson announced plans to buy American military aircraft to replace British aircraft, an action he said would save more than $840 million over a 10-yr. period. The two U.K. projects being dropped were the P–1154 vertical take-off supersonic strike aircraft and the HS–681 short take-off military transport. Both were made by the Hawker Siddeley group. American Phantom II's, made by McDonnell Aircraft, would be ordered to replace the P–1154. Phantoms were already on order to replace the Royal Navy's Sea Vixens. Lockheed's C–130's would replace the HS–681. The American planes would be equipped with British engines. On the question of the TSR–2, which the U.K. was considering replacing with General Dynamics' F–111, Mr. Wilson said there was not enough information yet to make a final decision. (Farnsworth, *NYT*, 2/3/65, 9; Clymer, *Balt. Sun*, 2/3/65)

- Soviet news agency Tass announced that firing of a new type of multistage rocket booster on Jan. 31 had been so successful that further tests in the Pacific series had been canceled. The rocket had travelled more than 8,000 mi. in the Pacific southwest of Hawaii. (UPI, Wash. *Daily News*, 2/2/65; UPI, *Wash. Post*, 2/3/65)

- A brightly illuminated object in the sky near Langley AFB, Va., was widely reported as a Ufo but identified by USAF as a weather balloon with the sun reflecting off its surface. (Newport News *Daily Press*, 2/3/65)

February 3: OSO II (Oso B2), NASA's Orbiting Solar Observatory, was successfully launched into orbit from Cape Kennedy by a three-stage Delta rocket. Preliminary orbital elements: apogee, 393 mi.; perigee, 343 mi.; period, 97 min.; inclination, 33°. The 545-lb. spacecraft included parts salvaged from the Oso B, damaged last April prior to launch, and components of a spacecraft built for prototype testing.

The second of eight spacecraft planned by NASA for direct observation of the sun, OSO II carried eight scientific experiments and had two main sections: the wheel (lower) section provided stability by gyroscopic spinning and housed the telemetry, command, batteries, control electronics and gas spin-control arms, and five experiment packages; the sail (upper) section was oriented toward the sun and contained solar cells and solar-pointing experiments. For the first time, the instruments, controlled by ground command, would scan the entire solar surface. Each scan required four minutes.

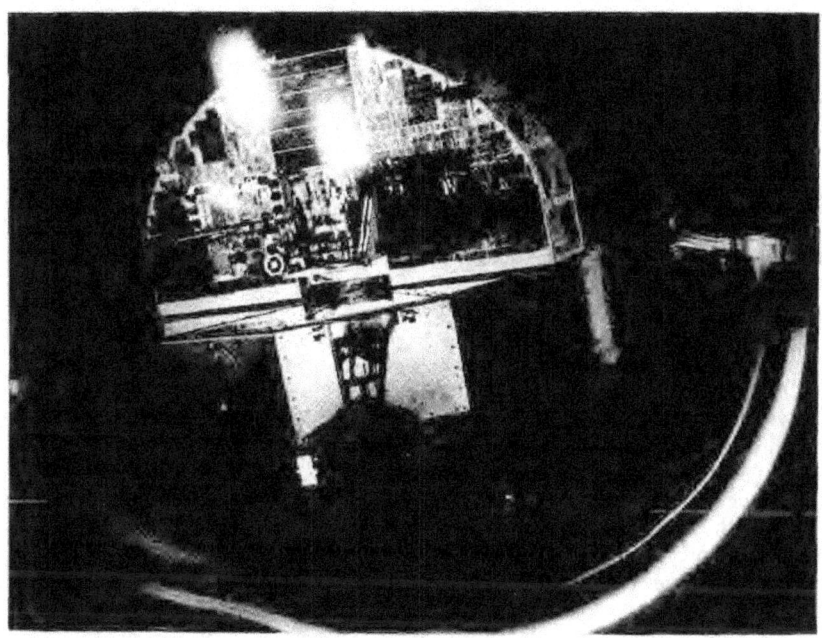

February 3: NASA's Orbiting Solar Observatory, OSO II, was launched from Cape Kennedy, Fla.

OSO II experiments were intended to map the frequency and energy of solar emissions and represented a joint Government-university-industry effort. NASA Goddard Space Flight Center managed the project. (NASA Press Kit Release 65-14; NASA Release 65-32; *Goddard News,* 2/8/65, 1-2)

- Nike-Apache sounding rocket was launched from Wallops Island, Va., to altitude of 87.7 mi. (141.1 km.) with experiments to measure the neutron intensity above the earth's atmosphere, the flux of solar x-rays, and Lyman-alpha radiation; and to determine ionospheric electron densities. All instruments functioned as predicted. (NASA Rpt. SRL)

February 3: NASA Administrator James E. Webb said during a panel discussion at the Military Electronics Convention in Los Angeles: "I think I can report that our ten-year aeronautical and space effort [beginning in 1961] has been well organized, it has stabilized at the $5\frac{1}{4}$ billion level, and has retained a well-worked-out balance among its various components. At the end of this ten-year period, we will have received back from our operating spacecraft the basic measurements of the space environment which will give us a much better scientific understanding of this environment and our engineers will have proved out the developmental concepts and engineering designs for effective operations of all kinds in space. Further, we will have a launch capability of six Saturn IB's and six Saturn V's per year, meaning that we could put almost two million pounds into orbit per year, if required. We will have logged more than five thousand hours of astronaut spaceflight time and learned a great deal about the relationship between man, equipment, the task assigned, and performance in the space environment. . . ." (Text)

- NASA Manned Spacecraft Center reported it had saved $12 million toward a $35 million cost reduction goal for FY 1965. Major portion of this saving was made possible by a suggestion from Dr. Robert R. Gilruth, MSC Director, that instead of spending the budgeted $7,873,000 for a static test stand for the Apollo spacecraft at Cape Kennedy, the reserve Titan Launch Complex 16 be modified for static test use. Cost of modifying the Titan launch complex would be $3,982,900, with a net saving of $3,890,100. (MSC *Roundup*, 2/3/65, 8)

- NASA Marshall Space Flight Center was readying its first Saturn V S–IC stage, designated S–IC–T, for static firing late this spring, NASA reported. The S–IC–T, a static test stage, would be hot-fired on a captive test stand in MSFC's West Test Area and would be ground tested repeatedly over a period of many months to prove out the propulsion system. (NASA Release 65–27)

- FCC vetoed a proposed Communications Satellite Corp. contract with Hughes Aircraft Co. for the design of satellites to be used by DOD. In a letter to ComSatCorp, the FCC took note of a protest by Philco Corp. (see Feb. 2), that it was as qualified as Hughes to bid on the proposed contract and said ComSatCorp must award the contract only after competitive bidding. ComSatCorp had asked the FCC to approve the proposed contract, waiving requirements for competitive bidding. (FCC Public Notice–C)

- USAF "ripple-launched" two Minutemen ICBM's from Vandenberg AFB. Both were launched from silos, the second within minutes of the first. (*M&R*, 2/15/65, 12)

February 4: Nike-Cajun sounding rocket with grenade payload to obtain temperature, wind, density, and pressure data was launched from Wallops Island, Va. to altitude 73.5 mi. (118.2 km.). Twelve grenades were to have exploded during rocket ascent, but two did not explode. All other instruments performed as anticipated. A similar experiment was launched from Point Barrow successfully. (NASA Rpt. SRL)

- USAF's XC–142A V/Stol, designed and built by Ling-Temco-Vought, made its first public flight at Grand Prairie, Tex. piloted by John Konrad. Designed to take off and land vertically, the experimental aircraft had a wing that could be moved in flight from the normal horizontal position to a vertical position, enabling it to hover, fly backwards, sideways, and rotate 360° in either direction over the same spot. It could fly forward as slowly as 25 mph without stalling and could be flown at maximum speed of 425 mph; cruising speed was 250 mph. Five of the airplanes would be delivered to Edwards AFB for further tests. (*Wash. Post*, 2/5/65; Clark, *Houston Post*, 2/5/65; *A&A*, 4/65, 8)

February 4: Scientists at Boeing Co., Seattle, had devised a "trampoline" bed designed to exercise the blood vessels in a weightless environment, it was reported. Compared in effect to a cocktail shaker, the device, by its to-and-fro motion, would send the blood surging from the head to the feet and back again. Some scientists had feared that days of inactivity in a weightless environment without exercising the blood vessels could result in death to an astronaut. (AP, Newport News *Daily Press*, 2/4/65; AP, *Huntsville Times*, 2/4/65; *Orl. Sen.*, 2/4/65)

- USAF presented a Lockheed Agena-B to the Smithsonian Institution, Washington, D.C., for permanent display in the National Air Museum. The Agena had performed as an orbital injection vehicle, space satellite (first to achieve circular and polar orbits), and as an intermediate stage booster for deep space probes. (Smithsonian Release)

February 5: MARINER IV was performing normally after nearly 10 weeks in space, NASA announced. At 9 a.m. EST the Mars probe was 14,421,246 mi. from earth and had traveled more than 117 million mi. in its sun-circling orbit. It was moving at a velocity of 14,478 mph relative to earth and 65,670 mph relative to the sun. Instruments aboard MARINER IV Mars probe detected a solar flare and the spacecraft telemetered data to a tracking station at Johannesburg, South Africa, for relay to the Jet Propulsion Laboratory. (NASA Release 65-30; L.A. Times, *Wash. Post*, 2/6/65)

- First major piece of flight-type hardware for the Apollo program, Service Module 001, successfully underwent a 10-sec. shakedown static test firing at NASA Manned Spacecraft Center's White Sands Operations. Service Module 001 was unlike previous boilerplate models in that it was constructed primarily of aluminum alloy and had an outside skin of honeycomb bonded between two aluminum sheets. Made by Aerojet-General, the service propulsion system engine had 22,000 lbs. of thrust. It would slow down the Apollo for entry into lunar orbit and speed up the spacecraft for escape from lunar orbit and the return to earth. (MSC *Roundup*, 2/17/65, 8; NAA *S&ID Skywriter*, 2/12/65, 1, 3)

- NASA Administrator James E. Webb, at Nebraska Wesleyan Univ. to receive an honorary doctorate, said in a speech: ". . . while our national policy is to maximize the peaceful uses of outer space . . . and to avoid the extension of weapons, we have no choice but to acquire a broadly-based total capability in space; a capability that can enable us to insure the protection of our national security interests while we actively seek cooperative peaceful development. . . .

"The Roman mastery of land and sea communications, the English mastery of the seas, the American mastery of the air and of nuclear energy were each accompanied by greatly enhanced prestige and followed by vast increases in power and position, new knowledge, the establishment of strategic international economic advantages, the wide use of new resources, great advances in military capability, and a quickening of national pride and vigor. Portentous realignments among nations were inevitable. These are the advantages the Russians now seek from their enormous investments in space. These are the advantages we cannot permit them to acquire and use against the non-Communist world.

"In these lessons of history lies the real challenge of space. The portents for our own time are clear enough in the early lead of the Soviet Union with the first Sputnik, Vostok, and Voskhod. The spurt in Soviet prestige brought a new assurance and weight in the international political arena, a new pride and confidence in Soviet national purpose. We have reacted quickly and with ever-increasing success, but the challenge of the mastery of space remains to be accomplished for us as a nation and for you as a member of the new generation. We are meeting this challenge, and in doing so enhancing the broadest values for our society and our world.

"Our power to survive as a great and vigorous Society is in the process of being proven again through our space efforts. Your own involvement in the actions and consequences will be far greater than you or I can fully appreciate today. . . ." (Text; *NYT*, 2/5/65)

February 5: NASA announced it had approved a Rice Univ. proposal for a satellite to measure radiation and radiation loss in the Van Allen belts, aurorae and airglow, bombardment of the upper atmosphere by energetic particles from space, and galactic and solar cosmic rays. The 125-lb. scientific satellite, to be known as Owl, would be designed, developed, and built by a Rice group headed by Dr. Brian J. O'Brien, and would be injected into a near-circular orbit at about 400 mi. altitude by Scout launch vehicle. After achieving orbit, the satellite would be oriented by a large permanent bar magnet so that one axis would be continuously aligned with the earth's magnetic lines of force. The Rice project would be part of the NASA University Explorers Program. Spacecraft and experiments would be tested at NASA facilities under the direction of NASA's Wallops Station, which also was assigned project management of Owl. (NASA Release 65-29)

- First S-II-S ground test stage in the Saturn S-II program was completed by North American Aviation at Seal Beach. The stage would be used for structural tests simulating critical thrust and pressure loads anticipated during Saturn V/Apollo flight missions. This stage would not be fired—it would have no engine. (NAA *S&ID Skywriter*, 2/5/65, 1)
- A new alloy known as NASA Modified TaZ-8 had been developed by NASA Lewis Research Center scientists John C. Freche and William J. Waters for use in modified X-15 nose sensors. The new material, which contained tantalum and zirconium, was necessary because the increased speed of the modified aircraft (X-15 No. 2)—up to 5,000 mph—would cause greater dynamic heating. (*Lewis News*, 2/5/65, 1)
- Menu released by NASA Manned Spacecraft Center in a request for bids from industrial firms interested in furnishing the Apollo astronauts with food for the journey to the moon included bacon and eggs, frosted flakes, toast, fruit juice, and a strawberry cereal bar. Food allowance of 8 lbs. would be dehydrated to reduce its weight. Astronauts would add water to the food from their drinking water supply. (Schefter, *Houston Chron.*, 2/5/65)
- National Science Foundation announced that a new radio technique might make it possible to study Antarctica's ice depth. The technique in-

volved sending radio waves down through the ice and measuring the time it took them to bounce back from the underlying ground. This would provide a measure of ice depth. The equipment was checked out at the South Pole where earlier seismic soundings had shown the ice to be 9,100 ft. deep. (UPI, *NYT*, 2/7/65, 77)

February 5: Deactivation of 129 obsolete intercontinental ballistic missile launch sites was underway. The Thor, Atlas, and Titan I missiles had been superseded by more modern weapons, including Titan II, Minuteman, and Polaris. Nearly $2 billion of property in 12 states was involved. Government agencies had been advised that equipment was available as military surplus. (Hill, *NYT*, 2/7/65, 64)

- A tentative plan of the Center for European Nuclear Research (CERN) to build a 300-billion-electron-volt particle accelerator in Bavaria was being opposed by residents of Munich, it was reported. CERN had stated that no final decision had been reached. (*NYT*, 2/7/65, 24)

February 6: Tabulations prepared by NASA Goddard Space Flight Center showed that more than 1,000 man-made objects—satellites, spacecraft, capsules, and assorted bits and pieces of them—had been placed in orbit since Oct. 4, 1957. Of these objects 243 were satellites launched by the United States or its allies and 94 were Soviet-launched satellites. 103 U.S.-sponsored satellites and 16 Soviet satellites were still in orbit. Of those no longer in orbit, 140 were U.S. and 78 Soviet. Many satellites had separated into two or more space objects or had broken apart accidentally or by design to produce space junk. GSFC records identified 469 hunks of junk of U.S. origin and 182 of Soviet as having orbited the earth at one time or another. Of these, 372 U.S. and 16 Soviet objects were still in orbit. (AP, *NYT*, 2/7/65, 80)

- Among 1965 recipients of the Arthur S. Flemming Award to outstanding young men in Federal Government were: Leonard Jaffe, Director of NASA Communication and Navigation Programs, for his work in communication satellite projects; Dr. Robert Jastrow, for his work in nuclear theory at the Goddard Institute for Space Studies in New York; Dr. Joseph F. Shea, Manager of the Apollo Spacecraft Program Office at NASA Manned Spacecraft Center, for his work in U.S. manned lunar landing program; and Wesley L. Hjornevik, Assistant Director for Administration at NASA Manned Spacecraft Center, for his work in construction of space environment simulator. (*Wash. Post*, 2/7/65; NASA Notice)

- Dr. Frank J. Low, research associate in the Lunar and Planetary Laboratory at the Univ. of Arizona, discovered three halved stars with halos around them which may be clues to stellar evolution. He said he believed these were stars throwing out material that would become building blocks of future stars. He identified the stars as Betelgeuse, Aldebaran, and Mu Cephei. (AP, *Phil. Eve. Bul.*, 2/6/65)

- Over 300 Government- and space industries-employed engineers were studying for master's and doctor's degrees utilizing closed circuit television with two-way communication in a program at the Univ. of Florida's College of Engineering. The system had been activated in September 1964. TV classrooms were at Orlando, Daytona Beach, Cape Kennedy, Melbourne, Patrick AFB, and NASA Merritt Island. (*NYT*, 2/7/65, 80)

- Marshal Nikolai I. Krylov, Soviet commander of the strategic rocket forces, said in *Krasnaya Zvezda*: "Representatives of the aggressive imperialist circles often brag about their rocket-nuclear weapons. In answer to this we can state with assurance that in respect to the quality and quantity of rocket-nuclear weapons, we not only do not lag behind those who threaten us with war, but far surpass them." (*Krasnaya Zvezda*, 2/6/65, 2, ATSS-T Trans.)

February 7: FAA Administrator Najeeb E. Halaby, questioned about the supersonic transport in New Orleans, cited the following advantages: the 220-plus passenger Sst in one year would carry as many passengers as does the *Queen Elizabeth* with a crew of 1,500; the Sst would effect obvious economies by decreasing air transportation time to a third of present levels; the program would provide approximately 15,000 skilled jobs a year that would otherwise not be filled; the Sst project would advance the technology of titanium as much as World War II aircraft production advanced that of aluminum. (*Wash. Post*, 2/8/65)

- A full-scale aluminum model of a 1,400-lb. telescoping space structure had been fabricated to verify design theory and manufacturing techniques, AFSC announced. Built by Martin-Denver, the 15-by-8-ft. expandable structure could be launched into space in a compact package and then, like a telescope, opened to full size after reaching orbit. Several of the expandable structures stacked on a booster's upper stage could be sent into space and expanded to form a rotating space station. (AFSC Release 4.64)

February 8: NASA conducted high-altitude grenade experiments almost simultaneously from launch sites in Alaska, Canada, and at Wallops Island, using a two-stage Nike-Cajun in each case. Grenades were ejected and detonated at intervals from about 25- to 56-mi. altitude. This was the third and final set in the current series to obtain upper-atmosphere wind, temperature, density, and pressure data at the three widely-separated geographic locations. The series marked the first time that such measurements had been made with sounding rockets within the Arctic Circle. (NASA Release 65–8)

- NASA Manned Spacecraft Center announced selection of Maj. L. Gordon Cooper (USAF) and Lt. Cdr. Charles Conrad, Jr. (USN), to make the seven-day Gemini V space flight. Gemini V would be the third manned Gemini space flight and would be made in 1965. Backup crewmen were two civilians, Neil A. Armstrong and Elliot M. See, Jr. (*MSC Roundup*, 2/17/65, 1; AP, *Wash. Post*, 2/8/65; AP, *Balt. Sun*, 2/8/65; Witkin, *NYT*, 2/9/65)

- Among the 11 scientists and engineers presented the National Medal of Science by President Johnson at a White House ceremony were Dr. Charles S. Draper, professor of aeronautics and astronautics at MIT, and Dr. Harold C. Urey, professor of astronomy at Univ. of Calif. and consultant on NASA Space Science Steering Committee. (*Wash. Post*, 2/9/65, 9; CR, 2/10/65, A590)

- 18 additional countries were applying for ownership in the $200-million international consortium to operate a global communications satellite system, it was reported. Eighteen nations and Vatican City had originally participated in forming the consortium in July 1964. U.S.'s ComSatCorp acquired 61% ownership and would serve as manager for the consortium.

  The new nations applying for ownership participation were Monaco, South Africa, New Zealand, Syria, Kuwait, Libya, Yemen, Brazil,

Morocco, United Arab Republic, Sudan, Iraq, Lebanon, Tunisia, Argentina, Jordan, Indonesia, and Ceylon. (*Av. Wk.*, 2/8/65, 25)

February 8: The world's first nuclear-powered weather station, designated Navy Oceanographic and Meteorological Automatic Device (Nomad), began its second year of successful operation 300 mi. out of New Orleans in the Gulf of Mexico. Developed by the Navy, the unattended station was powered by the AEC generator, Snap-7D. (*NYT*, 2/14/65, 90)

- A Polaris A-3 launched from the nuclear submarine *Sam Rayburn* represented the 16th consecutive success for that missile. The *Rayburn* was submerged off the coast of Cape Kennedy. (*M&R*, 2/15/65, 12)

During the week of February 8: In an interview, C. R. Smith, chairman of American Airlines, backed the Boeing Co.'s entry, one of two basic designs under study, in the Government's design competition for a supersonic transport aircraft: "I think the SST will have to have variable-sweep wings."

In variable sweep, the angle at which the wings meet the fuselage could be changed for efficiency at different speeds. At slow speed, the wings would be outstretched for greater lift; at high speed, they would be swept back sharply to reduce airflow drag. This principle was used on the F-111. (*NYT*, 2/14/65, 90)

February 8-10: American Astronautical Society presented a Symposium on Unmanned Exploration of the Solar System in Denver, Colo. Speaking about the Biosatellite Program, Dale W. Jenkins, NASA Office of Space Science and Applications said:

" . . . The Biosatellite Program is a second-generation series of carefully planned and selected experiments, including some highly sophisticated experiments which have required several years of baseline study and development. These orbiting recoverable Biosatellites provide an opportunity to test critically major biological hypotheses in the areas of genetics, evolution, and physiology. The Biosatellite studies will help delineate hazards to astronauts and assist in determining and defining effects on degradation of human performance. Prolonged manned flights may involve, for example, physiological changes such as decalcification of bones (particularly the vertebrae), loss of muscle tone and physical capability, and certain cardiovascular changes. Also, the effect of continued sensory deprivation on behavior and performance is unknown.

"Twenty experiments have been selected for flight to study the effects of weightlessness and decreased gravity during 3- to 30-day orbital periods. The experiments include a wide variety of plants and animals from single-cell organisms to higher plants and animals. The effects of weightlessness will be studied on the primate, especially the central nervous, the cardiovascular, and the skeletal systems during orbits of 30 days' duration.

"Experiments have been selected to study the effects of weightlessness combined with a known source of radiation to determine if there are any antagonistic or synergistic genetic or somatic effects on various organisms.

"Experiments are included to study the effects of the unique environment of the Earth-orbiting satellite and removal from the Earth's rotation in relation to biological rhythms of plants and animals. . . ." (Text)

Dr. Homer E. Newell, NASA Associate Administrator for Space Science and Applications, outlining progress made toward the objective of solar system exploration, said: " . . . It would appear . . . that enough experience and know-how has been accumulated to make the move to a five-ton Voyager spacecraft on the Saturn IB Centaur launch vehicle a reasonable next step in the unmanned exploration of the solar system. There is no question but that the size and weight of Voyager, plus the increased demands that will be placed upon it, will make the development of the Voyager a complex and difficult undertaking. But certainly, the Orbiting Geophysical Observatory and Mariner have shown us that we can deal successfully with complexity. As a matter of fact, the increased weight and size may afford considerable relief from the need to tailor every last function to a gnat's eyebrow in order to achieve the intended mission, as has been the case hitherto. . . ."

Dr. Newell said that reliability was probably the most difficult problem for deep-space long-duration missions and that ". . . the development of adequate spacecraft *systems* will not be the problem. The most serious threat to long life operation will lie in the potential random failure of one or more [spacecraft] components."

He added that this would probably be an ever-decreasing problem as we gained experience with launch vehicles and that ". . . launch vehicle reliability is far less a difficult program than that of long spacecraft lifetime for very deep-space missions."

In conclusion, Dr. Newell said: " . . . Nevertheless, the time has arrived when many thoughtful people urge a vigorous program of solar system exploration. The President has in his Fiscal Year 1966 budget request included funds to support initial conceptual and design studies of a Voyager spacecraft. Funds are also included for the development of a launch vehicle consisting of the Saturn IB plus the Centaur. Final decision as to whether to move ahead with the development of the Voyager spacecraft would come a little over a year from now." (Text)

Missiles and Rockets reported that scientists at the Symposium had differing opinions on Mars goals. Gilbert V. Levin of Hazleton Laboratories opposed the 1971 scheduled landing of the Voyager: ". . . although we insist that Mars should not be contaminated by terrestrial life before we search for Martian life, we fail to recognize that this is tantamount to saying that the U.S. must get there first, because the U.S. appears to be the only nation willing and able to sterilize its spacecraft.

"I'm all for Voyager, but an initial step in the Voyager program should be some early landers at the earliest opportunities. We should go ahead and devote efforts to develop a program to land on Mars in 1969."

Temple Neumann, Automated Biological Laboratory program engineer with Philco's Aeronutronic Div., agreed with Levin: "If the planetary biological exploration task is to be performed in a sound scientific manner, the U.S. must do it—first."

Lawrence B. Hall, NASA Special Assistant for Planetary Quarantine, re-emphasized "the disastrous effects that an unsterilized spacecraft could have on Mars. . . . If a single micro-organism should land on Mars and have a replication time of 30 days, it could grow to the bacterial population of Earth in eight years. This could not only compete with Martian life but could result in drastic changes in the geochemical and atmospheric characteristics of the planet."

JPL's Gerald A. Soffen said that "since the decision between Mariner landers and Voyager missions has not been made yet, scientific experimenters have to think in terms both of small payloads and large ones. Numerous small missions would provide a good opportunity to perform experiments in different locations and during different planetary seasons."

Bruce C. Murray, of Cal Tech, said: ". . . finding the right location, getting there, and interpreting the biological experiment results in a way that would indicate definitively whether life was or was not present would call for at least 100 times more photography than was currently assumed.

". . . atmospheric effects, color, seasonal changes, and the large number of locations of interest will make Martian pictures 10–50 times more difficult to interpret than lunar pictures."

Robert L. Sohn, TRW Space Technology Laboratories, stressed value of earth-return missions and recommended serious consideration of round-trip missions and multi-plan round trips using DSIF as guide. He suggested that an 800-lb. spacecraft could make a fly-by of Mars, using the Venus swing-by technique and return to earth to enable recovery of a 50-lb. capsule. Use of swing-by techniques for round trips reduces earth launch velocities to those of the favorable years and also reduces earth reentry velocities. Additional advantages were closer passage with Mars at encounter and the opportunity to gather data on two planets.

Elie A. Shneour, of Stanford University, said he ". . . could not say whether it was possible to draw up a set of experiments that would definitively search for life." He maintained, however, that "discovery of any form of extraterrestrial life will be tantamount to a basic determination of the nature of all life on that planet." (M&R, 2/22/65, 39, 41)

February 9: Joint Congressional Atomic Energy Committee, in hearing on AEC's FY 1966 authorization, asked NASA Administrator James E. Webb why NASA was dropping funds for the Snap–8 spacecraft nuclear auxiliary power project. In the joint NASA–AEC project, AEC was working on the reactor and NASA was working on the power conversion machinery. Snap–8 was one of three projects that had been deleted from the NASA FY 1966 budget request. Mr. Webb said: " . . . in the over-all budgeting . . . the President has a hard problem of adjusting resources to the needs of the Government. In this case, it was decided that these systems, these three systems, could not be financed within the resources available for allocation to NASA and therefore they were eliminated in the final decision relating to the President's budget, but not on our recommendation. . . .

"Because we are on the verge of significant technical milestones with our power conversion equipment, we believe we should phase out the program in an orderly way and provide the maximum amount of experience and data for future use.

"Therefore, we plan to continue current testing of components combined in the test loop to achieve at least 1,000 hours of operating time on each of the major components, by reprogramming our remaining fiscal year 1965 funds into these specific task areas.

". . . we expect to present to you and the committees, if you will permit us to do so, an orderly plan for using the resources we now have. This gives the Congressional Committees an opportunity to look at and plan and decide whether it does really fit what they believe is in the national interest rather than to take a sudden action. . . ."

Sen. Clinton P. Anderson (D-N.Mex.), Chairman of the Senate Aeronautical and Space Sciences Committee, expressed his belief "that it is too bad that the Bureau of the Budget trimmed you down on this work. I wish you had gone ahead with it through the test period, we would have learned some very significant things. I disassociate it from the other two [canceled projects] but Snap–8 should have gone on priority." (Transcript)

February 9: Six of eight OSO II experiments had been turned on and gave "excellent" data, NASA reported. The two experiments not yet operating were the ultraviolet scanning spectrometer provided by Harvard Univ. and the ultraviolet spectrophotometer provided by NASA Goddard Space Flight Center. Both had been turned on but were turned off to prevent damage to themselves or to the satellite when irregularities in the data received were noted. All other functions of the satellite—such as solar power supply, telemetry system, tape recorder, temperature control, and command system—were normal. (NASA Release 65–37)

- At a press conference during the Symposium on Unmanned Exploration of the Solar System, presented in Denver, Colo., by the American Astronautical Society, Univ. of California chemist Harold C. Urey said that he hoped the moon was "interesting enough to make the $20 billion exploration program 'worthwhile.'" He added, "If it turns out that the moon escaped from the earth, it will be just another incident and I will be disappointed. If, however, it was captured by the earth it will be an outstanding link in history."

  Urey backed the U.S. program designed to land men on the moon by 1970 and said he did not consider the cost excessive. (*Denver Post*, 2/10/65)

- During a luncheon speech at the Symposium on Unmanned Exploration of the Solar System, presented in Denver, Colo., by the American Astronautical Society, Maj. Gen. Don R. Ostrander, Commander of USAF Office of Aerospace Research, formerly NASA Director of Launch Vehicle Programs, said it was now generally agreed that the near-earth space area "looks more promising from the standpoint of potential military applications" than lunar bases, Mars flights, and other projects suggested earlier. USAF was seeking refinement of its knowledge in astronomy, geophysics, geodesy, and other areas. More pressing, he said, was to study the space environment as related to weapon systems and orbiting satellites. (Partner, *Denver Post*, 2/29/65)

February 9: In a report presented at the 55th national meeting of the American Institute of Chemical Engineers in Houston, NASA Lewis Research Center engineer E. W. Ott said that moisture in an astronaut's breath could escape into his space capsule, accumulate and float at zero gravity, and short out electrical systems it might come into contact with. He said something like this was believed to have happened when Astronaut Gordon Cooper had had to bring his space capsule in under manual control in May 1963. "There is good evidence that water found its way into automatic control equipment and caused malfunctioning." (Justice, *Houston Post,* 2/10/65)

- NASA Manned Spacecraft Center engineer John H. Kimzey told a meeting of American Institute of Chemical Engineers that fires during simulated spaceflight had the puzzling habit of burning fiercely initially, dying out so the flame disappeared, but flaring to life when force of gravity took over. Kimzey speculated that in weightless conditions, carbon monoxide, carbon dioxide, and water vapor created by the fire might surround the flame and cut off both oxygen and fuel. Motion pictures of the "dead" fires had shown no indication of either light or infrared heat energy coming from fire locations. (Burkett, *Houston Chron.,* 2/10/65)
- DOD announced that U.S. would sell United Kingdom: (1) F-4 (Phantom II) fighter/close-support aircraft and (2) C-130E combat assault transport aircraft. It had also been agreed that the two countries would expand existing program of cooperation in defense research and development. Serious consideration would be given to joint development of advanced life engine for vertical- and short-takeoff aircraft. (DOD Release 80-65)
- Douglas Aircraft Co. Missile and Space Systems Div. reported that tests conducted for USAF had indicated that a spin in a centrifuge might recondition astronauts living for weeks or months in a state of weightlessness. Previous research had indicated that long stays in the weightless state could have a debilitating effect on the body and cause the heart and circulatory system to lose their tone. (NYT News Service, St. Louis *Post-Dispatch,* 2/9/65)
- A USAF Strategic Air Command crew successfully launched a Minuteman ICBM from Vandenberg AFB. (*M&R,* 2/15/65, 12)

February 10: NASA Manned Spacecraft Center announced four women were among the applicants being considered for the new scientist-astronaut program. They would receive the same consideration as the male applicants. In the past, because of the requirement that applicants have either a test pilot rating or at least 1,000 hrs. in jet aircraft, women were not seriously considered. MSC had forwarded the names of just over 400 applicants to the National Academy of Sciences, which would make recommendations on selection of 10–15 scientist-astronauts. (Maloney, *Houston Post,* 2/11/65)

- At an AEC FY 1966 authorization hearing, Rep. Melvin Price (D-Ill.), acting chairman of the Joint Congressional Committee on Atomic Energy, attacked "wasteful, irresponsible vacillation" in developing nuclear power systems for use in space. He cited the Snap-50 project which "in 1962 had a development objective through flight test. About a year ago, the flight test objective was dropped for a complete flight system ground test. This year, we have had another change in objective, dropping the complete system ground test and cutting back to component test objectives. It thus appears we are moving rapidly backward in this program. . . ."

Rep. Price said millions of dollars were being spent to develop these power sources for DOD and NASA but that the projects were often stopped short of flight testing because of lack of funds. He mentioned "several specific construction items which were not approved by the Budget . . . [because] there was no indication of a user for the finished product. . . .

"The concern of the Committee is that this seems to be a pattern that happens on so many of these projects. Usually when it starts, you put several more millions in for a few years and then finally cut it off completely. . . . We are just worried about this pattern. If we thought it was going to end with the same result, it might be wise to cut it off earlier rather than later. . . . I have a pretty deep feeling we are back on this requirements merry-go-round." (Transcript)

February 10: Detection of the existence of life on Mars could be accomplished by a manned Mars-orbit mission without the necessity of a Mars landing, according to two NASA Ames Research Center officials in a Copley News Service interview. Alvin Seiff, Chief of Ames' Vehicle Environment Div., and David E. Reese, Jr., Assistant Chief of that division, said life on Mars could be detected from as far away as "several hundred thousand feet" from the planet's surface. "We think we could get good accuracy during even hypersonic flight around Mars. . . . We don't need to land men on Mars to find out what goes on there. We can find out about its atmosphere and whether life exists there through the use of a variety of instruments we now have at hand," Seiff said. Seiff and Reese were in Denver attending AAS Symposium on Unmanned Exploration of the Solar System. (Macomber, CNS, *San Diego Union*, 2/10/65)

- Hughes Space Systems Div. at El Segundo, Calif., signed a contract with NASA to propose designs of a beacon that could be placed on the moon as a guide for safe landing for moonbound Apollo astronauts. The beacon would be landed on the moon with a Surveyor spacecraft. (UPI, Phil. *Eve. Bul.*, 2/10/65)

February 11: MARINER IV received 12 commands from JPL to check out spacecraft equipment that would be used if the spacecraft was still operating normally when it reached Mars next July 14. These commands dropped a lens cover off the television camera, turned on a scanning platform that carried the camera and two Mars sensors, turned on portions of television system and checked out the capability of MARINER IV to perform the encounter sequence. It was not planned to take television pictures during this sequence. The lens cover was dropped at this time rather than at planet encounter to shake loose any possible dust particles that might interfere with the Canopus sensor, a light sensing device that locked on the star Canopus to prevent the spacecraft from rolling. (NASA Release 65-43)

- USAF Titan III-A rocket was launched from Cape Kennedy, hurled its third stage (transtage) and two satellites into orbit in a maneuverability test involving three different orbits. Primary goal of the mission was triple ignition of the transtage's engine. First firing, about five minutes after launch, injected the 7,000-lb. rocket-payload assembly into near-earth orbit of 128-mi. apogee and 108-mi. perigee. After traveling once around the earth, during which the rocket performed a deliberate somersault, the transtage ignited again, burned 37 sec., and shifted the rocket into an elliptical orbit of 1,766-mi. apogee and 116-mi. perigee. During one and one half orbits around the earth, the rocket performed a second deliberate somersault; a third firing put it in circular orbit with parameters of apogee, 1,737 mi.; perigee, 1,721 mi.; period, 145.6 min.; and inclination, 32.15°.

Titan III–A's transtage then ejected a 69-lb. experimental communications satellite (LES I), and ejected a 1,000-lb. metal chunk to demonstrate its ability to launch more than one payload.

LES I was to have fired a solid-propellant motor to move to an elliptical orbit with an apogee of 11,500 mi. and perigee of 1,725 mi., but the motor failed to fire. LES I continued to orbit near the transtage and the metal chunk. LES I (Lincoln Experimental Satellite) had been built by MIT's Lincoln Laboratory to test advanced components, materials, and techniques which might apply to future communications satellites. Radio signals were to be exchanged between LES I and ground stations. (UPI, *NYT*, 2/12/65; AP, *Houston Post*, 2/12/65; AP, Balt. *Sun*, 2/12/65; *U.S. Aeron. & Space Act., 1965*, 133)

February 11: Flight testing of the parachute landing system for two-man Gemini spacecraft was completed. The test simulated an emergency in which a stabilizing drogue chute failed to deploy from the capsule. Dropped from a plane at 17,000 ft., the unmanned, two-ton capsule landed safely after a pilot chute and the main 84-ft.-dia. chute deployed on schedule. This was the tenth straight successful test. (AP, *Houston Post*, 2/12/65)

- NASA announced that it would negotiate with Space Technology Laboratories and Thiokol Chemical Corp. for six-month, fixed-price contracts of approximately $1.5 million for definition of a program to develop and produce a 100-lb.-thrust rocket engine. The multipurpose engine, designated C–1, would be designed for spacecraft attitude control and maneuvering systems and also for launch vehicle ullage and attitude control systems. It would be powered by the hypergolic, storable liquid propellants monomethylhydrazine (MMH) and nitrogen tetroxide. (NASA Release 65–41)

- FAA released the first two volumes of a comprehensive five-part report on the sonic boom public-reaction study conducted in Oklahoma City in 1964. Their main conclusion: weather had a greater effect in determining the strength of booms than suspected, but the effect was within a corrective capability.

The first volume, "Sonic Boom Exposures During FAA Community-Response Studies Over a Six-Month Period in the Oklahoma City Area," prepared by NASA Langley Research Center, said measurements taken directly under the flight path of the supersonic fighters showed that about 80% of the booms were lower in intensity than scientists expected. About 20% equalled or exceeded the anticipated levels.

"Meteorological Aspects of the Sonic Boom," prepared by the Boeing Co., revealed that: weather had a greater effect on booms generated by planes flying less than mach 1.3 than those produced by aircraft exceeding mach 1.3; overpressures were increased by headwinds but decreased by tailwinds and crosswinds, with variations as much as 20%, particularly in the mach 1.3 range; under some atmospheric conditions, including such factors as wind, temperature, and even the time of day, sonic booms may vary from a complete cut-off with no boom heard to heavy overpressures concentrated over a small area or spread almost unlimited over a wide lateral area; turbulence had the effect of distorting booms and increasing or decreasing intensity and distribution. (FAA Release 65–15; UPI, *Minn. Trib.*, 2/12/65)

February 11: In a luncheon address to the National Security Industrial Association in Washington, D.C., Lt. Gen. W. A. Davis (USAF), Vice Cdr. of Air Force Systems Command, discussed AFSC accomplishments in 1964: ". . . Important strides were also made in the area of space. We carried out intensive studies on the Manned Orbiting Laboratory (MOL). Last month the Secretary of Defense announced that proposals are being requested from industry for design studies to assist in developing the cost and technical information needed to proceed with full scale development of the MOL. Titan III, the Standard Launch Vehicle 5A, completed two highly successful test launches. Systems Command also provided support to the National Aeronautics and Space Administration. This included the use of the Atlas-Agena combination to launch Ranger 7 on its successful photographic mission to the moon and Mariner 4 on its way to Mars. We also conducted the first launch of the man-rated Titan II. It was very successful.

"There are a number of tasks ahead of us in space. One of our most promising present programs is the Titan III space launching system. The Titan III will be used to launch the 24 satellites for the interim Defense Satellite Communications System. It will also be used to launch the Manned Orbiting Laboratory (MOL), which is designed to determine man's capability to perform military functions in space. The MOL will have an important bearing on our future space capabilities." (AFSC Release)

- In an isolation test in caves 330 ft. below ground near the French Riviera, two volunteers were reported to be "steadily losing time." Antoine Senni was about three weeks behind the actual date, observing New Year's Day on Jan. 20; Josiane Laures thought the date was Jan. 4, when it was actually Jan. 20. Scientists were conducting an experiment on man's ability to function in an environment where there was no day or night. (Reuters, *NYT*, 2/11/65, 54)
- Lockheed Missiles and Space Co. had been awarded $8,052,000 cost-plus-incentive contract for Agena D launch services at Eastern and Western Test Ranges during calendar year 1965, DOD announced. (DOD Release 87–65)
- Maj. Gen. George P. Sampson (USA), recently retired as Deputy Director of the Defense Communications Agency, was appointed Director of Operations for ComSatCorp. (ComSatCorp)
- *Moscovsky Komsomolets* reported that the wife of Valery Bykovsky, Russian cosmonaut, was expecting a second baby. (Reuters, *Chic. Trib.*, 2/12/65)

February 12: Escape system for the two-man Gemini spacecraft was successfully tested by NASA at the U.S. Naval Ordnance Test Station, China Lake, Calif. Simulating a pad abort condition, the test vehicle was mounted atop a 150-ft. tower equal in height to the Titan II launch vehicle; the side-by-side ejection seats were thrust out and away from the test vehicle to an altitude of about 350 ft. The mannequins landed by parachutes approximately 850 ft. downrange. The Gemini escape system was designed and built for NASA by Weber Aircraft Co. (MSC *Roundup*, 2/17/65, 2)

February 12: After almost 7 years, VANGUARD I appeared to be silenced. Its radio signals had weakened to a point where NASA tracking engineers thought the satellite might never be heard from again, according to NASA announcement. The six-inch, 3.25-lb. sphere was the second U.S. satellite, launched by USN as part of the International Geophysical Year program. For more than six years, it had transmitted radio signals from space with power from only six solar cells. Officially known internationally as 1958 Beta II, VANGUARD I was circling the globe every 134 min. and had an apogee of 2,442 mi. and a perigee of 402 mi. (NASA Release 65-45)

- USAF scientists at Hanscom Field, Mass., said they had hit EXPLORER XXII with a ground-based laser gun and had photographed the spot of reflected light and recorded it photoelectrically in relation to surrounding stars. (AP, L.A. *Herald-Examiner*, 2/13/65)

February 13: USAF Athena missile was fired from the Army's launch complex at Green River, Utah, to impact at White Sands Missile Range; a second firing was postponed because of technical difficulties. (AP, St. Louis *Post-Dispatch*, 2/14/65, 13A)

- California Institute of Technology received a $1,645,000 grant from the National Science Foundation to build the first of eight 130-ft.-dia. dish antennas to be trained on distant, recently discovered sources of radio energy, quasi-stellar radio sources, called "quasars"—the most distant objects yet discovered. (AP, *Wash. Post*, 2/14/65; Sci. Serv., *NYT*, 2/24/65, 5)

February 14: Dr. Fred Whipple, director of the Smithsonian Astrophysical Observatory, Cambridge, had suggested landing on a comet. Dr. Whipple also speculated that if a space vehicle were sent near a comet scientists could use a low-velocity probe that could be put into an orbit in the comet's vicinity for a week or more to study the velocities of gas and dust particles boiled off the comet by solar radiation. The probe would also be able to take core samples of the comet to give direct measurement of one of the oldest physical processes in the solar system, Dr. Whipple said. (*NYT*, 2/14/65, 50)

- JPL scientists had sent notices to professional and amateur astronomers asking them to keep the strip of Mars over which MARINER IV would fly next July 14 under surveillance from now on, with special emphasis on photography in March.

  "We don't know what we may learn through this procedure," a JPL spokesman said, "but we want all the information we can get. Suppose, for example, MARINER photographs what looks like a dust storm. We'll have a better chance of determining that fact if we have pictures of the same phenomenon taken through earth telescopes, even though it's a different storm months earlier." (AP, Seattle *Post-Intelligencer*, 2/15/65)

- The Royal Astronomical Society of London had awarded gold medal to Gerald Maurice Clemence, senior research associate and lecturer in the department of astronomy at Yale Univ., for his "application of celestial mechanics to the motions in the solar system" and for his "fundamental contributions to the study of time and the system of astronomical constants." (*NYT*, 2/15/65, 17)

February 15: NASA announced it had asked astronomers and scientists in 38 countries to help analyze and interpret the closeup photographs of the moon taken by RANGER VII in July 1964. The scientists would first receive a set of 199 high-quality pictures taken by RANGER VII's "A" camera; photographs taken by other cameras would be sent later.

NASA had also sent RANGER VII photographs to the European Space Research Organization, the European Launcher Development Organization, the International Committee on Space Research, and the United Nations. (UPI, Phil. *Eve. Bull.*, 2/15/65)

- Christopher C. Kraft, NASA Manned Spacecraft Center director of flight operations, said the three-orbit Gemini GT–3 flight with astronauts Virgil I. Grissom (USAF) and John W. Young (USN) would be much safer than Project Mercury orbital space flights. The astronauts would not depend solely on the braking rockets to bring them back to earth. They would make maneuvers during the first and third orbits to bring the spacecraft back through the atmosphere even if retrofiring braking rockets failed. Toward the end of the third orbit, near Hawaii, Grissom would fire the rockets for about two minutes, sending the Gemini spacecraft into a 54-mi. orbit which would be a reentry path. Over Los Angeles, the main braking rockets would be fired to drive the spacecraft down to a landing about 70 mi. east of Grand Turk Island in the Atlantic Ocean. Kraft said if the braking rockets did not fire, the GT–3 craft would land about 1,000 mi. due west of Ascension Island. (*Galveston News-Tribune*, 2/16/65)

- NASA announced it had determined the areas of Mars to be photographed by MARINER IV's TV camera during the July 14 fly-by. Recording of the first picture would occur when the spacecraft was approximately 8,400 miles above the Martian surface. MARINER's camera would be pointing at the northern Martian desert, Amazonis. The camera would then sweep southeast below the Martian equator covering the Mare Sirenum, the southern desert Phaethontis, Aonius Sinus, and into the terminator or shadow line. The spacecraft would be about 6,300 mi. above Mars for the final picture. (NASA Release 65–42)

- First successful flight test of a miniature mass spectrometer specifically for biomedical and environmental use was made at NASA's Flight Research Center. The system weighed 46 lbs., measured 10 x 10 x 20 in. with vacuum system, and could monitor and chemically analyze samples of gases that might be encountered in either the cockpit environment of the spacecraft or in the pilot's respiratory system. It could detect buildup of harmful gas or absence of necessary life support gas. The mass spectrometer was built by the Consolidated Systems Corporation, Monrovia, California. (FRC Release 6–65)

- NASA established an Office of Industry Affairs at the Pentagon by arrangement with the Assistant Secretary of Defense (Installations and Logistics), to coordinate DOD–NASA mutual interest procurement and contract management matters, including quality assurance. Clyde Bothmer, who formerly directed management operations in NASA's Office of Manned Space Flight, became Director. (NASA Release 65–55; NASA Ann. 65–35)

February 15: Clarence A. Syvertson had been named Director of NASA's new Mission Analysis Division of the Hq. Office of Advanced Research and Technology, NASA announced. He would be responsible for study of future missions for research and technology programs. The Mission Analysis Division, to be located at NASA Ames Research Center, would be staffed by outstanding scientists drawn from all NASA Centers and would be organized along aeronautical and space mission lines. (NASA Release 65–46; NASA Ann. 65–34)

- President Johnson sent to Congress his annual reports on the National Science Foundation, the ComSatCorp, and U.S. participation in the International Atomic Energy Agency. In message accompanying the NSF report, President Johnson said: "Close and understanding accord between science and public affairs is an imperative for free societies today." Science would be looked to for use in technology and industry, health programs, exploration, and, "most especially for the guidance that will permit us to proceed with greater security and greater confidence toward our goals of peace and justice in a free world."

    In a message accompanying the report on the ComSatCorp, the President said the goal of the U.S. was "to provide orbital messengers, not only of word, speech and pictures, but of thought and hope" for the world.

    "The past year has seen important advances in the program to develop a global communications satellite system. The first launch of a commercial satellite is to take place in the early months of this year.

    "Through the initiative of the United States an international joint venture has been established. Under the law I have designated the Communications Satellite Corp. as the U.S. participant. The corporation is to be the manager on behalf of all participants.

    "The corporation has now been financed, has constituted its first board of directors to replace the original incorporators and has moved forward with its program. All agencies of the Government with responsibilities under the act have made important and faithful contributions with the sympathetic assistance of the congressional committees concerned.

    "The new and extraordinary satellite telecommunications medium bringing peoples around the globe into closer relationship is nearer to fulfillment, heralding a new day in world communications."

    In its second annual report ComSatCorp noted that it had ended 1964 with about $190 million in short-term holdings and more than 137,000 shareholders. It reported it had agreements with 18 countries to join in a single global system with ComSatCorp as manager and said that a satellite was being readied for launching in March.

    The report on the Nation's participation in the International Atomic Energy Agency was accompanied by a covering letter which said 1963 "will possibly be marked in I.A.E.A. history as the year in which a firm foundation was laid for its system of safeguards against the diversion of materials to military use." (Text, *CR*, 2/15/65, 2605; *NYT*, 2/16/65, 1; AP, *NYT*, 2/17/65, 64)

- NASA and U.S. Army Materiel Command adopted an agreement for joint participation in low-speed and Vtol aeronautical research. The research program was centered at NASA Ames Research Center. (NMI 1052.7)

February 15: NASA had selected the Bendix Field Engineering Corp., Owings Mills, Md., to negotiate a cost-plus-award-fee contract for continued operation, maintenance, and support services of the NASA Manned Space Flight Network of tracking stations. Contract was valued at about $36 million over two years. (NASA Release 65–48)

- In National Science Foundation's annual report to the President and the Congress, NSF Director Leland Haworth said the Foundation was "attempting to formulate an approach . . . to interfield priority assessment which would take into account the probable contributions of NSF-supported basic research to the solution of a variety of national problems. Thus, for example, it is possible that a whole cluster of basic research activities might justifiably be supported in several fields of the behavioral and environmental sciences, all of which would in one way or another shed light on what is now called the 'transportation-urbanization' problem. . . ."

  Discovery of what may be the first real baby star—one apparently much smaller than the moon—was described in the NSF report. NSF credited the find to Dr. Willem J. Luyten, a University of Minnesota astronomer doing research aided by an NSF grant. Having roughly one-thousandth the diameter of the sun, the new-found dwarf apparently contained 300 tons of material per cubic inch of volume, more than 100 million times the density of water. There was no question about the discovery of the star, the report said. The only possible question was whether the distance to it had been figured accurately, because that would have a bearing on computing its actual size. (Carey, Wash. *Eve. Star,* 2/16/65; *Science,* 2/25/65)

- Lt. Gen. Frank A. Bogart (USAF, Ret.) was appointed Director of Manned Space Flight Management Operations. Since joining NASA on December 1, 1964, General Bogart had served as Special Assistant to the Associate Administrator for Manned Space Flight. (NASA Announcement 65–30)

- Dr. Eugene Konecci of the National Aeronautics and Space Council staff reported to the Medical Society of the State of New York at its annual convention that the semicircular canals of the inner ear had been demonstrated to play a key role in motion sickness that astronauts might experience in a rotating, orbiting spacecraft. Capt. Ashton Graybiel at the USN School of Aviation Medicine, Pensacola, expressed optimism that astronauts could be taught to overcome the effects of motion sickness. One way, he said, was to precondition selected persons by teaching them how to avoid movements that would invariably upset them. Another promising development, Dr. Graybiel said, was drug research. (Simons, *Wash. Post,* 2/16/65)

- Dr. Karl G. Harr, Jr., President of Aerospace Industries of America, Inc., addressed the Economic Club of Detroit:

  ". . . the aerospace industry of today does indeed represent a truly unique phenomenon in industrial history in almost all of its aspects. . . . it is that industry which places at the disposal of the nation—both its public and its private sectors—the capacity to manage the research, development and production of the most technologically advanced product that is possible—for whatever purpose desired.

". . . it is essential that we all understand the principal factors—historical, present and future—that have produced and will continue to produce this uniqueness.

"First, the genesis and evolution of what is today's aerospace industry is a direct product of the nation's post-World War II history and is inextricably linked thereto. . . . World War II unleashed for the world, but particularly for the United States, two revolutions which have been gaining momentum ever since. The first of these was a form of economic revolution which saw the economy of the United States surge into new dimensions. The second was a scientific/technological revolution which saw all that had gone before in man's scientific history fade into a pale background. . . .

"World War II itself provided an extreme example of the explosive expansibility of the industrial base of the United States. This expanded industrial base remained after the war to serve as a foundation for a general economic upsurge.

". . . the aerospace industry has become and remains, in a very real sense, an instrument of national policy, not only in terms of the hardware directly provided the government, but also as it underpins the economic/technological advances in the private sector of our economy. . . ."

Discussing the future of the industry, Harr noted that "the size and viability of this industry is not tied to defense and space programs, important as these have been and will continue to be in shaping its destiny. It is tied, rather, to the total technological progress of the nation, meaning the application of advanced technology to whatever purposes may be desired. Programs now well underway in such diverse fields as air freight, urban transportation, desalination, oceanography, 2000-mph aircraft and hundreds of others serve to illustrate this fact. . . ." (Text)

February 15: In an editorial headed "Space Racing After Seven Years," the *Miami Herald* said: "Fast starters don't always win. The match race in space between the United States and the Soviet Union is shaping up satisfactorily, from our viewpoint. The start was easy to identify, but the finish line is nowhere in sight." (*Miami Her.,* 2/15/65)

- Among Weather Bureau employees honored at the 17th Annual Dept. of Commerce Awards Program were: Dr. Sigmund Fritz, for outstanding contributions to meteorological research in the fields of solar radiation, ozone, and meteorological satellites, for highly distinguished authorship, and for exceptional leadership as Director of the Weather Bureau's Meteorological Satellite Laboratory; Louis P. Harrison, for highly distinguished authorship and outstanding contributions to the fields of barometry and psychrometry; David S. Johnson and Dr. S. Fred Singer, a joint award in recognition of unusual ingenuity, leadership, and guidance in the development and implementation of a National Operational Meteorological Satellite System; Jay S. Winston, for valuable contributions to meteorology in the areas of general circulation studies, the interpretation of weather satellite data, and the heat budget of the earth-atmosphere system. (Commerce Dept. Release WB 65-1)

February 16: Saturn I (SA-9) two-stage launch vehicle, launched by NASA from Cape Kennedy, orbited a 33,000-lb. multiple payload, of which

3,200 lbs. was the PEGASUS I meteoroid detection satellite. Orbital data: apogee, 745 km. (462 mi.); perigee, 496 km. (308 mi.); period, 97 min.; and inclination, 31.7°. This was the eighth successful test in eight flights for Saturn I; PEGASUS I was the first active payload launched in the Saturn tests.

At launch, an Apollo command and service module boilerplate (BP-16) and launch escape system (Les) tower were atop Saturn I, with PEGASUS I folded inside the service module. After first-stage separation and second-stage ignition, Les was jettisoned. When second stage (S-IV) attained orbit, the 10,000-lb. Apollo boilerplate command and service modules were jettisoned into a separate orbit. Then a motor-driven device extended winglike panels on the Pegasus satellite to a span of 96 ft. Each wing consisted of seven frames hinged together and made up of 208 panels. PEGASUS I remained attached to Saturn I's second stage as planned. A television camera, mounted on the interior of the service module adapter, provided pictures of the satellite deploying in space.

PEGASUS I exposed more than 2,300 sq. ft. of instrumented surface, with thicknesses varying up to 16/1000 in. As meteoroid particles collided with the surface of the panels, they would be registered electronically and reported to earth. Exposure of the large panel area over a long period would give designers of manned and unmanned spacecraft a good sample of meteoroid data.

PEGASUS I would be visible from the earth without the aid of telescope on clear nights. (NASA Release 65-38; *Marshall Star*, 2/24/65, 1, 5; AP, *Houston Chron.*, 2/16/65; Clark, *NYT*, 2/17/65; AP, Benedict, *Wash. Post*, 2/17/65; Hoffman, N.Y. *Her. Trib.*, 2/17/65; Sehlstedt, Balt. *Sun*, 2/17/65; *U.S. Aeron. & Space Act., 1965*, 133–134)

February 16: On the floor of the House of Representatives, Congressman George P. Miller (D–Calif.) commented upon the successful Saturn I launch: " . . . this morning the United States took another giant stride in the exploration of space. At 9:37 a.m. a Saturn rocket . . . with its 1,500,000 pounds of thrust, lifted off the launch pad at Cape Kennedy, Fla., on a mission to place in orbit around the earth the Pegasus satellite.

"This was the eighth launch of the Saturn rocket out of eight attempts, a truly outstanding scientific and engineering accomplishment of the men of the National Aeronautics and Space Administration and of the many contractors who worked so long and hard to make this event a success.

"I may say that the gratifying success of the Saturn booster has been matched in other programs as well.

"I need only point to the Tiros weather satellite.

"Nine have been launched out of nine attempts.

"I think we have every right to be proud of our space team on this day of outstanding achievement." (NASA LAR IV/30–32; *CR*, 2/16/65, 2630)

- North American Aviation's XB–70A made its fifth flight from Palmdale, Calif. Maximum speed was mach 1.6; maximum altitude 45,000 ft.; duration of flight, 1 hr. 10 min. During the flight the wingtips were folded to 25° and then to the full-down position of 65°. It was the first time this total deflection had been attempted. Flutter and stability characteristics, the inlet control system, and the air inlet bypass door system were investigated during the flight. Although the emergency landing parachute system did not function during landing at Edwards AFB, the aircraft completed a normal landing with normal braking. The drag chute had deployed, but the three-chute pack did not deploy. (*Av. Wk.*, 2/22/65, 22; UPI, *NYT*, 2/17/65, 74)

February 16: NASA's Flight Research Center engineers had made direct comparison of the noise levels generated by the XB-70 and a 707-120B commercial jet transport under the same atmospheric conditions. NASA made the measurements as part of its general study of runway noise conditions for use in the design of a supersonic transport. (FRC Release 8-65)

- NASA awarded a $8,879,832 fixed-price contract to the Univac Division of Sperry Rand Corp., for digital data processors to be used in Project Apollo. The contract also called for computer programming assistance in modifying present computer programs or developing new ones for Project Apollo requirements. (NASA Release 65-50)
- Dr. Hugh L. Dryden, NASA Deputy Administrator, received an honorary Doctor of Science degree from the Swiss Federal Institute of Technology in Zurich. The presentation was made by the Swiss ambassador, Dr. Alfred Zehnder, at the Embassy residence in Washington, D.C. (NASA Release 65-47)
- Dr. Charles S. Sheldon of the National Aeronautics and Space Council staff told MIT students: "The Russians are pretty cautious about disclosing much about their launch vehicles, but we know pretty well what vehicles they're using." Dr. Sheldon noted that ZOND II, the Soviet Mars probe, would pass near the planet perhaps one month after MARINER IV. The U.S. Mars probe was due to come within 5,400 mi. of the planet on July 14. (*Boston Globe*, 2/17/65)
- FAA Administrator Najeeb Halaby, British Aviation Minister Roy Jenkins, and French Aviation Minister Marc Jacquet met in London and agreed to work for joint establishment of operating conditions for supersonic jet transports. The British and French ministers arranged for the next stage in the development of their joint Concorde supersonic transport, which the British Labor government reportedly had wanted to cancel. The ministers also discussed a new Anglo-French project for a subsonic transport, an "air bus" that could take 200 to 300 passengers on short interurban hops. (*Wash. Post*, 2/17/65)
- Progress in developing the laser for communications use was evidenced by U.S. Army report that it had transmitted all seven of New York's standard television channels simultaneously on a laser. Although the seven TV channels had been transmitted over a distance only the width of a room, the Army said they could have been received at a range of several miles. Research described had been carried out at the Army Electronics Command's laboratories in Fort Monmouth, N.J. The Army was interested in laser communications because the narrow beams could be transmitted between specific points, making enemy interception difficult. (Sullivan, *NYT*, 2/17/65, 19)

February 17: NASA's RANGER VIII spacecraft, equipped with six television cameras to photograph part of the moon's surface, was successfully launched from Cape Kennedy by an Atlas-Agena B. Seven minutes after lift-off, the spacecraft and the Agena stage went into a parking orbit some 115 mi. above Africa; the Agena engines were cut off.

For the next 14 min., the combination coasted at 17,500 mph. Second burn of the Agena lasted 90 sec., increased the velocity to 24,476 mph, and freed the 808.8-lb. RANGER VIII from the major pull of the earth's gravity. Several minutes after injection, RANGER VIII was separated from Agena, which entered an elliptical orbit. About an hour after launch, RANGER VIII received and obeyed the command to deploy the solar panels that would convert solar energy to electrical power for its equipment. About $3\frac{1}{2}$ hrs. after launch, RANGER VIII completed its orientation maneuvering, achieved attitude stabilization, and pointed a high-powered antenna toward earth.

The projected impact area was the Sea of Tranquility, a dark area relatively free of crater rays, near the shadow line on the three-quarter moon. Lower-angle lighting was expected to give more contrast and better definition of detail than was in the photographs made by RANGER VII.

A small rocket aboard the craft would be fired later to correct a moon-miss error on either side of the target; tracking calculations showed that the path of the vehicle would miss the edge of the moon by only 1,136 mi., well within the correction capability. (NASA Release 65–25; AP, Benedict, *Wash. Post*, 2/18/65; UPI, *Chic. Trib.*, 2/18/65; AP, *Houston Post*, 2/18/65; Appel, *NYT*, 2/18/65, 1)

February 17: X-15 No. 2 was flown by Maj. Robert Rushworth (USAF) to 95,100 ft. altitude at a maximum speed of 3,511 mph (mach 5.27) to obtain data for several research programs. (NASA X–15 Proj. Off.; *X–15 Flight Log*)

- NASA and DOD announced a memorandum of agreement to establish a Delta launch capability at the Western Test Range (WTR). Costs would be shared, based on the estimated use of the vehicle by each agency. Existing USAF Thor-Able-Star launch sites would be adapted for Delta use wherever practicable. Launch pads and blockhouses would be used on a shared basis, with each agency responsible for its own missions. NASA would exercise launch vehicle control over all WTR Delta launches. NASA Goddard Space Flight Center and USAF Space Systems Division would be responsible for carrying out the agreement.

  NASA would be responsible for developing an improved Delta launch vehicle to meet both agencies' mission requirements for use at both WTR and ETR. DOD was planning to phase out the Thor-Able-Star and use the improved Delta for payloads carried by this vehicle class. (NASA Release 65–51)

- Hearings on NASA budget authorization for FY 1966 began before the House Committee on Science and Astronautics. Of the $5.260 billion requested for FY 1966, $4.576 billion was for research and development; $74.7 million was for construction of facilities; and $609.4 million was for administrative operations.

  Administrator James E. Webb testified: "This budget . . . supports an on-going successful research and development effort and the use of this knowledge to develop and test operating systems designed to give us what we need to know for national security, for applications in meteorology, communications, and other working satellite systems, and from which to make any decisions which may be called for in the future.

  "An important fact that underlies the President's 1966 budget decisions is that the program is now operating at a level of $5\frac{1}{4}$ billion dollars instead of at the optimum level of 5¾ billion originally recommended by President Kennedy or the 'fighting chance' level recommended last year by President Johnson. This means that we cannot accomplish the 15 Saturn V-Apollo flights now included in the program within the period of this decade. If all 15 flights are required to succeed in the lunar landing, then this will not be done before 1970. However, our overall major milestones are being met and we still have a reasonable opportunity for success on a flight earlier than the 15th and thus within this decade. In effect we will be launching toward the moon on earlier flights than we thought a year ago would be possible, but we simply cannot predict which flight will be the first either to orbit the moon or to land there.

"What we can say is this: the systems of equipment for the utilization of men for flights of all kinds out as far as the moon are now rapidly proceeding toward tests that will work out any imperfections; and our fast-developing knowledge of both the space environment and the capabilities of this equipment gives us more confidence than we had a year ago that we are on the right track and proceeding on a reasonable basis for the development of machines of this size and power. . . .

"In preparation of this budget, the President has faced two important facts. The first of these is that the central core of NASA activities as planned in 1961 is proceeding with excellent results. . . .

"The second major fact faced by the President was the necessity for a continued emphasis on supporting research and development. . . .

"Bearing these two facts in mind, this budget and this request for authorization call for an operating level which is approximately the same as that approved by Congress for fiscal years 1964 and 1965. This means that the work planned in 1961 for accomplishment in this decade must be spread out over a longer period, and the cost for the total will be increased. . . .

"The essential funds to give us some opportunity to make the lunar landing within this decade are included, as are funds for studies toward further use of the Saturn launch vehicles and the Appollo-LEM manned space flight systems in the period following the lunar landing. By 1969, we will have the capability to launch 6 Saturn-IB's and 6 Saturn V's per year. In the unmanned area, we have begun planning for a Voyager-Mars mission in 1971 with the possibility of a test flight in 1969; funds are included in the budget for expansion of this design effort during fiscal year 1966. Development and hardware procurement could then be initiated in fiscal year 1967 if appropriate. . . ." (Testimony; *1966 NASA Auth. Hearings*, 5-14)

February 17: Dr. Hugh L. Dryden, NASA Deputy Administrator, testified before the House Committee on Science and Astronautics on NASA's activities in the field of international cooperation: "Nineteen sixty-four was a year in which other nations emerged clearly as friendly competitors and valuable collaborators in space science and engineering and demonstrated by emulation their endorsement of our view that energetic efforts in these fields are essential contributors to the betterment of human society. I am thinking of such things as the energetic space programs of France, the demonstrated competence of Canada and Italy, the entry of the British aircraft industry into spacecraft engineering, the formal establishment of the European Space Research

Organization (ESRO) and the European Launcher Development Organization (ELDO), and finally of the successful first test flight of a large new booster by ELDO. . . .

"A year ago I reported our plans for including foreign experiments on NASA satellites under arrangements by which foreign experiments, selected in open competition with domestic proposals, are sponsored and financed by the experimenter's national space authority. One such experiment flew in 1964—a British ion mass spectrometer on EXPLORER XX, the U.S. fixed-frequency topside sounder. Six additional experiments were selected for flight, bringing the total to thirteen, with ten more under active consideration. We now have opened virtually all categories of NASA spacecraft, including Gemini and Apollo, to foreign participation on this cooperative basis. Indeed, we are now inviting foreign biomedic experts to a working conference in Houston, next April, to learn directly of the opportunities and constraints which apply to this program. . . .

"A noteworthy development was the fact that ESRO became the first foreign space agency to seek a ground station on American territory. After conducting a site survey and finding a suitable location near Fairbanks, Alaska, ESRO has formally requested the Department of State to begin negotiations for an agreement. The ESRO station is projected as an element in a network of tracking and data acquisition facilities. . . . France is establishing a North/South fence from France through the Canary Islands, Algeria, Upper Volta, Congo Brazzaville, and South Africa, with an injection-monitoring station in Lebanon. This chain will also serve ESRO, which expects to have additional stations at Spitsbergen, Brussels, and in Australia and the South Atlantic. Both the ESRO and French networks will be entirely compatible with NASA's, to maximize possibilities for mutual assistance. This is to our advantage, and we encourage it. . . .

"Let me bring you up to date on the status of our cooperation with the Soviet Union. You will recall that we have a series of agreements with the Soviet Academy of Sciences providing for three coordinated projects—in meteorology, in surveying the geomagnetic field, and use of ECHO II for communications tests. The project involving the observation and use of ECHO II is completed. The Soviet side observed the critical inflation phase of the satellite optically and forwarded the data to us; although not including radar data, which would have been most desirable. Communications via ECHO II between the U.K. and the U.S.S.R. were carried out in only one direction instead of two, at less interesting frequencies than we would have liked, and with some technical limitations at the ground terminals used. On the other hand, the Soviets provided very complete recordings and other data of their reception of the transmissions." (Testimony; *NASA Auth. Hearings*, 15–37)

February 17: Dr. Robert C. Seamans, Jr., NASA Associate Administrator, told the House Committee on Science and Astronautics: "I feel that our record over the past calendar year is evidence of the success we have had in building a team that is dedicated to effective management. The space flight record for 1964 is impressive by several standards: number of flights, percent of success, and variety of missions. The graph (spaceflight mission record) shows that, in terms of percentage, our

1964 success record is nearly the same as the high achieved in 1963; 83 percent compared with 85 percent in 1963. However, we more than doubled the number of successful missions from 1963 to 1964, from 11 to 25. . . .

"Other significant measures of the pace and rate of change in our space program have been our performance in terms of spacecraft operations and data return. These performance indicators are at the heart of a successful space program.

"For example, in 1964 an average of 54 experiments were functioning in space throughout the year; this is an improvement of over 35 percent from 1963, when we averaged 40 working experiments.

"The volume of information brought back from space, measured in millions of data points per day, shows a tenfold increase over previous years: in 1964 we were collecting about 57 million bits of information each day from our flight missions. . . ."

Dr. Seamans listed NASA management accomplishments during the past year "to find new techniques and new methods to carry out our jobs." Among them: establishment of Mission Analysis Div. from the Hq. Office of Advanced Research and Technology at Ames Research Center; conducting the joint DOD–NASA Launch Vehicle Cost Study; growth of incentive contract program ("over $1 billion are under active incentive contracts"); and application of phased project planning.

"The budget presented here has already undergone critical review by NASA's management, the Bureau of the Budget, and the President. It does not provide any contingency funds for the approved missions; it is predicated on a cost reduction program that will require us to operate more efficiently; it represents a carefully pared program priority list. In summary:

"First, NASA is dedicated to the accomplishment of the present approved missions and projects in terms of cost, schedule, and technical performance.

"Second, new effort is needed to maintain a position of leadership in aeronautics and space. This includes the definition of a new program for exploration of the planets commencing with Mars in 1971; the research and design necessary for effective extension of present Apollo and Saturn capabilities for manned flight; integration of the Centaur stage with the Saturn launch vehicle for planetary and other unmanned payloads; initiation of an advanced solar observatory satellite; and utilization of a prototype XB–70 aircraft for aeronautical research.

"Third, an aggressive research and advanced development effort must be maintained in many fields, including chemical and nuclear propulsion, to assure the nation meaningful options and alternatives in the selection of future aeronautical and space goals and the ability to react decisively to external pressures and opportunities. . . ." (Testimony; *NASA Auth. Hearings*, 37–51.)

February 17: Experimental solar still stations were being tested by Dr. Everett D. Howe, director of the Univ. of California at Berkeley's Sea Water Conversion Laboratory, it was reported. The small stills were located on islands in the South Pacific, where climate was favorable for testing solar distillation. The stills, made of light plastic and concrete or of metal and glass, produced two to five gallons of pure water a day.

Knowledge gained from the operation might eventually help the thousands of persons in nonindustrial countries where water was in short supply and fuel and energy for such things as distillation of water was scarce. (*NYT*, 2/17/65, 28)

February 17: A new extraterrestrial mineral, previously unknown in nature and christened sinoite (silicon oxynitride, Si_2N_2O), had been discovered, NASA Ames Research Center announced. Discovered in a meteorite which had fallen near the village of Jajh deh Kot Lalu in Pakistan in 1926, the mineral was grey in color and occurred in rough rectangular crystals. Scientists who made the discovery were Dr. Klaus Keil of NASA Ames Research Center; C. A. Anderson, Hasler Research Center, Goleta, Calif.; and Dr. B. H. Mason, American Museum of Natural History, New York. The meteorite had been made available for study by New York's Museum of Natural History. (ARC Release 65-5)

- Joan Merriam Smith, who flew solo around the world last year, was killed when a private plane she was piloting crashed in the mountains 45 mi. from Los Angeles. (UPI, *NYT*, 2/18/65, 42)
- The largest balloon ever constructed completed a successful 26-hr. flight over western U.S. Launched at Chico, Calif., by AFCRL, the 450-ft.-long polyethylene balloon lifted a 450-lb. instrumented payload to a record 142,000-ft. altitude. At float altitude, the balloon became oblate, or pumpkin-shaped, with dimensions of 330-ft. diameter and 270-ft. height. Payload consisted of instruments to measure atmospheric temperature, density, and pressures: telemetry equipment; and a command receiver for control of the flight. It was parachuted to earth and recovered near Logandale, Nev. (OAR Release 2-65-6)
- Soviet Marshal Vasily Sokolovsky said during a Moscow press conference held in connection with the 47th anniversary of the Red Army that the U.S.S.R. was armed with intercontinental and global rockets whose nuclear warheads were equal to 100 million tons of TNT. He claimed strategic rocket troops now formed the backbone of the Soviet armed forces so that the effectiveness of the Soviet air force had been sharply increased by supersonic planes equipped with nuclear-tipped, long-range rockets. He said the Soviet navy was now built around atomic submarines of virtually unlimited range that were equipped with powerful missiles. New Soviet antiaircraft defenses were capable of reaching targets flying at any speed at any altitude. He claimed that the Soviet Union had undertaken several measures toward the relaxation of international tension, among them a reduction of $555 million in military spending for the current year. (*Sovietskaya Moldaviya*, 2/18/65, 1, ATSS-T Trans.; AP, Wash. *Eve. Star*, 2/17/65; *Wash. Post*, 2/18/65)
- The Jodrell Bank Observatory, British tracking station, was visited by a group of six Soviet scientists led by M. V. Keldysh, president of the Soviet Academy of Sciences. (AP, *Boston Globe*, 2/17/65)

February 18: RANGER VIII lunar probe successfully executed a midcourse maneuver that corrected the path established at launch and aimed it for impact on the moon in the Sea of Tranquillity, an area centered 2.6° north of the lunar equator and 24.8° west of the moon's north-south line. Signal for the maneuver was radioed from earth to activate commands previously stored in the spacecraft's computer. At that time, the 808-lb. photo probe was 99,281 mi. from earth, traveling toward the moon at 4,100 mph. First command ordered the spacecraft to roll 11.6°; the second ordered the pitch maneuver of 151.7°; the third commanded the motor to burn for 59 sec. Then, after the spacecraft was ordered to break its attitude stabilization locks on the sun and earth, a "go" command was transmitted and RANGER VIII executed the maneuver in about 27 min. The correction completed, the spacecraft reacquired its stabilization locks and continued on its course. One measure of the accuracy of the maneuver was the current expected impact time on the moon: Feb. 20 at 4:57:30 a.m. plus or minus 60 sec. The original planned time was 4:57:30 a.m.

The photo probe's initial course would have missed the trailing edge of the moon by 1,136 mi. (NASA Transcript; L.A. Times, Miles, *Wash. Post*, 2/19/65; Appel, *NYT*, 2/19/65; UPI, Phil. *Eve. Bull.*, 2/18/65; AP, *Chic. Trib.*, 2/19/65; *Av. Wk.*, 2/22/65, 34)

February 18: Dr. George E. Mueller, NASA Associate Administrator for Manned Space Flight, reviewing overall manned space flight objectives and reporting on the Gemini and Apollo programs in testimony before the House Committee on Science and Astronautics, said:

"Chronologically, the first objective of manned space flight is to establish man's capabilities in space. Next is the establishment of a national competence for manned space flight, including the industrial base, trained personnel, ground facilities, flight hardware, and operational experience. Next, we use this capability for further space explorations and for other purposes. Finally, accomplishment of all these objectives brings about United States leadership in space.

". . . In 1964, we concentrated our efforts on Gemini ground tests and accomplished the first flight test. Filling the pipeline with hardware and carrying out development testing of subsystems were the major Apollo activities. Now in 1965, we have entered a year that will be devoted to Gemini flight test operations and the conduct of Apollo system development tests.

"Looking at the remainder of the decade, 1966 will be the year when we learn new space flight techniques in the Gemini Program, and conduct unmanned earth-orbital flight tests of the Apollo/Saturn IB space vehicle. In 1967, Gemini will be available as an operational system and we will carry out manned earth-orbital flights of the Apollo/Saturn V space vehicle. Manned flights of the Apollo/Saturn V space vehicle are scheduled for 1968, leading to the beginning of manned lunar missions before the end of the decade.

". . . I want to emphasize again that Apollo is an orderly program. The buildup of the Apollo effort has proceeded over more than three and a half years to its full strength. It is not a crash program.

"The duration of Apollo, as we reported to the Congress last year, is one of the longer United States research and development programs, resulting in a schedule that permits rapid, orderly progress. The Apollo priority is high but not overriding. Parallel and backup development efforts are limited. Flight testing is being carried out on a logical basis, and only after all possible tests are conducted on the ground.

"Efficient use of available resources is a major consideration in the conduct of the Apollo program, whereas a crash program follows the most expeditious course regardless of cost.

"Finally, crash programs typically have goals beyond the existing state of technology and pursue these goals under the pressure of having to achieve a technological breakthrough. Apollo, on the other hand, harnesses current technology in the development of launch vehicles, spacecraft and facilities to permit effective space exploration. The greatest challenges in Apollo, in fact, are in the integration of those systems and the men who must fly them as well as the provision of ground operational support, and the overall management of this enterprise."

Dr. Mueller said that data received and analysis continued in 1964 regarding radiation and the lunar surface indicated that these matters were of less importance than had been previously deduced: ". . . First, the chance of a significant solar event occurring during a mission is very low. Second, if the worst solar flare previously observed had occurred during an Apollo mission, the maximum dose that could have been received at the bloodforming organs by astronauts in the command module would have been about 10 per cent of the allowable safe dose, rather than 15 per cent as estimated last year.

"Regarding the lunar surface, the data from RANGER VII have been very helpful. The large area photography has indicated the probability that there are many areas of the moon's surface where the design of the lunar excursion module is adequate with respect to surface slope and roughness. . . ." (Testimony; *NASA Auth. Hearings*, 53–134)

February 18: NASA Langley Research Center announced it would negotiate with Ling-Temco-Vought, Inc., an $8 million incentive contract to provide complete system management for the Scout launch vehicle. The contract would continue support services and materials LTV had provided under several contracts. It would include systems engineering, logistic support, operational support, test program support, payload coordination, preflight planning, data reduction and analysis, standardization and configuration control, reliability and quality assurance, vehicle modification, checkout, and delivery. (NASA Release 65–54)

- NASA awarded a $10,940,000 contract to Douglas Aircraft Co., Inc., for mission integration and launch services of Delta launch vehicles at Cape Kennedy. The cost-plus-fixed-fee contract covered the calendar year 1965. (NASA Release 65–52)
- Saturn V launch vehicle retro-motors developed 100,000 lbs. of thrust in test of the solid-propellant motors at USAF Arnold Engineering Development Center for NASA. (AEDC)
- ComSatCorp met with representatives of aerospace companies it had invited to discuss specifications for 24 communications satellites for a proposed DOD satellite system. Previously ComSatCorp had indicated it would contract with the Hughes Aircraft Co. to build the satellites. But when the Philco Corp. protested to the FCC, the FCC required ComSatCorp to give Philco and other competitors a chance to show their capabilities. (ComSatCorp; Weekley, *Wash. Post*, 2/17/65)
- U.S.S.R. formally protested to Norway plans of the European Space Research Organization to establish a satellite tracking station in

Spitsbergen. The U.S.S.R. said such a ground station for tracking space satellites could be used for "military purposes" in violation of the 1920 Spitsbergen treaty. Norway later rejected the Soviet charge. (*NYT*, 2/20/65, 2)

February 18: ComSatCorp filed with FCC a notice of a proposed $300,000 contract with AT&T for research data and consultant services on ground stations for the global communications satellite network. FCC was notified that ComSatCorp had analyzed and evaluated other proposals and had held subsequent discussions with those making proposals. (ComSatCorp)

- The U.S. was pressing the U.S.S.R. for clarification of its view that it did not violate the nuclear test ban treaty with an underground explosion that released radioactivity. The large underground test took place Jan. 15 in the Semipalatinsk region of Soviet Central Asia. Four days later the U.S. announced that it had detected radioactive debris from the explosion over the Sea of Japan. (*NYT*, 2/19/65, 17)

February 19: In testimony before the House Committee on Science and Astronautics, Dr. Homer E. Newell, NASA Associate Administrator for Space Science and Applications, summarized significant mission results: "U.S. scientific satellites achieved the following firsts: discovery of the radiation belt, determination of the earth's irregular geoid, ionospheric topside sounding (with Canada), solar spectroscopy, x-ray and ultraviolet satellite astronomy, polar orbits, and highly eccentric orbits to map the earth's magnetosphere. Our deep space probes achieved the first successful direct monitoring of the interplanetary environment, the first lunar surface detail, and the first successful flight to Venus. We may soon achieve the first successful flight to Mars, if MARINER IV completes its mission. In launch vehicle development, this program has yielded the first rocket stage using the high energy propellant combination of hydrogen and oxygen. It has also yielded the first and only all solid propellant space booster. U.S. meteorological satellites have yielded the following firsts: daylight cloud photography, night cloud observations including surface and cloud top temperatures, world cloud coverage, global heat balance and stratospheric temperature measurements, and direct cloud picture transmission to local users. Our communication satellites have been first in the following achievements: erection of large structures in space and their use as passive reflectors of radio signals; active repeating of radio signals at various altitudes and orbits of interest to system designers; transoceanic and intercontinental relay of teletype, facsimile, voice, data, and television; and achievement of the first true geostationary orbit.

"The specific record of 1964 space missions of the Space Science and Applications Program . . . is particularly informative because most major program areas achieved at least one highly significant success in 1964. Of the 10 scientific satellite missions attempted, 7 achieved full success and 2 partial success."

Dr. Newell observed that SYNCOM III communications satellite had been placed in a "virtually perfect circular equatorial orbit," then maneuvered "to within about 1 mile of its station over the western Pacific where it successfully performed all its planned experiments.

... The Syncom maneuver was comparable to the feat of Ranger VII in flying to within 6 miles of its target on the Moon.

"Having completed our experiments with Syncoms II and III, we are turning them over to the Department of Defense. . . . If required, full-time communications could be provided between the United States and southeast Asia by Syncom III. . . ." (Testimony; *NASA Auth. Hearings*, 136–62)

February 19: NASA selected Philco Corp.'s Aeronutronic Div. for negotiations leading to a nine-month, $1 million contract for research, development, and preliminary design of a lunar penetrometer system applicable to the Apollo program. The penetrometer, an instrumented package capable of assessing the hardness, penetrability, and bearing strength of a surface upon which it is ejected, could furnish lunar surface information to an orbiting Apollo spacecraft for scientific assessment of remote sites inaccessible to manned spacecraft or unmanned earth-launch probes. NASA Langley Research Center would negotiate and manager the contract. (NASA Release 65–59; LaRC Release)

- NASA approved inclusion of three x-ray and gamma ray telescopes on the first Oao (Orbiting Astronomical Observatory) and rescheduled the Smithsonian Astrophysical Observatory's celescope experiment for the third Oao. The three x-ray and gamma ray telescopes, already fabricated, contained experiments for surveying the sky proposed by: MIT, to detect high-energy gamma rays that did not originate from earth; Lockheed Missiles and Space Div., to seek new sources of low-energy (soft) x-rays and to study those recently developed; and NASA Goddard Space Flight Center, to detect low-energy gamma rays. The celescope experiment was designed to map the stars and nebulae through observations in the ultraviolet region of the spectrum but had encountered development problems.

  Unaffected by the change and proceeding on schedule for a 1965 launch was the Univ. of Wisconsin's photometer-telescope system to measure the energy distribution and emission intensities of stars. (NASA Release 65–49)

- NASA's MARINER IV, en route to Mars, passed the 20-million-mile mark in its distance from earth. The spacecraft was functioning normally and was transmitting data on scientific measurements taken in interplanetary space. (NASA Release 65–58)

- Twenty Llrv Program personnel at NASA Flight Research Center were honored at an informal ceremony. Six employees received plaques for special and outstanding contributions to the Lunar Landing Research Vehicle project. (FRC *X-Press*, 2/19/65, 1)

- Col. John H. Glenn, Jr. (USMC Ret.), speaking at a National Space Club luncheon, said: "Looking back over the 3 years since the flight of Friendship 7, I am impressed most of all by the tremendous progress the United States has made in space science and technology.

  "I am proud of the determination the American people have shown to become the world's leading spacefaring nation. . . .

  "Three years ago, Mercury spacecraft were limited in weight to about 3,000 pounds. Today, the Saturn I booster is operational and can put 22,500 pounds into Earth orbit, or seven times the weight of Mercury. Saturn I–B, which will begin flying next year, will be able to orbit a payload equal to 11 Mercury spacecraft. Saturn V, which will fly in 1967, will be able to orbit a payload equal to more than 80 Mercuries. That's some jump—but it is no more than typical of the great gains we are making in national space capabilities.

"The accuracy requirements for all these missions are almost incomprehensible and are one of the least understood parts of the space program. A good case in point was the launch of Friendship 7 3 years ago. The speed at insertion into orbit was over 25,000 feet per second or 5 miles per second, and the booster and spacecraft were accelerating at approximately 240 feet per second at a steadily increasing rate. At booster cutoff, each error of 1.4 feet per second resulted in a difference of approximately 1 mile in apogee on the far side of the Earth. When you consider that before the onboard signal for cutoff was received, information had to be obtained by radar at the Cape, transmitted by landline to Goddard, run through the computers, returned by landline to the Cape, checked against Cape data and then transmitted 350 miles down range to the spacecraft, still allowing time for onboard delays in operation of relays, valves, and thrust termination, it looks like an almost impossibly accurate requirement.

"Those accuracies, however, are rather crude compared with some now being obtained on the deep space probes.

"Now we have Mariner IV which at 6 o'clock tonight will be 20,194,023 miles out from Earth on an extremely difficult and significant mission. As you know, there was a checkout of equipment aboard the spacecraft last week that indicates the chances are still good that we will get revealing pictures of Mars next July, when Mariner IV will be 134 million miles from Earth, so far it will take $12\frac{1}{2}$ seconds to get a radio signal back.

"To wax philosophical for a moment, we might liken our space program to one of Plato's allegories. He told of prisoners chained in a cave for so long they had lost touch with reality and felt that their whole existence was wrapped up in the shadows they could see on the wall ahead of them. He goes on to say that if one escaped to the outside world and returned to tell the others of what really lay outside the dark cave, they would probably think him completely crazy.

"Even though we have seen such tremendous break-throughs in scientific knowledge in recent years, our knowledge has necessarily been limited to such a cave, for practically all we know has been limited to this one tiny speck of earth in a much larger environment. But that is in the process of becoming changed and with a rapidity no one can forecast." (Text, *CR*, 2/22/65, A751–53)

February 19: Sen. A. S. (Mike) Monroney (D–Okla.) said in an interview that if commercial airlines would voluntarily join the Federal Government in subsidizing helicopter lines, these services might be saved. Commercial helicopter lines operating in New York, Chicago, Los Angeles, and San Francisco had received Federal subsidies since 1947, but President Johnson proposed in his budget message that they be cut off after Dec. 31.

Sen. Monroney said: "Congress isn't going to appropriate any more money. We haven't a chance of selling them or continuing the subsidy without added help from the airlines." His plan involved artificially lowering helicopter fares so that a greater portion of the flying public would use them. This higher load factor, combined with the use of larger, more economical aircraft and hoped-for improvements in helicopter technology might lead to self-sustaining helicopter service in a few years, the Senator said. (*NYT*, 2/19/65, 69)

February 19: On the floor of the Senate, Sen. John Stennis (D–Miss.) argued for development of an advanced manned strategic aircraft, pointing out that "for the first time in the history of American strategic air power, there is no follow-on manned bomber under development." He cited Gen. Curtis E. LeMay who testified last year: "I am afraid the B-52 is going to fall apart on us before we can get a replacement for it. There is a serious danger this may happen." (*CR*, 2/19/65, 3176)

- AFSC announced that an airborne jet engine analyzing system, designated to improve flight safety and operational readiness of USAF tactical aircraft, would be flight-tested for one year at Nellis AFB, Nev., and Davis-Monthan AFB, Ariz. The jet engine analyzer system would be used to monitor, analyze, and assess engine performance of turbojets; to assist in predicting required maintenance; and to indicate engine failures before they occurred. (AFSC Release 57.64)

- AFSC Aeronautical Systems Div. had awarded to North American Aviation, Inc., an $8,150,000 increment to previously awarded contract for the XB-70 aircraft, DOD announced. (DOD Release 100–65)

- Boeing Co. announced it had ordered its 737 model into production and that it had already received an order for 21 of the short-range jets from Lufthansa German Airlines. The 737 would be a twin-engine jetliner designed for short-haul routes. (UPI, *NYT*, 2/20/65, 52)

February 20: RANGER VIII lunar photography probe struck its target on the moon at 4:57:36.8 EST, after radioing to earth about 7,000 close-up pictures of the lunar surface during the last 23 min. of flight. The point of impact was 2.59° north latitude, 24.77° east longitude, in the Sea of Tranquillity, an area slightly east-northeast of the center of a full moon. The spacecraft impacted at slightly less than 6,000 mph. Total distance of travel along its trajectory from lift-off had been calculated as 248,766 mi. Accuracy of the shot was reflected in the fact that impact had been planned for 4:57:30, and at 3° north latitude and 24° east longitude.

RANGER VIII had been programed to execute a "terminal sequence" just before impact to point the six TV cameras more in the direction of flight; this sequence was omitted to allow the cameras to cover a larger area than planned and to provide greater continuity with the pictures transmitted by RANGER VII last July 31. A second change in the flight was to turn on cameras 23 min. before impact instead of 13 min. and 40 sec. as planned. The new time had been chosen so that initial pictures would be about equal to earth-based resolution and then continue into impact.

Two small anomalies: one part of the spacecraft had registered a higher temperature than had been anticipated and more telemetry data had been lost during midcourse maneuver than had been expected. RANGER VIII had been launched Feb. 17 from Cape Kennedy. (NASA Transcript; Appel, *NYT*, 2/21/65, 1, 65)

- Dr. Gerard P. Kuiper of the Univ. of Arizona, heading the panel for scientific evaluation of RANGER VIII photographs, said at a press conference that one of the most important results of this flight had been the discovery that the Sea of Clouds and the Sea of Tranquillity were similar in structure. He noted several "odd depressed regions" and said that they could be areas where collapse had occurred, which might suggest the presence of lava fields. He believed the lunar surface was composed of very light, frothy material such as would be formed when rock was melted and allowed to resolidify within a high vacuum, like that on the moon. The material envisioned by Dr. Kuiper might be considered similar to certain volcanic rocks found on earth: while it would probably be lighter than water, it could still have substantial strength. This theory was based on laboratory attempts several years ago to simulate conditions existing when the moon was formed.

Ewen A. Whitaker of the Univ. of Arizona said he felt the lunar material, which he thought had a consistency of crunchy snow, would support a manned spacecraft. He said color lines and sharp boundaries tended to show that the surface was some sort of frothy, lava-like material and definitely not dust.

Another member of the Panel, Dr. Harold C. Urey, of the Univ. of California, noted dimples on the moon's surface and said their curved walls indicated material must have been thrown out of their centers when comparatively soft terrain was gouged by heavy masses of rock. He thought he saw spots in the center of some dimples into which soft material might be draining and estimated the depth of some dimples at 50 to 60 ft. Dr. Urey also suggested the surface material might have the consistency of crunchy snow. (NASA Transcript; Appel, *NYT*, 2/21/65; Miles, *L. A. Times*, 2/21/65; AP, *Indianapolis Star*, 2/22/65)

February 20: No evidence of lunar origin had been found in rock samples from western Iowa tested at NASA Goddard Space Flight Center, said Donald E. Perry, GSFC Information Officer: "We had not . . . found anything in Iowa which could be identified as meteoritic or of the nature of a tektite."

NASA had requested a six-county area of western Iowa to submit rocks for analysis since GSFC astronomer, Dr. Walter O'Keefe, had had the theory that tektites came from the moon. Western Iowa had been chosen as a likely spot for the search for tektite fragments because of its heavy deposits of loose soil and near absence of natural rock formations. (Barton, Omaha *Sunday World Herald*, 2/21/65)

- Sir John Eccles, professor of physiology at Australian National University, cautioned Australia and New Zealand against overconcentration of scientific energy on space. At a scientific congress in New Zealand, he warned: ". . . we are spending too much of our resources, especially our intellectual resources, on the exploration of space when we have the much more important problem of life, and of man and his brain." (*NYT*, 2/21/65, 9)

- The Soviet Union was considering sending weather observers into outer space in manned meteorological satellites, Prof. K. I. Kondratief, Univ. of Leningrad, said at a meeting in Geneva of the World Meteorological Organization's Scientific Advisory Committee. (*NYT*, 2/21/65, 24)

February 21: Vice President Hubert Humphrey, Chairman of the National Aeronautics and Space Council, said in remarks taped for the NBC-TV program, "The Sunday Show," that the U.S. would extend its national strength into the space dimension. "We expect to explore the moon, not just to photograph it or visit it," he said. "We plan to explore and chart the planets as well. We shall expand our earth laboratories into space laboratories."

Mr. Humphrey praised the Nation's space effort, saying that such activities had encouraged economic development, stimulated new products and processes, and furthered the cause of peace. This was Mr. Humphrey's first public statement on the space program as Vice President. (NBC–TV)

February 21: U.S.S.R. launched COSMOS LIV, COSMOS LV, and COSMOS LVI on one rocket booster. All three satellites were moving in close initial orbits: apogee, 1,856 km. (1,141 mi.); perigee, 279.7 km. (170 mi.); period, 106.2 min.; inclination, 56°4'. Equipment aboard "for the further investigation of outer space" was functioning normally. (Tass, *Pravda*, 2/22/65, ATSS–T Trans.; *NYT*, 2/22/65, 12)

February 22: Vice President Hubert H. Humphrey spent six hours visiting Cape Kennedy launching pads, talking to space experts, and looking over NASA Kennedy Space Center's Merritt Island Launch Area (MILA). "I'm sure the American people can feel this program is in good hands," he said before returning to Washington. At one point, Humphrey rode to the top of the 100-ft. launching vehicle to be used in next month's manned orbital flight and exclaimed: "Man, oh man, what a fantastic job!" (NASA Release 65–57; AP, Wash. *Eve. Star*, 2/23/65)

- COSMOS LVII earth satellite was launched by the U.S.S.R. Orbital data: apogee, 512 km. (318 mi.); perigee, 175 km. (109 mi.); period, 91.1 min.; inclination, 64°46'. Tass said the satellite carried scientific apparatus "intended for the further investigation of outer space." Equipment was functioning normally. (Tass, *Pravda*, 2/23/65, ATSS–T Trans.)

- DOD Secretary Robert S. McNamara told the House Armed Services Committee that deferral of the decision on Nike–X production from FY 1966 to FY 1967 "should not delay an initial operational capability by many months beyond what we would expect to achieve if we were to start production in Fiscal 1966." He said this was primarily because of the development, test, and evaluation work already under way. He added that the FY 1966 requests included $400 million for the continued development of Nike–X "on an urgent basis." Of this, $10 million had been programed for preliminary production engineering. McNamara confirmed ". . . a broadening of the objectives of Air Force's Manned Orbiting Laboratory (MOL) program, including: 1) development of technology contributing to both manned and unmanned space operation; 2) development of manned capability to assemble and service large orbiting structures; and 3) other manned military space experimentation. In addition, MOL will be used to investigate servicing and assembly of non-military structures . . . and will progress to study man's biological responses during periods in orbit of as long as 30 days." (*Av. Wk.*, 2/22/65, 26; *M&R*, 2/22/65, 18)

- U.S.S.R. had kept the U.S. under relatively continuous surveillance with photo reconnaissance satellites launched as part of the Cosmos program, said Edward H. Kolcum in an article in *Aviation Week and Space Technology*. In 1964 14 such satellites were launched, he asserted. The article continued: "Soviet photo reconnaissance payload is believed to be an unmanned version of the Vostok spacecraft, which successfully carried cosmonauts into orbit six times. The unmanned camera mission uses the same launch facilities and same recovery techniques developed from Russian manned satellites. The recoverable section is the pressurized cabin, which weighs about 5,000 lb. when it is ejected from the main spacecraft for a parachute descent on land. . . .

"Six of these payloads were recovered after eight days; two after seven days, one after six days and another after five days. One came down after 24 hr. in orbit; one remained up five weeks, another eight and a half weeks, and another, launched Aug. 29, is still in orbit. Departures from the norm are believed to indicate retrofire malfunctions or failures. Most recent Soviet reconnaissance satellite was Cosmos 52, launched Jan. 11 and recovered Jan. 19. . . ."

He stated that 11 of the 14 reconnaissance satellites orbited in 1964 were orbited at 65° inclination. The remaining three orbited at 51° inclination—"an inclination that also permits the payload to sweep over the entire continental United States." The other Cosmos satellites, orbited at 49° inclination, had remained in orbit until they decayed naturally. They were "believed to be scientific applications and military development payloads." (*Av. Wk.*, 2/22/65, 22)

February 22: TSR-2, Britain's tactical and reconnaissance bomber, broke the sound barrier for the first time. The aircraft was flown to more than 1,400 mph over the Irish Sea, to a landing at Wharton. This was the 14th test flight of TSR-2 by its manufacturer, British Aircraft Corp. (Reuters, *NYT*, 2/23/65, 53)

- Martin Co. had delivered the first pair of prototype nuclear generators for use in space to NASA, it was reported. The units contained no radioactive fuel and would be heated electrically for their qualification tests. Each generator was designed to deliver 30 watts of direct current to the weather satellite Nimbus B, first NASA satellite to use a nuclear power source. (AP, *NYT*, 2/23/65, 21)

- John F. Mason outlined in *Electronics* the dramatic changes underway in U.S. tracking and communications stations around the world. "Before the end of the year, 85% of the telemetry gear on the Atlantic missile range will be replaced. Everywhere, new communications equipment is going in, new pulse radars are being installed and continuous-wave radar networks are being expanded. Slowly, the separate ranges are becoming an integrated global network. . . .

  "Besides the work going on at the ranges, research and development effort for programs of the future continues at an active pace at the various government and industry centers throughout the United States.

  "The reason for this general overhaul of the missile ranges is to support Apollo, the manned lunar mission, and approximately 70 other ambitious missile and space programs already under way. . . ." (*Electronics*, 2/22/65, 94–105)

- Esso Research and Engineering Co. announced development of a fuel cell that could convert methanol, a petroleum derivative, into electricity. The most immediate practical uses of the cell would be military, the company said. (*NYT*, 2/23/65, 48)

- Leonid Seliakov, a deputy to the Soviet aircraft designer Andrei N. Tupolev, said fundamental breakthroughs would be made in civil aircraft between 1970 and 1975. Seliakov was quoted in *Vechernaya Moskva* as saying: "Airliners will be designed for flight speeds up to 2,500 kilometers (1,550 miles) an hour. I am sure that in 10 years from now, Muscovites will be able to fly to Khabarovsk in three instead of eight hours." Khabarovsk is about 4,000 miles from Moscow. (AP, *NYT*, 2/23/65, 53)

February 23: USAF 1,175-lb. Project Asset reentry glider, last in a test series of six, was launched from Cape Kennedy by a Thor-Delta rocket booster on a 13,300 mph suborbital flight; the spacecraft then was lost at sea. The experiment was to test materials for future lifting body reentry designs; it consisted of 2,000 tiny heat-sensitive spots in ten different colors designed to change hue as the glider came back through the atmosphere from an altitude of 40 mi. USAF said that most of the information sought had been radioed to the ground during the 30 min. flight, but that visual examination of the glider was necessary for data on heat distribution.

Officials presumed the glider sank into the sea, although intermittent radio signals received had indicated it had been at least partially afloat for some hours. Planes and ships were combing an area in the Atlantic Ocean about 2,750 mi. southeast of Cape Kennedy for the missing craft. (AP, *Wash. Post*, 2/24/65; *U.S. Aeron. & Space Act., 1965*, 134)

- Addressing the U.S. Military Academy at West Point, NASA Administrator James E. Webb said: "I would like to acknowledge the debt that NASA owes to the armed forces for early and continuing work that has contributed to some of our most successful space projects. As you know, NASA works closely with the Air Force in the development of launch vehicles and in the general technology of space flight. . . .

  "As an example—among many—of cooperative NASA–DOD activities: on January 1, NASA transferred control of its operating SYNCOM II and SYNCOM III communications satellites to the Department of Defense. . . .

  "When the great DOD missile site construction program—which ran at one time to $2.8 billion a year—subsided, the Corps of Engineers brought its tremendous engineering capability to the construction of NASA facilities. As the Corps moved toward completion of its work for NASA, its abilities will have been enhanced by the experience of building these great new national resources. The Corps, with new skills, will be able to move to new national requirements with assurance that it has performed extraordinarily well in engineering fields never before attempted.

  "Today there are detailed to NASA 254 active duty military personnel. Five of our astronauts are graduates of this great Academy. Nothing could be more fitting. For the debt modern American science and technology owes to West Point is too large to be repaid. It can visibly be traced back to Sylvanus Thayer who not only is the father of this Academy but who had a tremendous influence for half a century in the field of technical and scientific education throughout the United States." (Text)

- Statement of Edmond C. Buckley, NASA Director of Tracking and Data Acquisition, was presented by Gerald M. Truszynski, NASA Deputy Director of Tracking and Data Acquisition, in testimony before the House Committee on Science and Astronautics: "Since 1958, NASA has provided tracking and data acquisition support by means of the NASA Satellite Tracking and Data Acquisition Network for approximately 50 DOD earth satellites. During 1965 we expect to provide continuing telemetry support for as many as 10 DOD earth satellites, and limited tracking support for several more. Also, our station at Carnarvon, Australia, being in an excellent geographical relation to the launch facilities at Cape Kennedy, will be used to support a number of DOD spacecraft, as well as NASA spacecraft, where data and flight control after one-half orbit is required. For example, the DOD TITAN III launch vehicle development program is vitally dependent on support by this station. We anticipate support of approximately 15 TITAN III operations per year for the next two years.

"In a similar fashion, the tracking and data acquisition support which the DOD provides for NASA is extensive. At Cape Kennedy, for example, DOD has supported the launch phase of each NASA space flight mission. The extensive support provided by DOD for the Mercury program is well known. The Gemini program requires continuation of this DOD support."

Referring in his testimony to current trends affecting planning for support networks, Mr. Buckley said: ". . . equally significant requirement, is the increase in the number of spacecraft which will have highly elliptical or synchronous orbits. Spacecraft in highly elliptical orbits must be supported by a particular ground station, much in the same way as done for deep space missions, i.e., a particular station is required to provide as much as 8–10 hours per day of its available time for support of one satellite. Spacecraft in synchronous orbits require support of a particular station for 24 hours per day. As a result, tracking and data acquisition links are being committed to longer support periods which means not only that more equipment is required, but additional personnel, ground communications, and other operations expenditures are needed to meet this upcoming satellite support workload." (Testimony; *1966 NASA Auth. Hearings*, 187–212)

February 23: On the floor of the House, Rep. George P. Miller (D–Calif.), Chairman of the House Committee on Science and Astronautics, commented on the successful RANGER VIII spacecraft as "yet another major American space achievement. For the second time in less than a year a Ranger spacecraft has successfully taken closeup pictures of the lunar surface and returned them to earth from a distance of a quarter of a million miles. . . ." (NASA LAR IV/33–35)

• Rep. John R. Schmidhauser (D–Iowa) commented upon and inserted in the *Congressional Record* an article which appeared in the Davenport *Times-Democrat* about the Davenport Alcoa plant and the construction of Pegasus satellites. It said: "A unique arrangement of special equipment that senses infrared energy—thus indicating which part of the satellite is facing earth—enables scientists to determine the direction each meteoroid is traveling when it strikes Pegasus. Such information will tell spacecraft designers the extent of possible damage from hits, enabling them to build manned craft which will be relatively unaffected by meteoroids." (*CR*, 2/23/65, A77374)

February 23: Scientists at MIT's new Center for Sensory Aids Evaluation and Development were screening and testing new items that could potentially help the blind and deaf-blind, it was reported. Scheduled for early testing was an inertial navigation system that could sense movement off a straight line. It was about the size of a cigaret package and would be held flat between the thumb and index finger. If a blind person holding this battery-operated device in his right hand veered off course to the left, a projection would hit him in the thumb; if he veered off to the right he would be hit in the index finger. The system of gyroscopes and accelerometers that would operate this instrument was also found in missile and spacecraft guidance systems. (Sci. Serv., *NYT*, 2/23/65, 31)

February 24: Dr. Raymond L. Bisplinghoff, NASA Associate Administrator for Advanced Research and Technology, told the House Committee on Science and Astronautics that "although space research, development, and operations have absorbed much of our resources within the past few years, the NASA has been and will continue to be dedicated to a strong program of aeronautical research keyed to the Nation's needs." He testified: ". . . The [aeronautics] program embraces the entire spectrum of flight from lowspeed private and V/STOL aircraft to hypersonic vehicles. I have already described . . . our part in the evolution of the XV-5A and F-111 aircraft, in the evaluation of supersonic transport proposals, and in the solution of jet transport rough air problems. Looking ahead to FY 1966, we are requesting $42.2 million in Research and Development for aeronautics. This figure can be separated into two categories: one funds a broad and continuing effort in the scientific disciplines underlying advances in all areas of air transportation, civil and military; the other funds a more concentrated attack on specific advances in air transportation whose potential is identified by research in the various scientific disciplines. . . .

"Throughout the aeronautics program budget, provision has been made to support the direct requests of the Department of Defense and the Federal Aviation Agency. It can be pointed out that although the Research and Development request in FY 1966 for aeronautics is $42.2 million, we expect to spend a total sum of $106.2 million in this field. The difference is accounted for by Administrative Operations and Construction of Facilities funds as well as supporting research and technology directly applicable to aeronautics in fields such as electronics, human factors, basic research, and others. Approximately 1600 direct personnel will be engaged in aeronautical research in FY 1966. . . ." (Testimony; *NASA Auth. Hearings*, 213–269)

- In testimony before the House Committee on Science and Astronautics, George Friedl, Jr., NASA Deputy Associate Administrator for Industry Affairs, said: "NASA spends about 93 percent of its dollars on contracts with industry, universities and private research organizations. These procurements during fiscal year 1964 amounted to $4.6 billion. Approximately 96 percent of this amount or $4.4 billion was awarded by our field installations in accordance with program and project research and development requirements. NASA contracts support our in-house research and development activities and establish our coupling with industry and the private scientific community. . . ."

Reporting on contract administration, Mr. Friedl testified: "The cost reimbursement contract has been the basic instrument for procuring research and development because the Government has had to risk the uncertainties and assume the high cost involved. No other type of contract provides the Government and the contractor the latitude and flexibility needed to relate scientific and technical requirements, schedules and use of resources to mission objectives. By adding incentive provisions pertaining to time, quality and cost to this type of contract, it is our intention to offset some of its deficiencies and strengthen the purpose of the Government-contractor relationship.

"NASA has made a concerted effort to introduce suitable incentive arrangements in our cost reimbursement type contracts whenever practicable. In each case, the objective is to encourage the contractor to manage better and improve his performance; adhere to schedules; and hold down costs.

"As a consequence of our incentive contracting drive, there has been a marked increase in this activity in the past $4\frac{1}{2}$ years. In fiscal year 1961 we had one contract worth about $100,000. By December 31, 1964, we had awarded 75 contracts with a target value of over $751 million, 7 of these have been completed leaving 68 contracts totalling over $724 million currently being administered. . . .

"In view of the undesirable features of letter contracts, NASA Headquarters began a concerted effort, early in 1963, to curtail the issuance of new letter contracts and to assure the timely definitization of all outstanding letter contracts. Headquarters issued instructions to all centers directing program and project managers to plan ahead and allow adequate lead time for the initial negotiation of definitive contracts. . . . At the end of January only 3 letter contracts having a total value of $4 million were outstanding. We expect that these contracts will be definitized in March 1965."

Mr. Friedl said that NASA had "structured a sound practicable managerial technique to direct the planning, approval and execution of . . . future programs. We believe that adoption of what we have termed 'phased project planning' will materially assist us in achieving this goal.

"Phased project planning represents an orderly sequential progression in the execution of NASA major projects. It provides for formulating proposed work goals and missions, and allows for decisions, reappraisal points for management consideration to advance or replan such proposals, as well as the resources to implement them.

"Specifically, phased project planning provides for four distinct phases as follows:

"Phase A Conceptual/Feasibility Phase . . .

"Phase B Preliminary Definition Phase . . .

"Phase C Final Definition . . .

and Phase D Development/Operation. . . ." (Testimony; *NASA Auth. Hearings*, 269–88)

February 24: Canada's National Defence Research Council said it would negotiate an agreement with NASA for Canadian operation of the rocket launching range at Churchill, Manitoba. The announcement said such an agreement would open the way for a new partnership between the two countries in research. (*NYT*, 2/26/65, 13)

February 24: NASA Kennedy Space Center announced it had awarded three contracts for equipment used on launch complexes at both Cape Kennedy and Merritt Island Launch Area (MILA). American Machine and Foundry Co. received $1,198,923 for umbilical devices that would provide fuel, liquid oxygen, and air conditioning to the fin section of Saturn V's first stage. $745,601.15 was awarded Kaiser Aluminum and Chemical Sales for the fabrication of bulk electrical cable for Complex 39. Spaco Inc. received $596,356 to fabricate interconnect cables for joining terminal boards in the umbilical towers of Complexes 34, 37, and 39. (KSC Release 35–65)

- NASA Marshall Space Flight Center awarded a $8,774,000 research and development contract modification to North American Aviation's Rocketdyne Div. for continued uprating of the H–1 rocket engine from 188,000 to 200,000 lbs. Uprated engines would be used in clusters of eight to provide a total thrust of 1.6 million lbs. in first stage of Saturn IB launch vehicle. Modification brought H–1 contract total to $20,648,500. (*Marshall Star*, 2/24/65, 6)

- NASA had contracted with Collins Radio Co., Dallas (Tex.) Div., to procure Unified S-Band Telemetry Systems for three 85-ft.-diameter antennas in support of Project Apollo. Under the fixed-price type contract worth $2,740,000, Collins would install the three systems at antennas to be built at Goldstone, Calif.; Canberra, Australia; and Madrid, Spain. (NASA Release 65–63)

- USAF abandoned the search in the Atlantic for the Project Asset glider launched Feb. 23. The 6-ft. spacecraft, which had just completed an otherwise successful 2,700-mi. experimental flight at 13,300 mph, was never sighted visually after impact in the Atlantic. The only guide was a weakening signal from its radio beacon that faded out yesterday afternoon. Although the glider had radioed valuable data, engineers had wanted to examine the skin of the spacecraft to determine the ability of its exotic metals and superalloys to withstand prolonged heat of reentry. (*Wash. Post*, 2/25/65; AP, Balt. *Sun*, 2/25/65)

- The number of women earning more than $10,000 annually in scientific government jobs had increased dramatically from 1959 because of interest in space programs, Mrs. Catherine Dryden Hock, NASA systems engineer, informed the New York Section of Society of Women Engineers. Between 1959 and 1963, number of women in Government grades of GS–12 and above in computer fields rose 790%; in mathematics and mathematical statistics, 137%; and in physical sciences, 122%. NASA's engineering force was 3% women. (NASA Release 65–60)

February 25: President Johnson visited NASA Headquarters, accompanied by Vice President Humphrey, for a briefing on the MARINER IV project and to congratulate and express appreciation to NASA officials and members of the Mariner and Ranger project team. The President recalled that he had sponsored legislation in 1958 that had created NASA: "I think it is really incredible that we have come so far. It was only seven years ago this month that we were deliberating and debating and still seeking to come to grips with the realities of the space age." Mr. Johnson told NASA officials that the people of America and the whole

February 25: President Lyndon B. Johnson is briefed on the Mariner mission at NASA Headquarters. Left to right, James E. Webb, NASA Administrator, President Johnson, Vice President Hubert H. Humphrey, and Dr. Hugh L. Dryden, NASA Deputy Administrator.

world were "deep in your debt." (NASA Announcement 65-43; Simons, *Wash. Post,* 2/26/65; Sehlstedt, *Balt. Sun.* 2/26/65; Young, *Chic. Trib.,* 2/26/65; Mohr, *NYT,* 2/26/65, 10)

February 25: USAF launched Thor-Agena D launch vehicle from Western Test Range with unidentified satellite payload. (*U.S. Aeron. & Space Act., 1965,* 135)

* ComSatCorp announced decision of DOD that continuation of its present program to secure satellite services, presumably with Ford Motor Co.'s Philco Corp. Div., was superior to that proposed by ComSatCorp. The satellites involved would make up "initial" DOD system; ComSatCorp might bid to supply advanced satellites. ComSatCorp hoped its separate commercial system would be afforded some DOD nonsecret traffic. Some military men had argued that, since the Government would build its own system for secret communications, it should also use these facilities for nonsecret transmissions. This had caused ComSatCorp to raise the question of the degree to which the Government should enter the communications business in competition with private enterprise. President Johnson had established policy in a report to Congress, ". . . a system tailored for the military's exclusive use, does not alter the policy under which . . . the Government will use the commercial satellite system for the transmission of the bulk of its traffic between the United States and various overseas areas." (*WSJ,* 2/25/65)

* NASA had granted an exclusive patent license, the second it ever issued, to Exactel Instrument Co. for a "line-following servo-system." The device, which "remembers" a given graph curve, could measure one characteristic of a physical situation and read out resulting characteristics in specific quantities. President of Exactel, Eugene A. Glassey had invented the servo-system while an employee at NASA Ames Re-

search Center. Issuance of the exclusive license on a Government-owned patent to a private individual was part of NASA's continuing effort to make aeronautical and space inventions available for commercial development as rapidly as possible. The only previous exclusive licensing was to Union Carbide in 1963 for a nickel-based alloy invented by a NASA scientist. (ARC Release 65-6)

February 25: NASA Kennedy Space Center awarded a $41 million cost-plus-award-fee supplement to the Chrysler Corp. for support services on the Saturn I and Saturn IB space programs. Chrysler would provide prelaunch, launch, and post-launch services at Complexes 34 and 37 through June 30, 1968. (KSC Release 43-650)

- DOD's Hibex, the high acceleration experimental booster, was successfully tested at White Sands Missile Range, N.Mex. (*M&R*, 3/8/65, 11)
- Douglas DC-9, a twin-jet airliner, made its maiden flight. The short-to-medium-range transport, expected to benefit smaller airports, flew from Long Beach, Calif., to Edwards AFB in two hours and 13 min. The plane had a wing span of 87 ft. and used about 3,500 ft. of the runway in taking off. Its cabin could accommodate up to 90 passengers. The DC-9 was expected to go into passenger service early next year. Orders or options for 121 of the planes had been received by Douglas, of which 24 were placed by Eastern Air Lines. (UPI, *NYT*, 2/26/65, 58; 2/26/65, 37)
- Dr. C. Stark Draper, head of the Dept. of Aeronautics and Astronautics at MIT, and Theodore C. Achilles, a former ambassador to Peru and presently vice chairman of the executive committee of the Atlantic Council of the U.S., were sworn in as consultants to NASA. Dr. Draper would be a technical consultant on a part-time basis; Achilles would be available for consultation on NASA's university program. (NASA Release 65-66)

February 26: PEGASUS I satellite, launched by NASA Feb. 16, was functioning normally and recording information to ground stations on the size and frequency of meteoroid "strikes" or impacts on all three sensor panel groups. Scientists at NASA Marshall Space Flight Center said the number of penetrations of the panels was not greatly different from the expected level. PEGASUS I had a wing-like structure 96 ft. long and 14 ft. in width, offering more than 2,300 square ft. of area instrumented to detect collisions with meteoritic particles. The basic information on the penetrating power and frequency of meteoroids was needed for the design of future spacecraft. In addition, data on temperature, power levels, and the intensity of radiation were being received. The latter were also as predicted. (MSFC Release 65-45)

- COSMOS LVIII satellite, containing "scientific equipment," was orbited by the U.S.S.R. Initial orbital data: apogee, 659 km. (409 mi.); perigee, 581 km. (360 mi.); period, 96.8 min.; inclination, 65°. Equipment was said to be functioning normally. (*Krasnaya Zvezda*, 2/27/65, 1, ATSS-T Trans.)
- X-15 No. 1 was flown by pilot John McKay (NASA) to 153,600-ft. altitude at a maximum speed of 3,750 mph (mach 5.40). Purpose was to check out landing gear revised recently, give pilot experience at higher altitude, and get apparatus data. (NASA X-15 Proj. Off.; *X-15 Flight Log*)

February 26: Dr. Frank K. Edmonson, chairman of the Astronomy Dept. at the Univ. of Indiana, said RANGER VIII photographs had suggested that the moon might have features in common with the Karst limestone formation in southern Indiana and that a request for aerial photographs of the Karst region had been made. RANGER VIII's pictures showed that the Sea of Tranquillity on the moon was pocked and mottled by innumerable depressions. Surface of the Karst limestone layers was similarly pocked with sink holes. Dr. Gerard Kuiper, chief experimenter for the RANGER VIII project, and Dr. Harold C. Urey of the Univ. of California at La Jolla proposed that these "dimples" were produced by drainage of material through holes in their bottoms. (*NYT*, 2/26/65, 10)

- Col. John H. Glenn, Jr., was sworn in as a consultant to NASA by Administrator James E. Webb. His duties would include taking part in conferences, making speeches in the U.S. and abroad, and checking on projects under way. (NASA Release 65–67)
- Joseph Campbell, Comptroller General, reported to Congress that the decision of NASA's Goddard Space Flight Center to lease rather than buy two electric substations from the Potomac Electric Power Co. had resulted in $174,000 of unnecessary costs thus far. Campbell explained, "We believe that the Center failed to make this determination because of the Administration's failure to provide guidelines to its employees, setting forth pertinent factors necessary for consideration in making decisions whether to lease or purchase property." He added, however, that NASA had agreed with GAO findings and would purchase substations as provided for in contract. The matter was nevertheless reported to Congress because it "further illustrates that significant unnecessary costs can be and are being incurred," when agencies do not make complete lease-versus-purchase studies. (*Wash. Post*, 2/28/65)
- Use of ComSatCorp's Early Bird communications satellite was subject of a London meeting between U.S. and European participants in the program. A general understanding was reached that once commercial service started, television networks could use the satellite system outside peak transatlantic telephone hours. The peak traffic hours were generally considered from about 9 a.m. to 3 p.m. EST. Exceptions could be made if major news stories broke in Europe during this period. (Farnsworth, *NYT*, 2/27/65, 51)
- Report of experiments by the European Organization for Nuclear Research indicated there was no fifth force in nature as had been proposed, independently, by two groups of American physicists to explain some unexpected experimental results. The four forces in nature were gravity, electromagnetism, and weak and strong nuclear forces. (Reuters, *NYT*, 2/28/65, 69)

February 27: Thiokol Chemical Corp. successfully static-fired its 156-in.-dia., 100-ft.-long solid propellant rocket motor—the largest yet fired—for 64 sec. The 900,000-lb. motor developed over three million lbs. of thrust, consumed over 800,000 lbs. of propellant, and generated temperatures up to 6,000°F. The solid propellant was encased in a half-inch-thick steel and nickel casing which apparently escaped damage. Also left intact was the 10-ton, 20-ft.-tall nozzle which rested on top of the 12-stories-deep testing pit. The motor was fired below ground level. Primary objective of the test was to validate design of the nozzle for use later on the 260-in. motor. A secondary mission was to check out the propellant processing system which would be used in the larger motor. The test was part of the large-solid demonstration program currently managed by NASA's Lewis Research Center. (Shipp, *Atlanta J/Const.*, 2/28/65; *M&R*, 3/8/65, 16)

February 27: NASA announced it had approved a grant of $100,000 for establishment of a Technical Utilization Program at the Univ. of Minnesota. Along with funds to be provided by private business concerns, the NASA grant would support the development and experimental testing of new ways in which developments in space science and technology could be rapidly transferred to and assimilated by business and industry. North Star Research and Development Institute would participate in part of the program. (NASA Release 65-69)

February 28: The first industry-produced Saturn I first stage (S-I-8) arrived at Cape Kennedy aboard the NASA barge *Promise* after a six-day trip from NASA Marshall Space Flight Center. The stage, which was 80 ft. long and 21.5 ft. in diameter, had been built by the Chrysler Corp. (MSFC Release 65-46; AP, *NYT*, 3/1/65, 12)

- Louis Walter, GSFC geochemist, told AP reporter his research with tektites indicated lunar surface may be sand-like. The key to this conclusion lay in Walter's discovery of the presence of coesite in tektites, believed to be particles of the moon sent into space when meteorites impact the lunar surface. Coesite, also found around the world at known meteorite craters and sites believed to have sustained meteoritic impacts, is a form of silicon dioxide—a major constituent of sand—produced under high pressure. "If we accept the lunar origin of tektites, this would prove or indicate that the parent material on the moon is something like the welded tuft that we find in Yellowstone Park, Iceland, New Zealand, and elsewhere," according to Walter. Welded tuft was said to have some of the qualities of beach sand. (AP, *Chic. Trib.*, 3/1/65)

- Three Univ. of California (Berkeley) scientists concluded on the basis of their laboratory studies that Dr. William M. Sinton's spectroscopic evidence of organic matter on Mars was not valid. Dr. Sinton of Lowell Observatory had made spectroscopic studies of Mars in 1959 that suggested infrared radiation from dark portions of Mars was comparable to that produced by some terrestrial plant life. The California chemists—James S. Shirk, William A. Haseltine, and George C. Pimentel—concluded Dr. Sinton had detected vaporized "deuterated water" (H_2O plus heavy hydrogen—deuterium) rather than plant-produced molecules. (UPI, *S.F. Chron.*, 2/28/65)

- NATO officials were examining preliminary bids for a $310 million NATO Air Defense Ground Environment (Nadge) system that would be used to protect continental Europe from enemy aircraft. Nadge was expected to be an improved version of the Sage system that had been used to defend the United States. At last December's NATO ministerial meeting, it was agreed that each country be guaranteed Nadge subcontracts equal to the amount the country was contributing to the program. The cost sharing formula for Nadge was based on the contributive capacity of the member countries; the advantage accruing to the user country; and the economic benefit to the countries in which the installations would be placed. Under this formula the U.S. was expected to contribute 30.85% of the cost of the program. (Smith, *NYT*, 2/28/65, 12F)

February 28: Prospecting for high-grade silver could be done, according to Thor H. Kiilsgaard, Chief of the Resources Research Branch of the U.S. Geological Survey, by using an infrared system mounted on aircraft. He explained that deposits of silver in the earth were associated with hot water and that areas of heat flow could be detected by the infrared devices. If the heat zones conformed with mineral zones or faults, silver might be present. (Sci. Serv., *NYT*, 2/28/65, 64)

During February: The prime and backup crews for the upcoming GT-3 three-orbital mission underwent parachute and egress training exercises. Parachute training, with the astronauts in space suits, was conducted in Galveston Bay, Tex. Egress training from a submerged Gemini boilerplate spacecraft was conducted in a large tank at Ellington, AFB. (MSC *Roundup*, 2/17/65, 2)

- Atlantic Research Corp. announced the Frangible Arcas meteorological sounding rocket, developed for USAF, had successfully passed flight tests at the Western Test Range. (*M&R*, 3/8/65, 11)

- *New York Times* continued its editorial opposition to the national objective for Project Apollo of landing a man on the moon in this decade. On Feb. 19, an editorial drew from the two successful major launchings of the week (RANGER VIII and SATURN I SA-9) the lesson that the kinds of experiments on these flights (lunar photography of RANGER VIII and PEGASUS I micrometeoroid detection satellite on Saturn I) proved there were many unmeasured perils in space and that "In the face of these uncertainties, the American space program ought to retain maximum flexibility of timing, rather than try at all costs to achieve the artificial goal of a manned lunar landing by 1970."

  On Feb. 22, following the successful conclusion of the RANGER VIII lunar photography mission, another editorial praised the accomplishment, then noted that the Ranger series was not providing all of the answers to lunar questions critical to the Apollo program, and concluded: "The two successful Ranger shots, however, make clear that much valuable information can be gathered about the earth's natural satellite by relatively cheap instrument-carrying rockets that do not risk human lives. This demonstration, and the continuing uncertainties about matters essential for a safe manned round trip to the moon, strengthen still more the case for making progress slowly, without any arbitrary deadline, on Project Apollo." (*NYT*, 2/19/65, 34; 2/22/65, 20)

- A warning that "In looking for life on Mars we could establish for ourselves the reputation of being the greatest Simple Simons of all time" came from Dr. Philip H. Abelson in an editorial in *Science*. Dr. Abelson was editor of the magazine and director of the Carnegie Institution's Geophysical Laboratory. He said he did not believe that life, particularly life resembling that on earth, would be found on Mars and proposed "a few inexpensive experiments" on earth to save years, billions of dollars, and the possibility of "considerable eventual disappointment" if the search for life on Mars should prove fruitless.

  Attempts to sterilize spacecraft to prevent them from carrying earth organisms to Mars might add "many years and billions of dollars" to the cost, the editorial said. It suggested, instead, careful selection of experiments to be sent to Mars and "relatively inexpensive studies here on earth" to determine whether sterilization were really necessary. (Clark, *NYT*, 2/13/65; *Wash. Post*, 2/13/65)

During February: Dr. Leo Steg, manager of General Electric Co.'s Space Sciences Laboratory, Missile and Space Div., was named Engineer of the Year—1964. He was cited for outstanding contributions to the advancement of space science and the engineering profession. The award was presented by an amalgamation of 41 societies during the 1965 National Engineers' Week in Philadelphia. (*Av. Wk.*, 2/1/65, 13)

- NASA's contributions to the technology of inorganic coatings were described in a new technology survey (NASA SP-5014) published by the NASA Technology Utilization Division. They were thermophototropic coatings; thermal control coatings for space vehicles; solid lubrication coatings; thermal insulation coatings; methods of applying coatings to substrates; measurement of coating optical properties; and refractory metal oxidation resistant coatings. (NASA Release 65-39, 65-44, and 65-61)

- GAO saved the military services a total of $254.7 million, AEC $3 million, and NASA $727,000 last year. This information was released in a 251-page document released by GAO in addition to the GAO Administrator's Annual Report to Congress. (*M&R*, 2/15/65, 9)

- Nikita Khrushchev, in his first known public appearance in Moscow since his removal from power in October 1964, visited the cosmonauts' monument on the outskirts of the city, Reuters reported. A militiaman on duty at the monument said: "Yes, it's quite true. Nikita Sergeyevich visited the monument and spent about 30 minutes." After the Soviet Union's three-man orbital mission, VOSKHOD I, last October, Khrushchev had been scheduled to welcome the cosmonauts to Moscow and to dedicate the monument, but his sudden retirement intervened. (Reuters, Waller, *Wash. Post*, 2/22/65, 1)

- France's Emeraude rocket, first stage of the Diamant booster, was successfully launched from Hammaguir Range, Algeria, after three failures. Its liquid-fueled engine provided 62,000-lbs. thrust for 88 sec. Twelve Emeraude launchings were originally scheduled. Second and third stages of the Diamant launch vehicle, both solid fueled, had already been successfully tested. No attempt had been made to launch the three stages linked together. (*Av. Wk.*, 3/22/65, 18; *M&R*, 3/22/65, 9)

- William Cohen, Chief of Solid Propulsion Experimental Motors in NASA's OART, discussed the great strides in large solid-propellant rocket motors taken in the past few years, in *Astronautics & Aeronautics* article. Among the new technologies he mentioned were maraging steels, ablative nozzles, vector control, and the cast-cure-test facility. Looking toward the future, Cohen said among the advanced concepts associated with large solids showing promise of success were reusable motor cases, insulation, and nozzle component; and failure-warning systems. (*A&A*, 2/65, 42–46)

- Cost and performance comparability of large solid-propellant rocket motors was topic of article by G.W.G. Van Winkle, Boeing Co., in *Astronautics & Aeronautics*. The information was based on research obtained in study made by Boeing for MSFC. In the same issue, Dr. Walter G. Berl of Johns Hopkins Univ. Applied Physic Laboratory discussed combustion instability in solid-propellant rocket motors. Four types of instability were listed, and the status of solutions to these problems was discussed. Dr. Berl concluded that it was "too much to expect that the always latent instability problem has been banished from the new propellants of the future. It is more likely that the most obvious troubles can be eliminated, partly through analysis, partly through recognition and exploitation of past trends. . . ." (*A&A*, 2/65, 48–61)

During February: In a report titled "Federal Funds for Research, Development, and other Scientific Activities," National Science Foundation said DOD obligations for R&D increased each year from $2.3 billion in 1956 to an estimated $7.5 billion in 1964. Although survey predicted a small decrease to $7.2 billion for 1965, the 10-yr. period showed a 200% gain. Support to applied research accounted for about 22% of 1965 R&D Defense funds, with 2 or 3% used for basic research; about 75% went for development. Profit-making organizations had done most of DOD's R&D during the 10 yrs., increasing from about 50% in 1956 to 65% in 1965. The report added: "On the other hand, the relative share of Defense research and development performed intramurally decreased each year from about 40% in 1956 to 21% in 1963, but an increase to 25% was expected in 1965. . . ." (NSF Rpt. 65–13)

- Experimental model isotopic thruster was tested at AEC Mound Laboratory, using heat from radioactive decay of polonium 210. Mound Laboratory was continuing development of polonium 210 fuel forms and fuel encapsulating techniques for specific space applications. The isotopic small rocket engine, or thruster, concept envisions use of a radioisotope to heat hydrogen, which is expelled through a nozzle to produce low thrust. (*Atomic Energy Programs, 1965,* 149)

March 1965

March 1: NASA Administrator James E. Webb sent a letter to the House and Senate space committees outlining major reprograming of funds planned by NASA during the remainder of FY 1965. Webb said $13 million had been allocated to large solid rockets in FY 1965 and nothing in the following year "due to the President's decision not to include funds in the NASA '66 budget." Close-out costs for the large solids would amount to $8.5 million in addition to the $13 million already earmarked "and would yield no technical confirmation of the planned objectives." By adding another $5.3 million, "bringing the FY '65 funding to $26.8 million," NASA would "carry the Phase I program through to completion." Phase I called for the manufacture and firing of two "half-length" rockets 78 ft. long and 260 in. in diameter. Additional close-out funds were also granted to the other programs not included in the FY '66 budget: $2.15 million for Snap-8; $3 million for the M–1 engine. (Text; *NASA Auth. Hearings* [Part 4], 279–88)

- First Saturn V booster (S–IC–T) had been moved to static test stand at NASA Marshall Space Flight Center to prove out its propulsion system. The 280,000-lb. stage, developed jointly by MSFC and Boeing Co., had two tanks with total capacity of 4,400,000 lbs. of liquid oxygen and kerosene, and five F–1 engines, each weighing ten tons, which provided total thrust of 7.5 million lbs. (MSFC Release 65–47; *Marshall Star*, 3/3/65, 1, 6)

- Louis Walter, geochemist at NASA Goddard Space Flight Center, reported that of the 400 specimens received from "Operation Moon Harvest" none analyzed was a meteor or other non-earth fragment. It had been theorized that because of the low gravity of the moon, meteoroids striking the moon might dislodge fragments which would be attracted by earth's gravity, and that analysis of the fragments would provide important clues to composition of the moon. (*Des Moines Register*, 3/2/65)

- Dr. Mose L. Harvey, Director of the Univ. of Miami Center for Advanced International Studies and history professor, was sworn in by NASA as part-time consultant in international affairs. Dr. Harvey also was a consultant to U.S. State Department's Policy Planning Council. (NASA Release 65–71)

- JPL's Dr. Robert Nathan had developed computer system that was doubling resolution of Ranger lunar photographs. Picture data were taken directly from magnetic tape and digitized for insertion into an IBM 7094, thereby bypassing kinescope response that had contributed to distortion of published Ranger pictures. Calibration data obtained before Ranger flight were used to remove noise and distortion which brought "a dramatic increase in resolution." Craters became visible that were not seen in original pictures. (*M&R*, 3/1/65, 8)

March 1: ComSatCorp announced delay of Early Bird synchronous satellite launch, previously targeted for end of March, because of decision to replace defective transistors and retest replacements. (ComSatCorp Release)

- David Sarnoff, Chairman of RCA, accepting National Commander's Award for Distinguished Service, said at the American Legion's fifth annual Washington conference: "The same sense of mission that ignited our young nation's Westward expansion a century ago should now be brought to bear in support of the President's space objectives. . . .

  "Leadership in space and in the communications art which is the key to mastery in space, translates itself today into political, military, economic and social leadership among the nations of the world. Technological leadership resembles a magnet which attracts other forces. When it is weakened, these forces are drawn into other orbits."

  President Johnson sent a message endorsing the award and praising Mr. Sarnoff's achievements "on behalf of a grateful nation." (*NYT*, 3/2/65)

- Editorializing in *Missiles and Rockets*, William Coughlin suggested a "useful mission" for which RANGER IX might be adapted: "Our unsolicited proposal to NASA is that Ranger be employed to return to Earth photographs of Earth from space. Satellites have told us the Earth is 'pear-shaped' rather than round and that it draws a perhaps invisible but comet-like tail after it through space. Photographs of the entire Earth globe as seen from space would have high scientific value. As a propaganda triumph, it would be unequalled. . . ." *M&R*, 3/1/65, 16)

- President Johnson, addressing 40 winners of annual Westinghouse science talent search, said science and politics should strive to "serve humanity." He added that this country was "very anxious to produce all the scientists that we can," and expressed hope that scientists would learn about government and politics. Larry Dean Howard of Canoga Park, Calif., won first prize for having developed a method of accurately defining the orbits of earth satellites through the use of differential calculus. Prize was a $7,500 Westinghouse scholarship. (AP, *NYT*, 3/1/65; Loftus, *NYT*, 3/2/65, 14)

- USAF conducted first inland Minuteman ICBM flight test, launching the missile from a silo near Newell, N. Dak. (*M&R*, 3/8/65)

March 1–3: The AIAA Unmanned Spacecraft Meeting was held in Los Angeles.

Maj. Gen. O. J. Ritland, AFSC's Deputy Commander for Space, said in address that focus on manned space events often caused us to lose sight of the numerous space missions adequately performed by unmanned spacecraft. Although much of unmanned spacecraft activity had directly supported manned missions, "unmanned space technology has benefitted only indirectly from manned space effort." He predicted that future manned missions might reverse this relationship and cited the objectives of the USAF MOL program as an effort to strengthen and expand technology for all space programs. MOL program would develop technology to improve manned space capability; demonstrate servicing by man of large structures in space; conduct basic scientific and general technological manned experimentation; determine biological response of man in space for extended periods. General Ritland said: "With a laboratory in space, astronaut-scientists or engineers can assemble, test, and observe the operation of many subsystems or components in the actual space environment. They can observe equipment failures on the spot and will be able to make necessary replacements or repairs. I have spent many hours . . . looking over space flight data—attempting to determine exactly what failed and why it failed. The time is near when we can overcome many of these frustrations and uncertainties by use of the astronaut to answer such questions or to relay data to the ground for detailed analysis." (AFSC Release)

Discussing future requirements for military satellite communications systems, Samuel P. Brown, Technical Director, U.S. Army Satellite Communications Agency, said: "The feasibility of gaining significant improvements in this area appears very good based on the lessons learned from the SYNCOM spin stabilized satellite and the approaches planned for the DOD's Initial Defense Communications Satellite Project and NASA's Applications Technology Satellite Program. From these and other programs is expected to evolve techniques for spacecraft stabilization which will permit the increase of satellite antenna gains by an order of magnitude." (Text)

NASA Deputy Associate Administrator for Space Sciences and Applications Edgar M. Cortright said the reason for a civilian and a military space program lay in fundamental differences in the respective roles of NASA and DOD: "NASA's role is to explore and exploit space for peaceful purposes. The DOD's role is to stay prepared to defend the United States and its allies, operating in any medium that furthers this end. The present space program with its great breadth would never have evolved under the DOD, which must necessarily devote its full attention to its awesome military responsibilities. . . .

"Fortunately, the two space programs are mutually supporting and blend together quite well. They use common equipment . . . and draw on the same scientific and industrial base . . . numerous projects are of great mutual interest. Top management in both agencies devotes substantial effort to insure close cooperation and to minimize duplication." (Text)

JPL's Dan Schneiderman, project manager for NASA's Mariner program, told AIAA delegates that data from MARINER IV's solar plasma probe, which ceased normal functioning ten days after the Nov. 28, 1964, launching, had become understandable to scientists through analysis of a component failure in the plasma probe. Telemetry from a second instrument indicated that a portion of the ionization chamber experiment, which measured radiation in space, was not operating properly. Schneiderman said the new failure was in the Geiger-Mueller tube. Schneiderman estimated that based on nitrogen consumption to date, there was enough gas available to keep MARINER IV stablized for about six years. He said there had been no loss of lock with Canopus since a special command was transmitted to the probe Dec. 17, 1964. (NASA Release 65-73)

March 1–3: The first NASA University Program Review Conference in Kansas City, Mo., assembled over 400 university representatives interested in learning how their institutions could qualify for NASA grants for space-related research or expand present programs. Dr. Thomas L. K. Smull, Director of the NASA Office of Grants and Research Contracts, reported that 200 universities were participating in the program, that some of the grants were for specific projects, some in university sustaining programs, and others for the support of predoctoral candidates. He said that while NASA was "mission oriented," its job was not limited to putting a man on the moon: "Its objective is the expansion of human knowledge of phenomena in the atmosphere and space. One problem is how the academic community can communicate with NASA." (McCoy, *Kansas City Star*, 3/1/65)

In a luncheon address, Dr. Raymond L. Bisplinghoff, NASA Associate Administrator for Advanced Research and Technology, urged educators not to strangle "the holy curiosity of inquiry." He said the success of the U.S. space program depended largely on "formation of ideas by individuals working as individuals in universities." (Text)

Sen. Stuart Symington (D–Mo.) told the Conference that the U.S. must "widen the scope of man's imagination, trample rough-shod over intellectually inhibiting barriers and stimulate to their fullest potential the mental powers of young and reasonably young Americans if the United States were to achieve and maintain preeminence in space." Symington emphasized the need for communication of new knowledge. (*Kansas City Times*, 3/3/65)

Dr. Willard F. Libby, Director of Univ. of California's Institute of Geophysics and Planetary Physics, reviewed activities supported by NASA multidisciplinary grant: "In the three years UCLA has administered [the . . . NASA grant, we have aided in bringing thirty-seven visiting scientists to this campus for short periods of time. This grant has supported fourteen visiting researchers for periods of up to one year. Through the use of these funds and program enrichment funds from the NASA Predoctoral Traineeship grant, we have aided in bringing seven new faculty members to this campus to augment the existing faculty in space-related fields. . . . Finally, we have made over fifty sub-grants to faculty for new starts on space-related research in various areas—Biology and Medicine, Physical Sciences, Engineering, and Business Administration." (Text)

• More than 250 scholars and theologians met in New York to discuss means of attaining world peace and "to lay groundwork for a theology for the dawning age of cybernation." Moral and technological implications of Pope John XXIII's encyclical *Pacem in Terris* were studied. Meeting was sponsored by Center for the Study of Democratic Institutions and the Fellowship of Reconciliation. (*NYT*, 3/2/65, 28; *NYT*, 3/3/65, 24)

March 2: Two seconds after lift-off, NASA's Atlas-Centaur 5, carrying a dummy Surveyor spacecraft, exploded and burned on Launch Pad 36–B at Cape Kennedy. Failure occurred when two of the three Atlas engines shut off simultaneously due to closing of a fuel-line valve. The 150-ton, 108-ft. rocket rose three ft. from the pad, then fell back to the ground and exploded. Although burning propellant covered most of the launch complex, no injuries to personnel were reported. Damage to the launching pad was estimated at $5 million.

Objectives of the Atlas-Centaur test had been to test the ability of its guidance system and hydrogen-powered second stage to send a payload the size of the 2,150-lb. Surveyor on a precise path to the moon and to evaluate how well the mock-up Surveyor spacecraft would withstand the stresses of launching. (AP, Phil. *Eve. Bull.*, 3/2/65; UPI, *Chic. Trib.*, 3/2/65; AP, *NYT*, 3/3/65; *Av. Wk.*, 3/8/65)

March 2: NASA invited international scientific community to propose research experiments and design studies for upcoming missions, primarily those scheduled between 1967 and 1970, and to propose space investigations not presently scheduled. In addition, they were invited to suggest experiments (1) involving the design and construction of entire spacecraft and (2) involving special characteristics or requirements calling for the development of a new Explorer spacecraft or for scheduling of additional missions for Explorers already developed. Proposals would be evaluated on scientific merit, technological feasibility, competence and experience of investigator, assurance of institutional support, and scientific adequacy of apparatus suggested. Proposal deadline: Jan. 1, 1966. (NASA Release 65–70)

- In a letter of explanation to Congress, NASA discussed priorities in the FY 1966 budget: ". . . As the President pointed out when he submitted the budget to the Congress, 'It is a budget of priorities. It provides for what we must do, but not for all we would like to do.' In assessing priorities and the most urgent national needs, the 260-inch solid propellant rocket program, the M–1 liquid hydrogen-oxygen rocket engine capable of providing 1½ million pounds of thrust, and the SNAP–8 nuclear electric power generating system to provide 35 kilowatts of electrical power in space could not be supported in the Fiscal Year 1966 budget.

    ". . . [NASA] is, therefore, preparing plans for reprogramming Fiscal Year 1965 funds so as to logically phase out these program activities in such a way as to obtain as much technical information as is possible for future use. . . . Every effort is being made to achieve the greatest possible benefit from the funds already invested." (Text)

- Prof. Thomas Gold, Cornell Univ. astronomer, discussing RANGER VIII photographs in an interview with John Lear, World Book Encyclopedia Science Service, Inc., suggested that long, narrow rills and irregular depressions could be caused by moon's surface collapsing into crevasses opened by the movement of a glacier hidden beneath lunar dust. He attributed gently rounded shapes to a shifting of small particles by electrical forces which, on earth, were inhibited by atmosphere. Concerning the manned expedition, Gold indicated: "The presence of ice oceans could give rise to many problems. But once these were solved, the ice itself could be mined and used to make hydrogen for fuel for rockets returning to Earth." Referring to the electrically-charged particles: "Many particles would be dislodged mechanically by the landing of a spacecraft or the footstep of a man. Once loose, the dust would jump in response to electrical attraction or repulsion. If particles landed on the astronaut's visor, brushing wouldn't remove them but would instead intensify the electrical charges affecting their behavior." Dr. Gold recommended more research on possible control of these electrical forces. (Lear, *Houston Chron.*, 3/2/65; *Ind. Star*, 3/7/65; WBE Sci. Serv.)

March 2: A $1,366,511 contract for construction of a high temperature heat load testing facility at NASA Flight Research Center was awarded to Santa Fe Engineers, Inc. The facility would be capable of producing temperatures up to 3,000°F on small isolated areas of aircraft; larger areas could be heated up to about 600°. Contract was awarded by Army Corps of Engineers, which was administering it for NASA. (FRC Release 9–65)

- A $1,260,000 contract to build an addition to Central Computer Facility at Slidell, La., had been awarded by NASA Marshall Space Flight Center to Quinn Construction Co., New Orleans, La. The computer facility was used to support the Michoud Operations in New Orleans and the Mississippi Test Operations, Hancock County, Miss. (MSFC Release 65–48)

- In a *Christian Science Monitor* editorial, Leonard Schwartz posed the problem of "how the capability represented by manned orbiting space stations can be used to enhance national security and promote peaceful-scientific uses of outer space." Schwartz suggested formation of an inspection agency—International Space Patrol—to neutralize military potential represented by manned space stations and to ensure usage of outer space for peaceful purposes only. He pointed out that Vice President Humphrey, one of the first proponents of an arms control agency and an international space agency, was now Chairman of National Aeronautics and Space Council, on which sat the Secretaries of Defense and State and administrators of AEC and NASA. This position provided him with "an appropriate vantage to supervise their arms control capabilities in order to reach a national decision which would reconcile control with security and scientific use of outer space." (*CSM*, 3/2/65, 4)

March 2–3: In testimony before the House Committee on Science and Astronautics' Subcommittee on Advanced Research and Technology, Edmond C. Buckley, Director of NASA's Office of Tracking and Data Acquisition, said that during 1965, data processing facilities would handle 70,000,000 data points per day and that there would be an increase to 200,000,000 data points per day in 1966. He continued: "In fiscal year 1966, effort under this category will be directed toward developing and evaluating techniques for building up the existing telemetry data reduction capability to match the increasing requirements. In order to reduce this tremendous amount of data in an efficient and reliable manner, new techniques must be evaluated for obtaining this additional capability.

"The heart of this prototype system is the Satellite Telemetry Automatic Reduction System (Stars). The development program for this system was initiated in prior years and is planned to continue through fiscal year 1968. The Stars equipment presently includes automatic editing, decommutation, and calibration of the telemetry data. Functions in addition to these will be included in the prototype equipment as the developmental subsystems become available. . . ." (Testimony; *NASA Auth. Hearings*, 1–87)

March 3: NASA's PEGASUS I meteoroid-detecting satellite recorded 19 wing punctures in its 3 to 4 million mi. travels. Earth-transmitted electronic signals might have been the cause of several recorded hits, but some were definitely meteoroid particles. PEGASUS I orbited the earth every 97 min. (AP, *NYT*, 3/4/65, 50; AP, *Wash. Post*, 3/4/65)

- NASA, at DOD's request, had halted SYNCOM II's westward drift at 68° east longitude over the Indian Ocean. Under the direction of project managers at NASA's Goddard Space Flight Center, the command signals had been sent from the Syncom station in Salisbury, Australia, beginning Feb. 20 and ending Feb. 24. No future major locational corrections were anticipated; SYNCOM II should remain in same general area indefinitely. (NASA Release 65–72)

- NASA Administrator James E. Webb said at a press conference held in conjunction with the NASA University Program Review Conference in Kansas City, Mo., that the space research program would cost $35 billion over a ten-year period. At the end of that time, NASA expected to have accomplished (1) 12 to 15 flights of the Saturn V, (2) 5,000 hrs. of astronaut flight time, and (3) the capability of lifting 240,000 lbs. from the earth and orbiting 90,000 lbs. (*Kansas City Times*, 3/3/65)

- Gen. Bernard A. Schriever (USAF), addressing American Management Assn. conference in New York City, announced recent approval and initation of USAF Spacecraft Technology and Advanced Reentry Tests program (Start), "a four-fold research spacecraft program to develop unmanned test vehicles capable of maneuvering to a precision recovery site after reentering from orbit." In a Baltimore *Sun* editorial, Albert Sehlstedt, Jr., said that the Martin Co. had designed for this program a new, wingless V-shaped plane, maneuverable in atmosphere because its shape would provide aerodynamic lift. The program would: (1) launch the SV–5 by Atlas booster, (2) continue Asset experiments to test vehicles entering atmosphere at very high speeds, (3) study effects of vehicles passing through atmosphere at slower speeds, and (4) relate to allied studies that had not yet led to specific designs for identifiable reentry vehicles. (Text; AFSC Release 31.65; Sehlstedt, Balt. *Sun*, 3/4/65)

- Rep. J. Edward Roush (D–Ind.), addressing the House, cited the 1965 National Science Foundation Report to the House Subcommittee on Science, Research and Development of the House Committee on Science and Astronautics, which pointed out the heavy concentration of Government research contracts in New York, California, and Massachusetts. "One-half of the 50 states have 96.78 percent of all the funds listed for the various States. The remaining 3.22 percent is shared by the other 25 states," the report continued. Roush maintained that more equitable distribution of Federal funds would alleviate economic depression in many areas. (*CR*, 3/3/65, 3895)

March 4: NASA's OSO II satellite, which completed its first month in orbit at 11:36 a.m. EST, had circled earth 419 times and daily returned about 7 mi. of tape-recorded data, NASA reported. Designed to provide detailed information on solar x-rays, gamma rays, and ultraviolet rays, OSO II was functioning normally except for failure of the Harvard College Observatory ultraviolet scanning spectrometer and for sporadic return of data from the spectroheliograph portion of Naval Research

Laboratory coronagraph. Earlier problems with data transmissions from GSFC ultraviolet spectrometer had been resolved. (NASA Release 65-74)

March 4: NASA's OGO I had received ground-administered "shock treatments" to correct faulty inverter. Continued malfunctioning of inverter, which supplied power for rotation of solar panels to maintain proper angle to the sun, would have shortened OGO I's lifetime for lack of electric power. All other systems were functioning normally except attitude control. OGO I was still spin-stabilized in orbit; apparently horizon scanners were obscured by experiment boom only partially deployed. 19 of the 20 scientific experiments were returning usable scientific data. (NASA Release 65-75)

- U.S.-Mexican agreement for operation of NASA tracking station at Guaymas, Mexico, had been extended to 1970, NASA announced. The station would be used to track Project Gemini and Project Apollo. The two Governments also agreed to cooperate on meteorological sounding programs. (NASA Release 65-76)
- Milton B. Ames, Jr., Director of NASA Space Vehicle Research and Technical Div., told the House Committee on Science and Astronautics' Subcommittee on Space Sciences and Applications that lightweight, flexible plastic baffles had proved more efficient for controlling fuel "sloshing" in launch vehicle's propellant tanks than heavy metal baffles. He said plastic baffles could also serve to prevent leakage of propellant gas used in fuel-pumping during weightlessness. (Text; *NASA Auth. Hearings*, 133-50)
- Dr. Maurice Goldhaber, Director of AEC's Brookhaven (N.Y.) Laboratory, testifying before a subcommittee of the Joint Senate-House Atomic Energy Committee, announced discovery of the "antideuteron," largest particle of antimatter yet known to be produced on earth. Antimatter consisted of various subatomic particles which could annihilate their particular opposite number if they struck them. Goldhaber later told newsmen that scientists had reported observing occasional particles of antimatter running earthward from outer space. "It could be that somewhere else in the universe there is an 'anticosmos' that occasionally leaks particles to the earth." (AP, *Louisville Courier-Journal*, 3/4/65)
- Basing his judgment on successful Feb. 27 firing of Thiokol's 156-in. solid propellant rocket motor, Harold W. Ritchey, President of Thiokol Chemical Corp., predicted U.S. could produce within 30 months a flyable rocket capable of generating 7 million lbs. of thrust. Brig. Gen. Joseph J. Bleymaier (USAF), Deputy Commander (Manned Systems) of USAF Space Systems Div., commented: ". . . this firing provides us with final proof that we can configure an all-solid space booster of tremendous capability when the requirement presents itself." (Appel, *NYT*, 3/5/65)
- Senate passed House-passed bill designating March 16 of each year as Dr. Robert Hutchings Goddard Day. (*CR*, 3/4/65, 4009, 4010)
- Firefly, a new life detection instrument containing an extract of common firefly's lamp, had been developed by NASA Goddard Space Flight Center to help determine how far out and how much life existed in earth's atmosphere. This information would be essential to prevent contamination of sterilized probes enroute through earth's atmo-

sphere. Firefly, containing luciferin, luciferase, and oxygen, would glow whenever it encountered adenosine triphosphate, a chemical essential to all life known on earth. Report of any encounter with live microorganisms would be immediate, precluding need for recovering detector. (GSFC Release G-5-65)

March 4: Columbia, Harvard, and Yale Universities' medical libraries, aided by a National Science Foundation grant, were linked by a network of computers and telephone lines, thereby giving students instant access to medical literature in all collections. Frederick G. Kilgour, Yale medical librarian, foresaw elimination of duplicate material when telecommunication and photographic reproducing devices were added to the network. Pages from a book in one city could be furnished to student in another city and even reproduced for him to check out. (Phillips, *NYT*, 3/5/65, 1)

March 5: NASA's MARINER IV spacecraft, at 8:02 a.m., EST, automatically switched from its omnidirectional antenna to fixed narrow beam antenna to communicate with earth, thereby becoming radio-ready for the remaining 130 days of its Mars flight. JPL received report from tracking station at Canberra, Australia, of a prompt increase in signal strength. (NASA Release 65-78)

- NASA Aerobee 150 sounding rocket launched from White Sands, N. Mex., went to a peak altitude of 188.5 km. (117 mi.). Primary experimental objective was to study the group of stars of Orion in the ultraviolet. Because of a failure with the attitude control system the experiment had no chance to operate. Experiment instrumentation was provided by Princeton Univ. Observatory. (NASA Rpt. SRL)

- In summary of activities of the NASA Office of Lunar and Planetary Programs in testimony before the House Committee on Science and Astronautics' Subcommittee on Space Sciences and Applications, Dr. Homer E. Newell, NASA Associate Administrator for Space Science and Applications, said: "The Ranger pictures represent our major scientific achievement in 1964. In addition to their direct value as new information, the subtle significance of these pictures toward increasing the value of other astronomical data is perhaps worthy of mention, as it may not be recognized generally. It is interesting to note how the information presented in the high resolution Ranger pictures has sent scientists scurrying back to the files of photographic plates taken years before to discover features which have remained unnoticed throughout the years. Some new interpretations of long recognized features have also been made possible by the close-up look obtained by Ranger." (Testimony; *NASA Auth. Hearings*, 56-111)

- At a House Science and Astronautics Committee budget hearing at NASA Manned Spacecraft Center, Rep. Olin Teague (D-Tex.) said he thought U.S. had about a 50-50 chance of landing a man on the moon by 1970 "if we get the money for our space team." Teague felt that America was ahead of Russia in development of scientific programs in space, but Russia was ahead in development of large boosters. Rep. Robert Casey (D-Tex.) stressed that the program would be considered a success even if 1970 schedules were not met. Rep. George Miller (D-Calif.), Chairman of House Science and Astronautics Committee, said that in 50 to 100 years, "people won't care if we made it in this decade, if the program itself is successful." Dr. Robert R. Gilruth,

MSC Director, said that the Apollo and Gemini spacecraft would not be limited to manned space program but also would be useful in other scientific programs. Teague expressed his disappointment at the military's failure to make greater use of NASA-developed spacecraft and boosters. He predicted that both Gemini and Apollo would be used some day as weapon carriers. (Maloney, *Houston Post*, 3/6/65)

March 5: NASA had awarded one-year, cost-plus-incentive-award fee contracts to nine firms for engineering, fabrication, and institutional support services to six laboratories and three offices of Marshall Space Flight Center. Cost of work was estimated at $58.5 million for one year and was primarily in support of the Saturn/Apollo launch vehicle program. The firms were Sperry Rand Corp., Brown Engineering Corp., Vitro Corp., Hayes International Corp., Northrop Corp., Spaco, Rust Engineering Co., RCA Service Co., and Management Services, Inc. (NASA Release 65-77; MSFC Release)

- USAF launched a Titan I ICBM from Vandenberg AFB, Calif., as one of a series of tests to determine compatibility of the missile with various payloads. (AP, *NYT*, 3/6/65, 9; *M&R*, 3/15/65, 11)
- General Dynamics Corp.'s F-111 fighter jet broke the sound barrier for the first time in a 1 hr. 32 min. flight test. Afterward, in quick-stop braking test, both tires in main landing gear blew out. (AP, *NYT*, 3/6/65, 11)
- Dr. Clyde W. Tombaugh of the New Mexico State Univ. Research Center was quoted in an editorial in the *Kansas City Times* as saying that the "canals" seen on Mars through telescopes were fractures of the planet's crust. He said: "The origin may be due to asteroids impacting on the surface, much as what happens when a stone hits the windshield of a car. I think I have the right answer. . . ." (McCoy, *Kansas City Times*, 3/5/65)
- West Germany was waging vigorous campaign by letter, circular, and word-of-mouth to persuade German technicians to leave their jobs in Egyptian aircraft and rocket industry. The campaign could be result of recent arrests in Cairo of several West German citizens on espionage charges. (Olson, *NYT*, 3/6/65, 7)
- Fred P. Strother, in charge of requirements for Boy Scout merit badges, announced that details of a space exploration merit badge were being worked out with NASA. (*NYT*, 3/6/65, 27)

March 6: John W. Findlay, Deputy Director at the National Radio Astronomy Observatory, Green Bank, W. Va., had been named Director of Arecibo Ionospheric Observatory in Puerto Rico for a one-year term beginning in the fall. He would succeed William E. Gordon. (*NYT*, 3/7/65, 75)

March 7: COSMOS LIX satellite, containing "scientific apparatus," was orbited by the U.S.S.R. Initial orbital data: apogee, 339 km. (210.9 mi.); perigee, 209 km. (129.6 mi.); period, 89.7 min.; inclination, 65°. Equipment was said to be functioning normally. (*Krasnaya Zvezda*, 3/9/65, 1, ATSS-T Trans.)

- Commercial aviation's first nonstop crossing of the Pacific was made by Qantas Airlines Boeing 707: San Francisco to Sydney in 14 hrs., 33 min. (Wash. *Daily News*, 3/8/65)
- DMS, Inc., aerospace market intelligence operation that published annual analysis of DOD and NASA budget requests submitted to Congress, fore-

cast a $106.57 billion market for the aerospace industry from 1966–1970, an increase of 13% for the five-year period. DMS preferred this "generally favorable market climate," to the "glorified major growth period of fiscal 1962–1964, when Government spending skyrocketed, inevitably producing an influx of hopeful though unusually ill-equipped competitors, followed by over-capacity as the market tapered off, and finally a retrenchment still under way." (*NYT*, 3/7/65)

March 7: A "Dictionary of Scientific Biography" containing essays on careers of scientists and mathematicians would be published by Scribner with a National Science Foundation grant of more than $250,000. Dr. Charles C. Gillispie, Princeton professor of History of Science, had been named Editor-in-Chief. (*NYT*, 3/7/65, Book Review Sec., 8)

• Professor Fred Hoyle, British astronomer, might accept U.S. position if U.K. Department of Scientific and Industrial Research determined not to build new Institute of Theoretical Astronomy which would house an American computer essential to his work. Hoyle complained last year that he had been prevented from using the only American-built computer in Britain that would do his work properly. (Feron, *NYT*, 3/8/65, 9; *Wash. Post*, 3/9/65)

Week of March 7: Drop tests at North American Aviation's Downey, Calif., plant demonstrated that substructure of Apollo spacecraft could withstand maximum Apollo water-landing conditions. A series of 18 more drop tests was planned. (*M&R*, 3/15/65, 7)

March 8: In first Pacific Ocean sounding rocket experiment from NASA's Mobile Range Facility, two two-stage Nike-Apaches were launched from USNS *Croatan* about one mile north of the equator at 84° west longitude. Conducted by NASA Goddard Space Flight Center, the mission of first rocket was to measure ionospheric currents and magnetic fields in "equatorial electrojet," a system of electrical current circulating in ionosphere in the region of magnetic equator which could be responsible for intensifying equatorial magnetic field at about local noon. Second Nike-Apache, conducting an experiment for Univ. of Michigan, was launched about 2 hrs. later carrying Pitot-static probe to measure pressure, temperature, and density in the region of 20 to 75 mi. altitude. (NASA Release 65–82; Wallops Release 65–12; NASA Rpts. SRL)

• The countdown rehearsal for the Gemini (GT-3) flight, conducted at Cape Kennedy, was delayed two hours because of (1) a propellant leak in Titan II rocket, (2) crossed wires in ground support equipment, (3) failure of some of the batteries to reach peak power immediately, and (4) faulty reading in control center. Project Gemini officials said none of these problems had been serious, but the combination would have caused a postponement on launch day. (*NYT*, 3/9/65; *N.Y. Her. Trib.*, 3/9/65; *Balt. Sun*, 3/9/65)

• In testimony before the Senate Committee on Aeronautical and Space Sciences, on "the status, management, and prospects of the aeronautical and space program," NASA Administrator James E. Webb said: "The progress during this period in the space program has been made possible by the cooperative efforts of many organizations and people. Ninety-four per cent of our work during Fiscal Year 1964 was conducted by American industry and involved a total of about 380,000 people in industry, universities, research institutes, and government installations. Almost 250,000 separate procurement transactions were initiated during this time.

". . . the past year saw the continued strengthening of the coordination and the mutual support between NASA and the DOD in space and aeronautics. The Aeronautics and Astronautics Coordinating Board has continued to be an effective medium for formal coordination. During 1964 NASA and the DOD developed procedures for the coordination of the space science programs; a national program in satellite geodesy was established by the DOD, NASA, and the Department of Commerce; a standardized basis for reporting space and aeronautical sciences research and technology information has been adopted; a joint NASA–DOD study was conducted to determine the launch vehicles needed to meet projected requirements during the next decade; a joint study was conducted of the current and planned lifting reentry vehicle research and development programs; the needs of NASA, the Air Force, and the Federal Aviation Agency were incorporated into an expanded flight research program utilizing the XB–70 aircraft to confirm theoretical and wind tunnel data on supersonic flight vehicles.

"All of this, Mr. Chairman, of course, is under an umbrella of policy followed closely by the [National Aeronautics and] Space Council. . . ."

Commenting on the Soviet space program, Webb said: "Our rapid rate of advance and the success we have achieved already has, we believe, denied the USSR many of the benefits and many of the options which the Soviets expected their space program to provide as a part of their forward thrust toward world domination. However, there is every evidence, on the basis of their activity during the past three years, that the Russians intend to maintain a vigorous effort in space, and, in fact, that their activities may be further increased. During 1963 and 1964 more Soviet spacecraft were put in earth orbit or deep space than in the six previous years combined. The number placed in orbit last year was double that of the year before. . . ." (Testimony; *NASA Auth. Hearings*, 13–50)

March 8: NASA Deputy Administrator Dr. Hugh L. Dryden reported on the status of NASA cooperation with the Soviet Union in testimony before the Senate Committee on Aeronautical and Space Sciences: "Let me review where we stand. Of three projects agreed to in 1962, the only one completed is that which involved communications tests with Echo II. . . .

"The second project—joint mapping of the geomagnetic field—is at the stage of exchanging ground-based magnetic observations . . . we are now acquiring data that was not previously available in the United States. . . .

"In the third project—for the coordination of meteorological satellite launchings and the establishment of a link for exchange of data—our prime purpose was and remains a sharing of the cost of providing weather satellite service and the exchange of satellite data. . . . we are . . . exchanging conventional data over the link, which . . . is financed on a 50–50 basis. I look forward for a meeting soon with Academician Blagonravov which will afford opportunity to review this exchange and the prospects of satellite data exchange. . . . a recent check shows the U.S. sending surface data for more stations than it receives but receiving upper air data for more stations than it sends. In sum, the present exchange is considered by the Weather Bureau to improve the quality of forecasts by our national weather services since it makes more data available in time for such forecasts than was the case prior to establishment of the link." (Testimony; *NASA Auth. Hearings*, 50–76)

March 8: NASA Associate Administrator Dr. Robert C. Seamans, Jr., discussing the management of NASA's aeronautics and space effort before the Senate Committee on Aeronautical and Space Sciences, said: ". . . our performance in terms of data returned is perhaps the most succinct evidence of success. The volume of information brought back from space in 1964 averaged 57 million data points per day in comparison with a previous high of some 6 million data points per day. This indicates not only more advanced instrumentation but also more reliable functioning of flight experiments. We averaged 54 working experiments throughout 1964, which represents an improvement of 35% over the 1963 average.

"We have achieved significant results in ground based experimentation, testing, aeronautical flights, and sounding rocket launchings. Work conducted in our wind tunnels continues to refine aircraft configurations for vertical takeoff and landing, supersonic transportation, and hypersonic flight. We are continuing to experiment with materials, fuels, turbines, injectors, and nozzles in order to improve the efficiency of air-breathing and rocket propulsion systems. The 3 successful power tests of the KIWI reactor demonstrated the applicability of nuclear energy to rocket propulsion. In 1964 we conducted 27 more flights of the X–15 aircraft, 19 of them over Mach 5, amassing data important to supersonic and hypersonic flight. In addition, we launched 131 successful sounding rockets from stations around the world to test new instrumentation and to obtain important scientific data in geophysics, astronomy, and meteorology. In the areas of manned space flight, the Apollo escape system has been successfully tested, and a boilerplate spacecraft checked out and flown on the Saturn I. A mock-up of the lunar excursion module has been approved. The Saturn IB and Saturn V 'battleship' upper stages have been successfully fired. The F–1 engine has passed its flight rating test. This record was established by the hard work and careful attention to detail of the government-industry-university team charged with aeronautic and space exploration. This total team, numbering 380,000 people, is managed by the relatively small hard core NASA organization of less than 34,000." (Testimony; *NASA Auth. Hearings*, 76–114)

- Gemini astronaut parachute system was successfully tested in drops from a C–130 at 15,000 ft. altitude by USN Chief Warrant Officer Mitch Kanowski and USAF Maj. Dan Fulgham over the Naval Air Facility El Centro. Parachutes deployed at 9,000 ft. as they would on actual Gemini missions. Additional tests would be made in drops from altitudes up to 35,000 ft. (Miles, *L.A. Times*, 3/9/65)
- U.S.S.R.'s ZOND II would pass within 900 mi. of Mars on Aug. 6, according to Soviet space scientist Prof. Mstislav Keldysh. This an-

nouncement was relayed by Dr. Charles S. Sheldon of the National Aeronautics and Space Council, who quoted Keldysh as saying ZOND II weighed about 2,000 lbs. Dr. Sheldon speculated that the probe's considerable weight—four times more than MARINER IV—could mean "it may be doing something more than a simple fly-by" of Mars. (UPI, *Denver, Post* 3/10/65; UPI, *NYT*, 3/11/65, 42)

March 8: According to *Missiles and Rockets*, Dr. Joseph Shea, director of Apollo spacecraft program at NASA Manned Spacecraft Center, said Apollo spacecraft was having no weight problems. He explained that the current weight of 89,000 lbs. was under 90,000 lbs. goal and there was room for additional growth since Saturn V booster had increased estimated payload capability to 95,000 lbs. Weight of Lem was increasing in early development stage, but new evaluation stemming from NASA decision to make it as safe in terms of redundancy as command and service modules, could raise weight from 29,500 lbs. to 32,000 lbs. Shea commented that a stable baffled injector had been selected for service module's propulsion system and it was currently undergoing qualification test series. Recent tests of heatshield in reentry tests with Scout (Aug. 18, 1964) achieved high total heat of 250 Btu's per square foot—about 80% of the heat expected to be encountered during return from moon. (*M&R*, 3/8/65, 14)

- NASA had decided to replace Lem's fuel-cell power subsystem with a more conventional battery system, *Missiles and Rockets* reported. Motivation was concern for reliability. Decision would not affect use of fuel cell in the Apollo command module. (*M&R*, 3/8/65, 14)
- In a letter-to-the-editor in *Missiles and Rockets*, Thomas M. Morse said that since there were no indications that the Russians were building a bigger booster for their lunar program, they might be planning to use a libration orbit to reach the moon. He described the libration orbit as an almost stable earth orbit in which a spacecraft would always be on a direct line between earth and moon, about 33,000 mi. from the moon. Advantages offered over the U.S.-planned lunar orbit included easier rendezvous; pre-parking of unmanned freight, shielding, and modules; unlimited rendezvous and liftoff windows; continuous line-of-sight communications between earth, rendezvous craft, and lunar landing crew; better radiation protection; improved safety factor; reduced cost. (*M&R*, 3/8/65, 6)
- California's Gov. Edmund G. Brown, in his second *Annual Economic Report* to the legislature, warned that new cutbacks in defense and aerospace spending could dilute "the reservoir of scientific brainpower and skilled manpower that has made California the leader in the space age." Brown said that 200,000 new jobs a year would have to be created and that the state had already contracted with aerospace firms for studies that might provide solutions "in transferring manpower from defense and aerospace production to other areas." (AP, *L.A. Herald-Examiner*, 3/8/65)
- France announced successful launching in the Sahara of the Emeraude stage of the Diamant booster. (*M&R*, 3/15/65, 11)

March 8–9: President Johnson's proposal to cut Federal subsidies to helicopter carriers in New York, Los Angeles, and Chicago by Dec. 31, was opposed by the CAB, who suggested continuation of Federal subsidies on declining basis until 1970. Sen. A. S. Mike Monroney (D–Okla.), Chairman of the Senate Aviation Subcommittee investigating ways to keep helicopter lines alive, suggested increased support from major airlines as possible alternative to Federal aid. Stuart G. Tipton, President of Air Transport Association which represented almost every scheduled U.S. airline, testified that helicopter lines received about $1 million a year in indirect support and that "as a matter of principle, this is as far as the airlines should go or be expected to go." He added that experience and advances in helicopters and poor-weather landing equipment were about to make helicopters potentially profitable in all large cities. Withdrawal of Federal support now would be disastrous. (Clark, *NYT*, 3/9/65; Clark, *NYT*, 3/10/65, 69)

March 8–12: "Efficiency and Perfection through People" was objective of AFSC's Internal Zero Defects Program which encouraged people to "set their own immediate goals and devise measurement techniques." Results of the program would be analyzed and recognition awards would be given to individuals making significant achievements. (AFSC Release 19.65; *CR*, A1315–A1318)

March 9: Thor-Agena D launched from Vandenberg AFB orbited eight military satellites, the most in any single launch to date. Two satellites would measure solar radiation (GREB VI and SOLRAD); two would test stabilization methods for future spacecraft (GGSE II and GGSE III); one would be used in geodesy (SECOR III); two would help calibrate satellite tracking networks (SURCAL satellites); and one would transmit radio broadcasts for ham operators (OSCAR III).

OSCAR III would transmit signals from 25 amateur radio channels over a 4,000-mi. radius and was being tracked by ham radio operators at Foothill Jr. College, Calif. Amateur tracking stations in 30 foreign countries were informally participating in the project. (*U.S. Aeron. & Space Act., 1965*, 135–36; GSFC *SSR*, 4/15/65; Clark, *NYT*, 5/19/65; *Wash. Post*, 5/20/65, A12; AP, Omaha *Eve. World-Herald*, 3/10/65)

• Gemini astronaut parachute system for use in launch emergency failed to function properly during test at El Centro, Calif. When the jumper stepped from a C–130 aircraft at 23,000-ft. altitude, a "ballute" (combination balloon and parachute) device for stabilizing the fall failed to deploy; the chute was opened manually at 12,000 ft. (UPI, *Minneapolis Trib.*, 3/10/65)

• In testimony on NASA lunar and planetary programs before the House Committee on Science and Astronautics' Subcommittee on Space Sciences and Applications, NASA Associate Administrator Dr. Homer E. Newell said: "The Surveyor program to date has accomplished a number of significant advanced developments that have found or will find their way into other programs.

"The closed loop automatic landing system has other potential space and terrestrial applications.

"The planar-array high gain antenna has several significant advantages over the usual parabolic dish.

"The doppler and altimeter radars represent a significant advance in technology and have been adopted by the Apollo Lunar Excursion Module.

"The throttlable high performance vernier engines, which have an almost unlimited operating lifetime, represent another significant step forward.

"The high performance spherical main retro rocket has appreciably advanced the state of the art. Several launch vehicle programs are interested in this motor as a high-energy upper stage.

"The Surveyor landing gear design represents a new high in efficiency of impact energy absorption.

"Many of the miniaturized geophysical instruments developed for Surveyor may have terrestrial applications." (Testimony; *NASA Auth. Hearings*, 243ff)

March 9: World's longest antenna had been stretched on the top of the Antarctic icecap to study radio conditions in space beyond the earth, the National Science Foundation reported. The antenna was a 21-mi., plastic-coated, 3/4-in. copper cable that radiated low frequency waves that traveled far out into space along a line of force in the earth's magnetic field. The waves followed the line of force as it curved back toward the earth in the opposite hemisphere. (UPI, *NYT*, 3/10/65)

- Rep. Westen E. Vivian (D–Mich.) told the House Committee on Science and Astronautics' Subcommittee on Space Science and Applications that he intended to request adoption of a policy to award one half of all NASA Phase 1A contracts to companies in areas presently receiving less than one half of NASA business. Chairman Joseph E. Karth (D–Minn.) said that although fiscal expedience demanded that procurement contracts go to industralized areas, geographic distribution should be seriously considered in the allocation of research and development funds. He said the Subcommittee would consider Vivian's proposal. (Transcript, 3/9/65)

- NASA Marshall Space Flight Center hosted group of 61 Navy, Air Force, and civilian personnel from Navy Field Office for Manned Orbiting Laboratory at Los Angeles. The group received briefings on Apollo and Saturn programs and saw facilities at Marshall. They had previously visited NASA Manned Spacecraft Center and were scheduled to tour NASA Kennedy Space Center. (MSFC Release 65–52)

- A job classification dispute at the Chrysler facility of the Michoud Saturn plant in New Orleans caused over 200 United Auto Workers (UAW) to walk off the job. Chrysler was responsible for developing first stage of Saturn IB rocket for NASA. (UPI, *Wash. Post*, 3/10/65)

- NASA Marshall Space Flight Center awarded $1,059,000 contract to Aetron, a division of Aerojet-General Corp., for installation of equipment on a Saturn V second-stage test stand at Mississippi Test Operations. Equipment would include consoles to check out systems on the flight stages being tested as well as in the area's test control center. (MSFC Release 65–53)

- Prompted by results of the experiments of Dr. Frank A. Brown, Jr., Northwestern Univ. biology professor, NASA and Northwestern Univ. scientists were studying a plan to orbit a potato around the sun in an attempt to prove whether man could survive in long trips in space. Dr. Brown had concluded that rhythmic patterns of wakefulness and sleep, glandular activity, cellular respiration, and all other biological cycles of most live organisms were timed by biological clocks outside the organism, not inside. The three primary forces were day-night changes and temperature, atmospheric, and pressure fluctuations.

For ten years, Brown had kept potato tubers under constant pressure and temperature and in constantly dim light, yet they continued to fluctuate at same rate and time as potatoes planted in Illinois and they detected atmospheric pressure changes. Brown concluded that "something is getting thru to the isolated potatoes to tell them what the weather is outside. It could be the earth's magnetic or electric fields or radiation, since they all observe a 24-hour pattern geared to the rising and setting sun, but we are not sure.

". . . biological clocks are necessary to keep a living system a coordinate, living, functioning whole. If the clocks are stopped, the organism may go beserk and die." If the orbiting potato were to die within 90 days, it would indicate that a 24-hour rhythm was vital. (Kotulak, *Chic. Trib.*, 3/9/65)

March 10: A dummy model of the Gemini spacecraft, dropped from 11,000 ft. altitude by a c-119 aircraft, parachuted into the Atlantic Ocean and was recovered by three USAF pararescue men. This was a practice mission in case Astronauts Virgil I. Grissom (USAF) and John W. Young (USN) had to abort their GT-3 flight during the launching phase. (AP, *Orl. Sent.*, 3/11/65)

• Reviewing NASA's activities in manned space flight in the last year, Dr. George E. Mueller, NASA Associate Administrator for Manned Space Flight, testified before the Senate Committee on Aeronautical and Space Sciences: "It is a pleasure to report that . . . there has been substantial progress in the development and testing of flight vehicles and earth-based facilities, in the nearly complete marshalling of the government-industry manned space flight team, and in the consolidation of firm program-wide management.

"During the past year, the Gemini Program has advanced to the point that we are ready for manned flight operations. The Apollo Program is entering a year of comprehensive development testing of major systems prior to the 'all-up' unmanned earth–orbital flights, which will begin in 1966. And our study of advanced manned missions has established that it is feasible to return dividends from the current investment by applying the wide range of Apollo capabilities to a number of other potential missions." (Testimony; *NASA Auth. Hearings*, 143ff)

• Harold B. Finger, Director of NASA Nuclear Systems and Space Power and Manager of the AEC–NASA Space Nuclear Propulsion Office, discussed NASA's electric thrustor program in testimony before the House Committee on Science and Astronautics' Subcommittee on Advanced Research and Technology: "We conducted the first successful flight test of an electric rocket engine in July, using the SERT I spacecraft. This flight demonstrated that ion beam neutralization will take place satisfactorily in space and, therefore, eliminated the only uncertainty regarding the basic feasibility of successful space operation. A second major accomplishment was the design, fabrication, and test of a 30 kilowatt thrustor. This thrustor is 10 times larger than previous ion engines, and demonstrates that we are successfully developing the engineering relations required to build the mega-watt size thrustors needed for spacecraft prime propulsion."

Discussing nuclear propulsion programs, he said: "1964 was a year of significant progress in the Nuclear Rocket Program. It was marked by the successful completion of the KIWI series of reactor development experiments and the successful initiation of the NERVA reactor testing. These reactor experiments, coupled with work in other portions of our Nuclear Rocket Program, provide assurance that the graphite core nuclear rocket can be available to fulfill its role as the next major space propulsion system.

"Of particular significance in 1964 was the successful demonstration of the adequacy of the reactor structural design, the elimination of reactor structural vibrations, full power reactor operation for over ten minutes at an altitude equivalent specific impulse of about 750 seconds, a rapid automatic startup, the ability to restart the rocket reactor, the determination of the effect of a maximum reactor power excursion, and the neutronic investigation of two rocket reactors located side-by-side as would be necessary in clustered engine configurations. . . .

"During this year emphasis will be placed upon extending our reactor technology to higher temperature, longer duration, and higher power while we proceed as rapidly as possible to close coupled nuclear rocket engine system testing. We face this task of developing nuclear rocket technology including component, subsystem, and engine system work, with a confidence that is based on the solid accomplishments in our reactor development program and with the knowledge that the technology we are developing will provide the propulsion capability that will ultimately be required for extensive space exploration." (Testimony; *NASA Auth. Hearings*, 243–300)

March 10: According to USAF Cambridge Research Laboratories study, a continuous barrage of meteoroids was causing moon to lose up to 6,000 tons a day and earth to gain 10,000 lbs. a day. Because of its strong gravity, earth absorbed about four times as many impacts as moon. (OAR Release 3–65–3; *Chic. Trib.*, 3/11/65)

- NASA announced award of $3,713,400 contract to Raytheon Co. to provide digital systems for Project Apollo. Options for additional displays and consoles, if exercised, could add $400,000 to basic price. The equipment was for use at NASA control centers and critical tracking stations to give instantaneous display of information received by encoded radio signals during Apollo flight permitting immediate decisions concerning welfare of the astronauts and conduct of the mission. (NASA Release 65–79)

- Sen. Claiborne Pell (D–R.I.) introduced in the Senate S. 1483, a bill to provide for a National Foundation on the Arts and the Humanities. (*CR*, 4/26/65, 8122)

- Roy W. Jenkins, British Aviation Minister, presented a plan to House of Commons to give grants of up to £100 ($280) each to householders plagued by noise of jetliners to soundproof their homes. It was estimated that about 40% of 200,000 householders affected would accept the grant, bringing the total cost to $7 million. An Airport Authority would be established to underwrite the cost of the grants. Householders complained that money was insufficient for adequate sound-proofing. (*NYT*, 4/11/65, 57)

March 10: Dr. W. Randolph Lovelace II, NASA's Director of Space Medicine, said NASA physicians screening future astronauts were eliminating people with heart and spine defects so slight they would be insignificant on earth. "We are interested in finding minute defects between the left and right heart," he explained. "If you lose pressure and you have this defect, thousands of little air bubbles may find their way up to the brain. If there is no defect, they are removed from the lungs. We are also looking for congenital defects of the spine. When someone experiences the acceleration astronauts do, however, a small defect may be magnified in effect." (Kass, *Houston Post*, 3/11/65)

March 10–11: Stuart G. Tipton, President of the Air Transport Association, testified before Senate Aviation Subcommittee that "more joint fares, perhaps more guaranteed flights, certainly more sales campaigns" by larger airlines might be initiated to help helicopter lines which were facing end of Federal subsidies. Tipton stipulated, however, that CAB's five-year declining subsidy proposal still would be essential. Earlier in the week, he had testified that airlines would not increase their aid.

Senators and Representatives from all three states that had helicopter service testified in support of subsidies. Sen. Robert F. Kennedy (D–N.Y.) said: "I think that, really, if we don't do it now [continue subsidy] it won't be done and that will affect not just New York and the other two cities but the entire country." Opposition came from Sen. William Proxmire (D–Wis.) who argued for a "user tax on the people who use helicopters . . ." (Clark, *NYT*, 3/11/65, 55; Clark, *NYT*, 3/12/65, 66)

March 11: NASA announced that MARINER IV, scheduled to reach the vicinity of Mars July 14, had traveled over 168 million mi.—more than half way. (NASA Release 65–80)

- USAF launched Thor-Able-Star booster from Western Test Range, placing in orbit an unidentified satellite and U.S. Army's SECOR II geodetic satellite. SECOR II "failed to deploy properly from its piggyback container." (*U.S. Aeron. & Space Act., 1965*, 136; *M&R*, 4/5/65, 12)

- Dr. C. O. Bostrom and Dr. D. J. Williams of the Space Research Div. of Johns Hopkins' Applied Physics Laboratory said danger of radiation damage to satellites from the artificial radiation belt created in July 1962, following the nuclear detonation over Johnston Island, was "now significantly less severe." Results of measurements by instruments aboard Navy research satellite 1963 38C showed that the number of high-energy electrons in the artificial radiation belt decreased by 50% in from three months to one year in different parts of the belt. The decrease in intensity as time passed would continue until natural levels of intensity were reached. Dr. Bostrom said, ". . . the observed time decay does show that the satellite radiation damage problems have been reduced by a factor of ten from what they were two years ago." (Balt. *Sun*, 3/11/65)

- Despite U.S. Federal Court's issuance of two temporary restraining orders, building trade employees halted construction on Saturn IB Launch Complex 34 at Cape Kennedy, for second straight day. Dispute involved general contractor's use of non-union subcontractor.

NASA claimed work on Launch Complex 34 was "critical" to the nation's space effort. (*Cocoa Tribune*, 3/11/65)

March 11: NASA Kennedy Space Center had extended for the second year two of the major contracts under which the NASA Merritt Island Launch Area was being operated. Extensions were negotiated with Trans-World Airlines, for base support services, and Ling-Temco-Vought, information services. (KSC Release 58-65)

- Dr. Hugh L. Dryden, NASA Deputy Administrator, speaking in Minneapolis, Minn., before the Twin Cities Section of AIAA, said: "We believe that activities in the exploration of space, a modern social need recognizable from the passage of the National Aeronautics and Space Act and the appropriation of large sums of money by the Congress, provide the essential environment to accelerate greatly the growth of theoretical and experimental science in many areas. It is true that this accelerated growth in science and technology is essential to the on-going development of space capability, but of deeper significance is the complex dynamic interaction between science, technology, and space exploration, which is essential to the growth of science, technology, and space exploration. In this case, as in the cases previously cited, to use an analogy from bacteriology, there has to be a nutrient solution (money and employment opportunities) to feed the scientific and technological effort, and as soon as this environment is provided, many latent efforts in science and technology begin to assert themselves and move forward.

    "I believe that this interpretation of certain aspects of the space program is more significant and meaningful than the current concepts of technology utilization and technological spinoff as incidental or serendipitous benefits of space exploration." (Text)

- Gerald L. Smith, NASA Ames Research Center, had been awarded $1,000 special service award for his computer analysis which resulted in decision to give ground-based navigation a primary role during Apollo lunar missions. Smith explained that, although radar tracking from earth and visual tracking onboard spacecraft were almost equally reliable, earth-based system could be maintained more easily and was not restricted by weight and size considerations. (ARC Release 65-8)

- American Academy of Arts and Sciences awarded its Rumford Prize to Dr. William D. McElroy for his analysis and isolation of chemicals that cause bioluminescence in the firefly and other organisms. He identified luciferin, luciferase, and adenosine-5-triphosphate (ATP). From his research he concluded that bioluminescence had evolved "as an accidental consequence of chemical reactions" in the organisms as they adapted to changing conditions in the environment. Dr. McElroy was head of the McCollum-Pratt Institute of Johns Hopkins Univ. and a member of the President's Science Advisory Committee. (*SR*, 4/3/65, 45-47)

- Victor D. Lebedev, U.S.S.R. Council of the National Economy, announced plan to convert 119 major industrial plants to electronic computer system of production management within two years. Aimed at ensuring fast access to detailed operating information, the systems would be introduced in heavy industry and consumer-goods production. For the future: computers serving individual plants or groups of plants would process detailed data and transmit generalized information to central agencies to aid in planning economy. (Shabad, *NYT*, 3/12/65, 8)

March 12: COSMOS LX satellite, containing "scientific apparatus," was orbited by the U.S.S.R. Initial orbital data: apogee, 287 km. (177.9 mi.); perigee, 201 km. (124.6 mi.); period, 89.1 min.; inclination 64° 42'. Equipment was said to be functioning normally. (Tass, *Komsomolskaya Pravda*, 3/13/65, 1, ATSS-T Trans)

- NASA Aerobee 150 sounding rocket launched from White Sands, N. Mex., went to a peak altitude of 155.5 km. (96.6 mi.) The primary experimental objective was to obtain ultraviolet spectra of Mars and Orion by the use of four spectrographs, provided by NASA Goddard Space Flight Center. Because of an attitude control system failure no experimental results were obtained. (NASA Rpt. SRL)
- USAF launched Atlas-Agena D booster with unidentified satellite payloads from Western Test Range. (*U.S. Aeron. & Space Act., 1965*, 136)
- Month-long experiment for NASA to test man's ability to withstand rotational stress ended at U.S. Navy School of Aviation Medicine. Capt. Ashton Graybiel, Research Director, expressed satisfaction with results of the test which confined four U.S. Navy men in a windowless, circular room, equipped with all necessary living accommodations. The room began rotating at 2 rpm's and in 16 days built up to 10 rpm's, stopping three times daily for meals. This pattern of speed build-up had no adverse affect on the men and produced no nausea or significant discomfort. This test, one of a series conducted by Naval School of Aviation Medicine, was to check new procedure for conditioning men for space flight. Since long space voyages could require rotating spacecraft to create artificial gravity, scientists wanted to determine spinning rate man could endure without discomfort. (NASA Release 65-84)
- Launching pad damage caused by the Mar. 2 explosion of an Atlas-Centaur rocket at Cape Kennedy amounted to $2 million and would take three to four months to repair, NASA reported. To avoid delay in the Atlas-Centaur launching scheduled for mid-summer, NASA was speeding completion of a new launching pad that was 90% completed and that could be ready in two months. (UPI, St. Louis *Post-Dispatch*, 3/14/65)
- DOD announced new type of defense contract for C–5A, 700-passenger supertransport and cargo plane: competitors must bid not only for initial development contracts, but for production and "lifetime" support of proposed aircraft. Lifetime support, estimated to be at least 10 yrs., would cover spare parts and ground maintenance equipment. The plane, expected to be biggest jet transport ever built, would have a gross take-off weight of 725,000 lbs. and a payload capacity of 250,000 lbs.

    Boeing, Douglas, and Lockheed were competitors for airframe contract; General Electric and Pratt & Whitney were competitors for engine contract. Contracts would be awarded this summer. $2.5 billion was estimated cost for a 58-plane program. (DOD Release 915-64)

March 12: AFSC announced award of five letter contracts totaling $3.8 million for conceptual phase of Mark II Avionics System being considered for use on F-111A aircraft. General Dynamics, Hughes Aircraft Co., Sperry Gyroscope Co., Westinghouse Electric Corp., and Autonetics Div. of North American Aviation, Inc., would perform analyses leading to system design recommendations integrating many subsystems. (AFSC Release 30-65)

- A 23-year old French nurse, after three months in a 240-ft. deep cave in Grasse, France, emerged thinking it was Feb. 25. Josie Laures had had no clock and a white mouse had been her sole companion in this experiment to test effects of solitude. She was flown immediately to Paris for three weeks of medical examinations. (AP, *NYT*, 3/13/65, 6)

March 13: President Johnson signed a bill and proclamation declaring March 16 "Robert H. Goddard Day." Dr. Robert Hutchings Goddard of Clark University had launched world's first liquid-fuel rocket at Auburn, Mass., on March 16, 1926. (Text, *NYT*, 3/15/65, 8)

- USNS *Croatan*, which had left Balboa, Panama Canal Zone, on Mar. 6, arrived at Lima, Peru. During the interval, ten two-stage sounding rockets had been launched from the deck of the ship, carrying upper atmosphere and ionosphere experiments for the Univ. of Michigan, the Univ. of New Hampshire, and NASA Goddard Space Flight Center. Three single-stage Arcas meteorological rockets, two of which carried experiments to measure ozone in the atmosphere, were also launched. (NASA Release 65-13)

- Seventy paintings and drawings rendered by 15 contemporary American artists at rocket and satellite launching stations were exhibited at the National Gallery, Washington, D.C. According to the National Gallery, the purpose of the NASA-sponsored art programs was to "record the strange new world which space technology is creating" and "to probe for the inner meaning and emotional impact of events of fateful significance to mankind." Accompanying the exhibit was a film "The World Was There" which contrasted secrecy of some nations' space programs with the openness of the American effort. (National Gallery Release, 3/14/65)

- In *New York Times* Richard Witkin said F-111 variable-sweep-wing plane, intended as the mainstay of U.S. fighter forces before 1970, had developed problems with engines and with the inlets that feed air to the engines. Officials maintained that problems were normal in any development, but conceded that fewer difficulties had been anticipated because of record number of wind tunnel tests. The two prototypes tested, one of which was supersonic, had continuously run into two main difficulties: (1) air flow through compressors of engines had become disturbed, causing erratic power output; and (2) combustion in afterburner section had been suddenly stopping. Otherwise, officials contended that flight tests, including tucking wings far back for high-speed runs, had been going better than anticipated. (Witkin, *NYT*, 3/14/65, 58; *Chic. Trib.*, 3/15/65)

- An article in *The Economist* questioned the political wisdom of the State Dept.'s ban on exchange of communications satellite information between Hughes Aircraft Co. and the British Aircraft Co.; it suggested that unfavorable repercussions to Anglo-American relations could result: "The reason had little to do with military security. The State Dept. appeared to think that American industry has a valuable monopoly in commercial satellites which should be exploited for maximum profit, which means keeping the know-how in America. . . . The first commercial satellite of Comsat happens to be the Hughes-built Early Bird due to be launched in the first half of April. Comsat is obliged . . . to distribute its orders among member countries on a basis proportional to the shares they hold. . . . Britain is the largest shareholder after the United States. So the less satellite know-how there is, particularly in Britain, the more work goes to the United States . . . this . . . is precisely what some people have been declaring the Americans would do whenever they found themselves in a position of technical superiority. . . ." (*Economist*, 3/13/65)

March 13: Two Russian airmen had set a world altitude record by flying M1-4 helicopter with a load of nearly two tons to 20,894 ft., Tass reported. (Reuters, *N.Y. News*, 3/14/65)

• Israel, to reassure U.S. of her peaceful intentions for use of atomic energy, had permitted two AEC commissioners to inspect Dimona reactor, a natural uranium, heavy-water-moderated type, capable of producing enough plutonium for several relatively small atomic weapons. Israel had imposed strict secrecy on the inspections, one a year ago and a second last month. U.S. tentatively concluded that Dimona was not being used to produce plutonium for atomic weapons but suggested that reactor be placed under inspection by International Atomic Energy Agency. Israel refused, explaining: (1) she should not be forced to place her national development under agency inspection until international inspection had been accepted by all nations, and (2) the Agency had discriminated against her in favor of Arab states in membership of its board and location of research centers.

Many American and British specialists feared that Israel could be "keeping the option open" to develop atomic deterrent against Arab nations. (Finney, *NYT*, 3/14/65, 1)

March 14: Writing about the visit of President Johnson and Vice President Humphrey to NASA Hq. for a briefing on NASA programs on Feb. 25, Lt. Gen. Ira C. Eaker (USAF, Ret.) said in an article for the *San Diego Union:* "I was particularly pleased at the deserved tribute the President paid Jim Webb and Hugh Dryden. I have known them both since 1937. They are extremely modest men. They avoid personal publicity. They are not jealous of subordinates, but prefer that the publicity and credit for NASA successes carry the pictures and headline the names of those members of the NASA team most directly responsible. For this reason they can attract and hold able people. . . .

"While we are giving out the space medals, it is only fair to say that without the vision and tenacity of Lyndon Johnson, the first man on the moon could not be an American.

"To have man's most dramatic and significant adventure become the achievement of a slave state instead of a free society would be intolerable." (Eaker, CNS, *San Diego Union*, 3/14/65)

• Soviet scientists announced development of compact, light-weight nuclear power system, similar to U.S. Snap program, to meet relatively low power requirements of up to several hundred watts by using heat from decay of radioisotopes as energy source for generation of electricity. Tass described system as a power-generating package weighing 150 kg. (330 lbs.) with a capacity of up to 200 watts and a lifetime of 10 yrs. Known as Beta, the installation was designed to ensure continuous operation of automatic weather stations in remote areas. (NYT, 3/15/65, 5)

March 14: U.S.S.R. reported that number of Soviet science teachers and scientific researchers had doubled between 1958 and 1963. At the end of 1963, 565,958 workers were engaged in scientific research and teaching, compared to 284,038 at end of 1958. Most of this increase was in persons having only basic undergraduate scientific and technical training, with women increasing more rapidly than men. Engineering sciences accounted for more than half the total of all Soviet scientific workers; physicians and mathematicians comprised second largest group; persons in medicine and pharmacy, the third. (Schwartz, NYT, 3/14/65, 18)

March 15: U.S.S.R. launched into orbit three earth satellites—COSMOS LXI, COSMOS LXII, and COSMOS LXIII—with a single booster rocket. Tass said three satellites were orbiting in close initial orbits: apogee, 1,837 km. (1,141 mi.); perigee, 273 km. (170 mi.); period, 106 min.; inclination, 56°. It was reported that the scientific apparatus onboard was functioning normally. (Krasnaya Zvezda, 3/17/65, 1, ATSS-T Trans.)

- The S-IB-1, Chrysler-built first stage of NASA's Saturn IB, arrived at NASA Marshall Space Flight Center from Michoud Operations in New Orleans for static-firing tests. The stage was 21 ft. in diameter, 80 ft. in length, and weighed 90,000 lbs. For Saturn IB program, its eight engines had been uprated to 200,000 lbs. thrust each and weight had been reduced by some 16,000 lbs. It would be returned to Michoud Operations in about six weeks for post-firing checks. Saturn IB vehicles would be used for earth-orbital missions of Apollo spacecraft. (MSFC Release 65-60)

- NASA and DOD had approved first phase of a General Dynamics proposal for 30%-uprated Atlas SLV X3 booster. This phase covered only reliability improvement by introduction of new components. Order to proceed on actual uprating was expected this month. (M&R, 3/15/65, 7)

- The House Committee on Science and Astronautics reviewed the master planning standards of ten major NASA centers and concluded that, considering the permanence of the space program, ". . . the installations and facilities required by NASA to implement the program should be planned . . . on a long-range basis, in recognition of permanency." In addition, they suggested that NASA: (1) develop "consistency of planning policy," (2) invest in master plans to prevent situations similar to "confused and congested layouts of Lewis and Wallops," (3) invest in facility planning, and (4) avoid procrastination and expediency. The Committee concluded that "NASA has achieved substantial success in master planning at many of its installations . . ." but that attempts should be made to succeed at all NASA installations. (House Report No. 167, 3/15/65)

March 15: A Benedictine nun, Sister M. Margaret Bealmear, said she had declined an invitation to apply for astronaut training and that she assumed the letter from NASA Manned Spacecraft Center had been a mistake. Sister Bealmear, a candidate for a doctorate in biology from the Univ. of Notre Dame, said she had received the invitation in December 1964. Invitations had been extended by NASA to select names appearing on a list provided by the National Academy of Sciences. (AP, *NYT*, 3/16/65, 5; MSC Historian)

- Dr. John T. F. Kuo, associate professor at Columbia Univ. Henry Krumb School of Mines, was studying the earth's gravity from each of the Empire State Building's 102 floors. Kuo was using a gravimeter sensitive to weight differences of one-billionth of a pound to measure gravitational acceleration on each floor. He felt that extrapolations from his figures might help in the "design of instrument measuring the gravitational acceleration on space vehicles as they hurtle through the universe." (*NYT*, 3/15/65, 29)

- Dr. Thomas F. Bates, professor of mineralogy and director of the Science and Engineering Institute at Pennsylvania State Univ., had been named science advisor to Interior Secretary Stewart L. Udall, it was announced. He would succeed Dr. John C. Calhoun, Jr., who was returning to his post at Texas A&M. (UPI, *NYT*, 3/16/65, 4)

- Despite boasts of increased Government volume by Westinghouse Electric Corp. and Sylvania Electric Products, Inc., most major companies complained of decline in defense, Government, and aerospace contracts. The Electronic Industries Association offered solutions to the problem: (1) look for new fields and products, (2) work harder to find Government contracts, or (3) continue complaining. (Smith, *NYT*, 3/15/65)

- "Project Stormy Spring," a meteorological study by the Air Force Cambridge Research Laboratories to develop more precise forecasting techniques for specific local areas, began. AFCRL scientists would investigate mesoscale structures and weather system dynamics in New England, particularly within a mesoscale. A varying distance measure, a mesoscale in New England in March was an area about 100 mi. sq. Major storm systems would be observed and analyzed for continuous periods of 24 to 36 hrs. each. A weather satellite, U–2, and C–130 aircraft would provide cloud photographs. The U–2 would also measure ozone distributions, temperature, wind, and radiometric data; the C–130 would contribute cloud physics, temperature, and wind data. Permanent and mobile radiosonde sites 60 mi. apart would comprise one aspect of the data-gathering network; special surface linkage of 25 sites spaced 20 mi. apart would gather wind, temperature, pressure, humidity, and precipitation data. The study would continue through April 30. (USAF OAR Release 3–65–5)

- *Aviation Week* reported theory of many U.S. officials that COSMOS LVII, launched by U.S.S.R. on Feb. 22, 1965, had strayed from its programed flight path and been deliberately destroyed the day after it was launched. The alleged reason was to prevent COSMOS LVII from falling into foreign hands. U.S. officials were said to have assumed that COSMOS LVII was a trial run for VOSKHOD II flight because of similar orbits: VOSKHOD II had 308 mi. (496.7 km.) apogee, 108 mi. (174.4

km.) perigee, 65° inclination; COSMOS LVII had 317 mi. (511.3 km.) apogee, 107 mi. (172.6 km.) perigee, and 65° inclination.

According to NASA Goddard Space Flight Center's *Satellite Situation Report*, 51 pieces of COSMOS LVII were in orbit Feb. 28, 1965; 39 pieces on March 15, 1965. The *Report* also listed COSMOS L, launched by U.S.S.R. October 28, 1964, in 88 pieces. (GFSC *SSR*, 3/15/65, 33, 37; *Av. Wk.*, 4/12/65, 34)

March 15: B. F. Goodrich Corp. had been selected by Hamilton Standard to replace International Latex Corp. as subcontractor for garment portion of the Apollo spacesuit. Change followed problems with certain portions of garment. (*M&R*, 3/15/65, 7)

• Astronaut R. Walter Cunningham suffered a simple compression fracture of a neck vertebra during exercise unrelated to astronaut training. Cunningham would be grounded during the three months he would wear a neck brace but would continue other phases of astronaut training. (AP, Wash. *Eve. Star*, 3/17/65)

• Opening the annual meeting of the National Research Council, Harvey Brooks, professor of applied physics at Harvard Univ., discussed recent trends in Federal support of research and development. Of the $14.5 billion R&D budget for FY 1966, he observed, nearly half was for space activities—expended principally through DOD and NASA. Scientific satellite programs accounted for 36% of all basic research expenditures. He noted the steady trend toward greater diversity in sources of Federal support for academic research. One indication is the fact that in 1954 DOD accounted for 70% of academic research but in FY 1966 for only 27%. (NAS-NRC *News Report*, 3/65, 1)

• Theory held by Soviet astrophysicists Vitaly Ginzburg and Leonid Ozernoi that intergalactic space is hot was reported by Tass. Scientists generally believed the hydrogen gas in intergalactic space to be cold (−273°C). Ginzburg and Ozernoi considered it "incomprehensible" that the gas could be cold yet neutral—no emissions in the 21-cm. wavelength had been detected from the intergalactic hydrogen. They theorized that the gas was heated by galactic explosions and likewise ionized by them, making impossible any 21-cm.-wavelength emissions. (Tass, 3/15/65)

• Lance battlefield missile was successfully test fired at the White Sands Missile Range. Built for the Army by Ling-Temco-Vought, Inc., the Lance was said to combine guided missile accuracy and range with the low cost and high reliability potential of a free rocket. It would complement division artillery and expand the capability for nuclear and non-nuclear fire. (AP, *NYT*, 3/18/65, 57)

March 16: Dr. Homer E. Newell, NASA Associate Administrator for Space Science and Applications, told the House Committee on Science and Astronautics' Subcommittee on Space Science and Applications that NASA had an obligation to make information gained from space exploration available to the public. He continued: "To help achieve this, a National Space Science Data Center was established at GSFC in April 1964. . . .

"The Data Center is responsible for the collection, organization, indexing, storage, retrieval, and dissemination of all scientific data resulting from experiments in space and the upper atmosphere. Since its establishment the Data Center has begun: (1) to maintain a continuing inventory of data from sounding rockets and spacecraft; (2) to acquire data generated by spacecraft previously launched; (3) to collect selected ground correlative data; and (4) to produce the announcement publications which support its functions.

"In anticipation of the need for this facility, NASA has established a line item in the Physics and Astronomy budget, Data Analysis, of three million dollars in FY 1966. Of this, 600 thousand dollars is for the operation of the Data Center and 2.4 million dollars is for analysis of data from a flight experiment under the flight project. After the initial results have been published by the Principal Investigator and the data are placed in the data center, the additional analyses of these data will be funded from Data Analysis funds on the basis of proposals from competent scientists throughout the Nation . . . This approach is expected to . . . encourage them to use all of the available information in their theoretical research."

Dr. Newell discussed NASA's orbiting observatory program: "The primary reason for . . . solar studies is to meet the overall NASA objective to expand human knowledge of space phenomena. . . .

"OSO-C [Orbiting Solar Observatory-C] is the next spacecraft to be launched and it is undergoing final testing at this time. On 30 May a solar eclipse of unusually long duration will occur. Every effort is being made to launch OSO-C prior to this event so that two OSO's, with complementary payloads, can be operating and transmitting unique data on the solar radiation at the time of the eclipse."

He said that the Orbiting Geophysical Observatory (Ogo) program would make a major contribution to our understanding of earth-sun-environment relationships and that although OGO I had not functioned as planned "it has proven that the basic spacecraft design is adequate and that large numbers of experiments can be integrated and operated from a single satellite. Furthermore, should OGO I continue to transmit data for a reasonable period, it is expected that the results will contribute substantially to studies of the Earth-Sun relationships.

"Investigation of the OGO I failure indicated there was no common cause for failure, but as a result of the investigation, design modifications and additional tests are planned for future OGO spacecraft. The modifications include: (1) relocation of the horizon scanner and certain boom appendages to assure a clear field of view for the horizon scanners; (2) the use of a new type development spring and the addition of separate appendage 'kick-off' springs; and (3) the relocation of the omnidirectional antenna." (Testimony; *NASA Auth. Hearings*, 461–580)

March 16: The communications blackout problem was discussed by Dr. Hermann H. Kurzweg, NASA Director of Research, Office of Advanced Research and Technology, in testimony before the House Committee on Science and Astronautics' Subcommittee on Advanced Research and Technology: "One of the phenomena that occurs in gases at high temperatures is ionization, that is, electrons are torn away by the high-speed collisions of the gas atoms and molecules. . . . The free electrons, produced by the high temperatures in the shock layer around a reentry vehicle, interfere with and block the propagation of radio signals. . . . This effect produces the communications-blackout problem. To understand what is going on and to eliminate, or at least minimize this communication difficulty, one must be able to calculate the distribution of free electrons about the body in order to predict when the plasma sheath will become opaque for certain radio frequencies. This calculation cannot be made until the flow field (temperature, density, pressure and velocity) about the body is known. A significant part of the fluid physics program is concerned with the investigations of flow fields. The results of these studies also give us better information on the heat transfer to reentry bodies.

"As a possible remedy for the communications blackout, we are studying the characteristics of various gases, called electrophylic gases, which have the unique property of capturing free electrons. Such a gas, which effectively reduces the electron concentration when injected into the flow, might solve the problem. . . . This work is tied closely with the work on radio attenuation going on at the Langley Research Center and the technique is being adapted to test a variety of fluids suggested by the work at Langley." (Testimony; *1966 NASA Auth. Hearings,* 447–62)

March 16: First observance of Robert H. Goddard Day. On the floor of the Senate, Sen. Stuart Symington (D–Mo.) spoke of Dr. Goddard's achievements as summarized by G. Edward Pendray in *Technology and Culture* (Fall 1963):

"Dr. Goddard—

"Was the first to develop a rocket motor using liquid propellants (liquid oxygen and gasoline) during the years 1920–25.

"Was first to design, construct, and launch successfully a liquid-fuel rocket—the event we mark today.

"First developed a gyrostabilization apparatus for rockets in 1932.

"First used deflector vanes in the blast of a rocket motor as a means of stabilizing and guiding rockets, also in 1932.

"Obtained the first U.S. patent on the idea of multistage rockets, in 1914.

"First explored mathematically the practicality of using rocket power to reach high altitude and escape velocity, in 1912.

"Was first to publish in the United States a basic mathematical theory underlying rocket propulsion and rocket flight, in 1919.

"First proved experimentally that a rocket would provide thrust in a vacuum, in 1915.

"Developed and demonstrated the basic idea of the bazooka near the end of World War I, although his plans lay unused until finally put to use in World War II.

"First developed self-cooling rocket motors, variable thrust rocket motors, practical rocket landing devices, pumps suitable for liquid rocket fuels.

"Forecast jet-driven airplanes, and travel in space." (*CR,* 3/16/65, 5051–52)

- In commemoration of Goddard Day, Dr. Hugh L. Dryden, NASA Deputy Administrator and other Washington officials telephoned greetings via RELAY II to Dr. Goddard's widow in Worcester, Mass. The call had been arranged by Vice President Humphrey, Chairman of the National Aeronautics and Space Council. Other events commemorating Goddard Day: At NASA Goddard Space Flight Center, a film on Dr. Goddard's life and work was premiered; at NASA Manned Space Flight Center, Astronaut Scott Carpenter spoke to several hundred science students about Dr. Goddard and rocketry; at NASA Marshall Space Flight Center, special recognition was shown, and at Smithsonian Air and Space Museum an original Goddard rocket was displayed. (NASA Release 65–87)

March 16: At the dedication of a new laboratory at Worcester Polytechnic Institute in memory of rocketry pioneer Dr. Robert H. Goddard, AFSC Commander General Bernard A. Schriever said that Dr. Goddard's writings still provided guidance to 1965 rocket men. General Schriever said the nation had made significant strides since Goddard conducted his first successful rocket launch 39 years ago. "His booster and its payload reached an altitude of 41 feet and traveled 184 feet before it impacted after a flight lasting about $2\frac{1}{2}$ seconds. By contrast, the first two-man Gemini orbital space shot scheduled for later this month will reach several hundred miles into space for three orbits. . . .

"The Air Force Titan II booster and the Gemini capsule stand almost 110 feet—over twice the altitude achieved by Dr. Goddard's historic rocket."

Mrs. Esther G. Goddard, Dr. Goddard's widow, attended the ceremonies. (AP, Balt. *Sun*, 3/17/65)

- A low-temperature, primary, non-rechargeable battery had been successfully tested over a range from $-100°$ C to $68°$ C, NASA Lewis Research Center engineers reported. Designed by the Livingston Electronic Co., the battery delivered constant power and, when fully developed, could be used on Mars where the nighttime temperatures were $-100°$ C and the average daytime temperature $-30°$ C. (LRC Release 65–20)

- North American and European television broadcasters met at ComSatCorp headquarters in Washington, D.C., and announced outline of inaugural broadcast between the two continents to demonstrate possibilities of Early Bird communications satellite for television use. Plans called for major part of telecast to be live transmissions of events in various countries. It would include live broadcasts from participating ground stations in Europe and North America, a short documentary history of past events carried on satellite television, and a brief explanation of how Early Bird worked and what it would mean to communications in the future. (ComSatCorp Release)

- Sen. Ralph Yarborough (D–Tex.) said on the floor of the Senate that results of Government-sponsored research should be "freely available to the American public" and that he "viewed with . . . skepticism any proposal to create a private monopoly" over this information. (*CR*, 3/16/65, 5051)

- A NASA-sponsored, 34-day spacecraft atmosphere test began as six Navy and Marine Corps fliers entered a space capsule at the Naval Air Engineering Center's Bioastronautics Test Facility in Philadelphia. The fliers would wear a full pressure space suit during three weeks of the period, eat a dehydrated menu, and breathe 100% oxygen while exposed to a simulated altitude of 27,000 ft. Investigators would conduct periodic tests to determine the overall effects, physiological and psychological, upon each of the men. (AP, Balt. *Sun*, 3/17/65)

March 16: Dr. Robert Gilruth, Director of NASA Manned Spacecraft Center, told a press conference that "there is a question whether astronauts can stand long confinement, let alone weightlessness." Dr. Gilruth was in Los Angeles to accept the 1964 Spirit of St. Louis Medal from the ASME at the Aviation and Space Conference. (Miles, *L. A. Times,* 3/17/65; NAA *S&ID Skywriter,* 3/19/65, 1)

- Abraham Hyatt, a former NASA Director of Plans and Program Evaluation, delivered the 9th Minta Martin lecture at the Conference on Aerospace Engineering at the Univ. of Maryland. He said that while much had been learned about the space environment since 1958, we still had only meager knowledge of the processes that operated on the sun; the sun-earth relationship; the sources of energy of the observed physical phenomena in space; the planets; and of many other properties of space. For a better understanding of the origin and space environment of the solar system, the origin and characteristics of the universe, or the possibility of life on other planets, measurements and experiments in space would be necessary for a long time to come, he said. (Program Notes)

- Dr. Wernher von Braun, Director of NASA Marshall Space Flight Center at Huntsville, Ala., received an honorary doctorate of laws from Iona College. (*NYT,* 3/17/65, 38)

- Dr. Athelstan Spilhaus, dean of Minnesota Univ.'s Institute of Technology and past chairman of the National Academy of Sciences, urged the Senate Commerce Committee to establish sea-grant colleges that could exploit ocean resources. He said that land-grant colleges had done a magnificent job in furthering agriculture and the mechanical arts and that sea-grant colleges could do the same in the field of oceanography. Dr. Spilhaus also spoke in support of a bill to provide for expanded research in the oceans and Great Lakes by creation of a national oceanography council. (AP, *NYT,* 3/17/65, 52)

- Yevgeny Artemyev, vice chairman of the Soviet Union's State Committee of Inventions, announced Moscow's intention to ratify the 82-yr.-old Paris Convention for the Protection of Industrial Property. The agreement required that each member state grant the citizens of other member countries in the matter of patents, trademarks, and other industrial property rights the same treatment it accorded its own nationals. The Soviet Union would be the 68th country to adhere to the convention. (*NYT,* 3/17/65)

March 17: MARINER IV's ion chamber experiment failed completely, Jet Propulsion Laboratory officials reported. Count-rate of the Geiger-Mueller tube portion of the experiment had become abnormal in February. The experiment had been designed to measure proton and electron radiation. Otherwise the spacecraft was operating normally; all other high-energy radiation detectors aboard were continuing their interplanetary measurements. In its 110th day of flight, MARINER IV was traveling 27,743 mph relative to earth and was 35,000,004 mi. from earth. It had traveled more than 178,000,000 mi. (NASA Release 65-90)

- First Saturn IB booster, the S-IB-1, was placed into a static test stand at NASA Marshall Space Flight Center for scheduled static firings. Built by Chrysler Corp., the 1.6 million-lb.-thrust, 90,000-lb. booster contained eight engines, was 21 ft. in dia. and 80 ft. long. The stage would be shipped to NASA Michoud Operations for post-firing checks. (*Marshall Star*, 3/17/65, 1, 2)

March 17: Discussing the need for sustaining engineering funds for Centaur starting in FY 1966, NASA Associate Administrator for Space Science and Applications Dr. Homer E. Newell testified before the House Committee on Science and Astronautics, Subcommittee on Space Sciences and Applications: "A preliminary study and design phase is being initiated by NASA this Fiscal Year [for adaptation of the Centaur to the Saturn IB]. The primary mission for this vehicle is the Voyager. Initial studies indicate this vehicle is capable of launching a payload to Mars in excess of 8000 pounds during all of the opportunities in the 1970's. Generally, the modifications necessary to create this stage combination are not particularly difficult. They do represent a large engineering effort, but there is nothing apparent at this time which indicates that new technologies will be required. The Centaur will be mounted, along with the Voyager, inside a fairing the size of the Saturn (260-inch diameter). By constructing this size fairing the technical problems associated with adaptation of the Centaur to this new booster are significantly reduced and the diameter required for all of the Voyager missions is obtained."

Dr. Newell described NASA's sustaining university program as an effort "to broaden the national research base in areas of importance to the national space effort and increase our capability to replenish continually the reservoir of basic knowledge. . . .

"In response to the continuing manpower requirements, NASA conducts a predoctoral training program, under which grants are made to universities to select and train outstanding students in space-related fields. Specialized training for selected students offers them identification with the national space effort, and involves them directly in the new programs of the space age. . . .

"At the present time, about 1,957 students are in training at 131 institutions. The disciplines represented by these 1,957 students are distributed as follows: physical sciences, 51 percent; engineering, 37 percent; life sciences, 8 percent; behavioral sciences, 4 percent . . . In September 1965, about 1,275 new students will begin their three years of study and research as NASA predoctoral trainees. At that time, 142 institutions, located in every state in the union, will be participating. With the proposed budget of $25 million for fiscal year 1966, about 1,300 new students would enter the program. Consequently, the NASA goal of an output of 1,000 Ph.D.'s per year will not be reached before fiscal year 1968 or fiscal year 1969 . . . Of the students participating to date, 40 trainees have received their Ph.D. degrees. . . ." (Testimony; *NASA Auth. Hearings*, 634-35)

- Astronauts Virgil I. Grissom and John W. Young gave the official name "Gemini 3" and the nickname "Molly Brown" to the spacecraft they would ride into orbit Mar. 23. (MSC Historian; AP, *Miami Her.*, 3/17/65)

- First six ships of a 20-vessel fleet that would participate in recovery of the Gemini GT-3 spacecraft following the two-man orbital flight scheduled for Mar. 23 left Cape Kennedy. Ships would be positioned from the mid-Atlantic to the Canary Islands. (*Wash. Post*, 3/17/65)

March 17: A strike was under way at the $256-million Mississippi Test Operations under construction in Gainesville, NASA announced. The dispute apparently concerned NASA's contracting policies. (AP, *St. Louis Post-Dispatch*, 3/18/65)

- DOD attracted more than 1,000 industrial representatives to its "regional unclassified briefing" in New York. It outlined the nation's military needs for the next decade and offered guidance in planning defense contracts. This was one of five meetings DOD had called throughout the country to provide industry, business, and labor with an idea of the military research, development, and production requirements. (Wilcke, *NYT*, 3/17/65, 65)

- FAA granted an air worthiness certificate for an automatic landing system developed jointly by the Boeing Co. and the Bendix Corp. It was the first system in the world to be so certified by FAA for operation in the U.S. and would enable users to apply to FAA for "Category II" certification under which a pilot could land with only 100 ft. downward visibility and 1,300 ft., or a quarter mile, forward visibility. Most airliners must land under "Category I" conditions under which the pilot must be able to see the last 200 ft. to the ground and must have at least a half mile forward visibility before he could land. First Boeing 707 or 720 jetliners equipped to land by computer would be available about Jan. 1966. (Appel, *NYT*, 3/18/65, 1, 14)

- FAA Administrator Najeeb E. Halaby announced that four Government agencies had joined forces to establish a national data bank for interagency exchange of information on civil manpower resources. The agencies were Dept. of Labor, Dept. of Health, Education, and Welfare, Civil Aeronautics Board, and Federal Aviation Agency. Halaby said availability of such a bank would make it possible to obtain more information on status of aviation manpower than FAA maintained. (FAA Release 65-20)

- Speaking on safety in the Space Age, John L. Sloop, NASA Assistant Associate Administrator for Advanced Research and Technology, told the 22nd Annual Greater Akron Safety Conference that "for the past ten years, the NACA and NASA have had a frequency rate (injuries per million man hours work) ranging from 3.2 to 2.1. The national industrial frequency average, I am told, is 6.12 for 1963 and the average for all of Federal government is 7.9." (Text)

- Dr. Robert H. Goddard was posthumously awarded the Daniel Guggenheim Medal by the American Society of Mechanical Engineers. Mrs. Goddard accepted the medal. (*Av. Wk.*, 3/22/65, 13)

- Brig. Gen. Charles A. Lindbergh (USAFR) was elected to the Board of Pan American World Airways. During his 36-year association with the airline, he had helped develop several aircraft from the Fokker and Sikorsky to the Boeing and Douglas jets. Recently he had worked on the supersonic transport and the fanjet Falcon. He was also a member of the NACA from 1931 to 1939. (*NYT*, 3/18/65, 47)

March 18: U.S.S.R.'s VOSKHOD II, manned by pilot Col. Pavel Belyayev and co-pilot Lt. Col. Aleksey Leonov, was launched from Baikonur Cosmodrome in Kazakhstan, Tass reported. The spacecraft set an altitude record, reaching an apogee of 495 km. (309 mi.)—higher than any manned spacecraft had flown. Other orbital data: perigee, 173 km. (108 mi.); inclination, 65°; period, 91 min.

During the second orbit, Lt. Col. Leonov, clad in a spacesuit with "autonomous life support system," stepped into space, moved about five meters from the spacecraft (tethered by a cable), and successfully carried out prescribed studies and observations: he examined the outer surface of the spacecraft; turned on a film camera; carried out visual observations of the earth and outer space; took horizontal, vertical, and somersaulting positions; and returned safely to the spacecraft. Tass said: "Outside the ship and after returning, Leonov feels well." He spent about 20 min. in conditions of outer space, including 10 min. free-floating in space. Entire procedure was carried out under control of Col. Belyayev, with whom continuous communication was maintained. A television camera fixed to the side of VOSKHOD II relayed pictures of the maneuver to Soviet ground stations.

Biotelemetric data indicated that both cosmonauts had satisfactorily withstood the orbiting and the transition to weightlessness: the pulse rate of Belyayev and Leonov was 70–72 beats a minute and the respiration rate 18–20 a minute. All spacecraft systems were functioning normally. Tass said VOSKHOD II would complete at least 13 orbits of the earth. (Tass, AP, *NYT*, 3/19/65; *Komsomolskaya Pravda*, 3/19/65, 1, ATSS–T Trans.; Haseltine, *Wash. Post*, 3/19/65)

March 18: Atlas launch vehicle sustainer engine system had been successfully fired for the first time using flox, a combination of liquid fluorine and liquid oxygen, as the oxidizer. This was the first time a complete engine system had been fired using this high-energy oxidizer. Approximately 20 firings would be conducted in the series using the standard concentration of 30% liquid fluorine to 70% liquid oxygen. Conditions involving thrust level, oxidant fuel ratio, and other engine variables would be run to establish engine performance limitations. The tests were being conducted under LRC contract, by North American Aviation's Rocketdyne Div., Canoga Park, Calif. (LRC Release 65–21)

- NASA launched a Nike–Apache sounding rocket with a 63-lb. payload from Wallops Station, Va., to peak altitude of 98 mi. The experiment was conducted for the Graduate Research Center of the Southwest, Dallas, Tex., and was designed to measure ion composition and neutral composition of the upper atmosphere as functions of altitude. Impact occurred 89 mi. downrange in Atlantic Ocean; no recovery was attempted. (Wallops Release 65–14)
- NASA Aerobee 150 sounding rocket was successfully launched from White Sands, N. Mex., to a peak altitude of 154.5 km. (96 mi.). The primary experimental objective was to obtain ultraviolet spectra of Mars and Orion by the use of four spectrographs. GSFC provided the payload instrumentation. (NASA Rpt. SRL)
- USAF launched Thor-Altair booster from Western Test Range with unidentified satellite payload. (*U.S. Aeron. & Space Act., 1965*, 136)
- NASA bioscience programs were discussed in testimony before the House Committee on Science and Astronautics' Subcommittee on Space Sciences and Applications by NASA Associate Administrator Dr. Homer E. Newell: "Results recently submitted by the U.S. and U.S.S.R. from flights up to five days in length indicate that long term space flight may

have several important and serious physiological and behavior effects upon the performance and well being of man that need to be investigated further. There were changes in the circulation system, in the biochemical characteristics of the blood and urine, and in the electroencephalogram indices, all pointing to a need for more detailed investigations. The results from the Biosatellite studies will have broad application to long term, manned space flight, including manned space stations and lunar and planetary bases.

"Prolonged manned flights may involve changes similar to those observed after 10 days of strict bed rest on the ground. These are moderate losses of bone minerals such as calcium, particularly in the vertebrae; loss of muscle tone and physical capability; certain cardiovascular changes; and metabolism in general. The effect of continued sensory deprivation on behavior and performance is unknown.

"Biosatellite experiments are of both scientific and practical importance and extremely profitable to investigate. We do not presently have sound theoretical bases for making precise quantitative (and in some cases qualitative) predictions of what we expect to happen. It is, therefore, important to carry out Biosatellite studies of suitable duration to critically demonstrate and test the effects of weightlessness on living organisms."

Outlining approaches to the search for extra-terrestrial life in NASA's bioscience programs, Dr. Newell testified: "(a) An attempt is being made to synthesize models of primitive single-celled organisms in the laboratory. . . .

"(b) The physical environments of the planets are being studied and characterized by instruments from the Earth, from high altitude balloons and from planetary fly-bys. . . .

"(c) Living Earth organisms are being grown under simulated planetary environmental conditions. . . .

"(d) Plans are being made for both unmanned and manned direct exploration of planets. . . ." (Testimony; *1966 NASA Auth. Hearings*, 806–41)

March 18: NASA Deputy Administrator Dr. Hugh L. Dryden told the annual meeting of the American Astronautical Society in Washington, D.C., that NASA planned to select 10 to 20 scientists to begin astronaut flight training this summer from over 900 applicants. Dr. Dryden said the Mercury astronauts had demonstrated man's ability as a sensor and manipulator, and to some extent as an evaluator, in orbit. "Early Gemini and Apollo flights will further examine these capabilities so that, in the future, man's full potential can be exploited." (AP, *NYT*, 3/19/65)

• In an article in the San Diego *Evening Tribune* deploring the strikes and labor unrest at Cape Kennedy and Merritt Island, Victor Riesel said: "Well over $100 million had been lost in strikes.

"NASA officials report 78 walkouts between Dec. 1, 1962, and Feb. 15, 1965. Total work loss has been 63,784 man days. This means there has been an average of more than five vital strikes a month. At least 35 of them have been illegal and have cost 49,596 man days." (Riesel, San Diego *Eve. Trib.*, 3/18/65)

• Tokyo Univ. Aeronautical Institute announced successful firing of a three-stage Lambda research rocket from Uchinoura in southern

Japan. The rocket reached an altitude of 680 mi. and landed in the Pacific northwest of the Marianas. (Reuters, *NYT*, 3/19/65)

March 18: Among the aerospace pioneers selected for San Diego's new International Aerospace Hall of Fame were Scott Crossfield, Charles A. Lindbergh, Gen. James H. Doolittle, Astronauts John Glenn and Alan Shepard, Dr. Wernher von Braun, Orville and Wilbur Wright, Robert H. Goddard, Jacqueline Cochran, and Amelia Earhart. Representatives of 287 organizations from throughout the world were on the nominating committee. Oil paintings of the honorees were unveiled at a dinner given in conjunction with San Diego's Space Fair 65 observance. (NAA *S&ID Skywriter*, 3/19/65, 1)

- Catholic Univ. was the first school in the Nation to offer undergraduate study in space science, said Dr. C. C. Chang, head of the Dept. of Space Science and Applied Physics established two years ago. In addition to space science, the department offered specialization in aerospace engineering, applied physics, and fluid mechanics and heat transfer. (Hoffman, *Wash. Post*, 3/18/65)

- Soviet VOSKHOD II Cosmonauts Pavel Belyayev and Aleksey Leonov talked with Cuban Defense Minister Raul Castro, who was in Moscow, and told him they had seen his island from space, Tass reported. "It was very beautiful, and her green colors were lovely," they said. (AP, 3/18/65)

- Rep. J. Edward Roush (D–Ind.), speaking on the floor of the House, compared the states in distribution of Federal research and development funds per scientist employed: "Of the seven states of Ohio, Indiana, Illinois, Iowa, Michigan, Minnesota, and Wisconsin only Illinois exceeds the national average of approximately $25,000 in research and development funds per scientist employed in educational institutions in this area. Even then this one state exceeds the average distribution by only $4,600. The shares of other states range from a high of $15,000 per scientist in Michigan down to only $9,000 in my own state of Indiana. In between these we find Minnesota, $13,000; Ohio, $11,000; and Wisconsin, $10,900.

  "Leading the national list is New Mexico with $163,000 per scientist followed by Nevada with $109,000 and California with $63,000 per scientist. At the very bottom of the list is Maine with only $4,000 per scientist.

  "... this matter of the uneven geographic distribution of Federal research and development funds is involving our national interest." (*CR*, 3/18/65, 5186)

- A spacesuit that would enable man to leave his spacecraft was discussed by Soviet doctor Vladimir Krichagin in a commentary for Tass written before the VOSKHOD II flight: "It is in fact a miniature hermetic cabin which consists of a metal helmet with a transparent visor, a multi-layer hermetic suit, gloves, and specially designed footwear. The spacesuit has its own power circuitry feeding communications, and a system of pickups of physiological functions . . . It is impossible to create atmospheric pressure within the suit because it would then inflate as a football . . . and the man would turn into a statue unable to bend his legs and arms . . . the air pressure inside the spacesuit should be at least 0.4 atmospheres . . . It was established that prolonged (over one hour) respiration in pure oxygen literally washes nitrogen out of the tissues of the body and then the pressure can be safely reduced. It was . . . possible to free a man in the spacesuit from . . . the immoblizing effect of an 'inflated football'. . . .

"There must be a steady supply of pure oxygen for the cosmonaut in spacesuit . . . his body has to 'breathe' and . . . give off up to 300 kilo-calories [every hour].

". . . the spacesuit has a special airconditioning system through which room temperature air is pumped into the spacesuit. This air carries away excess heat of the organism and skin-exuded moisture.

"To protect man in space from . . . heat . . . and cold . . ., the spacesuit is covered by thermal insulation layer and coated with a light color that deflects heat rays . . . In these spacesuits of the ventilation type . . . used air is injected into the environment.

"[In] spacesuits of the . . . regenerating type . . . the available air and hydrogen supply circulates from the spacesuit to a generating device and back. This device on the suit's surface removes carbon dioxide and excess moisture from the 'spent' air . . . replenishes oxygen supply and cools off gases to a preset temperature.

"This spacesuit may be used for prolonged work in space and for landing on the lunar surface." (UPI, Rosenfeld, *Wash. Post*, 3/19/65, 1, 2; Tanner, *NYT*, 3/20/65, 1, 3)

March 18: Soviet Cosmonaut Col. Pavel Popovich, who orbited the earth 48 times in August 1962, said during a televised news conference in Moscow: "In the future, we shall be able to discard the cord connecting the cosmonaut with his craft. A small rocket engine will help the man to return to his ship."

Vasily Seleznev, Soviet doctor of technology, told the news conference he thought the significance for further space research of Leonov's leaving his craft was that "in [the] future cosmonauts will take part in assembling spaceships. There may also arise the need for repairing the craft and, what is most important, there is the prospect of travel to other planets." Seleznev said the Russians hoped to reach the moon in the not too distant future. (Rosenfeld, *Wash. Post*, 3/19/65, 1, 2)

- Vice Adm. Hyman G. Rickover (USN) urged Congress to approve the construction of a new type of nuclear reactor that he said was vital to the welfare of the United States and perhaps the whole world. Adm. Rickover said the reactor—which he himself conceived—was called a "seed-blanket" reactor, would employ thorium as the major fuel and would produce more fuel than it consumed. It would run about nine years on one fuel charge. Reactors of this type—costing more than $263 million for the initial one—could extend the fuel resources of the United States by several hundred years and also produce electricity economically, he said.

Dr. Glenn T. Seaborg, chairman of the Atomic Energy Commission, testified that AEC had signed a memorandum of understanding with the state of California for the development and construction of the proposed $263 million prototype, and that whereas present "lightwater" reactors tapped only 1 to 2 per cent of the energy available in either uranium or thorium, the proposed reactor "will demonstrate technology which is expected to provide means for ultimately making available for power production about 50 per cent of the potential energy in thorium—which represents an energy source many times larger than that of the known fossil-fuel [coal and oil] reserves." Admiral Rickover said the proposed power device would have more than twice the electrical-generating capacity of any United States central power station. (AP, *NYT*, 3/19/65, 12)

March 18–19: Scientific research papers were presented by high school students at regional Youth Science Congress contests conducted by National Science Teachers Association in cooperation with NASA. Regional winners would compete at the National Youth Science Congress to be held in Washington, D.C., later this year. (LaRC Release; GSFC Release G-7-65)

March 19: After 26 hrs. of flight, Col. Pavel I. Belyayev landed VOSKHOD II manually near Perm, Russia. Tass announced. The two-man spacecraft had completed 17 orbits of the earth, one more orbit than planned, and had traveled 447,000 mi. This was the first time landing of a Soviet spacecraft had been described as manual. Impact of VOSKHOD II on the ground, later revealed as snow bank, was described as "soft." (Tanner, *NYT*, 3/20/65, 1, 3; Shabad, *NYT*, 3/21/65, 3)

- NASA plan for use of SYNCOM II in the communications link between the Gemini 3 spacecraft and Cape Kennedy was successfully tested in a GT-3 mission simulation. Telemetry signals and voice messages would come from the spacecraft to a surface ship, the USNS *Coastal Sentry*, in the Indian Ocean. The *Coastal Sentry* would transmit the signals to the Syncom surface station, USNS *Kingsport*, which would then be a few miles away. From there the signals would be transmitted to SYNCOM II, 22,300 mi. above the Indian Ocean, down to a ground station at Clark AFB in the Philippines, and by cable to a Nascom (NASA Communications Network) station near Honolulu. From Honolulu the transmission would go by cable to the U.S. and then by landline to NASA Goddard Space Flight Center and on down to Cape Kennedy.

  Simultaneously the signals would be transmitted from the *Coastal Sentry* via high frequency radio to a Nascom station near Perth, Australia. Cable would carry it to the Nascom station at Honolulu. There, the better reception of the two transmissions would be sent to the Cape. (NASA Release 65-93)

- NASA launched a scientific payload for the Univ. of Michigan from Wallops Station using a two-stage Nike-Tomahawk. The 122-lb. payload, consisting primarily of a thermosphere probe in the form of a 32-in. ejectable cylinder, was boosted to a peak altitude of 315 km. (196 mi.). Purpose of the experiment, a joint project of the Univ. of Michigan and Goddard Space Flight Center, was to measure density and temperature of electrons and neutral particles at 75–200 mi. altitude and to test a solar aspect sensor. This was the first firing of Nike-Tomahawk configuration from Wallops Island. (NASA Rpt. SRL; Wallops Release 65-16)

- President Lyndon B. Johnson sent a message of congratulations to Australian Prime Minister Sir Robert Menzies, on the occasion of the dedication of a new NASA lunar and planetary spacecraft tracking station at Tidbinbilla near Canberra, Australia. The station would be operated entirely by Australians, as are the two other NASA facilities in Australia. (NASA Release 65-89)

March 19: Vice President Hubert H. Humphrey, Chairman of the National Aeronautics and Space Council, made his first address on the U.S. space program at the Goddard Memorial Dinner sponsored by the National Space Club in Washington, D.C. He said: "I intend to be an advocate of a dynamic space program—a program which will succeed in reaching to goals we have set—and one which will see new goals—one that can see beyond the moon and into fields where we can only speculate about the knowledge awaiting us."

The Vice President spoke briefly about the Soviet Union's VOSKHOD II flight: "It is well for us from time to time to take stock—to take a careful look—in order to see how we are making out in comparison with our main competitor. The facts are that we do have very strong competition and hence we have another big reason for a major space effort—namely, prudence. Our national security alone would suggest reason enough for us to strive for absolute leadership in space exploration."

Humphrey pointed out that the Soviets remained ahead in propulsion for their rockets, while the U.S. continued to lead "in the directly useful fields of weather reporting, navigation, and communications." He continued: "The Soviets clearly have an advantage in studying the effects of space environment on human beings. . . . We can salute the Russian achievements . . . but we would be foolish if we did not understand the military implications of Soviet space science as well as our own.

"Each Russian shock has produced action here. But a mature nation should not need shock treatment. We are a peaceful nation . . . but we would ignore the real interests of the free world if we diminished our military efforts in space."

In the principal presentation, the widow of the scientist presented the Robert H. Goddard Memorial Trophy to Dr. William H. Pickering, Director of the Jet Propulsion Laboratory and leader of the RANGER VII team that obtained the first close-up pictures of the moon's surface.

The National Space Club Press Award for "an outstanding role in adding significantly to public understanding and appreciation of astronautics" went to *Aviation Week and Space Technology;* Nelson P. Jackson Aerospace Award for "an outstanding contribution to the missile, aircraft, and space field" was presented to Florida Research and Development Center, Pratt & Whitney Aircraft Div. of United Aircraft; Robert H. Goddard Historical Essay Award was made to John Tascher, Case Institute of Technology, for *U.S. Rocket Society Number Two: the Story of the Cleveland Rocket Society;* Robert H. Goddard Scholarship ($1,500 to the university of the recipient's choice) for "the purpose of stimulating the interest of talented students in space research and exploration" was awarded Willard M. Cronyn, a graduate student in Maryland Univ.'s Dept. of Physics and Astronomy. (Text; Program; Carmody, *Wash. Post*, 3/20/65)

- "Present-day Americans are thinking, working, and risking to find ways, first to explore, and then to use, the new environment of outer space," said NASA Administrator James E. Webb in an address to the New Mexico Chapter of the American Institute of Industrial Engineers in

Albuquerque. He continued: ". . . the exploration of space has brought a new force into the affairs and life of this Nation. Once more the American people confront a new environment—harsher, more demanding, more inspiring than any man has ever tried to enter before. . . . We cannot yet foresee all the consequences of man's entry into space. But the record of history is clear, that the mastery by one nation of a new environment, or of a major new technology, or the combination of the two as we now see in space, has always in the past had the most profound effects on all nations and on all the peoples of the earth." (Text)

March 19: In an interview with *Izvestia*, one of the two directors of the Soviet space program, the "chief designer," whose identity had never been revealed, said the VOSKHOD II program had called for Lt. Col. Leonov to spend "10 minutes outside the cabin" but that he could have stayed much longer. He said the weight and space saved by having two men aboard VOSKHOD II instead of three men, as on VOSKHOD I, had been used to install a decompression chamber and related equipment. The designer said Leonov's spacesuit was equipped with "duplicate systems" to ensure a high degree of reliability and that a bellows had been installed to allow bending of the torso, arms, and legs. *Izvestia* said in another article that Leonov's spacesuit consisted of five layers: a heat reflecting layer outside; material for strength; airtight material; heat insulating material; and an inside layer containing a ventilation system.

The "chief theoretician," joint director of the Soviet space program, told *Izvestia* that Col. Leonov's venture into space had shown that future astronauts might find it easier to work in space than on earth. He said that "we shall yet live to see the day when orbiting platforms appear in space—resembling scientific research institutes in the earth's upper atmosphere." The theoretician was also quoted as saying that Leonov's principal assignment had been to determine man's reaction to "weightlessness in free space." He told Tass: "We obtained in practice what we had visualized theoretically before." (Tanner, *NYT*, 3/20/65, 1, 3)

- President Johnson sent congratulations on the Mar. 18 VOSKHOD II space achievement to Anastas Mikoyan, Chairman of the Praesidium of the Soviet Union. The message said: "All of us have been deeply impressed by Lt. Col. Aleksei Leonov's feat in becoming the first man to leave a space ship in outer space and return safely. I take pleasure . . . in offering on behalf of the people of the United States sincere congratulations and best wishes to the cosmonauts and the scientists and all the others responsible for this outstanding accomplishment." (NYT, 3/20/65, 3)

- Pope Paul VI, speaking to the "workers of the world" on St. Joseph's Day, expressed the hope that the "great and marvelous" Soviet space achievement would "serve to render men better, more united and intent to serve ideals of peace and common good." (*NYT*, 3/20/65, 3)

- Charles A. Wilson, an expert in management and development of space and other advanced systems, had been named Project Manager for NASA's Project Biosatellite at the Ames Research Center. He succeeded Carlton Bioletti, who had retired. (ARC Release 65-9)

March 19: NASA signed a five-year $235-million incentive contract with the AC Spark Plug Div. of General Motors Corp. for manufacture, testing, and delivery of primary navigation and guidance systems for Apollo's three-man command module and the two-man lunar excursion module (Lem). The systems were being designed by MIT. (MSC *Roundup*, 3/19/65, 8)

March 20: President Johnson, asked during a press conference, "where does our space program stand in relation to the Soviets' in the wake of their latest feat?" replied: "The Soviet accomplishment and our own scheduled efforts demonstrate, I think dramatically and convincingly, the important role that man himself will play in the exploration of the space frontier. The continuing efforts of both our program and the Russian program will steadily produce capability and new space activity. This capability, in my judgment, will help each nation achieve broader confidence to do what they consider they ought to do in space.

"I have felt since the days when I introduced the Space Act and sat studying Sputnik 1 and Sputnik 2 that it was really a mistake to regard space exploration as a contest which can be tallied on any box score.

"Judgments can be made only by considering all the objectives of the two national programs, and they will vary and they will differ. Our own program is very broadly based. We believe very confidently in the United States that we will produce contributions that we need at the time we need them. For that reason, I gave Mr. Webb and his group every dollar in the Budget that they asked for a manned space flight.

"Now the progress of our program is very satisfactory to me in every respect. We are committed to peaceful purposes for the benefit of all mankind. We stressed that in our hearings and our legislation when we passed the bill, and while the Soviet is ahead of us in some aspects of space, U.S. leadership is clear and decisive and we are ahead of them in other realms on which we have particularly concentrated." (Transcript; *Wash. Post*, 3/21/65)

‡ NASA Aerobee 300A sounding rocket was successfully launched from Wallops Station, Va., to a peak altitude of 326.2 km. (203.6 mi.). Primary objective was the nighttime measurement of the density and temperature of neutral N_2 using an omegatron mass spectrometer, and the simultaneous measurement of electron temperature and density using a small cylindrical electrostatic probe. A secondary objective was the testing of a lunar optical sensor especially developed for thermosphere probe application. Univ. of Michigan provided the experiment instrumentation. (NASA Rpt. SRL)

‡ VOSKHOD II's two-man crew, Col. Pavel Belyayev and Lt. Col. Aleksey Leonov, rested under medical supervision at an undisclosed place in the northern Ural mountains, Tass reported.

Soviet space flight headquarters at Baikonur Cosmodrome in Kazakhstan reported that VOSKHOD II's antennas had burned away as the spacecraft reentered the earth's atmosphere. The descent had been tracked by radar units.

Lt. Col. Andrian G. Nikolayev, Soviet Cosmonaut, said the order to use manual controls in landing VOSKHOD II was given by a Soviet ground station, *Izvestia* reported. It was not known whether the manual landing was part of the original program or was made necessary by a malfunction of the automatic controls. (Shabad, *NYT*, 3/21/65, 3)

March 20: Soviet Cosmonaut Andrian Nikolayev said in the press that "all these operations—the orientation of Voskhod 2 and switching on of the braking engine—were performed by my colleague cosmonauts by hand, without the help of automation. They performed this task brilliantly. They carried out this landing excellently." He did not say if they had landed in their target landing area. (Balt. *Sun*, 3/20/65)

March 21: NASA's RANGER IX, equipped with six television cameras, was successfully launched toward the moon from Cape Kennedy by an Atlas-Agena B. After the Agena had carried the 800-lb. RANGER IX into 115-mi.-altitude parking orbit with 17,500 mph orbital speed, the Agena engines were cut off. Second burn of the Agena lasted about 90 sec., increasing the velocity to about 24,525 mph and freeing RANGER IX from the major pull of the earth's gravity. RANGER IX then continued on its 2½-day, 245,000-mi. trip to the moon. About 70 min. after launch, NASA announced the spacecraft had been commanded to deploy its solar panels that would convert solar energy to electrical power for its equipment.

Projected target was the crater Alphonsus, about 12° south of the moon's equator, where gaseous emissions had been reported. On the day of impact, Alphonsus would be illuminated by slanting sunlight, producing long shadows and bringing out subtle surface features. The terminator—dividing line between the dark and sunlit portions of the moon—would be only 11° from Alphonsus.

Five hours after lift-off, NASA announced that RANGER IX's course was so accurate it would hit the moon only 400 mi. north of the crater target; an inflight maneuver would be executed later to correct this small course error. (NASA Release 65–25; *Wash. Post*, 3/22/65; Sehlstedt, Balt. *Sun*, 3/22/65; Sullivan, *NYT*, 3/22/65; *WSJ*, 3/22/65)

* Leonid I. Brezhnev, Soviet Communist Party First Secretary, talked by telephone to Cosmonauts Pavel Belyayev and Aleksey Leonov and promised them a fitting reception when they arrived in Moscow. He thanked them for the successful fulfillment of their mission. They said they felt well. Congratulations on the VOSKHOD II flight were sent to Brezhnev by Mao Tze-tung and other Chinese leaders, Peking Radio reported. (Loory, *N.Y. Her. Trib.*, 3/22/65; AP, *N.Y. Her. Trib.*, 3/22/65)

* Soviet Cosmonauts Col. Pavel I. Belyayev and Lt. Col. Aleksey Leonov appeared in public for the first time since they landed VOSKHOD II in the Perm region Mar. 19. They were en route to Baikonur Cosmodrome in Kazakhstan where they were expected to undergo detailed medical checkups and debriefings by scientists and technicians before being welcomed in Moscow in Red Square. (Shabad, *NYT*, 3/22/65, 1, 3)

* At a news conference reported by Soviet press, Col. Pavel Belyayev and Lt. Col. Aleksey Leonov, the two-man crew of the Soviet spacecraft

VOSKHOD II, said they had sighted an artificial satellite during their Mar. 18 flight: "We shouted with surprise when we saw it slowly rotating about 800 meters [900 yards] from our ship." Neither the satellite nor the orbit in which it was traveling was identified.

The cosmonauts related the part that each had played. Col. Belyayev had operated the controls of the decompression chamber through which his companion left the spacecraft, recorded Leonov's pulse and respiration rate, and oriented the spacecraft so that Leonov was always in sunlight during the televised sequence transmitted to earth. Col. Leonov said that when he opened the hatch of the air lock after decompression, he was "struck by a flow of blindingly bright sunlight like an arc of electric welding." The spacecraft was in its second orbit, passing over Kerch Strait. Space had an unexpected aspect, he said: "Ahead of me was black sky, very black. The sun was not radiant, just a smooth disc without an aureole. Below was the smooth-level earth. You could not tell it was a sphere, only by the fact that the round edge showed on the horizon." The acrobatics tired Leonov, especially because of the effort required to move. He said that although the program required that he carefully wind the rope that had tethered him to the craft, he found it "a waste of time" and simply pulled it into the hatch. "The commander quickly closed the hatch cover and injected pressure into the air lock," Leonov said.

Describing the manually controlled landing, Col. Belyayev said the controls were switched on in time and all systems had "worked without a hitch." He said the spacecraft landed in the northern Ural mountains between two big spruce trees in snow 5–10 ft. deep. (Shabad, *NYT*, 3/23/65, 1, 23)

March 21: Over 500 contractors shared the work in NASA's $1.35 billion Gemini manned space flight project, it was reported. The biggest contractors were aircraft companies, but computer manufacturers, major airlines, telephone companies, and small businesses, manufacturing highly specialized items were included. (Hines, Wash. *Sun. Star*, 3/21/65)

March 22: In NASA FY 1966 authorization hearings before the Senate Committee on Aeronautical and Space Sciences, NASA Administrator James E. Webb testified: "Among the hard decisions and difficult choices which had to be made in the preparation of this budget was the decision to terminate the programs to develop the M-1 large liquid hydrogen fueled engine, the large 260-inch solid propellant motor, and the SNAP-8 nuclear electric power supply. The reduction in the requests for space technology activities amounting to about $48 million when compared with fiscal year 1965, results mostly from these terminations. However, as this Committee knows, there is pending before it notification of a plan to reprogram $16,950,000 of 1965 funds so that these projects can be carried forward into 1966 to appropriate developmental points at which important segments of the engineering data for which the projects were originally planned can be obtained for incorporation in our total bank of technological and engineering knowledge."

Mr. Webb was questioned by Sen. Walter F. Mondale (D–Minn.) on when the first U.S. extravehicular activity was planned, and he replied: "Within the next year. We are not sure on which GEMINI flight we will do it as yet." Senator Mondale asked: "When do we plan our first rendezvous maneuver?" and Mr. Webb replied: "Within the next year, maybe the latter part of this year." (Testimony; *NASA Auth. Hearings,* 623, 663)

March 22: Testifying before the House Committee on Science and Astronautic's Subcommittee on Space Sciences, NASA Associate Administrator Dr. Homer E. Newell said that since success of any program was measured by the nature of the data provided, NIMBUS I had more than achieved design objectives: ". . . during its three and one-half weeks of life, Nimbus took 12,137 individual frames of AVCS pictures, an estimated 1,930 APT cycles, and over 6,880 minutes of HRIR data. Hurricanes Cleo, Dora, Ethel, and Florence were observed and Typhoons Ruby and Sally in the Pacific were located by this spacecraft. . . .

"The launch and successful operation of Nimbus I has proved the success of the basic Nimbus spacecraft design. It has also given NASA a better insight as to what additional modifications will be required in the system design for the next Nimbus flight. As mentioned previously, the primary limitation of the first Nimbus flight was the result of the failure of the Agena B vehicle to inject the spacecraft in the proper polar, near-circular orbit and the failure in the spacecraft solar paddle rotation mechanism. The first of these failures resulted in less than complete global cloud coverage and the second reduced spacecraft lifetime. . . ." (Testimony; *NASA Auth. Hearings,* 928–35)

- Telemetry data from RANGER IX indicated that the probe was on such an accurate course toward the moon that JPL engineers decided to delay for one day a planned mid-course correction. RANGER IX began its 245,500-mi. trip to the moon Mar. 21, and was 144,488 mi. from earth at 9 p.m. EST. (UPI, Wash. *Daily News,* 3/22/65; Hines, Wash. *Eve. Star,* 3/22/65; AP, Phil. *Eve. Bull.,* 3/22/65)

- More than 900 representatives of news media had been accredited, making the GT-3 mission of Astronauts Virgil I. Grissom and John W. Young the most intensely covered event in the history of space exploration. Nearly 1,200 newsmen had requested credentials from NASA. (Wash. *Eve. Star,* 3/23/65)

- In an editorial in *Aviation Week and Space Technology,* editor Robert Hotz said: "The trail-blazing mission of the Soviet Voskhod 2 still is continuing as these lines are written, but it has already opened a new chapter in the history of man's conquest of space. It also has emphasized again that, unless some drastic changes are made, this history will be written primarily in the Russian Cyrillic alphabet with only an occasional U.S. footnote technically necessary. . . .

"All of this Soviet progress again emphasizes strongly the ultra-conservatism of the U.S. manned space flight program and the utter inadequacy of the tiny step-by-step approach that sounds so convincing when defending under-funded programs. This approach is sounding more and more idiotic in the face of Soviet space achievements. . . .

"Each Soviet manned space flight makes it clearer that the Russians are widening their lead over the U.S. in this vital area. It also makes it clear that the many billions the American people have poured willingly into our national space program for the purpose of wresting this leadership from the Soviets are not going to achieve that goal under the present management. . . ." (Hotz, *Av. Wk.,* 3/22/65, 11)

March 22: Reporting on public reactions to the two-man Gemini flight scheduled for Mar. 23, Samuel Lubell said in an editorial in the *Washington Daily News:* "In recent weeks more than half of the persons interviewed said funds for moon trips would be the part of the Federal budget they would cut first. Another third named space exploration in general. This interviewing took place before Russia's space exploit of last Thursday." (Lubell, *Wash. Daily News*, 3/22/65)

- NASA Langley Research Center scientists Arthur L. Newcomb, Jr., Nelson J. Groom, and Norman M. Hatcher reported their work on an infrared sensing instrument to help a spacecraft determine which way was up, at the IEEE national convention. The device described was sensitive to the difference between infrared radiation in space and that emitted by a planetary or lunar body; it employed a mechanically-driven system of mirrors to scan the region of space in which it was operating. Radiation gathered by the mirrors was focused into four germanium lenses, each containing a thermistor sensitive to infrared. When the scanning mirror crossed the horizon of a planet, the increase or decrease registered on the thermistor and generated an electronic signal that could be processed through a series of special circuits to provide a stabilizing or control command to the spacecraft.

    The new sensor concept was expected to be useful for weather and communications satellites, as well as for space probes and other types of spacecraft. (LaRC Release)

- Britain's Blue Streak Rocket, first stage of the European Launcher Development Organization's (ELDO) satellite project, was successfully launched to an altitude of 150 mi. from Woomera, Australia. (Reuters, *Wash. Post*, 3/23/65)

- Reasons for choosing the moon crater Alphonsus as the target for RANGER IX were given by David Hoffman in an article in the *New York Herald Tribune:* "First, they are just plain curious. Rangers 7 and 8 photographed two lunar seas and taught scientists that all such 'maria' are pretty much the same. Now scientists want pictorial coverage of the moon's rugged highlands.

    "Alphonsus' walls rise 7,000 to 10,000 feet above its crater floor, and in the basin thus formed astronomers have observed reddish gas seeping from the surface. The question, then, is whether Alphonsus is really a lunar equivalent of a live volcano.

    "Second, some space experts believe Apollo astronauts, as they descend on the moon, may encounter an emergency. That emergency might force them down in the moon mountains instead of onto a flat lunar plain. Accordingly, NASA wants to know surface roughness of the smoothest part of the moon mountains.

    "Third, there are some who believe the smoothest areas on the moon actually lie within the great craters (Alphonsus' diameter is 70 miles). If this proves true, astronauts might select a crater floor as their touchdown point, assuming there is no volcanic activity." (Hoffman, *N.Y. Her. Trib.*, 3/23/65)

- Theo E. Sims, Manager of NASA Langley Research Center's Project Ram, reported results of reentry communications blackout research before the IEEE national convention in New York. Sims said significant progress had been made toward understanding the fundamental nature of the blackout problem and suggested that vehicle shape selection, signal frequency choice, use of static magnetic fields, and material addition to the flow field were all possible solutions.

Flight experiments, he indicated, had shown the materials addition technique to be useful at speeds up to 12,000 mph, and an experiment to be flown on the first manned Gemini spacecraft would attempt to demonstrate the effectiveness of water addition at even higher speeds. (LaRC Release)

March 22: NASA's actions in releasing foreign satellite information were criticized in a report by the House Committee on Government Operations, based on study by its Foreign Operations and Government Information Subcommittee. Committee stated NASA had deleted from its biweekly *Satellite Situation Report* certain Soviet launches because they were designated as secret information by Norad. "NASA has not once challenged these security classifications, blindly accepting the military decision. . . ." Compounding the problem, NASA had "publicized the facts about Soviet failures [Sept. 15, 1962, letter from Administrator Webb to Senate and House space committees] after those facts had been carefully deleted from its routine report of satellite information.

". . . NASA has ignored two clear requirements of law—the requirement for civilian control over nonmilitary space activities and the requirement for the fullest possible flow of public information. By yielding, automatically, to the military judgment on what the American people shall know about Soviet space activities, NASA fails to implement its legal mandate. By playing an on-again, off-again secrecy game, NASA tends to confuse the American public. . . .

"Therefore, the committee recommends that, in every possible instance consistent with the dictates of national security, NASA exercise its right to challenge military-imposed restrictions by requiring justification and, thus, carry out the mandate to keep the American people informed. . . ." (House Rpt. 197)

- FAA issued a special regulation banning unauthorized aircraft of U.S. registry from the designated recovery and associated areas "during the time determined to be necessary for the safe conduct of the Gemini flight and recovery operations." (FAA Release 65–21)
- AFSC's 6595th Aerospace Test Wing assumed responsibility for Atlas launches into the Air Force Western Test Range in support of the Army Nike antimissile program and the USAF Advanced Ballistic Reentry Systems (Abres) program. (AFSC Release 46.65)
- *Newsweek* reported that plans to capture world's speed record with YF–12A "mystery plane" had been blocked by Defense Secretary McNamara because he felt Congress might press for mass production of the jet—a move he opposed. Present record was held by U.S.S.R. (*Newsweek*, 3/22/65)
- "[Dr. Robert H.] Goddard's dream was the object of derision 39 years ago. Who, we must wonder, is the dreamer today who is being ignored? Where is he? What is he working on that will change this world so vastly 39 years from now? . . ." These were queries in an editorial by William J. Coughlin in *Missiles and Rockets*. Coughlin lamented the fact that much of the U.S. technological progress in the

missile/space field was directly keyed to a race with the Soviet Union. He said that "if we do not provide the atmosphere and support required for the acceptance of bold new challenges, the onward pace of U.S. science and technology will falter, then stop." (*M&R*, 3/22/65, 46)

March 23: GEMINI III Astronauts John W. Young (foreground) and Virgil I. Grissom in spacecraft immediately prior to launch.

March 23: NASA's GEMINI III spacecraft ("Molly Brown"), with Astronaut Virgil I. Grissom (Maj., USAF) as command pilot and Astronaut John W. Young (LCdr., USN) as pilot, was successfully launched from Eastern Test Range on three-orbit GT-3 mission by a two-stage Titan II.

Within six minutes after lift-off, GEMINI III and its two astronauts were injected into elliptical orbit with apogee, 224 km. (139 mi.); perigee, 161 km. (100 mi.); period, 88 min. Speed of spacecraft was 16,600 mph. Toward the end of the first orbit, 93 min. after launching, the first maneuver was performed: Grissom fired two small thruster rockets that pushed "backward" on the spacecraft, slowing it down by about 45 mph. Lessened velocity caused GEMINI III to drop in altitude to a near-circular orbit with apogee, 169 km. (105 mi.); perigee, 158 km. (98 mi.). Second maneuver occurred during second orbit: Astronaut Grissom used the thrusters to turn the spacecraft broadside to its flight path. Then he gave a burst that pushed the craft about 1/50th of a degree from the original course; short bursts, fired rapidly, slowed the craft and he turned it into a course nearly parallel to his original one. Third maneuver came in the third orbit: Grissom fired the spacecraft thruster rockets, dropping into an orbit with perigee of 82 km. (52 mi.). Manually controlling reentry, the astronauts turned the spacecraft's blunt end forward, ejected the section carrying the retrorockets. Four hours and 53 min. after launching, GEMINI III safely landed in the Atlantic Ocean off Grand Turk Island, considerably off target and some 50–60 mi. away from the recovery ship, *Intrepid*. Navy frogmen from hovering aircraft fastened a float around GEMINI III. Original plans had called for the spacecraft, with the astronauts still inside, to be hoisted aboard the recovery ship and immediate medical checks made. When Grissom became seasick the men were picked up by helicopter and landed on the *Intrepid*; the spacecraft was recovered later.

The astronauts helped perform two experiments. One was the irradiation of human blood to test the combined effects on it of weightlessness and irradiation. The other was to squirt small jets of water into the plasma sheath that surrounded the spacecraft as it reentered the earth's atmosphere, testing a theory that a fluid flowing through the ionized layer of atoms would permit radio signals to penetrate the communications blackout common to reentry.

Gemini officials said that, so far as was known, this was the first time a manned spacecraft had maneuvered in orbit, changing its orbital path. (NASA Release 65–81; NASA Transcript; Clark, *NYT*, 3/24/65, 1, 22; Simons, *Wash. Post*, 3/24/65; Bishop, *WSJ*, 3/24/65)

March 23: RANGER IX underwent a midcourse correction maneuver at 7:03 a.m. EST that would aim the spacecraft more accurately for impact on the moon crater Alphonsus on Mar. 24. The maneuver consisted of a series of radio signals that changed the spacecraft's attitude and then, through a 31-sec. burn of a small jet engine, speeded up its flight by 40.6 mph. RANGER IX was then 175,416 mi. from earth, traveling at 2,943 mph.

Newly estimated impact point was 12.9° south latitude and 2.3° west longitude—only four miles from the original target point of 13° south latitude and 2.5° west longitude. Before the correction maneuver, RANGER IX was headed for a point about 400 mi. north of Alphonsus.

JPL Director Dr. William H. Pickering said during a press conference that the landing should be well out of the shadow of the towering peak in the center of Alphonsus—a possibility that had caused JPL scientists some concern since light was needed for the picture-taking. (*L.A. Times*, West, *Wash. Post*, 3/24/65; Hill, *NYT*, 3/23/65, 1)

- President Johnson told Astronauts Virgil I. Grissom and John W. Young during a telephone call: "Your mission . . . confirms once again the vital role that man has to play in space exploration, and particularly in the peaceful use of the frontier of space. I am sure you would be the first to say that on this flight, as well as on our other manned flights in space, there were heroes on the ground as well as in space, and the record made by men like Jim Webb, Dr. Dryden, and Dr. Seamans, as well as all of those at the Cape, Cape Kennedy, and around the world, is a very proud record under Project Mercury and now on Project Gemini. And to all of those who have helped to make our space flights safe and successful, I want to . . . say 'Well done'." (*Wash. Eve. Star*, 3/24/65)

- Vice President (and NASC Chairman) Hubert H. Humphrey, visiting Cape Kennedy for the day, congratulated Astronauts Grissom and

Young and commended all participants throughout the world for "this tremendous flight of three orbits.

"... this step forward commits us to the next project. Once we have completed the Gemini series, we move on to the Apollo Project and we move on even beyond that. ... Let me say that the American economy is better because of the space program. American education is better because of the space program. American industry is better because of the space program and Americans are better because of the space program. We are emphasizing here one great character of American life—excellence, performance, achievement. . . . These are efforts well made and money well spent. . . ." (Transcript)

March 23: Following the GT-3 space flight, Dr. George E. Mueller, NASA Associate Administrator for Manned Space Flight, said at a press conference: "This particular flight is noteworthy for many reasons. Perhaps most importantly it is the first manned flight of a Gemini vehicle and it represents, then, the first step in the remaining twelve Gemini flights. In this flight . . . we did for the first time carry out an orbital maneuver in space. Another first was the first demonstration of reentry control. We did control reentry landing point on this mission. Another first was the use of Syncom for communications with the Coastal Sentry Quebec during the course of the flight." (NASA Transcript)

- AFSC Commander Gen. Bernard A. Schriever said in the keynote address at the Air Force/Industry Planning Seminar in Dayton that "we need a broader perspective and greater vision in our conceptual planning . . . we need to be more farsighted." He continued: "The Soviet Union is making a major effort to surpass us in science and technology. The Soviets now have approximately the same number of scientists and engineers that we have. But every year they graduate an average of 200,000 scientific and technical students as compared with about 120,000 a year in this country. It is also worth noting that the number of scientific institutions in Russia has grown from about 3000 in 1957 to about 5000 in 1965.

  "Both of these facts indicate that the Soviets are deadly serious when they talk about the importance of science and technology to their global ambitions. We must more than match their effort, not only too maintain our national security but also to keep our world markets." (Text)

- World Meteorological Day was celebrated by the 125 member nations of the World Meteorological Organization, a specialized agency of the United Nations. (Commerce Dept. Release WB)

- An editorial in *Red Star*, the Soviet Defense Ministry newspaper, revealed that the booster that had launched VOSKHOD II had developed 1.43 million lbs. of thrust. The article said Soviet rockets were "unmatched" and that the VOSKHOD II flight "expedites the appearance of orbital stations and the landing of people in the heavenly bodies." (Loory, *N.Y. Her. Trib.*, 3/24/65)

- Cape Kennedy and Moscow's Red Square were linked in a British television program, "East Meets West," marking U.S. and Soviet space achievements. First part of the program showed the triumphant return to Moscow of Cosmonauts Pavel Belyayev and Aleksey Leonov. Then the scene switched to Cape Kennedy to show preparations for the GT-3 flight of Astronauts Virgil Grissom and John Young. Both parts were screened "live"—the Moscow scenes via Eurovision and the Cape Kennedy one via communications satellite. (AP, Wash. *Eve. Star*, 3/24/65)

March 24: RANGER IX photograph of moon, 38.8 seconds before impact and 58 miles above lunar surface.

March 24: After transmitting 5,814 close-up lunar pictures to earth, RANGER IX, traveling at 5,977 mph, impacted the moon at 9:08 a.m. EST at 12.9° south latitude and 2.1° west longitude in the crater Alphonsus. The 10-ft., 800-lb. spacecraft, last in the Ranger series, was only four miles off target. NASA had made real-time TV coverage available and the three major networks broadcast "live" pictures during the last ten minutes of RANGER IX's flight. First pictures, taken as the photographic probe was 1,300 mi. from the moon, had about the same degree of detail as telescopic views from earth. Those taken a few seconds before impact defined objects as small as 10 in. across, including close-ups of canal-like rilles on the floor of the crater and dimple-like depressions at points along the rilles.

Photographs shown on television were taken by the "B" camera, one of two wide-angle cameras used. Four narrow-angle cameras took other shots. Pictures were received on 85-ft. antennas at Jet Propulsion Laboratory's Goldstone Tracking Station in the Mojave Desert and recorded on both 35mm film and magnetic tape for detailed analysis. Simultaneously signals were relayed by microwave to the JPL laboratory in Pasadena where an electronic scan converter "translated" electronic impulses from the 1,152-lines-per-picture of the RANGER IX signal system to the standard 500 lines of commercial television.

The Ranger program had begun inauspiciously in 1961 with a series of failures and near-misses. Rangers 1 and 2 had been designed to test the spacecraft and launch vehicle but were not injected into the desired orbit. RANGER III, IV, and V were to rough-land a seismometer package on the moon to record moon quakes, and to transmit closeup photos of the moon to earth by radio. None of the missions was successful. RANGER VI, first of the reworked and redesigned spacecraft, impacted within 17 mi. of its point of aim—but its television system failed. On July 31, 1964, RANGER VII successfully relayed to earth 4,316 high-quality close-up photos of the lunar surface. RANGER VIII, launched on Feb. 20, 1965, transmitted 7,137 pictures. Total number of photographs from RANGER VII, VIII, and IX was 17,267. (NASA Release 65-96; Sullivan, *NYT*, 3/22/65, 1; AP, Dighton, *Wash. Post*, 3/25/65, A1, A12, A16; Hill, *NYT*, 3/25/65, 1, 23; NASA Proj. Off.)

March 24: A panel of scientists analyzed slides of the RANGER IX lunar pictures at a post-impact press conference and noted that crater rims—some with level areas—and ridges inside the walls seemed harder than the plains but that floors of the craters appeared to be solidified volcanic froth that would not support a landing vehicle. Volcanic activity was inferred from indications that the moon had at least three types of craters not caused by meteorite impact.

Dr. Ewan A. Whitaker of the Lunar and Planetary Laboratory of the Univ. of Arizona said parts of the highlands around the crater Alphonsus and ridges within it seemed harder and smoother than the dusty lunar plains. Dr. Gerard P. Kuiper of the same laboratory said of the crater: "It might well be better to make landings there."

Most significant finding of RANGER IX's photographs, according to Dr. Eugene Shoemaker of the U.S. Geological Survey, was the smoothness of the crater walls and of the long ridges on the floor of the crater.

Dr. Harold Urey of the Univ. of California referred to black patches in the pictures which he said might be composed of graphite: ". . . these dark halo craters are due to some sort of plutonic activity beneath the surface of the moon. They do not look to me like terrestrial volcanoes. . . . They look like a unique lunar type of object." Dr. Urey said a Soviet scientist had reported a red flare near a peak in Alphonsus and that analysis had indicated presence of a molecule with two carbon atoms. He said this was "a very curious situation because this molecule . . . does not escape from any known volcano" on earth. (NASA Transcript)

March 24: After watching televised pictures of the moon's surface transmitted by RANGER IX, President Johnson issued a congratulatory statement that said: "Ranger 9 showed the world further evidence of the dramatic accomplishments of the United States space team. Coming so close after yesterday's Gemini success, this far-out photography reveals the balance of the United States space program.

"Steps toward the manned flight to the moon have become rapid and coordinated strides, as manned space maneuvers of one day are followed by detailed pictures of the moon on the next.

"I congratulate the scientists, the engineers, the managers—private contractors as well as Government—all who made this Ranger shot and the successes of its predecessors the great space advances that they have been." (Text, *NYT*, 3/25/65)

- First Biosatellite nose-cone test was conducted at White Sands Missile Range to evaluate aerodynamic and reentry characteristics of the spacecraft designed to carry biological specimens into—and back from—space. AFCRL's balloon-launch group was assisting NASA in conducting the tests, which involved carrying the nose cones by balloons to 88,000–100,000-ft. altitudes, releasing them, then studying their behavior during descent. Evaluated were the drogue ejection mechanism, deployment of parachute systems, descent rate, and vehicle oscillation and impact velocity. A second successful test was conducted April 29. (OAR *Research Review*, 7/65, 30)

- An editorial in the Baltimore *Sun* said: "Yesterday's Gemini flight is described as 'historic' and so it was. So too is each successful new space exploration, launched by whatever country, manned or unmanned. . . . What is happening is that a body of knowledge is being accumulated through increasingly accurate photographs and increasingly sophisticated exercises and experiences on the part of the adventurers of our age, the astronauts. . . ." (Balt. *Sun*, 3/24/65)

- XB–70A experimental supersonic bomber broke world aviation weight and speed endurance records during a one-hour 40 min. flight. It took off weighing 500,000 lbs., the heaviest at which any aircraft had been flown, and flew at continuous supersonic speeds for 80 min., longer than any other aircraft had. It cruised at a top speed of 1,400 mph and was piloted by Al White and Van Shepard. (AP, *Wash. Post*, 3/25/65; *NYT*, 3/25/65; AP, *Wash. Eve. Star*, 3/26/65)

- An editorial in the *Washington Evening Star* said: ". . . judging from Soviet cosmonaut Leonov's spectacular 'walk' in the high heavens last week, the Russians seem to be well ahead of us at the moment. Interestingly enough, however, in marked contrast to the wide-open American procedure, they do not let the outside world have any look at either the launching or the landing of their spacemen. This furtiveness makes one wonder about the nature of their program and whether they're really accomplishing as much as they claim to be.

"In any case, regardless of what the Russians are hiding, there can be no doubt that the Grissom-Young flight represents an important advance for the United States in the race to the moon. Technically, we are ahead of the Reds in many respects, and it is entirely possible that we'll make lunar landings before them." (Wash. *Eve. Star*, 3/24/65)

March 24: An editorial in the *New York Herald Tribune* referred to the U.S.–U.S.S.R. race for the moon: "The moon remains an elusive target, but it gets closer all the time. . . .

"Ideally, this should be a cooperative venture, enlisting the common efforts of the peoples of all nations; instead, so far at least, it is a race between the United States and the Soviet Union. Because it is a race; because space technology is, in major part, inseparable from military technology; because space prestige is, however illogically, a factor in the struggle to keep the earth free, we have to compete. NASA's ambitious program of a manned Gemini flight every three months promises a vigorous competitive effort. But the American effort does not parallel that of the Soviets; each is giving priority to different techniques, and the comparative standings in the race are hard to measure. What is clear, however, is that the Grissom-Young flight has carried the American program a long way forward—and beyond that, and more importantly in the long perspective of history, it has brought closer the day when man, not American man or Soviet man, finally breaks the terrestrial bonds that hold him to his native planet." (*N.Y. Her. Trib.*, 3/24/65)

- "The three-orbit flight by Virgil I. Grissom and John W. Young was in some ways the most remarkable space trip yet accomplished by this country's astronauts," said an editorial in the *New York Times*. "Particularly impressive was the apparent success of a series of maneuvers to change the Gemini's orbit—maneuvers that will be required to join two spacecraft in orbit, notably on the return leg of the projected manned flight to the moon." (*NYT*, 3/24/65, 44)

- In a speech before the National Association of Broadcasters in Washington, D.C., Gen. Bernard A. Schriever (USAF) remarked that the Soviet's space science timetable "always seems to put them one step ahead of us." He said: "It is still true that we lead in some aspects of space exploration, such as the total number of space shots, number of scientific probes, and practical applications of space satellites for such purposes as communications and weather observation. On the other hand, the Soviets lead in a number of areas with both propaganda and practical implications.

  ". . . Thus, they have put into space the first satellite, the first living creature, the first man, the first woman, the first multi-man space ship and now the first man to step out of the capsule and into space itself. They also hold the world record for time in orbit, orbital distance, orbital weight lifted, and highest orbital altitude. . . .

  "How will the Soviets use their space capabilities? . . . we are interested. . . ."

  Gen. Schriever said ground tests would begin shortly for a collapsible and expandable space laboratory for possible use as a space station: "The structure can be compressed into a small package and expanded to a cylinder 10 ft. in dia. and 25 ft. long." (Text)

- U.S.S.R. announced that Cosmonaut Valentina Nikolayeva-Tereshkova would arrive in Algiers Mar. 26 at the invitation of Algerian President Ahmed Ben Bella. (UPI, *Wash. Post*, 3/25/65, D10)

March 24: Both U.S. and U.S.S.R. space research were criticized by a Vatican weekly magazine, *L'Osservatore della Domenica*, which said they were using it as a "political instrument."

In an editorial, the publication's deputy director, Federico Alessandrini, said space competition was "beneficial because it widens man's understanding and offers new methods of observations which tomorrow will allow man to attain other goals.

"But, as one can see, the political instrument made of it limits its results and reveals . . . an obstacle to progress." (AP, *NYT*, 3/25/65)

- Aircraft operations in the U.S. increased 10% for the second consecutive year, according to statistics reported in *FAA Air Traffic Activity, Calendar Year 1964*. Ten percent gains were made in each of three major categories: total aircraft operations (takeoffs and landings at 278 airports with FAA airport traffic control towers)—34.2 million; instrument approaches at Air Route Traffic Control Center (ARTCC) areas—1.005 million; and IFR (Instrument Flight Rule) aircraft handled at ARTCCs—11.7 million. (FAA Release 65–22)

March 25: MARINER IV was nearly 40 million mi. from earth, traveling 30,000 mph relative to the sun. It had covered 188 million mi. in its orbit around the sun. The Mars probe had transmitted to earth more than 160 million bits of engineering and scientific information about planetary space. (NASA Release 65–95)

- Soviet Union launched COSMOS LXIV with scientific instruments aboard for investigation of outer space, Tass announced. Orbital data: apogee, 271 km. (167 mi.); perigee, 206 km. (127 mi.); period, 89.2 min.; inclination, 65°. All systems were functioning normally. (*Pravda*, 3/26/65, 1, ATSS–T Trans.)

- USAF launched an unidentified satellite from Vandenberg AFB on a Thor-Agena D booster. It also fired its 85th Minuteman ICBM. (UPI, *Phil. Inq.*, 3/26/65)

- NASA Administrator James E. Webb reported to President Johnson and the Cabinet on both the two-man GT–3 flight and the RANGER IX photographic mission. Mr. Webb made these points: "The most significant accomplishment of the GT–3 flight was that . . . it provided verification of the basic design, development, test and operations procedures NASA is using to develop manned spacecraft, man-rated launch vehicles and a world-wide operational network. . . .

"We now know that at least two spots, and perhaps three, when we look more carefully at the RANGER IX pictures, are at least smooth enough for the Lem [manned moon landing]. . . ."

An American astronaut probably would be able to open his spacecraft and partly emerge from the cabin during the GT–5 flight. Mr. Webb said under questioning that there might be some possibility of achieving this in the next Gemini flight, but that GT–5 was more likely.

He regarded a Russian cosmonaut's leaving a space vehicle briefly as spectacular but said the U.S. was more intent on developing a space suit that would enable American astronauts to work outside on space vehicles and develop or put together space centers. (Text; UPI, *N.Y. Her. Trib.*, 3/26/65)

March 25: At a press conference, Maj. Virgil I. Grissom (USAF) and LCdr. John W. Young (USN) described the three-orbit GT-3 flight of Mar. 23, as busy, exhilarating, near-perfect, and short on surprises. They said it was highly significant for future flight in space since it proved that a spacecraft could be maneuvered precisely, at will, and more independently of the ground than before. They said it also proved that man can eat and safely dispose of wastes as they will need to do on long flights.

Major Grissom suggested two possible reasons that the "Molly Brown" had undershot the target landing area: one was that something might have gone wrong during the final orbit change or when subsequently the braking rockets were fired to start the spacecraft's descent; the other was that there might have been a miscalculation of the craft's center of gravity. (NASA Transcript)

- Soviet President Anastas Mikoyan sent President Johnson congratulations on the Gemini GT-3 space flight. (AP, Wash. *Eve. Star*, 3/25/65)
- Use of a special airlock through which Lt. Col. Aleksei Leonov passed from the spacecraft cabin into space and back again was a major factor in the success of the VOSKHOD II flight Mar. 18, it was reported. According to Soviet sources, the preservation of normal pressure inside the spacecraft throughout flight had had an important psychological effect on both Col. Belyayev and Col. Leonov. Findings were to be discussed at a press conference to be held by the cosmonauts Mar. 26. (Shabad, *NYT*, 3/26/65)
- Tass reported that the Soviet Union was making extensive use of RANGER VII photographs presented to the Pulkovo Observatory: "Prof. Alexander Markov, who supervised the study of the photos, told a Tass correspondent that the materials received from the United States would be used to study the size and distribution of moon craters, to ascertain the origin and development of the entire lunar relief. He emphasized the particular topicality of these problems 'in view of the landing of spacecraft on the lunar surface planned for near future.'" (Loory, *N.Y. Her. Trib.*, 3/26/65)
- Gen. Curtis E. LeMay (USAF, Ret.) said in a speech at a dinner meeting of the National Security Industrial Assn. where he received the James Forrestal Memorial Award that the "United States should observe with great care any tendency of the Soviet Union to develop space weapons. Already there is considerable reason for concern about Soviet capabilities in space. Many of the techniques the Soviet Union has developed so far point strongly toward a military space effort. The development of a capability by the Soviet Union to deliver strategic weapons from near space or to deny to the United States the opportunity to continue its present programs in space would amount to a serious threat and would negate our present favorable balance of military power." General LeMay criticized "current conservatism in the Department of Defense growing out of economic considerations" and said responsible officials should reappraise existing military R&D policies. (Sehlstedt, Balt. *Sun*, 3/26/65; Raymond, *NYT*, 3/29/65, 36)

March 25: Dr. Wolfgang B. Klemperer, pioneer in glider and missile design, died of pneumonia. A fellow of the AIAA, the AAS, and the British Interplanetary Society, Dr. Klemperer had been active in preparations for a NASA project to photograph a solar eclipse on May 30 from a jet airliner over the South Pacific. (*NYT*, 3/27/65, 27)

- Kenneth Gatland, Vice-President of the British Interplanetary Society, urged U.S.-U.S.S.R. cooperation in manned lunar exploration in *New Scientist* article. ". . . it seems we are faced with the ludicrous situation of the world's two most powerful nations, each with massively expensive rival programmes, heading for a common objective which each proclaims is being pursued in the highest interests of peaceful scientific exploration." A joint venture would have the advantage of providing for contingencies such as rescue of astronauts possibly stranded on the moon or in lunar orbit—a capability not included in Project Apollo. "This situation can only be satisfactorily resolved by the provision in lunar orbit of a second soft-landing vehicle and back-up crew capable of mounting an emergency rescue operation. To achieve this would require a specially adapted version of the craft already designed to soft-land astronauts.

    "This is where the merit of US-Soviet cooperation lies for, as an international venture, a project to land men on the Moon would surely not be undertaken as envisaged in project Apollo; and certainly not with such rigid constraints on time. In all probability it would be planned as an operation rather than a solo mission, with logistic support from a second space vehicle placed in lunar orbit ahead of the main expedition. . . .

    "The essential requirement in terms of the eventual lunar expedition is that launchings should be coordinated so that expedition components arrive in lunar orbit together. By the mid-1970's, orbital rendezvous techniques should be well established with the ability of men to move between orbiting vehicles. An agreed crew could then descend to the lunar surface while another ship remains in reserve orbiting the Moon in case of need. Alternatively, a reserve vehicle might be landed, unmanned, in advance. . . .

    "Although at this stage such . . . [a combined lunar expedition] would have little influence on overall costs, it could mean a great deal to the safety of initial manned missions.

    "Such a move would demand concessions on both sides. It would mean America abandoning her 1970 target date for placing men on the Moon, and while allowing Russia to keep her rocket secrets she would have to be prepared to reveal her programme for manned spaceflight. . . ." (*New Scientist*, 3/25/65, 774–76)

March 26: X-15 No. 1 was flown by Maj. Robert Rushworth (USAF) to 101,900-ft. altitude at a maximum speed of 3,580 mph (mach 5.2) to obtain data using infrared scanner and to check the Honeywell inertial guidance system. (NASA X-15 Proj. Off., *X-15 Flight Log*)

- NASA postponed indefinitely the launching of a beacon Explorer satellite from Wallops Island. The launching had been scheduled for March 30. (*NYT*, 3/27/65)
- Soviet Cosmonaut Pavel Belyayev told a Moscow news conference that VOSKHOD II had been scheduled to land after 16 orbits, but that there was an inaccuracy in "the solar system of orientation" that

prevented use of the automatic landing system. He said he then had to obtain radioed permission from the Soviet space center to land by manual control after the 17th orbit. The landing site was overshot "by a certain distance" Belyayev said without disclosing how much.

Belyayev said success of the GT-3 flight of Astronauts Virgil I. Grissom and John W. Young "was a national achievement of the United States." He congratulated "the courageous American cosmonauts," and said: "May the flights of both ours and American cosmonauts be dedicated to unraveling the mysteries of the universe in the interests of science and for the good of all mankind."

Belyayev said VOSKHOD II was capable of maneuvering in space as did the U.S. GEMINI III but that this was not in the Soviet flight plan.

Leonov described time outside the ship saying "it is too early to call it a pleasant walk. It could not have been done without hard training." He reported his small push on VOSKHOD II to move away from it after going out of the hatch started the spacecraft into slow rotation. In pulling himself back to the VOSKHOD II by his cable, Leonov disclosed he had yanked rather vigorously and had to put his hand out to avoid collision with the spacecraft.

Belyayev said he and Leonov were found by a helicopter 2½ hrs. after a soft landing in snowy woods near Perm. He said VOSKHOD II was airlifted back to the launch site at Baikonur in Soviet Central Asia and could be used again if necessary. (AP, Wash. *Eve. Star*, 3/26/65; Shabad, *NYT*, 3/27/65; *Flight International*, 4/8/65, 542–44)

March 26: Astronauts Virgil I. Grissom and John W. Young were honored in a White House ceremony where President Johnson conferred NASA Exceptional Service Medals on both men and pinned a cluster on the NASA Distinguished Service Medal awarded Major Grissom for his July 21, 1961, suborbital Mercury flight. He was the first man to make two space flights.

NASA Associate Administrator Dr. Robert C. Seamans received the NASA Distinguished Service Medal for his direction of space efforts. Harris M. Schurmeier received an Exceptional Scientific Achievement Medal for his direction of the Ranger program.

President Johnson said: "A sense of history is present strongly here today. All of us are conscious that we have crossed over the threshold of man's first tentative and experimental ventures in space. . . .

"Since we gave our program direction and purpose seven years ago, many successes have been achieved through the efforts of a great American team, which now numbers 400 thousand men and women in industry, on campuses, and in government. And this team is inspired and stimulated and led by a former Marine and a great public servant —Jim Webb."

Following the ceremony, a motorcade bearing Vice President Humphrey, the astronauts, and their party took the Pennsylvania Ave. parade route, where thousands had gathered to cheer them, to the Capitol; a luncheon in their honor was jointly sponsored by Sen. Clinton Anderson (D–N.Mex.) and Rep. George P. Miller (D–Calif.), chairmen of the Senate and House space committees.

At 5 p.m. the group returned to Capitol Hill for a Congressional reception hosted by House Speaker John McCormack

(D-Mass.). (NASA Release 65-98; Text; Carmody, *Wash. Post*, 3/26/65; *NYT*, 3/27/65, 1)

March 26: Propulsion system and structure of the hypersonic Sprint antimissile missile was successfully tested by the Army at White Sands Missile Range. Although the missile was being designed for launchings from underground cells, the Sprint was launched from an aboveground launcher for the test. (DOD Release 137-65)

- Smithsonian Institution's National Air Museum placed on display a quarter-scale model of the GT-3 spacecraft, a full-scale model of RANGER IX along with some of the photos it took, and a model of MARINER IV Mars probe. The spacecraft were part of an exhibit depicting NASA's broad program of space research. (NASA Release 65-100)

- It was announced that the special magnetic actuator which worked shutters on RANGERS VII, VIII, and IX, that photographed the moon, and on all nine Tiros weather satellites would be granted a patent. The device moved the shutter at a constant velocity so that the exposure was uniform. It was invented by RCA engineers Langdon H. Fulton and Thomas D. Tilton. (Jones, *NYT*, 3/7/65, 35)

March 27: M. V. Keldysh, President of the U.S.S.R. Academy of Sciences, commented on the VOSKHOD II flight in an article in *Izvestia*: "One of the most significant accomplishments in the conquest of space was the experiment dealing with man's emergence into space. New, grandiose perspectives are now open for the construction of orbital stations, the docking of spacecraft in orbit and the carrying out of astronomical and geophysical investigations in space. In the near future it will be possible to create, in orbit around the earth, a Space Scientific Research Institute in which scientists representing the most diversified fields will be able to work. The results obtained as a result of the flight of 'Voskhod-2' are most important steps on the way toward carrying out flights to the moon and on to other celestial bodies." (*Izvestia*, 3/27/65, 5, ATSS-T Trans.)

- Astronauts Virgil I. Grissom and John W. Young had congratulated Soviet Cosmonauts Pavel Belyayev and Aleksey Leonov on the VOSKHOD II flight, *Izvestia* disclosed. (UPI, *Wash. Post*, 3/28/65)

March 28: Robert J. Schwinghamer, NASA Marshall Space Flight Center, received American Society of Tool and Manufacturing Engineers' Research Medal. Schwinghamer was cited for his research "leading to a better understanding of materials, facilities, principles, and operations, and their application to better manufacturing." (MSFC Release 65-58; *Marshall Star*, 3/17/65, 1, 6)

March 29: Gemini GT-3 Astronauts Maj. Virgil I. Grissom and LCdr. John W. Young were given traditional heroes' welcome from New Yorkers at a parade given in their honor. Honored with the astronauts was Dr. Robert C. Seamans, Jr., Associate Administrator of NASA. They were met by Mayor Wagner and the city's official greeter, Commissioner Richard C. Patterson of the Department of Public Events. Mayor Wagner presented gold keys to the city to the astronauts and Dr. Seamans at a ceremony at City Hall. He also presented the city's Gold Medal of Honor to Major Grissom and Dr. Seamans and the Silver Medal of Honor to Commander Young. At the United Nations, Secretary General U Thant presented medals and two auto-graph sets of U. N. outer space commemorative stamps to the astronauts. (Sibley, *NYT*, 3/29/65, 36; Talese, *NYT*, 3/20/65, 1; Orl. *Sent.*, 3/30/65)

March 29: *Pravda* described Lt. Col. Leonov's exit and return to VOSKHOD II in giving the first detailed description of the inside of the spacecraft. The airlock was apparently built into the place occupied by a third astronaut during the VOSKHOD I flight Oct. 12. After Col. Leonov moved into the airlock, his companion, Col. Belyayev pressed a button that closed the inside door and created a vacuum inside the lock chamber. At the prescribed moment, Col. Belyayev pressed a second button that opened the hatch between the airlock and space, allowing Col. Leonov to climb out. The procedure was apparently reversed for the astronaut's return. (AP, *NYT*, 3/30/65)

- USAF announced successful test firing of a simplified rocket engine called Scorpio. The engine had eight combusters in a ring around a nozzle and an injector that sprayed fuel into the combusters through several ports. Scorpio developed 200,000 lbs. thrust and would be modified to produce greater power. (AFSC Release 44.65; AP, Balt. *Sun*, 3/30/65)

- Construction work at Cape Kennedy and Merritt Island Launch Area was halted when an Orlando union local set up picket lines to protest a contractor use of non-union labor. USCE estimated that more than 4,500 of about 5,000 building trades workers refused to cross the lines. NASA had advised the National Labor Relations Board. This marked the sixth time in 14 mos. that a labor dispute had crippled construction work on Merritt Island where launching facilities were being built. (AP, *Chic. Trib.*, 3/30/65)

- DOD Advanced Research Projects Agency had selected three contractors for research programs in the materials field: Martin Co., awarded $1 million, subcontract with the Univ. of Denver and conduct a three-year program on the high energy rate of forming metals; Union Carbide Corp., with $2.5 million, would subcontract with Case Institute of Technology and the Bell Aerospace Corp. and conduct a three-year research program on carbon composite materials; Monsanto Research Corp., awarded approximately $2 million, would subcontract with Washington Univ. of St. Louis, Mo., and conduct a two-year research program on high-performance composites. (DOD Release 193–65)

March 30: A copper-plated 46½-lb. "minilab," instrumented to measure radiation variations in the earth's magnetic field, was launched to 8,700-mi. altitude from Cape Kennedy on a four-stage Blue Scout Jr. rocket. It carried three sensing devices designed to produce a radiation profile during its two-hour climb into the Van Allen radiation belt and the two-hour plunge back through the earth's atmosphere to the Indian Ocean. (UPI, *NYT*, 3/31/65; *U.S. Aeron. & Space Act.*, 1965, 138)

- Emergency landing of VOSKHOD II was the third such failure in the Soviet space program, according to an unidentified Czechoslovak scientist, member of the Astronautic Commission of the Czechoslovak Academy of Sciences, during a panel discussion on Radio Prague. He said there had been two earlier failures in the unmanned Vostoks. The disclosure was made in reply to a listener's letter. (*NYT*, 4/1/65, 6)

March 30: Gemini GT–3 Astronauts Grissom and Young were feted as heroes in Chicago, where they motorcaded from O'Hara International Airport through the city to City Hall. An estimated one million thronged the streets shouting joyous ovations and flinging a deluge of tickertape and confetti. At luncheon with city officials the astronauts were given honorary Chicago citizenship medallions, and later a reception was given in their honor. Accompanying the astronauts were members of their families and NASA Deputy Administrator Dr. Hugh L. Dryden and Mrs. Dryden. (Wiedrich, *Chic. Trib.*, 3/31/65)

- Dr. Harold Brown, Director of Defense Research and Engineering, appeared before the House Appropriations Committee's Subcommittee on DOD Appropriations, in testimony supporting DOD's request for $6.709 billion new obligational authority for FY 1966 research, development, testing, and evaluation.

  He discussed the Vela nuclear detection satellites, orbiting in nearly circular orbits. "All four satellites remain in operation, providing data on the radiation background and the operation of detectors in space." He outlined the AACB's 1964 launch vehicle study, which "was intended to identify overall effects and provide a data base for, rather than to resolve, individual user program booster selections or near-term booster improvement questions." The study "confirmed earlier estimates" of launch vehicle needs for the near future. [See Jan. 26, Jan. 27] (*DOD Appropriations Hearings* [Part 5], 1–30)

March 31: Nike-Apache sounding rocket was launched from Wallops Island with NASA Lewis Research Center experiment to study three wavelengths of light in the airglow: one in the red part of the spectrum, another in the yellow, and a third in the green. Altitude of the airglow was measured with phototubes mounted on the rocket. A 26-in.-dia. mylar balloon helped scientists correlate measured light intensity and altitude with density of the atmosphere. (Wallops Release 65–19; LRC Release 65–26)

- U.S. Army disclosed it had orbited a three-satellite earth-mapping system, with two of the spacecraft circling the earth from west to east and the third traveling from pole to pole. The satellites were of the Secor type. Two were fired into orbit earlier this month; the other was launched Jan. 11, 1964. The three spacecraft, each with a radio receiver and transmitter, were helping pinpoint locations on earth that were widely separated by large bodies of water. (AP, *NYT*, 4/1/65, 11; *M&R*, 4/5/65, 12)

- Studies carried out under NASA contract by the Union Carbide Research Institute had demonstrated the ability of many life forms to adjust to at least partial Martian conditions. It had also been demonstrated that lack of oxygen produced surprising results: turtles with little or no blood; plants that could endure lower temperatures than plants raised in normal air. Such temperature resistance would be an advantage on a cold planet like Mars. Dr. Sanford M. Siegel disclosed these findings during a press tour of Union Carbide and said that if earth life could withstand Martian conditions so well, Martian life, if there ever had been any, must have been able to evolve to cope with the situation there. (Sullivan, *NYT*, 4/1/65)

March 31: Discussing NASA Kennedy Space Center's evaluation measurement program for cost-plus-award-fee contracting before the GE Annual Method and Work Measurement Conference in Gainesville, Fla., John E. Thomas of KSC's Support Operations listed eight points designed to give a thorough profile of the contractor: (1) quality of work; (2) personnel profile; (3) care and control of Government property; (4) effectiveness of the contractor's training programs; (5) speed of compliance with work requests; (6) contractor attitude; (7) cost-control practices; (8) business management practices. He said that from these data the KSC Evaluation Board determined how much of the fee the contractor had earned. (Text)

- Maj. Virgil I. Grissom and LCdr. John W. Young, the Gemini astronauts, returned home to Houston and to an enthusiastic welcome by a crowd of some 12,000 persons. The astronauts walked by much of the crowd, shaking hands. "We've had a pretty tough week, then came a couple of days of debriefing, then three parades, but today is the best of all—when we get to come back home," Major Grissom said. Commander Young said, "We're sure happy to see all you smiling Texans." (UPI, *NYT*, 4/2/65, 12)

- USAF sonic boom series over Chicago, which had begun Jan. 4, ended. (*Chic. Trib.*, 3/31/65)

- Senate Armed Services Committee approved a $15,284,000,000 military authorization bill for DOD; an unrequested $82 million was added for development of a new manned bomber to replace the B-52 and B-58, no longer in production. (Raymond, *NYT*, 3/31/65)

- NASA Administrator James E. Webb told the American Society of Photogrammetry and the American Congress on Surveying and Mapping, convening in Washington: ". . . since the dawn of the Space Age—in less than eight years—one of our most important tasks has been that of mapping—mapping the surface of the world and its geodetic figure; mapping the world's weather, as revealed in its cloud patterns as seen from above; mapping the earth's outermost atmosphere in three dimensions, and exploring its interaction with the newly-discovered solar wind; seeing and mapping astronomical sources for the first time in ultraviolet and X-radiation from outside the earth's atmosphere; and mapping areas of our moon to an accuracy 2,000 times better than that now achievable from earth, and preparing to map areas of Mars to an accuracy as much as 100 times better than that attainable from earth. . . ." (Text)

- All but six of the 170 pieces into which Soviet satellite COSMOS LVII had shattered after being orbited Feb. 22 had fallen to earth, according to GSFC's *Satellite Situation Report.* Another disclosure of the report was that a U.S. satellite orbited March 9 from WTR was orbiting in eight pieces, four of which were transmitting signals. (GSFC *SSR*, 3/31/65)

- Construction workers at NASA Kennedy Space Center returned to work, ending a two-day walkout which NASA spokesman said cost the government $200,000 a day. Pickets of United Association of Plumbers and Pipefitters were withdrawn when Assistant Secretary of Labor James Reynolds agreed to meet with union representatives Apr. 5. (UPI, *Cocoa Trib.*, 3/31/65)

March 31: Lt. Gen. James Ferguson, USAF Deputy Chief of Staff (R&D), stated in FY 1966 appropriations hearings of House Appropriations Committee's Subcommittee on DOD Appropriations: "I cannot help but believe, if we take a look at the last 40 years of Russian national development, that they are watching for an opportunity to gain a major military advantage over us. I cannot help but feel that they are examining opportunities in space very thoroughly for this particular purpose.

"In order to be able to offset any advantage which they may discover, I feel we must move as rapidly as we can in this area, and take full advantage of any other national space programs such as the NASA activity.

"The big program that we hope to get a go-ahead on here shortly is the Manned Orbital Laboratory. Here we think we will achieve a number of answers in the next 2 or three years. . . ." (*DOD Appropriations Hearings* [Part 5], 148)

- USAF announced a high vacuum test chamber that would simulate space environment and altitudes up to 990,000 ft. was being constructed at Wright-Patterson AFB. Liquid metal system components such as space radiators, and expandable structures such as solar reflectors, would be tested in the chamber. Chicago Bridge and Iron Co. was constructing the facility, which would be completed in Sept. 1965, under a $699,780 contract awarded in Nov. 1964. (AFSC Release 1.65)

During March: Asked in an interview for the *San Diego Union* if the U.S. would succeed in landing a man on the moon in this decade, Dr. Donald F. Hornig, special assistant to President Johnson for science and technology, said: "When you lay down a schedule, it says that if everything goes as I see it, making allowances for reasonable difficulties, this is what I'll do. It's a tight schedule and will take a lot of doing. We also have to acknowledge that unforeseen problems may arise. . . . When we started in 1961 on a nine-year program it was not wishful thinking but it was a purely paper exercise. We have slipped some on our schedules, but in a sense we have gained ground in that we have not run into any serious difficulties yet. We are now entering the hardest period of all, when the pieces begin to come out of the factory and have to be put together and tested."

Answering a query if there would be a manned expedition to Mars one day, he said: "It would be harder than going to the moon. I don't anticipate he will go soon. But we have started the unmanned exploration. The results may whet our appetite or may prove that conditions are so inhospitable that it isn't worth the effort." (*San Diego Union*, 3/7/65)

- JPL scientists W. L. Sjogren and D. W. Trask reported that as a result of RANGER VI and RANGER VII tracking data, DSIF station locations could be determined to within 10 meters in the radial direction normal to the earth's spin axis. Differences in the longitude between stations could be calculated to within 20 meters. The moon's radius had been found to be 3 km. less than was thought, and knowledge of its mass had been improved by an order of magnitude. (*M&R*, 3/22/65, 23)
- NASA Manned Spacecraft Center analysis showed that radiation shielding offered by the Apollo Lunar Excursion Module (Lem) was negligible:

a particle flux producing a 1-rem dose in the Apollo command module would produce a 17-rem dose in the Lem. The Apollo space radiation warning system would provide advance indication of need for astronauts to return from the Lem to the command-service modules. (*M&R*, 3/22/65, 23)

During March: USAF San Bernardino Air Materiel Area reported that Atlas and Titan ICBM's scheduled for phase-out by summer would be used in antimissile and space booster research and development assignments. Requests had been received to use the silos as civil defense shelters and for storage of petroleum, gas, and grain. (*M&R*, 3/22/65, 12)

- NASA's Office of Technology Utilization published a technology survey on advanced valve technology growing out of space research. (NASA Release 65-92)
- A land exchange between the U.S. Government and New Mexico was nearing completion, clearing the way for construction of a $20 million rocket testing complex to be built by Bell Aerosystems Co. near the White Sands Missile Range. (AP, *Houston Chron.*, 3/24/65)
- Republican minority of the Joint Congressional Economic Committee said, after reviewing the President's *Annual Economic Report*, that the U.S. emphasis on defense, space, and other Federal research was giving the other industrial nations the opportunity to concentrate on civilian-oriented research, which might enable them to build superior economies. (*Av. Wk.*, 3/29/65, 78)
- The theory that temperature change of 3.5° C or more in 5 min. of horizontal jet flight was a true indicator of clear air turbulence (Cat) had been disproved by George McLean, AFCRL. He explained that Cat did not always occur near jet streams and that when it did, the angle at which the plane hit the jet stream was a determining factor. (OAR Release 365-6)
- British Meteorological Office's Skua solid-propellant sounding rocket was described by Kenneth Owen in *Indian Aviation*. The eight-foot-long, five-inch-diameter rocket had been in use since the beginning of the year as a tool for weather observations and other research. A series of Skuas would be launched as part of IQSY; launchings were planned at the rate of three a week during the nine two-week periods of IQSY known as "World Geophysical Intervals." (*Indian Aviation*, 3/65, 73-74)
- Interview of Dr. Boris Yegorov, Soviet physician-cosmonaut and member of the three-man VOSKHOD I spaceflight crew, by Novosti Press, appeared in *Space World*. Yegorov mentioned nothing about any ill effects of spaceflight conditions, but did say:

    "Several times we tried to break away from the chair and hang a bit in the cabin. I must tell you that it's far from a pleasant sensation. It's also entirely inconvenient to sleep thus. One tries rather to lean on something: either with his head against ceiling or with his feet against the chair. During weightlessness it's much more pleasant to be tied to the chair. . . .

    "During the time we worked none of us had any unpleasant sensations because of weightlessness: we felt fine." (*Space World*, 3/65, 37-38)

April 1965

April 1: The S-IB-1 stage of the Saturn IB booster was successfully static-fired by Chrysler for the first time at NASA Marshall Space Flight Center; the test lasted about 30 sec. Powered by 8 Rocketdyne uprated H-1 engines, each developing 200,000 lbs. of thrust, S-IB-1 stage would be fired at least one more time before being returned to Michoud Operations in New Orleans for checkout. It would then be shipped to Cape Kennedy for launch early next year. (MSFC Release 65-75)

- A prototype Tiros weather satellite was donated to the Smithsonian Institution's National Air Museum by Dr. Hugh L. Dryden, NASA Deputy Administrator, on behalf of NASA, in commemoration of the fifth anniversary of NASA's TIROS I launch.

  Dr. Dryden said: ". . . nine experimental meteorological satellites of the Tiros series have been successfully launched and operated.

  "Seldom, if ever, has a complex technological effort in its early phases returned such valuable dividends as this project. In the early stage Tiros was an Army project. When the National Aeronautics and Space Administration was created in 1958 it took over the development of the spacecraft.

  "The United States Weather Bureau utilized the data from the very first experimental flight. The first Tiros had been in orbit only a few hours when it began transmitting to NASA ground stations cloud photographs of good quality. The Weather Bureau was quickly able to apply the pictures to its day-by-day forecasting. During the years since then, Tiros satellites have literally been working their way around the world, benefitting men everywhere by supplying previously unobtainable weather data. At this stage, it is impossible to estimate how many lives have been saved and how much property loss avoided through use of Tiros information, but the totals must already be substantial."

  David Arthur Davies, Secretary-General of the World Meteorological Organization, discussed international reaction to meteorological satellite developments, listing three main points: (1) ". . . the tremendous impact which this new means of observing the atmosphere has had upon the world scientific community. . . . [For instance,] it was the realization that the meteorological satellite was . . . a turning point in the long history of man's endeavors to improve his knowledge and understanding of his environment—the atmosphere" that led to the establishment of the World Weather Watch. (2) The impact of the meteorological satellite upon the United Nations. The ". . . impact of the TIROS satellites was so great as to inspire the General Assembly of the United Nations to take the very unusual step of adopting a resolution on a scientific question of this kind [Resolution 1721 on International Cooperation in the Peaceful Uses of Outer Space] and to maintain its interest from that time." And, (3) ". . . the general feeling of gratitude and admiration towards the United States which the launching of TIROS I and which the decision to distribute the data to all countries throughout the whole world engendered."

Speaking at the ceremony, Dr. Robert M. White, Chief of the U.S. Weather Bureau, praised the Tiros program and said that the NASA-Weather Bureau Tiros Operational Satellite System (TOS), expected to be operational early next year, would modify a Tiros satellite similar to TIROS IX to permit daily observation of clouds in the earth's atmosphere. He added: "And one day we may even be using the moon as a base for establishing a weather station to monitor and study terrestrial weather." Dr. White predicted continued NASA-Weather Bureau cooperation: (1) to further develop "satellite visual and infrared sensing devices for the indirect probing of the atmosphere"; (2) to "broaden the meteorological satellite system as a means of data collection"; and (3) to "pursue the use of synchronous satellites for weather observations."

Dr. Morris Tepper, Director of Meteorological Programs in NASA's Office of Space Sciences, recalled the launching of TIROS I: "It was a very exciting morning—waiting for my first countdown . . . someone fixed a leaky lox line at the launching pad by wrapping a wet rag around the leak and freezing it solid . . . The launch vehicle, the Thor-Able, performed exceptionally well. The spacecraft was placed into an exceptional orbit. The next question was—what would we see? . . . And finally we had our picture—this first picture from TIROS I. Yes, there were clouds in it . . . The first three pictures were . . . carried to Dr. Glennan, the first Administrator of NASA, and finally we all trekked over to the White House and interrupted a Cabinet Meeting to show President Eisenhower the results of this remarkable space capability." (Texts; NASA Release 65-102)

April 1: To date, 46 sounding rocket launchings had been made from the USNS *Croatan* operating at sea off South America's west coast, NASA announced. 32 of the firings were two-stage sounding rockets carrying upper atmosphere and ionosphere experiments; 14 were single-stage vehicles to obtain high-altitude meteorological data. Launchings were part of NASA's sounding rocket program for the 1964–65 International Quiet Sun Year (IQSY) when solar flare and sunspot activity were at a minimum. Expedition data would be correlated with findings of scientists throughout the world conducting experiments to study IQSY phenomena. (NASA Release 65-104)

- Vice President Hubert H. Humphrey visited NASA Flight Research Center. (FRC *X-Press*, 4/9/65, 1, 2)
- FAA announced one-month extensions, through April 1965, of design contracts with Boeing Co. and Lockheed Aircraft Corp., airframe contractors; and General Electric Co. and Pratt & Whitney Div. of United Aircraft Corp., engine contractors, for U.S. supersonic transport program. Extensions applied to design contracts awarded to four companies for period Jan. 1, through Feb. 28, 1965, with provisions for one-month extensions from Feb. 28, through June 30. Dollar amount

of each one-month airframe contract extension was $1 million ($750,000 Government, $250,000 contractor); dollar amount of each one-month engine contract extension was $835,000 ($626,250 Government, $208,750 contractor). (FAA Release 65–24)

April 1: NASA awarded a $1,307,347 firm-price contract to Space Corp. to fabricate, test, assemble, install, and check out engine servicing platforms at Kennedy Space Center's Launch Complex 39 on Merritt Island. (KSC Release 72–65)

• Members of Southern Interstate Nuclear Board, official agency of the 17 states of the Southern Governor's Conference for service and assistance in nuclear energy and space technology, toured Cape Kennedy and received briefing on NASA activities there. (KSC Release 73–65)

• Najeeb Halaby, FAA Administrator, announced that he would ask Congress for enabling legislation authorizing a ten-day, federally-sponsored International Aerospace and Science Exposition, to be held the summer of 1966 at Dulles International Airport, Washington, D.C. The Exposition, approved by President Johnson March 31, 1965, would attempt to stimulate export sales of U.S. products and to demonstrate U.S. accomplishments in aerospace and related sciences. (FAA Release 65–25)

• A proposal was made that Great Britain streamline its space and scientific research efforts by dissolving the Dept. of Scientific and Industrial Research and transferring its activities to the Ministry of Technology and the Science Research Council. Control of British scientific attaches in embassies abroad would be transferred to Dept. of Education and Science which would coordinate its activities with Science Research Council and the Ministry. (*Av. Wk.*, 4/12/65, 33)

April 2: Summary report of NASA's Future Programs Task Group, directed by Francis B. Smith of LaRC, was sent by NASA Administrator James E. Webb to the chairmen of the Senate Committee on Aeronautical and Space Sciences and House Committee on Science and Astronautics. Report presented "the results of studies made during 1964 to answer inquiries made by President Johnson as to criteria and priorities for space missions to follow those now approved for the decade of the 1960's. . . ." It examined (1) conditions and constraints for future planning, (2) major capabilities existing and under development, (3) intermediate missions, and (4) long-range aeronautical and space developments. Report concluded:

". . . The details of these new missions such as specific spacecraft designs and exact mission plans will, of course, be the subject of continued study. . . . Continued space exploration will be an evolutionary process in which the next step is based largely on what was learned from the experience of preceding research and flight missions. The pace at which these new programs will be carried out will necessarily depend upon many other factors, such as the allocation of budgetary and manpower resources and the changing National needs of the future.

"This study has not revealed any single area of space development which appears to require an overriding emphasis or a crash effort. Rather, it appears that a continued balanced program, steadily pursuing continued advancement in aeronautics, space sciences, manned space flight, and lunar and planetary exploration, adequately supported by a broad basic research and technology development program, still represents the wisest course. Further, it is believed that such a balanced program will not impose unreasonably large demands upon the Nation's resources and that such a program will lead to a pre-eminent role in aeronautics and space." (Text; *NASA Auth. Hearings* [Part 3], Senate Comm. on Aeronautical and Space Sciences, 1015–1102)

April 2: Fifty years ago President Woodrow Wilson appointed the first members of the National Advisory Committee for Aeronautics. The first meeting of the NACA was held on April 23, 1915, in the office of Secretary of War Lindley M. Garrison. Brig. Gen. George P. Scriven, Chief Signal Officer, was elected temporary chairman. (Hunsaker, *40 Years,* 247; *A&A, 1915–60,* 3)

- MARINER IV's star-tracking guidance system was updated to compensate for changing angular relationship between spacecraft and the star Canopus. (NASA Release 65–111)
- Landing pads that might be used on unmanned or manned vehicles in NASA's Project Apollo were patented for NASA. Bowl-shaped, the pads would be attached to the spacecraft's struts by ball joints and would be braced inside by collapsible ribs to absorb lateral shock. The underside of the bowl would be covered by material similar to sheet aluminum designed to shear away if the pads should slide. The inventor, Josef F. Blumrich of NASA Marshall Space Flight Center, said the pads would support a vertical landing on level terrain and would not dig in or transmit undue shock if they should slip against rocks; they were designed to settle on rock or dust or a combination of the two. (Jones, *NYT,* 4/3/65, 34)
- USAF designated Textron's Bell Aerosystems Co. an associate prime contractor to supply rocket engines for the Agena space vehicle, it was announced. Change would enable the AFSC Space Systems Div. to procure Agena rocket engines directly from Bell Aerosystems. Bell had designed, manufactured, and tested the Agena rocket engine since 1956 under subcontracts from Lockheed Missiles and Space Co. Agena had orbited more than 80 percent of the USAF and NASA satellites and had placed approximately 60 per cent of the free world's functional unmanned payloads in space. The Bell Agena engine, which had contributed largely to that percentage, had been fired in space approximately 200 times and had achieved a record exceeding 99.3 per cent. (Bell Release)
- Canadian Defence Minister Paul Hellyer announced the Mar. 31 shutdown of the $227-million Mid-Canada Warning Line, an electronic aircraft-detection device. Mr. Hellyer said that the shutdown would save $13 million annually and that improvements in the Pinetree radar system had made coverage by the Mid-Canada Line unnecessary. (AP, *NYT,* 4/4/65, 12)
- Hsinhua, official Chinese Communist press agency, announced public display in Peking military museum of a pilotless U.S. reconnaissance plane, shot down over central south China, Jan. 2, 1965, "by the Air Force." (*NYT,* 4/3/65, 2)

April 3: Check out of AEC's SNAPSHOT satellite.

April 3: AEC's 970-lb. SNAPSHOT spacecraft carrying Snap-10A nuclear reactor was successfully launched from Vandenberg AFB by an Atlas-Agena booster into nearly circular polar orbit; 820-mi. (1,320 km.) apogee; 788-mi. (1,269 km.) perigee; 112 min. period; 90.17° inclination. Four hours after injection into orbit, radio command from earth activated the 250-lb. nuclear reactor by moving internal shielding that had kept the emission of electrons from the uranium-235 fuel element from reaching the chain reaction stage. The reactor would provide electric power for a 2.2-lb. ion engine. This was the first attempt to test a reactor-ion system in orbit.

Twelve hours after launch, radio signals from the Agena vehicle carrying the reactor indicated it was producing 620–668 watts of electricity—some 20% over its designed power. Electricity generated by the reactor would be stored in a 480-lb. bank of batteries and released as the ion engine was put through start-stop tests during a three-month period. The engine would manufacture its own power by electrically vaporizing the 3½ oz. of the metal cesium in its fuel tank into atomic particles and expelling them at high speed through a nozzle to provide thrust of two-thousandths of a pound.

AEC said the satellite would stay aloft more than 3,000 yrs.—far beyond the 100 yrs. it would take for the reactor's radioactive elements to decay to a safe level. The reactor would be shut down after a year, the ion engine after about three months. If successful, the test would signal the first operation in space of a light, compact, propulsion system that would produce power over long periods on small amounts of fuel for (1) surveillance and patrol satellites functioning in orbit for years, and (2) manned spaceships capable of speeds of 100,000 mph on trips to distant planets now beyond the reach of conventionally-fuelled rockets.

Also orbited was U.S. Army SECOR IV geodetic satellite. (Hill, *NYT*, 4/5/65; AP, *Wash. Post*, 4/4/65; UPI, *Chic. Trib.*, 4/5/65; AEC Release H–60; *U.S. Aeron. & Space Act., 1965*, 139; *Atomic Energy Programs, 1965*, 151)

April 3: NASA Nike-Apache sounding rocket was launched from Ft. Churchill, Canada, to altitude of 204.67 km. (127.2 mi.) with Rice University experiment to make time resolution measurements of electron fluxes within an aurora for use in determining transit times of these electrons from their sources. Performance was satisfactory. (NASA Rpt. SRL)

• USAF School of Aerospace Medicine was conducting experiments on 13 rhesus monkeys at Oak Ridge National Laboratory to discover how nuclear radiation would affect auditory, visual, and motor systems. Studies might ultimately reveal how man would be affected under similar conditions. Each monkey was conditioned to respond to a visual or auditory cue; by measuring the time required for animal to respond before and after radiation exposure, scientists could determine the effect of radiation on monkey's ability to perform. Preliminary results had confirmed that "animals exposed to radiation undergo a period shortly after irradiation in which they are totally unable to function." (*NYT*, 4/4/65, 68)

• In *Saturday Review*, Science Editor John Lear reviewed GSFC's Project Firefly as "an epic experiment that will at least track the essential spark of life wherever it can be found beyond the earth."

He reviewed Dr. William D. McElroy's pioneering research in bioluminescence [see March 11] and noted that Norman E. MacLeod, head of GSFC Bioscience Group, emphasized in interviews the contribution of the Johns Hopkins scientist. He also reviewed the flight of "robot photographer named Ranger 8," concluding "The Russians tend to be more practical about small but crucial obstacles than Americans do. Although they are years ahead in rocketry (having now demonstrated the ability to move a man out through the hatchway of a spaceship in flight and safely back again—a preliminary step to using the hatchway to link the two spaceships that will travel as one to the moon), they have not yet been so brash as to announce a date by which they will make a manned landing on the moon. Before we become still more acutely embarrassed by our lunar braggadocio, it would seem wise for Washington to abandon the virtually impossible 1970 deadline for putting an American on the moon." (*SR*, 4/3/65, 45–48)

• Sen. J. W. Fullbright (D–Ark.), speaking at Virginia Polytechnic Institute, criticized the U.S. "crash program aimed at landing on the moon by 1970 at a cost of $20-to-$30 billion." He said that ". . . the moon is only one of our aspirations, a distant one at that, and in the meantime we have children to educate and cities to rebuild." Fulbright cited education as the nation's paramount deficit and advocated orienting "our space program to our own needs instead of letting the Russians determine for us what we will do and how much we will spend." (UPI, *Boston Sun. Globe*, 4/4/65)

April 3: "Our military space program is a wall decoration," said James J. Hagerty, Jr., in an editorial in the *Journal of the Armed Forces*. He continued: "The technology is there, but we are not exploiting it. Our DOD civilian leadership is content to drift along with the idea that someday we'll get around to it if we need it. This attitude seems to be based on the theory often advanced by Secretary McNamara and echoed by [NASA Administrator] Mr. Webb in his Hill testimony, that there is 'little chance that the Russians can develop a surprise military [space] capability' . . . If there is any chance at all, we should be doing something more than we're doing." (Haggerty, *J/Armed Forces*, 4/3/65, 8)

- Walter Henry Barling, Sr., who built the Barling bomber in 1923 for Gen. Billy Mitchell, died at 75. Mr. Barling was one of aviation's first test pilots and his Barling bomber was the world's largest airplane at the time. (AP, *NYT*, 4/5/65, 31)

April 4: Gemini spacecraft, scheduled for a four-day manned flight this summer, was delivered to Cape Kennedy. It was flown by cargo plane from McDonnell Aircraft Corp., prime contractor for manufacture of the craft, where it had undergone simulated flights. Astronauts James A. McDivitt and Edward H. White II, who would pilot the Gemini 4, also had made simulated flights at McDonnell. (AP, *Wash. Post*, 4/5/65)

- Dr. Edmund Klein of Roswell Park Memorial Institute for Cancer Research and Dr. Samuel Fine, Northeastern Univ. professor, in a report prepared for the 149th national meeting of the American Chemical Society, disclosed that laser beams may cause damage to the eyes, brain, and other organs in a way that may not be immediately apparent. Klein recommended that researchers "err on the side of safety in precautionary measures."

    Lasers are devices for concentrating light into extremely powerful beams; researchers were exploring their usage in fields of communications, eye surgery, cancer treatment, and in chemical and other industrial applications. (AP, *Houston Post*, 4/5/65)

- Dr. Krister Stendahl, Harvard Divinity School, replying to the question of how the discovery of intelligent creatures on other planets would affect religions on earth, said: ". . . it would be a refreshing shock to our faith if there were something like intelligent life elsewhere in the Universe. It would force us to enlarge our image of God and find our more humble and proper place within his creation." (*Boston Sun. Globe*, 4/4/65)

April 5: One of TIROS IX's two cameras had stopped returning useful photographs, NASA announced, possibly because of malfunction of a diode. Second camera was taking about 250 pictures daily of the earth's cover. Project engineers at NASA Goddard Space Flight Center had begun a "turnabout" maneuver to prevent the meteorological satellite from overheating and to ensure continued solar power. Maneuver would not affect satellite's picture-taking ability.

Launched into polar orbit Jan. 22, 1965, TIROS IX had apogee of 1,605 mi. and perigee of 435 mi. The "cartwheel satellite," so called because it was moving through space like a rolling wheel with the cameras mounted opposite each other on the perimeter, had taken more than 32,000 pictures, 92% of them useful to weather forecasters. (NASA Release 65-120)

April 5: The White House announced scientists appointed by President Johnson to his Science Advisory Committee: Dr. Lewis Branscomb, chairman of the joint Institute for Laboratory Astrophysics of the National Bureau of Standards; Marvin L. Goldberger, professor at Princeton Univ.; Kenneth Pitzer, president of Rice Univ.; Dr. George Pake, professor at Washington Univ.; and Dr. Gordon McDonald, Univ. of California at Los Angeles' Institute of Geophysics and Planetary Physics. Also announced was the nomination of Frederick G. Donner, chief executive officer of General Motors Corp., for reappointment to ComSatCorp's board of directors. (*Wash. Post,* 4/5/65)

- NASA selected three aerospace firms to develop a concept and prepare preliminary designs for hypersonic ramjet research engine: Garrett Corp., General Electric Co., and Marquardt Corp. Total value of first phase of contract would be about $1.5 million. During 9-mo. parallel studies, opening phase of NASA's Hypersonic Ramjet Experiment Project, the companies would prepare engine development plans that would serve as technical proposals for the second phase of the program. The ramjet engine, because of its relative fuel economy at hypersonic speeds, was expected to be useful for hypersonic transport aircraft, boosters, and spacecraft flying within the atmosphere. Flight research with the engine mounted on the X-15 aircraft was planned. Hypersonic Ramjet Experiment Project would be under the technical direction of NASA Langley Research Center, with the assistance of NASA Ames, Lewis, and Flight Research Centers. (NASA Release 65-110)

- NASA Nike-Apache sounding rocket was successfully launched from USNS *Croatan* carrying an instrumented payload to provide data on the neutron intensity, solar x-ray flux, Lyman-alpha radiation, and ionosphere electron density at different latitudes. Experiment was conducted for the Univ. of New Hampshire. (NASA Rpt. SRL)

- NASA Administrator James E. Webb appeared before House Committee on Appropriations' Subcommittee on Independent Offices, in support of the $5.26 billion NASA appropriation requested by President Johnson for FY 1966. He said: "The budget submitted to the Congress by the President provides for activities that are essential to continuing the progress that we have made towards our goal of preeminence in space sciences, application satellites, manned space flight, and advanced research and technological development necessary for aircraft improvements and for future space activities. It does not provide for everything that we could do or would like to do. In fact, it has been necessary within the strict budget requirements imposed by the President that certain desirable project activities started in previous years be omitted from the 1966 budget. . . .

"Within the confines of this limited budget, the President has provided the funds necessary to preserve the opportunity that we still believe we have to accomplish a manned lunar landing and exploration within this decade. The margin for insurance that had been built into our original program plan has largely disappeared. However, we now estimate this may be possible if we can maintain our current successful development efforts and make the all-up systems testing procedure work on the very large Saturn V-Apollo combination to launch men toward the Moon on earlier flights than we had originally planned. There is, therefore, still an opportunity to accomplish this national space objective within the time specified. Our work to date gives us somewhat more confidence than we had a year ago that we can still achieve the objectives that were planned in 1961 in spite of a limit on resources that will not fund all the flights planned at that time. It is important, however, to keep in mind that in Gemini we are just now in a position to find out by flight experiments how men can live, work, remain efficient, and make important contributions in space for extended periods. . . ." (Testimony; *Ind. Off. Approp. Hearings* [Part 2], 846–96)

April 5: Announcement was made at NASA Manned Spacecraft Center that Astronauts Walter M. Schirra, Jr. (Cdr., USN) and Thomas P. Stafford (Maj., USAF) had been selected for the first Gemini docking and rendezvous mission, scheduled for launch "the first quarter of 1966." Virgil I. Grissom (Maj., USAF) and John Young (Cdr., USN) would be the backup crew. (Transcript)

- An equipment modification to permit opening of the hatch on Gemini 4 had been successfully tested, William Normyle reported in *Aviation Week & Space Technology*. Hoses connecting the spacesuits to the spacecraft's environmental control system were lengthened to permit the astronaut to stand and partially emerge through the hatch. NASA had not yet approved a spacecraft-depressurization and hatch-opening exercise for the two-man spaceflight. (Normyle, *Av. Wk.*, 4/5/65, 27)
- NASA had published 110-page illustrated report containing ten papers on diversified utilization of space-research knowledge delivered at NASA and Univ. of California-sponsored workshop held in Los Angeles, June 2, 1964. (NASA Release 65–109; NASA SP–5018)
- Danish satellite tracking station official reported what he believed to have been the explosion of a U.S. satellite launched by USAF Mar. 25. About ten brilliantly lighted objects crossing the sky were at first assumed to have been meteors. (*M&R*, 4/26/65, 11)
- Antoine Senni emerged from a cave 333 ft. below ground near Cannes, France. Senni had entered the cave Nov. 30, 1964, to test effects of isolation on human system. (Reuters, *Wash. Post*, 4/6/65)
- Gen. Bernard A. Schriever, AFSC Commander, spoke in a luncheon address on military technology at the World Affairs Council in Los Angeles: ". . . we can expect substantial improvements in materials with respect to their strength, stiffness, and ability to operate at high temperatures.

  "One such material, a composite formed from boron fibers in a plastic binder, has been demonstrated in the laboratory to have approximately five times the specific strength of today's aircraft alloys. . . . This . . . will give increased strength at greatly reduced weight. Another material is oxide dispersed nickel, which can make possible an increase of several hundred degrees in turbine operating temperatures, enough to double the thrust of today's jet engines, with no increase in weight. . . .

"In propulsion, these advances in materials and component technology can make available engines for vertical takeoff and landing aircraft with more than double our present thrust-to-weight ratios and transport engines with half of today's fuel consumption. The use of hydrogen would make feasible engines for long range hypersonic craft flying at 7,000 miles per hour—almost four times as fast as the most sophisticated supersonic transport now proposed. And the aircraft will be of smaller size to do the same job.

"New technologies in flight dynamics, such as laminar flow control, can materially increase the ranges of transport aircraft. If laboratory boron composite structures pan out, we could build aircraft that could carry twice the payload at the same weight and range of present models. With further understanding of variable geometry wings we can alleviate the difficulties of operating at a variety of combinations of speed and altitude." (Text)

April 5: "Within a decade . . . space could be as vital to defense as nuclear weapons are today," postulated an article in *U.S. News and World Report.* It continued: "The deep conviction of top U.S. Air Force leaders is that Russia is directing its main energies and resources not to the moon, but to mastery of space nearer earth. Some are convinced that Russia, far behind in the missile race, is now striving to leapfrog the U.S. and move ahead with manned satellite weapons." (*U.S. News,* 4/5/65)

April 5–7: The Second Space Congress of the Canaveral Council of Technical Societies was held in Cocoa Beach, Fla. Rep. Olin Teague (D–Tex.) reportedly said in a speech that the House Committee on Science and Astronautics supported a military man-in-space effort and "almost unanimously" favored restoring $30 million to the Apollo program. Rep. Teague revealed that the Committee had written to President Johnson to stress the need for a decision on the proposed USAF Manned Orbiting Laboratory program and to urge him to "take a careful look as soon as possible and make a decision" as to whether or not the Gemini spacecraft would be used in the MOL program. (*M&R,* 4/12/65, 16)

In answer to the question of what man could do in space to contribute to the military mission, Maj. Gen. Don R. Ostrander, Commander of USAF Office of Aerospace Research, said at the Space Congress: "I believe that the MOL will enable us to come up with some of the answers." (Text)

Dr. George E. Mueller, NASA Associate Administrator for Manned Space Flight, speaking before the Space Congress, said: ". . . extravehicular activity, as accomplished by the Soviets, and orbital changes, as accomplished by Gus Grissom and John Young . . . are essential to future progress in space exploration. Both are objectives of our Gemini Program and both are techniques that we must learn in order to carry out the Apollo Program. We have long assumed that both were objectives of the Soviet Program.

"Given these assumptions, the difference between the scheduling of these experiments in the Soviet program and ours is a detail of relatively minor importance. It has been our judgment that maneuvering and changing orbits are more important than extravehicular activity for the progress of our program. For this reason, we scheduled the conduct of such maneuvers for the first manned flight in the Gemini Program. We must assume that the Soviets had their good reason for scheduling extravehicular activity on an earlier flight in their program." (Text)

E. Z. Gray, also of NASA's Office of Manned Space Flight, discussed future programs. He stressed that one of the cardinal rules guiding the planning was that maximum use must be made of hardware either already developed or currently in development. (M&R, 4/12/65, 16)

April 6: ComSatCorp's 85-lb. EARLY BIRD I, the first commercial communications satellite, was successfully launched from Cape Kennedy with a three-stage Thrust-Augmented Delta (Tad) booster. An hour after launching, flight control center confirmed that the satellite had entered an elliptical transfer orbit with apogee, 22,677 mi. (36,510 km.); perigee, 908 mi. (1,463 km.); period, 11 hrs. 10 min., and was sending clear radio signals. NASA handled the launching under a contract with ComSatCorp.

About 40 hrs. after launching, a kick motor aboard EARLY BIRD I would be fired to adjust the path of the satellite to a synchronous circular orbit at 22,300 mi. altitude above the Atlantic. EARLY BIRD I would become the first link in ComSatCorp's proposed worldwide satellite communications system and would relay radio, television, teletype, and telephone messages between North America and Europe. (Clark, *NYT*, 4/7/65; AP, Balt. *Sun*, 4/7/65; ComSatCorp)

- Subcommittee Chairman Albert Thomas (D–Tex.) and the House Independent Offices Appropriations Subcommittee were highly critical of Astronaut Virgil Grissom's deviation from flight plan instructions during the GEMINI III flight and eating a sandwich instead of fasting. According to published reports, one Subcommittee member referred to a "$30 million corned beef sandwich," and another asked NASA Administrator James E. Webb how he could control a multi-million dollar budget if he could not control two astronauts. (*Av. Wk.*, 4/12/65, 25; Hines, Wash. *Eve. Star*, 4/15/65)

- Defense Secretary Robert S. McNamara confirmed that the U.S. had given Great Britain option to purchase F–111 aircraft and spare parts totaling more than $1 billion for its Royal Air Force. Delivery orders for the F–111 were expected to be placed after completion of the British defense review. (DOD Release 210–65)

April 7: ComSatCorp's EARLY BIRD I communications satellite successfully received, amplified, and returned a television signal to Andover, Me., ground station in an unscheduled communications test. ComSatCorp Vice President Siegfried H. Reiger said that "the picture quality of the test pattern was excellent." (Clark, *NYT*, 4/8/65; AP, Balt. *Sun*, 4/8/65)

- USAF announced that data from AEC's Snap–10A satellite indicated "an extremely high noise factor" when the ion engine was turned on, making it impossible to determine whether it was operating properly. Scientists said the engine, which on Apr. 2 had operated normally for an hour, would not be tested further until additional analyses were made. The difficulty had not interfered with the major experiment—operation of the Snap–10A nuclear reactor. (UPI, *NYT*, 4/8/65)

April 7: Four airmen emerged with high voices and a hunger for meat after five weeks of confinement in a simulated space cabin at the USAF School of Aerospace Medicine. Scientists were studying a helium-oxygen atmosphere for possible future space cabin work because it did not produce decompression sickness in astronauts and was less hazardous in terms of spacecraft fires. (*Chic. Trib.*, 4/8/65; *M&R*, 4/12/65, 10)

- NASA Marshall Space Flight Center awarded IBM a 5-yr. $175,125,000 contract for integration and checkout of instrument units for Saturn IB and Saturn V programs. Initially announced in 1964, the contract would give IBM the additional responsibility for structural and environmental control systems and integration of all systems. (MSFC Release 65-79)

- NASA Administrator James E. Webb was asked by Rep. Charles R. Jonas (R-N.C.) in NASA appropriations hearing of the Subcommittee on Independent Offices, House Committee on Appropriations, to "set to rest" the rumor that NASA was planning to phase out MSFC in Huntsville, Ala. Mr. Webb explained that during his recent visit to Alabama leading Alabama businessmen had asked "questions about the future and whether the budget was going to be larger, and whether more would come to Alabama. Perhaps injudiciously, I said, 'Unless we can recruit better and more able people for the new phase of our program, you are not going to keep what you have.' . . .

  "We have a real problem in recruiting the kind of people needed to manage these contracts with American industry to go and live in Alabama, and the image of the State has been one of the problems that we have had. I pointed this out to the businessmen, and pointed out to them also that not only the problem of our recruitment was involved, that the State itself, in my opinion, was missing a valuable opportunity to use these kinds of people to build up its own economy, because the very existence of them there in the various areas could be of great benefit to the State. . . ." (*Ind. Off. Approp. Hearings* [Part 2], 1264-65)

- Soviet cosmonaut commander Air Force Lt. Gen. Nikolai Kaminin denied foreign newspaper reports that some of his men had died in unannounced space shots. Kaminin, writing in *Krasnaya Zvezda*, said: "The names of people who have allegedly died listed in foreign papers are mostly names of nonexistent cosmonauts." He said the aim of the reports "is to weaken the tremendous impressions made by the achievements of Soviet science and technology in space." (AP, *Huntsville Times*, 4/7/65)

- New York World's Fair opened for its second season. It featured NASA-DOD U.S. Space Park, containing two and one half acres of full-scale rockets and spacecraft. Among the exhibits were a full-scale Gemini model, an X-15 model, full-scale reproductions of Tiros, Nimbus, Relay, Telstar, and Syncom satellites, and AURORA 7 Mercury spacecraft.

  An honorary astronaut card signed by Astronaut Alan B. Shepard, Jr. (Cdr., USN), the first American in space, and Astronaut Virgil I. Grissom (Maj., USAF), the first astronaut to make two trips into space, was available at the U.S. Space Park to young visitors taking a ride in the full-scale animated Mercury spacecraft on display there. (Press Release)

April 7: Dr. Franklin P. Dixon, NASA Director of Manned Lunar and Planetary Mission Studies, told Twin Cities AIAA Chapter in Minneapolis that NASA was "investigating and planning manned missions and experiments beyond the presently approved Gemini and Apollo programs. . . .

"A logical sequence for future NASA manned space flight programs . . . begins with the Gemini and Apollo program base. The next logical development is the Apollo Extension System (AES) which is a stepping stone to advanced Earth-orbital operations, to lunar-orbital surveys, and to lunar surface exploration. The AES Earth-orbital activities are a development phase for an orbital research laboratory or early space station as well as a lunar exploration station. Based on the Apollo Command and Service module technology, we can also develop advanced logistic systems for larger orbiting space stations of indefinite life or for greater expansion of lunar exploration if desired. The advanced orbiting space station can likewise lead to an orbiting launch complex for planetary missions such as Martian flyby and exploration shelters or a lunar base for potential exploitation of the lunar environment. . . . In Earth orbit, the AES can provide for experimental operations in the three major fundamental areas . . . : (1) flights to conduct scientific research in space requiring man's presence; (2) Earth-oriented applications to increase the nation's strength, and (3) development of advanced technology for support of both manned and unmanned space operations. . . . In the field of Earth-oriented applications of manned space operations, NASA has been conducting studies and investigations jointly with the Departments of Commerce, Agriculture, Interior and Defense to determine how we might apply Apollo's unique capabilities to improve our ability to forecast weather, to communicate globally at high data rates, to make an up-to-date inventory of the world's resources, to monitor air and sea traffic on a global scale, to support a world-wide air-sea rescue service, to make better forecasts of food production and to provide a data-gathering system on a global scale. Experiments are also being evaluated to enhance over-all development of space operations. Biomedical, behavioral and other medical studies would be conducted as well as the development of advanced subsystems and technology for spacecraft. . . ." (Text)

- National Science Foundation reported that three New Mexico State Univ. engineers were studying satellites' radio signals in an attempt to determine exact shape of the earth. Under an NSF grant, the engineers had set up and were manning a special tracking unit at U.S. McMurdo Station in Antarctica and were tuned in on three spacecraft in polar orbit that passed near McMurdo 42 times daily. Stanford Univ. scientists had established a unit at Byrd Station to receive information from NASA's Pogo, to be launched later this year. (UPI, *NYT*, 4/11/65, 2)

April 8: MARINER IV, 49,373,799 mi. from earth and traveling 34,738 mph relative to earth, had covered 206,868,340 mi. in its journey toward Mars at 9:00 a.m. EST. (NASA Release 65–111)

- NASA Goddard Space Flight Center awarded RCA a $4.6-million contract to provide a real-time deep space tracking and data acquisition system for support of Project Apollo missions. Contract called for installation, checkout, and documentation of RCA's long-range (32,000 mi.) FPQ radar on land made available near a NASA site on Cooper's Island, Bermuda, through a land-lease agreement with DOD. The "Q–6" radar would have a flexible capability to support NASA programs other than manned flight. (GSFC Release G–9–65; GSFC Release G–10–65)

April 8: Army Corps of Engineers awarded Fisher Construction Co. a NASA-funded $1,497,728 fixed-price contract for construction of Lunar Mission and Space Exploration Facility at Manned Spacecraft Center. (DOD Release 220–65)

• NASA Administrator James E. Webb said at the U.S. Naval Academy: ". . . we are on the verge of another major breakthrough—the capability to forecast weather at least five days in advance with better accuracy than we can now predict 24 to 36 hours ahead. Atmospheric systems such as weather balloons and ground and seabased instruments which are already developed, together with satellite systems and high speed computers, should make it practicable in the next few years to establish a global observation system. As distinguished from the satellites whose main mission is cloud cover photographs, the more advanced future system will be able to map the structure of the earth's atmosphere in terms of wind, temperature, and pressure at various altitudes."

He continued: "We foresee the possibility of carrying sensors in satellites that will give us the thermal patterns of the ocean's surface which, when compared with the atmospheric conditions in any area, may give us the ability to predict the formation of fog. Similarly, ocean currents can be mapped and studied to advance the science of oceanography. We can even measure sea state—roughness of the sea —from a satellite." (Text)

• Dr. Eugene Shoemaker, head of the astrogeological branch of the U.S. Geological Survey, said in an interview with the *Houston Post* while at Rice Univ. as a speaker in the President's Lecture Series that the Ranger program had cost a total of about $200 million. He estimated that each Ranger shot had cost just under $30 million and said that although four of the seven Ranger missions had failed, it would have been foolish to settle for one success: "Just imagine that the Martians sent a Ranger-like camera to take pictures of the earth. With just one shot, they'd end up with pictures of a space no bigger than the size of an urban lot, or of the peak of the Alps, or of the sand dunes in Arabia. Could they tell anything about the earth from pictures of just one of these?" The Ranger program, just concluded with the success of RANGER IX, gave U.S. scientists good pictures of three different areas of the moon, Shoemaker said. "A Ranger picture is worth a million computer words." (Perez, *Houston Post,* 4/8/65)

• Panel on Science and Technology of the House Committee on Science and Technology reported on its sixth meeting (aeronautics), Jan. 26–27. Report was a comprehensive summary of views by the Committee and Panel members and the more than 150 scientists and engineers attending as representatives of Government, industry, and the scientific and academic communities. In its general conclusion, report stated three objectives for future improvement of U.S. civil aeronautics: "Insure that our economy continued to have the best air transportation system to give it a continuing advantage in world competition"; "Insure that U.S. aeronautical development is immediately responsive to the demand, and sufficiently great to continue leadership in the domestic and world markets"; and "Maintain recognized world leadership in technical matters to insure a favorable image and stature of the U.S. technological competence in aeronautical development."

Some of its general observations on the future of aeronautics:

"There is a need for more centralized direction, control, and procedure . . . [of the] widely dispersed . . . technical competence and expertise behind aeronautical development in the United States. . . .

"The aircraft industry in general is willing to contribute to any program designed to further aviation advancement, but the degree of their contribution will depend upon the extent of Government support, and the availability of a market. The extent is also dictated by the extent of their earnings on marketable products for which the Government is usually the principal customer.

"There are indications that an insufficient amount of research effort is being put forth in the hypersonic regime of the flight spectrum, particularly in the field of propulsion.

"The aeronautical research and development capability of NASA is not being used to its maximum capacity." (House Rpt. 227, 32–34)

April 8: In address on "The Early History of the Space Age" at the Univ. of Wisconsin, Eugene M. Emme, the NASA Historian, said: "The Space Age clocks on. Never before have basic alterations in fundamental knowledge, in practical engineering, and for an universal perspective been thrust so quickly upon mankind. . . .

"Few serious thoughts, whether associated with the physical or social sciences, or humanities, can ignore some aspect of the space venture. Like it or not, man's time for space mobility is here." (Text)

- The Flight Safety Foundation, under FAA contract, conducted day and night tests in the purposely-wrecked Constellation aircraft at Deer Valley, Ariz., to obtain data on emergency evacuation of passengers in survivable accidents. "Passengers" were local volunteers; airline stewardesses were provided by several air carriers. Evacuation duplicated obstacles passengers would face in real situations. Passenger reactions were recorded with remotely-controlled motion picture cameras; certain phases of the operation were timed with precision clocks. Test results would aid in planning advanced studies which would explore seat spacing, aisle widths, and other related factors. (FAA Release 65-27)

April 9: ComSatCorp's EARLY BIRD I communications satellite, launched April 6 by NASA, was placed into a "near letter perfect" synchronous orbit, with apogee, 36,637.1 km. (22,765 mi.); perigee, 35,041.9 km. (21,774 mi.) The five-day-early maneuver was accomplished by firing small retrorocket onboard satellite 19.7 sec. The satellite would be allowed to drift about 5°—over 300 mi.—to the exact point over the Atlantic where it would remain for its expected three to five year lifetime. (*NYT*, 4/10/65)

- USAF launched Blue Scout Jr. space probe from Eastern Test Range with instrumented payload to measure space environment effects on biological samples. The probe reached altitude of about 18,000 mi., reentered over the South Atlantic Ocean. Telemetry was received for only 15 min. (*U.S. Aeron. & Space Act., 1965*, 140)

April 9: Dr. George E. Mueller, NASA Associate Administrator for Manned Space Flight, announced change of primary control of manned flight missions from Cape Kennedy to Manned Spacecraft Center Mission Control Center. Christopher Kraft, mission flight director for GT-3 flight, completed Mar. 28, would serve as mission director for GT-4 flight scheduled for later this year. MSC Mission Control Center would provide centralized control of manned spaceflight programs from launch through recovery; computer-driven time and data displays would report instantly the status of astronauts, spacecraft, and supporting operations to mission/flight director. Most information would travel over land lines. (Transcript; NASA Release 65-119)

- NASA awarded MIT separate cost reimbursement contract, with no fee, to cover further work on guidance and navigation of Apollo command and lunar excursion modules. The new contract, running from March 1 through November 4, 1965, totaled $15,529,000, including $1.4 million to support research activities in the guidance and navigation field. (NASA Release 65-116)

- NASA was negotiating with Grumman Aircraft Engineering Corp., prime contractor to NASA Goddard Space Flight Center for Oao program, to convert prototype Oao into flight-ready spacecraft. The contract was expected to exceed $8 million. The converted prototype, to be designated Oao A-2, would be the third spacecraft scheduled for launch in Oao program. First planned launch in the series was scheduled for late this year or early next year at Cape Kennedy. (NASA Release 65-115)

- The *Christian Science Monitor* asked Dr. Homer E. Newell, NASA Associate Administrator for Space Science and Applications, and Dr. Philip H. Abelson, Director of the Geophysical Laboratory of the Carnegie Institution of Washington and editor of *Science* magazine, to present elements of the debate on the question "Man in space: is it worth $40 billion?"

    Dr. Newell presented the case for manned space flight: "The manned space flight effort serves to round out the total program. Its primary aim is to develop a broad space capability that will secure to this nation strength, security, flexibility, and freedom of choice in space. Landing men on the moon and returning them to earth has been chosen as the means to this broader, more substantive end, and it is not to be considered as the only justification for our manned space effort."

    Dr. Abelson, speaking for the critics, said: "The unmanned program has been a substantial contributor to our international prestige. Moreover, prestige based on science and technology tends to be enduring. . . .

    "Our Apollo program was launched for reasons of international prestige. The yield has not been very good or very lasting. How many citizens can now recall the names of the astronauts and of their capsules? We can expect much the same reaction when we finally accomplish a moon landing." (*CSM*, 4/9/65)

- NASA announced publication of a summary of research results of the joint NASA–USAF–USN, 10-yr. X-15 flight program. (NASA Release 65-114; NASA SP-60, *X-15 Research Results*)

April 9: After two days of discussion with West Germany's Minister of Defense Kai Uwe von Hassel, Britain's Minister of Defence Denis Healey told a news conference in Bonn, Germany, that the two countries had agreed to develop by the 1970's a light combat Vtol fighter and possibly a heavy aircraft to succeed the F-104 Starfighter. Healey added that studies were being conducted on other weapon projects, including tanks and tank equipment. (*NYT*, 4/10/65, 46)

- Editorial in *Life* put into perspective the "break-throughs" and spectacular "firsts" recently achieved in space exploration—U.S.S.R.'s VOSKHOD II, U.S.'s GEMINI III, RANGER IX, ComSatCorp's EARLY BIRD. "The first Sputnik was less than eight years ago, but already the space age has reached what President Johnson calls an 'early maturity.' Each technical advance is a planned and measured consequence of the previous one; Mercury fed Gemini and Gemini feeds Apollo; each hero stands on the shoulders of predecessors who are also his contemporaries. . . .

  "Our space program is, as Johnson puts it, 'a national asset of proven worth and incalculable potential.' Its cost is leveling off at about $7 billion a year. One hopes this includes enough to land us on the moon before the Russians—and what's wrong with wanting to be first? . . .

  "Our program, which may or may not be overtaking the Russian, is well past its own first period of jumpy desperation. We can stick to it in confidence." (*Life*, 4/9/65)

- U.S.S.R. was building a spaceship designed not for space flight, but for exhibition in a new space museum to be built at the site of Moscow's Space Monument. Inside the model cabin, which would have a seating capacity of 100, a movie showing the earth as it appeared from space would be shown to visitors. (AP, San Diego *Eve. Trib.*, 4/9/65, 22)

- Communist China's failure to conduct a scheduled second nuclear test in March was reported by an unidentified U.S. researcher in an interview with AP. He said reasons for the delay might be technical or political. (AP, *NYT*, 4/11/65, 94)

April 10: One of the five F-1 engines on the Saturn V booster was successfully static fired at NASA Marshall Space Flight Center for $16\frac{3}{4}$ sec. (*Marshall Star*, 4/14/65, 1)

- In a speech to the Interact Conference of First Rotary District 696 in Orlando, Fla., KSC's Richard E. Dutton, said: ". . . NASA's major launch facility for space vehicles and unmanned and manned spacecraft [is] the John F. Kennedy Space Center and its new Merritt Island Spaceport. I hope you noticed that I used the term Spaceport, instead of Moonport, as it is often referred to in the news media. We call it a Spaceport because its basic concept is not to exist as a research and development facility for any *one* mission only; it is being created to function as an actual *port*, with a space vehicle launch rate that may be some day as high as one manned launch per month.

  "However, just as important to consider is the spaceport's capacity for growth. It can accommodate launch vehicles with up to 40 million pounds of thrust, 32.5 million pounds more than the Saturn V here can deliver. Because of this, the United States has not invested three quarters of a billion dollars in a facility which will serve only to launch a manned lunar mission. It has acquired a permanent installation which will serve the requirements of the National Space Program for years to come.

"But these facilities, like the lunar landing mission, are themselves only a manifestation of a greater entity—people. At present, 2,500 NASA and 6,300 contractor employees work at the Center. By 1967, when the spaceport becomes operational, 3,000 government employees and 10,000 contractor employees will be employed." (Text)

April 10: First General Dynamics F-111A developmental aircraft, in its 13th flight, reached 40,000 ft., its highest altitude so far, USAF announced. (*Av. Wk.,* 4/19/65, 27)

- 17-yr.-old John J. Breaux, who exhibited a "soundovac" that could "solve any mathematical problem when a formula was available," and 17-yr.-old Douglas A. Whithaus, who based his exhibit on development of a liquid-gaseous-propellant rocket engine, were entrants in the Greater St. Louis Science Fair selected to compete in the National Science Fair, May 6–8. (St. Louis *Post Dispatch,* 4/10/65).
- Fred Callahan, 16, of Ft. Benning, Ga., prepared to launch Zeus 2, possibly the largest rocket built by an amateur. Zeus 2, nine ft. long with 2,000-lb. thrust, could reach peak altitude of 64 mi. Zeus 1 was launched by Callahan three years ago. (*Wash. Daily News,* 4/10/65)

April 11: NASA Marshall Space Flight Center had awarded a ten-month, $10,934,377, cost-plus-award fee contract to Mason-Rust Co. to continue support services at Michoud Operations, New Orleans, and at its Computer Operations Office in Slidell, La. (MSFC Release 65–84)

- The case of Thiokol's 260-in.-dia. solid motor ruptured during initial hydrotest of the Newport News Shipbuilding & Drydock Co., builders of the case. Cause of the failure had not been determined. (*M&R,* 4/19/65, 14; *Av. Wk.,* 4/19/65, 29)
- Commenting that contributions to science made by the space probes and satellites had been "interesting, all of it useful, none of it genuinely, eye-poppingly unexpected," an editorial in the *San Francisco Sunday Chronicle* continued: "Surprisingly enough, space research has produced several by-products with a practical end.

"The most significant to the world as a whole are the reconnaissance satellites with which Russia and the U.S. are now mutually inspecting each other's and everyone else's military installations with the kind of accuracy that has given Washington excellent pictures of the tower on top of which the Chinese atom bomb was exploded. They can prevent any significant military move from going undetected; a by-product of them are the weather satellites.

"Less is heard about the progress of early warning satellites designed to pick up the flaming tails of enemy missiles; this could be either because they have run into trouble or, like the satellites the Polaris submarines steer by, they are too successful to be mentioned. The possibility of putting H-bombs into satellites is not mentioned either in these days, but this time because the Russians and the Americans seem to have decided by mutual consent to forget it: the risks of an unmanned satellite going wrong were too great, and the risk of a manned one going berserk was even greater." (*S. F. Sun. Chron.,* 4/11/65)

April 12: Aerobee 150 sounding rocket launched from White Sands, N. Mex., carried instrumented payload to 125 mi. (200 km.) altitude. Payload was a spectroheliograph to obtain a monochromatic picture of the sun. Experiment was conducted by NASA Goddard Space Flight Center. (NASA Rpt. SRL)

April 12: ComSatCorp announced that clear test signals transmitted via EARLY BIRD between Andover, Me., and stations in Goonhilly Downs, England; Pleumeur Bodou, France; and Raisting, W. Germany, had demonstrated that communications satellite's equipment to receive messages from the European stations was functioning properly, as was its receiver tuned to the Andover station. (AP, *Chic. Trib.*, 4/13/65)

- AEC granted a full-term, ten-year operating license to NASA's Plum Brook Reactor Facility, NASA announced. The Plum Brook reactor, which produced 60,000 kw. of thermal power at peak operation, was being used in basic research relating to development of a nuclear rocket and of systems and components for space nuclear auxiliary power. The Facility is part of NASA Lewis Research Center. (LRC Release 65–27)

- USAF had awarded to General Dynamics a fixed-price-incentive-fee contract covering initial procurement of 431 F–111 aircraft, DOD announced. The contract was expected to exceed $1.5 billion. (DOD Release 228–65)

- Tass announced: "Scientists of the Sternberg Astronomical Institute believe they have received perhaps the first evidence that we are not alone in the universe." The report referred to a strange pattern in signals emanating from a radio source believed being beamed at earth from another civilization.

  During the past year, the Soviet announcement continued, Soviet scientific listeners have noted that the signals come and go like the radio equivalent of a revolving beacon. Every hundred days the signals get strong and then fade out again.

  The Tass announcement quoted Dr. Nikolai Kardashev as saying: "A super civilization has been discovered."

  Dr. Kardashev had first announced a year ago that he thought the radio signals from a source known as CTA–102 came from intelligent beings. Tass indicated that radio astronomers at Britain's Jodrell Bank station had also observed CTA–102. (Loory, *N.Y. Her. Trib.*, 4/13/65; Simons, *Wash. Post*, 4/13/65)

- A spokesman for Britain's Jodrell Bank Radio Telescope Observatory said concerning the Tass report that radio signals from CTA–102 might come from intelligent beings in outer space: "We have made measurements on these sources and confirmed that they are very weak and very small. But there is no observational evidence at Jodrell Bank to show any variation in the signal strength received. We would have to scrutinize carefully the Russian evidence before making any further statements." (AP, *Balt. Sun*, 4/13/65)

- Fourth flight of General Dynamics' second USAF F–111A developmental aircraft lasted 1 hr. 40 min. Speeds ranged from 138 to 354 kt., with wings swept at 16°, 26°, and 70°. Landing gear and flaps were worked up and down during the flight.

  Fifth flight of the aircraft lasted 2 hr. 10 min. and attained a speed of mach 0.8 and an altitude of 27,000 ft. Wings were swept at 16°, 26°, and 70°. (*Av. Wk.*, 4/19/65, 27)

- In a *Missiles and Rockets* editorial, William J. Coughlin questioned NASA's wisdom in drawing up mission requirements for lunar exploration. The article said: "Dr. Homer E. Newell, NASA associate administrator, told Congress last year: 'Ranger will play an important role in the support of Project Apollo.' . . .

"Not a single change has been made in any part of the Apollo system or in the program's operational plan as a result of the Ranger findings. None is contemplated. The reason for this is simple. The Block III Rangers were incapable of producing any such data. . . .

"The case for Surveyor and Lunar Orbiter as supports for the Apollo program . . . is not a very strong one. . . .

"Dr. Newell sees them as part of what he calls the 'total program for exploring the Moon.' . . . [He said] in the following statement to Congress: 'You will have a lunar landing. That lunar landing will involve a few hours of stay on the Moon, a look that the astronauts can make, a few collections of samples, maybe some simple tests, and maybe the implacement by the astronauts of monitors to be left on the lunar surface.'

"After their departure, Dr. Newell sees the instruments carrying on. Lunar Orbiter wheels overhead. Surveyor explores areas on the moon which Man would have difficulty in reaching. . . .

"We suggest that if anyone proposed exploring the Antarctic in such a manner, he would be clapped in the pokey as a nut. Man is going to the Moon and he is going to explore it. Expenditure of billions of dollars on instruments remotely controlled from Earth to do the same job is folly." (*M&R*, 4/12/65, 46)

April 12: Chickens exposed to one half to three times the earth's gravity had contracted chronic acceleration sickness in tests conducted at the Univ. of California. Dr. Russell R. Burton, a veterinarian at the University conducting the experiments as part of a program supported by NASA and the Office of Naval Research, said there was great variation among the chickens in susceptibility to the sickness: "Some chickens will show symptoms after a few days at 1.5g, but others not until many months at 3g, and, of course, some never exhibit any of the symptoms. However, once the sickness develops, symptoms are the same." Sick fowl developed enlarged adrenal glands and their digestive functions became abnormal. Some chickens' legs were paralyzed as a result of increased gravity forces.

Objective of the tests was to determine effects of artificially altering body weight. Interest in increased gravity fields stemmed from greater fields present on other planets such as Jupiter, which has gravity 2½ times that on earth. (*Av. Wk.*, 4/12/65, 79)

• Robert Hotz, editorializing in *Aviation Week and Space Technology*, said that it could be a "dangerous mistake" to defer development of earth-orbital operational capabilities until financial and technical peak loads of Apollo had been passed: "The Soviets obviously have chosen the earth-orbital approach to their lunar landing mission. Therefore, they necessarily must develop rather fully their hardware and operational techniques in this area as a vital prelude to their lunar landing attempts and not as a postlude, in the manner of current U.S. planning. They also have made little attempt to conceal their primary military interest in the development of manned spacecraft operations in the earth-orbital area.

"Thus, it is entirely possible that unless U.S. policy is drastically changed soon, the Soviets may have an opportunity to achieve the technical surprise in space that they so narrowly missed in the race to an intercontinental ballistic missile." (*Av. Wk.*, 4/12/65, 21)

April 12: *Pravda* announced the birth of Russia's third "space baby": a son to Cosmonaut Valery F. Bykovsky and his wife Valentina. (UPI, *Wash. Daily News*, 4/13/65)

Week of April 12: European Space Research Organization (ESRO) selected Laboratoire Central de Telecommunications (LCT), a wholly-owned French subsidiary of International Telephone and Telegraph Corp., as prime contractor for development of Esro 1 polar ionosphere satellite. The $3-million contract awarded called for development and production of one prototype and two flying satellites—one a backup—to gather information on ionospheric and particle conditions in the northern polar region. (*Av. Wk.*, 4/12/65, 37; *Av. Wk.*, 4/19/65, 30)

April 13: Establishment of a Joint Meteorological Satellite Program Office (JMSPO), to identify, compile, and coordinate requirements from the military services and the Joint Chiefs of Staff for use of meteorological satellites, was announced by DOD. JMSPO would continually review the NASA meteorological satellite program and would define military applications of the national system and the DOD technical efforts to support the national program. (DOD Release 229-65)

- At a news conference, astronomers at Moscow's Sternberg Institute of Astronomy repudiated the Tass report that radio signals had been received from a "super civilization" in outer space. The astronomers explained that their studies had been based on a radio signal from a point in space called CTA-102—a designation of the California Institute of Technology for a quasi-stellar radio source. Signals had been picked up from CTA-102 systematically in fluctuating strength that followed a regular 100-day pattern. They said that although no other radio emission from outer space had the same periodicity, it was too early to tell whether the radio signals were artificially made by intelligent beings or whether they came from a natural source.

  The Soviet astronomers appealed to their Western counterparts to help study CTA-102 to determine whether the signals were artificially or naturally made. (AP, Balt. *Sun*, 4/14/65; Post News Service, *Houston Post*, 4/14/65)

- NASA had awarded Douglas Aircraft Co. $2,697,546 contract modification to test Saturn V instrument unit and S-IVB stage instrumentation in a space environment. The test program would be conducted in Douglas' 39-ft.-dia. space simulator at Huntington Beach, Calif., and would simulate a typical Saturn V flight from launch to earth orbit and injection into lunar path. Tests would begin in early 1966. (MSFC Release 65-88)

- Reported that Dr. William I. Donn of Columbia Univ.'s Lamont Geological Observatory, Dr. Wilbur G. Valentine of Brooklyn College, and Dr. Bertram D. Donn of NASA Goddard Space Flight Center had challenged presently accepted ages of the earth (4.5 billion yrs.) and the sun (5 billion yrs.). They had asserted that the oldest of continental rocks were so very ancient that the sun's and the earth's ages allowed too little time for continent formation by earthly processes and from earthly materials. Two alternative explanations were proposed: (1) either the sun and the earth must be much older, perhaps by a half-billion years or more; or, (2) the original continents were thrown down upon the planet's surface when objects from space—hundreds of miles across in size—crashed into the earth. Research results had been published in the *Bulletin of the Geological Society of America*. (Abraham, *Phil. Eve. Bull.*, 4/13/65)

April 13: Brig. Gen. Joseph S. Bleymaier (USAF), Deputy Commander for Manned Space Systems of AFSC's Space Systems Div., announced at a Washington, D.C., luncheon for Aviation/Space Writers that two used Gemini spacecraft would be flown by USAF in tests for a Manned Orbiting Laboratory (MOL). This would be the first time that a Mercury or Gemini spacecraft had been flown twice. Both Air Force flights would be unmanned and would test the effect of cutting a hatch into the heat shield on the capsule's blunt end. (*NYT*, 4/15/65, 8)

- Second General Dynamics-USAF F-111A developmental aircraft made its sixth flight, lasting 1 hr. 30 min. Wings were swept at 16°, 26°, and 70°. (*Av. Wk.*, 4/19/65, 27)

- Soviet astronomers were seeking increased research funds. At a meeting of the Presidium of the Academy of Sciences, physicist Lev A. Artsimovich reportedly assailed what he called the inadequacy of the observational equipment available to Soviet astronomers and noted that the U.S. had more large telescopes than did the U.S.S.R. He accused those charged with making appropriations of underestimating the importance of astronomy, while overestimating the importance of and being overly generous to nuclear physics: "At the present time, expenditures on astronomical work in our country are no more than a few percent of the investments in elementary particle physics. Our progeny will probably be surprised that we divided in such strange proportions the efforts directed to investigate the great world of stars and the artificial world of elementary interactions [of nuclear particles]." (*NYT*, 4/13/65)

- "Award of the [$40 million] contract [for 28 Atlas SLV-3s] reflects plans by the Air Force and the National Aeronautics and Space Administration to use the Atlas in a variety of future space missions," Robert Zimmerman said in an article in the *San Diego Union*. He continued: "Its versatility as a launching vehicle lies in the 'plug-in' concept which allows electronic instruments for various missions to be installed on the basic booster as requirements for the mission may dictate.

    "Before the Atlas was standardized into the SLV-3 it would take a year to 18 months to equip one booster for a particular mission. Now, an SLV-3 can be outfitted for any mission in three to four months." (Zimmerman, *San Diego Union*, 4/13/65)

- According to official sources, both the U.S. and U.S.S.R. had exploded certain of their own satellites in orbit to prevent their falling into other hands, but neither nation was known to have attempted to knock down a spacecraft belonging to the other. (Clark, *NYT*, 4/4/65, 1)

- Commenting on blockade to prevent Negroes from using North Merritt Island ocean beach—federally-owned property released for public use by NASA—Dr. Kurt H. Debus, KSC Director, said: "If difficulty should continue to arise in implementing a basic public policy of non-discrimination, the Kennedy Space Center would be obligated to withdraw the beach from public use." (*Miami Her.*, 4/13/65)

April 14: MARINER IV set a distance record for communications from American spacecraft. The Mars probe transmitted data from 54 million miles out, exceeding the record of 53.9 million miles set by MARINER II in 1963. (AP, San Diego *Eve. Trib.*, 4/14/65; NASA Releases 65-111, 65-117)

- ComSatCorp's EARLY BIRD I communications satellite reached its permanent station over the Atlantic Ocean: apogee, 22,243 mi. (35,811 km.); perigee, 22,224 mi. (35,780 km.); period, 23 hrs. 56 min. 57 sec.; inclination, .085°; location, 28.0° west longitude. (ComSatCorp)

- In a "topping out" ceremony, signifying that the Vehicle Assembly Building at NASA's Merritt Island Launch Area had reached its maximum height of 525 ft., a 38-ft., four-ton steel beam inscribed with emblems of the companies and Government agencies participating in the building's construction and autographed by contractor and Government personnel, was hoisted into place in the upper reaches of VAB's steel skeleton. Scheduled for completion in 1966 as an integral part of Launch Complex 39, VAB would have 7.5 acres of floor area, would be 525 ft. tall, 518 ft. wide, and 716 ft. long. Within the 129 million cu. ft. of the structure, Apollo-Saturn V launch vehicles would be assembled in an upright position in a controlled environment. (KSC Release 86-65)

- NASA launched from Wallops Island a four-stage Journeyman (Argo D-8) sounding rocket with 130-lb. Univ. of Minnesota payload. Firing was timed to correspond closely with passage of the OGO I satellite in an unsuccessful attempt to compare and correlate radiation belt electron and proton measurements. Sounding rocket reached peak altitude of 1,031 mi.; experiment package impacted in the Atlantic Ocean about 1,200 mi. downrange.

   Telemetry indicated proper functioning of instrumentation during the 26-min. flight, but no useful data were returned because the nose cone covering the payload failed to eject and the experiment package was not exposed to energetic particles in the radiation belt. (Wallops Release 65-21; NASA Rpt. SRL)

- First of four Stellar Acquisition Flight Feasibility (Staff) flights planned by USAF failed 73 sec. after launch of the experiment aboard a Polaris A-1 booster. The experiment's Stellar Inertial Guidance System (Stings) was operating open-loop and was not guiding the missile, which had to be destroyed when it veered off course. Stings had been locked onto the star Polaris and had tracked properly through the first 54 sec. of flight until time of second-stage ignition, when the trouble with the launch vehicle apparently developed. Period during which the Stings operated was time of highest dynamic pressure; data received were termed excellent.

   Main purpose of the Staff flight was to test a telescope-like device intended to allow a Stings to take a reading from Polaris after piercing the earth's cloud cover and to plot an exact trajectory to a target area. (*M&R*, 4/19/65, 9)

- NASA Administrator James E. Webb told the Harvard Business School Club of New York: "The impact of the space program cannot be described just by a recital of the flow of technology to industry. The NASA system of management, for example, has efficiently mobilized for research and development in aeronautics and space some 400,000 men and women and is utilizing some 20,000 industrial companies under prime and subcontract arrangements. We are handling about 250,000 procurement actions a year, and over 150 universities are involved in the scientific, engineering, and training programs required for the rapid solutions and high standards the program requires."

He continued: "It should be emphasized that our space program is not a crash effort. It is a planned, deliberate development over a ten-year period.

"Through our programs at NASA, we are proving out important new mechanisms through which investments made in science and technology can pay substantial dividends. The social, economic, and political forces at work in our society today are dependent, as never before, on developments in science and technology." (Text)

April 14: NASA Administrator James E. Webb said at the Boy Scout Launch-O-Ree in New York City that the "future will be determined in large measure by the kind of talented and dedicated youth found in the Boy Scouts. Science and technology, which form the basis for the national space program, are pioneering areas within which many of you can find opportunities for satisfaction and service." (Text)

- Dr. Frederick Seitz, President of the National Academy of Sciences, speaking at the end of Purdue Univ.'s three-day symposium on "Science and Public Policy—Evolving Institutions," warned that the present system of Federal grants might be "disastrous" to some areas of science if not modified. "The man with the big, obvious project tends to get his Federal grant today, but the lonely individual with an off-beat idea does not fare so well," he said. Dr. Seitz favored a large-scale, supplementary system of Federal grants for research in science and the humanities that would permit the individual university to determine how the grant would be disposed. "Block grants would enable a university administration to draw upon talents of its faculty and administrators in deciding how funds for a certain area of research are allocated," he argued. Dr. Seitz said that such a Federal grant-giving agency would be patterned after the National Science Foundation and might fulfill the role envisioned for the National Humanities Foundations proposed in bills currently before Congress. (Sullivan, *NYT*, 4/15/65, 30)

- Dr. Joseph F. Shea, Apollo Program manager at NASA Manned Spacecraft Center, announced at a press conference at North American Aviation's Tulsa facilities that the Tulsa plant would build 16 Apollo service modules. Apollo contract work there totaled more than $61 million. (Leslie, *Tulsa Daily World*, 4/15/65)

- "Positive action must soon replace delay and procrastination" on the development of an American supersonic airliner, Sen. A. S. (Mike) Monroney (D–Okla.), Chairman of the Senate Aviation Subcommittee, told a Washington, D.C., meeting of the Society of Automotive Engineers. Monroney said that U.S. failure to build the plane could "choke off" 375,000 jobs within several years. Sen. Monroney added that if U.S. carriers did not fly supersonic planes as early as foreign airlines, it could mean a loss of $1 billion a year in passenger revenues. "If we capitulate, it would mean the eventual loss of technical superiority and a second class airline industry," he said. (*NYT*, 4/15/65, 15)

April 14: Supreme Court Justice William O. Douglas told Philadelphia Rotary Club members that money being spent to put a man on the moon could be better spent ending water pollution in the United States. He claimed that costs for equipping the Nation with adequate sewage disposal was about equal to that of sending a man to the moon in the Apollo project. (AP, *Galveston News-Tribune*, 4/15/65)

- Arthur E. Jenks, retired FAA official, received the Laura Taber Barbour Air Safety Award for 1965 at a luncheon in Washington, D.C., given by the Society of Automotive Engineers in conjunction with its annual meeting. The award, sponsored by the Flight Safety Foundation, was presented to Jenks because of his "contributions to improving the techniques for flight checking the accuracy of air navigational aids and improvement of landing aids on and around airports." (FAA Release 65-30)

April 15: Lunar Excursion Module (Lem) ascent engine underwent a 5-sec. test firing under ground level conditions at White Sands Missile Range. Initial indications were that the test had been successful. The 3,500-lb.-thrust hypergolic engine was built by Bell Aerosystems and used a 50-50 mixture of Udmh and hydrazine for fuel and nitrogen tetroxide for the oxidizer. (MSC *Roundup*, 4/30/65, 1)

- Vice President Hubert Humphrey wrote to Cape Kennedy technician Richard Tennis: "I understand that you are the gentleman who corrected the problem of the oxidizer leak on the Gemini-Titan [GT-3].

  "I simply wanted to express to you the thanks of all of us here in Washington who have watched so carefully the success of this program. It is the excellent and quick efforts of people like yourself that have made this program so successful." (KSC *Spaceport News*, 4/15/65, 2)

- Federal Urban Renewal Administration would approve location of the NASA Electronics Research Center in the Kendall Sq. area of Cambridge by declaring the area eligible for an urban renewal project, the *Boston Globe* reported. According to an unidentified Federal official, an eligibility report prepared by the Cambridge Redevelopment Authority had been approved by the New York regional office and approval from Washington, D.C., was expected soon. (*Boston Globe*, 4/15/65)

- The Associated Press applied to FCC for recognition as "an authorized entity for the purpose of buying service from the Communications Satellite Corporation." AP was the first organization to take advantage formally of the clause in the Communications Satellite Act of 1962 that authorized ComSatCorp to furnish circuits "to the carriers and to other authorized entities, foreign and domestic." The law, however, did not define an authorized entity, also known as "authorized user" (in contrast to an "authorized carrier"). (Gould, *NYT*, 4/27/65, 1, 25)

- A. J. Hayes, president of the International Association of Machinists, said at a Dallas briefing of industry sponsored by DOD and the National Security Industrial Association that Federal procurement officers were meddling in negotiations of labor and the aerospace industry to the extent that free collective bargaining was being eroded away. He said the affected unions would not settle this year for less than the 57-cent package in wage increases and fringe benefits recently worked out for the United Auto Workers. (AP, *Denver Post*, 4/15/65)

April 15: Battelle Memorial Institute reported reasons the sweet potato would be the best vegetable for a space garden: (1) it would yield a large number of calories per pound and would have a high count of vitamin A; (2) its leaves are edible, either cooked or raw; (3) under simulated space conditions, it would grow in 90 to 120 days; (4) it would give off oxygen and absorb carbon dioxide, aiding air conditions inside a spacecraft. The plan, Battelle said, would be to grow the sweet potato in a spacecraft in a soilless culture to provide fresh vegetables for astronauts. (AP, *Wash. Post*, 4/16/65)

• A Cairo newspaper revealed that the United Arab Republic was training men for space flight. No date for a possible launching was given. (UPI, *Milwaukee J.*, 4/16/65)

April 15–16: World scientists met in a special conference on the lunar surface sponsored by the International Astronomical Union and NASA Goddard Space Flight Center at Greenbelt, Md.

Noting areas of disagreement among scientists, theoretical astrophysicist Thomas Gold of Cornell Univ. tried to explain why the Ranger pictures resolved so little: "The Ranger pictures are like a mirror. Everyone sees his own theories reflected in them." Prof. Gold saw a moon covered with dust; young craters composed of solid rock while older craters had somehow gone soft; and vast sheets of ice locked under compacted sediment beneath much of the lunar surface.

Dr. Harold C. Urey, Nobel prize-winning chemist from the Univ. of California, referred to evidence of widespread collapse of the lunar surface, probably due to underground movement: "The RANGER IX's pictures scared me more than anything. There's all sorts of evidence that some of these craters are sinking."

Dr. Eugene Shoemaker of the U.S. Geological Survey said that chances that the moon's surface was too soft for the 15-ton Lem were "almost vanishingly remote." (Simons, *Wash. Post*, 4/16/65; Clark, *NYT*, 4/16/65)

Dr. Ewan A. Whitaker agreed with findings in the paper he presented for his colleague, Dr. Gerald P. Kuiper of the Univ. of Arizona's Lunar and Planetary Laboratory. Dr. Kuiper concluded that the lunar surface had a bearing strength of between one and two tons per square foot. His calculations, made from data extracted from RANGER IX photographs, was based on the size of rocks ejected from a given impact crater and the distance they traveled. Other tentative findings were that the dark portions of the maria were due to some unknown fluid flows and not lava or ash flows; that the maria were not completely covered with lunar dust; and that the moon's surface exhibited a remarkable series of fracture patterns which could be due to polar contraction, tidal effects, or some other force. (Clark, *NYT*, 4/16/65; Simons, *Wash. Post*, 4/16/65; *Av. Wk.*, 4/26/65, 34)

Boris J. Levin, section chief of the Institute of Earth Physics, U.S.S.R. Academy of Sciences, said studies based on radioactive emissions from meteorites and on lunar data indicated that the interior of the moon partially melted some two million years after the formation of that body began: "If you assume the moon is of the same material as meteorites, it is necessary to assume that the interior at one time was partly molten." Prof. Levin said the moon was formed simultaneously with the earth and was not originally part of it. It was about 10 earth-radii distant and later shifted to the present position. He added: "We believe that there is a lava flow not covered by dust." (*Wash. Post*, 4/17/65; *Milwaukee J.*, 4/17/65; *CSM*, 4/26/65; *Av. Wk.*, 4/26/65, 34)

Dr. John Clark, NASA Director of Space Sciences, said that a year ago NASA officials had hoped that Ranger would tell something about the topography of the moon: "That in turn would tell something about the geometry needed for the landing vehicle. Ranger has done this and now we look to the Surveyor spacecraft to tell us the bearing strength of the moon's surface." (AP, *Houston Post*, 4/17/65; Clark, *NYT*, 4/17/65)

Dr. Fred Whipple of the Smithsonian Astrophysical Observatory said the moon's surface might be lower than had been calculated: "The data indicates that RANGER VII and VIII, and maybe RANGER IX, landed one second late because the moon was one mile small. The moon's surface at the point of landing was lower by two kilometers (a mile and a quarter) than the average lunar radius." (AP, *Houston Post*, 4/17/65; *Milwaukee J.*, 4/17/65)

April 16: Saturn V launch vehicle (S–IC stage) was static-fired for the first time, at NASA Marshall Space Flight Center. The five F–1 engines were ignited in a test which lasted 6½ sec. during which they generated a thrust of 7.5 million lbs. (160,000,000 hp.). This was the first full cluster test and was made on a recently completed 400-ft.-tall test stand. The S–IC was the first stage of 364-ft.-tall Saturn V-Apollo combination that would ultimately take astronauts and equipment to the moon. Associate Administrator for Manned Space Flight Dr. George E. Mueller congratulated MSFC personnel on the successful test: ". . . As this was one of the key milestones in the whole lunar landing program, its successful performance, 12 weeks ahead of schedule, has a great bearing on our program." (MFSC Release 65–92; *Marshall Star*, 5/5/65, 5)

* NASA had signed a $9.6-million contract with Ball Brothers Research Corp. to build, integrate, and test two Orbiting Solar Observatories. The spacecraft, designated Oso–D and Oso–E, would contain experiments designed to advance understanding of the sun's structure and behavior and the physical processes by which the sun influenced the near-earth environment and interplanetary space. The amount included $800,000 obligated by letter contract signed Feb. 17, 1964. (NASA Release 65–129)

* Following a six-hour visit to Cape Kennedy and the Merritt Island spaceport, Mayor Willie Brandt of West Berlin said: "The space challenge is not only the responsibility of young Americans and Russians, but also that of young Europeans." Mayor Brandt said the European space effort should be a combined effort and that Germany would welcome any cooperation. (AP, *Orlando Star*, 4/17/65; AP, *Miami Her.*, 4/17/65)

* FAA approved the British-built BAC 111, a new short-haul jet airliner, for passenger-carrying operations in the U.S. FAA's airworthiness

certificate was awarded after a 3½-yr. evaluation program. (UPI, *NYT*, 4/18/65, 49)

April 17: U.S.S.R. launched COSMOS LXV from the Baikonur launch complex 200 mi. northeast of Tyuratam, Tass announced. The satellite carried scientific instruments for continuing the Soviet space exploration program. Orbital data: apogee, 342 km. (212.4 mi.); perigee, 210 km. (130.4 mi.); period, 89.8 min.; inclination, 65°. All systems were functioning normally. (Tass, *Krasnaya Zvezda*, 4/18/65, 1, ATSS–T Trans.; *SBD*, 4/22/65, 290)

- In an article discussing major American testing sites, Howard Simons and Chalmers M. Roberts of the *Washington Post* said: "Indeed it is from Vandenberg and not Cape Kennedy, Fla., that the majority of American satellites are launched. Between Jan. 1, 1964, and Oct. 31, 1964, for example, 33 or three times as many satellites were successfully put into space from Vandenberg as from Cape Kennedy.

  "The great majority of the satellites launched from Vandenberg, the hub of what is officially called the Air Force Western Test Range, are military satellites with secret payloads or reconnaissance cameras capable of peering down on Russia and China." (Simons and Roberts, *Wash. Post*, 4/17/65)

April 18: United Airlines and Eastern Airlines had placed the first orders for Douglas Aircraft Co.'s new DC–8–61 jetliner, seating 251. The aircraft would be the largest commercial jet in existence, having a total length of 187 ft. 4 in., and would cost about $8 million. United would buy five of the Model 61's and take options on two more; Eastern had ordered four aircraft for delivery late next year. (*NYT*, 4/18/65, 84)

- Soviet Union announced that pilot A. V. Fedotov had established a new world speed record for 1,000-km. closed route. He flew an E–266 aircraft with 2,000-kg. (4,409 lbs.) cargo at average speed of 2,320 kph. This exceeded by 253 kph the world speed record for that class held by U.S. pilot Harold E. Confer in a B–58 Hustler aircraft. (*Krasnaya Zvezda*, 4/18/65, 1, ATSS–T Trans.)

- A shipment of American Hawk missiles was unloaded recently at the Israeli port of Haifa, Israel announced. (UPI, *Wash. Daily News*, 4/19/65)

April 19: A detailed report on the progress of the Mars-bound MARINER IV spacecraft was presented at annual meeting of the American Geophysical Union in Washington, D.C.:

MARINER IV, launched Nov. 28, 1964, was on course to fly by Mars shortly after 9 p.m. EDT on July 14. Four of MARINER IV's six experiments were still working well. The ionization experiment had ceased to function and data from the solar plasma probe were only partially interpretable. At 3 p.m., MARINER IV was 58,176,037 mi. from the earth. It had traveled 221,330,000 mi. on its journey of 325 million miles.

MARINER IV had returned a considerable amount of scientific data. A cosmic ray telescope aboard the 575-lb. spacecraft had, for example, "observed" more solar protons than alpha particles from the sun. John A. Simpson of Univ. of Chicago said this indicated there was a "different kind of mechanism operating on the sun for accelerating these particles in space."

A report from a team of scientists from NASA GSFC and Temple Univ. indicated that MARINER IV was encountering increasing amounts of cosmic dust as it moved further away from the sun. MARINER IV's cosmic dust detector had been hit 95 times.

Dr. James A. Van Allen predicted that if Mars had a magnetic field no stronger than $\frac{1}{30}$th the intensity of the earth's, MARINER IV would detect it in July.

Richard Sloan of JPL said he and his colleagues planned to try to establish a radio lock with MARINER IV in September 1967 after it had journeyed through space and come back to within 40–50 million miles of earth. (NASA Releases 65–117, 65–117–A, 65–117–B, 65–117–C, 65–117–D, 65–117–E, 65–117–F; Transcript)

April 19: Six Navy and Marine flyers emerged from a cylindrical chamber at Philadelphia's naval air engineering center where they had spent 34 days in a simulated journey into space in an experiment sponsored by NASA. The project was designed to collect and analyze information on long confinement in a space atmosphere, specifically, how pure oxygen would affect the blood, the lungs, thinking, and eating. Cdr. Kenneth R. Coburn, project manager, called it "a major success," noting that "we find that man can live for long periods of time—for a month anyway—without any bad effects." (AP, *Chic. Trib.*, 4/20/65)

- DOD announced award to Lockheed Missiles and Space Co. of $3,000,000 increment to existing contract for engineering support for Agena system. (DOD Release 246–65)

- Edward L. Hays, chief of crew systems at NASA Manned Spacecraft Center, announced that the crew of the Gemini GT–4 flight would wear the qualified Extravehicular Activity (Eva) spacesuit during their flight. (AP, *Wash. Eve. Star*, 4/19/65; *M&R*, 4/26/65, 7)

- Excerpts from comments on management of research and development activity by Dr. L. R. Hafstad, director of General Motors Research and Defense Research Laboratories, appeared in *Aviation Week and Space Technology:* "In the modern laboratory the basic research activity is essentially an information-gathering intelligence operation. The operatives must be trained to speak, and allowed to speak, the language of the area on which they are expected to keep informed, and to interact with other researchers in the same area. It is this apparently excessive freedom of action on the part of employees which makes for the concern of students of administration about the management of research, or the lack thereof. My conclusion is that most of this problem evaporates once it is realized that a director of research directs the research program—but certainly not the individual researchers.

    "The partnership of science, engineering and industry is really a rather new concept developed since the turn of the century and only now reaching maturity. An even newer concept is the partnership of science, engineering, and government. A problem we must face up to —whether we represent industry, government or science—is the effective use of research in creating a better future for everyone.

    "There is never a dearth of projects—the difficulty is to pick worthwhile projects. It is here that I feel that the discipline of the profit and loss statement is essential. . . ." (*Av Wk.*, 4/19/65, 21)

April 19: Two teams of scientists collecting dust from Greenland and Antarctic icecaps presented their findings to the American Geophysical Union, meeting in Washington. The scientists were collecting particles by "core sampling"—boring through the ice with a thermal drill and analyzing particles to determine their origin. Team studying Greenland samples—E. L. Fireman, J. Defelice, and C. C. Langway, Jr., of the Smithsonian Astrophysical Observatory and the U.S. Army—believed their dust samples to be nonterrestrial in origin. Team studying Antarctic samples—M. B. Giovinetto of the Univ. of Wisconsin and R. A. Schmidt of NASA—was not certain of the origin of these particles. They reported a high concentration of spherules in the core samplings; the amount of these particles, which closely matched those found in volcanic eruptions, made identification of dust origin more difficult. They had collected dust from 165-ft. core of ice, representing 400 yrs. accumulation. Greenland team had drilled to depth of 1,800 ft. and expected to continue to 5,000 ft. National Science Foundation would use the same thermal drill—beginning in summer of 1967 or 1968—to drill to 8,000-ft. depth through the south polar ice. (Simons, *Wash. Post,* 4/20/65, 1)

Week of April 19: Cryogenic propellants were loaded for the first time into a ground test model of the NASA Saturn S–IVB upper stage to verify the design of the stage and fabrication techniques, and to demonstrate operational procedures. The S–IVB, 58 ft. long and 21.5 ft. in diameter, was being built for Saturn IB and Saturn V by Douglas Missile and Space Systems Div. for NASA MSFC. (MSFC Release 65–98)

April 20: NASA had awarded a $3,135,977 contract modification to the Boeing Co. for preparatory work leading to dynamic testing of the Saturn V moon rocket at NASA MSFC. Boeing would perform engineering services for the Saturn V dynamic testing program and would supply instrumentation equipment for the test stand. (MSFC Release 65–94)

- NASA Ames Research Center had let a $1,382,000 contract to the American Machine and Foundry Co. for fabrication of an advanced flight simulator which could simulate nearly all flight situations for aircraft and spacecraft except cases involving either high acceleration forces on the pilot or aerobatics.

  Designed by the Research Facilities and Equipment Div. at Ames, the simulator would have "six degrees of freedom," the capability to move in all possible axes of motion: fore and aft, vertical, and side-to-side; also pitch, roll, and yaw. It would be unique in having 100 ft. of lateral motion. This would be needed to simulate supersonic transport (Sst) flight since the crew would be far forward of the center of rotation of the aircraft. (ARC Release 65–12)

- North American Aviation Co.'s XB–70A experimental bomber reached altitude of 59,000 ft. and speed of 1,500 mph on its tenth flight from Edwards AFB. Duration of flight was 1 hr. 39 min., of which 1 hr. 14 min. was at supersonic speed, boosting its total supersonic flight time to 5 hrs. 5 min. (AP, Wash. *Eve. Star,* 4/21/65, A7)

- The X–15 research aircraft was praised by William Hines in an article in the Washington *Evening Star:* "The United States spent nearly a quarter-billion dollars to produce three copies of the X–15, unromantically known as 'No. 1,' 'No. 2,' and 'No. 3.' Modifications, maintenance and operation charges have by now pushed the bill close to a third of a billion.

"By any rational standard, the X-15 has been worth every penny. It has given the United States far more than mere supremacy in the flight record books; it has provided a foundation for advanced aeronautical technology that could have been obtained in no other way." (Hines, *Wash. Eve. Star*, 4/20/65)

April 20: Dr. Werner R. Kirchner, vice president and manager of Aerojet-General Corp.'s solid rocket operations, announced that a new solid fuel multipulse rocket engine containing several charges of propellant that could be separately fired by electrical signal had successfully completed its first series of test firings. The rocket could zip, glide, and dart about much like a bird, he said, or could lie dormant in space a year and then be restarted on command. Key to multipulse firings was described as a lightweight thermal barrier separating each charge. Aerojet had conducted demonstration firings of six flight-weight configurations in the company-funded program. (AP, *Denver Post*, 4/21/65; *Av. Wk.*, 4/19/65, 30)

- Donald E. Crabhill of the Bureau of the Budget discussed "Space Programs and the Federal Budget" before the National Space Club: "What are some of the significant factors to be pointed out in the relationship between the space program and the budget?

"The first is, of course, the matter of growth in the funding for space and the current absolute amount of funds allocated to space programs, including not only NASA, but also DOD, AEC, and activities in this area by other agencies. In FY 1957, approximately $150 million was expended by the Federal Government on space programs. In FY 1960, the total was still below $900 million. In FY 1966, the tenth year of the space age, the President's budget provides for space expenditures of $6.9 billion.

"Where does this amount stand in relation to amounts in the administrative budget for other programs? It is less than the total amounts to be spent in 1966 on national defense; on health, labor, and welfare programs; and on interest on the national debt. But it is greater than that to be spent for any other function of Government. Space expenditures of all agencies will be greater in 1966 than those for international affairs and finance, for agriculture, for natural resources, for commerce and transportation, for housing and community development, for veterans benefits and services, or for other general Government.

"The space program has not been, since it was initiated, and is not today, a budgetary underdog.

"The second specific point to be made is that the budget process by its very nature is an exercise in priorities. . . . A great many merely desirable projects get deferred throughout the Government every year under the press of the budgetary process.

"In the past, this pressure has not been felt as severely in the space area as it has in most others because of the emphasis that has been given to creating in a hurry a vast capability to operate in space. The space program has been very successful in meeting this aim. In fact, it has been so successful that space is now coming of age with other Government programs. We will soon have a technical capability to do a great many more space missions than we as a nation will probably want to pay for. . . .

"There is one other point that, as a budget examiner, I feel I must mention. Funding and schedule estimates for space programs have been historically quite unreliable. Cost estimates have tended not so much merely to *grow*, but to *multiply!* At the same time, schedules have tended to slip, slip, slip.

"This was an understandable situation while the space program was new, but we have had enough experience that there will be considerable resistance from now on to escalation in price and radical slips in schedule of the next generation of space projects. The more detailed planning we are doing now, the phased project procurement processes, and the experience we have gained in the technology and the techniques of space operations must be expected to show returns in better ability to make good cost and schedule estimates in the first place, and then to meet the cost and schedule targets that are approved." (Text)

April 20: Three American scientists were honored by the American Geophysical Union during an honors meeting at the National Academy of Sciences in Washington, D.C.: Norman F. Ness of NASA Goddard Space Flight Center received the John Adam Fleming Award for research done by means of instruments aboard NASA's EXPLORER XVIII satellite; Gordon J. F. MacDonald of the Univ. of California at Los Angeles was given the James B. Macelwane Award for work on a variety of subjects ranging from the center of the earth to the solar corona; Hugo Benioff, professor emeritus at the California Institute of Technology, was awarded the William Bowie Medal for "unselfish cooperation in research." (Wash. *Eve. Star*, 4/21/65)

April 21: Pegasus B, second of the "winged" micrometeoroid detection satellites, arrived at Cape Kennedy to be readied for launch during the next two months. Similar to PEGASUS I, Pegasus B would occupy a simulated Apollo service module aboard the SA-8 vehicle. A boilerplate model of the Apollo command module would be placed above the Pegasus; in orbit, the Apollo modules would be jettisoned and the satellite exposed. Preliminary data from PEGASUS I indicated it was confirming current theory on micrometeoroid density. (MSFC Release 65-85; *Marshall Star*, 4/21/65, 1, 2)

- NASA absolved Astronaut Virgil I. Grissom (Maj., USAF) of any blame in the 58-mi.-off-target landing of the GEMINI III spacecraft following the three-orbit flight Mar. 23, according to MSC spokesman. The mishap was attributed to the fact that the spacecraft did not develop as much lift as expected. The possibility that Major Grissom might have banked GEMINI III improperly as a result of misunderstanding instructions from ground stations had been investigated. (UPI, *NYT*, 4/21/65, 11, MSC GEMINI III Fact Sheet)

- EARLY BIRD communications satellite would relay a sampling of scientific, cultural, and entertainment events televised live at 35 sites in North America and Europe during an hour-long inaugural program, "This is Early Bird," scheduled for 1 p.m. EST, May 2, ComSatCorp announced. (ComSatCorp Release; Adams, *NYT*, 4/21/65, 91)

- Thomas W. Thompson of Cornell Univ. said in a paper presented at the meeting of the American Geophysical Union that half the moon's surface had been mapped in a lunar mapping program using the radio-radar telescope at Cornell's Arecibo Observatory, Puerto Rico. From the radar signal returns, the hardest areas of the moon were the rim and floor of the relatively new craters. The floors of the older craters and the surface of the maria were covered by a three-to-four-meter-thick layer of highly porous material often referred to as "lunar dust." (Simons, *Wash. Post*, 4/22/65; *NYT*, 4/22/65)

April 21: Dr. Gordon H. Pettingill, Dr. Rolf H. Dyce, and Dr. Thomas Gold of Cornell Univ., reported to the meeting of the American Geophysical Union that through radar studies with Cornell's 1,000-ft.-diameter radiotelescope at Arecibo, Puerto Rico, they had found an apparent "flat spot" on the planet Mars that seemed to correspond to markings seen there through telescopes. They also reported that radar observations indicated the planet Mercury rotated on its own axis once each 54 to 64 days, exposing all sides to the sun in a year. Its full day, corresponding to a 24-hr. earth cycle, would be about 180 earth days long. It was inconclusive whether Mercury rotated in the opposite direction from its orbit—a retrograde rotation—or in the same direction as its orbit—a direct rotation.

Dr. Gold also speculated that Mercury could not have been in its present orbit for much longer than 400 million years. Otherwise, he postulated, the sun would have held the planet over a long enough period of time to force it into a synchronous or 88-day rotation. This suggested to Gold that Mercury might once have been a moon of Venus but broke away or was tugged away to establish its own orbit around the sun. (Hines, Wash. *Eve. Star*, 4/21/65; Clark, *NYT*, 4/21/65, 17; Simons, *Wash. Post*, 4/21/65)

- In a statement of FAA policy outlined by FAA Administrator Najeeb E. Halaby, FAA's obligation was affirmed to regulate private conduct of pilots but only to the extent required in the public interest; to recognize the right of the general public to be informed and to be heard; to apply the regulatory hand evenly in similar situations, while also recognizing the different rights, duties, and operational requirements of the various segments of the aviation community; and to manage the airspace as a national resource in a manner best serving the requirements of all users while also recognizing the interests of people on the ground. (Text)
- Dr. Homer E. Newell, NASA Associate Administrator for Space Science and Applications, was among the ten outstanding Federal Government employees chosen by the National Civil Service League to receive Career Service Awards May 19. (*Wash. Post*, 4/22/65)

April 21–23: A Technology Status and Trends Symposium was held at NASA Marshall Space Flight Center for industry and university officials and invited guests. Purpose of the conference was to make available for general use in everyday life the results of research and engineering carried out in connection with the U.S. space program. (*Marshall Star*, 4/21/65, 1, 5; NASA SP-5030)

- At AIAA/AFLC/ASD Support for Manned Flight Conference in Dayton, Ohio, Temple W. Neumann of Philco Corp. reviewed studies of manned Mars missions and discussed the importance of "early biological precursor missions" to Mars. He concluded:

    "It has been shown that the lack of biological, as well as critical environmental, data about Mars can have important ramifications in not only the cost, but possibly even in the feasibility of performing early manned missions to Mars. The importance of preliminary knowledge about the interaction of possible Martian organisms with man and his equipment has been shown to significantly affect surface operations, decontamination requirements, and equipment reliability. Further, the need for some preliminary data about the nature of Martian organisms is necessary in order to intelligently design an experimental program for use by the first manned landing expedition. The conclusion can therefore be supported that a precursor biological mission, such as that represented by the current ABL studies, is mandatory in the early 1970 time period if manned missions are to make effective use of the mid-1980 launch opportunities." (Text, AIAA Paper 65-249)

April 22: With arrival of the sea-going launch platform USNS *Croatan* at Valparaiso, Chile, NASA completed a successful expedition of launching scientific experiments off the west coast of South America. A total of 77 sounding rockets were fired, 45 of them Nike-Cajun and Nike-Apaches, and 32 of them single-stage meteorological rockets. Firings occurred at various position from 5° north to 60° south of the equator. Five experiments were conducted at or near the 60th parallel at about 78° west longitude. The project was part of the NASA sounding rocket program being conducted during the 1964-65 International Quiet Sun Year. Expedition data would be correlated with findings of scientists throughout the world conducting experiments on IQSY phenomena. (Wallops Release 65-22)

• Two NASA sounding rockets, a Nike-Cajun and a Nike-Apache, were launched at Wallops Station after dark and about one hour apart. Both rockets released chemiluminescent gas clouds, which observers on the ground used to measure atmospheric winds, shears, turbulence, and vertical motions. Nike-Cajun reached altitude of 128 km. (79.5 mi.) and the Nike-Apache, 145 km. (90.1 mi.) (NASA Rpt. SRL)

• NASA selected Ling-Temco-Vought and Lockheed Electronics Co. for competitive negotiation of contract covering operational support services for laboratories and test facilities at NASA Manned Spacecraft Center. The support contract would be cost-plus-award-fee for one year with options to extend for four additional one-year periods. First year costs were expected to exceed $2 million. (NASA Release 65-133)

• NASA selected three industrial firms with which to negotiate similar preliminary design contracts for a Voyager spacecraft to undertake unmanned scientific exploration of the planets: the Boeing Co., General Electric Co., and TRW Space Technology Labs. The three-month, fixed-price contracts would each be worth about $500,000. (NASA Release 65-135)

• At Purdue Univ., Dr. George E. Mueller, NASA Associate Administrator for Manned Space Flight, discussed in a speech NASA's emphasis on man's part in future planned space experiments: "The role of man in space is basic to any discussion of our planned space experiments. . . . We have always recognized his inherent characteristics as a sensor, manipulator, evaluator and investigator.

"As a sensor, man adds little to automatic equipment in space—sometimes nothing at all. . . . instruments can measure . . . phenomena that man cannot perceive at all.

"But instruments are limited by the knowledge we now have on earth; they cannot cope with the unexpected or the unknown. Man, on the other hand, can operate in any unprogrammed situation and reap full benefits of the true objective of manned operations. He can explore the unknown.

"The second function of man in space is manipulation. Gus Grissom demonstrated superbly last month that a man can operate the spacecraft controls for delicate maneuvering. . . .

"In the conduct of space research also, man as a manipulator can probe into his environment. He can make use of motor responses and verbal skills to carry out procedures and to assemble, operate and repair equipment. . . .

"With the capacity to evaluate, man achieves a substantial degree of self-reliance in controlling what he perceives and how he reacts. When a man remembers, analyzes, compares, and induces—using a solid foundation of knowledge—he has improved the degree to which meaningful data can be translated into useful knowledge. . . .

"The most advanced role of man in space is that of an investigator who responds creatively to unexpected situations. He is able to postulate theories and hypotheses, and to devise and use systematic measurements. In this role, the astronaut is a full-fledged scientist." (Text)

April 22: NASA Manned Spacecraft Center's Public Affairs Officer, Paul Haney, announced that daily newspapers might have ½-hr. interviews with the crew of the GT-4 flight on the same basis as television networks and wire news services. Astronauts James A. McDivitt (Maj., USAF) and Edward H. White (Maj., USAF) would spend two full days in personal interviews at MSC early in May. There would be a mass press conference for all news media in Washington, D.C., on April 30. Without such an arrangement, the only newspapers that would have had personal interviews would be those that subscribed to the service that paid astronauts for their stories. (*Houston Post,* 4/23/65)

- New sunspots heralding the start of a new 11-yr. cycle were discussed at sessions on the International Years of the Quiet Sun held in Washington, D.C., under auspices of the American Geophysical Union and the International Scientific Radio Union. Scientists said the asymmetrical birth of the new cycle suggested it might not reach as intense a maximum as usual.

  The cycle was of vital interest to planners of a manned moon landing since it had been discovered that some solar eruptions shoot out protons at so close to the speed of light they could kill an astronaut. While astronauts were on the moon, or inside the Lem, they would be poorly protected against such a proton shower. Dr. Herbert Friedman, of the Naval Research Laboratory, said during the symposium they would be comparatively safe if they could return to their orbiting command capsule. The goal, therefore, he said, was to learn enough about these events so that astronauts could have sufficient warning to take refuge in their spacecraft. (Sullivan, *NYT,* 4/23/65)

- A two-day conference began at NASA Manned Spacecraft Center on international participation in space biomedical experiments on U.S. manned spaceflights. About 50 doctors from 17 countries attended. (*Houston Chron.,* 4/21/65; NASA Release 65-31)

April 22: First of four Canadian Stol CV-7A transport planes was accepted by the USA. The aircraft would undergo extensive service, engineering, and climatic tests in the next year. (DOD Release 253-65)

- DOD announced award to Thiokol Chemical Corp. of $2,300,000 increment to existing contract for production of Minuteman Stage I operational and flight test rocket motors. (DOD Release 255-65)

April 23: U.S.S.R. launched its first communications satellite MOLNIYA I into orbit: apogee, 39,380 km. (24,459 mi.); perigee, 497 km. (309 mi.); period, 11 hrs. 48 min.; inclination, 65°. *Krasnaya Zvezda* reported that the "basic purpose of launching the Molniya-1 communications satellite is to accomplish the transmission of TV programs and to perform two-way multichannel telephone, phototelegraphic and telegraphic communication. All the onboard equipment on the satellite and the ground radio network are operating normally, and the first transmission of TV programs between Vladivostok and Moscow were successfully completed." (Tass, *Krasnaya Zvezda*, 4/24/65, 1, ATSS-T Trans.)

- Successful simultaneous two-way transmission of television tests via EARLY BIRD communications satellite between the U.S. ground station at Andover, Me., and European ground stations at Pleumeur-Bodou, France; Goonhilly Downs, England; and Raisting, W. Germany, was announced by ComSatCorp. The pictures were of good quality. (ComSatCorp Release)

- X-15 No. 3 was flown by Capt. Joe Engle (USAF) 79,700 ft. altitude at a maximum speed of 3,657 mph (mach 5.48) to obtain data for heat transfer experiment with surface distortion panel ablative test. (NASA X-15 Proj. Off.; FRC Release; *X-15 Flight Log*)

- Successful completion of formal flight qualification tests of the uprated H-1 rocket engine for use in the Saturn IB space vehicle was announced jointly by NASA Marshall Space Flight Center, under whose technical direction the engine was being developed, and by Rocketdyne Div. of North American Aviation, Inc., its manufacturer. Two engines were used for the qualification program. In 51 firings they operated successfully for 4,581 seconds—more than 75 min.—and produced 200,000 lbs. of thrust (188,000 lbs. was previous power rating). (MSFC Release 65-96)

- NRX-A3, experimental Nerva nuclear reactor engine fueled with liquid hydrogen, was successfully hotfired for about 8 min., including 3½ min. at full power. A loose circuit connection caused the engine to shut off prematurely after the 3½ min. of full power. (Nerva Proj. Off.; *Wash. Post*, 4/25/65; *Rover Chron.*)

- Thompson-Ramo-Wooldridge had proposed to NASA a design for a Deep Space Planetary Probe System to be used in a flyby of Jupiter, Saturn, or Pluto. The spacecraft would consist of a large dish antenna, possibly as large as 16 ft. in diameter, which would telemeter data back to earth. Power would be supplied by 10-watt Snap-19 generator. The spacecraft could be boosted by either Atlas-Centaur or Saturn IB-Centaur with upper-stage assist from available solid rockets or from the Poodle, a low-thrust radioisotope rocket engine. Flyby missions for the probe could be made in 1970 to Jupiter, to Saturn in 1972, and ultimately to Pluto. (SBD, 4/23/65, 297)

April 23: Addressing a citizen's seminar at Boston College sponsored by the College of Business Administration, Rep. George P. Miller (D–Calif.), Chairman of the House Committee on Science and Astronautics, gave examples of the potential down-to-earth benefits of space research:

"—Automated highspeed, urban and interurban rail transportation, such as the four-hour trip between Boston and Washington mentioned recently by President Johnson.

"—Better communications systems, more reliable radios and television sets, improved home appliances.

"—Reduction of rust and corrosion by controlling bacteria which space researchers found to thrive by eating and digesting metal.

"—Prevention of muscular atrophy and new methods of treating Paget's disease, osteo-porosis and kidney stones. All this springing from the studies of weightlessness.

". . . also new knowledge about the processes of aging, and cancer." (White, *Boston Globe*, 4/23/65)

- Prof. Hannes Alfven of the division of plasma physics at the Royal Institute of Technology in Stockholm, Sweden, revived a theory that the moon was once an independent planet. In an article written for *Science,* he said that "many if not all of the craters of the moon were produced" by an "intense bombardment of fragments of itself" when the moon swept too close to the earth and partly disintegrated under the tremendous tidal forces that were generated. "It is also possible," the theory suggested, "that so much of the lunar matter fell down [on this planet] that the upper layer of the earth—the crust—originally derives from the moon." Prof. Alfven wrote that this theory was first stated by H. Gerstenkorn of Hanover, Germany, and published in 1954 in *Zeitschrift fur Astrophysik* under the title "Uber die Gezeitenreibung beim Zweikorperproblem" ("About Tidal Friction in a Two-Body Problem"). (Osmundsen, *NYT*, 4/24/65, 31; Myler, *Wash. Post*, 4/24/65)
- USAF received at Travis AFB, Calif., the first of 65 C–141 Starlifter cargo jets to be delivered this year, DOD announced. The aircraft were capable of carrying 30 tons of cargo or 123 combat troops 6,000 mi. nonstop at a speed of about 500 mph. (UPI, *NYT*, 4/24/65, 15)
- FAA announced that U.S. airports known to FAA numbered 9,490 at the end of 1964, an increase of 676 over previous years. Over the past five years, the annual increase in landing facilities reported to FAA had averaged 623. (FAA Release 65–36)

April 24: Second major Saturn V milestone this month: First five-engine ignition test of the Saturn V second stage, the S–II, was conducted at the Santa Susana, Calif., static test laboratory of North American Aviation, Inc., NASA Marshall Space Flight Center announced. The five J–2 engines, built by NAA's Rocketdyne Div., would produce one million pounds thrust. Short-duration firings leading to full-duration tests of nearly 400 sec. would follow the ignition firing. (MSFC Release 65–99)

- In an address at Duke University, Vice President Hubert H. Humphrey said: "How fortunate we are to live in this dramatic and creative period of change, of challenge, of opportunity. How great is our responsibility to achieve excellence of mind and spirit to do the tasks that must be done.

"I appeal, therefore, to you the generation of 1965.
"Make no little plans.
"Have not little dreams.
"Do not set your standards and goals by those of your mother and father.
"Do not set your standards and goals by those of this time.
"Challenge the impossible. Do what cannot be done.
"Thirty years ago it was 'Brother, can you spare a dime.'
"Today we reach the stars." (Text, *CR*, 4/26/65, 8179–80)

April 24: The John Young Award, a medal specially struck by the citizens of Orlando, Fla., was presented to the astronaut as a highlight of the John Young Day celebration. The medal would be used in future years to honor Orlando residents for outstanding achievements, but would not necessarily be awarded annually. (*Orl. Sent.*, 4/18/65)

- Soviet astronomer Dr. I. S. Shklovsky of the Sternberg State Astronomical Society in Moscow suggested 100 years as the age for a source of radio waves known only as 1934 minus 63. These figures pinpoint its position in the southern sky. 1934 minus 63 would be the youngest known natural object in the sky. (Sci. Serv., *NYT*, 4/24/65, 9)

- Employees of NASA Kennedy Space Center began moving into the new headquarters building on Merritt Island. The move of more than 1,700 employees would be completed by mid-August. (KSC *Spaceport News*, 4/22/65, 1)

April 25: FAA Administrator Najeeb E. Halaby said that FAA's sonic boom tests over Oklahoma City last year had shown that construction of a supersonic airliner prototype was clearly warranted: "My current conclusion is that a supersonic airplane can be designed in terms of configuration, operating attitudes and flight paths so as to achieve public acceptance in the early 1970s." Halaby's statement was in conjunction with release of a three-volume final report on the Oklahoma City experiment.

The FAA report, which included preliminary results from boom tests at the White Sands Missile Range, concluded that only abnormally massive booms would create serious problems. A principal finding in the "community reactions" study stated: "Substantial numbers of residents reported interference with ordinary living activities and annoyance with such interruptions, but the overwhelming majority felt they could learn to live with the numbers and kinds of booms experienced during the six-month study."

The three volumes just released were "Structural Response to Sonic Booms," "Community Reactions to Sonic Booms in the Oklahoma City Area, February–July 1964," and "Final Program Summary, Oklahoma City Sonic Boom Study, February 3–July 30, 1964."

Publication of these three documents completed the five-part Oklahoma City report. Two volumes had been made public in February 1965. (FAA Release 65–34)

- Expansion of the role of the National Science Foundation and expenditure by Federal mission-oriented research agencies of more money on basic research were two major recommendations of a special panel of the National Academy of Sciences to Congress. Recommendations were in a report, *Basic Research and National Goals*, submitted to the House Committee on Science and Astronautics.

The panel, headed by Dr. George B. Kistiakowsky of Harvard Univ., former science adviser to the President, was comprised of 15 scientists, engineers, and economists. The panel held that the National Science Foundation, as the sole agency of Government whose purpose was support of science across the board without regard for immediate practical gains, should be expanded to serve as a "balance wheel" to soften the impact of variable research policies of mission-oriented agencies on "little science." The recommendation that agencies should devote greater portions of their budgets to basic research was based on the view that in many cases these budgets were becoming stationary while the capacity for scientific growth was expanding. The panel also recommended that in some cases the Congress should extend the mission of the agency to include the pursuit of certain branches of basic research.

Three general opinions were widely held by the panel regarding the balance of science support today: first, Federal funds should be allocated with some consideration to the geographical-social effects of their expenditure; second, biological sciences had been under-supported and should receive support to expand them faster than the physical sciences; third, there was an impending crisis in the physical sciences because mission-oriented agencies, faced with stationary budgets, would probably not expand their support of basic physical research as fast as capacity to do basic research expanded. (Clark, *NYT*, 4/26/65, 55; *SBD*, 4/30/65, 330)

April 25: Dr. Hideo Itokawa, professor at Tokyo Univ. and deputy director of Japan's Institute of Space and Aeronautical Science, was quoted on Japan's role in space activity by Peter Temm in an article in the Washington *Sunday Star:* "Space research is not a competition. It should be a cooperative undertaking among all countries, to explore and study the universe.

"Both America and Russia appear to be chiefly interested in artificial satellites and manned space vehicles. I see Japan's role as filling some of the gaps skipped over by these two nations.

"I believe it is possible that Russia may be preparing to abandon its project of putting a man on the moon in favor of assembling a satellite space station; at least, this how I interpret the recent Voskhod flight and its emphasis on carrying out tasks outside of the capsule.

"I sincerely hope if this is so, that American space scientists will not swerve from their intentions of getting to the moon. There are many sides to space research, and the ideal approach is for all nations engaged in the new science to tackle different areas.

"That way, we will all progress at a faster rate." (Temm, *Wash. Sun. Star,* 4/25/65)

April 26: A 37-man study group chaired by Dr. Colin Pittendrigh of Princeton Univ. and convened by the Space Science Board of the National Academy of Sciences at the request of NASA had reconfirmed the Academy's appeal for exploration of Mars to receive "the highest priority among all objectives in space science—indeed in the space program as a whole" and endorsed NASA plans to use the 1969–73 favorable Mars window for intensive study of the planet with the Voyager spacecraft. In its final report transmitted to NASA Administrator Webb, the group said that "given all evidence presently available, we believe it entirely reasonable that Mars is inhabited with living organisms and that life independently originated there," and thus that "the biological exploration of Mars is a scientific undertaking of the greatest validity and significance."

The panel noted, however, that "while we are eager to press Martian exploration as expeditiously as the technology and other factors permit, we insist that our recommendation to proceed is subject to one rigorous qualification: that no viable terrestrial microorganisms reach the Martian surface until we can make a confident assessment of the consequences."

The group made seven basic recommendations: (1) "every opportunity for remote observation of Mars by earth-bound or ballon-and satellite-borne instruments should be exploited"; (2) "... An adequate program for Martian exploration cannot be achieved without using scientific payloads substantially larger than those currently employed in outer unmanned space research program.... We see very substantial advantages in the use, from the onset, of the new generation of large boosters which are expected to become operational toward the end of the decade"; (3) since flyby missions "yield at best a fleeting glimpse of the planet" and carry a relatively small array of instruments, "we deliberately omit an explicit recommendation in favor of any flyby missions additional to those already executed or planned"; (4) "Every effort should be made to achieve a large orbiting mission by 1971 at the latest. This mission should precede the first lander.... By 'large' we mean a scientific payload that would include instrumentation for infrared and television mapping, microwave radiometry and bistatic radar, infrared spectrometry, and optical polarimetry"; (5) "The first landing mission should be scheduled no later than 1973 and by 1971 if possible" and will "ultimately demand a large lander" like Abl (Automated Biological Lab); (6) "The task of designing an Abl should be initiated immediately as a continuing project"; and (7) to maintain "a continuing dialogue among all potential investigators and the engineers responsible for implementing their scientific goals," the Academy's Space Science Board should have a standing committee. (NAS Release; Abraham, Phil. *Eve. Bull.*, 4/26/65; Hines, Wash. *Eve. Star*, 4/26/65, 2; Sullivan, *NYT*, 4/27/65, 1)

April 26: The Federal Communications Commission confirmed it expected to rule soon on who should own the initial American ground stations providing access to communications satellites. The established international carriers, including AT&T, RCA Communications, Western Union International, and ITT World Communications, had accused ComSatCorp of exceeding its statutory authority and demanded the right to share in the ownership of the ground stations. (Gould, *NYT*, 4/27/65, 1, 25)

• Dr. Charles A. Berry, chief of medical operations at NASA Manned Spacecraft Center, had said that new body sensor equipment developed for astronauts had "stretched the doctor's stethoscope to reach 100 miles," reported Norm Spray in an article in the *Houston Chronicle:* "This could open the door for new types of medical research and treatment potentially as important to the family physician as to space scientists, Dr. Berry believes.

"'Right now,' he said, 'we think our sensing and monitoring system would be a tremendous tool in hospital recovery and intensive care rooms.

" 'Basic medical data that is reliable and distortion-free could be fed from each patient to a central computer or console. Each patient could be watched as closely as if a nurse or even a doctor were constantly at his bedside.' " (Spray, WBE Sci. Serv., *Houston Chron.*, 4/26/65)

April 26: Sen. Claiborne Pell (D-R.I.) discussed in the Senate S. 1483, bill which he had introduced to establish a National Foundation on the Arts and the Humanities. He cited article by Frank Getlein on the recent "Eyewitness to Space" exhibition at the National Gallery of Art. Getlein's article showed "how cooperation between our Government and the arts can illuminate some of the most exciting moments in our important explorations in space." The exhibition contained some 70 paintings and drawings by 15 artists under the NASA art program.

". . . The work shows total freedom and a wide variety, ranging from the superb illustrationist's style of Paul Calle to the highly individual abstraction of Washington artist Alfred McAdams.

". . . The space effort, therefore, from Huntsville to the launching apparatus at Cape Kennedy, to the pickup system in the Pacific, is covered at once as a set of visual phenomena and an immensely varied set of artistic responses to those phenomena. . . .

"The NASA art program is a modest step but a carefully made one in the gradually reemerging relationship between American art and the American Government. It deserves study by those interested in the larger problem." (Getlein, Wash. *Eve. Star*; *CR*, 4/26/65, 8122–23)

- Groundbreaking ceremony for Univ. of Maryland's $1.5-million Space Science Building was held at the College Park, Md., campus. Dr. Homer E. Newell, NASA Associate Administrator for Space Science and Applications, and Edward F. Holter, Vice Chairman of the Univ. of Maryland Regents, shoveled the first spadefuls of dirt. (*Wash. Post*, 4/27/65, A12)

- Dr. Roman Smoluchowski of Princeton Univ. said at the American Physical Society's meeting in Washington, D.C., that there was no life on Mars. All seasonal changes in the color of the planet could be traced to bombardment of minerals with energetic radiation under varying temperatures.

  Dr. Jane Blizard of Boulder, Colo., also speaking at the APS meeting, suggested that any astronaut braving a 400-day journey to Mars would be likely to get a fatal dose of radiation. Maybe, she said, long range forecasting of solar storms can be perfected in time. Or maybe "superconductive magnetic doughnuts" could be devised to shield spacecraft from barrages of protons spewed out in solar storms. (Hines, Wash. *Eve. Star*, 4/27/65)

- Jean Delormé, president of France's L'Aire Liquide and head of Eurospace, said he believed there could be no large-scale European space program without formation of a European equivalent to the U.S. National Aeronautics and Space Administration. He called for establishment of a central coordinating body that would be the supranational European NASA, with the power to make financial decisions. Delormé was addressing opening of 12-day U.S.-Eurospace conference in Philadelphia. (AP, *NYT*, 2/27/65, 17)

- Dr. Paul Herget, professor and director of the Univ. of Cincinnati

Observatory, was awarded the James Craig Watson Medal of the National Academy of Sciences "for important contributions to the field of celestial mechanics, and particularly his application of electronic-computer techniques to calculations of the orbits of comets, earth satellites, and asteroids." He previously had responsibility for developing operations of the Vanguard Computing Center, which provided tracking information on early scientific satellites. Henry Draper Medal for original investigation in astronomical physics was awarded in absentia to British radioastronomer Martin Ryle. (NAS Release; NAS-NRC *News Report*, 4/65, 4)

April 26: Speaking before the Fourth Symposium on Advanced Propulsion Concepts in Palo Alto, Calif., Gen. Bernard A. Schriever, AFSC Commander, said that the Air Force was studying the possibility of using hydrogen-burning, supersonic combustion ramjet engines, known as Scramjet, to power hypersonic aircraft: "The Scramjet is the most promising approach we have today for sustained hypersonic flight . . . . it could be used effectively on hypersonic aircraft with both military and commercial applications." He said experience gained with the research airplane might lead to the hypersonc aircraft and could make feasible the development of recoverable launch vehicles for flight speeds up to about 8,000 mph. This would permit delivery of very large payloads into space at far greater economy than is presently possible. (Text, AFSC News Release 65.65)

• DOD had asked NASA to consider using Minuteman I missiles scheduled to be removed from their silos, as launch vehicles, *Missiles and Rockets* reported. NASA Hq. transferred study to Langley Research Center. LaRC was expected to complete its feasibility investigation in three to four weeks. (*M&R*, 4/26/65, 7; LaRC)

• Maj. Gen. Don R. Ostrander. Commander of USAF's Office of Aerospace Research, said at the Fourth Symposium on Advanced Propulsion Concepts in Palo Alto, Calif., that important changes in America's research and development posture during the past few years "are the result of the more stringent requirements that must be met before increasingly complex and expensive systems can be approved for development." He continued: "These changes have placed more emphasis on research and exploratory development.

"Coupling—or reducing the time lag between discovery and application—is the proposed solution for accomplishing this tremendous task. The problem of coupling is the problem of time. . . ." (OAR Release 4-65-3)

• Passage of bills concerned with freedom of information was urged by William J. Coughlin in an editorial in *Missiles and Rockets:* "Intent of the bills is to establish a Federal public records law and to permit court enforcement of the people's right to know the facts of government. Providing for sensible exceptions in the case of sensitive and classified information, the proposed law would require every agency of the government to make all its records promptly available to any person. . . .

"The onus for restrictive news management usually falls on the Dept. of Defense, and rightly so, but there are a number of individuals in the National Aeronautics and Space Administration who are inclined to regard the agency as a preserve which should be off limits to the press. It is to the credit of Administrator James E. Webb that he has a consistent record of correcting abuses of press freedom that are called to his attention. The same cannot be said of his counterpart in the Defense Department." (M&R, 4/25/65)

April 27: The House Committee on Science and Astronautics unanimously approved a $5.2 billion NASA authorization for FY 1966, cutting only $75 million from the President's request. An unrequested $27.2 million was included for the 260-in. solid propellant program, the M–1 liquid hydrogen engine, and the Snap 8. Biggest single reduction was $42 million cut in $3.6 billion request for Apollo. Other programs affected by the cut included Oao, Ogo, Surveyor, Rover, Lunar Orbiter, and Centaur. (*WSJ*, 4/28/65)

- NASA Marshall Space Flight Center announced $40 million modification to contract held by General Electric Co. for the design of electrical equipment for Saturn vehicle launch support. Modification would cover the design portion of the work involved in providing electrical support equipment for Saturn IB and Saturn V launches. (MSFC Release 65–100)

- W. L. Everett, chief test pilot for the Ryan Aeronautics Co., was catapulted to his death from an XV–5A experimental plane after the vertical take-off aircraft developed mechanical difficulties.

    A witness said the XV–5A was at only 800 ft. and upside down when Mr. Everett ejected: "When he ejected, he ejected straight into the ground." The parachute did not have time to open. (*N.Y. Her. Trib.*, 4/28/65; Miles, *Wash. Post*, 4/28/65)

- Gen. William F. McKee (USAF, Ret.), NASA Assistant Administrator for Management Development, was named to succeed Najeeb E. Halaby as Administrator of the Federal Aviation Agency. (*Wash. Post*, 4/28/65)

- Frederick G. Donner, chief executive of General Motors, appeared before a Senate Commerce Committee on his renomination by President Johnson as a director of the Communications Satellite Corporation. Asked about rivalry from the Soviet Union in view of their recently-launched comsat, Donner said he regarded this about the same way he did Soviet automobiles as far as competition was concerned. (AP, *Wash. Post*, 4/28/65)

- "Self-organizing flight controller," featuring device that could cope with unexpected flight conditions of satellites and aircraft, was being developed by AFSC Research and Technology Div. Applying "probability state variable devices," bionics researchers had recreated function of a living nerve cell in a device called "Artron" (artificial neuron). Networks of Artrons in electronic cluster functioned like living neurons: they became self-organizing, achieving problem-solving, and learning new ways to capitalize on their mistakes and find new ways of performing a given task. AFSC stressed that flyable self-organizing flight controller was 5–10 yrs away. (AFSC Release 50.65)

- Dr. Geoffrey Bennett, chief medical officer of the British Ministry of Aviation, gave a progress report on the Anglo-French supersonic transport, the Concorde, at the annual meeting of the Aerospace Medical Association in New York. He said problems of designing a supersonic aircraft safe enough for commercial use were proving less difficult than had been expected: "It is quite heartening to find that the further one goes along, the less difficult things seem to be."

After his talk, Dr. Bennett said in an interview that such potential hazards as loss of air pressure in the cabin, accumulation of ozone, radiation, and problems of acceleration aroused much worry and discussion a few years ago. Overcoming these problems by proper designs had proved less difficult than many expected, he said. (Schmeck, *NYT*, 4/28/65, 89)

April 27: President Charles de Gaulle said in address delivered over French radio and television: "In the economic, scientific and technical domain, . . . we must see that our activities, for the essential part, remain under French management and control. We must also meet, at whatever cost, the competition in advanced sectors. . . . Finally, when it is opportune, in order to combine our inventions, our capabilities and our resources in a given branch with those of another country, we must often choose one of those which is closest to us and whose weight we do not think will overwhelm us.

"That is why we are imposing a financial, economic, and monetary stability upon ourselves which frees us from resorting to outside aid; we are converting into gold the dollar surpluses imported into our country as a result of the American balance of payments; we have over the past six years multiplied by six the funds devoted to research; . . . we are joining with England to build the world's first supersonic passenger aircraft; we are ready to extend this French-British collaboration to other types of civil and military aircraft; we have just concluded an agreement with Soviet Russia concerning the perfection and use of our color television process. In sum, however large may be the glass offered to us, we prefer to drink from our own, while touching glasses round about. . . ." (Text, *Atlantic Comm. Qtrly.*, 6/22/65)

- Rep. Emilio Q. Daddario (D-Conn.) disclosed in speech before Washington Section, National Association of Science Writers, that the House Subcommittee on Science, Research and Development (of which he was chairman) was planning an investigation of the National Science Foundation. He said: "For some years, there has been the need to review its work and to determine if it were, in fact, thoroughly successful in promoting the progress of science, the national health, prosperity and welfare and for other purposes." (*Wash. Post*, 4/28/65, C9)
- In a speech before the Aero Club of Washington, Dr. Raymond L. Bisplinghoff, NASA Associate Administrator for Advanced Research and Technology, ventured predictions for the next 20 years in aeronautics and astronautics. He noted the steady increase in civil aircraft output and the expansion in air travel "at a rate better than 12 percent per year for more than 15 years." He predicted "a 20-fold rise in air traffic volume over the next 25 years," but said that in order to reach its full potential the aircraft must be improved "in at least three important respects": reduction of minimum speeds for safe controlled flight; increase of maximum flight speed; and greater simplicity and economy of operation. He cited NASA research in these vital areas. (Text)
- Dr. Erhard Loewe, vice president of the German company Telefunken, outlined Eurospace's long-range goals at Eurospace Conference in Philadelphia: ". . . we want to avoid errors as far as possible and derive the greatest possible profit from experience gained in the U.S. . . ."

Loewe said that Eurospace would urge support of the Aerospace Transporter, conceived as a two-stage vehicle—both piloted—able to carry a 5,000-lb. payload into a 180-mi. (300-km.) altitude orbit and capable of rendezvous with an orbiting satellite. Loewe said that the Aerospace Transporter "signified as much to Europe as the trip to the moon does to the U.S. and the U.S.S.R."

Other specific projects in the Eurospace recommendations: space stations, because long-term platforms were believed necessary to exploit space scientifically and economically; communications satellites in a system that would be integrated with the worldwide system of the U.S. ComSatCorp; applications and scientific satellites, for high-capacity commercial television, weather forecasting, and data collecting; ground facilities for basic R&D. (*Av. Wk.*, 5/10/65, 74–81)

April 27: Dr. George B. Kistiakowsky, professor of chemistry at Harvard Univ. and former special assistant to President Eisenhower for science and technology, was selected Vice President of the National Academy of Sciences for four-year term beginning July 1, 1965. NAS also elected 35 new members in recognition of their distinguished and continuing achievements in original research.

National Academy of Engineering, holding its first annual meeting in coordination with NAS, elected 19 new members including Dr. Raymond L. Bisplinghoff, NASA Associate Administrator for Advanced Research and Technology. (NAS Releases; NAS-NRC *News Report*, 4/65)

April 28: X-15 No. 2 was flown by pilot John McKay (NASA) to 92,600-ft. altitude at a maximum speed of 3,260 mph (mach 4.80) to obtain data on the landing gear modification and on stability and control. (NASA X-15 Proj. Off.; FRC Release)

- USAF orbited two unidentified satellites with a single Atlas-Agena D launch vehicle launched from Vandenberg AFB. (AP, *NYT*, 4/30/65, 40; *U.S. Aeron. & Space Act.*, 1965, 140)
- In its 11th test flight, the XB-70 aircraft reached a speed of 1,630 mph and an altitude of 62,000 ft.—both records for the XB-70. The aircraft's total time in the air was 14 hrs. 41 min. Flight was made from Edwards AFB with NAA pilots Alvin S. White and Van Shepard. (Clark, *NYT*, 4/29/65)
- Quasi-stellar radio sources ("quasars"), cosmic x-ray sources, and neutron stars were discussed at NAS meeting in Washington. Jesse L. Greenstein of Mt. Wilson and Palomar Observatories suggested quasars were signs of galaxies forming—"the first condensation" of intergalactic gases. William A. Fowler of Cal Tech revived his earlier theory (proposed with Fred Hoyle) that these sources were huge energy masses created by explosive contractions of gigantic stars. Herbert Friedman of NRL presented new evidence that cosmic x-ray sources and neutron stars were not the same things. (Scientists at NRL had earlier suggested that some cosmic x-ray sources were the theoretical neutron stars.) Edwin E. Salpeter of Cornell Univ. reiterated the neutron star hypothesis. He suggested neutron stars could be oscillating stars which generate such great amounts of gravitational energy that the x-rays are produced. (Simons, *Wash. Post*, 4/29/65, A5)
- Dr. Harold C. Urey, Univ. of California physicist, told members of the Overseas Writers Club in Washington, D.C., that Communist China

could produce hydrogen bombs by a comparatively simple process and could possibly develop a nuclear delivery system in five years. Dr. Urey said Communist China had surprised world scientists, including himself, when it produced a nuclear bomb last fall with uranium 235—one of the technically most difficult ways to produce the nuclear bomb. (AP, *NYT*, 4/30/65, 3)

April 29: NASA's EXPLORER XXVII (BE–C) satellite was successfully launched into orbit from Wallops Island aboard a four-stage Scout rocket. Orbital parameters were: apogee, 796.5 mi. (1,162.4 km.); perigee, 579.7 mi. (921.3 km.); period, 108 min.; inclination to the equator, 41°. Primary mission of the 132-lb., windmill-shaped satellite was geodetic measurement: irregularities in the earth's gravitational field would be mapped by analysis of the Doppler shift of radio signals from the spacecraft. As a secondary mission, EXPLORER XXVII would provide data related to ionospheric studies and would evaluate further the use of laser techniques in deriving orbital and geodetic information and for deep space communication.

All systems were operating as planned. (Wallops Release 65–24; NASA Release 65–147; NASA Proj. Off.)

- MARINER IV set world space communications distance record shortly after 3:00 a.m. EST when it reached a straight-line distance from earth of 66 million mi. Soviet scientists reported two years ago that they lost radio contact with their MARS I spacecraft March 21, 1963, after 149 days of flight at more than 65 million mi. (NASA Release 65–141)

- USAF launched Thor-Agena D from Vandenberg AFB with unidentified satellite payload. (*U.S. Aeron. & Space Act., 1965*, 141)

- Second successful Biosatellite nose cone test at White Sands Missile Range was conducted by AFCRL, which was assisting NASA in evaluating reentry of the spacecraft after being released by balloons at altitudes of from 88,000 ft. to 100,000 ft. First such test had been conducted March 24. (OAR *Research Review*, 7/65, 30)

- An accelerated reservoir light-gas gun had set a world speed record of 25,300 mph for controlled flight of a visible object, of known mass and shape, and over a known distance in a ground facility in tests at NASA Ames Research Center, Ames announced. The shot was 3,200 mph faster than the previous record. In the light-gas gun used, an explosive charge was set off in a cylinder behind a plastic piston. The explosion pushed the piston into a chamber of hydrogen gas, compressing it, and the gas in turn pushed the projectile out of the firing tube. A light gas must be used because it has low mass and would expand at the highest speed after compression.

With the ability to move objects this fast, researchers could extend their knowledge of space flight problems. (ARC Release 65–13; ARC *Astrogram*, 4/29/65, 1, 2)

- DOD announced interagency agreement whereby Defense Supply Agency would furnish NASA about $500,000 worth of electronic items annually on a reimbursable basis. The agreement would involve approximately 12,000 centrally-managed items at DSA's Defense Electronics Supply Center in Dayton, Ohio. (DOD Release 272–65)

April 29: At the Spring Meeting of the American Physical Society, Dr. Homer E. Newell, NASA Associate Administrator for Space Science and Applications, attempted to answer the question "What can the space program do for experimental sciences like physics": ". . . the impact that space techniques are having and have already had on geophysics . . . is three-fold in character. First, the geophysicist finds in the space program powerful tools to use in a new approach to solving old problems. Secondly, the application of space techniques to geophysics has already turned up a number of exciting new problems, greatly broadening the scope of the discipline. Thirdly, as space probes, and eventually men, reach other bodies of the solar system such as the moon and planets, the domain of geophysics grows beyond the confines of a single body of the solar system. Let us consider each of these extensions to geophysics a little further.

". . . space techniques have provided new tools for studying old problems of geophysics. Geodesy, meteorology, upper atmospheric physics, ionospheric research, and sun-earth relationships have all benefited from the application of space techniques. In the case of geodesy, the influence of the earth upon the orbits of various artificial satellites has been measured by careful radio, radar, and optical tracking and used to obtain quantitative measures of the various harmonics in the expansion of the earth's gravitational potential. As a consequence of such measurements it has been found that the earth's equatorial bulge is some 70 meters greater than one would expect. . . . Other departures of the geoid from the figure of hydrostatic equilibrium have also been determined from these satellite measurements. . . . These measurements in turn have important implications for the distribution of matter within the earth, and for the internal strength of the earth's mantle." (Text)

- At a news conference in Washington, D.C., Dr. George E. Mueller, NASA Associate Administrator for Manned Space Flight, said that although "extravehicular activity" was not planned for Gemini astronauts until GT-5, "we are working hard at trying to qualify the space suit and the hatch itself to see whether we can accelerate that date."

  If their spacesuits and the spacecraft's hatch passed tests in time, Astronaut Edward H. White (Maj., USAF) would lean halfway out of the capsule for perhaps 15 min. on flight GT-4, scheduled for early June. He and Astronaut James A. McDivitt would attempt to orbit the earth 63 times in 98 hours, taking off from Cape Kennedy and landing in the Atlantic near Grand Turk Island.

  Maj. White and Maj. McDivitt appeared at the news conference with their backup crew—Lt. Cdr. James A. Lovell, Jr. (USN), and Maj. Frank Borman (USAF). (Transcript)

- Dr. Winston E. Kock, Director of NASA Electronics Research Center, was guest of Dr. Robert R. Gilruth, Director of NASA Manned Spacecraft Center, on a tour of MSC facilities. While in Houston, Dr. Kock addressed the annual banquet of the Institute of Navigation. In his speech he revealed ERC would investigate possibilities of new guidance techniques for future ion-propelled (or other low-thrust) spacecraft, employing Mossbauer radiation as an accelerometer to monitor systems performance on the spacecraft. He termed Mossbauer radiation "the most precise electromagnetic frequency yet known" in guidance applications. (MSC *Roundup*, 5/14/65, 7)

April 29: National Academy of Engineering's first award, the Charles Proteus Steinmetz Centennial Medal, was presented to RCA President Elmer W. Engstrom, for his outstanding leadership in electrical engineering for more than 30 years. (NAS–NRC *News Report*, 4/65, 4)

- Dr. Charles H. Townes, provost of MIT, reported at the meeting of the American Physical Society in Washington, D.C., that a laser beam had been used to produce sound waves more than a million times higher in pitch than those audible to the human ear. Dr. Townes explained that the laser beams at MIT had been used to produce oscillations constituting sound waves in solids and liquids. Sound had been produced by means of the laser at 3,000 mc in a fluid and 60,000 mc in a solid. It should be possible, Dr. Townes said, to generate sound at 300,000 mc in diamond. (Sullivan, *NYT*, 4/30/65)
- John G. Lee, pioneer aircraft designer and former director of research for United Aircraft Corp., had joined NASA as a part-time consultant to NASA Administrator James E. Webb on aeronautical research. (NASA Release 65–143)
- A full-size model of the Soviet Union's Vostok spacecraft was placed on public view for the first time. The spherical, silvery capsule, mounted on a model of the last stage of its launch vehicle, was on display in Moscow's permanent Exhibition of National Economic Achievement. The 4.6-ton Vostok had a diameter of 7½ ft. (*NYT*, 4/30/65, 8)

April 30: S–IVB stage of the first Saturn IB launch vehicle—first piece of flight hardware from Douglas Space Systems Center at Huntington Beach—had been shipped aboard NASA barge Orion to Douglas Sacramento Test Flight Center for flight readiness testing. The stage, 58 ft. long and 21.5 ft. in dia., had single Rocketdyne J–2 engine, developing 200,000 lbs. thrust. (MSFC Release 65–104)

- NASA had awarded $300,000 grant to the Dept. of Interior's Bureau of Mines for a three-year research program on the potential use of lunar materials to support manned exploration of the moon. The research team, utilizing data from NASA's unmanned lunar programs, would study the possible production, processing, and uses of materials on the moon for the construction, supply, and operation of manned lunar bases. Faculty consultants and graduate students from Univ. of Minnesota would assist as part of the Bureau's program to develop future capabilities at educational institutions. (NASA Release 65–144)
- NASA Flight Research Center awarded separate lifting body study contracts to McDonnell Aircraft Co. and Northrop Norair. The two separate six-month studies would investigate a vehicle concept whose sole mission would be the basic research involved with reentry of a manned lifting body from orbital flight. Preliminary objectives included determination of problem areas and their influence on design. Both contracts were fixed price; McDonnell received $152,496 and Norair $150,000. (FRC Release 11–65)
- James E. Webb, NASA Administrator, addressed meeting of Eurospace in Washington, D.C.

    "Launch vehicle and propulsion requirements for more distant applications have led us to establish the feasibility of nuclear reactors for space propulsion purposes, and continuing attention will be given to this field. Data obtained in 10 years of extensive technical effort have now experimentally verified the analytical predictions of performance for this type of propulsion. And, of course, the supporting technologies which would be necessary for difficult and distant future missions must also be considered, the power sources, including fuel cells, radio isotope sources, reactor power plants, vastly improved communications technology, pointing and orientation technology, highly reliable and long-lived componentry, and life support systems, including closed ecological systems. In this wide range of prospects for the more distant future, we are not committed to a particular line of development nor to given systems. We are too early in the space age to make such commitments. . . ." (Text)

April 30: C. Leo De Orsey, financial advisor and attorney for the seven original astronauts and acting president of the Washington Redskins football team, died. (UPI, *Houston Chron.*, 5/1/65; AP, *NYT*, 5/2/65, 89)

- Operational control of U.S. weapons to intercept and destroy armed satellites had been assigned to the Space Defense Center at Colorado Springs, *Denver Post* reported. The Space Defense Center included the Space Detection and Tracking Systems (Spadats), which recorded the launches of all space vehicles, foreign and domestic, and logged precise orbital data until they decayed in the earth's atmosphere. (Partner, *Denver Post*, 4/30/65)

During April: More than 100 delegates from Eurospace toured U.S. aerospace installations, including NASA Kennedy Space Center, Goddard Space Flight Center, and facilities of U.S. firms corresponding to Eurospace member companies. Purpose of the U.S. European Space Conference was to bring together top industrial leaders from European and American aerospace companies to review problems posed for the industry by evolution of space technology. (*M&R*, 4/26/65, 9)

- A $2.3-million test facility expected to improve space storability of liquid and solid rocket propulsion systems would be completed at the Air Force Rocket Propulsion Laboratory at Edwards AFB, *Missiles and Rockets* reported. (*M&R*, 4/26/65, 10)
- The 2,000th full-scale solid rocket motor of the Polaris A–3 model was shipped to Navy's Pacific Missile Facility where it would be integrated into an operational missile. (*J/Armed Forces*, 4/24/65, 15)
- Walter R. Dornberger, vice president in charge of research for Bell Aerosystem Co., wrote in the company's bimonthly magazine, *Rendezvous*, the United States was spending too much for space exploration. As a start to cutting costs, Dornberger proposed developing space boosters that could be recovered and reused. (AP, *Milwaukee J.*, 4/14/65)
- AFCRL experiment proved that a radio signal transmitted by an orbiting satellite could be trapped between two layers of the ionosphere and, upon emergence, channeled to ground stations half way around the world. Scientists had been aware of the ionospheric ducting capability for a number of years, but it had not been fully explored before the orbiting satellite experiment. (OAR Release 4–65–1)
- Dr. Willard F. Libby, chemist and Director of UCLA Institute of Geophysics and Planetary Physics, advocated emphasis on manned scientific missions in the U.S. space program. "In my opinion, space is a

great unknown from which we will obtain *many* new scientific discoveries." He approved of the use of scientist-astronauts, "but they must be backed much more wholeheartedly by the entire scientific community, particularly the academic community, than is at present the case. Education will help to accomplish this eventually, but there is a particular urgency to determine the post-Apollo objectives in the near future." A solution to the immediate problem, which had been proposed to and adopted by NASA: formation of a "Scientific Task Force," made up of scientists about the same age as the astronauts, to work and live at MSC and be closely connected with the astronauts, MSC Director Dr. Robert R. Gilruth, Director of the Office of Space Sciences Dr. Homer E. Newell, and the Advisory Committee for Science and Technology. The Scientific Task Force would educate the scientific community with the manned space flight program and thereby acquire its ideas on the subject, and acquaint the NASA directors with the ideas for scientific experiments suggested by the academic community. (*A&A*, 4/65, 70–75)

During April: Martin Summerfield, Princeton Univ., said in AIAA editorial that most of the critics of the U.S. space program were erecting and knocking down "straw men." Some of the attacks on the space program were designed to divert space funds "to other, supposedly more important purposes," and these viewpoints are "pushed too hard and can lead the nation in dangerous directions." The more significant criticism on scientific grounds was that ground-based instruments (supplemented by unmanned probes) can gather data about space, the moon and other celestial bodies more effectively than rocket-launched exploration. This criticism, he said, "misses the mark completely because it takes for granted that the national space program—or at least the NASA part of it—was conceived simply as a scientific venture. . . ." He recalled the words of the National Aeronautics and Space Act of 1958, which provided "clearly . . . the overriding intent to develop the technology of space flight as an extension of the former NACA's commitment to aeronautical flight. . . .

"The real issue is whether the nation should continue to develop the technology for flight in space, capitalizing on such useful applications as seem practical from time to time. The answer can only be 'yes,' and nothing less than a vigorous program will do. It makes no sense to insist that so broad a program be evaluated in competition with telescopes or unmanned scientific probes. Advances in space science will not substitute for flying capability. Each of these efforts is important in its own right. . . ." (*A&A*, 4/65, 23)

- Orville H. Daniel discussed small rockets—chiefly meteorological sounding rockets—in *International Science and Technology*. "Forty years ago, when Dr. Goddard was performing his first experiments, all rockets were small rockets. Today, with thrusts nearing 10 million pounds and rocket vehicles approaching the size of small skyscrapers, a 500-pound-thrust rocket seems like a relic of the past. Nevertheless, such small rockets remain as important to science and as challenging to technology as Dr. Goddard's fledglings were in his day. About 1500 of them were fired last year for various scientific purposes. . . ." (*Int. Sci. & Tech.*, 4/65, 32–37)
- Cosmic x-ray detection experiment carried aloft by an Aerobee sounding

rocket discovered the first two extragalactic x-ray sources and identified a variable x-ray source within the Milky Way galaxy. The two extragalactic sources—Sygnus A and M–87—were found to emit x-radiation 10 to 100 times their radio and light energy. The variable x-ray source was Casiopeia A. Details of the experimental results were announced March 2, 1966, by Dr. Herbert Friedman, Naval Research Laboratory physicist. Project was conducted by Dr. Friedman, E. T. Byram, and T. A. Chubb of NRL under sponsorship of NRL and National Science Foundation. (Clark, *NYT*, 3/3/66; *A&A*, 4/66, 98, 100)

May 1965

May 1: NASA Administrator James E. Webb, speaking at Rose Polytechnic Institute, Terre Haute, Ind.: "Indeed, the success of the national space program depends to a very large degree on the quality and the extent of involvement by the universities. Their most important contribution would naturally be in doing the jobs they are uniquely qualified to do, that is, in research and in educating and training at both the undergraduate and graduate levels the scientists, engineers, and other professional personnel required by the space program. . . .

"With its university program, the National Aeronautics and Space Administration is approaching a goal established early in its history. That goal, when achieved, will provide a substantial increment to those trained men who are capable of guiding this country's undertakings in science and technology confidently toward future needs that are only partially visible to us now. That goal is being pursued in institutions of higher learning where men teach and practice their specialties in the context of other highly refined fields of interest. Surely, this concept is broader than the space program itself." (Text)

- YF-12A, USAF's twin-jet, delta-winged interceptor prototype, established four speed and altitude records at Edwards AFB: (1) 2,062 mph straight-away speed record, breaking the 1,655.9 mph previous record held by the Soviet Union's E-166; (2) 80,000-ft. record for sustained altitude in horizontal flight, exceeding the E-166's 74,376-ft. record; (3) 1,688 mph record for 1,000-km. closed-course event with 2000-kg. (4,409-lb.) cargo, surpassing the 1,441 mph record set by the E-166 in April 1965; and (4) 1,642 mph record for 500-km. closed-course event, topping Soviet performance of 1,452 mph. USAF pilots Col. Robert L. Stephens and Lt. Col. Daniel Andre set the first two records; Maj. Walter F. Daniel and Capt. James Cooney, the others. YF-12A performed under requirements of the Fédération Aéronautique Internationale, world authority for verification of flight records. (DOD Release 281-65; *NYT*, 5/9/65, 88)
- Possibility that the wake of ice crystals—contrails—produced by supersonic jets would persist and spread into a thin, semipermanent haze layer at about 14-mi. altitude, increasing temperature of the air mass below, altering global wind patterns, and effecting unpredictable climate changes had been suggested by several weather specialists, reported Walter Sullivan. (Sullivan, *NYT*, 5/1/65, 1)

May 2: The recommendation to NASA by NAS-convened study group [see Apr. 26] that Mars receive "the highest priority among all objectives in space science," evoked editorial comment from the *New York Times:*

"The biological exploration of Mars will not be cheap, and available funds for scientific research and development are limited.

"The likely costs and returns of the search for Martian life must be compared with those from, say, programs for stepped-up research into cancer or for building giant accelerators that would permit physicists to peer more deeply into the recesses of the atomic nucleus. That broader consideration may well suggest a less concentrated program than the scientists had recommended.

"Such a decision would have the added advantage of allowing more time for an effort to make the search for Martian life a cooperative international project and not, . . . merely one more arena for the wasteful duplication that is the essence of Soviet-American space competition." (NYT, 5/2/65)

May 3: Nike-Cajun sounding rocket was launched from Wallops Station, Va., to obtain temperature, wind, density, and pressure at a time of minimum zonal wind flow by exploding twelve grenades during the ascent of the rocket. Two grenades did not eject and a third exploded before complete ejection, causing complete failure of experiment. Coordinated firings did not occur simultaneously at Ft. Churchill or Pt. Barrow due to weather conditions and payload problems. (NASA Rpt. SRL)

- FAA announced one-month extensions, through May 1965, of design contracts with Boeing Co. and Lockheed Aircraft Corp., airframe contractors; and General Electric Co. and Pratt & Whitney Div. of United Aircraft Corp., engine contractors, for U.S. supersonic transport program. Extensions applied to design contracts awarded to four companies for period Jan. 1 through Feb. 28, 1965, with provisions for one-month extensions from Feb. 28 through June 30. Dollar amount of each one-month airframe contract extension was $1 million ($750,000 Government, $250,000 contractor); dollar amount of each one-month engine contract extension was $835,000 ($626,250 Government, $208,750 contractor). (FAA Release 65-40)

- Gemini Astronaut John W. Young (LCdr., USN) was presented the Navy's astronaut wings by Secretary of the Navy Paul H. Nitze. (AP, *Wash. Post*, 5/4/65)

- EARLY BIRD I transmitted clear pictures and sound of live television programs between Europe and North America for 14 hrs. demonstrating its usefulness in regularly scheduled television. For three weeks, television's use of EARLY BIRD I would be restricted to Mondays; daily commercial use would not begin until fall when rates had been fixed. The satellite would be used on other days for telephone purposes and transmission of recorded information. (ComSatCorp; Gould, *NYT*, 5/4/65, 75)

- A GEMINI III experiment in which blood cells subjected to a known dosage of radiation were allowed to float around weightless in a container showed that weightlessness had no effect on irradiated human blood cells, according to Charles W. Mathews, Gemini program manager. He also explained why GEMINI III landed about 60 mi. short of predicted spot: The pilots were instructed to fly a bank angle based on wind-tunnel data of Gemini spacecraft's lift characteristics. But in actual reentry, the spacecraft's "lift was only about ⅔ of what we had expected it to be." Onboard instrumentation showed the discrepancy, but the command pilot followed ground instructions. When he ultimately changed the angle, based on the onboard display, it was too late to achieve the spacecraft target. (Transcript)

May 3: Editorializing in *Aviation Week and Space Technology*, Robert Hotz said that during the Eurospace meeting in Philadelphia, European members had made significant points of interest: "Europe needs a technically strong, economically beneficial and politically imaginative space program of its own if it is to remain a powerful economic entity and maintain its present standard of general prosperity . . . Europe must organize its technical and political resources on an over-all European level to be successful in space technology. . . . European industry faces a formidable task in selling the economic and political benefits of space technology to its people and governments. . . . European industry must develop its own space technology and cannot remain technically dependent on the U.S. regardless of how much support this country is willing to provide." Hotz concluded that "the fact that the discussions were so blunt and realistic proved the value of an organization such as Eurospace where these admittedly knotty problems can be aired. . . ." (Hotz, *Av. Wk.*, 5/3/65, 11)

- Discussions at last week's Eurospace meeting in Philadelphia indicated that "Europeans are eagerly seeking means to acquire U.S. technical know-how and systems management capability without buying hardware," wrote William J. Coughlin in a *Missiles and Rockets* editorial. He continued: "This was recognized in a blunt statement by Lockheed vice president Elmer P. Wheaton:

" 'As we see the situation, the real reason today for joint U.S.-European industrial cooperation is to facilitate acquisition by Europe of the technical capability the United States has been fortunate enough to develop. If we objectively appraise the existing circumstances, we all recognize that U.S. cooperation will often simply strengthen the European ability to compete more effectively with U.S. firms. With these facts in mind, it is obvious that the purchase of U.S. hardware does not best fulfill Europe's aims'. . . .

"As Lord Caldecote, managing director of the guided weapons division of British Aircraft Corp., put it: 'I cannot believe European taxpayers will be prepared to put forward money for programs on which American firms are prime contractors'. . . .

"The most hopeful route to European space collaboration probably lies in the proposals put forward for navigation, meteorological and television satellites." (Coughlin, *M&R*, 5/3/65, 46)

May 4: Aerobee sounding rocket successfully launched from NASA Wallops Station, Va. carried 317-lb. payload to 90-mi. altitude and impacted about 54 mi. downrange in the Atlantic. Conducted by NASA Goddard Space Flight Center, the stellar spectroscopy experiment measured special radiation of two stars, Spica and Alkaid, utilizing an ultraviolet stellar spectrometer and an input telescope with a 13-in. aperture. Performance of a gimbaled star tracker and modified attitude control (Strap) was also tested. Data were telemetered to ground station during flight. (Wallops Release 65-26)

May 4: Sen. Russell B. Long (D–La.) introduced a bill (S. 1899) in the Senate "to prescribe a national policy with respect to the acquisition, disposition, and use of proprietary rights in inventions made, and in scientific and technical information obtained, through the expenditure of public funds." Senator Long said in introducing the bill: "New discoveries derived from research supported by public funds belong to the people and constitute a part of the public domain to which all citizens should have access on terms of equality." (CR, 5/4/65, 9023–9027)

- Aerospace Corp.'s $22 million expenditure to construct buildings in California when space was available in nearby U.S. facilities was criticized by Comptroller General Joseph Campbell in his testimony before the House Armed Services Special Investigations Subcommittee investigating Aerospace. Campbell said that Aerospace had also incurred "certain questionable costs which appear to be of interest." (AP, *NYT*, 5/5/65)

- Dr. Eugene B. Konecci of the National Aeronautics and Space Council staff discussed future manned aerospace flight before the American Astronautical Society meeting in Chicago:
    "A great deal of lifting-body research is being performed by NASA and the USAF. In the not too distant future we will enter into the truly second generation manned spacecraft era by relying more on a higher L/D (lift-drag ratio) such as a hypersonic L/D of about 1.3. . . . The lifting body second generation manned spacecraft gives operational versatility for reentry from a number of orbit planes and gives a recovery capability at a number of landing sites within the United States. This versatility also increases the margin of safety for the astronauts. . . ." (Text)

- Orbit of MOLNIYA I Soviet communications satellite was slightly corrected to increase its usefulness for relaying telecasts between Moscow and Vladivostok. Soviet Communications Minister Nikolai D. Psurtsev told *Izvestia* that the firing of a special rocket motor aboard the satellite had raised the apogee to 40,045.2 km. (24,872.8 mi.); perigee to 548.4 km. (340.6 mi.); and the period to 12 hrs. Previous orbital parameters: apogee, 39,467.7 km. (24,514.1 mi.); perigee, 498.4 km. (309.5 mi.); period, 11 hrs. 48 min. The corrected high elliptical orbit put MOLNIYA I within the visibility of Russia's ground stations for the greater part of its period. (*NYT*, 5/5/65, 6)

May 4–6: U.S.S.R. Mars probe ZOND II had stopped transmitting data to earth, Russian physicist, Gennadii Skuridin, told the AAS–IIT Research Institute Symposium on Post-Apollo Space Exploration in Chicago. Cutoff apparently resulted from a failure in the probe's solar panels caused by meteoroid impact or solar radiation, he said. Other facts about the Soviet space program made public for the first time: (1) pressure in Cosmonaut Aleksey Leonov's spacesuit during his walk in space on March 18, was about 5.9 lbs. psi; (2) Leonov had trouble with his vision and in orienting himself while in space, but was capable of performing useful work; (3) Soviet scientists have the technological know-how to perform orbit-changing spacecraft maneuvers. A thirty-minute movie of Leonov's walk in space gave closeup views of construction of VOSKHOD II's airlock, Leonov's spacesuit, gloves, footwear, and life-support equipment back pack.

Discussing investigation of space by the U.S.S.R., Skuridin said that from 1962, the problem of going to the moon had been studied with the Cosmos series of spacecraft. He said COSMOS III and COSMOS IV had studied solor plasma, its energy and location in earth areas, and during the period April 24 to May 2, 1962, had transmitted to earth more than 50 million measurements; 20 million more had been stored in a data-storage system. COSMOS XLI had investigated charged particles at 40,000 mi. altitude. The Elektron series, he continued, had made important measurements of the atmosphere up to an altitude of 3,000 km. The ions of hydrogen, carbon, and oxygen had also been measured.

Discussing future flights, Skuridin said the Soviet Union would like to study Saturn, Pluto, and the sun, but added that a satellite was needed that could be launched to far-off planets and the sun and return to earth. (UPI, *NYT*, 5/6/65, 2; Kotulak, *Chic. Trib.*, 5/7/65; *M&R*, 5/10/65, 12, 13)

May 4–6: Preliminary plans for Apollo Extension System (Aes) development required selection of three major spacecraft contractors, NASA official told *Missiles and Rockets* during Symposium on Post-Apollo Space Exploration in Chicago: one to devise single payload plan, one to cover physical installation of experimental payloads and checkout systems, and one to translate Apollo spacecraft into Apollo extension vehicle with a six-week manned orbiting capability. NASA official attributed this decision to a reluctance to depend on a single contractor and a desire for broad-based readily available industrial capability. (*M&R*, 5/10/65, 13)

May 5: Soviet Cosmonaut Aleksey Leonov, first man to walk in space, had received 1/230th of the permissible radiation dose, proving that space travel is radiation-safe, Tass announced. (Reuters, *NYT*, 5/6/65, 2)

• NASA Assistant Deputy Administrator Dr. George L. Simpson, Jr., was named Chancellor of the University System of Georgia and would assume the duties of the new post July 15. Simpson, who had joined NASA in 1962 as Assistant Administrator of Public Affairs, later became Assistant Administrator for Technology Utilization and Policy Planning. In July 1964 he assumed the additional duties of Assistant Deputy Administrator. A native of North Carolina, he had been a professor at the Univ. of North Carolina and a planner of the Research Triangle, cooperative endeavor of the Univ. of North Carolina, Duke Univ., and North Carolina State College. (AP, Wash. *Eve. Star*, 5/5/65)

• USAF Chief of Staff General John P. McConnell, speaking at a meeting of the National Press Club in Washington, D.C., said: ". . . As airmen, all of us in the Air Force look at space with real concern. Will it someday become an area of military operation? If so, what will be the U.S. posture? In military language, what is our readiness? . . .

"Space exploration is the responsibility of the National Aeronautics and Space Administration. . . . The act which created the National Aeronautics and Space Administration gave NASA broad responsibilities for meeting many of the broad needs of the nation. It also stated that the Department of Defense should be responsible for and direct those space activities peculiar to or primarily associated with the development of weapon systems, military operations or the defense of the United States.

"So we have both NASA and Air Force assigned specific responsibilities. We have the basis of a partnership. And a partnership it is in carrying out the national program as recommended by the President and authorized and funded by Congress. The intent of Congress is very clear. The members wanted the broad space capabilities of the nation to be built up as rapidly as possible without unnecessary duplication of effort or of waste. This we are attempting to do. And while I would not ordinarily try to speak for Jim Webb, the NASA Administrator, I think I can speak for him today on this subject, in saying, that it is a very well understood mutual objective between the Air Force and NASA."

Asked his opinion about spending of $20 billion to reach the moon, McConnell said: "I think it is necessary for us to get everything we can out of space. And I think we should get it as rapidly as we can at as reasonable cost as we can. But you can't get it rapidly and at the same time cheaply . . . going to the moon is just the end product of what we are getting out of it. If we were just going to the moon, I wouldn't think it would be worth 20 billion dollars to go to the moon. But I don't hesitate to say that all of the other things which we have to do, the preliminaries, and the things that we're going to learn in the process of achieving that goal is well worth the expenditure of whatever money is required to attain the knowledge which we will attain as a result of this project." (Text)

May 5: Boeing Co. unveiled to the public a mockup of its Molab (Mobile Laboratory), a six-wheeled vehicle being studied by NASA for use in manned exploration of the moon. (AP, *Tulsa Daily World*, 5/6/65)

May 6: MARINER IV, after 159 days in space, was 72 million mi. from earth, had travelled 243 million mi. The spacecraft continued to return scientific and engineering data to ground stations daily and to set new records for distance of communications. (NASA Release 65–148)

• U.S. House of Representatives passed a bill authorizing appropriations to NASA for FY 1966 totaling $5,183,844,850, as follows: $4,537,121,000 for research and development; $60,675,000 for construction of facilities; and $586,048,850 for administrative operations. NASA had requested $5.26 billion.

During the debate preceding passage of the bill, Rep. James G. Fulton (R–Pa.) said: "We have moved quickly. But we are not in a crash program. We are now conducting a reasonable program . . . it is a well-planned program.

"It is impossible to believe that in the fiscal year 1959 only $48,354,000 was authorized for space [NASA]. In fiscal year 1960 it went up ten times to $485,550,000. It doubled again in fiscal year 1961 to $915 million.

"In fiscal year 1962 it went to $1,361,900,000.

"In fiscal year 1963, it went to $3,742,162,000 and in fiscal year 1964 to $5,238,119,400.

"In fiscal year 1965 it went to $5,193,810,500.

"For this fiscal year, the committee has recommended $5,183,844,-850, which is down from last year's level.

"I want the House to know we have gone over these programs thoroughly. We have made cuts in the committee, and the cuts were worthwhile. They are responsible, and they are substantial. They are not small."

Regarding funds restored to the NASA budget, Rep. Fulton said: "The M-1 engine development, the 260-inch engine development, and the SNAP-8 development, were ongoing programs of research that were approved by the committee and authorized by the House over the past several years. However, for reasons of economy, the Administrator cut these three programs entirely from the NASA budget. The committee on the other hand, believed that such actions in the long run would be extremely wasteful and later result in very high costs when it would become necessary to reactivate these programs.

"Consequently, the committee restored $15 million to the M-1 program to continue it on a technological development level, $6.2 million to the 260-inch solid rocket program to carry it through the test firing of two full length rockets, and $6 million to the SNAP-8 to continue it at the scheduled level of effort."

Rep. Olin E. Teague (D-Tex.) discussed changes made by the Manned Space Flight Subcommittee: "The total request by NASA for manned space flight for fiscal year 1966 is $3,567,052,000. . . . The subcommittee is recommending a total reduction of $42,825,000.

"NASA requested $3,249,485,000 for research and development in manned space flights. Total reduction in research and development amounts to $30 million. All of this reduction comes from the Apollo program. It is the view of the subcommittee that in the areas of Apollo mission support and engine development that program improvements could be made. However, the reduction was made in the total request to allow NASA to make program alterations with a broad management latitude of choice without adversely affecting the total program. It was recognized by the subcommittee that NASA, prior to coming before the committee, had made substantial reductions in their total research and development program. A further reduction was also made by the Bureau of the Budget. Based on this, the $30 million reduction is considered a maximum amount that could be taken without jeopardizing the pace and progress of the Apollo program." (CR, 5/6/65, 9291, 9296, 9301)

May 6: Saturn V booster (S-IC stage) was static-fired for the second time at NASA Marshall Space Flight Center. The five F-1 engines were ignited in a 15-sec. test during which they generated 7.5 million lbs. thrust. Tests of this stage would gradually increase in duration until full-length firing of 2½ min. was reached in late spring or early summer. (MSFC Release 65-117)

• USAF Titan III-A rocket was fired from Eastern Test Range in a maneuverability test in which the third stage (transtage), carrying two satellites, executed a series of consecutive and intricate maneuvers. Primary goal of the mission was four separate ignitions of the transtage's engines—a feat never before attempted.

First firing, after burnout of the first two stages, lasted 296 sec. and injected the 7,000-lb. rocket-payload assembly into near-earth orbit of 125-mi. (201.3 km.) apogee, 108-mi. (173.9-km.) perigee, and 88.1-min. period. After one earth orbit, about 90 min. after launch, the third stage ignited a second time, for 37 sec., driving the stage upward into an elliptical orbit of apogee, 1,757 mi. (2,828.8 km.) and perigee, 115 mi. (185.2 km.). Two and one-half hours later, transtage's two 8,000-lb.-thrust engines burned a third time, for 27 sec., to circularize the orbit at 1,743-mi. (2,806.2-km.) apogee and 1,729-mi. (2,783.7-km.) perigee. Thirty seconds after shutdown of the transtage, an 82-lb. Lincoln Laboratory experimental communications satellite (LES II), equipped with its own rocket motor to shoot itself into a higher elliptical orbit, was spring-ejected from the stage. LES II attained orbit of 9,364-mi. (15,076-km.) apogee; 1,753-mi. (2,822-km.) perigee; 315-min. period; and 31.35° inclination. Then, 42 sec. after LES II was released, a 44.5-in.-dia., 75-lb. hollow aluminum radar calibration sphere (LCS I) was ejected from the transtage. LCS I was to remain in near-circular orbit with 1,743-mi. (2,806.2-km.) apogee, 1,729-mi. (2,783.7-km.) perigee. Seven hours after launch, the transtage was fired a fourth time, driving it into a final elliptical orbit of 2,317-mi. (3,730.4-km.) apogee; 1,725-mi. (2,777.3-km.) perigee; 157-min. period; and 32.07° inclination. (UPI, *NYT*, 5/7/65, 12; *Av. Wk.*, 5/10/65, 33; USAF Proj. Off.; *U.S. Aeron. & Space Act., 1965*, 141)

May 6: NASA announced its agreement with the Brazilian Space Commission (CNAE) to cooperate in scientific sounding rocket program to investigate the lower regions of the ionosphere, emphasizing the effects of cosmic rays. NASA would provide and CNAE would launch two sounding rockets from Natal, Brazil; scientific payloads would be constructed by Brazilian technicians at NASA Goddard Space Flight Center. NASA and CNAE would combine to provide ground support equipment, to analyze data, and to publish the results of the experiment. In addition, NASA would launch one instrumented sounding rocket from Wallops Station, Va., in a complementary experiment. The project would contribute to observance of 1965 as International Cooperation Year. (NASA Release 65–149)

• To assure expeditious completion of NASA's Mississippi Test Facility—permanent national center for ground testing of large launch vehicle stages—Marshall Space Flight Center announced two changes in preparation of the installation: (1) buildup in personnel would start immediately; (2) MSFC planning, construction, and activation elements would be grouped into a new Mississippi Test Facility Task Force. Jackson Balch, until now MSFC's assistant deputy director, technical, would have the dual titles of Mississippi Test Facility site manager and head of the MTF Task Force. A permanent organization to operate MTF once it was activated would be formed later. (MSFC Release 65–114)

• Techniques for weather predictions reliable up to two weeks were discussed at Geophysics Corp. of America in Bedford, Mass., by Dr. D. Q. Wark of the U.S. Weather Bureau, Dr. William Nordberg of NASA Goddard Space Flight Center, and Dr. Jean I. F. King of GCA. These scientists had successfully utilized radio waves to collect weather data and were planning to build a new weather satellite which could log greater amounts of data and provide constant coverage. They proposed placing weather buoys in the oceans and weather balloons in the atmosphere equipped to relay data to the orbiting satellite which, in turn, would relay data to ground stations. (Hughes, *CSM*, 5/6/65)

May 6: Russian communications satellite MOLNIYA I, because of its higher and sharply elongated orbit, could transmit continuously several hours longer than American Telstars, reported Tass. Tass claimed that Telstars could transmit uninterruptedly for only 30 min. (Reuters, *NYT*, 5/7/65, 3)

- Chairman of the UCLA Astronomy Dept. Dr. L. H. Aller believed the moon might be as solid as metal below the top few inches of surface, reported George Getze in the *Los Angeles Times*. According to Getze, Aller said chances were good that the chemical composition of the moon was more like the sun's than the earth's and that elements in the sun as gases would be found in the moon as solids. "The first few inches of the moon's surface may have been changed a good deal by meteor hits and solar radiation, but if we go down a few feet we will probably find that the composition is like the sun's," he said. (Getze, *L.A. Times*, 5/6/65)

- Editorializing in the *Evening Star*, Richard Fryklund said: "It is a pity . . . that the hot, new plane, called the YF–12A, has almost no chance to be used by the Air Force for anything except tests and speed records. . . .

  "The reason: Secretary of Defense McNamara doubts that any new interceptor is needed or that the Air Force's nomination is the right plane even if one is needed. . . .

  "Three of the records set by the YF–12A on May Day are considered to be the most important performance checks on any airplane: Speed over a straight course (2,062 miles an hour, or about mach 3.2), altitude (80,000 feet, though it can go higher) and speed around a circular course (1,688 miles an hour)." (Fryklund, *Wash. Eve. Star*, 5/6/65, 7)

May 7: U.S.S.R. launched COSMOS LXVI with scientific instruments aboard for investigation of outer space, Tass announced. Orbital data: apogee, 291 km. (180.7 mi.); perigee, 197 km. (122.3 mi.); period, 89.3 min.; inclination, 65°. All systems were functioning normally. (Tass, 5/7/65)

- The President of Aerospace Corp., Dr. Ivan A. Getting, replying to Comptroller General Joseph Campbell's charges that Aerospace had spent $22 million to build new facilities in California when Government space was available, told the House Armed Services Special Investigations Subcommittee that the separate buildings assured "the financial independence and stability to enable the corporation to perform its mission," and that the construction would be paid for with earnings from Government contracts and fees. Chairman of the Subcommittee Rep. Porter Hardy, Jr. (D–Va.), said that the hearings had revealed "startling deficiencies in the control of public funds made available to Aerospace." (AP, *NYT*, 5/9/65, 76)

- Civil Aeronautics Board approved a United Air Lines plan to lease eight Boeing 727–22 jet airliners from a group of 22 banks rather than buy them directly from Boeing. United told the CAB, in applying for approval of the new agreement, that the lease arrangement would give it the use of the planes on a cost basis substantially more favorable than if it leased the planes some other way or financed their purchase through commercial bank borrowings. Each plane would be leased to United for 13 yrs. from date of delivery. (*WSJ*, 5/7/65, 6)

May 7: Civil Aeronautics Board request for $2.1 million in subsidies for commercial helicopter lines in New York, Chicago, and Los Angeles, to be paid during July–December period, was denied by House Appropriations Committee. (*WSJ*, 5/7/65, 2)

- Pan American Airways announced it would purchase four additional Boeing 727 jet aircraft, bringing its Boeing purchase program total to 19. (*WSJ*, 5/7/65, 3)

May 8: "We are already getting ready for the next manned flight," Soviet Cosmonaut Pavel Belyayev's backup pilot wrote in a Soviet air force journal. "We are getting acquainted with the construction of a new ship" and "planning new flights on new courses with more complicated assignments." (*Wash. Post*, 5/8/65)

May 9: LUNA V, a 3,254-lb. instrumented moon probe, was successfully launched by U.S.S.R. on an undisclosed mission. According to Tass announcement, the probe was launched by multi-stage rocket into a parking orbit and then fired toward the moon. All onboard equipment was said to be functioning normally and a U.S.S.R. station tracking the probe was receiving "scientific information." Tass reported that LUNA V was "moving along a trajectory close to the planned one." At 10:00 p.m. Moscow time, the probe was 110,000 km. (68,323 mi.) from earth. (Tass, 5/9/65)

- Sir Bernard Lovell, director of the radiotelescope facility at Jodrell Bank, England, said that the telescope would try to track Soviet lunar probe LUNA V on May 10. "We have been expecting the Russians to make an attempt to achieve a soft landing of an instrumented package on the moon for some time now," he said. "This may possibly be the attempt." (*NYT*, 5/10/65)

- Studies on flight handling qualities of a manned lifting body reentry vehicle during the later stages of reentry and during the landing approach were being jointly conducted by NASA and Cornell Aeronautical Laboratory at NASA's Flight Research Center using a T-33 jet aircraft specifically modified for AF Systems Command by Cornell. Cornell was working under a NASA-funded $231,000 contract which also included human transfer–function studies and ground simulation of the lifting body. (FRC Release 12–65)

- Recently released photograph of the recoverable capsule of the U.S.S.R. Vostok spacecraft revealed that the craft was spherical and that one third of it was covered with an unidentified material marked by concentric rings. In a *New York Times* article, Frederic Appel said that the U.S. had rejected a spherical design for U.S. spacecraft because of its lack of dynamic stability and because, during reentry, too much surface was exposed to hot gases deflected by the heat shield raising the internal temperature above allowable limits. Appel speculated that the Soviets might have solved the problem with greater heat insulation or a more powerful cooling system and that the material marked by concentric rings could be the remains of a heat shield that had burned away. (Appel, *NYT*, 5/9/65, 14)

- U.S.S.R displayed some of its newest, most powerful missiles during a parade across Red Square in Moscow commemorating 20th anniversary of victory over Hitler's Army. Missiles never before displayed included two three-stage missiles about 110 ft. long and 10 ft. in diameter which Tass described as of "unlimited" range and as similar to the rockets that orbited the Vostok and Voskhod spacecraft; two missiles of similar construction—about 65 ft. long—described by Tass as "intercontinental rockets" using solid fuel; a massive self-propelled missile consisting of a tracked carrier topped by a stubby rocket resting as if in a pod and described by Tass as a solid-fuel medium-range missile of "tremendous destructive power." This was the first time the Soviet Union had officially reported it possessed a solid-fuel rocket of the intercontinental, or orbital, type.

Also in the parade were a Polaris-type missile used by submarines and what Tass described as an "antimissile missile." These types of weapons had been displayed before. (Tanner, *NYT*, 5/10/65)

May 9: Dr. Richard L. Lesher, consultant to NASA since June 1964 and a special assistant to Breene M. Kerr, NASA Assistant Administrator for Technology Utilization since Nov. 1964, became NASA Deputy Assistant Administrator for Technology Utilization. (NASA Release 65-161)

- Sixty college science and engineering students selected in a nationwide competition were awarded NASA grants to participate in a summer space science program at Columbia Univ. (*NYT*, 5/9/65, 34)

May 9-12: During NASA Conference on Aircraft Operating Problems, NASA scientists reported to Government and industry technical experts on research accomplishments leading to improved aircraft usefulness and safety. Held at NASA Langley Research Center, the technical sessions were under the sponsorship of NASA's Office of Advanced Research and Technology and included 34 papers. (NASA Release 65-160; NASA SP-83.)

May 10: Tass announced that LUNA V probe had undergone a planned midcourse maneuver to change its trajectory. (Tass, 5/11/65)

- Evidence of life on earth 2.7 billion yrs. ago was reported by Univ. of California professor and Nobel prize winner Melvin Calvin. The evidence was in the form of two chemicals, phytane and pristane, extracted from the Soudan Formation, a carbon-rich and precisely-dated geological stratum in Minnesota. Both are carbon-hydrogen compounds; both are manufactured only by living systems; both are stable enough to have survived unaltered. As Calvin reconstructed it, both chemicals were synthesized by chlorophyll-containing plants—fairly high forms of life requiring long ancestry. First signs of earthly life must therefore have existed 800 million yrs. prior to the date currently accepted. (*Newsweek*, 5/10/65)

- Rep. James C. Corman (D–Calif.) announced that a poll taken among his constituents showed that 68.7% supported a program to land an American on the moon by 1970; 14% felt the program should be slowed down; 17.3% disapproved of the program. (*CR*, 5/10/65, A2275)

- Newest U.S. telescope, a 24-in. reflector for photographing stars, was operating at Univ. of Rochester under direction of Dr. Stewart Sharpless. It would be used to study the structure of the galaxies, the gas and dust between stars, and the evolution of variable stars. (Sci. Serv., *Wash. Daily News*, 5/10/65)

- USAF scientist Dr. John W. Evans received DOD's Distinguished Civilian Service Award for his research on the physical processes of solar magnetic fields, mass motions of the solar photosphere, and growth and development of solar flares. (OAR Release 5-65-1)

May 10: Second stage (S-IV) for the tenth and last Saturn I launch vehicle was delivered to Kennedy Space Center, NASA, aboard "Pregnant Guppy" aircraft. The stage was flown from Douglas Aircraft Co.'s Sacramento, Calif., facility. (MSFC Release 65-135)

- U.S.S.R.'s antimissile missile and other powerful rockets were shown in action for the first time in a film on Moscow television, "Rockets in Defense of Peace." Included were test firings of surface-to-air, air-to-surface, and underwater missiles as well as launchings of intermediate and intercontinental surface-to-surface ballistic missiles, some from underground silos. Also displayed were installations of the Soviet antimissile defense, including testing stations, computer centers supplying data for interceptions, and launching sites for interceptor missiles. One sequence showed firing of an antimissile missile and the interception of an intercontinental ballistic missile at an unspecified altitude. (Shabad, *NYT*, 3/11/65, 4)

- In a *New York Times* article, Jack Gould suggested that statesmen planning EARLY BIRD I telecasts prepare their speeches well in advance and consider time differences in their scheduling. He noted that President Johnson's speech had received limited European coverage because it was hastily arranged and that the address of West Germany's Chancellor Ludwig Erhard had suffered because of an unusually poor simultaneous English translation. (Gould, *NYT*, 5/10/65, 59)

May 11: NASA Administrator James E. Webb, speaking to the Washington Board of Trade, said: "In 1959, when NASA attempted 14 space flights, we had 37 percent success in missions and launch vehicles. Last year we attempted 30 missions, more than twice as many as in 1959, and the percentage of success in missions went up to 83, with 93 percent success in vehicles. So far this year, the percentages are holding close to those of 1964." (Text)

- Successful 75-min. test of USAF F-111A supersonic fighter bomber was conducted at Edwards AFB by Lt. Col. James W. Wood (USAF) who flew at 760 mph and to 30,000 ft. (AP, *NYT*, 5/12/65)

- F-111B, USN version of the F-111 multipurpose fighter designed for use by both USAF and USN, was displayed for the first time during a rollout ceremony at the Grumman Aircraft Engineering Corp. plant at Peconic, L.I. Secretary of the Navy Paul H. Nitze was the principal speaker. Test pilots demonstrated the variable-sweep wing which could extend almost perpendicular to the fuselage for take-offs, landings, and slow flight, and then pivot back sharply for supersonic flight. In a news conference, Brig. Gen. John L. Zoeckler (USAF), F-111 project manager, acknowledged that F-111B was 500-600 lbs. "overweight," but said that "very substantial strides" had been made in weight reduction. He added that "some compensation" in performance would be achieved in later USN models by addition of high-lift devices. The two-man, all-weather, supersonic aircraft was designed to fly at about 1,600 mph. (DOD Release 285-65; Hudson, *NYT*, 5/12/65, 18)

- A third solid-fuel Pershing ballistic missile unit would be moved to Europe this month, DOD announced. The Pershing could reach 400 mi.

with either a nuclear or conventional warhead and would replace the slower-firing liquid-fuel Redstone missile. (*Wash. Post,* 5/11/65)

May 11: NASA announced closing of its Santa Monica and Dallas Area Professional Staffing Offices and moving of its New York office to Boston to assist in recruitment program for new Electronics Research Center in Cambridge. (NASA Release 65-156)

- Dr. Raymond L. Bisplinghoff, NASA Associate Administrator for Advanced Research and Technology, announced the appointment of Francis J. Sullivan as Director of the Electronics and Control Div. of NASA's Office of Advanced Research and Technology. Mr. Sullivan had been serving as Acting Director since Sept. 1, 1964. (NASA Release 65-152)

- Decision to narrow the choice of type of broadcast satellite it would consider from three to two was announced by ComSatCorp president Joseph V. Charyk at a stockholder's meeting in Washington, D.C. Two of the approaches under consideration involved satellites that would operate about 6,000 mi. above the earth: one would have 18 satellites, orbiting in random positions; the other would have 12 satellites, orbiting at "phased" or controlled positions. The third approach, being tested in EARLY BIRD I, had satellites placed at an altitude of 22,000 mi. in synchronous orbit. In his speech, Charyk revealed that the corporation had decided to drop from consideration the 6,000-mi. random version. It had been discovered, he reported, that a 6,000-mi.-high satellite could be controlled more easily than ComSatCorp had believed when it first started studying random satellites as one alternative approach. Moreover, Charyk said, it now appeared that a satellite could be designed that would operate either at 6,000 mi. in controlled positions or at the 22,000-mi.-high, synchronous position.

  ComSatCorp therefore would invite satellite designers to offer bids to build this type of satellite. (Denniston, *Wash. Eve. Star,* 5/11/65, 12)

May 11-21: 1965 COSPAR (Committee on Space Research) meeting was held in Mar del Plata, Argentina, where it was moved from Buenos Aires because of student demonstrations.

A new working group was formed, with Morris Tepper (Chief of Meteorological Programs, NASA) as chairman. Called Working Group VI, for Scientific Space Experiments Concerned with Properties and Dynamics of the Troposphere and Stratosphere, it was formed to "further international understanding of, and cooperation in, the use of rocket and satellite systems and techniques for meteorological research, and to promote international discussions involving meteorologists with scientists of other disciplines in order to provide a good climate for the development of imaginative new approaches to the use of rockets and satellites for meteorological research." (NAS-NRC *News Report,* Vol. xv, 5/6/65, 6)

Dr. O. Z. Gazenko, physiologist and member of the Soviet Academy of Sciences, said that cosmonauts had no difficulty knowing the orientation of their bodies and experienced no nervous disorders if they were given visual cues. He based his remarks on experiences of Soviet cosmonauts, especially those of Lt. Col. Aleksei Leonov in VOSKHOD II: "When he saw the spacecraft, he had no problem knowing his orientation, but it was different when he didn't see the spacecraft."

During the launch phase, the cosmonauts' pulse and breathing rates were greater than noted during centrifugal tests in ground laboratories. In VOSKHOD II, it took a comparatively long time for the normal levels to be reached, according to Gazenko. Comparison of the data of the Voskhod flights with other space flights showed fewer cardiovascular variations and better responses to stress. He recommended crews of several people since "the feeling of togetherness of cosmonauts is very important." (Text)

Success of the fully stabilized British Skylark rocket in obtaining new astrophysical data was described. Skylark was a single-stage, solid-fuel vehicle designed to carry 150–200-lb. payload to 200 km. (124 mi.) altitude. Using the sun as a reference, Skylark could achieve pointing accuracy of between two and three minutes of arc in pitch and yaw, reported Kenneth Pounds, lecturer at Leicester College, England, and one of its users: "The new Skylark has revolutionized the whole field of rocket research as far as we're concerned." He pointed out that many scientific experiments, such as taking x-ray photographs of the sun, could not be done by an unstabilized rocket: "You need 100 sec. or more exposure time, plus roll stabilization, or the photographs will be blurred." (COSPAR Rpt.)

M. S. V. Rao, reporting on the Thumba, India, experiments conducted on World Days during the 1964–1965 IQSY, said east northeasterly winds with speeds of 60–90 knots were observed in the stratosphere during the monsoon. In the mesosphere, data revealed a region of unusually strong winds with high shear. Rao reported that radar observations of the rate of dispersion of chaff confirmed existence of complex pattern of high shears and pronounced turbulence in the equatorial mesosphere in the monsoon season. (Text)

NASA scientists at Wallops Station, Va., had made a similar rocket launching during the Thumba experiments to get a synoptic picture. Arnold Frutkin, NASA Director of International Programs, said at the COSPAR meeting that "these data were the first relating to the monsoon problem on a global scale. It shows what very important work less advanced countries can do." (M&R, 5/24/65, 17)

Activities in the 1964 U.S. space program were summarized by Dr. Richard W. Porter, National Academy of Sciences delegate to COSPAR: ". . . Satellite storm warnings, intercontinental television, voice and data transmissions via satellite, all-weather navigational 'fixes' for ships at sea, and precise map making by means of satellite observations have become almost commonplace events. Space launchings at frequent intervals are providing a continuous stream of new information of value to science and mankind. . . . In total, the various agencies of the United States carried out sixty-one successful satellite and space probe launchings; however, because of the occasional practice of launching several satellites at a time . . . the total number of useful discrete payloads in Earth orbit or escape trajectory was seventy-seven. . . .

"In addition, the United States launched seven large high-altitude rocket probes, in the range from 700 to 1100 km., and well over one hundred other scientific sounding rockets, most of which reached altitudes between 110 and 250 km. Twelve hundred twenty-three small meteorological rockets, having a payload of about 5 kg. and a maximum altitude of about 60 km. were fired on regular schedules by the meteorological rocket network and more than one hundred scientifically instrumented large balloon flights were made during the period.

"Technological advances made during this period which will contribute significantly to the space-research capability of the United States include the launching of three SATURN I booster rockets, capable of putting about 7500 kg. of useful weight into Earth orbit, one TITAN IIIA booster, and successful tests of the CENTAUR liquid hydrogen rocket. Electrostatic ion accelerator rocket propulsion devices were tested in space during 1964, and more recently a nuclear reactor with thermoelectric energy conversion devices successfully began an endurance run in space which is still continuing. Passive gravity-gradient stabilization techniques have been perfected by means of additional satellite tests to the point where this technique is ready for useful employment in a variety of space applications. The highly directional properties of a lasar beam were successfully used in tracking a satellite. Significant improvements were also made in sounding rocket, high altitude balloon design and in data conversion facilities." (Text)

K. Maeda, chief Japanese delegate to the sixth international space symposium of COSPAR, told *Missiles and Rockets* Japan would launch its own satellite with its own launch vehicle within the next three years. A four-stage, solid-fueled rocket would be used, with the Mu rocket as first stage. The satellite, to be used solely for scientific research, would weigh between 50–100 kgs. and be sent into a 500–1,000-km. (311–621-mi.) orbit. (*M&R*, 5/17/65, 9)

At a news conference during the COSPAR meeting, A. A. Blagonravov, chief Soviet delegate, said that in view of the difficulties of soft landing on the moon, the Soviet Union would probably try to soft land another Lunik before attempting to land cosmonauts. He said the lunar surface must be known in detail and "should be examined by automatic stations." Because of the problems involved, he added, "it is not possible to set any date for a lunar landing." (*M&R*, 5/24/65, 17)

May 12: LUNA V "hit the moon in the area of the Sea of Clouds" at 10:10 p.m. Moscow time [3:10 p.m. EDT], Tass announced. The release continued: "During the flight and the approach of the station to the moon a great deal of information was obtained which is necessary for the further elaboration of a system for soft landing on the moon's surface." The announcement revealed no further details of the landing. Western experts saw evidence that the Soviets had attempted a soft landing and failed. (Tass, 5/12/65; Shabad, *NYT*, 5/13/65, 1, 24)

- USAF launched Blue Scout Jr. space probe from Eastern Test Range with instrumented payload to measure pitch angle and magnetic field intensity in space. Probe attained 8,536-mi. altitude in its 3-hr. 50-min. flight and returned useful data to earth before falling into Indian Ocean. (*U.S. Aeron. & Space Act., 1965*, 141)
- First developmental test of a possible landing system for the Apollo Spacecraft was successfully performed at NASA Manned Spacecraft

Center with the drop of a boilerplate spacecraft from a crane into a 700,000 gallon water tank. The boilerplate was fitted with two pairs of rockets and an 8-ft.-long altitude sensor. Rockets were mounted in the outer rim of the heat shield; thrust vector of the rockets was aligned with the gravity vector of the spacecraft.

Structural reinforcement of the heat shield area was current solution for preventing damage to the spacecraft in a rough water landing. If the landing rocket system proved desirable, it would cut several hundred pounds from the weight of the Apollo command module in addition to providing an improved emergency and landing capability. (MSC *Roundup*, 5/28/65, 8)

May 12: USN would build new stations at Raymondville and Roma, Tex., as part of its SPADATS (Space Detection and Tracking System) surveillance network for detecting satellites passing over the U.S., reported Warren Burkett in the *Houston Chronicle*. (Burkett, *Houston Chron.*, 5/12/65)

- NASA announced award of $15 million contract to Grumman Aircraft Engineering Corp. to build an additional Orbiting Astronomical Observatory. Grumman already was building three Oao spacecraft under a previously awarded contract. (NASA Release 65–154)
- At Bell Telephone Laboratories a two-mile-long laser beam was folded into a ten-foot-long space by reflecting the beam back and forth more than 1,000 times between two mirrors. By distorting the shape of the mirrors to enable the beam spots to form a pattern of slowly changing ellipses, scientists kept the reflections separate. Bell predicted that a computer utilizing this effect could store 1,000 bits of information which could be read out serially one bit every billionth of a second. (*NYT*, 5/12/65)
- DOD awarded Smith and Sapp Construction Co. a $1,616,970, NASA-funded, fixed-price contract for construction alterations to existing spacecraft facilities at Cape Kennedy. (DOD Release 323–65)
- Soviet's first communications satellite MOLNIYA I maintained direct radiotelephone communications between Vladivostok and Sofia, Warsaw, and Prague for almost three hours. (Tass, 5/12/65)
- In interim decision, FCC awarded ComSatCorp for two years "sole responsibility" for design, construction, and operation of three ground stations for a global communications network. Future of AT&T-owned Andover, Me., station was not discussed. (ComSatCorp)
- XB–70 and Boeing 707 noise comparison results were reported by FRC engineers Carol S. Tanner and Norman J. McLeod at Aircraft Operating Problems Committee meeting at LaRC. During takeoffs both aircraft reached maximum noise level in the frequency range of about 125 cps. Data from tests would aid in prediction of runway noise levels for the proposed supersonic transport. (FRC Release 13–65)
- Capt. Robert F. Freitag (USN, Ret.), Director of NASA Manned Space Flight Field Center Development, told Theodore von Kármán Memorial Seminar in Los Angeles that solutions to air and water pollution "could very well develop out of the research now being undertaken to develop self-sustaining life support systems for astronauts on missions of long duration." (West, *L.A. Times*, 5/13/65)

May 13: MARINER IV, 78,277,013 mi. from earth at 9 a.m. EST, had covered 251,691,170 mi. along its orbit. The Mars probe was travelling 46,214 mph relative to earth and was returning data and scientific information continuously. (NASA Release 65-159)

- President Johnson transmitted to Congress a plan to merge the Weather Bureau, the Coast and Geodetic Survey, and the Central Radio Propagation Laboratory of the National Bureau of Standards into an Environmental Science Services Administration. "The new administration will then provide a single national focus for our efforts to describe, understand, and predict the state of the oceans, the state of the lower and upper atmospheres and the size and shape of the earth . . . as well as enhance our ability to develop an adequate warning system for the severe hazards of nature . . . which have proved so disastrous to the Nation in recent years." He added that Federal agencies "concerned with the national defense [and the] exploration of outer space" would receive improved services and that combining of offices and technical facilities would save money. (White House Release)

- Gemini GT-4 countdown rehearsal at KSC with Astronauts James A. McDivitt (Maj., USAF) and Edward H. White II (Maj., USAF) was delayed because of a minor fueling problem. Launch of the GT-4 mission was scheduled for June 3. (AP, Galveston News-Tribune, 5/14/65)

- An Emeraude rocket was successfully fired by France from the Hammaguir range, Algerian Sahara, to a planned altitude of 112 mi. It was topped by a mockup of the Topaze rocket which was to be the second stage of the Diamant launcher that France was developing. (Reuters, NYT, 5/18/65; Root, Wash. Post, 5/18/65; M&R, 5/31/65, 11)

- NASA Goddard Space Flight Center researchers Dr. John B. Schutt and Charles M. Shai announced development of a new series of inorganic spacecraft paints with promise of commercial application; report given at a meeting of the Philadelphia Society for Paint Technology and the Philadelphia Section of the American Chemical Society. The paints would utilize an alkali-metal silicate as a binder and an inorganic phosphate as a wetting agent. They would adhere to most metals and non-metals; would not crack, peel, chalk, flake, or fade when subjected to temperatures between 1,800°F and −320°F; would be washable; could be made in any color; and would have a long shelf life. (GSFC Release G-13-65)

- American Broadcasting Co. notified the FCC it was preparing plans for a domestic communications satellite to relay network television programs to affiliated stations for rebroadcast, thereby raising for the first time the question of a company other than ComSatCorp owning and operating a comsat. (Gould, NYT, 5/14/65, 1)

- NBC announced it would televise the June 3 Gemini GT-4 spaceflight in color. It would be the first live-color coverage of a space flight. (Doan, N.Y. Her. Trib., 5/14/65)

- Cornell Univ. astronomers at Arecibo radiotelescope facility revealed that their radar observations of the planet Mercury April 25 indicated that Mercury rotated on its axis once every 59 days, rotating in the same direction as its orbit. This new study confirmed clearly that Mercury did not have a retrograde rotation and laid to rest the classic view that Mercury did not rotate on its axis at all. The astronomers had reported their findings on Mercury's rotation in Washington last

April 21 but at that time were not sure whether the rotation was retrograde or direct. (Hines, Wash. *Eve. Star*, 5/13/65)

May 13: The Sofar (Sound Fixing and Ranging Device), used to locate ICBM's through a small explosive charge set off as the missile sank, would be transformed into a rescue device for aircraft and ships in the Pacific, Capt. John M. Waters, Jr. (USCG), told a U.S. Coast Guard-sponsored North Atlantic Search and Rescue Seminar in New York. A pressure switch mechanism would fire the explosive at 2,500 ft. below the surface—depth at which sound waves encountered least resistance; sound of the explosion would be picked up by four hydrophone listening stations and the exact disaster site plotted. Capt. Waters said the device was "practically foolproof" and had been endorsed by the Naval Aviation Center. Each Sofar locator would cost about $75. (Bamberger, *NYT*, 5/14/65, 65)

- Soviet engineer T. Borisov suggested that cause of LUNA V's apparent failure to soft land on the moon might have been failure of the braking rockets to fire "precisely when needed," the *New York Times* reported. Borisov pointed out that earth stations could not help the automatic equipment during this phase because it takes $2\frac{1}{2}$ sec. for radio signals to make round-trip between earth and moon. (*NYT*, 5/14/65, 3)

- A descriptive report on the three generations of Soviet manned spacecraft—VOSTOK I through VOSTOK VI, first generation; VOSKHOD I, second generation; and VOSKHOD II, third generation—was prepared by *Space Daily*, in collaboration with Soviet space officials and the Novosti Press Agency: "The Soviet's first three generations of manned spacecraft are injected into orbit within a standard cone-cylinder configuration with a maximum length of 30.3 feet and a maximum diameter of 8.7 feet . . .

  "The launch-to-orbit vehicle is comprised of four major components: the last stage of the rocket; the instrument and service module; the cosmonaut cabin and re-entry capsule; and the nose cone and fairing. . . .

  "The cabin for the first two generation spacecraft remained in external configuration essentially the same. The major modification . . . was the internal arrangement providing a capability for two astronauts instead of one which included the requirement for an additional hatch. The third generation spacecraft has required not only a major modification for the internal arrangement, for the third cosmonaut, but has forced a configuration addition to the 7.5 foot sphere with the attachment of the airlock. . . .

  "For the first generation spacecraft the cosmonaut was seated in the center of the sphere with his back to re-entry portion of the sphere. The capsule had three hatches: the egress hatch, the parachute compartment hatch and an equipment access hatch . . . The parachute compartment was located to the left and rear of the cosmonaut . . . Antennas for the radio system of the re-entry capsule were located 180 degrees from the stagnation point of the heat shield. Even in that location it is possible that heat build-up destroyed all protruding systems as evidenced by the landed VOSTOK. . . .

"The first generation vehicles weighed about 10,430 pounds, after ejection of the nose cone and fairing and separation of the third stage of the booster. . . . The cabin for the first and second generation missions weighed about 5300 pounds.

"In addition, VOSKHOD II represents an advancement to a more operational type of vehicle with an arrangement indicating its role for extensive Earth-orbital operations.

"The airlock for the VOSKHOD II mission would represent the farthest evolution of the Soviet manned spacecraft program . . . a cylindrical projection to the basic vehicle, positioned within the nose cone and fairing above or forward of the cabin in the antenna region. With respect to the cosmonauts the airlock would be above and to the front as they remained in their seats; its position would be 180 degrees from the stagnation point of the heat shield." (*SBD*, 5/13/65, 68–70)

May 13-14: Executives of four competing companies briefed the USAF Space Systems Div. source selection board on their Manned Orbiting Laboratory (Mol) entries. Represented were the Boeing Co., General Electric Co., Lockheed Aircraft Corp., and Douglas Aircraft Co. (*Av. Wk.*, 5/31/65, 22)

May 14: NASA and FAA announced formation of a joint 12-member coordinating board to strengthen joint planning and facilitate exchange of information between the two agencies. The board would focus its attention on aeronautical research, development, and testing activities to gain the greatest return from available resources and to avoid duplication. Co-chairman would be Dr. Raymond L. Bisplinghoff, NASA Associate Administrator for Advanced Research and Technology, and Robert J. Shank, FAA Associate Administrator for Development. (NASA Release 65–155)

• Sen. A. S. Monroney (D–Okla.) told a meeting of the American Helicopter Society in Washington, D.C., that NASA was spending too small a share of its budget on aviation research.

Senator Monroney, the chairman of the Senate Aviation Subcommittee, said he became angry when he compared the $43 million earmarked for aeronautics next year with the space agency's total budget of $5.2 billion.

He said that although the agency allocated less than 2 percent of its budget to solving the many flight mysteries it acknowledges still exist, the agency's working-level scientists wanted to do more in this area.

Monroney also said he disagreed with those who contended that subsidy for the helicopter airlines was wasteful and unwarranted.

He said that while helicopters might not have made the progress many wished for and some had promised, commercial revenues had increased, costs had declined, equipment had improved, and capability to operate on instruments had been developed. (AP, *NYT*, 5/15/65)

• NASA Administrator James E. Webb announced during a ceremony at Western Reserve Univ. honoring retiring Dr. T. Keith Glennan, president of Case Institute of Technology, that Glennan had been asked to return to NASA as an adviser. Webb said Glennan would be asked to review NASA spending plans for the next ten years.

Glennan, Webb's predecessor as NASA Administrator, was appointed by President Eisenhower to head the agency when it was formed in 1958. (Ludwigson, Cleveland *Plain Dealer*, 5/15/65)

May 14: Secretary of Defense Robert S. McNamara told the House Appropriations Committee that about $1.2 billion—80 per cent of the allocated money—had been wasted on the abortive B-70 bomber project, Howard Margolis reported in the *Washington Post*. The question of how much of the money spent was wasted arose when McNamara was asked whether knowledge from the B-70 work would be valuable to other military and civilian projects. McNamara suggested that at least 80 per cent of the money had been wasted, Margolis said. McNamara's general view had been that substantial "waste" of this sort was unavoidable in the defense program since it was rarely possible to know how valuable a development project would be before large sums had been spent. Margolis added that McNamara suggested minimizing such waste by insisting on good evidence of probable value before allocating large expenditures and, even then, by limiting spending as much as possible until the value of a project was proven. (Margolis, *Wash. Post*, 5/15/65)

- A special educational television satellite station to carry color or black-and-white TV direct to home receivers was proposed to NASA by Hughes Aircraft engineer Dr. Harold Rosen. (*Time*, 5/14/65; *CR*, 5/20/65, A2549)
- Sen. Henry M. Jackson (D-Wash.) reported that ComSatCorp would construct a $6 million ground station at Brewster, Wash., and that FCC had approved ownership of the station by ComSatCorp. (AP, *Oregonian*, 5/14/65)
- A mouse-size "algatron," life-support system designed to make outer space habitable for astronauts on prolonged missions, was demonstrated by Univ. of California scientists Dr. William J. Oswald and Dr. Clarence G. Golueke. In the system bacteria break down animal wastes, algae live off the result, and emit oxygen while absorbing carbon dioxide. According to the scientists' report, the algatron, in which a mouse lived for six weeks and could have stayed indefinitely, would weigh about 1,000 lbs. in a man-sized version. (*Wash. Post*, 5/14/65)
- A lunar dust cloud produced by braking rockets of Soviet probe LUNA V as it attempted a soft landing on the moon May 12 was photographed by the observatory at Rodewisch, E. Germany, said the observatory's director in an interview with ADN, E. German press agency. The tracking station had made photographs of the lunar approach of the spacecraft at 15-sec. intervals. At the moment of best visibility—10:15 p.m. Moscow time—the dust cloud was 140 mi. long and 50 mi. wide. It had disappeared by 10:21 p.m. Moscow time. This was the first indication that braking rockets aboard the spacecraft had been operative. Soviet announcement had given landing time for LUNA V as 10:10 p.m. Pictures of the dust cloud were published in *Izvestia*. (*NYT*, 5/16/65, 6; AP, Wash. *Sun. Star*, 5/16/65)
- Communist China exploded its second atomic bomb "over its western areas" at 10 a.m. Peking time, according to Hsinhua, the Chinese Communist press agency. (Reuters, *NYT*, 5/15/65, 2)

May 15: NASA Administrator James E. Webb, speaking to the University of Alabama Alumni in Washington, D.C.: "During the five years ending this month, NASA will have awarded to the University general-purpose grants, project contracts in support of research, and traineeships amounting to over $4.8 million.

"This sum has supported 68 research projects and renewals and the training of 30 graduate students.

"In the last academic year 63 faculty members, 67 graduate students, 51 undergraduate students, and 25 others were supported through NASA research and predoctoral training programs.

"Over the past five years 49 faculty members, 61 graduate students and 73 undergraduate students participated in engineering research sponsored by NASA.

"In addition to this support—and in addition to support for the physics, mathematics and chemistry departments—the Marshall Space Flight Center has guaranteed support for the graduate training program at Huntsville to a total of $750,000 in five years. This Huntsville program permitted the establishment two years ago of resident master's degree programs in five disciplines. A sixth was added last year. In two years, 2,729 students have participated." (Text)

May 15: "There are 593 objects in earth orbit today," said Maj. Gen. Horace A. Hanes (USAF), Commander of the 9th Aerospace Defense Div., at an Armed Forces Week celebration at Selfridge AFB, Mich. He said these ranged from the six-in.-dia. Vanguard satellite through the 90-ft.-dia. Echo satellite. Hanes said the primary mission of his division was to detect and warn the U.S. of a mass ballistic missile attack: "We use radar stations in Alaska, Greenland, and England for this. But to detect satellites and other objects in earth orbit we use these radars plus a variety of other equipment including special optical cameras eleven feet high that weigh 3,000 pounds." (Pipp, *Detroit News,* 5/16/65)

• A newspaper article summarizing a report of the International Civil Aviation Organization on the safety record of the non-Communist world's airlines in 1964 said: ". . . its more than 100 member airlines, which include U.S. carriers, ended the year with the lowest fatality rate on record, 0.61 deaths to 100 million passenger-miles flown, 22.5 percent below 1963, the best previous year." (CR, 5/19/65, 10592)

Week of May 16: A $300,000 telescope produced through gifts of parts and money was put on display at the Stamford (Conn.) Museum and Nature Center. The 22-in. photo-visual telescope, designed to track even man-made satellites, was the result of a project compared to a "barn raising." A spokesman said that at dedication on June 13, plaques would be distributed to 51 major contributors of equipment and labor, and certificates to 81 other cooperators. (Devlin, *NYT,* 5/23/65)

May 16: EXPLORER XXIII and PEGASUS I meteoroid technology satellites continued to transmit useful information after months of operation in the space environment, reported Milton B. Ames, Jr., NASA Director of Space Vehicles Research and Technology.

EXPLORER XXIII, launched by Scout rocket from Wallops Station, Va., Nov. 16, 1964, was last of three S-55 series satellites which were the first spacecraft orbited specifically to measure meteoroid penetrations through spacecraft structures. Performance of EXPLORER XXIII had been entirely satisfactory, and indications were that it would have a useful life of more than a year, Ames said. Orbital parameters were: apogee, 615 mi. (990 km.); perigee, 290 mi. (467 km.); inclination to the equator, 51.95°.

PEGASUS I, launched Feb. 16, 1965, was first of a series of three satellites intended to measure meteoroid penetrations of greater structural thicknesses and contained a meteoroid penetration area of almost 2,300 sq. ft. Ames said that although useful results had been obtained with .0015-in.-thick panels, the data obtained with .008-in.-thick and .016-in.-thick panels had not been fully satisfactory because of difficulties in the operation of the detection system. Still, PEGASUS I had provided significant information leading to improvement of detection systems on the remaining two Pegasus spacecraft. Orbital parameters for PEGASUS I were: apogee, 451 mi. (726 km.); perigee, 311 mi. (500.7 km.); inclination, 31.75°. (NASA Release 65–157)

May 16: Editorializing, the *Hartford Courant* said: ". . . Lunik V's purpose was openly said to be a soft landing on the moon, an experiment that might have sent back the first pictures of the moon from the actual lunar surface, and information about the physical nature of that surface. Possibly the Russians were sure they had the problems of a soft landing solved. But just possibly they decided this time to be frank and out in the open about the whole business. After all, why be scared? Look at all the failures the United States has admitted. And right now the Russians have something to console themselves with. It's called honesty, and its just as good to be distinguished for this as it is for technology." (*Hartford Courant,* 5/16/65)

May 17: Britain and France signed an agreement to jointly build two supersonic military aircraft for the 1970s: (1) a strike trainer; and (2) a pivoting-wing attack plane. The strike trainer, to be based on France's twin-engine Breguet 121, would be built by the British Aircraft Corp. in cooperation with the Société des Ateliers d'Aviation Louis Breguet. Rolls Royce, Ltd., and Turbomeca, a French engine concern, would supply the engines. The variable sweep wing, aircraft similar to the American F-111 fighter-bomber, but smaller, would be based on the concept of British aircraft designer Dr. Barnes Wallis and built by the British Aircraft Corp. and the Société Générale Aéronautique Marcelle Dassault.

The agreement committed each country to an initial expenditure of $56 million, most of which would be spent on a prototype for the strike trainer. (Farnsworth, *NYT,* 5/18/65, 8)

- NASA Marshall Space Flight Center had awarded a $1,600,000 contract to Aero Spacelines, Inc., to transport Saturn upper stages and outsize rocket components in its modified Boeing Stratocruiser, Pregnant Guppy. The contract would run through June 1966. (MSFC Release 65–123)
- *Aviation Week* reported: "NASA is considering the possibility of launching two manned Gemini spacecraft within a few days of each other so that the two would operate concurrently in space for a day or two." Noting that the plan was not yet approved, the item speculated that such action would probably not take place until late in the Gemini program. Since only one Gemini launch stand existed, it was most

likely that the second vehicle would be erected and checked out first, then stored until the first had flown. (*Av. Wk.*, 5/17/65, 23)

May 17: Robert Hotz said in an editorial in *Aviation Week and Space Technology:* "With each passing year it becomes more and more apparent that the Soviets agreed to the partial nuclear test ban treaty at a time most advantageous to them and most disadvantageous to us. The Soviets already had tested their nuclear warheads over the entire spectrum—from underwater devices to 50-megaton air bursts including live ICBM warheads. The U.S. had not tested any of its nuclear warheads in strategic systems and can only theorize about the effects that high-altitude nuclear blasts in the 50-megaton-and-up range will have on communications and control networks of silo-based ICBMs and other strategic systems.

"Mr. McNamara has based his defense policy on the belief that he will be able to detect any new Soviet weapons development in time to develop a U.S. counter-measure before the Russians can become operational with their new force. Since several of the new Soviet ICBMs and an anti-ICBM shown in recent Red Square parades came as a complete surprise to the Western intelligence community, it would appear that this assumption by Mr. McNamara is open to serious challenge. History may prove that Mr. McNamara's view of the time span available for the U.S. to counter-develop weapons to thwart a Soviet challenge is as wrong as his forecasts of the war in Vietnam." (*Av. Wk.*, 5/17/65, 21)

- "The Soviet Union, with its May 9 display of missile and space might, has dealt a major blow to the complacency of those persons in the United States who consistently have underestimated the competence of the Russians in these fields," wrote William J. Coughlin in an editorial in *Missiles and Rockets.* He continued: "The appearance of Soviet solid-fuel missiles of a type similar to the U.S. Minuteman ICBM indicates that the Soviets finally have overcome the chemical roadblock which until now has made possible the U.S. lead in solids. . . .

"In a film which the Moscow correspondent of the *New York Times* estimated to be at least three years old, the Soviets also displayed launchings from an underground silo. The combination of these events suggests the Soviets now are in a position to rapidly close the missile gap with the United States to the point where it is of no consequence in military calculations. . . .

"In the film release, the Russians for the first time showed their anti-missile missile in action. One sequence was of intercept of an ICBM. . . .

"The increasing Soviet confidence also is indicated in the space field. The Soviets let it be known more than a month in advance that their next space spectacular could be expected May 9. . . . The launch of LUNIK V obviously was right on schedule. The acknowledgement after launch but in advance of impact that its goal was a lunar soft landing also is a more realistic approach to space developments than previously shown.

"This shift toward a franker attitude is supported by the open admission of the Zond II Mars probe failure by Soviet scientists attending the Space Exploration Symposium in Chicago on May 4.

"At the same time, the Russians released more information on the Soviet space program at the Chicago meeting than heretofore.

"All of this points toward greater maturity in both Soviet missile programs and Soviet space programs. The competition therefore is far keener than many persons in the U.S. have been willing to admit.

"The conclusion is clear. The U.S. cannot afford to let down or it will be far outdistanced in areas which will continue to be vital in its national security and well-being for many, many years." (Coughlin, M&R, 5/17/65, 74)

May 17: Communist China's second nuclear bomb was the warhead on a missile launched from a military base and detonated in the air after traveling an undisclosed distance, asserted the Japanese newspaper *Asahi Shimbum*. The bomb was exploded May 14 over Western China. (UPI, *Wash. Daily News*, 5/17/65, 18)

May 18: X-15 No. 2 flown by pilot John McKay (NASA) to 102,100 ft. altitude at maximum speed of 3,541 mph (mach 5.17) to obtain data for stability and control evaluation, star tracker checkout, advanced X-15 landing dynamics, and landing gear modification checkout. (NASA X-15 Proj. Off.; *X-15 Flight Log*)

- USAF launched an unidentified satellite from Vandenberg AFB with a Thor-Agena D booster combination. (UPI, *NYT*, 5/19/65, 2)

- TELSTAR II had successfully turned off its tracking beacon as scheduled after two years and nine days of service and 4,736 orbits of the earth, Bell Telephone System engineers announced. This would not affect the comsat's usefulness, but would conserve energy and permit other satellites to use the channel that was cut off. TELSTAR II was expected to remain usable for at least three more years. (UPI, *NYT*, 5/20/65, 18)

- Memorandum of Understanding for a cooperative Argentina-U.S. program of meteorological sounding rocket research was signed by Teofilo Tabanera for the Comision Nacional de Investigaciones Espaciales (CNIE) and Hugh L. Dryden for NASA. Specific purpose of this experimental program was to obtain high-altitude meteorological data in the vicinity of Chamical, Argentina, by Boosted-Dart and Arcas sounding rockets and to evaluate Argentine ground support equipment in conjunction with the payloads. General purpose of the experimental program was "to develop a basis for future meteorological rocket soundings on an operational basis." The program was contemplated as "one element in a projected inter-American, experimental, meterological sounding rocket research network (EXAMETNET)." (Memo of Understanding)

- 3C-9, a quasar (quasi-stellar radio source) receding from the earth at 149,000 mps or 80% of the speed of light, had been discovered with the 200-in. telescope at Mt. Palomar Observatory, Walter Sullivan reported in the *New York Times*. It was the most distant of a new generation of five quasars which included CTA-102, the object Soviet astronomers had suggested might be transmitting signals under intelligent control. All appeared to be so distant that their life had probably ended during the billions of years required for their light to reach earth. Dr. Allan R. Sandage of Mt. Palomar Observatory said his studies of brightness and velocities of these five quasars and four others previously calculated resulted in evidence supporting the "oscillating universe" theory. Data on the nine quasars' velocities largely was the work of Dr. Maarten Schmidt, Mt. Palomar Observatory. (Sullivan, *NYT*, 5/18/65, 1, 2; 5/23/65, 6E)

May 18: Stanley R. Reinartz, previously deputy manager of NASA Marshall Space Flight Center's Saturn I/IB Program Office, had been named program manager of the newly established Saturn IB/Centaur office, MSFC announced. The office would manage the program definition and design phase of the three-stage Saturn IB/Centaur space vehicle system. (MSFC Release 65–124; *Marshall Star*, 5/26/65, 1)

: NASA Lewis Research Center planned to buy enough ⅛-in.-dia. pingpong balls to fill a bucket-like device 12 ft. in dia. and 19 ft. deep. The miniature pingpong balls would be used to cushion experiments in LRC's 500-ft.-deep zero-gravity shaft. Experiments would be recovered intact for evaluation and later reuse. The pingpong balls, it was hoped, could cushion up to 6,000 lbs. (LRC Release 65–34)

: Four Ohio college students ended a six-week isolation test at Wright-Patterson AFB, Ohio, to study diets, effect of continuous wearing of a spacesuit, and microbiology of the human body. The four, comprising the eighth group to take part in space tests conducted by the Aerospace Research Labs., spent the first three weeks on a balanced but monotonous diet and the last three weeks on a liquid diet with the same nutrients as their earlier meals. All agreed that astronauts would probably be able to wear spacesuits for long missions but that "something would have to be done" about the proposed liquid diet. (AP, *NYT*, 5/19/65; AP, Cleveland *Plain Dealer*, 5/19/65)

: Najeeb E. Halaby, retiring FAA Administrator, speaking at the annual news conference of the Aviation-Space Writers Association in Albuquerque, urged President Johnson to make the "tough decision" to develop 2,000 mph airliners to handle expanding travel in the 1970s. He said opponents of the supersonic transport project had "seriously overstated" the safety and other problems involved.

Mr. Halaby received the Monsanto Chemical Co.'s aviation safety award for the "most significant and lasting contribution to aircraft operating safety in 1964." President Johnson sent him a congratulatory telegram hailing his "outstanding performance" as aviation administrator. (UPI, *NYT*, 5/19/65)

: Representatives of companies planning to buy the supersonic Concorde airliner were told in a report prepared by the joint builders, British Aircraft Corp. and Sud Aviation France, that the makers were confident, following extensive wind tunnel tests, that the Concorde represented "the best possible compromise for a supersonic transport" and would be "safe and easy to fly." A special report on the problem of sonic boom said tests had shown that the calculated extent of these sharp detonations had been "generally pessimistic." It said that climb and acceleration techniques were being developed that would keep the shock waves of air causing these booms as slight as possible.

The experts present for the three-day talks on the airliner's progress were from Air France, British Overseas Airways Corp., Pan American World Airways, American Airlines, Continental Airlines, Qantas, Air India, and Middle East Airlines, which had together ordered or taken options on 45 of the aircraft, valued at $560 million. (Reuters, *NYT*, 5/19/65, 94C)

May 18: Sen. Margaret Chase Smith (R–Me.), interviewed by a group of women correspondents, was critical of the Administration's failure to "pinpoint" objectives beyond its 1970 goal to put a man on the moon. She said she found it "hard to believe" the Administration wasn't thinking beyond the moon to Mars and Venus but that "it's difficult to get the answers." (Dean, Wash. *Eve. Star*, 5/19/65)

- A fuel cell system had successfully operated for more than 1,300 hrs.—the time it would take a spacecraft to make nine trips to the moon and back—producing electricity and drinking water from hydrogen and oxygen, John L. Platner of the Allis-Chalmers Research Div. told the 19th annual Power Source Conference in Atlantic City. Platner gave details of the cell's performance in reporting on an advanced 2,000-watt fuel system being built by Allis-Chambers for NASA. (UPI, *Wash. Post*, 5/19/65, A21)

May 19: A 71-ton Little Joe II rocket fired from White Sands Missile Range, N. Mex., to test the Apollo spacecraft escape system split into fragments three miles above ground following a series of excessive rolls occurring about 25 sec. after launch. The escape rocket fired immediately, however, and carried the 14-ton Apollo boilerplate free of the debris; the parachute recovery system operated normally, lowering the command module to the ground. Apollo program manager Dr. Joseph F. Shea said: "Although the prime objectives of the high altitude abort test were not met, the launch escape system proved its mettle in an actual emergency, which is the purpose for which it was designed." The launch escape subsystem would be used to propel the spacecraft and its crew to safety in the event of a Saturn launch vehicle failure either on the pad or during powered flight.

Little Joe II had been programed to carry the test vehicle, Boilerplate 22, to 22-mi.-altitude in 89 sec.; an escape motor would propel the spacecraft to a peak altitude of about 35 mi. Finally, the three 84-ft.-wide parachutes would lower the command module to earth. (NASA Release 65–145; *N.Y. Her. Trib.*, 5/20/65; NAA *S&ID Skywriter*, 5/21/65, 1, 2; *NYT*, 5/20/65, 42; MSC *Roundup*, 5/28/65, 8)

- U.S. launched eight military satellites into orbit from Vandenberg AFB March 9 with a Thor-Agena D booster, NASA disclosed. This was the greatest number of payloads the U.S. had ever orbited with a single launch vehicle and was believed to exceed any multiple launching made by the Soviet Union. Orbital parameters: apogee, 585 mi. (942 km.); perigee, 561 mi. (903 km.); inclination to the equator, 70°. Two payloads would measure solar radiation; two would test stabilization methods for future spacecraft; one would map the earth's surface; another, Surcal (Space Surveillance Calibration), would help improve precision of satellite tracking networks; another, Oscar (Orbiting Satellite Carrying Amateur Radio), would broadcast on frequencies that amateur radio operators could track. The satellites had been unidentified until NASA listed them in its periodic satellite summary.

    The summary also showed that unmanned COSMOS LXI, COSMOS LXII, and COSMOS LXIII, launched by U.S.S.R. March 15 with a single launch vehicle, had become 26 satellites or pieces of satellites. COSMOS LXVI, and two companions, launched May 7, had fallen out of orbit. (GSFC SSR, 4/15/65; Clark, NYT, 5/19/65; Wash. Post, 5/20/65, A12)

May 19: NASA launched a two-part 104-lb. sounding rocket payload from NASA Wallops Station, Va., to measure electron densities and ion composition of the upper atmosphere. Designed as a mother-daughter experiment—with radio signals to be sent from daughter to mother—the payload separated as planned at about 170-mi. altitude and the two sections reached peak altitude at 605 mi. The sections were programed to rise separately for about 8 min. and reach a distance apart of about 3 mi. Experimental information was radioed to ground stations and no recovery of the sections was required; they impacted in the Atlantic Ocean. Measurement of the differences between the signals of the two devices, monitored by ground stations, was expected to provide more accurate profiles of upper atmosphere electron density. The launching was timed to occur while Canadian satellite ALOUETTE was passing nearby. ALOUETTE's instruments would provide a horizontal profile of ionospheric and ion densities and temperatures to be correlated with findings of the mother-daughter experiment. (Wallops Release 65-30)

• The Gemini 2 spacecraft which made a suborbital unmanned flight from Cape Kennedy Jan. 19, 1965, would be reworked by the McDonnell Aircraft Corp. and delivered to USAF in July 1966 for a preliminary unmanned flight in the USAF Manned Orbiting Laboratory Program, NASA Manned Spacecraft Center announced. USAF would launch the spacecraft in an unmanned suborbital flight to test the Gemini B heat shield design. The heat shield would have a hatch to allow crew transfer from the Gemini to the Orbital Laboratory. (NASA Release 65-166)

• NASA successfully launched Argo D-4 sounding rocket from Wallops Station, Va., to peak altitude of 588 mi. Objective of 17½-min. test was the measurement of phase differences to determine electron density along the rocket trajectory. Experiment was provided by Pennsylvania State Univ. (NASA Rpt. SRL)

• Dr. Homer E. Newell, NASA Associate Administrator for Space Science and Applications, was among the ten outstanding Federal Government employees who received a career service award from the National Civil Service League. (Mohr, NYT, 5/20/65)

• A $784,600 contract had been awarded to Mechling Barge Lines, Inc., for towing three Saturn space vehicle barges, NASA MSFC announced. Two of the barges, *Promise* and *Palaemon*, would be used to carry the Saturn I and IB boosters. A third, being readied, would transport the larger Saturn V booster. The contract covered a one-year period. (MSFC Release 65-128)

• Dr. Wernher von Braun, Director of NASA Marshall Space Flight Center, was named chairman of the International Sponsors Committee for Clark Univ.'s $5.4 million Robert Hutchings Goddard memorial library scheduled for completion by 1968.

Several nuclear-powered, self-supporting lunar bases and a wide variety of space stations would be in operation by the year 2000, Dr. von Braun told the luncheon meeting of the National Space Club in Washington, D.C. He made his predictions during the question and answer period following his speech on Dr. Robert H. Goddard's contributions to American rocketry.

The greatest activity in space 35 yrs. hence would be in earth orbits, von Braun felt, and space would provide a "tremendous military field." This field would not be the science fiction concept of orbiting hydrogen bombs, but rather a broad program of military reconnaissance. Photography and direct observation of foreign military developments were cited.

Space stations would be in a variety of orbits and many would be manned by scientists and repairmen shuttling back and forth in reusable vehicles. Scientists would spend up to six weeks at a time in the stations to make their observations. The use of reusable boosters would cut the cost of delivering payloads to orbit down to some 10% of today's costs, von Braun added. (NSC *Newsletter*, 5/65, 6/65)

May 19: "Early Bird should not be construed by any government as just another door to be opened when there is a self-serving point to be made, and a door to be slammed when that point is in danger of being questioned," said Dr. Frank Stanton, president of the Columbia Broadcasting System, in a speech at the Career Services Awards dinner of the National Civil Service League in Washington, D.C. Dr. Stanton said it was agreed the peoples of the world should have an opportunity to hear foreign leaders, but that this must be done in an atmosphere of freedom "with openness and in candid discussion." He added: "Early Bird must not be transformed from the unprecedented opportunity into the most universal and pervasive censorship—both affirmative and negative—ever known." (*NYT*, 5/20/65, 75)

- Dr. Johannes H. Klystra, interviewed in his laboratory at the State Univ. of New York in Buffalo, revealed that laboratory mice and dogs had survived completely submerged in heavily oxygenated salt water; the lungs had extracted oxygen from the pressurized liquid. Dr. Kylstra said that man might one day find it useful to develop techniques for breathing liquids as an aid in the exploration of the two new realms that are just opening up to him: space and the ocean depths. A space flier, for example, could be protected from the destructive forces of a less-than-soft landing on another planet if he were in a cockpit filled with oxygenated liquid that he could also breathe; a free-swimming underwater explorer with liquid-filled lungs could go deeper, stay longer and ascend faster and more safely than a diver breathing a gaseous mixture of nitrogen and oxygen. (Osmundsen, *NYT*, 5/19/65, 49C)

- Bendix Corp. would receive from USAF a $2,666,840 initial increment to a $22,123,000 fixed-price contract for modification and improvement of the AN/FPS-85 space track radar. Work would be done in Towson, Md., and at Eglin AFB, Fla. (DOD Release 343-65)

May 20: NASA-AEC successfully performed a restart of the NRX A-3 Nerva experimental engine at Jackass Flats, Nev. The firing lasted for 18 min., including 13 min. at the engine's full power rating. The engine was the same one that had run for four minutes Apr. 23 before being shut down prematurely due to spurious malfunction. (SNPO-N-65-9; *Wash. Eve. Star*, 5/21/65; *Rover Chron.*)

- USAF launched unidentified satellite payload with Thor FW4s booster from WTR. (*U.S. Aeron. & Space Act., 1965*, 142)

May 20: Ground test version of the Saturn V booster (s–ic–t) was fired by NASA Marshall Space Flight Center for 41 sec., MSFC announced. It was the third and longest firing of the five engines, which developed 7.5 million lbs. thrust. The firing seemed entirely satisfactory, based on preliminary evaluation of data. (MSFC Release 65–131)
- USN's F–111B fighter aircraft, originally designated TFX, was given its first test flight over Long Island at 2,000-mph for an hour and 18 min., during which the variable wing-sweep of the craft was tested. (*Wash. Post*, 5/19/65, 11; UPI, *Wash. Post*, 5/20/65, 2)
- NASA engineers Harry Carlson and Francis E. McLean believed the sonic boom problem in the operation of the supersonic transport could be solved by fattening the fuselage just forward of the wing, thereby altering the air flow in such a way as to cut the boom to an acceptable level, reported Richard P. Cooke in the *Wall Street Journal*. Fattening the Sst fuselage forward of the wings, said the NASA engineers, would also help the lift and might permit room for more seats. (Cooke, *WSJ*, 5/20/65)
- AFSC announced that an airspace surveillance and weapons control system had been proposed for installation in the Ryukyu islands, southwest of Japan. Through use of semi-automatic data processing, the Ryukyu Air Defense System (Rads) would pick up airspace intruders in its area almost instantly, enhancing defense capabilities of the Pacific Air Force in that area. The system would consist of radars, ultra-fast communications, data processors, display consoles and command posts where decision makers could direct manned or unmanned weapon interception. Returning aircraft could be directed home or to alternate bases through the system. (AFSC Release 54.65)
- Newest Soviet aircraft, including the 186-passenger, four engine Il–62, designed for nonstop intercontinental service, were displayed at an exhibition of airliners and helicopters at Moscow's Vnukovo Airport. The Il–62, whose engines were mounted on the tail section of its fuselage, had a cruising speed of 500 to 550 mph and a range of 5,500 mi. Boris Kharchenko, chairman of the Soviet aircraft export organization, said the Soviet Union was seeking orders this year for both the Il–62 and the Tu–134, a medium-range, two-engine jetliner. Delivery would be in 1967. (*NYT*, 5/21/65)
- Secrets unearthed by MARINER II and just made public were reported by Frank Macomber in the *San Diego Union:* "Venus is no lush sea-and-swamp world, possibly teeming with primitive life, as some astronomers have speculated. Under its eternal cloud cover, the planet's surface must be like fuming slag or lava. The surface temperature is about 800°F.—hotter than molten lead.

    "The clouds surrounding Venus are a dense, unbroken pall of hydrocarbon smog, boiling up to at least 60 miles from the planet's surface."

    Macomber said MARINER II was regarded as one of the most successful of U.S. spacecraft. (Copley News Serv., Macomber, *San Diego Union*, 5/20/65)
- General Bernard A. Schriever, AFSC Commander, said in an address to the Aviation-Space Writers' Association Conference in Albuquerque: "The Air Force responsibility for our nation's military developments in space is clearly established. This morning I would like to review

our current progress in the areas of unmanned space programs, boosters and propulsion, and finally, manned space programs. . . .

"In the late 1950s, a small group of Air Force officers began a program to develop a space-based missile detection and warning system. To obtain information on the background as observed from space and on the signature of ballistic missile rocket motors, the Air Force initiated a series of measurement programs. Instrumented aircraft were used to obtain data on our missile target, from many aspects and in various weather conditions. Concurrently, a spacecraft 'piggyback' program for background measurements was instituted. This program has resulted in information of great value and is still collecting valuable data. . . .

"The second area of interest is anti-satellite defense. Last September, President Johnson announced the existence of operational U.S. anti-satellite defense systems. . . .

"The third area of interest is the detection of nuclear detonations in space. The original effort was formerly known as 'Vela Hotel,' and has now emerged as the present Vela Satellite Program. . . .

"In 1963 the first pair of Vela Satellites was launched from Cape Kennedy; the second launch occurred in 1964. Both launches were completely successful, and the four satellites are still functioning. . . .

"The last area that I would like to consider in unmanned military space systems is communication satellites—commonly called COMSAT. Our current philosophy of controlled response has placed an additional emphasis upon communications between field commanders and the highest level of our nation. . . .

"In summary, space is a new environment of activity. We need to exploit it effectively for our own purposes to prevent it from being used against us. We are aware of the many problems confronting us and do not pretend to have all the solutions. But much has been done, and we are building a broad technological base to meet the even greater challenges of the future." (Text)

May 20: NASA MARINER IV was 85 million miles from earth and traveling faster than 48,000 mph, NASA announced. A radio signal from the spacecraft, traveling at the speed of light, would take more than $7\frac{1}{2}$ min. to reach a ground station. The Mars probe was returning scientific measurements and engineering data continuously and daily setting a new record for distance of communications. (NASA Release 65–167)

- A NASA report on its Aircraft Noise Research Program to the House Committee on Science and Astronautics and the Senate Committee on Aeronautical and Space Sciences said: ". . . there is a growing understanding that efforts at a practicable and mutually effective solution will need to be evolutionary in nature, and involve a dedicated attack on all major aspects of the problem. These include the acquisition of definitive information on the manner in which aircraft noises are generated and propagated, and the associated development of efficient methods for the reduction of adverse aircraft noise at its source; the establishment of safe and efficient aircraft operating procedures that minimize and control the exposure of airport community property to undesirable aircraft noise; and the provision of a rational understanding of the specific aircraft noise factors which produce subjective annoyance for various activities and environments of a community population, and of optimum methods for the control and adjustment of community property usage in critical noise areas in the vicinity of the airport." (*CR*, 5/27/65)

May 20: AEC's Snap–10A nuclear reactor, aboard SNAPSHOT satellite, launched by USAF into circular polar orbit Apr. 3, automatically shut down on May 16 for unknown reasons, AEC announced. Snap–10A had been producing power for its own telemetry; first indications of malfunction came when telemetry ceased. Telemetry resumed about 40 hrs. later, powered by stand-by batteries, and indicated the reactor had shut itself down and was no longer producing power.

The prototype of future auxiliary power systems, planned to operate at least 90 days, had been operating successfully although the ion engine experiment included in the spacecraft had been shut down when it developed electronic noise. The spacecraft containing the defunct power system would remain in orbit more than 3,000 yrs.; it would take 100 yrs. for the reactor's radioactive elements to decay to a safe level.

AEC said Snap–10A had provided valuable information for design of future nuclear propulsion systems. (AP, *Wash. Post*, 5/21/65; UPI, *NYT*, 5/22/65, 5; *Atomic Energy Programs, 1965*, 151)

- Enriched uranium of U–235 was the fuel used by the Chinese May 14 in their second nuclear explosion, according to preliminary analysis of airborne radioactive debris, AEC announced. It found "implausible" reports that the nuclear device had been carried by a missile although the detonation took place "above ground." AEC said the May 14 test was somewhat larger than China's first explosion of Oct. 16, 1964, which was equal to 20 kilotons or the Hiroshima bomb. (*NYT*, 5/21/65; *Wash. Post*, 5/21/65, A27)

- Dr. Jeanette Piccard's 1934 balloon flight, establishing the still current women's world altitude record for a balloon, was celebrated in Dearborn, Mich., by a ceremony and placing of a marker near the takeoff site. The balloon had a 600,000 cu. ft. volume, reached 57,559 ft. altitude, and took Dr. Piccard from Dearborn, Mich., to Cadiz, Ohio. (*CR*, 5/18/65, A2465)

May 21: U.S. and Argentina jointly announced plans to collaborate in launching weather rockets to gain information about hemispheric weather patterns. Under terms of an agreement, Argentina would provide launching facilities, would transport rockets and equipment from the U.S. where they would be manufactured, and would assemble and launch the rockets. U.S. launchings would be made from Wallops Station; launching pads in Argentina would be at Chamical. Other Latin American countries had been invited to participate in the program. (AP, *NYT*, 5/22/65)

- Vice President Humphrey, Chairman of the National Aeronautics and Space Council, said at the 16th annual luncheon for Albert Lasker Medical Journalism Awards in Washington, D.C.: "The most important race is not the space race or the arms race. It is the human race. If America can get excited about putting a man on the moon in 1970, why can't we get excited about putting a lot of people on their feet by the same date? . . . some day we will be able to tell the world that science has discovered the secrets of aging or of cancer or of muscular dystrophy or multiple sclerosis or mental retardation. That news will outrank in importance even the wonderful tidings that man has landed on the moon." (Text)

May 21: David N. Buell of Chrysler Corp. told the Aviation-Space Writers' Association Conference that an unmanned spacecraft could be launched to the sun by 1975 or 1980 with a modified Saturn IB/Centaur booster and that it could obtain information vital to space exploration and a better understanding of the universe. Buell envisioned the solar spacecraft as a bi-conal structure with the forward cone pointing toward the sun and acting as a sunshade, bolstered by refrigerants inside the craft. (UPI, *Wash. Post,* 5/23/65)

- "MARINER IV, speeding toward Mars for a rendezvous in July, has knocked out the romantic notion that the ruddy planet is the site of a dying civilization millions of years older than ours and far wiser," wrote David Dietz in the *Knoxville News-Sentinel*. Continuing: "This theory holds that the planet is drying out, losing its atmosphere and its water supply and that the inhabitants have taken refuge in underground cities.

  "Well, if this is the case, one thing is certain. The Martians forgot to take their radios with them. For the past five months, Mariner 4 has been sending a steady stream of radio chatter back to earth . . . If little Mariner 4 can do that, there is no apparent reason why the Martians couldn't do the same, providing, of course, that there are Martians of superior intelligence." (Dietz, *Knoxville News-Sentinel,* 5/21/65)

- David H. Hoffman, aviation editor of the *New York Herald Tribune* and Arthur C. Clarke, British science writer, were cited by the Aviation-Space Writers' Association for outstanding articles in 1964. Mr. Hoffman received the James J. Strebig memorial award for his series on air safety. Mr. Clarke was honored for an article published in *Life* magazine on communications satellites. (*N.Y. Her. Trib.,* 5/23/65)

May 22: NASA's 200-lb. Project Fire II spacecraft—similar in shape to an Apollo command module—was launched into a ballistic trajectory from ETR by an Atlas D booster that sent it over 500 mi. into space in test of reentry heating of spacecraft returning from the moon. Some 26 min. later, when the ballistic path of the payload turned it toward earth, a solid-fueled Antares rocket fired for 30 sec., accelerating the payload into the atmosphere at 25,400 mph. As a fireball estimated at 20,000°F formed a shock wave in front of the spacecraft, instruments in its interior radioed information to tracking stations. Tracking reports indicated that the heat probe impacted 32 min. after launch in the south Atlantic about 5,130 mi. southeast of Cape Kennedy. The spacecraft had been dubbed a "flying thermometer" because it was to radio more than 100,000 temperature readings.

First Project Fire flight took place from Cape Kennedy April 14, 1964, and was the fastest controlled in-flight reentry experiment ever conducted. The spacecraft reached a speed of more than 25,800 mph and telemetered many important direct measurements of reentry heating. (NASA Release 65–131; AP, *Wash. Sun. Star,* 5/23/65)

May 22: Jack N. James of the Jet Propulsion Laboratory, responsible for MARINER IV's cameras during the July 14 Mars flyby, told the Aviation-Space Writers' Association Conference that photographs taken by the probe were not expected to show signs of life that might exist on the planet since surface detail in the photographs would not be great. James said MARINER IV's cameras probably would be fixed on the planet by command from earth; previous plans had called for this to be done automatically by equipment in the spacecraft. (AP, Wash. *Sun. Star*, 5/23/65; Wash. *Eve. Star*, 5/24/65)

- The Gemini 4 manned spaceflight had been scheduled for June 3, NASA announced. The four-day flight would last about 97 hrs. 50 min., and would increase the U.S.'s hours of manned space flight to about 257 hrs. No decision had been made about opening the two-man spacecraft and letting one astronaut stand exposed to space. (Clark, *NYT*, 5/22/65, 8)

- President Chung Hee Park of the Republic of Korea, his wife, and members of his official party visited Kennedy Space Center where they were briefed on NASA programs and toured facilities at Cape Kennedy and on Merritt Island.

   In a luncheon statement, President Park said: "You are now engaged in a breath-taking race with Moscow for the conquest of space. . . . I should like to invite your attention to the stark reality that there are some fools engaged in utilizing space power politically, psychologically and militarily for sinister and dangerous purposes.

   "They are absorbed in developing space power not for the true purpose envisaged by mankind but for making it an instrument with which to conquer the world.

   "Needless to say, they are Communists. I believe you [Americans] have the responsibility of causing the Communists to desist from this dangerous play and of well preparing yourselves to douse a fire if it breaks out of that play. . . ." (NASA Off. Int. Aff.; KSC *Spaceport News*, 5/20/65, 1, 5; AP, *Miami Her.*, 5/23/65)

- In an interview at Reed College, Dr. John A. Simpson, professor of physics at Univ. of Chicago's Enrico Fermi Institute for Nuclear Studies, said that the U.S.'s present space policy was based on scientific achievement "and this has been diverse, thorough and deep and has led to wondrous discoveries." He lauded U.S. developments in weather and communications satellites which he termed an "outstanding example" of peaceful developments in space exploration. "Russia is mainly concerned with putting a man on the moon and has ignored, for the most part, the U.S. goals of achieving a better physical understanding of our solar system—and contributing to civilization's use of it." (*Sun. Oregonian*, 5/23/65)

- European Broadcasting Union's administrative council issued a statement saying it was concerned by the possibility that "prohibitive" charges might make it impossible to transmit television programs over EARLY BIRD I communications satellite. The council expressed the hope that the first three experimental years of intercontinental satellite television would not be "cut off at the start of commercial satellite operation." (Reuters, *NYT*, 5/23/65, 19)

- Soviet pilot Natasha Prokhanova, flying an E-22 supersonic jet trainer, climbed to 79,000-ft. altitude, exceeding the world altitude record for women of 56,073 ft. set by U.S. pilot Jacqueline Cochran in 1961 at Edwards AFB in a Northrop T-38 Talon supersonic jet trainer. (*NYT*, 5/31/65, 24)

May 23: The Life Sciences Committee of the National Academy of Sciences' Space Science Board recommended to NASA that American astronauts returning from the moon and planets be kept in quarantine for at least three weeks to prevent possible contamination of the earth by extra-terrestrial organisms, Howard Simons reported in the *Washington Post*. Recommendation was in a report entitled "Potential Hazards of Back Contamination from the Planets."

Other recommendations included the need to avoid decontamination of returning equipment until it had been subjected to biological study; the possible need for the astronauts to shed their outer garments on the moon and Mars before returning home; the need to conduct immediate research on any samples of extraterrestrial life brought to earth; and trial runs to acquaint astronauts with methods for minimizing chance of contamination. (Simons, *Wash. Post*, 5/23/65)

- United Press International had announced it would seek to establish a worldwide satellite communications system, either on its own or in partnership with others, if the governments concerned granted the necessary permission. (*NYT*, 5/23/65)

May 24: AFSC had selected nine graduates of its Aerospace Research Pilot School, Edwards AFB, Calif., to participate in crew performance studies for manned space flight to be conducted by NASA at the Martin Co. Three seven-day lunar landing simulations would be made using a simulated Apollo lunar landing mission. Each would utilize a three-man crew. (AFSC Release)

- President Johnson said, in transmitting NSF's sixth annual report to Congress on weather modification programs that control of weather was not beyond the reach of man: "The development of methods for altering weather and climate is a subject of quickening interest in the Congress and the Executive Branch . . . as, indeed, it is to all of the human race. We must recognize that the achievement of such a capability would mean vast economic and social gains for human life on this earth." (House Doc. 188)

- EARLY BIRD I linked audiences at the Parke-Bernet Galleries in New York City and at Sotheby's in London for the first trans-Atlantic art auction, ComSatCorp reported. BBC broadcasted a portion of the auction for British TV viewers. The telecast marked the fourth successive Monday on which the satellite had carried a commercial program free of charge to show its potential. (ComSatCorp Release; Esterow, *NYT*, 5/25/65, 1)

- The British Government announced plans for conversion of weights and measures to the metric system over the next ten years. The announcement meant the U.S. would be the only major power using nonmetric units.

Sen. Claiborne Pell (D-R.I.) said on the floor of the Senate: "The United States finds itself in the odd position of having inherited our anachronistic system of quarts, pounds, and inches from the British, only to find that the parent of the system has recognized its impracticality and is moving over to the metric system. This leaves us virtually alone in the world in our insistence upon our system of weights and measures, which originated in medieval times." (Farnsworth, *NYT*, 5/25/65, 6; *CR*, 5/24/65, 11023)

May 25: Saturn I (SA-8) launch vehicle, launched from Eastern Test Range, orbited a 23,000-lb. payload of which 3,200 lbs. was the PEGASUS II meteoroid detection satellite and 9,700 lbs. was Apollo boilerplate command and service modules (BP-26). This was the ninth successful test in nine flights for Saturn I.

At launch, Apollo command and service module boilerplate spacecraft and launch escape system (LES) tower were atop Saturn I, with PEGASUS II folded inside the service module. After second-stage ignition, LES was jettisoned. After injection into orbit, the Apollo boilerplate was jettisoned into a separate orbit and a motor-driven device extended 96 × 14-ft. winglike panels on PEGASUS II, exposing 2,300 sq. ft. of instrumented surface. The satellite was attached to the Saturn's S-IV second stage and would remain so during its lifetime. Each wing consisted of seven frames hinged together and providing mountings for a total of 208 detector panels. As particles collided with this surface, the penetrations would be registered and reported to earth. Orbital data: apogee, 460 mi. (741 km.); perigee, 316 mi. (509 km.); period, 97 min.; inclination, 31.8°.

Primary purpose of the flight was to gather information on frequency of meteoroids encountered in the near-earth environment for use in design of future manned and unmanned spacecraft.

PEGASUS II, an improved version of PEGASUS I launched Feb. 16, 1965, would be visible to the naked eye under favorable conditions near dawn and dusk. (NASA Release 65-151; MSFC Release 65-121; *Marshall Star*, 5/26/65, 1; AP, *NYT*, 5/26/65, 10; *U.S. Aeron. & Space Act.*, 1965, 143)

- X-15 No. 1 was flown by NASA's Milton O. Thompson to 179,800-ft. altitude at a maximum speed of 3,418 mph (mach 4.87) to obtain data on the Honeywell inertial system checkout, MIT horizon photometer, Pace transducer, RAS (Reaction Augmentation System) modification

May 25: Saturn I launch of PEGASUS II from Cape Kennedy.

checkout, and pilot altitude buildup. (NASA X-15 Proj. Off.; *X-15 Flight Log*)

May 25: U.S.S.R. launched COSMOS LXVII containing scientific equipment for investigation of outer space. Orbital parameters: apogee, 350 km. (217 mi.); perigee, 207 km. (128 mi.); inclination to earth, 51.8°. Onboard equipment was functioning normally. (*Krasnaya Zvezda*, 5/27/65, 1, ATSS-T Trans.)

- During the planned 4-day flight June 3, Astronaut Edward H. White (Maj., USAF) would leave the Gemini 4 spacecraft for 12 min. "if conditions are favorable," MSC officials announced at press conference. He would be secured to the craft by a 25-ft. safety line.

  NASA said the decision had been delayed "so final qualification tests could be completed on the spacecraft, spacesuit, secondary life support pack and umbilical."

  The 12-layer protective suit that Astronaut White would wear had been worn for more than 200 hrs. and White himself had worn it during more than 60 hrs. of tests. Among other things, it had had pellets fired at it at a speed of 30,000 fps to simulate the impact of small meteoroids.

  The flight's command pilot, Astronaut James A. McDivitt (Maj., USAF) would not open his hatch but would take movies of White through a spacecraft window. Astronaut White would take a 35-mm. still camera loaded with color film on his "walk" in space. Although he had practiced acrobatics, White had no planned program and would "use his own judgment as to what to do while outside the ship." Exit from the spacecraft was planned for the second orbit. (Transcript; Clark, *NYT*, 5/26/65, 1, 11; UPI, *Wash. Post*, 5/26/65, A3)

- Al J. Hayes, International Association of Machinists president, said union negotiations with Aerojet-General Corp. would not halt the scheduled two-man Gemini shot at Cape Kennedy on June 3. He said union members would continue work at Cape Kennedy even if a walkout were called against Aerojet General. (UPI, *Wash. Post*, 5/26/65, A3)

- X-22A vertical/short take-off and landing aircraft (V/Stol) was inspected by Government and military representatives at the Bell Aerosystems plant in Niagara Falls. Its unique characteristic was the ducted fan concept of propulsion consisting of four shrouded propellers—two forward and two on the tips of the 39-ft. wing aft—driven by four T-58 turbine engines. The four engines, expected to propel the aircraft at a cruising speed of 300 mph, were run for about five minutes.

  X-22A was constructed for the Army, Navy, and Air Force under a Navy-administered contract for $25 million. First flight test would be made in September 1965. (DOD Release 341-65; AP, *NYT*, 5/26/65, 94)

- Minute amounts of fresh radioactive debris from detonation of Communist China's second nuclear bomb were registered over the U.S. by the Division of Radiological Health of the U.S. Public Health Service. Pollution was far below the hazard level. (AP, *Wash. Eve. Star*, 5/26/65, 5)

May 25-26: More than 300 representatives of NASA and industry attended the 1965 Cost Reduction and Management Improvement meeting at NASA Marshall Space Flight Center. (*Marshall Star*, 6/2/65, 2)

May 26: NASA launched an ionosphere experiment from Wallops Station, Va., on a four-stage Javelin (Argo D-4) sounding rocket. Primary objectives of the flight were to measure ion and electron densities and temperatures and the ionic composition in the upper atmosphere. A malfunction in the launch vehicle caused the 140-lb. payload to reach an altitude of only 200 mi. instead of the planned 520 mi. Telemetry data were received for about nine minutes. Project officials termed the flight a partial success despite the failure to achieve peak altitude. (Wallops Release 65-31)

- PEGASUS II had reported two meteoroid punctures, NASA announced. The hits were recorded on the .0015-in. and .008-in.-thick aluminum-covered detection panels. (NASA Release 65-175)
- On the floor of the Senate, Sen. Ralph W. Yarborough (D-Tex.) advocated that rights to patents from Government-sponsored research should belong to the Government: "In this struggle between the public interest and those who seek a public subsidy to enrich private coffers, the stakes are immense. The Federal Government every year becomes more involved in the financing of scientific research. This being the case, it is the responsibility of Congress to protect the public purse, rather than to construct private pipelines from the Public Treasury to private recipients." (*CR*, 5/26/65)
- "Establishment of a communications satellite system for commercial purposes is a matter entrusted to the Corporation under the Communications Satellite Act," was the reply of ComSatCorp President Joseph Charyk to the FCC regarding the American Broadcasting Co.'s proposal to launch its own satellite. The FCC had requested the views of ComSatCorp on the proposal. (ComSatCorp Release)
- In Second Annual Sight Lecture to the Wings Club, Dr. Jerome C. Hunsaker, former Chairman of the NACA and MIT professor of aeronautics, said:

  "We cannot return to the time when the century was young, yet we still need the ingenuity and luck of gifted individuals. It is important to establish an environment with incentives to bring new ideas forward.

  "I think of the British Admiralty's prize for a ship's chronometer. The chronometer appeared, and changed the entire art of navigation. Lilienthal's gliding experiments, the Wrights' flights and Sikorsky's helicopter were individual contributions, not in government programs.

  "Scientists have a favorable climate for their own research provided by the Universities and Foundations, with opportunity for publication and recognition through the learned societies. Could we not devise a plan to bring ideas of individuals before sensitive and wise people who would select wheat from chaff and arrange for development testing of some of the harvest. We must be patient. I am reminded of Dr. Paul Foote's remark that, for a new chemical, it is usually seven years from test tube to tank car.

  "What we must avoid is centralized control of the exploration of ideas by the people responsible for immediate needs. There is nothing more discouraging to an engineer than the statement: 'We have no requirement for what you are thinking of.'

"Today, U.S. military power is supreme, but our intent and resolve are more in question than our strength. General LeMay says, 'We must make more determined and longer range plans and commitments. . . . We must look further into the future to foresee the threats that lie ahead.'

"Quantum advances in technology follow availability of scientific knowledge plus creative imagination and financial risk taking. International cooperative effort has been valuable in the past in research, and could be valuable in development work when the threat of destructive purpose becomes less.

"Let us never think we have no requirement for men with new ideas." (Text)

May 26: First stage (s–i–10) for the tenth and last Saturn I launch vehicle left MSFC's Michoud Operations aboard the barge *Promise,* to arrive at KSC May 31. This was the second s–i stage built at Michoud by Chrysler Corp. Space Div. (MSFC Release 65–135)

- ComSatCorp may well face competition from foreign satellite communications systems in the next few years, David Sarnoff, chairman of the Board of Radio Corporation of America predicted at the convention banquet of the Armed Forces Communications and Electronics Assn. in Washington, D.C.: "We can expect that ultimately Russia will set up a satellite communications system competitive to our own and offer it to other nations on favorable terms determined more by political than economic considerations."

  Mr. Sarnoff advocated creation of "a single, privately owned American company" to handle all international communications currently handled by six private carriers. He argued that, among other benefits, a single "unified carrier" was the only way the U.S. could "deal on equal terms with foreign government [communications] monopolies.

  The RCA chairman warned that in only five years the interim agreement between ComSatCorp and the 45 participating nations would be up for re-evaluation. The U.S., he said, "will have to negotiate a new contract under different circumstances and possibly vastly altered bargaining conditions."

  It was technically feasible, Mr. Sarnoff said, that direct radio/TV broadcasting by satellite could be undertaken by 1975. Three equatorial, synchronous orbit satellites powered by nuclear energy, each equipped with a three-TV-channel capability, would be able, he said, to broadcast programs to the entire United States and parts of Canada. He estimated that the three satellites, exclusive of ground stations, would cost $30 million and compared this with the $50 million annual cost to the three major networks for leasing circuits to transmit programs to their affiliated stations or to the $30 million cost of a single large city television station. (*WSJ,* 5/27/65, 6; Wash. *Eve. Star,* 5/27/65)

- Missile lead of the U.S. was put at three to one in an article by Richard Fryklund in the Washington *Evening Star:* "U.S. intelligence estimates are that the Soviet Union has 245 to 295 intercontinental ballistic missiles on launchers ready to be fired.

  "The United States has 900." (Fryklund, Wash. *Eve. Star,* 5/26/65, 2)

May 26-28: NASA-sponsored Fifth National Conference on the Peaceful Uses of Space and St. Louis Bicentennial Space Symposium was held in St. Louis with participants from Government, education, industry, and the scientific community. (NASA SP-82)

NASA Deputy Administrator Hugh L. Dryden, delivering the keynote address, said: "The rate of growth of space activities in the first six years of the space age has been unprecedented in the history of a new field of science and technology but there are signs of attainment of a certain degree of maturity. The most obvious is the establishment, following several years in which available funds nearly doubled each year, of a level of five to five and a quarter billions for congressional appropriations to NASA, or about seven billions for space activities of all agencies at the suitable level. . . .

"Maturity is also indicated by the drastic reduction in the number of unsuccessful missions, the result of increased knowledge and experience in the previously unknown field of space. Thus in calendar year 1958 in the first three months of NASA, four missions were attempted without a single success. In the following year eight of fourteen were successful, whereas in 1964 twenty-five of thirty more difficult missions were successful, a percentage of 83 which has been maintained now for three years. . . ." (NASA Release 65-83; Text, NASA Release 65-165)

Answering the query "What does the future hold in store?" NASA Associate Administrator for Manned Space Flight Dr. George E. Mueller told the Symposium about future manned flight options: "In near-earth space, missions could include low and high inclination, polar, or synchronous orbits to accomplish research, technological, and applications objectives. . . .

"In a low inclination orbit, below the Van Allen belts, the basic problems of keeping men in space for extended periods can be studied, rendezvous and resupply problems could be worked out, and scientific experiments conducted.

"In synchronous orbit, where the spacecraft hovers over a fixed area of the earth all the time, experiments could be carried out which involve manned observations over a given portion of the earth or which use man to assist in the operation of various experimental systems.

"In polar orbit, scientist-astronauts could monitor and observe the entire surface of the earth as it passes beneath the spacecraft, mapping it and surveying most of the world's resources. . . .

"In earth orbit . . . a medium-size manned orbiting research laboratory might be developed. Such a space station would accommodate six to nine men and remain in orbit for up to five years. . . . Resupply vehicles, or space shuttles, could be used for crew rotation and for delivery of equipment and supplies. The laboratory would provide roomy quarters with a shirt-sleeve environment for conducting a wide variety of experiments in space. It would also contain a centrifuge, should it be found essential for reconditioning crew members to withstand the effects of gravity after periods of weightlessness.

"Following this a larger permanent manned orbiting research laboratory accommodating 20 to 30 men, might then be developed, by assembling three or four of the medium-size laboratories in space. Artificial gravity could be provided in the laboratories by rotating them about their axes.

"Possibly the most challenging long-term goal of the entire space program is manned exploration of the planets—especially of Mars." (Text)

Comparing the space programs of the U.S. and the U.S.S.R. Dr. Edward C. Welsh, Executive Secretary of the National Aeronautics and Space Council, said:

"1. In number of earth-orbiting payloads the United States has launched almost three times as many as has the USSR, although the 1965 rate is less than two to one.

"2. In the weight of such payloads, the USSR has put up almost three times as much as has the United States.

"3. In propulsion, the Soviets have from the beginning enjoyed an operational advantage over the United States. However, we are currently making great strides in this regard and it is hoped that we will keep moving up the propulsion ladder so as not to be overtaken again.

"4. In manned space flight, the USSR is ahead of the U.S., not only in hours of flight but also in multi-manned flight and extravehicular activity. So far, the U.S. astronauts have completed 40 orbits of the earth, the Soviet cosmonauts have completed 342 such orbits. Moreover, as our Gemini schedule proceeds and contributes continued progress, we must look for much more activity on the part of the Soviets.

"5. In the application of space developments to directly useful purposes, the United States is well ahead, particularly in such fields as weather observations, navigation, and communications. However, the Soviets have potential capabilities of these types and have already begun to show some actual experience in space communications.

"6. In lunar and interplanetary activity, the U.S. may have an edge with the spectacular success of the Rangers and Mariners. We have developed this advantage, even though the Soviets have made a greater relative commitment in this regard, both from the view of absolute numbers of launches and also in regard to weight of payloads.

"7. Based upon clear knowledge of our own program and upon assertions by the Soviets about theirs, one can reasonably conclude that both countries have manned lunar landing projects under way. It would be impossible to state definitely who is ahead in this regard but I am hopeful that we will turn out to be.

"8. As regards the collection of scientific data from space, both countries have made impressive strides, resulting in a possible advantage to the USSR regarding the effects of space environment on human beings.

"9. Both countries are in a position to make many observations from space, but both countries have pledged not to orbit weapons of mass destruction and have stressed that their programs are dedicated to peaceful uses. I can only speak for this country in regard to our intent and do state that we will maintain our defenses while pledging not to use space for aggressive purposes."

Dr. Welsh warned: "Let us not expect our space program to proceed indefinitely without some tragedy involving our astronauts." (Text)

Dr. Raymond L. Bisplinghoff, NASA Associate Administrator for Advanced Research and Technology, said: "In assessing our growth in space capability in terms of three steps from earth to earth orbit, from earth orbit to moon, and from moon to planets, it is important to recognize that the first two steps rest on essentially the same technologies. These are technologies which have evolved for decades and which are familiar: chemical energy conversion, relatively common engineering materials, measurement and control systems generally consistent with aircraft and ground technology and microwave communications. However, the third step will demand performance and efficiency well beyond the first two. An entirely new level of technology is needed; nuclear energy conversion, new refractory materials, accuracy of sensors—improved by orders of magnitude—and laser communications. There are the underlying requirements of higher reliability and longer lifetimes than have yet been demonstrated, together with low specific weight.

"The requirement for improvement in this spectrum of space-related technologies will drive them well beyond their present level. The presence of difficult goals can have a profound influence on earthbound consumer products through the advancement of common fields of technology in addition to opening the gateway to deep space. The NASA program of advanced research and technology embraces most of these elements at least in their fundamental forms. Without this research the space program would soon wither and die. With it, by the year 2000, an enormous influence can be exerted on national prestige and strength." (Text)

Discussing space projects of the future at the Conference on the Peaceful Uses of Space, Dr. Wernher von Braun, Director of NASA Marshall Space Flight Center, said: "The reusable vehicle seems to be the key to development of an economical earth-to-orbit transportation system. Passenger conveniences must be improved so that scientists, engineers, technicians, military personnel—and even politicians and journalists—can make the trip.

"One of the methods we have been studying several years combines the experience gained in the X–15 rocket plane program with present Saturn know-how, for building a high performance two-stage rocket "plane"—called the Re-Usable Orbital Transport. It appears entirely practical to develop a vehicle that would not subject passengers to more than three g's in ascent or descent.

"In the orbital transport under study, the first stage would fly mission paths similar to the X–15, with the second stage, carrying passengers and cargo, launched from a piggy-back position. The second stage would fly into and out of orbit, gliding to a power-off landing after re-entry in the same manner the X–15 does now as routine procedure.

"It would offer passengers who are in a hurry transportation over global ranges with about one-hour flight time. If we can develop a single or two-stage chemical rocket aerospace vehicle and learn to fly it over and over before it is worn out, the high-income traveler should find the operational cost acceptable. But, of course, the thing we must have is the demand—the traffic, cargo, and passengers to make the system economical.

"After we have tried our wings in the immediate earth environment, our next major step in exploring and utilizing the solar system is the moon. And after that, the planets." (Text)

Dr. Joseph V. Charyk, ComSatCorp president, announced at the Space Symposium that the corporation might invite the aerospace industry to submit detailed proposals for satellites that would connect the television networks to their affiliated stations and would provide new facilities for airplane companies to communicate with aircraft in flight.

Dr. Charyk's disclosure was a consequence of the American Broadcasting Company's recent proposal to put up its own comsat to relay TV shows to affiliated stations for rebroadcast to home viewers.

Dr. Charyk said a satellite to relay television programs to affiliated stations involved no new basic engineering problems and offered "real potential, sound economic basis."

He envisioned a television satellite equipped with 12 channels, three of which would serve each of the four time zones. A satellite of essentially the same design could serve the airplane companies, he noted. (Gould, *NYT*, 5/29/65, 55)

May 27: All test phases of the Project Fire II reentry heating experiment conducted at Cape Kennedy May 22 were satisfactory, NASA announced. Preliminary examination of telemetry data indicated that heating information was received throughout reentry and that all test sequences occurred as scheduled. (NASA Release 65-179)

- USAF launched Atlas-Agena D from WTR with unidentified satellite payload. (*U.S. Aeron. & Space Act., 1965*, 143)
- Army Lockheed XH-51A, fastest helicopter in the world, demonstrated its rigid rotor system and auxiliary jet engine in a successful test flight. It had a top speed of 272 mph. Without thrust from the engine, the XH-51A could be operated as a helicopter. (*Wash. Post*, 5/27/65, A7)
- An explosion two minutes after launch ruined a USAF attempt to send a plastic replica of an astronaut's body into space aboard an Atlas missile from Vandenberg AFB. The dummy was instrumented to measure space radiation at various depths of the body. Cause of the explosion was not immediately determined. (AP, *Wash. Post*, 5/29/65)
- First experimental color television transmissions through the Soviet comsat MOLNIYA I were reported by Tass. Programs were transmitted continuously for more than nine hours from the Moscow television center via MOLNIYA I to an unidentified ground station about 1,000 mi. from Moscow and by land lines back to the Soviet capital. Tass said the tests included color television systems developed in the U.S., France, and the Soviet Union. The U.S.S.R. and France had recently concluded an agreement to cooperate in development of a joint system. (*NYT*, 5/28/65, 2)
- Dr. Kurt Waldheim, Austrian U.N. delegate, was unanimously elected to head the U.N. Outer Space Committee. (*NYT*, 5/28/65)
- It was reported that President Johnson was disappointed that the two-man Gemini-Titan 4 spacecraft scheduled for June 3 launching, had no cameras aboard for simultaneous TV transmission of the space walk. The President had hoped that at completion of the four-day flight by Astronauts McDivitt and White, the U.S. would have pictures similar to those released by the Soviet Union after VOSKHOD II flight. TV cameras had been sacrificed for experimental instruments. (Humphrey, Phil. *Eve. Bull.*, 5/27/65)

May 27: "If Major Edward H. White leaves his space capsule during next Thursday's Gemini 4 flight, it will only be a 'space spectacular' stunt," said Rep. George P. Miller (D–Calif.) during a news conference in San Francisco. Rep. Miller, chairman of the House Committee on Science and Astronautics, had made the same comment at the time of a similar feat by the Soviet Union. (*NYT*, 5/28/65)

- NASA would hire 330 additional summer employees, ages 16 through 21, in support of the Youth Opportunity Campaign announced by President Johnson May 23, NASA disclosed. Instructions had been sent to 11 NASA field centers directing them to begin recruiting for work to begin as early in June as possible. (NASA Release 65–177)
- A working model of Electro-Optical Systems, Inc.'s new 100-lb., 15-in.-dia. ion engine, using accelerated ions to gain thrust, was presented to Smithsonian Institution in Washington, D.C., for display in the Arts and Industries building. (*Wash. Post*, 5/27/65, F3)

May 27–29: Forty educators from Alabama, Arkansas, Louisiana, Mississippi, Missouri, and Tennessee attended a NASA-Univ. of Alabama Educational Symposium whose prime purpose was to determine the impact upon the curriculum of secondary schools of new knowledge and developments in science, sociology, and human relations created by NASA MSFC activities. Symposium and workshop were conducted by the University under a MSFC contract. (MSFC Release 65–129)

May 28: X–15 No. 3 was flown by Capt. Joe Engle (USAF) to 209,600-ft. altitude at a maximum speed of 3,754 mph (mach 5.17) to obtain data on NSL radiometer, Langley scanner, and boundary-layer noise. (NASA X–15 Proj. Off.; *X–15 Flight Log*)

- NRX A–3 Nerva reactor, joint NASA–AEC project to develop a nuclear rocket, was restarted and operated for the third time at Jackass Flats, Nev. Total operating time was 45 min., including about 7 min. at more than 40% of its designated 55,000-lb. thrust capacity. Function of the test was to explore control system response characteristics in low and intermediate power ranges. (SNPO–N–65–9; UPI, *NYT*, 5/29/65, 8; *Rover Chron.*)
- NASA Administrator James E. Webb said in a statement to Pat Houtz of the *Huntsville Times*: ". . . it is extremely important that both the Legislature and the Governor fully understand the importance of the George C. Marshall Space Flight Center operation to the success of the current United States effort in space and, also, the importance of our ability to work in that state in an environment conducive to the most effective utilization of our ablest scientists, engineers, technicians, and industrial contractors." (Text; *Huntsville Times*, 5/28/65)
- Dr. George Mueller, NASA Associate Administrator for Manned Space Flight, said at a news conference in Cocoa Beach, Fla., that Astronauts McDivitt and White would attempt to steer the Gemini 4 spacecraft to a rendezvous with the spent second stage of its booster rocket. This plan was outlined for the flight: When the Gemini 4 spacecraft separated from the second stage six minutes after launching, Maj. James A. McDivitt, as command pilot, would fire jet thrusters to hold a tight formation with the spent stage, which would trail the astronauts by about 300 ft.

Throughout the first orbit, the astronauts would make a complete check of all their systems. At the start of the second orbit they would begin preparing for Maj. White's emergence by unpacking life support packs, the maneuvering unit, and the 25-ft. lifeline.

Sweeping over the Indian Ocean during this orbit, they would begin to depressurize the spacecraft cabin and pressurize their spacesuits. Over Hawaii, Maj. McDivitt would maneuver the spacecraft to within 25 ft. of the second stage.

Maj. White would open his hatch and at a point west of Guaymas, Mexico, he would leave the vehicle. That would be about three hours after launching.

Using the hand-gun maneuvering unit, Maj. White would slowly rotate toward the second stage, which is 27 ft. long and 10 ft. in dia. and would be equipped with two flashing lights. The astronaut would carry a 35-mm movie camera to take pictures of the earth, star background, the booster, and the spacecraft.

After 10 min., over Florida, Maj. White would begin returning to the spacecraft. The cabin would be repressurized and the suits depressurized.

Then Maj. McDivitt would fire thrusters so that the spacecraft would move about 16 mi. away from the booster. During the fifth orbit, about three hours later, the Gemini again would be maneuvered so that it would approach the second stage high over Africa.

The craft would close to within 10 ft. this time to determine how well they can approach an orbiting craft, sighting on the flashing lights. (Transcript; Appel, *NYT*, 5/29/65, 1)

May 28: Scientists at NASA Lewis Research Center had successfully operated a high-field-strength cryomagnet having a volume many times larger than any previously known, NASA announced. The cryogenic magnet would provide research facilities for magnetics, solid state physics, and plasma physics. Effects of high-strength magnetic fields on life could also be examined, using plant life, fruit flies, and small animals placed in the field. (NASA Release 65-170; LRC Release 65-38)

- Lt. Gen. Walter K. Wilson (USA) received NASA's Outstanding Leadership medal for his "outstanding leadership as Chief of Engineers, United States Army, in directing the effective application and utilization of the resources of the Corps of Engineers in the design and construction of facilities crucial to the successful exploration of space by the United States and the application of its space technology for the benefit of mankind." (NASA Release 65-180)

- A tariff for the use of Early Bird satellites for transmission and reception of voice, record, data, telephoto, facsimile, television, and other signals was filed by ComSatCorp with the FCC.

Beginning Sunday, June 27, voice channels would be available between 5 a.m. and 9 p.m. EDT on a daily basis. Minimum rental period, one month; rent, $4,200. Additional consecutive periods would be rented at $140 per day. Voice channels would be two-way. No refunds would be given for interruptions of less than 30 min. or for those caused by solar eclipse. Interruptions of 30 min. or more not the responsibility of the customer would be refunded at roughly $3 per 30-min. interruption. ComSatCorp could request temporary surrender of a voice channel for TV use and, in that event, would refund charges in amounts proportional to the surrender period if it falls between 5 a.m. and 8 a.m. or between 2 p.m. and 9 p.m., or in amounts twice proportional to the surrender period if it falls between 8 a.m. and 2 p.m.

Also beginning June 27, television channels would be available as frequently as feasible. Hours would be 5 a.m. to 8 a.m. and 2 p.m. to 9 p.m. (Schedule I) and 8 a.m. to 2 p.m. (Schedule II). Channels would accept standard TV signals. TV channel rentals must be made for at least a 30-min. period. Rent: $2,400 for first 30 min., $475 per immediately following 15 min. (Schedule I); $3,825 for first 30 min., $710 per immediately following 15 min. (Schedule II). Regular channels would be on one-way monochrome. Two-way monochrome and one-way color would rent for an additional 50%. Interruptions of more than 30 sec. and not the responsibility of the customer would be refunded in amounts proportional to the interrupted period. If a customer canceled his application for use of a TV channel, he would be billed as though he had not and would be required to pay any additional charges involved in acquiring temporary use of a voice channel if his application had made such acquisition necessary. TV channels would be rented on a first-come-first-serve basis.

Rates covered only transmissions between Andover, Me., and the satellite. The arrangement whereby refunds would be made for temporary TV use of voice channels was necessary because EARLY BIRD I cannot handle both kinds of transmissions simultaneously. (ComSatCorp Release)

May 28: Supersonic transport airframe and engine design contracts had been extended through the month of June, FAA announced. The airframe contractors were the Boeing Co. and Lockheed Aircraft Corp.; engine contractors were the General Electric Co. and the Pratt & Whitney Div. of United Aircraft Corp. (FAA Release 65–46)

- The House Armed Services Real Estate Subcommittee approved Air Force plans to dispose of Atlas and Titan missile sites representing an investment of $856,900,000. The 14 missile complexes, embracing 113 missile launching silos, were being declared surplus as a result of the obsolescence of the missiles they were built to accommodate. (AP, *NYT*, 5/29/65, 25)

May 28–29: Fifth Scientific Conference of the Polish Astronautical Society was held in Krakow. Reports were read on many important aspects of rocketry and space travel and on space physics, technology, and biology. Some 93 persons attended the conference which reviewed projects conducted in Poland and abroad. (*Skrzydlata polska*, 6/27/65, 9)

May 29: NASA successfully launched EXPLORER XXVIII Interplanetary Monitoring Probe (Imp–C) from ETR on a three-stage Thor-Delta booster. A slightly longer than planned burn by the third stage engines placed the 130-lb. probe into an orbit with 164,000 mi. (264,040 km.) apogee and 120 mi. (193 km.) perigee instead of the scheduled orbit of 130,000 mi. (209,300 km.) apogee and 120 mi. (193 km.) perigee. Inclination was 34°; period, 5 days, 22 hrs. The spacecraft was equipped with devices to report on the earth's magnetic field, cosmic rays, and the solar wind throughout its highly elliptical orbit.

Confused telemetry signals from the EXPLORER XXVIII for 3½ hrs. after launching made it seem the spacecraft had not separated from the third stage of the booster; however, later signals indicated that all spacecraft systems were operating normally, that separation had occurred.

The Imp series began with EXPLORER XVIII (Imp-A) launched Nov. 26, 1963. (NASA Release 65-164; Wash. *Sun. Star*, 5/30/65)

May 29: MARINER IV, NASA's Mars flyby and photographic probe, reached the distance of one AU (Astronomical Unit) from earth at 9 p.m. EST. An Astronomical Unit is the mean distance of the earth from the sun that had been established, partially from data received from MARINER II, as 92,956,000 mi. The probe had traveled over 271 million mi. in its orbit; its velocity relative to the earth was 51,442 mph. (NASA Release 65-171)

- An "antirock"—a meteorite composed of anti-matter—may have hit the earth in 1908, accounting for what was perhaps the most violent explosion ever observed on earth, said a report in *Nature* by Dr. Clyde Cowan of Catholic Univ. and C. R. Atluri and Dr. Williard F. Libby of the Univ. of California. The 1908 explosion, referred to as the Tunguska meteorite, took place in the air at a height estimated at three miles. Its effects were comparable to those of a nuclear weapon with a yield equivalent to that of 30 million tons of TNT.

  The hypothesis had been supported, to some extent, by an analysis of tree rings formed during, before, and after the year of the explosion. It was calculated that an anti-matter explosion would create enough additional atoms of carbon 14 to produce a worldwide enrichment of this radioactive substance. In the study, a 300-yr.-old Douglas fir from Arizona and an oak tree from near Los Angeles were analyzed. Wood was stripped from a number of annual rings from 1873 to 1933. In both trees, the highest content of carbon 14 was from wood formed in 1909, the year after the explosion. Another supporting fact was that the blast left no cloud such as that produced by an atomic or chemical explosion; a mass of anti-matter, plunged into the atmosphere, would be annihilated, leaving no cloud. (Sullivan, *NYT*, 5/30/65, 1, 50)

- NASA Lewis Research Center scientist Charles A. Low, Jr., was co-recipient (with William R. Mickelsen) of a patent for a radio-isotope generator with attached propulsion system. Low and Mickelsen would use a colloidal particle thrustor to provide the propellant. Research was underway at LRC on use of colloidal particles as propellant in various thrustor designs. The system could cut interplanetary flight durations by as much as one-half or increase interplanetary payloads by substantial amounts. (LRC Release 65-39)

May 30: A modified Convair 990A jet transport—NASA's new high-altitude research laboratory—carried 30 scientists from five countries and a million dollars worth of delicate instruments in a race with a total eclipse over the South Pacific. Path of the eclipse stretched from the northern tip of New Zealand 8,000 mi. east to the coast of Peru. Except for a few small islands, the eclipse was not visible in any heavily inhabited parts of the earth.

Taking off from Hilo, Hawaii, the jet flew at an altitude of 39,500 ft.; in its 9 min. 42 sec. race with the 1,700 mph eclipse, it reached a speed of 587 mph, doubling observation time possible from a ground-based station.

May 30: NASA's Convair 990 airborne research laboratory photograph of solar eclipse.

First indications were that the mission was a complete success; detailed analyses of data from 13 observation projects would be made. Most obvious phenomena were large prominences on the sun, Jupiter shining brightly in the sun's corona, and long corona streamers flashing with surprising brightness.

Scientists in the mission were from Belgium, The Netherlands, Italy, Switzerland, and the United States. (NASA Release 65-178; AP, *NYT*, 6/1/65, 20)

May 30: "A report recently prepared by the Science Policy Research Division of the Library of Congress . . . notes that seven of the agencies —the Weather Bureau, Air Force, Federal Aviation Agency, NASA, Army, Navy, and Treasury Department—ran through about $266 million last year in collecting and reporting identical weather information," said Fred Blumenthal in an article in the *Washinton Post*. He continued: "If the current structure of our weather efforts continues unchanged, untold millions of dollars will keep going down the drain. The obvious solution would be to establish one central national agency to handle the collection and reporting of all weather data, giving us the same service at a cheaper price or better service for the same price—or possibly even less. Then each of the agencies which are now duplicating each other's efforts can use the information for their own purposes." (Blumenthal, *Wash. Post,* 5/30/65)

* Maj. Virgil I. Grissom (USAF) was honored by a parade in his hometown of Mitchell, Ind. (*Indianapolis Star*, 5/21/65)
* Soviet press published first technical details and a sketch of the communications satellite MOLNIYA I, launched April 23. The satellite, which was visible nine hours a day from Soviet ground stations, had been used for experimental transmissions of television programs, including color, between Moscow and Vladivostok.

According to the drawing and the text description, MOLNIYA I had an airtight cylindrical body with conical ends, one of which contained a rocket engine used to correct the orbit, as well as other solar orientation devices.

Expanding like spokes of a wheel from the cylinder were six long panels of solar batteries to supply electrical power to the satellite for retransmission of signals received from earth. It was equipped with two parabolic antennas, one active, the other in reserve. These were pointed toward earth with a high-precision direction finder for earthbound transmissions. MOLNIYA I would make two 12-hr. revolutions around the earth every 24 hrs. One loop would take it over the Soviet Union during daytime periods, when it could be of most use for transmissions. The other loop would take it over the United States. (*NYT*, 5/31/65, 6)

May 30: A successful 60-sec. ground test of the solid-fuel rocket programed to launch Japan's first artificial satellite in 1968 was announced by Tokyo Univ. scientists. The rocket had a maximum thrust of 200,000 lbs. (*Wash. Post*, 5/30/65)

May 31: All technical problems threatening the scheduled launching of Gemini 4 on June 3 had been cleared up, NASA said. One problem involved what had been thought to be a malfunction in the water management system of the spacecraft which would provide water for drinking and cooling. Instead of a leak, technicians found that a valve had been left open in error. A second problem, rupture of an underseas communications cable 10 mi. south of San Salvador in the Bahamas, was sidestepped when USAF rented a commercial cable from Puerto Rico to West Palm Beach to replace the severed line. (UPI, Wash. *Daily News*, 5/31/65; Appel, *NYT*, 6/1/65, 16; AP, *Wash. Post.*, 6/1/65)

- Students at West Bend High School near Milwaukee and at Lycee Henri IV in Paris talked for 40 min. via a two-way circuit on EARLY BIRD I in the first transatlantic linkup of classrooms by live television.

  The exchange was conceived by the Univ. of Wisconsin's educational television station, WHA–TV, in cooperation with Radiodiffusion-Télévision Française. Sound and picture transmission were excellent. (*NYT*, 5/1/65, 32)

- Members of the International Association of Machinists struck Aerojet-General Corp.'s rocket-engine and torpedo plant at Azusa, Calif., because of a wage dispute. (AP, *NYT*, 6/1/65, 25)

- "Along with the development of a nuclear weapon, Communist China has also conducted a program to develop rockets," wrote Cheng Chu-yuan in *Military Review*. He continued: "Since early 1956, when Peking mapped out the 12-year plan for the development of science and technology, jet propulsion has been listed as one of the 12 major tasks, exceeded only by the use of nuclear energy. The project is under the supervision of the Institute of Mechanics in the CAS.

  "In 1958 several new institutes were set up within the Academy—an Institute of Upper Atmosphere Physics in Wuhan; and an Institute of Automation and Remote Control and an Institute of Mechanics and Electronics both in Peking. All of these institutes participate in the rocket program. . . .

"The Science and Technology University of the CAS, the Tsinghua University in Peking, and the Peking Aeronautical Engineering College are the three important centers for training engineers and technicians in rocketry. During the past ten years, more than 3,000 college students, specialists in aeronautical engineering, have been graduated. In 1963 the China Aeronautical Engineering Society was formally established, indicating the rapid growth of technical manpower in this field.

"Since Communist China has several capable men with long years of experience in the rocket field, and since China began her rocket project almost in the same period with the nuclear weapons program, the development of a rocket booster might soon be anticipated. It is quite possible that China may launch her first rocket within the next two or three years." (*Military Review*, 5/65, 10F)

May 31: Aerobee 150A sounding rocket, launched from Wallops Station, Va., attained peak altitude of 90.5 mi. (145 km.). Primary experimental objective was to measure spectral irradiance of the stars Spica and Alkaid in the wavelength interval from 1,100 A to 4,000 A. Instrumentation consisted of an ultraviolet stellar spectrometer with photometers and optical telescope. Attitude control was obtained with a modified Attitude Control System (ACS), a roll-stabilized gyro platform, and an optical tracker—the combination known as Strap. The Strap system performed correctly. Due to an incorrect gain setting, the star-tracker failed to lock-on throughout most of the flight. The telescope and spectrometer functioned properly, but obtained no data due to the failure to lock-on. Experiment was conducted by GSFC. (NASA Rpt. SRL)

During May: A camera capsule from the Saturn I SA-7, launched from Eastern Test Range Sept. 18, 1964, was found in shallow waters off San Salvador in the Bahamas. Color film in the capsule had deteriorated and was not usable. The capsule was the third one found of the eight flown on the SA-7. The first two were found in November 1964, near San Salvador and Eleuthera Islands. Film in these capsules was in good condition. (*Marshall Star*, 5/19/65, 6)

• Bell Telephone Laboratories astronomers detected radio waves that seemed to be "flying in all directions through the universe." Dr. Arno A. Penzias and Dr. Robert W. Wilson made the observations with the horn antenna developed for communications satellite research at Holmdel, N.J. Princeton Univ. scientists led by Dr. Robert H. Dicke, Prof. of Physics, unaware of the BTL observation, reached a prediction of the existence of such waves, which they theorized were remnants of light waves from the primordial explosion giving birth to the universe. In this theory of the universe's origin—the "big bang" theory—the galaxies all originated at a single point, shooting outward ever since the cataclysmic event. According to the theory, the light waves were stretched into radio waves by the expansion of the universe. (Sullivan, *NYT*, 5/21/65)

• Remarkable adaptability of some fungi and bacteria to life in atmosphere containing high concentrations of ammonia and methane was discovered by Dr. S. M. Siegel and Miss Constance Guimarro of the Union Carbide Research Institute and reported in *Icarus*. The report suggested there might therefore be life on Jupiter, where extremes of temperature and a dense atmosphere of these noxious gases would seem to make life-forms resembling those on earth unlikely. The research was supported by NASA contract. (Schmeck, *NYT*, 5/12/65; *Icarus*, IV/1965, 37–40)

During May: Carl Sagan, Harvard Univ. and Smithsonian Astrophysical Observatory, and Sidney Coleman, Lyman Laboratory of Physics, Harvard Univ., reviewed the need for sterilization of Mars-bound spacecraft to protect that planet from contamination. Using probability theory, Sagan and Coleman specified formulas to provide predictions of onboard-experiment and Mars-contamination relationships. (*A&A*, 5/65, 22–27)

- Research expenditures in 1963 totaled $5.9 billion in the national economy, according to National Science Foundation report. Of this amount, $3.4 billion (58%) was financed by the Federal Government and $2.1 billion (35%) by industry, with colleges and universities and other nonprofit institutions providing the remaining 7%. In performance of research, industry spent $3.2 billion (54%); colleges and universities, $1.4 billion (24%); Federal Government, $.9 billion (14%); and other nonprofit institutions, $.4 billion (7%).

  Of the $5.9 billion total, nearly $2 billion was expended for basic research and the rest for applied research. Predominant in basic research performance were the colleges and universities and other nonprofit institutions, spending more than $1 billion (57%) of the $2 billion total. (NSF *Reviews* . . ., Vol. I, 5/65, 1, 2)

- Article in *Soviet Life* by Academician Anatoli Blagonravov described the "three-directions of modern astronautics":

  "The first is the study of the earth's upper atmosphere and the portion of space adjoining our planet. . . .

  "The second is the study and exploration of . . . the moon.

  "The third is the study of solar space, including the nearest planets, Mars and Venus.

  "Soviet scientists are working in all three directions.

  "The first to be launched, always, are the automatic scouts, followed by men.

  ". . . The final stage in the solution of the first problem—exploration of near space—will probably be to set up a permanent manned space observatory-town, with bilateral contact maintained through rockets. Of course, long before this, reliable systems of meteorological sputniks, worldwide television sputniks, navigation sputniks, etc., will have been established. . . .

  "Several interesting moves have been taken in the second direction [lunar exploration]. . . . The automatic devices have not yet explored the moon's surface in detail, have not yet determined the conditions prevailing there. Presumably, they will be followed by animals. Only after the safe return of the ship to earth has been assured will man go to the moon.

  "Manned landing will be preceded by numerous earth-moon flights. In the course of these flights the conditions along the entire route will be studied, detailed maps of the lunar surface made, and lunar space investigated. Need I add that all these flights will be made by teams of scientists only?

"The first stage in the exploration of the moon will be to set up a permanent research station on its surface. . . .

"In the third direction—the exploration of near solar space and near planets—only the first steps have been taken, the first flights of automatic scouts. . . . The interplanetary routes will be explored again and again by automatic stations that will bring back much needed information on space and the nature of the planets to which they are sent. Only then will the first interplanetary expeditions take off. They will carry even larger teams than the lunar reconnaissance expeditions." (*Soviet Life*, 5/65, 48)

During May: Dr. Lee A. DuBridge, President of Cal Tech, discussed objectives of the space program and what men hoped to learn through the space program about the moon, the planets, the sun, interplanetary space, and the earth itself. He concluded:

"Man's growing understanding of the nature and constitution of the universe had led to new advances in our knowledge of physics and of chemistry; and these in turn have led to applications of this knowledge to the development of things which men have found useful. We have never been able to predict in advance what the usefulness would be of new knowledge about the nature of the physical universe. All we know is that, by and large, new knowledge always has proved useful—and often it has proved useful in the most unexpected and unforeseeable ways. No one would have predicted that Newton's enunciation of the laws of motion would lead to the age of machinery; that Faraday's experiments would lead to the age of electricity; or that Einstein's theory of relativity and Bohr's theory of the atom would lead to the atomic bomb and atomic power.

"We do know one thing: that scientific research which has been aimed at purely practical problems though it often has been of great value, has over a long run been of less value in producing wholly new things than has the research aimed solely at the extension of knowledge. The extension of man's knowledge is the basic and the overriding purpose of the space exploration program." (Text, JPL *Lab-Oratory*, 5/65, 10–12)

June 1965

June 1: Aerobee 150 sounding rocket launched from White Sands, N. Mex., went to peak altitude of 113 mi. (180 km.). Preliminary experimental objective was to obtain clear spectrograms of ultraviolet light from stars. Experimentation was provided by Princeton Univ. Observatory. (NASA Rpt. SRL; AP, *NYT*, 6/28/65, 2)

- North American Aviation, Inc., was awarded a $17 million increment to a previous contract for the XB–70 flight test program. USAF Aeronautical Systems Div. was the contracting agency. (DOD Release 374–65)

June 2: U.S. Senate passed a bill (H.R. 7717) authorizing appropriations to NASA for FY 1966 totaling $5,196,826,350, as follows: $4,533,350,000 for research and development; $67,376,350 for construction of facilities; and $596,100,000 for administrative expenses. (*CR*, 6/2/65, 11816)

- NASA Administrator James E. Webb, testifying before the Senate Judiciary Committee's Subcommittee on Patents, Trademarks, and Copyrights, said: "Fundamental to NASA's approach to the patent policy question and to technology utilization is our belief that active effort must be expended, and meaningful incentives provided, if the byproducts of the space efforts are to flow to the general public through entrepreneurs willing to risk investment capital. . . .

  "If NASA's experience has served to establish one principal, it is . . . that 'a single presumption of ownership does not provide a satisfactory basis for Government wide policy on the allocation of rights to inventions.' NASA's experience further establishes . . . that the 'Government has a responsibility to foster the fullest exploitation of the inventions for the public benefit.' " (Transcript)

- William B. Rieke, NASA Deputy Associate Administrator (Management) for Manned Space Flight, was appointed Deputy Associate Administrator for Industry Affairs, replacing George Friedl, Jr. Prior to his appointment to NASA in 1962, Rieke was president of Lockheed Aircraft International, Inc. Friedl would continue to serve NASA as a consultant. (NASA Ann.)

- Grove Webster had been appointed Director of NASA Hq. Personnel Div., NASA announced. He had previously served as deputy and acting director. (NASA Release 65–182)

- AFSC Commander Gen. Bernard A. Schriever predicted at a Retired Officers' Luncheon in Washington, D.C., that "the next major breakthrough in international commerce will be low cost, long haul air transportation, which could be derived in large part from prior military experience." Gen. Schriever said he recognized "that there are problems involved in translating military systems into commercial systems . . . but these problems can be successfully attacked and solved if there is adequate long range planning now." (Text)

June 3–7: NASA's GEMINI IV spacecraft was launched at 11:16 a.m. EDT with two-stage Titan II booster from the Eastern Test Range and began the four-day space flight of Astronauts James A. McDivitt (Maj., USAF) and Edward H. White, II (Maj., USAF), who would make 62 revolutions around the earth in 97 hrs. 56 min. Two minutes and 36 sec. after liftoff, the first stage of the booster separated. Six minutes later, traveling at 17,567 mph, the spacecraft was inserted into an orbit with apogee, 174.8 mi. (283.2 km.); perigee, 100 mi. (161 km.); period, 94 min. Original plans had been for GEMINI IV to be maneuvered within 25 ft. of the burned out second stage of the TITAN II booster rocket and for White to approach and possibly touch it during his extravehicular mission. Three hours into flight, ground stations reported that excessive tumbling of the second stage had increased atmospheric drag and that it was orbiting 32 mi. ahead of and 5 mi. below GEMINI IV. Mission Director Christopher Kraft confirmed Command Pilot McDivitt's suggestion to abandon further attempts at rendezvous because of a potential fuel shortage.

White's extravehicular activity, planned for the second orbit, but delayed until the third to allow astronauts more preparation time, began at 3:45 p.m. EDT. The cabin was depressurized; White, equipped with tether carrying oxygen and communication and with chest pack for emergency oxygen supply, emerged from the spacecraft just past Hawaii. Carrying a modified 35 mm. single-lens reflex camera loaded with color film and propelled by a hand-held, oxygen-jet gun, he went three times to the full length of his 25-ft. tether and then returned, using the gun to halt his motion and prevent his hitting the spaceship. When the gas supply in the gun was depleted, he returned to the spacecraft by gently tugging on the tether line. At one point, McDivitt exclaimed: "You smeared my windshield, you dirty dog."

Flight plans had called for a ten-minute walk in space but White remained outside the spacecraft for 22 min. He experienced no disorientation during his "walk." When he finally heeded commands to return to the capsule, he had difficulty closing the hatch and decided not to reopen it to jettison excess equipment. "It's the saddest moment of my life," White said as he reentered the spacecraft.

On June 5 during the 17th orbit, the astronauts spoke to their wives at MSC. During the 20th orbit, McDivitt spotted a satellite with "big arms sticking out." He was unable to identify it positively.

On June 6 during the 48th orbit, trouble developed with the spacecraft's computer and attempts to repair it with the aid of ground instructions failed. The malfunctioning computer made it necessary for GEMINI IV to reenter on a ballistic trajectory.

Throughout the flight the daily routine of the astronauts included eating, exercise, and performance of medical and scientific experiments. They alternated rest periods. During 12 of the 62 orbits, when GEMINI IV passed through a heavy radiation area called the South Atlantic anomaly, Astronauts McDivitt and White switched on radiation and magnetic field measuring devices to take readings inside and outside the spacecraft and near their bodies. They also attempted to improve the knowledge of the earth's terrain through high-quality color photographs; to measure with instruments the electrostatic

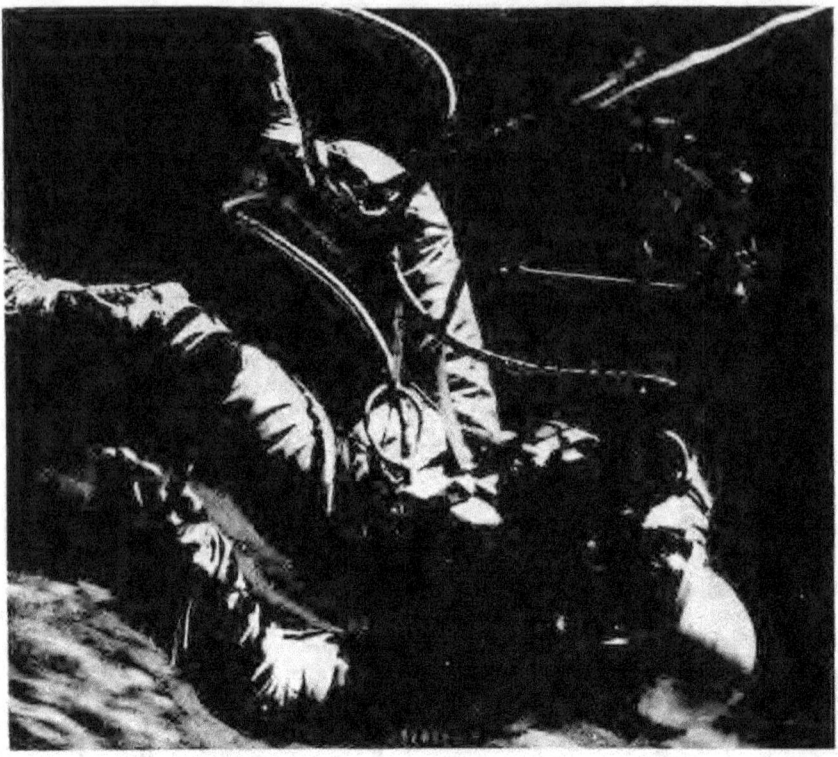

June 3: Space walk of GEMINI IV Astronaut Edward H. White.

charge that accumulates in space and on objects in space; to better define the magnitude and direction of the earth's geomagnetic field; to test the accuracy of part of a prototype navigational system for future space capsules; to measure heartbeats to try to determine the effects of prolonged weightlessness on functioning of the heart; to determine the effects of limited exercise in space through use of a tension cord; to take high-resolution photographs of cloud formations over the earth to aid weathermen in improved forecasting from weather satellites; to determine more exactly the elevation of the earth's atmosphere and its layers through use of filtered film; and to determine if long periods of weightlessness might make the bones brittle.

On June 7, to begin reentry, McDivitt fired a 100-lb. thrusting rocket for two minutes, 41 sec.—one second too long—to guide the capsule into an orbit with 156.2-mi (252-km.) apogee and 100-mi. (161-km. perigee. Twelve minutes later, McDivitt fired the retrorockets; 2½ min. later he placed the spacecraft into slow rotation to reduce reentry dispersion. Communications with the spacecraft then ceased for about 4 min. as ion-sheath blackout phase of reentry began. GEMINI IV entered the final stage of reentry at 1:07 p.m. EDT when the stabilizing chute emerged and damped the oscillations of the descending craft. The main parachute opened shortly afterward and slowed the spacecraft for its final landing at 1:13 p.m. EDT, about 450 mi. east of Cape Kennedy. The landing was 40 mi. off target because of the one-second error in the firing of the thrusting rockets and one-second delay in the firing of the retrorockets.

Fifteen minutes after splashdown, Navy frogmen, lowered into the water by helicopter, placed a flotation collar around the capsule. Before egressing from the spacecraft to be hoisted to helicopter, astronauts took biomedical data on themselves. About 40 min. later, Majors McDivitt and White were landed by helicopter on the flight deck of the carrier *Wasp.* They were taken immediately to the ship's sick bay for postflight examinations.

Dr. Charles Berry, Chief Flight Surgeon for the astronauts, said after the examinations: "We have knocked down an awful lot of straw men. We had been told that we would have an unconscious astronaut after four days of weightlessness. Well, they're not. We were told that the astronaut would experience vertigo, disorientation when he stepped out of that spaceship. We hit that one over the head." The most serious problem was fatigue. "Both men were bushed," said Dr. Berry. Maj. McDivitt had a few flecks of caked blood in his nostrils, caused by the dryness of the mucous membranes from inhaling pure oxygen for so long. X-rays taken of the astronauts' little fingers and heel bones both before and after the flight to determine if long exposure to weightlessness would cause a substantial loss of calcium were being studied.

Commenting on the historical significance of the Gemini flight, Dr. George E. Mueller, Associate Administrator of NASA's Office of Manned Space Flight, said the flight had included "at least two world firsts": a record length for two-man flights and the first time a pilot had maneuvered outside a ship with a propulsion gun, which gave him control over his movements.

American astronauts had flown a total of 259 hrs. 34 min. in space; Soviet astronauts had accumulated 507 hrs. and 16 min. (NASA Release 65–158; NASA Proj. Off.; *NYT,* 6/4/65, 1,15; Wash. *Eve. Star,* 6/4/65, A1,A6,A10; *Wash. Post,* 6/4/65, A1,A7,A8; *WSJ,* 6/4/65, 4; Clark, *NYT,* 6/5/65, 1,12; Wash. *Eve. Star,* 6/5/65, A3; *Wash. Post,* 6/5/65, A1,A7; *NYT,* 6/6/65, 1,70; *NYT,* 6/7/65, 22C; *Wash. Post,* 6/7/65, A1,A4,A6; Balt. *Sun,* 6/7/65; Justice, *Wash. Post,* 6/8/65; Simons, *Wash. Post,* 6/8/65; *Time,* 6/8/65, 20,25,25A; *NYT,* 6/8/65, 22C; *NYT,* 6/9/65, 1,22; Lee, *Houston Post,* 6/9/65; *Wash. Post,* 6/11/65, A1; Wash. *Eve. Star,* 6/13/65, A1; *Wash. Post,* 6/15/65, A1.)

June 3: Viewers in Great Britain and at least 11 Western European nations were able to watch the Gemini 4 lift-off live via EARLY BIRD I satellite. Picture reception was reportedly clear. ComSatCorp, cooperating with European ground station owners, arranged for free use of EARLY BIRD I between 8:00 a.m. and 12 noon EDT. (*Wash. Post,* 6/4/65; ComSatCorp Release)

• Department of Defense should, without further delay, commence full-scale development of a manned orbital laboratory (MOL) project, recommended a report by the Military Operations Subcommittee of the House Committee on Government Operations. The report also suggested that DOD "pursue a more realistic security policy in its program," and that "careful and intensive consideration be given to achieving future economies in space operations with greater standardization of systems and subsystems and through such techniques as recoverable and reusable boosters and spacecraft." Moreover, "NASA and DOD [should] use each other's facilities and resources to the greatest possible extent." (Text)

June 3: RCA Communications, Inc., asked FCC for the right to lease and operate 30 of EARLY BIRD I's voice-grade and television channels operating between the U.S. and Europe. FCC approval would enable RCA, owner of the National Broadcasting Co., to provide international television and closed-circuit transmission programs across the Atlantic. (*WSJ*, 6/4/65, 6)

- A home-made U.S. flag, carried by astronauts Virgil I. Grissom (Maj., USAF) and John W. Young (LCdr., USN) on the March 23 GEMINI III flight, was hoisted at NASA Manned Spacecraft Center the moment Astronauts James A. McDivitt, (Maj., USAF) and Edward H. White II (Maj., USAF) went into orbit. Flag would be flown only while astronauts were in spaceflight. (AP, Knoxville *News Sentinel*, 6/3/65)

- House voted to allow Gen. William F. McKee (USAF, Ret.) to serve as Administrator of the Federal Aviation Agency and retain his military status. Effect of the legislation, if it were to become law, would be to let McKee draw reduced military retirement pay of $8,404 annually in addition to the Administrator's salary of $30,000. Without the bill, he would have to give up his military status and retirement benefits. (*CR*, 6/3/65, 11961; AP, Wash. *Eve. Star*, 6/4/65)

- American Telephone and Telegraph Co. made the first formal bid to become a customer of ComSatCorp. In a filing with the FCC, AT&T asked for the right to lease 100 voice-grade channels that would operate through EARLY BIRD I comsat and connect with European telephone companies. AT&T also asked that it be allowed to acquire a television channel for use from time to time in providing TV service to and from Europe in ventures with various European communications carriers. (*WSJ*, 6/3/65, 4)

June 3: In an article suggesting the possibility of catastrophe during the Gemini GT-4 flight, William Hines had said in the Washington *Evening Star:* "The truth is that the only 'first' the United States has a chance to achieve in Gemini 4 is 'first casualties in space'—and nobody (including the Russians) seek this dubious record." Commenting on the planned "rendezvous," he said: ". . . the other and more questionable added spectacular on the present mission—is a risky business of unknown proportions . . . There are several purely technical objections to this exercise. First, it is not part of the original Gemini 4 mission, but an afterthought. Second, the spacecraft was not built for rendezvous. Third, the rocket was never envisioned as a target vehicle and is uncontrollable in space. . . .

"But if tragedy should strike as a result of something that happens on the first few orbits of Gemini 4, how will the responsibility be fixed: As pilot error . . . or politician's error?" (Hines, Wash. *Eve. Star*, 6/3/65)

- Lockheed Missiles and Space Co. was awarded a $10,789,000 increment to an existing contract by USAF for FY 1965 launch services for Agena-D program. (DOD Release 379-65)

- First Negro to receive USAF astronaut training, Capt. Edward J. Dwight, Jr. (USAF), denied *Ebony* magazine's charges that he had been eliminated from selection by the NASA program because of his race. Dwight said the charge "apparently had some information out of context." A statement released by NASA in response to queries about Dwight said: "A formal objective rating system based on flight experience, academic background and supervisory ratings was developed and used by a preliminary selection committee in rating the candidates to make sure that the best qualified were selected as finalists to be considered for the 14 available astronaut positions. Of the 136 candidates, 102, including Capt. Dwight, were eliminated by the primary selection committee, leaving a group of 34 finalists of whom the 14 best were chosen.

"Selection is made on a best qualified basis without regard to race, religion or sex.

"Capt. Dwight did possess the basic qualifications; he did not score sufficiently high to be selected under the rating system." (UPI, *NYT*, 6/3/65; *WSJ*, 6/3/65, 1; Wash. *Eve. Star*, 6/3/65)

• National Park Service delivered to NASA a comprehensive report outlining ways to make Kennedy Space Center a major tourist attraction. (Wash. *Eve. Star*, 6/2/65)

June 4: Man would someday control enough energy to maneuver planets in their solar systems to suit his own purposes, British astronomer Dr. Fred Hoyle predicted at the dedication ceremony of Brown Univ.'s new science building. He explained: "The only large and still mainly untapped reservoir of energy is from nuclear fuels. The conversion of deuterium in the waters of the ocean to helium is the biggest potential resource. . . There's just the possibility that Venus may possess an atmosphere of hot steam. If this is so, pushing Venus a little farther from the sun would cause the whole steam atmosphere to collapse into a much cooler ocean. In fact . . . with the energy availabilities I have been talking about, conversion of Venus to a planet like earth would be an entirely feasible proposition." (*NYT*, 6/6/65, 70)

June 5: Izvestia, official Soviet newspaper, said that the "walk" in space by Astronaut Edward H. White II (Maj., USAF) had given an impetus to U.S. plans for using outer space for military purposes and cited a recommendation by a Congressional committee for an orbital laboratory run by the Pentagon: "These are the evil 'uses' of man in space being thought up here by some influential people in Washington where, incidentally, not one official word has been said about the peaceful purposes of the flight of Gemini 4." (AP, *NYT*, 6/6/65, 82)

June 6: The 2 million people of Melbourne, Australia, saluted Astronauts McDivitt and White as GEMINI IV passed overhead by turning on all the lights in the city. McDivitt said to the Carnarvon tracking station: "Tell them I thank them for lighting the night for me." (AP, Wash. *Eve. Star*, 6/7/65, A6; *Houston Chron.*, 6/7/65)

• NASA terminated its contract with the Thiokol Chemical Corp., and its subcontractor, Newport News Shipbuilding and Drydock Co., for development of a 260-in.-dia. solid rocket motor. Reason for the decision was the difficulty encountered by the contractors in processing and welding special steel required for the rocket case and the time which would be required to develop an alternate method.

During a hydrostatic pressure test April 11 of the first rocket case produced under the Thiokol contract, failure had occurred at a pressure considerably below the normal operating firing pressure with the consequent destruction of the case. A second case, presently about 60% complete, had been constructed by the same method and was subject to the kinds of faults that resulted in destruction of the first case.

Aerojet-General Corp., a second contractor undertaking development of a 260-in. solid rocket motor, was not affected by termination of Thiokol contract. (NASA Release 65–187; *Marshall Star*, 6/16/65, 6)

June 6: Dr. Edward C. Welsh, Executive Secretary of the National Aeronautics and Space Council, told the graduating class of Clark University: "Your country has wisely although belatedly made a firm decision to conduct a vigorous national space program. It is a clear responsibility of each of us to learn what he can about that program, find out what it means to the country as a whole and to the individuals who live here." He continued: "As we attempt to carry out our responsibilities in educating the general public about space, there are a few concepts which need particular emphasis:

"1. *Education.* The space program has been a catalyst, a stimulus to education at all levels, with particular attention to science and engineering. . . .

"2. *National Security.* . . . How much more secure are we, due to improved communications, more accurate navigation, and more complete weather information? How much is it worth to be better informed about potential sources of danger? How can we assess the advantage of developing competence to detect and offset possible aggression from space? How important is it to know that we intend to keep peace and freedom in space so that all who would go there with peaceful intentions are free to do so? I cannot judge the worth of this national security. . . .

"3. *Innovations.* The space program stimulates the development of new materials, new products, new productive processes, and new managerial techniques. . . .

"4. *International Status.* A substantial difference in influence in world affairs evolves from whether a country is in a first position or a second position in power. In many respects, a nation's relative position depends on how it stands in advanced technology. Power and influence in world affairs depend to a great degree upon the technological capability of a nation. . . . The ideal picture is that of a nation strong in ideas, in technology, in freedom, in standard of living, and in military power to protect the viability of the other prestige ingredients. The space program, effectively and imaginatively conducted, contributes positively to all of those ingredients. Of even greater importance is the potential impact the space program can have on world peace through substituting competence in space exploration for competence in building implements of aggression. . . .

"5. *Economics.* Combining the best talents in management, in engineering, and in science, with the most modern facilities available, the net result of the space program is the production of progress. . . ." (Text)

• Russian Cosmonaut Lt. Col. Aleksey Leonov, first man to walk in space, was quoted as calling the GEMINI IV flight of Majors James A. McDivitt (USAF) and Edward H. White II (USAF) "a very interesting one." During an interview with the Bulgarian news agency, he congratulated the Gemini crew, wished them "a happy landing back on earth," and said "no doubt the experience from our flight must have helped the U.S. space program very much in preparing and executing the flight." (Reuters, *NYT*, 6/7/65)

June 6: ABC science editor Jules Bergman reported that the satellite sighted June 4 by Astronaut James A. McDivitt (Maj., USAF) was a secret U.S. military reconnaissance satellite with cameras. He said that space officials had been unable to identify it because DOD would not admit the existence of a U.S. reconnaissance satellite. (AP, Balt. *Sun*, 6/7/65)

- Pope Paul VI, in a mid-day Pentecostal message to a crowd in St. Peter's Square, blessed Astronauts McDivitt and White: "Our benediction goes to all on earth and rises also to the skies for those who are exploring astral paths." (*NYT*, 6/7/65, 33)
- The National Science Foundation's past activities were reviewed in a report by the Library of Congress Legislative Reference Service for the House Committee on Science and Astronautics. No judgments were made, but the 286-page report called attention to broad areas of concern: (1) Could the Foundation meaningfully promote the progress of American science on a very limited budget? (2) What was its relationship to the Office of Science and Technology which was developing national science policy? (3) What was its relationship to American universities conducting research? (4) Had it effectively gathered data and made statistical analyses essential to research planning and administration? (5) Should the Foundation continue to manage large-scale research projects? (Text)

June 6–26: 50 helium-filled tetrahedronal plastic balloons were released over New York City by the Weather Bureau to gather air pollution information at altitudes of 500 ft. or less. Each balloon carried a radar beacon which was tracked by weather radar to obtain a continuous record of its position; some were followed by helicopters to gather additional information on pressure, temperature, and air pollution values. (U.S. Weather Bureau Release)

June 7: President Johnson telephoned the Nation's thanks to Gemini Astronauts James McDivitt (Maj., USAF) and Edward H. White II (Maj., USAF) and told them that they had written their names "in history and in our hearts." He concluded: "What you've done will never be forgotten. We can hope and pray that the time will come when all men of all nations will join together to explore space together and walk side by side toward peace. And you two outstanding men have taken a long stride forward in mankind's progress, and everyone in this nation, and I think in the free world, feels in your debt." The astronauts, aboard the carrier *Wasp* in the Atlantic for medical tests, were invited to spend the week-end with the President at his Texas ranch. (Kilpatrick, *Wash. Post*, 6/8/65, A14)

- ITT asked the FCC for authority to lease 41 of EARLY BIRD I's 240 voice-grade channels. (Weekley, *Wash. Post*, 6/8/65, D8)
- Educators from various elements of NASA and a group of 65 lecturers employed in the NASA Spacemobile program began a week-long training session at NASA Langley Research Center. (LaRC Release)
- "Soviet spaceships make their landings on terra firma, and practically everybody in the space flight business agrees that this is much more desirable than splashing down in the ocean," wrote the Washington

Evening Star. The article said that water landings by U.S. spacecraft proved that "U.S. manned spacecraft lack the capability to come back to land." (AP, Wash. *Eve. Star*, 6/7/65, A6)

June 7: "U.S. space officials are in no particular hurry to develop a ground landing because water landings have worked so well," reported the *Washington Post.* The article said that the "Soviet landing system—partly due to the Russians' large and sparsely populated land area and their lack of sea forces—required Russian astronauts to parachute out of the descending spaceship." This put stress on the airman—especially after long periods of weightlessness. (AP, *Wash. Post*, 6/8/65, A14)

- Soviet cosmonaut Gherman Titov, who orbited the earth 17 times in August 1961, announced that his wife was expecting a baby in a few weeks. (UPI, Houston *Chron.*, 6/7/65)

June 8: LUNA VI, 3,179-lb. instrumented moon probe, was successfully launched by U.S.S.R. with a multi-stage rocket into a parking orbit and then fired on a trajectory toward the moon. All onboard equipment was said to be functioning normally and the trip was scheduled to last about three and a half days, according to Tass announcement. (Tass, 6/8/65)

- Snap 8, NASA–AEC experimental reactor, had run continuously at power for 209 days, thereby completing the longest known power operation of a nuclear reactor. Built by Atomics International Div. of North American Aviation, Inc., Snap 8 began power operation in November 1963, ran 91% of the total time available, and produced more than five million kw. hours of heat. It was operated in a shielded AEC test facility near Los Angeles. (Atomics International Release AI–18)

- Selection of Radiation, Inc., to negotiate a fixed-price contract for Pulse Code Modulation (Pcm) Data Handling Equipment Systems was announced by NASA Goddard Space Flight Center. The proposed $1,700,000 contract would call for design, manufacture, and spare parts for 11 PCM systems. (GSFC Release 6/14/65)

- ComSatCorp filed application with the FCC for licenses to operate commercially EARLY BIRD I and the ground station at Andover, Me. (ComSatCorp Release)

- France announced it had successfully completed tests of its three-stage Diamant rocket scheduled to orbit a French satellite in 1966. (Reuters, *Detroit News*, 6/10/65; Reuters, *NYT*, 6/10/65)

- Communist Hungary acquiesced to a U.S. State Dept. request and suspended a Budapest radio broadcast which could have interfered with base-to-ship communications during the GEMINI IV splashdown period. (Wash. *Eve. Star*, 6/8/65)

June 9: USAF launched Thor-Agena D booster rocket with unidentified satellite from Vandenberg AFB. (AP, Balt. *Sun*, 6/10/65)

- USAF launched Blue Scout Jr. space probe from ETR on a 10,897-mi. altitude flight to measure effects of space radiation on human tissue equivalents. Useful data were telemetered, and after the 4-hr., 32-min. flight the probe fell into the Indian Ocean. (*U.S. Aeron. & Space Act., 1965*, 145)

- LUNA VI, Russian probe, passed the halfway point on its journey to the moon. All systems continued to function normally, Tass announced. (Tass, 6/9/65)

June 9: Astronauts McDivitt and White, onboard the carrier *Wasp* for medical debriefing, received a congratulatory telegram from Soviet Cosmonaut Yuri Gagarin, first man in space: "We send you our congratulations after the success we witnessed of the spaceflight in ship Gemini 4. We express hope that spaceflights will be to serve the world and make progress for humanity." (AP, Haughland, *Wash. Post,* 6/10/65, A1, A11)

- NASA released a photograph of the GEMINI IV spacecraft taken by Astronaut Edward H. White II (Maj., USAF) during his walk in space. The picture was one of an unannounced number that Major White recorded with a 35 mm. camera mounted atop a space gun that he used for maneuverability. Films of Astronaut White floating in space would be made available for public showing "sometime within the next month." (AP, *NYT,* 6/11/65, 12)

- An analysis of worldwide press reactions to the successful GEMINI IV flight indicated that most newspapers took the occasion to call for U.S.-Soviet cooperation in space ventures. "That this [cooperation] is unlikely shows how far our politics lag behind our technology," Britain's *Manchester Guardian* said. Dutch newspapers joined in asking for cooperation in space and Trouw said the duplication of effort by the Soviet Union and the United States "is a waste of money and know-how." The West Berlin press highly praised the U.S. policy of reporting the Gemini flight as it was taking place. Swedish newspapers refrained from any comment amid a wave of anti-American sentiment over Viet Nam and the Dominican Republic. The *Hindustani Times* commented: "It is a pity that the two nations [U.S. and Soviet Union] are going about the job with such secrecy. The time has come for both nations to pool their resources and make the projected attempt on the moon a truly human adventure." (*Wash. Post,* 6/9/65, A3)

- Joint U.S.–U.S.S.R. space effort was urged by *L'Osservatore della Domenica,* Vatican magazine. An editorial said press comment on the GEMINI IV spaceflight had been slanted by "a competitive mentality for which, especially in this field, there is no longer any reason." It continued: "Let us ask ourselves . . . if those who on earth appear divided by unbridgeable distances may not meet and collaborate in space." (UPI, *NYT,* 6/11/65, 7)

- EARLY BIRD I communications satellite demonstrated its ability to relay commercial voice messages, photographs, and data between Europe and the U.S. During one exchange a news photograph was relayed from London to the N.Y. offices of the Associated Press; simultaneously, Pan American Airways was relaying international airline reservations and communications, including pilot and control messages. Other uses included transmission of pictures and layouts for a fashion magazine and relay of bank signatures from New York for verification of a check-signer in London. Except for a slight echo bounced back from Europe on the voice transmissions, the reception was excellent. (ComSatCorp Release; Dewar, *Wash. Post,* 6/10/65, A3)

- The earth has four vast bulges roughly in the design of a pyramid and four equally large depressions, reported scientists from Johns Hopkins Univ. Applied Physics Laboratory. "They are as big as the North

American continent," said Dr. Robert R. Newton, Supervisor of Space Research and Analysis at the Laboratory. Discoveries announced were credited to ANNA IB geodetic satellite, launched Oct. 31, 1962, and a number of USN research satellites. By studying rises and dips in their orbits, scientists were able to determine gravitational highs and lows. One high point centers over Ireland in the northern hemisphere and sprawls northward toward the pole. Another extends across the equator from New Guinea northward toward Japan. Another is south of Africa centered about half way to Antarctica, and the fourth is west of South America with its apex off Peru. (Johns Hopkins Univ. Applied Physics Lab. Release; Myler, UPI, *Wash. Post*, 6/9/65, A3; AP, *NYT*, 6/10/65, 17)

June 9: USAF awarded Aerojet General a $5,101,000 initial increment to a $28,294,800 fixed-price contract for first and second stage engines for the Titan III-X/Agena program. (DOD Release 392-65)

- Gen. Bernard A. Schriever (AFSC), in talk on "Materials and Tomorrow's Air Force" at the Air Force Materials Symposium in Miami Beach, said: "The rate of progress in materials will be measured by accomplishments in four areas. First, the acquisition of new fundamental knowledge which will lead either to new materials or to the utilization of existing materials to meet specific needs. Second, the exploratory development of materials in advance of specific system requirements. Third, the evaluation of new materials for potential applications. And fourth, the investigation of economical manufacturing techniques and equipment. The importance of vigorous efforts in all of these areas is obvious when we remember that it takes from 5 to 10 years to translate a laboratory result into hardware.

    ". . . the people of the Air Force Materials Laboratory are making significant contributions to present and future Air Force systems. Their recent accomplishments include development of methods for ultrasonic detection of corrosion in aircraft fuel tanks and for installation of corrosion-resistant rivets in aircraft skins. They have done important work with reinforced structural composites; with graphites for leading edges, nose cones, and rocket nozzles; and with chemically resistant seals for liquid rocket propulsion systems." (Text)

- "Much has been done in an effort to alleviate aircraft noise. Annual expenditures have risen from $100,000 in 1961 to $1.3 million to be spent by FAA alone in 1965," FAA Administrator Najeeb E. Halaby told a National Aircraft Noise Symposium in New York City. He warned: "If we fail to make progress in reducing noise . . . pressure may perhaps ultimately exclude the availability of aviation facilities to our centers of trade and commerce. I feel it is essential that we all recognize the price that we may be forced to pay for a tolerable environment next door to our airports." (Text)

June 10: LUNA VI, Soviet probe, would miss the moon by 160,000 km. (99,379 mi.) because of an unsuccessful midcourse maneuver, Tass announced. Engine used to adjust the spacecraft's trajectory could not be switched off, causing a deviation from the planned course. (Tass, 6/10/65)

- A grant of nearly $4 million to the Univ. of Minnesota to strengthen its scientific and technological capability, particularly in space research" was made jointly by NASA and National Science Foundation. (NASA Release 65-191)

June 10: Astronauts James A. McDivitt (Maj., USAF) and Edward H. White II (Maj., USAF) were flown from Mayport, Fla., to Ellington AFB near Houston for reunion with their families. (Clark, *NYT*, 6/10/65, 1,47)

• First computer landing of commercial airliner with fare-paying passengers was made in London by British European Airways' Trident; touchdown termed smooth by test crew and passengers. Trident was the first civil aircraft certified to employ the automatic landing system, Autoflare, developed by Smith & Sons, Ltd., a British aviation engineering company, in association with Hawker Siddeley Aviation Co. (*NYT*, 6/11/65)

• Formation of a program to achieve a more powerful Atlas booster for future Agena and Centaur missions was announced by NASA. Presently designated SLV-3X, the program would seek a 21,000-lb. propellant capacity increase in the standard Atlas booster by making the top of the vehicle cylindrical and would increase the thrust of three Atlas engines by using modified Saturn H-1 fuel injector and improved turbines. AFSC Space Systems Div. would act as NASA's procurement agent. NASA Lewis Research Center would supervise; General Dynamics Convair Div. was expected to receive the contract. Use of an uprated Atlas would increase the Surveyor mission capability by 600 lbs. and would permit similar payload increases for Lunar Orbiter, Ogo, Oao, and Applications Technology Satellite (Ats). (NASA Release 65-192)

• W. C. Fortune, manager of NASA Marshall Space Flight Center's Mississippi Test Facility since November 1962, had been selected to evaluate the cooperative efforts of the Government–industry Saturn rocket team and "to ascertain that maximum utilization is obtained from the giant new super rocket family now under development," announced MSFC. (MSFC Release 65-146)

• The computer aboard the GEMINI IV, launched June 4, that was to have controlled the landing of the spacecraft was returned to IBM for tests to determine the cause of failure. (*NYT*, 6/11/65)

• U.S.S.R. dominated the International Air Show at Le Bourget, France, with premiere showing of Il-62 186-passenger jetliner, powered by four turbofan engines, each developing 23,100 lbs. thrust, and display of the M-110 crane helicopter reported to have set an unofficial world's record last month by hoisting 25 tons more than 8,000 ft. into the air. Cosmonaut Yuri Gagarin answered questions in the Soviet pavilion. (UPI, *Miami Her.*, 6/11/65)

• Curtiss-Wright Corp. could participate in future naval weapons procurement, including the procurement of weapon system trainers, DOD announced. Last March when R/Adm. Allan M. Shinn, Chief of the Bureau of Weapons, testified before a closed session of the House Appropriations Committee he disclosed "inadequate performance" on the part of Curtiss-Wright in connection with the P3-A trainer program. (Text; *NYT*, 6/11/65, 12)

June 11: Saturn V booster (S-IC-T stage) was successfully static-fired for 90 sec. at NASA Marshall Space Flight Center. During the test, longest to date, the five F-1 engines developed 7.5 million lbs. thrust and all four outer engines were gimbaled to simulate the motion required to control the vehicle in flight. (MSFC Release 65-148; *Marshall Star*, 5/16/65, 1)

June 11: At a news conference at NASA Manned Spacecraft Center on the Gemini GT-4 spaceflight, Astronaut McDivitt said: "I saw three things that looked to me like they were satellites on the earth. I saw two over the Pacific, I guess. One . . . near Hawaii. . . . I saw a white object and it looked like it was cylindrical and it looked to me like there was a white arm sticking out of it. . . . We saw another one at night. It looked like just a pin point of light in the sky. . . . And I saw another one over the western Pacific again just shortly before I got into the sunlight on the windshield. . . . The only one I could even define the shape of at all was the first one and it looked a lot like an upper stage of a booster."

Astronaut White commented: ". . . we were looking to find out: Could man control himself in space? And the answer is yes, man can control himself in space." McDivitt continued: "The first thing we learned was that the Gemini 4 is a liveable spacecraft for at least four days." (Transcript; *NYT*, 6/12/65, 3)

- President Johnson announced during an impromptu visit to NASA Manned Spacecraft Center—his first—that he had nominated Astronauts James A. McDivitt (Maj., USAF) and Edward H. White II (Maj., USAF) for the rank of lieutenant colonel. In a speech before nearly 5,000 MSC employees, Mr. Johnson said: "The race in which we of all generations are determined to be first is the race for peace in the world.

   "In the labors of peace—as in the explorations of space—let no man doubt for the moment that we have the will, and the determination, and the talent, and the resources required to stay the course and see those labors through." (Text; Stern, *Wash. Post*, 6/12/65, A3; Semple, *NYT*, 6/21/65; MSC *Roundup*, 6/25/65, 1)

- Sen. E. L. Bartlett (D-Alaska) introduced a bill to extend privileges and immunities, including tax and customs granted international organizations, to the European Space Research Organization (ESRO). ESRO was considering building a satellite telemetry command station near Fairbanks, Alaska, and was seeking the same special treatment afforded NASA with regard to its tracking stations abroad and personnel abroad. (*CR*, 6/11/65, 12836)

- Rep. Albert Thomas (D-Tex.), Chairman of the House Appropriations Committee's Subcommittee on Independent Offices, paid tribute to NASA on the floor of the House: ". . . too much credit cannot be given the top management of the Space Agency . . .

   "To this group of distinguished gentlemen, must go the credit of spending some $17 billion without the slightest breath of scandal attached to the many thousands of transactions." (*CR*, 6/11/65, 12829)

- Laser beams could be used to track satellites, a group of scientists reported at a news briefing at NASA Goddard Space Flight Center. The briefing followed a two-day meeting discussing efforts to track EXPLORER XXII and EXPLORER XXVII satellites which were orbiting the earth at a height of about 700 mi. Dr. Henry Plotkin, Head of GSFC's Optical Systems Branch, said the experiments had indicated that: 1) a beam of laser light from the ground could be directed with sufficient accuracy to strike satellite reflectors; 2) turbulence in the atmosphere would not break the beam enough to interfere with its lighting up the satellite; 3) very short bursts of light from lasers could be used to measure the range of a satellite precisely by means of timing the flight and rebound of the pulse; 4) reflected light could be photographed against a stellar background to provide angles by which the satellite could be identified very accurately. (Transcript)

June 12: Kennedy Space Center's "lost time rate" because of strikes was 10 times that at all the other missile and space installations combined, reported Victor Riesel in the *Indianapolis Star*. He said that during the 15 months ending March 31—the deadline months preceding the GT-4 spaceflight—there were 93 work stoppages at all missile and space sites in the country. Of this number, 56 were at KSC. 65,144 man-days were lost. At all missile bases and space centers, there was a total of 13 major walkouts; 10 of these involved KSC. (Riesel, *Indianapolis Star*, 6/12/65)

- Canada's Black Brant research rocket underwent its first successful launch from Ft. Churchill, Manitoba. (*M&R*, 6/28/65, 11)
- Discovery of "quasi-stellar blue galaxies," termed a "major new constituent of the universe," was announced by Mt. Wilson and Mt. Palomar observatories. The blue galaxies resembled quasi–stellar radio sources ("quasars") except that they did not emit strong radio waves. They appeared to be 500 times more plentiful than quasars and numbered about one to every 100,000 normal galaxies. According to Dr. Allan Sandage of Mt. Palomar, the newly discovered blue galaxies substantiated the theory that the "universe is a finite, closed system originating in a 'big bang,' that the expanding universe is slowing down, and that it probably pulsates once every 82 billion years." (Sullivan, *NYT*, 6/13/65, 1, 81)
- British Broadcasting Co. filed a petition with the FCC to request halving suggested fee of $3,825 for a half-hour's television use of EARLY BIRD I communications satellite. In addition to ComSatCorp's charge, any television user must pay an identical fee to the group of 17 European countries that helped finance EARLY BIRD I; this combined fee of $7,650 would cover the cost of transmission only between Andover, Me., ground station and one European point and would not include costs for ground lines to broadcasting stations. One BBC expert, who estimated that at the proposed rates a half-hour transatlantic program would cost more than $11,000, concluded that British broadcasters would be able to use the satellite only for the "most compelling matters." (Lewis, *NYT*, 6/13/65, 1)

June 13: A 22-nation European Post and Telecommunications Congress ended its 12-day meeting in Lisbon after appointing a coordinating committee to deal rapidly with problems of international radio and television communications by satellite. Discussions in the congress, which was closed to the press and public, centered on improving postal services and satellite-relayed radio and television transmissions. (AP, *WSJ*, 6/14/65, 24; AP, Wash. *Eve. Star*, 6/13/65, A-11)

June 14: MARINER IV successfully performed a final tracking correction before its encounter with Mars on July 14. A preprogramed command electronically altered the look-angle of the star sensor to compensate for the changing relationship between the spacecraft, the sun, and

Canopus. The star sensor must be pointed at Canopus so that the Mars probe would be properly aligned and stabilized in attitude. (NASA Release 65-198)

June 14: A crowd of two million gathered in Chicago during the parade and motorcade honoring Astronauts James A. McDivitt (Maj., USAF) and Edward H. White II (Maj., USAF), accompanied by Vice President Hubert H. Humphrey. Honorary citizens medals were presented the astronauts at a special City Council meeting. Civil rights leaders postponed a demonstration protesting de facto school segregation in deference to the celebration. (AP, *Wash. Post.* 6/16/65; Wehrwein, *NYT,* 6/15/65)

- Radio station WTOP in Washington, D.C., assisted NASA in conducting a radio signal interaction experiment employing a Nike-Apache rocket launched from Wallops Sta., Va., with a 55-lb. instrumented payload. WTOP transmitted a steady modulated tone for several minutes during the flight to enable Univ. of Illinois scientists to measure interaction of the WTOP signal on a signal of a different frequency broadcast from Wallops Sta. Both were received by instruments in the payload as the rocket rose to peak altitude of 110 mi. (NASA Release 65-195; Wallops Release 65-35)

- EARLY BIRD I communications satellite experimentally transmitted to a Paris physician an electrocardiogram of a passenger on the S.S. *France,* 2,000 mi. at sea, the French Line reported. (*NYT,* 6/15/65, 70; AP, *Wash. Post,* 6/15/65, A14)

- AT&T and ITT asked the FCC to reverse its May 12 decision awarding ComSatCorp temporary control over the initial three U.S. ground stations which would comprise important segments of a global satellite communications system. ITT, in its petition, contended that the ruling supported an "unwarranted monopoly in international communications." AT&T argued that the licensing policy was not in the public interest. (*WSJ,* 6/14/65, 24)

- M2-F2 manned lifting body research vehicle was rolled out at Northrop Norair's Hawthorne, Calif., plant and accepted for NASA by Paul Bikle, Director of NASA Flight Research Center. The craft would be dropped from beneath the wing of a B-52 bomber at high speeds in tests to determine how this configuration would perform in the critical period during reentry if it were carrying astronauts. (AP, *Wash. Post,* 6/17/65, A3; ARC *Astrogram,* 6/24/65, 1)

- *U.S. News and World Report* suggested that the success of the Gemini GT-4 flight should prompt reassessment of the U.S. position in the race with the Soviet Union. Two conclusions were noted:

    "1. In the civilian space race, White's self-propelled 'space walk' and McDivitt's ability to maneuver the spaceship put the U.S. ahead in at least two key areas and gave the U.S. a fighting chance eventually to overtake the Russians in the race to the moon. . . .

    "2. In the military space race, maneuvering of the Gemini spacecraft demonstrated that the region just above the earth—the inner space belt—could soon become vital to American security." (*U.S. News.,* 6/14/65)

- Lt. Col. Aleksey Leonov's comments on extravehicular activity during the March 18 VOSKHOD II flight were quoted in a review report authored by Prof. N. M. Sissakian of the Soviet Academy of Sciences and delivered in Paris at the Second International Symposium on Basic Environmental Problems of Man in Space: "I found that the slightest shift in direction of the force of impact caused rotation in the corresponding plane. Those persons who will be working in space will obviously have much to do in securing their bodies in . . . [the weightless state]. As for the so-called psychological barrier that was supposed to be insurmountable by man preparing to confront the cosmic abyss alone, I not only did not sense any barrier, but even forgot that there could be one." (Wetmore, *Av. Wk.*, 6/21/65, 25)

June 14: That antimatter could exist in aggregations of particles, not only as isolated subatomic particles, was demonstrated by physicists studying under AEC funds at Columbia Univ. Nevis Cyclotron Laboratory and Brookhaven National Laboratory. Protons placed in Brookhaven's AGS synchrotron were hurled at almost the speed of light and energy of 30 billion electron volts at a target of beryllium; scientists used a high-transmission mass analyzer to detect anti-deuterons in the debris of collisions between high-energy protons and nuclei of atoms in the target. Research report appeared in *Physical Review Letters*. The existence of the antideuteron had been predicted theoretically, but its actual production indicated that properties of the nuclear force were closely mirrored in the world of antiparticles and that an antiworld would be conceivable in terms of contemporary nuclear physics. (Schmeck, Jr., *NYT*, 6/14/65, 1)

- A model of TU–144, proposed supersonic passenger plane, was displayed by the Soviet Union at the International Air Show at Le Bourget, France. TU–144 was designed for a capacity of 121 passengers, a speed of 1,550 mph, and a range of 4,000 mi. (Kamm, *NYT*, 6/16/65, 1, 9)

- An instrumented experiment package capable of recording lunar phenomena and relaying information to earth, would be installed on the moon by astronauts before their return to earth, reported Howard Simons in the *Washington Post*. Simons said that NASA officials explained that the package would contain "combination of instruments to measure the moon's gravity and atmosphere, heat flow and solar wind, proton activity and micrometeorite impacts for as long as a year." Such information would be helpful in planning the establishment of permanent lunar bases and in studying the history of the earth and the solar system. (Simons, *Wash. Post*, 6/14/65, A9)

- Discussing Russian-American cooperation in space in a letter to the editor in the *New York Times*, Donald Spero, a student at Columbia Univ. School of Engineering, said: ". . . technical integration of the U.S. and (assumed) Russian lunar programs is out of the question. Hardware for every phase of the Apollo program has already been designed and built. . . .

  "The integration of a Russian booster and an American capsule would be a technical impossibility. . . . The only plausible alternative for initial lunar exploration would be to include a Russian cosmonaut in the Apollo crew or one of our astronauts as a member of the Russian expedition. Even if problems of language and pilot training could be overcome, political and propaganda considerations eliminate this alternative.

  "Realistic possibilities for cooperation lie in the areas of unmanned probes, communication and weather satellites, and eventually manned planetary exploration and establishing of lunar bases." (*NYT*, 6/14/65)

June 14: A *New York Times* editorial by Harry Schwartz concerning Soviet-American cooperation in space: "The real issues relate to the advantages and disadvantages in the moon race itself—including, of course, its propaganda aspects.

"The argument that cooperation will not mean significant savings is strongest for the immediate future, but its force weakens rapidly as one extends the time horizon of both nations' future space efforts. Even in the next year or two both countries could gain from a full pooling of space technology and knowledge because this would reduce the number of Gemini-type flights each would have to engage in. . . .

"Major cost advantages can certainly be gained by agreement on a division of labor between the Soviet Union and the United States, if it is accompanied by a decision to send mixed crews on major missions. For example, a pooling of information and resources might permit one country to focus on the hardware needed for the moon trip, while the other concentrated on the equipment needed to send men to Mars. . . .

"But the major savings from real Soviet-American cooperation in space might come from another direction entirely. In both countries influential voices are urging major military efforts looking to the creation of armed national space fleets . . . The time is past due for decision between space cooperation, or the extension of a rivalry that could cost both Soviet and American peoples dearly—and perhaps not solely in terms of vast sums wasted." (Schwartz, *NYT*, 6/14/65, 31, 32)

June 14-21: PDP-5 and PDP-8 (Program Data Processing) computers, reported to simultaneously collect and analyze oceanographic data and to use data received by radio from artificial earth satellites to fix the position of ships, were displayed at the Ocean Science and Ocean Engineering Conference and Exhibit in Washington, D.C. PDP-5 had first been used by the U.S. Coast Guard during the 1964 International Ice Patrol season to predict the speed and course of icebergs drifting into major ship lanes. (Callahan, *NYT*, 6/14/65, 58M)

Dr. James H. Wakelin, Jr., president of the Scientific Engineering Institute of Boston, said in an address at the Conference: "We must look forward to undersea dwellings, laboratories and military installations in which men would live and work for the economic good and military defense of the United States." Dr. Wakelin advised President Johnson to appoint a National Advisory Commission on the Ocean to develop a 10-yr. program for study, exploration, and use of the seas. (UPI, *NYT*, 6/15/65, 6)

Capt. Jacques-Yves Cousteau urged the organization to "preserve and protect the sea from pollution." He also warned against conducting undersea explorations entirely with instruments and suggested: "Let us go down . . . and see for ourselves, with our eyes." (Casey, *Wash. Post*, 6/16/65, A14)

June 14–25: "Science in the Sixties," a seminar sponsored by the Air Force Office of Scientific Research, was held in Cloudcroft, N. Mex. In opening remarks, Maj. Gen. Don Ostrander, AFOSR Commander, said: "The purpose of these Cloudcroft meetings is to stimulate ideas—to act as an intellectual catalyst.... We all have a responsibility to try to understand the complex interrelationships between science and technology, and between technology and national defense; through understanding, to participate in the excitement and urgency of the creative turmoil which is such an inescapable part of the age in which we live!"

Historian A. Hunter Dupree, professor of history at the Univ. of California (Berkeley) and a consultant to NASA, said that scientists with a negative attitude regarding the Nation's manned space flight activities had a laboratory-limited view of scientific endeavor and had lost perspective of the contributions made to American and world science through exploration and survey expeditions in the field. He pointed out the relationship of the Pacific voyages of Capt. Cook to Darwin's later theory of evolution and said: "One can as little predict the results of space exploration as Captain Cook could have predicted Darwin's theory."

According to Dupree, it was the general expansion of knowledge that would lead to later fruitful developments. But to justify these developments immediately or to justify exploration in terms of predictable developments would be a mistake. (Simons, *Wash. Post,* 6/16/65; AFOSR Release 5-65-2; *Aerospace Historian,* 10/65, 106-110)

An artificial frog's eye which could be sent to the surface of Mars to detect living organisms was described by Warren McCullough and Louis Sutro. Research had revealed there were four varieties of ganglion cells in the eye of the frog—each processing different information. The MIT scientists had proposed the following scheme to NASA: the artificial eye would be coupled to a microscope in a tiny computer. Samples of Martian soil would be seen by the frog's eye through the microscope. When movement was detected, the eye would inform the computer, which would decide whether a picture of the moving organism should be taken for relay back to earth. (Simons, *Wash. Post,* 6/18/65, A1)

Theories on biological rhythms were proposed by Colin Pittendrigh of Princeton Univ. at the AFOSR seminar. He suggested that oscillations or biological rhythms were serving a fundamental function that was not yet fully identified and that all organisms undergo oscillations with a periodicity that matches that of the external world—roughly 24 hrs. Light, even in negligible amounts, could alter these oscillations. In Pittendrigh's view, once the true face of biological clocks—time measuring mechanisms innate in all living organisms—was seen, science would have vital clues to how life developed on earth and how biological rhythms determine what it is all living things do. (Simons, *Wash. Post,* 6/22/65, A6)

Star collisions were suggested by astrophysicist Thomas Gold of Cornell Univ. as one way that energy now associated with a host of new objects observed throughout the universe was released. He said a prime candidate for providing the right kind of environment for star collisions was elliptical galaxies. In their predeath condition, elliptical galaxies start to lose stars that comprise the galaxies. Star loss causes the galaxies to shrink and become denser. The remaining stars rush in and out through the heats of these galaxies at speeds possibly as high as 24 million mph—greatly enhancing the chance for star collisions. The effect of collisions at these speeds would be to release amounts of energy equivalent to that calculated to be stored in the quasi-stellar radio sources. Gold had not observed such star collisions, but dense regions on the "brink of destruction" had been detected. "We must inspect each in turn," Gold said, adding, "maybe we will learn that something totally different is involved, a new type of energy source that physics doesn't know about." (Simons, *Wash. Post*, 6/23/65)

June 15: U.S.S.R. launched COSMOS LXVIII containing scientific equipment for the investigation of outer space. Orbital parameters: apogee, 334 km. (207 mi.); perigee, 205 km. (127 mi.); inclination to earth, 65°. On board equipment was said to be functioning normally. (*Pravda*, 6/16/65, 1; *Izvestia*, 6/17/65, 4, ATSS–T Trans.)

- High-speed transmission of weather data between the U.S. and France was provided by EARLY BIRD I communications satellite. Information gathered by TIROS IX weather satellite during a 24-hr. period and assembled on a chart at the World Weather Center in Maryland, was relayed to the Andover, Me., ground station; then, via EARLY BIRD I, it was transmitted to the French ground station at Pleumeur-Bodou and on to the French National Weather Center in Paris. Conducted jointly by the Weather Bureau and ComSatCorp, in conjunction with Press Wireless, Inc., and Alden Electronics Corp., the demonstration illustrated a new and advanced forecast method which would include transmissions of facsimile charts and data at eight times the speed of present networks. (ComSatCorp Release; AP, Balt. *Sun*, 6/16/65)

- CBS became the first U.S. network to issue a formal statement about commercial rates proposed May 28, 1965, by ComSatCorp for EARLY BIRD I transmission: "We shall have to make future determinations as to the use of Early Bird on a case by case basis, depending on the importance or urgency of the news to be transmitted. Certainly, the cost structure proposed for the use of Early Bird militates against its use on a routine basis. Unless urgency requires transmission by Early Bird, we shall have to continue to rely upon air shipments of film and taped coverage of European news." (CBS News Div.; Adams, *NYT*, 6/16/65, 87)

- NASA announced completed negotiations with Aerojet-General Corp. for two-phase $11,163,051 contract to design, develop, and deliver three exhaust nozzles for use in testing the 5,000-mw Phoebus nuclear rocket reactor. Phase I would be a $1,837,971 cost-plus-fixed-fee contract and would include four-month preliminary design study of the nozzle, and evaluation of fabrication and testing methods as well as a joint design effort involving Aerojet, LASL and American Car and Foundry, Inc. Phase II would be a $9,325,080 incentive contract and, relying on the results of Phase I, Aerojet would be required to design, develop, test and deliver three nozzles to the Nuclear Rocket Development Sta., Jackass Flats, Nev., by the end of 1967. The contract would be under management of the joint AEC–NASA Space Nuclear Propulsion Office. (NASA Release 65-196)

- Dr. George E. Mueller, NASA Associate Administrator for Manned Space Flight, said at the National Space Club in Washington, D.C., that it would take "a great deal of effort over a number of years" for the United States to achieve first place in space and warned it would be "a mistake to believe" that the successful GEMINI IV spaceflight had "overcome a lead of several years" held by the Soviet Union. He said "the most important result" of the GEMINI IV flight might be the condition of the astronauts upon their return, based on the preliminary medical examinations of Astronauts McDivitt and White. The final medical report on the flight and their postflight condition would take about two months. (Text; Clark, *NYT*, 6/16/65, 13)

June 15: Defense Communications Agency had awarded contracts to six firms to conduct parallel systems design studies for the Advanced Defense Communications Satellite Project: ComSatCorp, General Electric Co., Hughes Aircraft Co., Philco Corp., RCA, Defense Electronic Products, and Space Technology Lab. The fixed price contracts ranged from $135,000 to $196,000.

The studies, to be completed in three or four months, would be used as a basis for design of any advanced operational satellite communications system. (DOD Release 402-65)

- Dr. Hugh L. Dryden, NASA Deputy Administrator, was awarded an honorary Doctor of Science degree from Princeton Univ. (Off. of Deputy Administrator)

- Honorary Doctor of Astronautical Science degrees were conferred on Astronauts James A. McDivitt (Maj., USAF) and Edward H. White II (Maj., USAF) by their alma mater, the Univ. of Michigan where Maj. White received his BS degree in 1959 and Maj. McDivitt, his MS degree in 1959. The astronauts then attended a ceremony dedicating the University's new $1.7 million space research building and rode in a motorcade through downtown Ann Arbor. NASA's official representative at the festivities was Dr. Floyd L. Thompson, Director of the NASA Langley Research Center, also a Michigan alumnus. (LaRC Release; AP, Balt. *Sun.* 6/16/65)

- G. Mervin Ault, Associate Chief of Material and Structures Div., NASA Lewis Research Center, discussed refractory metals in an honors lecture before American Society for Testing and Materials (ASTM) meeting at Purdue Univ. Refractory metals—such as tungsten, tantalum, molybdenum, columbium—have strength at high temperatures and corrosion resistance to alkali metals. "The past decade has resulted in greater progress in refractory metals than ever before achieved for any one class of structural materials," Ault said. The lecture commemorated metallurgist Horace W. Gillett and was sponsored jointly by ASTM and Battelle Memorial Institute. (LRC Release 65-44)

- In surprise move, U.S.S.R. landed the world's largest plane at the International Air Show, Le Bourget, France. Designated An-22, the aircraft could carry 720 passengers or 80 tons of cargo and would weight 250 tons with maximum cargo. Powered by four turboprop engines, each with twin propellers rotating in opposite directions, the aircraft, with maximum load, would have a range of 3,100 mi. at cruising speed of 420 mph. Maximum speed would be 460 mph; maximum altitude 36,000 ft. An-22 would require 4,300 ft. for takeoff but only 2,600 ft. for landing. It was designed by Oleg Antonov and was called "Antaeus" for the mythical Libryan giant wrestler who drew new strength every time he touched the ground.

USAF supersonic B-58 Hustler jet bomber crashed on landing at Le Bourget, killing the pilot, Lt. Col. Charles Q. Hubbs (USAF), and injuring the two crew members. The aircraft was arriving from Torrejon Air Base, Spain, to take part in the air show. A U.S. B-58 had crashed at the International Air Show in 1961. (AP, *Wash. Post*, 6/16/65, A3; Kamm, *NYT*, 6/16/65, 1, 9; *WSJ*, 6/16/65, 1; *Av. Wk.*, 6/21/65, 24)

June 15: Referring to the "real success of Luna 6," an article in the Philadelphia *Evening Bulletin* said: "This is not the first Russian failure in space. But it is the first open admission of failure. Americans, who have had their own failures, can't help but warm up a little in the glow of such non-Marxist honesty." (*Phil. Eve. Bull.*, 6/15/65)

- Carl L. Norden, inventor of the famous bombsight, died. Mr. Norden's device developed for USN was used by AAF B-17's and other bombers during World War II. (AP, *NYT*, 6/16/65, 43)

June 16: X-15 No. 3 flown by pilot Capt. Joseph Engle (USAF) to 244,700 ft. altitude at maximum speed of 3,404 mph (mach 4.66) to measure ultraviolet radiation and noise intensity of the boundary layer of air. (NASA X-15 Proj. Off.; *X-15 Flight Log*)

- Poland launched its first meteorological rocket. The two-stage vehicle was 2.5 m. (8.2 ft.) long and reached an altitude of 37,000 m. (121,360 ft.). (*M&R*, 6/28/65, 10)

- XB-70A research bomber, leaving Edwards AFB, flew 1,700 mph at 65,000-ft. alt. on its 13th test flight. It landed three minutes earlier than planned because of a possible leak in a hydraulic system. (AP, *Balt. Sun*, 6/17/65)

- An atomic clock so accurate it could help determine the position of a rocket hurtling at 238,000 mph toward the moon within three-quarters of an inch was in production at Varian Associates, UPI reported. The clock would be about the size of a hatbox. (UPI, *NYT*, 6/16/65, 31)

- NASA Administrator James E. Webb, in Subcommittee of the Senate ate Committee on Appropriations' hearings on the requested $5.26 billion appropriation request for NASA in FY 1966, said:

    "Recent events have clearly demonstrated two important facts about space activities. First, the United States has shown that it can successfully build and launch complex spacecraft to measure the space environment over large regions of our solar system and to extend our knowledge of our neighboring space bodies. We have developed a capability to produce large launch vehicles, to test them, and to launch them successfully. We are producing the space hardware for environmental testing that will prove out our concepts and engineering for the large launch vehicles and spacecraft that will be required to operate out to and on the moon and meet all the demands of our other difficult undertakings. We have successfully developed space technology for improved communications and weather reporting and forecasting systems. The Ranger program, completed with Ranger IX, provided 17,000 closeup pictures of the moon that have not only given us a better understanding of its topography but may reveal totally unexpected processes taking place below the surface. The first two manned flights of the Gemini program verified the system for using man in space, the capability of the Gemini spacecraft, the capability of an astronaut to operate outside of his spacecraft, and the utility of the ground net and mission control, and provided the first tests of some of the equipment designed to accomplish rendezvous and docking. They also served as an orbiting space laboratory with several experiments included on both flights.

"The second major fact demonstrated by recent space events is that the Soviet Union continues to make a major commitment to its aeronautical and space activity. In late 1964, they launched the first multi-manned mission with the three-man Voskhod I satellite. So far in this calendar year, they have launched 17 Cosmos satellites; in the Voskhod II flight they achieved the first extravehicular activities of man in space; in April they placed in orbit Molniya I, their first active communications satellite; in May they launched a Lunik spacecraft to the moon with a successful midcourse correction but apparent terminal failure; and only a few days ago they launched another Lunik spacecraft to the moon with an apparent unsuccessful midcourse correction. They, too, are expanding upon a sound basis for both manned and unmanned activities in space. The growth of their space activity is quite apparent. The exhibition in Paris yesterday afternoon of a new very large air transport indicates the same kind of emphasis on equipment to use the earth's envelope of air.

"In aeronautics, it is important to note the increasing tempo of our research in not only the aerodynamics, loads and structures, propulsion, and operating problems of supersonic flight, but hypersonic flight as well. There is a resurgence of interest in airbreathing propulsion in the form of advanced turbojet and ramjet engines to meet the requirements of supersonic and hypersonic transports and to make them competitive with transports operating in the subsonic range. And of course, we are also engaged at the other end of the speed spectrum in our work with vertical takeoff and landing aircraft. . . ."

In response to questioning, Mr. Webb said: "A substantial amount of time is now being put into aeronautics by our top people. Remember, we have to go through the air to get to space. The use of thin wall structures and the use of power delivered by engines all come out of the same research competence which we have.

". . . I have been asked once or twice to consider whether NASA should take on the management and development of prototypes and all other factors relating to the building of a supersonic transport.

"Each time I have pointed out that we spend a large number of our dollars through the military services because they have the procurement capability. They are the only people in the U.S. Government today who know how to let a contract, monitor a contract, and take delivery on large airplanes and large numbers of airplanes. We use them for that purpose in boosters where they have already developed the competence; and in new boosters like SATURN V, we also use their contract administration and their Project 60 for engines. . . ." (*Ind. Off. Approp. Hearings*, 1095–1195)

June 16: Max Quatinetz of NASA Lewis Research Center addressed International Powder Metallurgy Conference in New York. He discussed LRC research in adding fibered metals to tungsten to strengthen that metal, which has a high melting point but is brittle and difficult to work. Quatinetz described a new method of producing the fibered compounds

—extrusion of powdered metals. Researchers had formed tungsten composites containing high-temperature additives such as oxides, borides, nitrides, and carbides; they had noted increases in the metal's stress-rupture life of up to 50 times. Quatinetz observed that the new method of fibering would have wide potential application in materials research. (LRC Release 65-45)

June 16: Dr. Werner R. Kirchner, vice president and manager of Solid Rocket Operations, Aerojet-General Corp., received AIAA's James H. Wyld Propulsion Award during the Institute's Propulsion Joint Specialists' Conference in Colorado Springs. He was cited for "outstanding contributions to the field of solid rocketry for over 15 years, including the development of thrust-vector control and thrust-reverser systems that made possible the use of solid rocket motors in ballistic missiles." (*NYT*, 6/17/65, 54M)

June 17: Charles W. Mathews, manager of the Gemini program, and Astronauts James McDivitt (Maj., USAF) and Edward White II (Maj., USAF) received NASA's Exceptional Service Award from President Johnson in a special White House ceremony. Introducing the President, NASA Administrator James E. Webb said: ". . . we . . . should never forget that at the beginning of the space age, in 1957, the challenge of this new frontier which was laid down to us was first met by the man who is now the President of the United States and who has so graciously invited us here today to indicate again his interest in, and the importance he attaches to, the new systems we have developed for building our national competence in space and using the science and technologies acquired to work toward a peaceful world and a better world.

"This great leader of our nation, and of the Free World, is still pioneering, this time on an even more difficult frontier where we must learn to master the restrained but decisive use of the powers which technology gives our nation. Those of us who are responsible for the build-up of our new base of technology believe that power, based on advanced technology, can provide new means to hold back those ruthless forces which answer not to the need of all men for security, freedom, dignity, and opportunity. The pioneering which President Johnson is engaged in today on this new frontier is, if anything, more important than his pioneering actions in 1958 to create our national program in aeronautics and space.

"Seldom in the history of the world has one man had to play so vital a role in developing the tools of modern science and technology and then in the development of a national capability to use them to achieve cooperation toward a world consistent with our own ideals and those we have sought for others as well as ourselves."

Accompanying citations noted "outstanding contributions" and singled out Major White as "the first man to engage in self-propelled extra-vehicular activity." Terming the three "the Christopher Columbuses of the 20th century," Mr. Johnson said their work had nudged the world toward greater international cooperation. "Men who have worked together to reach the stars are not likely to descend together into the depths of war and desolation," he said. Later, the recipients were guests at a luncheon held by Vice President Humphrey and received accolades in both the House and Senate. A crowd estimated at

50,000 applauded the motorcade as Mathews and the astronauts rode to the Capitol.

In the evening, Majors White (USAF) and McDivitt (USAF) narrated a 20-min. film of the GEMINI IV flight for the chiefs of foreign diplomatic missions. President Johnson, in a surprise appearance, told the astronauts to "take the Presidential plane and travel outside this country again." He said: "Many people in many lands were thrilled by what you have done. I want you to join our delegation in Paris and go out among the friendly peoples of the earth to share with them the excitement and thrills of your experience."

Astronauts White and McDivitt then returned to the White House where provisions had been made for them and their families to remain overnight. (Text; Clopton, *Wash. Post*, 6/18/65, A1, A3; UPI, *N.Y. Her. Trib.*, 6/18/65; Semple, *NYT*, 6/18/65, 1, 13; Sehlstedt, *Balt. Sun*, 6/18/65)

June 17: X-15 No. 1 flown by pilot Milton Thompson (NASA) to 108,523 ft. altitude at maximum speed of 3,541 mph (mach 5.145) to measure and record infrared radiation and to conduct further flight checkouts on the new inertial guidance system. (NASA X-15 Proj. Off.; *X-15 Flight Log*)

- President Johnson said during a Washington press conference that "we are going to build" a supersonic passenger airliner to compete in the world market against a supersonic transport being developed jointly by British and French interests and one the Russians intended to enter in the competition. Mr. Johnson told reporters he wanted the best plane possible, one that the airlines would buy as an economically attractive investment. (Transcript, *NYT*, 6/18/65, 14)

- A Nike-Apache sounding rocket was launched from Wallops Sta., Va., to peak altitude of 109.9 mi. (176.8 km.) in an experiment to measure electron densities. Good signals were received on all telemetry channels throughout the flight; indications were that good data were obtained. Instrumentation was provided by GCA Corp. (NASA Rpt. SRL)

- USAF's attempt to launch Titan III-C was unsuccessful when a series of minor technical problems and then bad weather were encountered. Two of thee technical holds were attributed to faulty instrumentation. The third hold was caused by a drop in pressure in the second stage oxidizer tank. (Ubell, *N.Y. Her. Trib.*, 6/18/65)

- F-4C Phantom jet pilots downed two attacking Communist Korean War-vintage MiG-17's about 80 mi. south of Hanoi. (AP, *NYT*, 7/11/65)

- The Senate, in a 46-20 vote, gave final Congressional approval to a bill waiving the restriction barring military men from the post of Administrator of the Federal Aviation Agency, clearing the way for President Johnson to appoint Gen. William F. McKee (USAF, Ret.) of NASA. (*CR*, 6/17/65, 13541; *WSJ*, 6/18/65, 8)

- Sen. Wayne Morse (D-Ore.) introducing a bill to amend the NASA Space Act of 1958 regarding patent rights to inventions, said the bill had two objectives: "The first is to reestablish congressional control over the disposition of patent rights by the national aeronautics and space agency [sic], and the second is to provide that private companies desiring to acquire interests in such patents and processes repay the taxpayers of this country fair market value pursuant to the so-called Morse formula." (CR, 6/17/65)

June 17: Commenting on anxiety about the Soviet Union's capability of delivering strategic weapons from near space, Dr. S. Fred Singer said in an article in *Reporter:* "If a bomb is released from a satellite without giving it any propulsion, it will stay with the satellite and simply blow it up. For a bomb from a satellite to be directed to a point on earth, it must be propelled not only with a lot of rocket power but also with exceedingly fine guidance. In principle, this can be done from a satellite or from the moon or even from the planet Pluto; but the cost and complexity is enormously greater than that of an equally effective ICBM buried deep in the earth itself." (*Reporter*, 6/17/65, 14)

- In what he termed "Coming of Age in Houston," William Hines wrote in the Washington *Evening Star* of "a new, mature outlook on the part of NASA." He said: "There was a conscious effort to deglamorize (but not depersonalize) the [Gemini 4] astronauts, and to focus attention on the mission rather than on celebrities who would subsequently tell their stories in *Life*. The decision not to give the spacecraft a name and to use the radio call sign 'Gemini 4' was a step in this direction.

  "The determination of NASA to rid itself of what has been called the 'Hollywood syndrome' and handle space flights as transcendental news events instead of tawdry theatrical productions did not come easily, or without prodding from the outside. But once the decision had been made not to try any longer to fool all the people all the time, a new era in public understanding of space dawned." (Hines, Wash. Eve. Star, 6/17/65)

June 18: USAF Titan III-C, launched from the Eastern Test Range, became the most powerful rocket known to have been lofted and the first liquid-fuel spacecraft to be lifted from its pad with solid-fuel rockets. Two 120-in.-dia., 86-ft.-long solid strap-on boosters generated a peak thrust of 2.647 million lbs. ½ sec. after ignition to propel the liquid-fuel core vehicle to an altitude of 24 mi. Less than two minutes after lift-off, the boosters were jettisoned by firing of 16 small rockets. The liquid-fuel engines of the 127-ft., three-stage core vehicle then fired a 474,000-lb. thrust burst that injected the vehicle's third stage (transtage) with 29,285-lb. lead ballast and instrument payload into an orbit with apogee, 116.2 mi. (187 km.); perigee, 110.4 mi. (177.7 km.); period, 88.1 min; inclination to the equator, 32.175°. This was the heaviest payload ever orbited; insertion velocity was 25,584 fps.

The solid-fuel motors, made by United Technology Center, were formed in 10-ft.-dia. segments stacked inside metal casings. Adding two segments to each of the two five-segment boosters used would permit Titan III-C to put 32,000 lbs. into orbit. Payload could be increased to 40,000 lbs, using 13-ft.-dia. segments of another type already test-fired by UTC. Liquid stages used nitrogen tetroxide and Aerozene 50 (a 50/50 mixture of hydrazine and unsymmetrical dimethylhydrazine) as the oxidizer and fuel; the propellants ignited hypergolically. A series of twelve Titan III-C tests was projected.

Most powerful rocket previously launched was NASA's Saturn I, which produced 1.5 million lbs. thrust. The most powerful known Soviet rocket, which orbited VOSKHOD I and VOSKHOD II, had been rated by American experts at 800,000–900,000 lbs. thrust. The two-stage Titan II that launched GEMINI III and GEMINI IV into orbit produced 430,000 lbs. of thrust. (Clark, *NYT*, 6/19/65, 1; AP, Benedict, *Wash. Post*, 6/19/65, A3; Hines, Wash. *Eve. Star*, 6/19/65; *Av. Wk.*, 6/28/65, 16–19)

June 18: A briefing given by NASA Administrator James E. Webb to President Johnson and the Cabinet in session said that the success of the last two Gemini missions "has proved the design and confirmed the results of the ground tests, has increased our confidence in the reliability of the over-all Gemini systems, and has enabled the National Aeronautics and Space Administration to advance the Gemini program such that rendezvous and docking are now scheduled during calendar year 1965." (Text)

• NASA launched 3.5-ton Aerobee 350, new two-stage research rocket, from Wallops Station, Va., in first flight test. The 52-ft.-long, 22-in.-dia. rocket carried 367 lbs. of performance instrumentation to peak altitude of 235 mi. and impacted in Atlantic Ocean 160 mi. from launch site.

Designed and developed by Space-General Corp., Aerobee 350 had a main stage "sustainer" propelled by four liquid-fuel engines, each developing 4,100 lbs. thrust. Booster stage used a solid propellant Nike motor with 51,000 lbs. thrust. Booster and sustainer stages fired simultaneously, with the booster burning out and separating about 3.2 sec. later. The rocket would boost a minimum-weight payload of 150 lbs. to an altitude of 290 mi. and a maximum payload of 500 lbs. to 210 mi. (Wallops Release 65–37; NASA Rpt. SRL)

• An ionospheric sounding probe launched by NASA on a Nike-Apache vehicle from Wallops Station, Va., reached peak altitude of 116 mi. in an experiment to measure electron density in the E region of the ionosphere. Electron profile data were obtained during both the ascent and descent portions of the flight trajectory and were telemetered to ground receiving stations during the flight. Experiment was conducted for National Bureau of Standards' Central Radio Propagation Lab. and NASA Goddard Space Flight Center. (Wallops Release 65–36)

• NASA announced it would negotiate with Lockheed Missiles and Space Co. for mission modifications on seven Agena-D second stages for future missions. Total cost of the modification would be more than $13 million. Five of the Agenas would be used with Atlas boosters to launch the Applications Technology Satellites (Ats); the other two, also to be boosted by Atlas launch vehicles, would be used for the third and fourth Orbiting Astronomical Observatories (Oao). Lockheed would design, develop, and fabricate equipment and match the Agenas with the Atlas boosters and the spacecraft. (NASA Release 65–199)

• ComSatCorp was authorized by FCC to begin temporary commercial service at midnight via EARLY BIRD I to replace circuits lost due to a break in the Canada-to-England cable. The authorization would be in effect until midnight June 26.

A temporary tariff, approved by FCC, set a rate of $420 a day, per circuit, for the first two days of commercial operation and $210 per circuit for the remainder of the time. (ComSatCorp Release)

June 18: Canada's Black Brant research rocket underwent its second successful launch from Ft. Churchill, Manitoba. (*M&R*, 6/28/65, 11)

- Brig. Gen. Joseph S. Bleymaier (USAF), Deputy Commander (Manned Systems) of AFSC Space Systems Div. and head of the Titan III-C program, and project leaders Cols. David V. Miller (USAF) and Otto C. Ledford (USAF) were honored by Maj. Gen. Ben I. Funk (USAF) for the success of the Titan-III program. General Bleymaier's commendation for the Legion of Merit was for "outstanding service to the United States." Colonels Miller and Ledford were awarded Commendation Medals. The ceremony was conducted at Eastern Test Range one hour after launch of Titan III-C-1. (Clark, *NYT*, 6/19/65, 1, 11)

- USAF resumed tests of the Athena reentry program with two launches from Green River, Utah, into White Sands Missile Range. One was termed a complete success; the other 80% successful. (*M&R*, 6/28/65, 11)

- First use of USAF B-52 heavy jet bomber in anger when 28 U.S. Strategic Air Command B-52's dropped 750- and 1,000-lb. conventional bombs on Viet Cong units in South Vietnam. B-52 had been mainstay of U.S. global thermonuclear deterrent for almost a decade. (Margolis, *Wash. Post*, 6/18/65, A1, A18; EPH)

June 19: Vice President Humphrey and the GEMINI IV astronauts, Maj. James A. McDivitt (USAF) and Maj. Edward H. White II (USAF), won cheers and applause from visitors to the International Air Show at Le Bourget, France. Attendance at U.S. pavilion, which had been poor, picked up appreciably. Also present were NASA Administrator James E. Webb, and Charles Mathews, manager of the Gemini program.

A scheduled formal meeting between the American astronauts and Lt. Col. Yuri Gagarin, first man in space, fell through when the Soviets announced that Col. Gagarin would be "too busy." However, at an official luncheon, Gagarin stopped at White's table and the two astronauts shook hands and spoke briefly. (AP, *NYT*, 6/20/65, 38; UPI, Wash. *Daily News*, 6/19/65; AP, Hudgins, *Wash. Post*, 6/20/65, A28)

- TIROS VII meteorological satellite completed two years in orbit without a failure. NASA had orbited TIROS VII June 19, 1963, with a Thor-Delta launch vehicle launched from ETR (then called AMR). (NASA Proj. Off.)

- Gemini 5 spacecraft was flown by cargo carrier to Kennedy Space Center, NASA, for the seven-day flight scheduled for Aug. 9. (UPI, *NYT*, 6/20/65)

- British physicist Samuel Tolansky of London Univ.'s Royal Holloway College theorized that a carpet of black diamonds valuable for industrial purposes had formed on the lunar surface over the ages because of meteor impact. He cited diamonds found in the El Diablo meteorite crater in Arizona. (AP, Wash. *Eve. Star*, 6/19/65, 1)

- "It is time now to put the manned military control of space on a crash basis equal in priority to the Apollo program," said Rep. John W. Wydler (R-N.Y.), member of the House Science and Astronautics Committee, in a letter to the editor in the *New York Times*. He suggested the following steps be taken: "The first M.O.L. flight is scheduled from two and one half to three years from now. This should be speeded up at least a year and the necessary sacrifices made to achieve it. The Gemini capsules required for the M.O.L. project should be ordered at once. To achieve our goals effectively the manned earth orbiting program should be placed under military control. . . .

"I believe the only way the Department of Defense can meet its responsibilities in 'near space' is to assume direction of the manned earth orbiting program. It should reorganize the U.S. Air Force into the U.S. Aerospace Force and make it truly that. . . .

"The decision we must make is not whether there will be military control of space but rather whether that control will be Russian or our own. . . ." (*NYT*, 6/19/65, 28)

June 19: The Space Act of 1958 may have unwittingly provided competition that is getting results, said an editorial in the Cleveland *Plain Dealer*: "There was conjecture then, and there is conjecture now, that NASA and the Air Force duplicate efforts in the parallel development of rockets. There are rumors of smouldering controversy.

"But yesterday's dramatic blastoff of the triple-barrelled Titan 3-C by the Air Force, coming closely on the heels of NASA's sensational Gemini performance, indicates the competition, thus far is beneficial to both." (Cleveland *Plain Dealer*, 6/19/65, 23)

- USAF has come up with something for NASA to reckon with, commented William Hines, concerning the Titan III-C success, in an article in the Washington *Evening Star*: "It is the Air Force position—which NASA will now be forced to try to disprove—that anything Saturn IB can do, Titan III-C can do better.

"The Air Force, for its part, must now try to prove Titan III-C's reliability over the long haul . . . Gen. Joseph S. Bleymaier Jr., head of the Titan III-C program, said it is his goal to make every one of the 12 shots in the Titan III-C development series a 100% success.

"Equally important with reliability is cost. Bleymaier says the Titan III-C can be produced in quantity for $12.8 million, or just a little more than half the $22 million it is estimated NASA's Saturn IB will cost. Titan III-C's $800 million development cost is but a fraction of what NASA will have spent to get the first Saturn IB off the ground." (Hines, *Wash. Eve. Star*, 6/19/65)

June 20: Vice President Humphrey, in France for the International Air Show at Le Bourget with Astronauts White and McDivitt, met for 80 min. with President de Gaulle. (Tanner, *NYT*, 6/21/65; Newport News *Daily Press*, 6/22/65)

- Two single-engine propeller-driven Skyraider tactical bombers downed two Communist jet MiG-17's over North Vietnam. (AP, *NYT*, 7/11/65)

June 21: F-1 rocket engine completed its 1,000th test firing at NASA MSFC's Rocket Engine Test Site where it operated at its full thrust of 1,500,000 lbs. for 165.6 sec. Test was conducted by North American Aviation's Rocketdyne Div. In a cluster of five, F-1 would provide 7,500,000 lbs. thrust in the S-IC first stage of the Saturn V booster that would launch Apollo lunar missions. (MSFC Release 65-154; *Marshall Star*, 6/23/65, 1)

- NASA would negotiate with the Rocketdyne Div. of North American Aviation, Inc., for 22 H-1 rocket engines with 200,000-lb. thrust for use on the 12 Saturn IB launch vehicles presently planned, NASA Marshall Space Flight Center announced. Engines, with supporting services, would cost more than $6 million. (MSFC Release 65–155; *Marshall Star*, 6/23/65, 1)

June 21: Dr. Robert C. Seamans, NASA Associate Administrator, received the New England Aero Club's Godfrey L. Cabot Award in Boston. He was cited for "outstanding contributions to aeronautics." (NASA Release 65–193)

- West German satellite tracking station at Bochum monitored radio signals from a new Soviet space probe. The observatory said it had received the signals since 11:26 a.m. EDT. (Reuters, 6/21/65)

- AEC reorganized its space-related R&D activities and established a Division of Space Nuclear Systems. All AEC space-oriented work on Snap reactor and isotope electric power systems was transferred to newly-created Space Electric Power Office in that division. Isotopic thruster propulsion work, formerly under AEC Division of Isotopes Development, was transferred to NASA-AEC Space Nuclear Propulsion Office (SNPO). "A major advantage of the new organizational alignment [was] the improved communication and ease of coordination between AEC and NASA in the power area." (*Atomic Energy Programs, 1965*, 141)

- Immediate planning for a fourth jetport for the New York City area was advocated by Harold E. Gray, President of Pan American World Airways, at the Annual Aviation Luncheon of the Queens Chamber of Commerce: "Maybe the need is eight years from now. Maybe it is twelve years from now. But it would take ten years, as I understand it—starting today—to develop a fourth airport. So, tomorrow may be too late to make this start. Mañana is not soon enough for me!" This was the first time a Pan Am official had stated the airline's position on the jetport issue. (Hudson, *NYT*, 6/22/65, 58)

- Rep. Oren Harris (D–Ark.), in France for the International Air Show, endorsed the idea of an international exhibition at Dulles International Airport in 1966. Harris, who headed the House Committee on Interstate and Foreign Commerce which would handle any American counterpart of the Paris show, made his statement after comment by American aircraft builders participating at Le Bourget that the U.S. had let the Soviet Union "steal the show."

    During the first days of the exposition, the Russians had displayed the prototype of a 720-passenger commercial aircraft to be ready in two years. Also, Yuri Gagarin, first man in space, was on hand. The U.S. had recovered some lost ground during the final weekend by flying in Vice President Hubert Humphrey and Astronauts Edward White II (Maj., USAF) and James McDivitt (Maj., USAF). American jets had also put on spectacular acrobatic demonstrations.

    Rep. Harris said the project had been discussed at the White House but that further studies by the FAA and the Budget Bureau would be necessary. (UPI, Bruns, *Wash. Post*, 6/22/65)

- The U.S. showing at Le Bourget was discussed by Robert Hotz in an *Aviation Week and Space Technology* editorial: "The United States is presenting a sorry spectacle at the 26th Paris Air Show [June 11–21] that does grave injustice to its genuine aerospace capability in relation to its principal competitors, including the Soviet Union. As the show draws to a close, only a bold stroke of public relations, such as jetting Astronauts White and McDivitt with their Gemini capsule to Le Bourget for a weekend finale would offer any chance for the U.S. to recoup the prestige lost last week through the combination of top-level government indifference, official naivete and the stifling channels of interagency bureaucracy between the various government departments involved in the U.S. show effort. . . .

"For the past two Paris air shows, the NASA exhibit has been a major sensation, convincing Europeans of U.S. technological leadership. This year, NASA was conspicuously absent. Even frantic pleas from the French to have Astronaut John Glenn appear to match the Soviets' Yuri Gagarin were strangled in the maze of interdepartmental coordination. . . .

"The net result of this sad combination of government bungling at Le Bourget has been to spend large sums of the taxpayers' dollar to create the impression that the United States is resting smugly on its technical oars, complacently relying on rapidly obsolescing military hardware for the present and craftily confident this aging equipment can be foisted on its European allies in the near future." (Hotz, *Av. Wk.*, 6/21/65, 13)

June 21–July 16: Space Research Summer Study—1965, convened at Woods Hole, Mass., by NAS–NRC Space Science Board, reviewed the National space program. Various working panels were formed to examine in detail the three topics of particular focus: planetary exploration, astronomy requirements, and the role of man in space research. About 200 invited scientists participated in the multi-disciplinary review, which was under general chairmanship of George P. Woollard, Director, Hawaii Institute of Geophysics. NASA provided financial support. (NAS—NRC, *News Report*, 9/65, 5)

June 22: MARINER IV spacecraft was functioning well on its mission to Mars, but the public should not "expect too much" from the photographs of Mars it was scheduled to take July 14, NASA scientists said during a press briefing. It was conjectural whether the 21 photographs the Mars probe would take would be clear enough to disprove or verify the theories held by some scientists that there are canals and some form of life on Mars. Scientists who would study the photographs pointed out that 21 pictures would only enable them to see 1% of Mars. Nevertheless, any pictures of the surface would be far superior to the best observations now obtainable with earth-based telescopes.

The first few photographs might be made public immediately after being received, but the others probably would not be released until they had been studied for weeks or months. (Transcript, Sehlstedt, *Balt. Sun*, 6/23/65; Hill, *NYT*, 6/23/65, 7)

- X-15 No. 2 flown by pilot John McKay (NASA) to 155,900 ft. altitude at maximum speed of 3,938 mph (mach 5.64) to obtain data on star tracking cameras, landing gear modification checkout, stability and control and advanced X-15 landing dynamics. (NASA X-15 Proj. Off.; *X-15 Flight Log*)
- The GEMINI IV spacecraft's onboard computer that failed toward the end of the GT–4 flight was working well in ground tests. It would be

tested under simulated orbit conditions to determine if the problem might have been in the inertial guidance system. (UPI, *NYT*, 6/22/65, 36; Phil. *Eve. Bull.*, 6/22/65)

June 22: Two Nike-Apache sounding rockets launched two sodium vapor trail experiments from Wallops Station, Va., as part of NASA's upper atmosphere meteorological research program. Vapor trails were ejected through a region 40 to 124 mi. above earth. One rocket was fired on an azimuth of 90° (due east) and the other on a 130° azimuth (southeast) so that the sodium trails were in the same altitude region at about the same time but several miles apart to provide data on wind behavior variations over a lateral distance as well as at various altitudes. (Wallops Release 65-38)

- President Johnson nominated Astronauts Walter M. Schirra (USN) and John Young (USN) for promotions. Commander Schirra was nominated for captain and Lt. Commander Young for commander. Each had received his current rank in 1961, before the space flights. All the original astronaut team had now been promoted. (AP, *Wash. Post*, 6/23/65, A12)

June 23: Three Nike-Apache sounding rockets launched vapor trail experiments from Wallops Station, Va., as NASA completed a five-shot, two-day series of experiments to measure wind direction and velocity over the Atlantic coast as part of its meteorological research program. Payloads also contained instrumentation to compare electron densities with wind dispersion and to measure electron temperatures. Experiments were conducted for the GCA Corp. under contract to NASA. (Wallops Release 65-39)

- Pegasus C, third meteoroid detection satellite, arrived at NASA Kennedy Space Center aboard the aircraft "Pregnant Guppy." Pegasus C was scheduled for launch during summer of 1965 by SA-10, the last Saturn I vehicle. (MSFC Release 65-159; *Marshall Star*, 6/30/65, 2)

- President Johnson nominated Gen. William F. McKee to be Administrator of the Federal Aviation Agency. He made the appointment after signing a bill that exempted General McKee (USAF, Ret.) from the requirement that the FAA administrator be a civilian. (AP, *NYT*, 6/25/65)

- NASA Administrator James E. Webb said in an address to the Defense Supply Agency in Alexandria, Va.: "We are at a watershed of history—we are at a point where man has made reality of his wildest imagination—he has created machines which enable him to move outward from the earth into the new environment of space, not just in thought, but in actuality, taking long strides toward the stars. . . .

   "With the initial planning and procurement phases behind us, we are now approaching a period when decisions on the next generation of major activities in space will be made. We have created and tested a workable managerial capability to direct the planning, approval, and execution of future programs. We believe that adoption of what we have termed 'Phased Program Planning' will materially assist us in achieving this goal.

   "Phased Program Planning represents an orderly, sequential progression in the execution of major NASA projects. It provides for formulating proposed work goals and missions, and allows for reappraisal points for management consideration to advance or replan such proposals, as well as the resources to implement them." (Text)

June 23: NASA ordered new computer equipment that would make possible the processing of data sent back by spacecraft at a rate 40 times faster than in the systems currently used. The contract, ranging from an initial procurement of about $8 million to a possible total of $18 million if all contract options were exercised, was being negotiated with International Business Machines Corp. (NASA Release 65–205; AP, *NYT*, 6/24/65, 4; AP, *WSJ*, 6/24/65)

- Joint plans for higher wage and improved fringe benefits negotiations in 29 agreements with major aerospace companies were formulated at a conference in Washington, D.C., between leaders of the United Automobile Workers and the International Association of Machinists. Contracts discussed would expire within the next six months. (AP, *NYT*, 6/24/65, 21)

June 24: A wage dispute involving the International Alliance of Theatrical and Stage Employees caused a shutdown of much of the construction under way at Kennedy Space Center, NASA, as about 3,000 of some 5,700 construction workers stayed off the job because of the picket lines. (UPI, *NYT*, 6/25/65, 28)

- NASA and the U.S. Army had entered into an agreement which would establish a joint effort in the area of low-speed aeronautical research to be accomplished in facilities at NASA Ames Research Center, Ames reported. The program would be conducted in cooperation with personnel of the U.S. Army Materiel Command. (ARC *Astrogram*, 6/24/65, 1)

- USAF launched an unidentified satellite from Vandenberg AFB with a Thor-Able-Star booster. (UPI, *NYT*, 6/26/65, 6)

- NASA would stick with its Saturn launch vehicles for the Apollo program, Maj. Gen. Samuel C. Phillips (USAF), Apollo Program Director, reportedly said in Tulsa. Asked about the possibility of "leap-frogging" in the Apollo and making use of Titan III–C—because of its launch potential—instead of Saturn V, Phillips said "no," and commented: "The Saturn vehicle will do what we want insofar as the Apollo program is concerned, and leapfrogging now probably would mean some setbacks rather than advancing the Apollo project." He said launch vehicles such as Titan III–C, which had a potential of 11 to 15 million lbs. of thrust, most certainly would have major roles in future space operations. "But NASA is committed to the Saturn for the Apollo and we plan to stick with this vehicle." (Leslie, Tulsa *Daily World*, 6/24/65)

- Vice President Humphrey announced that the National Aeronautics and Space Council, of which he is chairman, would meet July 1 to consider a proposed USAF manned orbiting laboratory. Meeting was later postponed to July 9. (*NYT*, 6/26/65, 5; EPH)

- Dr. Raymond L. Bisplinghoff, NASA Associate Administrator for Advanced Research and Technology, submitted his resignation, effective at the end of August, to accept the presidency of Case Institute of Technology, Cleveland, Ohio. Dr. Bisplinghoff had come to NASA in 1962 from MIT where he was Deputy Head of the Department of Aeronautical Engineering. (NASA Release 65–208)

June 24: NASA Administrator James E. Webb presented NASA Exceptional Service Award to Dr. George F. Simpson, Assistant Administrator for Policy Planning, in an informal Headquarters ceremony. Having joined NASA in 1961, Dr. Simpson resigned to become Chancellor of the State of Georgia University System. (EPH)

- First Federal regulations specifically governing agricultural flying and related activities were announced by FAA. Effective Jan. 1, 1966, the rules would establish national standards and requirements for private and commercial agricultural operator certificates, operating rules, aircraft airworthiness, pilot qualifications, and record keeping. (FAA Release 65-50)

- Senate adopted a resolution requesting President Johnson to proclaim Sept. 17 and 18 as special days in honor of the memory of James Smithson, the Englishman who willed all his property for establishment of the Smithsonian Institution in Washington, D.C. 200th anniversary of his birth would be celebrated this year. (AP, *NYT*, 6/26/65, 20)

- Col. Jack Bollerud (USAF) was appointed Deputy Director of Space Medicine, Office of Manned Space Flight, NASA Hq. He would be deputy to Dr. W. Randolph Lovelace II. (NASA Release 65-207)

June 25: Soviet Union launched COSMOS LXIX artificial earth satellite containing scientific equipment for the study of outer space. Orbital parameters: apogee, 332 km. (206 mi.); perigee, 211 km. (131 mi.); period, 89.7 min.; inclination to the equator, 65°. All instruments were operating normally. (*Izvestia*, 6/26/65, 1, ATSS-T Trans.; *Pravda*, 6/26/65, 1, ATSS-T Trans.)

- USAF launched two unidentified satellites from Vandenberg AFB using a single Atlas-Agena D booster. (UPI, *Chic. Trib.*, 6/26/65; *U.S. Aeron. & Space Act., 1965*, 146)

- First Phoebus 1A nuclear reactor test was conducted at NRDS, the reactor operating successfully at full power for $10\frac{1}{2}$ min. The reactor was damaged during shutdown when the facility liquid hydrogen supply was unexpectedly exhausted. Test was part of NASA-AEC Phoebus program to extend graphite reactor technology developed under the Kiwi series to higher power and temperature, ultimately leading to high-thrust nuclear engine system for space exploration. (*Rover Chron.*; *Atomic Energy Programs, 1965*, 145)

- NASA Lewis Research Center announced successful test of a large-scale facility to investigate the boiling of liquid sodium. Test was just one milestone in the complex and extensive research being conducted on the properties and engineering performance of liquid metals. Liquid alkali metals, having excellent heat-transfer capabilities and large liquid range, were being considered as working fluids in future advance electric power systems in spacecraft. (LRC Release 65-46)

- NASA announced it would negotiate with Douglas Aircraft Co. for nine additional S-IVB flight stages to be used as the third stage of the Saturn V launch vehicle being developed at NASA Marshall Space Flight Center. Work would also include related spares and launch support services. Value of the S-IVB contract presently totaled some $312 million. The new work was expected to exceed $150 million. (NASA Release 65-209; MSFC Release 65-162; *Marshall Star*, 6/30/65, 1)

June 25: Thiokol Chemical Corp. received from USAF a $3,195,500 fixed-price contract for design, development, and firing of a 156-in. solid rocket motor. USAF also awarded United Technology Center a $10,500,000 increment to a previously issued contract for design, development, fabrication, delivery, and flight testing of large segmented solid propellant motors. (DOD Release 426–65)

- Ives, Whitehead & Co., Inc., a Washington, D.C., management and trade consultant company, proposed that the Nation's program to develop a supersonic airliner be financed by a special Government corporation similar to ComSatCorp. The plan would eliminate the necessity of asking Congress to appropriate the minimum $1 billion needed for development of the aircraft. The proposed company would be known as the SST Development Corp. and would be authorized to raise capital funds through private investment channels, by issuing and selling bonds or notes or both. The corporation would then enter into contracts with airframe and engine manufacturers whose designs had been selected by FAA. The corporation's obligations would be repaid by a predetermined percentage of the profits earned by industry by selling the planes to private airlines and by royalty payments added to the sale price of the plane by the airlines. (UPI, *NYT*, 6/26/65, 42)

- NASA announced the appointment of Robert F. Thompson as Mission Director for the future Gemini missions and Col. C. H. Bolender (USAF) as Mission Director for the first and second Apollo/Saturn IB flights. Thompson and Bolender were assigned to the Mission Operations Organization in the Office of Manned Space Flight, NASA Hq., and would have overall responsibility for directing assigned missions. Christopher Kraft would continue in his regular assignment as Flight Director for Gemini missions.

  Thompson was Chief, Landing and Recovery Div., NASA Manned Spacecraft Center, before receiving this assignment. Col. Bolender had directed a studies group in the office of the USAF Chief of Staff and had engaged in extensive guided missiles and aeronautical systems work. (NASA Release 65–211)

- Civil Aeronautics Board (CAB) authorized Pan American Airways and Trans World Airlines to subsidize New York Airways' helicopter service temporarily. CAB also announced that two stockholders in New York Airways, Robert G. Goelet and John Hay Whitney, had lent a total of $165,000 to the company. (AP, *NYT*, 6/26/65, 42)

- South African Prime Minister Hendrik F. Verwoerd said his government would not admit American Negroes if they were assigned to work in satellite tracking stations operating in South Africa. The Johannesburg *Sunday Times* commented: "The United States will have to decide whether it can afford morally to overlook Dr. Verwoerd's remarks." (Lelyveld, *NYT*, 6/27/65)

- A centennial meeting in Moscow paying homage to the work of Gregor Mendel marked the end of a 20-yr. period during which his work was discredited and Soviet biological science was based on the work of geneticist Trofim D. Lysenko. (*NYT*, 6/27/65, 2)

June 26: NASA Administrator James E. Webb announced at a press conference during the 23rd annual Hampton County (S.C.) Watermelon Festival that the Gemini V manned space flight would be an eight-day mission—the time required to fly to the moon, explore its surface, and return to earth. Webb also disclosed that NASA would announce next week the selection of six scientist astronauts.

Speaking at "Mendel Rivers Day" ceremony, Webb pointed out that Congressman Mendel Rivers, Chairman of the House Armed Services Committee, "has steadily supported the Nation's effort to build strength in space." Webb also said: "Thoughtful students of national power and its uses are increasingly aware that America's security as well as her leadership of the Free World, depends directly upon our progress in mastering and using space.

"The National Aeronautics and Space Administration cooperates closely with the Department of Defense, to ensure that the technological progress we make in developing the peaceful uses of space will be drawn upon as needed to help keep the peace—in space and on earth.

"As you can readily understand, the development of military space systems follows those basic research fields which provide the scientific understanding and technological capability to make such systems possible. It is NASA's job, as the space research and development agency, to provide this basic knowledge and know-how."

Astronauts Virgil I. Grissom and John W. Young accompanied Webb to this festival. (NASA Release; Text; *Aerospace Historian*, 10/65, 111–14)

June 26: 20th anniversary of the United Nations celebrated in San Francisco. President Lyndon B. Johnson said in an address to the General Assembly: "The movement of history is glacial. On two decades of experience none can presume to speak with certainty of the destiny of man's affairs. But this we do know and this we believe: Futility and failure are not the truths of this organization brought into being here 20 years ago.

"Where historically man has moved fitfully from war toward war, in these last two decades man has moved steadily away from war as either an instrument of national policy or a means of international decision. . . .

"The promise of the future lies in what science, the ever more productive industrial machine, the ever more productive, fertile and usable lands, the computer, the miracle drug and the man in space all spread before us. The promise of the future lies in what the religions and the philosophies, the cultures and the wisdoms of 5,000 years of civilization have finally distilled and confined to us—the promise of abundant life and the brotherhood of man." (Text, *NYT*, 6/26/65)

June 26: Thirteen NASA astronauts left NASA Manned Spacecraft Center for an area near King Salmon AFB, Alaska, for a week-long study of a large volcanic ash flow believed to be similar to the surface of the moon. The ash was deposited in 1912 from a volcanic eruption and is the largest flow of its type in the world. (AP, *NYT*, 6/22/65, 2)

• An S–IVB facility vehicle and an S–II simulator arrived in New Orleans aboard the USNS *Point Barrow* enroute to NASA Kennedy Space Center. Both the S–IVB and S–II were upper stages of the Saturn V launch vehicle. The S–IVB would also serve as the second stage of the Saturn IB booster. (MSFC Release 65–161; *Marshall Star*, 6/30/65, 2)

June 26: Sealab II, described as having some interesting "physiological" similarities to the proposed USAF Manned Orbiting Laboratory program, was scheduled to begin the middle of August by the Office of Naval Research. In the experiment, part of a long-range project to determine how effectively man could work under the sea, two diving teams of ten men each would descend to the Pacific Ocean bottom off La Jolla, Calif. Two of the divers were expected to stay on the bottom for 30 days, living and sleeping in specially-designed, 57-ft. long quarters about 210 ft. below the surface.

It was anticipated that Cdr. M. Scott Carpenter (USN), presently on loan from NASA, would be leader of the first team. (Anderson, *Chic. Trib.*, 6/27/65)

• The solar boat had been found feasible by Army engineers after extensive tests, NANA reported. It was a lightweight craft that operated solely by sunlight falling on power-generating cells. The Army said: "The solar propulsion boat may have potential military application where it is necessary to operate quietly and without using conventional fuel." (NANA, Detroit *News*, 6/27/65)

June 27: The six scientist–astronauts selected for the Apollo program were announced in the Nation's press: Owen K. Garriott, 34, associate professor of physics, Stanford Univ.; Edward G. Gibson, 29, senior research scientist, Applied Research Labs., Aeronutronic Div., Philco Corp.; Duane E. Graveline, 34, flight surgeon, NASA Manned Spacecraft Center; Lt. Cdr. Joseph P. Kerwin (USN), 33, staff flight surgeon, Air Wing, 4, Cecil Field Naval Air Sta., Pa.; Frank Curtis Michel, 31, assistant professor of space sciences, Rice Univ.; Harrison Schmitt, 29, astrogeologist, U.S. Geological Survey.

They were chosen from a group of 16 nominees submitted to NASA by the National Academy of Sciences. NAS had screened about 400 applications forwarded by NASA earlier this year.

The six new scientist–astronauts were to have been announced officially by NASA June 29 but NASA officials confirmed the six named on June 28, the day after the press stories. (Schefter, *Houston Chron.*, 6/27/65; UPI, *Wash. Post*, 6/27/65)

• First clear spectrograms of ultraviolet light from the stars were obtained on a Princeton Univ. rocket experiment originally thought to have been a failure, a spokesman for the university's rocket program announced.

The films from the June 1 flight from White Sands Missile Range had at first been feared to be fogged, but after being developed by special techniques, showed spectra of starlight with a fineness of detail never before achieved. A detailed report would be issued after the films had been studied further. (AP, *NYT*, 6/28/65; *Wash. Post*, 6/29/65)

• The U.S. should take the lead in establishing an international patent system, Dept. of Commerce Deputy Assistant Secretary for Science and Technology William W. Eaton said at a patent conference in Washington, D.C. At present, an inventor must take out separate patents in each of several foreign countries or run the risk of his idea being exploited. The new system would eliminate this problem by having one international patent cover each invention, Mr. Eaton said. (Sci. Serv., *NYT*, 6/27/65, 53)

June 27: In his column in the New York *Journal American,* Bob Considine cited an item written about Soviet failures in space by Julius Epstein, a research associate with the Hoover Institute of War, Revolution, and Peace, Stanford Univ.: "According to reliable reports in Washington, the Soviets have lost at least three cosmonauts on their way to the moon. My first publication of these assertions in 1962 met with no denial from our National Aeronautical and Space Agency when I forwarded a copy and asked for comment. A free-lance writer, researching the possibility of Soviet failures, tells me that NASA informed him that all such information had been classified as top secret. They recommended that he use my material! Isn't that a reasonable indication for the veracity of this record?" (Considine, N.Y. *J. Amer.,* 6/27/65)

- Soviet parachutist holding world and national records, Vyacheslav Zharikov, flew through the air at 119 mph at the end of an airplane-towed cable, Tass reported. He then dropped free, opened his parachute above 100,000 spectators at an air show at Tula, U.S.S.R., and landed safely. (AP, *Wash. Post,* 6/28/65, A17)

June 28: Pictures of Mars relayed to earth by MARINER IV would be released to the public within 36 to 48 hrs. after they were taken July 14, NASA announced. MARINER IV was expected to take more than 20 photographs in about 24 min. and radio them back to earth in digital form; while not as detailed as the Ranger photographs, they were expected to make a valuable contribution to space exploration.

MARINER IV would pass within 5,600 mi. of the planet. The pictures would be taken at somewhat greater distances while Mars was in sunlight relative to the spacecraft. (NASA Release 65-210; UPI, *NYT,* 6/29/65)

- NASA officially confirmed the six scientist-astronauts named in the Nation's press June 27. They were Owen K. Garriott; Edward G. Gibson; Duane E. Graveline; Lt. Cdr. Joseph P. Kerwin (USN); Frank Curtis Michel; and Harrison Schmitt. (NASA Release 65-212)

- USAF and NASA were finishing plans for joint lifting-body tests, reported William Normyle in *Aviation Week and Space Technology.* He said a joint USAF-NASA Flight Test Group was being formed at Edwards AFB and would include NASA and USAF pilots who would flight-test three lifting-body configurations in a program to be monitored by NASA. Two configurations, the M-2 and the HL-10, were NASA designs built by Northrop Corp's Norair Div. Third configuration, USAF's SV-5, was developed by Martin Co. (Normyle, *Av. Wk.,* 6/28/65, 19)

- In response to newsmen's questions, a spokesman for the State Dept. said the U.S. planned to continue operating its two satellite tracking stations in South Africa despite the warning by South African Prime Minister Hendrick Verwoerd that Negro Americans cannot be employed there. Of the two tracking stations, DOD operated one and NASA the other. The DOD station employed about 50 American technicians, almost all civilians, and about 25 South Africans. The NASA station had only one American, a liaison officer; other employees were South Africans—some of them non-white. (Halloran, *Wash. Post,* 6/29/65, A16)

- EARLY BIRD I began commercial operations when President Johnson formally inaugurated telephone service via communications satellite in a

25-min., 6-nation conference call with European officials, including British Prime Minister Harold Wilson, West German Chancellor Ludwig Erhard, and Swiss President Tschudi. Participants reported satisfaction with results. (ComSatCorp Release; Robertson, *NYT*, 6/29/65, 12)

June 28: At a Special Awards Ceremony at NASA Manned Spacecraft Center, the Presidential Citation was presented to John H. Robinson of Resources Management Div. Robinson was one of the three individuals responsible for organizing a control center, preparing all Pert networks, and operating this control center. Operation was so effective that USAF was able to complete a modernization program far ahead of schedule with a monetary saving of several million dollars. (MSC *Roundup*, 7/9/65, 3)

- President Johnson accepted the resignation of Eugene G. Fubini as Assistant Secretary of Defense (Deputy Director of Research and Engineering), effective July 15. (UPI, *Wash. Post*, 6/29/65)
- A strike by 86 members of the International Alliance of Theatrical and Stage Employees carried work stoppage at Kennedy Space Center into the fifth day. They had walked out in a wage dispute with a firm providing printing and reproduction services for NASA; about half the 5,300 construction workers at KSC had honored their picket lines, halting work on $178 million worth of projects. (*Wash. Post*, 6/29/65)

June 29: NASA successfully conducted an Apollo boilerplate pad abort test at White Sands Missile Range to check the launch escape system. The test simulated an abort from ground level, using the Apollo launch escape system for propulsion. This type of abort would be necessary in an actual mission if serious trouble developed with the Saturn launch vehicle just before or during ignition of the Saturn engines.

Boilerplate 23A, the command module used for this test, powered by the launch escape rocket's 155,000 lbs. of thrust, traveled 5,000 ft. above the range. Eleven sec. after ignition was signaled from the blockhouse, canards deployed near the top of the escape motor, causing the spacecraft to pitch aerodynamically to a blunt-end-forward position. Three seconds later, the tower jettison motor ignited, removing the tower and boost protective cover from the spacecraft. The forward (apex) heat shield was jettisoned .4 sec. later to uncover the parachute containers. Dual drogue parachutes were deployed by mortars from the upper deck two seconds after the LES was jettisoned. They slowed the spacecraft's descent, then disreefed to stabilize the module in a blunt-end-forward position. When the drogue parachutes were jettisoned, three pilot chutes were deployed to extract the three main chutes from their containers. The main parachutes were deployed in reefed condition, then disreefed to lower the spacecraft to a gentle landing about one mile from the launch site. The flight sequence took about one minute. This was the first Apollo boilerplate to be reused. (NASA Release 65–202; UPI, *Chic. Trib.*, 6/30/65; MSC *Roundup*, 6/25/65, 1)

- President Johnson signed the bill authorizing $5,190,396,200 for NASA during FY 1966. (AP, *Wash. Post*, 6/30/65)
- At Kennedy Space Center, NASA problems with radio frequencies forced at two-day postponement of an attempt to launch Tiros 10 weather

satellite, designed to study hurricanes and typhoons. (*Wash. Post*, 6/29/65)

June 29: Ten firms had submitted proposals to the NASA Marshall Space Flight Center for preliminary design study of the Optical Technology Satellite being considered as part of the Apollo Extension System (Aes). Two contracts would be awarded for parallel studies.

Objective of the proposed Ots program was to advance NASA's capability in space optics technology by performing several engineering and scientific experiments in space. Selected contractors would review existing conceptual designs for Ots and analyze the technical feasibility and justification for the performance in space of experiments being considered. The contractor would then create a conceptual design with onboard experiments.

In Phase II, the contractor would perform a conceptual design study of the Ots subsystem to establish a sound base for overall preliminary design. The program would be under NASA Office of Advanced Research and Technology. (NASA Release 65-213; MSFC Release 65-213)

- Capt. Joseph Engle (USAF) qualified for the military rating of astronaut by piloting the X-15 No. 3 research aircraft to an altitude of 280,600 ft. at a maximum speed of 3,432 mph (mach 4.94). By exceeding 264,000 ft. (50 mi.) Engle, 32, met the USAF astronaut requirements.

  Purpose of the flight was to use a scanning device to obtain measurements of the earth's horizon. The measurements would be used to establish the design criteria for navigational systems for future spacecraft. (NASA X-15 Proj. Off.; NASA Release 65-201; *X-15 Flight Log*)

- USAF awarded Hughes Aircraft Co. a $13,468,725 contract for modifications of missiles and space parts. (*WSJ*, 6/29/65, 9)

- Pickets were removed from entrances to Kennedy Space Center, NASA, when printing and reproduction workers voted to end their 6-day-old construction-crippling strike and submit the dispute to the President's Missile Sites Labor Commission. Removal of the pickets would enable 2,200 construction workers who had honored the lines to return to vital space projects. (AP, *Balt. Sun*, 6/30/65)

- The 10 crew members of a Pan American World Airways Boeing 707 jetliner that caught fire just after take-off and lost an engine and part of a wing would receive FAA's Exceptional Service Citation and medals, FAA announced. Capt. Charles H. Kimes and his crew landed the plane safely at Travis AFB; all passengers were clear of the plane within the two-minute goal set by FAA as a maximum safe evacuation time. (*NYT*, 6/30/65, 59)

June 30: A four-stage, 60-ft.-long Journeyman (Argo D-8) sounding rocket launched by NASA from Wallops Station, Va., carried a 137-lb. instrument package to an altitude of 1,060 mi. on a 25-min. flight that ended 1,700 mi. offshore in the Atlantic, east of Bermuda. After launch, the payload, unfurled two 35-ft. antennas to provide a single dipole antenna, measuring 70 ft. from tip to tip. Main objective of the mission was to measure the intensity of radio frequency energy originating largely from outside the solar system. This was done at three frequencies: 750, 1,125, and 2,000 kc. Secondary objectives were the investigation of previously detected radio noise on the top side of the ionosphere and measurement of electron density in that part of the ionosphere.

Information would be used in planning future space radio astronomy experiments to investigate certain characteristics of antennas in the ionospheric region. Experiment was conducted in cooperation with the Univ. of Michigan. (NASA Release 65–214; Wallops Release 65–40; AP, Balt. *Sun,* 7/1/65)

June 30: Lunar Landing Research Facility, a controlled laboratory for exploring and developing techniques for landing a rocket-powered vehicle on the moon, had been put into operation at NASA Langley Research Center. The $3.5 million facility included a rocket-powered flight test vehicle which would be operated while partially supported from a 250-ft. high, 400-ft.-long gantry structure to simulate the one-sixth earth gravity of the moon in research to obtain data on the problems of lunar landing. (LaRC Release 6/30/65)

- ComSatCorp asked the FCC for authority to assemble a third Early Bird-type satellite from existing parts for a future launch. A back-up satellite to EARLY BIRD I was already assembled; components for a third comsat were completed earlier this year. The application before the FCC requested permission to assemble these components. (ComSatCorp Release)
- Voice of NASA Administrator James E. Webb was relayed from Washington, D.C., to Tulsa, Okla., via RELAY II for the dedication of Tulsa's Central Library. Mr. Webb spoke briefly before triggering an electronic signal that turned on the lights in the new building. (AFP; Tulsa *Daily World,* 6/10/65)
- Senate confirmed the nominations of Gen. William F. McKee (USAF, Ret.) as Federal Aviation Agency Administrator and David D. Thomas as his deputy administrator.

President Johnson formally accepted the resignation of McKee's predecessor, Najeeb E. Halaby, with a "Dear Jeeb" letter praising Halaby's "vigorous and dynamic leadership," particularly in supersonic transport development. (*CR,* 6/30/65, 14824; UPI, *Wash. Post,* 7/1/65)
- NASA procurements for FY 1965 totaled $5,187 million, an amount 13% greater than the amount awarded during FY 1964. About 94% of NASA's procurement dollars was contracted directly or indirectly to private industry: About 79% of net dollar value was placed directly with business firms; 4% was placed with educational and other nonprofit institutions or organizations, 5% with Cal Tech for operation of JPL, and 12% with or through other Government agencies. 90% of the dollar value of procurement requests placed with other Government agencies resulted in contracts with industry (awarded on behalf of NASA); also, about 76% of NASA-placed funds under the JPL contract resulted in subcontracts or purchases with business firms. (NASA *FY 1965 Annual Procurement Rpt.,* 54)
- During FY 1965, NASA awarded $121,115,000 in grants and research contracts to 190 colleges and universities in 50 states, the District of Columbia, and 8 foreign countries. In addition, $25,527,000 was awarded to 53 nonprofit institutions in 20 states and the District of Columbia. (NASA Proj. Off.)

June 30: Strongly urging greater emphasis on U.S. aeronautics and aviation than now exists, Sen. A. S. (Mike) Monroney (D–Okla.), Chairman of Senate Aviation Sub-committee, addressed Air Force Association in Dallas, Tex.

"I am afraid some people, and I mean people of high stature and great influence, seem bent on relegating aeronautics to a secondary position. This I think is a crucial mistake. In aeronautics we stand on the threshold of a boundless upsurge which may well dwarf what has been achieved by aviation over the past 61 years.

"Also, the economic productivity of aeronautics makes dollars . . . and sense. It means money in the bank, or rather gold in Fort Knox. On the other hand, I don't know of any scheme . . . and I doubt that one will emerge within this century . . . for fighting the balance of payments deficit in outer space.

"There simply are no dollars floating around between here and the moon. For the time being, outer space *commerce* just doesn't seem to be in the cards. And as long as this is the situation, I don't think we should banish aeronautics to the back seat . . . or have a NASA budget which is better than 98 percent spacecraft and less than 2 percent aircraft. And this in spite of the fact that NASA and its forerunner, NACA, were ostensibly founded to serve aeronautics!

"I am not saying this to downgrade space, but to suggest that we strike a balance based on how the national interest and economy are served best. . . . This widening disparity between NASA's space and aeronautical efforts . . . this wall flower treatment of the airplane . . . is a reckless gamble, economically, politically and militarily, that will come back to haunt us in the future. My criticism, by the way, is not directed at the level of competence in that agency but at the lack of programming of aeronautical goals and the failure to mobilize this nation's brain power on behalf of aviation. It seems to me the real deficiency is that the well qualified aeronautical talent in NASA is so far down in the hierarchy that these men rarely ever get involved in the decision making. It's crystal clear that other nations, the Russians included, have cut this Gordian knot and are rapidly filling the vacuum which we so invitingly created for them. . . ." (Text; AP, *NYT*, 7/2/65)

- Dr. Albert J. Kelley, Deputy Director of NASA Electronics Research Center, retired as a U.S. naval commander. He had been assigned to detached duty with NASA by USN in 1960 and would continue with ERC as a civilian. (*Boston Globe*, 7/1/65)
- A Titan II ICBM was fired from Vandenberg AFB underground silo on a routine flight down the Western Test Range by a Strategic Air Command (SAC) crew. (UPI, *Boston Globe*, 6/30/65)
- The last of 800 USAF Minutemen I ICBMs became operational at Warren AFB, Wyo. Weighing about 65,000 lbs., the three-stage solid propellant missiles would have a range of more than 6,300 mi., a speed of over 15,000 mph, and would carry a nuclear warhead. The missiles were housed in individual blast-resistant underground launch sites and could be stored for long periods with a minimum of maintenance. (AFSC Release 97.65)

During June: First patent for production and separation of plutonium was granted to AEC Chairman Dr. Glenn T. Seaborg and co-inventors the late Dr. Joseph W. Kennedy and Dr. Arthur C. Wahl. The patented procedure included treatment of uranium in a reactor to produce plutonium—first synthetic element to be seen by man—and to separate and recover the plutonium by a method called oxidation reduction. (Jones, *NYT*, 6/26/65, 33)

During June: Interviewed in *Data*, NASA Associate Administrator for Advanced Research and Technology Dr. Raymond L. Bisplinghoff discussed basic research in the NASA program:

"The term basic is employed for two reasons. The basic research program involves elements which undergird everything we do in aeronautics and space. Such areas as materials, fluid physics, electrophysics and applied mathematics form a common base for all of our work. In addition, most of this program is carried on at a fundamental level, that is, at the atomic and molecular level. If we select, for example, the field of materials we can observe that nearly everything we do in atmospheric and space flight is limited in some way by materials. . . ."

Asked about the role of industry in NASA's advanced research and technology program, Dr. Bisplinghoff said:

"About 75 per cent of the OART research and development budget is spent outside the NASA organization. This money goes to university and research institutes and to industrial contractors capable of carrying out the advanced work making up this program." (*Data*, 6/65, 22–25)

- Five-week study of atmosphere contaminants in a closed environmental system was concluded at USAF School of Aerospace Medicine, Brooks AFB, Tex. The three phases of the test: determining leak rate of the main test cell; checking the unmanned cabin; and checking the manned, fully operational chamber (2 weeks). In the third phase, four volunteer airmen lived inside the cabin in the strictly controlled experiment. They were not allowed to shave or wash; they subsisted on liquid nutritional compounds; they kept logs of diet consumed and waste volume and time; they occupied themselves only by operating psychomotor test panels or, for recreation, watching television. Study was conducted in cooperation with NASA. (AFSC AMD Release 65-125)

- Harold B. Finger, Manager of NASA–AEC Space Nuclear Propulsion Office, discussed nuclear-rocket technology in *Astronautics & Aeronautics:*

"The recent reactor test experience shows that nuclear rockets can be made available to furnish thrust at high specific impulse for many possible post-Apollo missions. Whatever direction the future space program may take—whether toward extensive manned lunar exploration, unmanned solar-system exploration, or manned planetary exploration—the performance advantages of nuclear rockets will be available, valuable, and, certainly for the latter mission, necessary. Through this program the country will have the options it must have in selecting future missions. For the manned planetary missions, which they can perform for all planetary opportunities, nuclear rockets offer such substantial spacecraft weight reductions, and associated savings in cost, that no less-efficient form of spacecraft propulsion could be seriously considered. . . ." (*A&A*, 6/65, 34–35)

During June: A decade of nuclear-rocket research at Los Alamos Scientific Laboratory was described in *Astronautics & Aeronautics* by Roderick W. Spence and Franklin P. Durham, both of LASL. The 10-yr. effort reached new levels of attainment in 1964-65, with successful tests of the Kiwi-B-4E and NRX-A2 reactors, and initial strides in the more advanced Phoebus reactor program. "The entire operations of both Kiwi-B-4E and NRX-A2 gave very close to the desired results and met or exceeded all of the test objectives.

". . . the past decade of experience has given us confidence that nuclear-rocket engines can be built and that they will prove to give good performance with high reliability." (*A&A*, 6/65, 42–46)

- W. Y. Jordan, Jr., R. J. Harris, and D. R. Saxon, all of MSFC, said in *Astronautics & Aeronautics* article that clustering of nuclear rocket systems up to 10,000 mw. of power had been studied. These studies indicated the concept showed sufficient promise to warrant more detailed design studies, and these studies had been initiated. They stated that a "modular nuclear vehicle system concept, which offers a flexible multipurpose space-transportation capability, now appears possible through development of only one basic propulsion system and vehicle stage. . . ." (*A&A*, 6/65, 48–52)

- A. O. Tischler, Director of Chemical Propulsion, Office of Advanced Research and Technology, NASA Hq., said in *Astronautics & Aeronautics* article that the time was rapidly approaching "when space will no longer be something we throw darts into, but rather an environment in which working propulsion systems maneuver and transport spacecraft payloads, eventually bringing them back to Earth for reuse. . . .

"Present space systems lag the launch vehicles in both development status and sophistication. But we can expect considerable performance improvement by the use of more-sophisticated space propulsion systems. Moreover, ground facilities that simulate the space environment are becoming available. With these to examine new concepts in depth, we can anticipate greater and more certain progress in spacecraft propulsion technology in the future. . . ." (*A&A*, 6/65, 60–62)

July 1965

July 1: NASA and the Brazilian Space Commission (CNAE) signed an agreement in São José dos Campos, Brazil, providing for a cooperative project for studying hemispheric weather patterns by launching meteorological sounding rockets from Brazil. Project would be part of the Inter-American Experimental Meteorological Sounding Rocket Network (EXAMETNET).

Under the terms of the agreement, Brazil would transport the rockets and equipment from the U.S. to Brazil, assemble and launch the rockets from Brazilian launch facilities, and provide meteorological data to other participants in the network. NASA would lend ground support equipment such as radar to CNAE; train Brazilian personnel in the handling of sounding rockets and in reducing meteorological data; and provide data obtained at NASA's Wallops Station, Va., to other participants in the network.

Agreement was similar to one signed by the U.S. and Argentina on May 18, 1965. The project entailed no exchange of funds between the two countries. (NASA Release 65-258)

- NASA and India's Dept. of Atomic Energy (DAE) signed a memorandum of understanding for a joint space research program to be conducted in India by NASA and the Indian National Commission for Space Research. Program would include two sounding rocket launchings to investigate upper atmosphere wind shears, turbulence, and diffusion by chemical release payloads; six launchings to investigate relationship between wind shears and sporadic E; two launchings to investigate the equatorial electrojet; and two launchings to measure electron and ion densities and other phenomena in the D region.

  DAE would make available the scientific payloads; range and range support facilities; and personnel for conducting the scientific experiments, for range support, and launching operations. NASA would furnish the sounding rockets, two test payloads, cameras, and nose cones as well as necessary training for DAE personnel at NASA centers. No exchange of funds was provided for. All scientific results of experiments would be made freely available to the world scientific community. (NASA Release 65-259)

- USN F-111B variable-sweep-wing fighter flew supersonically for the first time, reaching speed of mach 1.2 and 30,000-ft. altitude. During the 54-min. flight, the F-111B for the first time changed the angle of its wings from a virtually straight 16° takeoff configuration to a maximum sweep supersonic configuration of 72.5°. The flight took place at Calverton, N.Y. (Gen. Dynamics Corp. Release; *NYT*, 7/26/65, 38M)

July 1: XB-70A research bomber reached a speed of 1,870 mph and 68,000-ft. altitude during a one-hour, 44-min. flight from Edwards AFB. The 185-ft., 500,000-lb. aircraft, expected to reach its design maximums of 2,000 mph and 70,000 ft.-altitude on its next flight, flew over Arizona, Nevada, Utah, Idaho, Wyoming, and California. (AP, Balt. *Sun,* 7/2/65)

* Astronauts Frank Borman (Maj., USAF) and James A. Lowell, Jr. (LCdr., USN), had been assigned as the prime flight crew for the Gemini 7 mission scheduled for the first quarter of 1966, NASA announced. Backup crew for the flight, which would last up to 14 days, would be Astronauts Edward H. White, II (Lt. Col., USAF) and Michael Collins (Maj., USAF). Borman and Lovell had been the backup crew and White the pilot for GEMINI IV. (NASA Release 65-218)
* General Electric Co. told a news conference in New York that it had successfully converted a space age component—SCR—to a device for home use that could dial speed or heat like tuning in a radio. A silicon-controlled rectifier, SCR had one of the highest power-amplification capacities of any semiconductor, no moving parts, and could control kilowatts in thousandths of a second. The unit would cost 35 to 50 cents for GE customers. GE did not comment on what GE appliances would initially incorporate SCR. (Smith, *NYT,* 7/2/65, 37)
* NASA announced changes in the names of NASA organizations: from Mississippi Test Operations to Mississippi Test Facility; from Michoud Operations to Michoud Assembly Facility; from MSC White Sands Operations to White Sands Test Facility. (MSFC Release 65-107; NASA Hq. Bull. 1-65, No. 13)
* NASA selected Federal Electric Corp., a subsidiary of ITT, for negotiation of an award-fee contract to provide logistical and technical information support services to NASA Manned Spacecraft Center. Contract would be for one year with renewal provisions for two additional years. Estimated cost for the first year was $1.5 million. (NASA Release 65-217)
* Over 50 security guards at NASA Goddard Space Flight Center—members of the International Union of United Plant Guard Workers of America—went on strike over wage and contract issues with their employer Wackenhut Services, Inc. Wackenhut, operating at GSFC under contract, provided additional guards; there was no breakdown in security.

  Some 250 guards at NASA test sites near Las Vegas, Nev., also went on strike against Wackenhut. They were members of a separate local union. (*Wash. Post,* 7/2/65, A3; DJNS, Balt. *Sun,* 7/2/65; Wash. *Eve. Star,* 7/2/65)
* Continuation of Weather Bureau-Navy Project Stormfury during the 1965 hurricane season was announced. Hurricanes and cumulus clouds would be seeded to investigate the feasibility of modifying clouds and tropical storms. The hurricane research program began in 1961, included experiments on hurricanes Esther (1961) and Beulah (1963) and on tropical clouds (1963). Objectives for 1965 were to intensify the hurricane-seeding experiments and to begin new experiments on hurricane rainbands. (Commerce Dept. Release WB 65-100)
* NASA Goddard Space Flight Center announced selection of six firms for contract negotiations to provide nonpersonal scientific and engineering

support services for GSFC. Final negotiations were expected to result in cost-plus-award-fee contracts for two years at a total cost of about $16 million. Contracts would cover requirements for ten divisions at GSFC; although requirements of a single division would be fulfilled through a single contract, one contractor might service more than one division in several cases. Firms selected: Fairchild Hiller Corp.; Vitro Corp. of America; Electro-Mechanical Research, Inc.; Consultants and Designers, Inc.; Lockheed Electronics Co.; and Melpar, Inc. (NASA Release 65-216; GSFC Release 15-65)

July 1: Gen. William F. McKee (USAF, Ret.) was sworn in as Federal Aviation Agency Administrator—less than 24 hrs. after confirmation by the Senate—in a White House ceremony. President Johnson, presiding, announced that General McKee's primary task would be development of a supersonic transport (Sst). Mr. Johnson had approved five recommendations of the President's Advisory Committee on Supersonic Transport: "Those five recommendations . . . are: first, the next phase of design covering an 18-month period beginning about August 1, 1965; second, the four manufacturers—Boeing Company, Lockheed Aircraft Company, General Electric Company, Pratt & Whitney Division of United Aircraft—be invited to continue this phase of the program; third, the FAA Administrator be authorized to enter into contracts with the airframe manufacturers to undertake detailed airframe design work and test them over the next 18 months; fourth, the FAA Administrator be authorized to enter into contracts with the engine manufacturers to construct and test over the next 18 months demonstrator engines to prove the basic features of the engines; and five, and finally—and very importantly—that the Congress be requested to appropriate—Senator Magnuson—the necessary funds to initiate the next phase of the program. And for this purpose I shall request an appropriation of $140 million."

Ira C. Eaker commented in the *San Antonio Express:* "When McKee succeeds in his SST mission, as I believe he will, any citizen will be able to buy a $50 ticket for a flight from Washington to Los Angeles in 55 minutes, or purchase for $55 air passage from New York to Paris in one hour. Of course, there will be cynics who will ask: 'Who wants to fly to Paris in an hour?' The answer is: Most of the people who want to go to Paris.

"There is one predictable human trait. People will always go for the fastest transport. Not long after man got on a train, he took his freight out of the covered wagons and off the oxcarts forever. When people took up with automobiles, they soon put their produce, their pigs and their poultry on trucks.

"There is also some algebraic or arithmetic relation between the speed of travel and the number of people who travel. About 10 times as many people fly the Atlantic now in five hours as crossed when the trip took five days. No doubt 10 times more people will cross the ocean when they can do it in an hour." (White House Release; Eaker, *San Antonio Express,* 7/8/65, 14)

• Sen. A. Willis Robertson (D-Va.) issued a statement opposing President Johnson's plan for a supersonic commercial airliner:

"I attended the conference at the White House at which the secretary of defense explained his plans for the development of a supersonic airplane that would be the fastest of any in the world, making mach 2.5 to 2.7, or about 1,800 miles per hour. The cost of developing this plane will be between 2 and 2.5 billion dollars, and when developed, the planes will probably cost 20 to 25 million each.

". . . if we succeed in the development of this plane any airline that agrees to use them will have to be largely subsidized both for the purchase of the plane, which will have a limited life, and likewise for its operation. Consequently, I feel towards this project like I do about the shot to the moon—a fine advertising scheme, but not worth what it is going to cost the taxpayers." (Text)

July 1: USAF launched an Atlas D ICBM from Vandenberg AFB in test of the Army's Nike-Zeus antimissile missile. (UPI, *NYT*, 7/2/65)

- Harvey Brooks, Dean of the Division of Engineering and Applied Physics, Harvard Univ., succeeded George B. Kistiakowsky as Chairman of the National Academy of Sciences' Committee on Science and Public Policy. (NAS-NRC *News Report*, 7/8/65, 1)
- American Institute of Physics established a Center for History and Philosophy of Physics, which was the merging of AIP's Niels Bohr Library of the History of Physics, History of Physics Archives, and Project on the History of Recent Physics in the United States. (AIP Release)

July 2: TIROS X meteorological satellite was launched from Eastern Test range by a Thrust-Augmented Delta booster into a near-perfect sun-synchronous orbit from which it would photograph tropical storm breeding areas. Orbital parameters: apogee, 517 mi. (832 km.); perigee, 458 mi. (737 km.); period, 100.6 min.; inclination to the equator, 81.4°. The spacecraft at first was spinning too fast for picture-taking because of inability to command activation of an automatic slowdown device. A more intense ground signal sent during the second orbit activated the device, slowing the satellite's spin rate from 138 rpm to about 13 rpm.

TIROS X would photograph the hurricane-breeding area between 70° north and 30° south latitude, about 60 to 80 percent of the earth. Its two vidicon cameras would take more than 400 pictures daily. TIROS IX in its "cartwheel" movement in orbit had been able to provide about 100 percent coverage of the earth daily.

The weather observation spacecraft was the 10th successful Tiros (Television Infrared Observation Satellite) to be launched by NASA in as many attempts. TIROS X was funded by the Weather Bureau, while the previous nine were NASA research and development spacecraft. It was one of three satellites purchased by Weather Bureau in 1963 "to be used to assure continuity of satellite observations for operational purposes." (NASA Release 65–229; NASA Release 65–197; AP, Wash. *Eve. Star*, 7/2/65, A3; KSC *Spaceport News*, 7/8/65, 4)

- PEGASUS II meteoroid detection satellite, orbited May 25 from Kennedy Space Center, had reported "hits" on panels of all three thicknesses of aluminum target material, NASA reported. 14 penetrations had been recorded on the .016-in.-thick detection panels; five on the .008-in. panels; and 34 on the .0015-in. panels.

In the first month report on PEGASUS II, Milton B. Ames, Jr., Director of NASA Space Vehicle Research and Technology, said that although the number of penetrations varied slightly from earlier predictions, "the figures follow closely our statistical computations based on results from PEGASUS I and EXPLORERS XVI and XXIII. Continued successful operation of PEGASUS II will give us a good base for use by spacecraft designers in predicting the frequency and size of meteoroids which constitute a hazard to space flight." Ames referred to a new capacitor fusing arrangement which would enable project engineers to disconnect a single malfunctioning capacitor while leaving other capacitors in the same group of panels working. If a malfunction serious enough to warrant disconnection of the entire panel group occurred, this could be done by ground command. 36 capacitors on PEGASUS II were working improperly during the first four weeks and had been disconnected by ground command to prevent a drain on the spacecraft's power supply. (NASA Release 65–219; MSFC Release 65–168)

July 2: NASA's MARINER IV spacecraft was in its 216th day of flight and had only 12 more days before it would pass within 5,700 mi. of the planet Mars. MARINER IV would take and record up to 21 pictures of Mars on July 14 for playback to earth. (NASA Release 65–215)

- U.S.S.R.'s COSMOS LXX artificial earth satellite was successfully launched containing scientific equipment "for continuing the outer space research in conformity with the program announced by Tass on March 16th, 1962," a radio system for precise measurement of the orbital elements, and a radio telemetry system for sending data to earth. Orbital parameters: apogee, 1,154 km. (716.7 mi.); perigee, 229 km. (142 mi.); period, 98.3 min.; inclination to the equator, 48.8°. Equipment was functioning normally.

  Bochum Observatory (W. Germany) had reported picking up signals different from those of the Cosmos series. Heinz Kaminski, head of the Institute for Satellite and Space Research at Bochum, West Germany, said he tracked two Soviet space satellites in orbit although the U.S.S.R. announced the launching of only one: "Apparently the Russians are preparing for a space rendezvous—perhaps between these two satellites or between a manned space ship and a satellite in parking orbit." No confirmation of the Bochum report was available from other sources. (Tass, *Krasnaya Zvezda*, 7/3/65, 1, ATSS–T Trans.; AP, Wash. *Eve. Star*, 7/2/65, 1; UPI, *Cocoa Trib.*, 7/2/65)

- NASA and the Univ. of Hawaii had signed a cost-sharing contract for design and installation of an 84-in. telescope at a site to be selected in Hawaii, NASA announced. An initial sum of $475,000 had been awarded to the University to initiate development and construction of the intermediate-size planetary telescope with fused-quartz optical system and coudé spectrograph. The University would assume costs for buildings, utilities, and supporting services. (NASA Release 65–220; *Marshall Star*, 7/14/65, 9)

- A $3,200,000 contract had been let to the E. A. Hathaway Co. for construction of the new 93,000-sq.-ft. Advanced Flight Simulation Laboratory at NASA Ames Research Center, ARC announced. It would contain the most advanced aircraft simulator known (for supersonic transport studies); the most powerful centrifuge yet built (up to 50 times the force of gravity); a mid-course navigation facility where stars and planets could be projected just as they would be seen by astronauts in spaceflight; and a "virtually frictionless" satellite attitude control facility. Total values of these facilities would be $10,068,000. (ARC Release 65-16)

July 2: FAA announced that in accordance with President Johnson's decision to move ahead with development of the U.S. supersonic transport, work under Sst design contracts would be accelerated during the month of July. Contracts were with the Boeing Co. and Lockheed Aircraft Corp. for the airframe and with General Electric Co. and the Pratt & Whitney Div. of United Aircraft Corp. for the engine. Dollar total of the contracts in July, including a Government cost-share of 75% and a contractor cost-share of 25%, would be $5,670,000—an increase of $2 million above the monthly level of funding for the four design contractors during the period June 1, 1964, through June 30, 1965. (FAA Release 65-56)

- NASA had awarded an $8,150,833 contract modification to General Electric Co. for supporting the Computation Laboratory at NASA Marshall Space Flight Center. (MSFC Release 65-169)
- Blount Brothers Corp. had been awarded a $6,745,000 fixed-price contract for modification of Saturn IB Launch Complex 37 at NASA Kennedy Space Center, DOD announced. (DOD Release 444-65)
- Five Japanese-made research rockets had been shipped to Indonesia, Reuters reported. The Japanese Trade Ministry reportedly said it authorized export of the rockets for "purely cosmic observation use." The rockets were said to have the capability for use as armed missiles. (Reuters, *Wash. Post,* 7/2/65)
- Statistics on intercontinental ballistic missiles were given by Robert R. Brunn in an article in the *Christian Science Monitor:* "In the Soviet Union 270 to 300 intercontinental ballistic missiles (ICBMs) are in place. Underground and overseas the United States has more than 1,300.

    "Within five years the Soviets may have emplaced 500 to 600 of these great birds with nuclear warheads. But by then the American missile arsenal may have the destructive power of 3,000 1965-type ICBMs."

    A multiple warhead for the Minuteman had been developed, Brunn said, which could triple the effectiveness of this missile: "Approaching enemy territory, it could separate into three warheads, each with its own guidance system. It would be steered to the target and report if it was going to miss or hit." (Brunn, *CSM,* 7/2/65, 9)

July 3: U.S. Bureau of Mines said its scientists and engineers had begun research on how to tap the mineral resources of the solar system. This attempt to develop "extraterrestrial mining techniques" was being made at the Bureau's research center in Minneapolis for NASA. The idea was not to mine the moon and planets for materials usable on earth but rather to develop means of exploiting celestial bodies for resources explorers could use to build bases or travel farther into space. The Bureau said that, because of its nearness, the moon "is likely to be the site of the first extraterrestrial mine." (Bur. Mines Release, 7/4/65; UPI, *NYT,* 7/4/65, 27)

- Gen. Charles P. Cabell (USAF, Ret.) was sworn in as consultant on organization and management development activities to NASA Administrator James E. Webb. Military assignments during his career included that of directing the Joint Staff, Joint Chiefs of Staff. He was deputy director of the Central Intelligence Agency from 1953 to 1962. (NASA Release 65-226)

July 3: Dr. N. M. Sissakian of the Soviet Academy of Sciences said in an article in *Krasnaya Zvezda* that the attention of Soviet biomedical researchers had been increasingly drawn to the complex effects of combinations of spaceflight factors. They were interested in the stability of cells and organisms to very low temperatures, and the effect of artificial cooling of animals on their ability to withstand oxygen deprivation, acceleration, radiation, and other factors. The ability of algae and lower animals to survive a vacuum was of interest to exobiologists, as was the ability of certain chemical compounds to screen out harmful effects of ultraviolet irradiation. As part of the continuing effort to simulate conditions existing on other planets and study their effects, soil infusoria were found able to adapt to the wide temperature range in a chamber simulating Martian atmosphere. (*Krasnaya Zvezda*, 7/3/65, 6)

- Leonid I. Brezhnev, First Secretary of the Soviet Communist Party, warned that Soviet nuclear missile strength was greater than Western intelligence estimates suggested. Speaking to graduates of the Soviet military academies during a Kremlin ceremony, Mr. Brezhnev said that the quantity of intercontinental and orbital missiles at Soviet disposal was "quite sufficient to finish off once and for all any aggressor or group of aggressors."

  Recent American intelligence reports had indicated U.S.S.R. had about 270 ICBM's, most of them slow-firing, liquid-fuel weapons that were unwieldy compared with Minuteman. Reference to orbital rockets had been made on at least two occasions in recent months: Moscow television made the claim in May; a June issue of the magazine *Ogonek* referred to them. Neither gave details. (Grose, *NYT*, 7/4/65, 1, 2; UPI, Shapiro, *Wash. Post*, 7/4/65)

July 4: Six out of ten Minnesotans said in a recent survey conducted by the *Minneapolis Tribune's* Minnesota Poll that the U.S. should at least maintain its $5 billion a year budget for space exploration. 59% felt the U.S. had outdistanced the U.S.S.R. 10% considered the "space race" about even. (*Minn. Trib.*, 7/4/65)

- Status report on MOLNIYA I comsat, launched by the Soviet Union April 23, was given to Tass by I. P. Petrushkin of the U.S.S.R. Ministry of Communication: "Tests have demonstrated that the combinations of the terrestrial and onboard communication equipment of the 'Molniya-1' satellite assure the possibility of the reliable transmission of black and white TV. Color TV sessions yielded encouraging results.

  "On Sundays, via the 'Molniya-1,' a regular exchange of TV programs is being conducted between Moscow and the Pacific coast. During May and June, tests were also run on the quality of the multichannel telephone system. The systems were simultaneously loaded with phone conversations, tonal telegraphy, phototelegraphy, and radio broadcasting programs. Even under such an 'optimal' regime, the communication channels function fairly reliably.

  "In June, there was put into operation the commercial long-distance phone link between Moscow and Vladivostok, functioning via the 'Molniya-1.'" (Tass, *Krasnaya Zvezda*, 7/4/65, 4, ATSS-T Trans.)

July 4: Robert F. Thompson, named mission director for NASA's Gemini program on June 25, had decided not to accept the job, UPI reported. Thompson, who had decided "for personal reasons" to turn down the assignment, would stay on instead as head of the landing and recovery division at NASA Manned Spacecraft Center. Christopher C. Kraft would temporarily serve as mission director and flight director for the Gemini program. (UPI, *Houston Chron.*, 7/4/65)

- In a newspaper interview, Soviet Cosmonaut Yuri Gagarin expressed surprise that the U.S. had shown primarily military aviation equipment at the Paris International Air Show. He said he regretted he could not meet longer with American astronauts McDivitt and White. (Tass, *Zarya Vostoka*, 7/4/65, 3)
- The Soviet Union's apparent effort to penetrate the world commercial aircraft market was reported by Evert Clark in the *New York Times*. He said the U.S.S.R. had formed a new organization called Aviaexport to direct commercial sales. It was expected to pick satellite countries as its first target, then so-called "dependent" nations, and eventually the Western countries. Showing of the Antonov 22 700-passenger aircraft and a model of a supersonic airliner scheduled to begin test flights in 1968 had enabled the Soviet Union to dominate the Paris International Air Show (June 11–20) until the arrival of Vice President Humphrey and Astronauts White and McDivitt. (Clark, *NYT*, 7/5/65, 30)

During week of July 4: Ambassador Tran Thien Khiem, South Viet Nam, visited NASA Langley Research Center. (*Langley Researcher*, 7/16/65, 8)

July 5: British Minister of Aviation Roy Jenkins said at a London conference on European cooperation in advanced technology that Britain might seek collaboration with France in developing the VC–10 aircraft into an "air bus" capable of carrying 265 passengers. This move was thought to be precipitated by activities of two American companies, Boeing and Douglas, both of which had approached BOAC with their versions of an air bus. (Farnsworth, *NYT*, 7/6/65, 54M)

- Seven French girls ended a 15-day isolation experiment in a cave 360 ft. below the surface. They were apparently in good condition but said they had lost their sense of time within the first 24 hrs. The experiment ended nearly a day before it should have according to their reckoning. (Reuters, *NYT*, 7/6/65, 24; UPI, *N.Y. Her. Trib.*, 7/6/65)

July 6: ComSatCorp, rejecting petitions by seven communications carriers to reduce its authority, strongly supported FCC's decision giving ComSatCorp "sole responsibility for the design, construction, and operation" of three ground stations to support the Corporation's global satellite system for commercial communications.

Answering the charge that station ownership and operation should be assumed by the carriers on a competitive basis, ComSatCorp said the operational date of the global system would be needlessly delayed if ownership of stations were decided on a case-by-case basis.

To the complaint that the FCC decision gave ComSatCorp control of "terrestrial facilities" for traffic-processing between interface points and the stations themselves, ComSatCorp replied that it would "look first to the carriers to provide" the facilities. Further, "should ComSatCorp determine, in any particular situation, to propose construction of its own communications links, the soundness of any such proposal would, of course, be subject to scrutiny by the Commission with full consideration of the views of all interested parties."

Against the allegation that ComSatCorp-constructed terrestrial facilities would introduce costly "backhauls," ComSatCorp said backhauling was "routine and unavoidable" throughout the communications industry. (ComSatCorp Release)

July 6: Existing contracts with the Boeing Co. and the Bendix Corp. had been extended to incorporate modification of the Apollo Lunar Excursion Module (Lem) as a shelter for use with a Local Scientific Survey Module (Lssm) for astronaut surface mobility, MSFC announced. Boeing would receive $565,000 and Bendix $570,000 for the new work. Both contracts would run for about nine months. The two firms had completed individual studies of a possible lunar mobile laboratory (Molab) under the original terms of the contracts. The Molab would be much larger than the Lssm now being studied. (MSFC Release 65-173)

- Lockheed Aircraft Corp. pronounced successful the first flight of its new helicopter, the 286. The Corp. said the five-place aircraft, designed to travel at 174 mph, was two weeks ahead of schedule, with FAA certification expected later in 1965.

  Lockheed hoped the utility helicopter would find a wide range of use in transport, rescue work, and various military missions, including antisubmarine warfare. (*WSJ*, 7/6/65, 2)

- Thiokol Chemical Corp.'s Reaction Motors Div. was awarded a $10,600,000 Navy contract to continue production of packaged liquid rocket engines for the Navy and Air Force Bull Pup missile. (Thiokol Release; *WSJ*, 7/6/65)

- Rep. Burt L. Talcott (R–Calif.) inserted in the *Congressional Record* a letter from E. J. Stecker, president of Holex, Inc., and an exhibitor at the Paris International Air Show (June 11–20): "At Paris the Russians made us look like idiots and we cooperated so beautifully that it almost looked as though our public relations program was being directed by the Kremlin. . . .

  "I walked through the aircraft park where the Russian and American exhibits were practically side by side. The Russian aircraft were exclusively commercial, the American almost 100 percent military. Think of the irony of the situation. . . .

  "There was a long line waiting to enter the . . . space exhibit of the U.S.S.R. . . . inside was a full scale model of the Vostok space capsule and its rocket motor suspended from the ceiling, what appeared to be an operational Vostok in a glass case and Mr. Gagarin, Russia's first man in space, who shook my hand . . . and gave me a Vostok lapel pin. This was easily the most popular exhibit at the show and I would estimate that 70 percent of the attendees had Vostok pins and were wearing them. . . .

  "The Russians then flew in their great misshapen 750-passenger transport aircraft which really impressed the public. I examined it and as a pilot and engineer of many years standing, I feel it economically and militarily foolish, but the publicity value was tremendous.

"Then there was the announcement of the proposed U.S.S.R. SST transport which looks like a retouched Concorde. This also hit all of the front pages.

"In short, the Russians stole the show with an obsolete space capsule much inferior to our Mercury, an artist's sketch of a supersonic transport and an overgrown, awkward monstrosity of an airplane. But they could do it because everything they did was aimed at and shown to the general public. . . .

"When White, McDivitt, and the Vice President finally arrived, it was a triumphant tour surrounded by Secret Service, press, public relations and photographers and the great mass of the unwashed public including the exhibitors were generally ignored and forgotten. . . .

"There were a few good points. James Webb, Administrator of the NASA, visited the U.S. exhibitors. He came unheralded and alone and had time to stop and talk for a few minutes without the aid of 50 photographers. In this, Jim was unique among the U.S. officials and should be commended. The USAF Thunderbirds and the U.S. Navy Blue Angels put on fantastic flying exhibitions on Thursday evening. But it occurred around 7:30 p.m. when most of the public had departed.

"The point is this: The Russians appealed to the people; we ignored the people and appealed to the press. As a result, the Russians only lost the front pages once and that was when we made them through the unfortunate crash of our 2-58." (CR, 7/6/65, A3555–6)

July 6: President Johnson's ordering 18 mo. of additional research on the supersonic airliner (Sst) program was assessed as a calculated risk by Robert J. Serling in *Washington Post.* The President was betting that the extra year and a half of development work would produce a plane so far superior to the British-French Concorde that the Concorde's far earlier introduction would mean little in terms of sales, he said.

Serling pointed out that while advance orders for a purely paper Sst had outnumbered those of the under-construction Concorde by more than 2 to 1, the U.S. lead was built on the airlines' belief that (1) U.S.'s plane would be far superior and (2) it would not come along too far behind the Concorde. However, present plans were for flight-testing the Concorde in 1968; the Soviet Union had announced it would have a supersonic airliner operational even before the Concorde. An inservice U.S. Sst was not likely before 1974 or 1975. (Serling, *Wash. Post,* 7/6/65, A8)

July 7: Some Presidential advisers who had once recoiled at the idea of a supersonic race were now wondering how long this country could delay without losing the bulk of the world market, wrote Evert Clark in the *New York Times:* "For a variety of reasons, most of the controversy over timetables has been kept out of the public forum, mainly because of Secretary of Defense Robert S. McNamara, the chief Presidential advisor on the plane . . . Mr. McNamara swore the Government officials and private citizens on the advisory committee to secrecy at their first meeting, and that secrecy has been maintained." (Clark, *NYT,* 7/7/65, 21)

* NASA's MARINER IV began feeling the gravitational pull of Mars a week before it was scheduled to take the first close-up pictures of the

planet. The tug was noticed at 5 p.m. EDT in a slight change in speed as the 575-lb. spacecraft, 128,054,720 mi. from earth and 1,721,770 mi. from Mars, neared the end of its 228-day trip.

At noon, the speed relative to Mars was 9,879 mph. Tracking engineers said the speed, which had been dropping two mph every six hours, would lessen because of Mars' gravity at a rate of one mph every six hours through July 10 when the speed would begin to increase. No further sensing of the planet's presence in space was expected until July 14 when instruments aboard MARINER IV might detect an increase in radiation. (AP, *NYT*, 7/8/65, 13; AP, *Orl. Sent.*, 7/7/65)

July 7: Paul Haney, Public Affairs Officer at NASA Manned Spacecraft Center, said no extravehicular activity (Eva) was planned for the eight-day Gemini V mission scheduled for August 17 or for October's two-day Gemini VI flight.

Astronauts Frank Borman (Maj., USAF) and James Lovell (LCdr., USN), primary crew for Gemini VII, said they would attempt to communicate with a ground station via laser beam during their 14-day mission planned for early 1966. The laser—a beam of intense light—can transmit voices or other messages. Ground station at White Sands, N. Mex., aiming by radar, would fire the first laser signal at Gemini 7 as it passed overhead. The Gemini 7 crew would attempt to answer by pointing their capsule downward, aiming, and shooting a 10-lb., hand-held transmitter toward the ground. (Transcript; AP, Balt. *Sun*, 7/7/65; UPI, *Wash. Post*, 7/7/65; UPI, *NYT*, 7/7/65, 20)

• The radar set that would guide Gemini astronauts on rendezvous missions in space, beginning with the August 19 Gemini V flight, was demonstrated by Westinghouse Corp.—working under an $18 million contract with NASA—at Friendship International Airport, Baltimore. Astronauts L. Gordon Cooper (Maj., USAF) and Charles Conrad (LCdr., USN) would carry a self-contained Rendezvous Evaluation Pod (Rep) into orbit on the back end of their spacecraft. An Agena rendezvous radar transponder and flashing beacons would be packaged in the Rep along with batteries and antenna. Midway through the second revolution, at 2 hrs. 25 min. after liftoff, command pilot Cooper would yaw the Gemini 5 spacecraft 90° to the right. Explosive charges would eject the Rep from its canister northward at 5 fps. Gemini 5 would then maneuver away from the Rep to attain a position six miles below and 14 mi. behind the Rep. Subsequent phase adjustment would place Gemini 5 in a co-elliptical orbit—that is, the spacecraft would be at constant altitude below the Rep but reducing the trailing distance, since the spacecraft in its lowest orbit would be traveling faster than the Rep.

Range and range-rate data would be displayed to the Gemini 5 crew by the rendezvous radar system. The radar system would continuously compute distance and angles from the spacecraft to the Pod, and calculate the maneuvers necessary to effect rendezvous.

At five hours and 36 min. after liftoff, if the mission went as planned, Gemini 5 should be closing in on the Rep just north of the Carnarvon, Australia, tracking station. Gemini 5 would not physically dock with Rep; the experiment would simply provide training for

Gemini VI and other rendezvous missions and evaluate the rendezvous radar hardware to be used on the Agena target vehicle in later missions. (Clark, *NYT*, 7/8/65, 12; Hines, Wash. *Eve. Star*, 7/8/65; MSC *Roundup*, 7/23/65, 1, 2)

July 7: The new Magnetic Field Components Test Facility at NASA Goddard Space Flight Center—which would create the precisely-controlled magnetic environment necessary for testing and calibrating spacecraft instruments intended to measure the low magnetic fields in outer space—became operational. The facility was also equipped to demagnetize the spacecraft carrying the magnetic measuring instruments. (GSFC Release G-16-65)

- Leo D. Welch, the first Chairman and Chief Executive Officer of ComSatCorp, announced his intention to retire from active direction of the Corporation. Mr. Welch said he had privately informed the ComSatCorp Board of Directors on May 21 that he wished to slow down the pace of his activity. (ComSatCorp Release)

- Tass released a picture of what was described as a working model of an orbiting space station. Six hermetically sealed compartments branched out from a central stem. The compartments included a control desk, a laboratory, a garden, an orientation system, radar section, and a heliostation with a system for carrying on conversation with incoming spaceships. (Tass, AP, *Wash. Post*, 7/8/65, D5)

- M. I. Kiselev and E. B. Galitskaya had worked out a method of controlling spaceships by means of solar pressure. They proposed a system of reflecting black and white blades which would work like a propeller or a "solar rotary mill." By regulating the inclination of the blades, one could obtain a torsional moment and change its direction and deceleration. Under conditions of weightlessness, the large size of the blades needed to provide the required torsional moment did not present any difficulties, and the blades would eliminate the need for additional energy sources for maneuvering in space. (*Kazakhstanskaya Pravda*, 7/7/65, 3

- Photos of the rim of the terrestrial globe and of the twilight aureole taken by VOSTOK VI revealed two turbid layers in the stratosphere, Prof. Georgiy Rozenberg was reported in *National Zeitung* as saying at a space physics conference. The two layers contained relatively dense water-and-ice-covered particles at heights of 11,500 and 19,500 m. (37,950 and 64,350 ft.). Prof. Rozenberg assumed that sulphuric oxide emitted by volcanoes played an important role in the formation of the layers. He concluded that the colored luminescense following major volcanic outbreaks was related to this phenomenon. (*National Zeitung*, 7/7/65, 6)

- Consensus of W. German observers at the International Air Show (June 11-20), as reported in *Der Spiegel*, was that the Soviet aircraft were obvious copies of Western models, but with inferior workmanship (often hand-made details) and instrumentation: the Ilyushin-62 was a poor copy of the British VC-10; the Tupolev-134 was a hybird of the French Caravelle and British BAC 111; the Antonov-24B was obviously copied from the Dutch "Friendship" and the British "Handley Page Herald"; the navigational instrument used on the Tu-134 was the American World War II Eyeball Mark One; the design model of the

Tu–144 so resembled the Concorde that it was humorously referred to as the Concordovich. (Heumann, *Der Spiegel*, 7/7/65, 86–87)

July 8: Transfer of control of SYNCOM II and SYNCOM III communications satellites from NASA to DOD was completed. Under direction of the Defense Communications Agency, the three telemetry and command stations for maintaining precise control and positioning of the satellites would be operated by USAF at locations in the Seychelles Islands, Hawaii, and Guam.

Army would be responsible for earth communications facilities used with the Syncom satellites except for two shipboard terminals owned and operated by USN. Army's Strategic Communications Command would continue to man and operate all ground terminals.

SYNCOM II—launched by NASA July 26, 1963—would be maintained at a position between 60° and 80° east longitude; SYNCOM III— launched by NASA Aug. 19, 1964—would be positioned between 170° and 174° east longitude. Both satellites were in orbit at 22,300-mi. altitude. NASA would continue to receive reports on the telemetry from the two satellites from DOD and would continuously evaluate their performance in space. (DOD Release 451–65)

- A new F–1 engine test stand was used for the first time at NASA Marshall Space Flight Center's West Test Area. The 10-sec. initial firing of the 1.5 million-lb.-thrust engine was primarily for checking out the new facility.

  On another test stand at MSFC, Chrysler Corp. fired the second Saturn IB booster, manufactured by Chrysler at MSFC's Michoud Assembly Facility. The test, scheduled to run for 30 sec., was terminated automatically after three seconds because of a faulty signal from an engine pressure switch. (MSFC Release 65–178; *Marshall Star*, 7/14/65, 1)

- X–15 No. 2 piloted by NASA research pilot John B. McKay attained a speed up to 3,659 mph (mach 5.19) and an altitude of 212,600 ft., photographed Gamma Cassiopeia with four 35-mm. cameras. Purpose was to verify theoretical data on the physical composition of the stars. (NASA X–15 Proj. Off.; *X–15 Flight Log*)

- NASA announced Pegasus C meteoroid detection satellite was equipped with small aluminum sub-panels that at some future date an astronaut could, if desired, detach and bring back to earth. The panels would provide the first actual samples of meteoroid impact and would have tested some 43 types of thermal coatings.

  Pegasus C would be launched July 30, 1965, into 332 mi.-altitude circular orbit at 28.9° inclination—close to a nominal manned flight path—rather than into an elliptical orbit like that of PEGASUS I and II. (NASA Release 65–228; MSFC Release 65–175)

- NASA had awarded Brown Engineering Co. a $3,630,000 contract for building nine discrete control equipment systems for use with Saturn V launch vehicle. Two of the systems would be installed in a Saturn V systems development facility at NASA Marshall Space Flight Center in Huntsville. The other seven would be delivered to Kennedy Space Center, NASA, Launch Complex 39: three systems would be placed in launch control centers; four would be installed on Saturn V mobile launchers. (MSFC Release 65–176)

July 8: Referring to a French rocket launching site, comparable to Kennedy Space Center, that would be operating in French Guiana by 1968, Pierre J. Huss said in the *New York Journal American:* "French experts claim the Guiana location makes it possible to fire toward the east at an angle that makes use of the earth's rotation speed. They say it also offers optimal conditions for launching vehicles to the moon or Mars." (Huss, *N. Y. J/Amer.*, 7/8/65, 9)

- MSFC Director Dr. Wernher von Braun, speaking before the International Christian Leadership World Conference in Seattle, said the two dominant forces shaping the course of human events in our revolutionary age were science and religion. And, Dr. von Braun said, "it is depressing to witness a growing misconception that these two powerful forces are not compatible." On the contrary: science was trying to harness the forces of nature around man, while through religion man sought to control the forces of nature within.

  Dr. von Braun said science and scientists had been blamed for the desperate dilemma today, because science had utterly failed to provide a practical answer on how to handle the powerful forces it had unleashed. He said the blame for the wrongful use of force could not be pinned on science: "Science, all by itself, has no moral dimension. The same drug that heals when taken in moderation will kill when taken in excess. Only when a society accepts and applies a scientific advance do we add a moral dimension to it." (*Marshall Star*, 7/14/65, 2)

- Gen. Bernard A. Schriever, AFSC Commander, predicted in a luncheon address to participants in the National Youth Science Camp "a revolution in aeronautics within the next 15 years" as a result of advances being made now.

  Breakthroughs being made in development of more powerful rockets, heat resistant metals, and more sophisticated engines would soon lead to vertical takeoff craft capable of undergoing sustained flight at speeds up to 12 times the speed of sound, General Schriever postulated. (Text; Wash. *Eve. Star*, 7/9/65)

- ComSatCorp requested 26 design-engineering companies to submit proposals by July 20 for architectural and engineering services for construction of a ground station site at Brewster Flat, Wash. The proposed station would provide communications services to Hawaii and nations of the Pacific as part of a global commercial satellite system. The RFP's asked for plans sufficiently detailed to enable interested contractors to bid on construction and to enable the station to be operational by September 1, 1966. (ComSatCorp Release)

- Sen. Strom Thurmond (R–S.C.) expressed concern on the floor of the Senate about advances in Soviet strategic weaponry and inserted in the *Congressional Record* an article which compared Soviet and American development of large solid rocket motors:

  "First, the development of rockets has received a high priority continuously in the Soviet Union for 20 years, since the end of World War II.

  "Second, the Soviet effort has been broadly based and produced several generations of vehicles with increasing performance capability.

"Third, solid-propellant rocketry is playing a major role in current Soviet operations and in their future plans.

"Fourth, U.S. intelligence has suffered a major failure if the Soviet missiles in the May 9 parade were not fake.

"An early review of U.S. intelligence and military planning operations by both the Congress and the administration definitely is in order if the Soviets are operating large solid rockets." (CR, 7/8/65, 15359-61)

July 8: Soviet astrophysicist Rolan Kiladze had introduced a new theory that every planet's rotation was caused initially by powerful bombardments by clusters of particles in its path. (*Sovetskaya Latviya*, 7/8/65, 14)

- Paul Mantz, veteran pilot who owned and flew a large collection of rebuilt vintage airplanes, was killed when a home-built aircraft he was flying for a film sequence crashed near Yuma, Arizona. (*Arizona Republic*, 7/9/65)
- Prof. Wolfgang Pilz, leader of a team of West German scientists helping the United Arab Republic build rockets, had quit the project and returned to West Germany, it was reported. (Smith, *NYT*, 7/9/65, 7)

July 9: Extraterrestrial dust particles would be collected by U.S. and foreign scientists using five rockets launched into the extreme upper atmosphere from Churchill Research Range, Canada, by the Air Force Cambridge Research Laboratories (AFCRL), NASA and USAF announced jointly. Each of the five Aerobee 150 rockets would carry a recoverable device called a "Venus Flytrap" designed to capture micrometeoroids and noctilucent cloud particles at various altitudes for laboratory inspection.

NASA had provided partial funding for the project and had arranged for participation of organizations from abroad: the Commonwealth Scientific and Industrial Research Organization, Radio Physics Laboratory, Sydney, Australia; the National Center for Radioactivity Research (CNRS), France; the Meteorological Institute, Stockholm Univ., Sweden; Max Planck Institute, Heidelberg, Germany; and the Univ. of Tel Aviv, Israel. (NASA Release 65-223; AFOAR Release 7-65-1)

- Eleven of America's 34 astronauts left NASA Manned Spacecraft Center for a 10-day field trip into the volcanic regions of Iceland. Making the expedition were Edwin Aldrin, William Anders, Charles Bassett, Alan Bean, Eugene Cernan, Roger Chaffee, R. Walter Cunningham, Donn Eisele, Russell Schweickart, David Scott, and C. C. Williams. The trip would be part of the astronauts' training in the type of geological formations they might find on the moon. (UPI, *Orl. Sent.*, 7/7/65)
- Freeman J. Dyson, professor of physics at the Institute for Advanced Study, Princeton, N.J., charged official Washington with the political "murder" of Project Orion, a nuclear-pulse-powered rocket, which he said would have been far better than any other kind of propulsion for rapid exploration of the solar system. The project was ended early this year after seven years of scientific and engineering studies that had cost $10.3 million and proved, Dyson said, the feasibility of nuclear-driven spacecraft. Its "killers," Dyson said, were

DOD, NASA, the promoters of the nuclear test-ban treaty, and U.S. scientists generally. Writing in *Science*, Dyson continued: "The story of Orion is significant because this is the first time in modern history that a major expansion of human technology had been suppressed for political reasons." (*Science*, 7/9/65)

July 9: A J-2 rocket engine had been retired from service after being test-fired a total of 10,686 sec. in 60 separate tests, MSFC announced.

J-2, developed for NASA by Rocketdyne Div. of North American Aviation, would burn liquid hydrogen and liquid oxygen and develop 200,000 lbs. thrust at altitude. Five clustered J-2's would provide a million lbs. thrust for the S-II (second) stage of the Saturn V booster. A single J-2 would power the S-IVB, top stage for both the Saturn IB and Saturn V launch vehicles. (MSFC Release 65-177)

- Robert V. Reynolds was named Assistant FAA Administrator for General Aviation Affairs by FAA Administrator William F. McKee. Reynolds would serve the growing needs of general aviation which had more than 88,700 out of a total of 90,935 active civil aircraft. FAA had forecast that the general aviation fleet would grow to 105,000 by 1969. (FAA Release 65-58)
- Yemen and Pakistan had signed agreements increasing to 46 the number of countries which had joined in the international joint venture for development of a global commercial communications satellite system, ComSatCorp announced. They would share with other nations in the financing and ownership of the space segment of the global system, consisting of the satellites themselves, tracking, control, and related functions, but not earth stations. (ComSatCorp Release)

July 10: Probable reasons for the GEMINI IV crew's failure to achieve rendezvous with the booster during the June 3-7 spaceflight—based on runs in a rendezvous simulator—were listed by Ron Simpson, Guidance and Control Branch, NASA Manned Spacecraft Center: (1) visual rendezvous requires extended ground practice; (2) maneuver is nearly impossible without radar if the spacecraft and target are more than 4,500 ft. apart; (3) more fuel than originally expected was required. (*Houston Chron.*, 7/10/65)

- XC-142A aircraft, designed to take off vertically and fly horizontally to a top speed of 430 mph, began 18 mo. of extensive testing at Edwards AFB, Calif. The four-engine, propeller-driven aircraft, first of five to undergo evaluation and performance tests, had been turned over to USA July 8 at Ling-Temco-Vought, Inc., Dallas. (*N.Y. Her. Trib.*, 7/11/65)
- Resignation of Eugene M. Zuckert as Secretary of the Air Force, effective Sept. 30, was announced by President Johnson. To succeed Zuckert, who had served since Jan. 23, 1961—longer than any previous Air Force secretary—the President named Dr. Harold Brown, DOD Director of Defense Research and Engineering since May 3, 1961. No replacement was named for Dr. Brown.

President Johnson also accepted the resignation of Dr. Brockway McMillan as Undersecretary of the Air Force, also effective Sept. 30. (White House Release; Horner, *Wash. Eve. Star*, 7/11/65, 12)

July 11: Reports by U.S. astronauts of seeing details on the earth's surface while orbiting the earth at altitudes of over 100 mi. did not surprise physicists at the U.S. Bureau of Standards' Central Radio Propagation

Laboratory in Indianapolis, reported the *Indianapolis Star*. Wind tunnel experiments there had proven that the closer a viewer is to turbulent air, the more it distorts an image: the image wavers in brightness, changes position, and shifts in and out of focus. All these effects diminish contrast—an essential factor in identifying objects in aerial reconnaissance. The effects are reduced and seeing improves as the viewer moves away from the turbulence. (Lewis, Sci. Serv., *Indianapolis Star*, 7/11/65)

July 11: *New York Times* editorial on July 14 MARINER IV flyby of Mars said: "On Wednesday, if all goes well, one of the epoch-making experiments in the history of science will take place as MARINER IV comes within 6,000 miles of Mars and takes pictures for transmission back to earth. Not since Galileo first trained his telescope on the moon has there been such a prospect for a quantum leap in man's knowledge of a nearby world.

"Whatever the results of the picture-taking experiment, the accomplishments of MARINER IV are already historic. Launched last Nov. 28, it has traversed a preassigned course for more than seven months. During this time it has flown roughly 350,000,000 mi. while communicating back to earth new scientific data as well as a steady flow of information on its flight and its internal condition. This performance far exceeds all similar earlier feats.

"For comparison we may note that MARINER II—justly famed for its flight past Venus in 1962—needed to fly only three and a half months and 180 million miles to reach its objective. And while MARINER II sets a record signalling to earth from a distance of 53.9 million miles away, MARINER IV has recently been transmitting from the neighborhood of Mars, now over 130,000,000 miles distant from this planet." (*NYT*, 7/11/65, E10)

July 12: Soviet cosmonaut Col. Pavel Belyayev disclosed in an article in the newspaper *Sovietsky Patriot* that the heat was so intense when he was forced to land VOSKHOD II by manual controls that drops of molten metal ran down the portholes. He said when he discovered that something was wrong with the automatic landing system of the two-man spacecraft, he asked ground control for permission to use the manual system—something that had never been done before during a Soviet manned spaceflight. (UPI, *Houston Chron.*, 7/12/65)

• Soviet astronomer A. Markov stated that a lunar crater would be the most suitable landing site inasmuch as the floor of the crater would be composed of solid material. Because of the steep inclines in certain lunar formations, it would be impossible for deep layer of dust to accumulate, he noted. (*Berliner Zeitung*, 7/11/65, 4–6)

• Dr. Cyril Ponnamperuma of NASA Ames Research Center's Exobiology Div. and ARC research assistant Ruth Mack had synthesized the five chemical building blocks of DNA and RNA in a simple laboratory model which duplicated conditions believed to exist on earth from three to four and a half billion years ago, NASA announced.

DNA and RNA, nucleic acids, form the core of and are the "prime movers" of all living cells in plants, animals, and man. The five DNA–RNA building blocks, known as nucleotides, are made up of a nitrogenous base, a sugar, and a phosphate. Their synthesis under laboratory conditions could be a major advance toward explaining the origin of life on earth; it could have retraced a critical phase in the chemical evolution of organic material which had to occur before the appearance of life itself.

Synthesis of nucleotides had been done before, but always by long, complex laboratory procedures. (NASA Release 65–221; ARC Release 65–17; *Marshall Star*, 7/14/65, 9)

July 12: Strange objects moving through space were reported sighted in two widely-separated areas of Portugal. The Azores Weather Bureau claimed interference from one which stopped its electromagnetic clocks. Descriptions of the objects were similar to official Argentine and Chilean military reports of sightings in the Antarctic recently. (*Orl. Sent.*, 7/12/65)

- Senate defeated by 61-to-16 the proposed Proxmire amendment to make an across-the-board reduction of 5% in the NASA appropriations bill. This left the Senate bill, still to receive its final vote, at $5.19 billion. The House had passed a $5.16 billion appropriation.

  Sen. William Proxmire (D–Wis.) had contended that the manned lunar landing program constituted an "excessive waste." Sen. Joseph S. Clark (D–Pa.) had felt the problem was not one of waste or inefficiency but felt that more of the nation's resources should be put into education, housing, pollution abatement, and other urban problems. Sen. Wayne Morse (D–Ore.) had supported the reduction but had proposed that the cut be 25% or 50%. "The American people have been thoroughly taken in by a TV spectacular," he said and charged that the space program "is all for the purpose of gratifying our national ego." Senator Morse had indicated that no other single agency conducted programs "with such vague objectives with such little return to the national Government."

  Proxmire had charged that NASA was the prime example of an agency whose expenditures Congress had difficulty in controlling, that certain project costs within the NASA program could be reduced, and that NASA used too great a proportion of our monetary, material, and manpower resources.

  The 16 Senators voting for the Proxmire amendment were Boggs, Clark, Cooper, Douglas, Fullbright, Gruening, Kennedy (N.Y.), McGovern Miller, Morse, Morton, Mundt, Nelson, Proxmire, Williams (N.J.), and Williams (Del.). (*CR*, 7/12/65, 15927–40)

- Brig. Gen. Julian H. Bowman (USAF, Ret.) was appointed Special Assistant to the NASA Associate Administrator for Manned Space Flight. General Bowman would handle special management problems in areas such as program manpower and organization, career management, and industry relationships. (NASA Hq. Bull., 7/20/65)

- USAF scientist Dr. Hubertus Strughold told the Aviation and Space Writers Meeting in Washington, D.C., that the possibility that frozen oceans beneath the surface of Mars may support life cannot be ignored. Men may even find enough moisture there to provide them water and oxygen for an expedition, "and thus a critical problem of extraterrestrial resources would no longer exist," he said. The idea that an underground water table exists on the otherwise arid planet was only a hypothesis with no evidence at present; but a combination of older theories by other scientists plus "common horse sense" led him to speculate the water was there, Dr. Strughold said. (Text)

July 12: AEC Administrator Dr. Glenn T. Seaborg, in an interview with *U.S. News and World Report* reviewing the 20th anniversary of the exploding of the first atomic device in New Mexico on July 16, 1945, said a dozen countries had the potential to "join the present five-member nuclear club." Japan, India, West Germany, Sweden, Italy, Canada, and Israel were capable of producing an atomic bomb soon; Spain, Brazil, Yugloslavia, Egypt, and Switzerland had the scientific talent and available resources to produce a bomb in a little longer time, he said. (*U.S. News*, 7/19/65, 13)

- Reviewing developments in the TFX aircraft program—now designated F-111, Richard Elliott wrote in *Barron's:* "Since it was launched in 1963, the TFX program has cost roughly $900 million. Another $700 million has been earmarked for it in fiscal 1966. If the Pentagon buys as many F-111's as it now plans, the price tag through the early 1970's will run to at least $8 billion. . . . (Elliott, *Barron's*, 7/12/65)

July 13: NASA invited scientists to propose, by August 15, the various experiments to be included in a continuing Voyager program of exploration of the planets. The Mars Voyager program would begin with a test flight in 1969 followed by an orbiter-lander flight in 1971 with other flights following in 1973, 1975, and beyond. The program would also include missions now under consideration for flights to Venus and other planets. The experiments to be included in the 1971 mission of the landing capsule would be subject to strict sterilization requirements. In following programs, there would be considerable latitude in the choice of experiments. (NASA Release 65-230)

- Senate passed the Independent Offices appropriation bill 84-2, with the NASA portion at $5.19 billion as reported out by the Appropriations Committee. Negative votes were cast by Senators Dirksen (R-Ill.) and Young (D-Ohio). (*CR*, 7/13/65, 16008-27)

- President Johnson, asked during a White House news conference if he could give a status report on USAF's Manned Orbiting Laboratory, said: "No, I am not in a position to make a statement on that at this time. The Space Council has had some briefings in connection with the matter. There is a study going on every day in that connection, but I would not want to go further than that now." (Transcript, *Wash. Post*, 7/14/65)

- Vice President Hubert Humphrey and his party, which included Rep. Carl Albert (D-Okla.) and Sen. Fred Harris (D-Okla.), visited NASA Manned Spacecraft Center and were briefed on the Gemini program by Charles W. Mathews; on the Apollo program by Dr. Joseph F. Shea; and on the life support systems by Richard Johnston of Crew Systems.

  Mr. Humphrey—it was his first visit to MSC—took a "ride" in the Gemini docking trainer with Astronaut Walter M. Schirra, Jr. (Cdr., USN) and then he and his party observed a simulated Gemini liftoff at the Mission Control Center and a portion of a mission. (MSC *Roundup*, 7/23/65, 1; *Houston Post*, 7/14/65)

- JPL Director Dr. William Pickering held up some fuzzy photographs of patches of the moon during a JPL press conference and said this was the best quality to be expected from the MARINER IV television system. The pictures should be better than any ever obtained from earth if the system operated as planned, he said. Each of the MARINER IV pictures would consist of 200 rows of 200 dots. Each dot would be sent as a number ranging from 0 to 63, designating the darkness of the dot. Reconstruction of the picture would be done by a computer at JPL. Because the first views would scan the planet obliquely, it might not be until the third image that one was received showing any obvious surface features. This, according to Dr. Pickering, might mean that no pictures would be displayed publicly until July 16. (Sullivan, *NYT*, 7/15/65, 1)

July 13: Rains forced Princeton Univ. scientists to postpone launching from Palestine, Tex., of a Stratoscope II balloon and its telescope designed to photograph Saturn and its rings. No new launch date had been set. (*Houston Chron.*, 7/14/65)

- A ground firing of the Saturn S-II battleship stage was conducted for 25 sec. at the Santa Susana test site of North American Aviation, NASA Marshall Space Flight Center announced. The firing—longest to date in the S-II program—was one of a series leading to full flight duration runs of nearly 400 sec. Next test, expected within a week, was planned for about two minutes. (MSFC Release 65-183)

- U.S. Weather Bureau, Coast and Geodetic Survey, and Central Radio Propagation Laboratory of the National Bureau of Standards were merged to form Environmental Science Services Administration (ESSA). Creation of ESSA came two months after President Johnson transmitted the reorganization plan to Congress. (30 *FR* 8819 (1965))

- ComSatCorp had filed applications with FCC for construction of two new earth stations and related facilities, and expansion of a third existing station. One new station would be located at Brewster Flat, Wash., and a second on the island of Oahu, Hawaii. Overall costs of these facilities, including land acquisition, construction, and establishment of related electronic equipment, was estimated at $6 million each. ComSatCorp said it was negotiating to purchase from AT&T the existing earth station at Andover, Me., and asked authority to modify and expand the facility, pending a purchase agreement. (ComSatCorp Release)

- Two Presidential citations were among 12 awards presented to NASA Marshall Space Flight Center employees recently, MSFC announced. Ralph Butler of the Aero-Astrodynamics Laboratory had suggested an improved method of measuring local atmospheric conditions by using radio-controlled model airplanes. Franklin Williams of the Saturn V Program Office had made two cost-reduction proposals, one concerning x-ray radiation protection and the other recommending an inexpensive covering for clean room insulation. Butler and Williams saved the Government an estimated $306,000. (MSFC Release 65-182)

- The rash of reports on strange visitations to earth received comment from Walter Sullivan in the *New York Times*: "Whether or not [flying saucer enthusiasts] are aware of it, this is the time when space probes would be arriving from Mars, if inhabitants of that planet were engaged in a similar effort [to the MARINER IV Mars flyby]. The earth overtakes Mars every two years and two months, coming within a few dozen million miles of it. There was such an approach this spring. The United States and the Soviet Union anticipated it by firing vehicles toward Mars last November. If there were any Martians equipped to do so, they would have fired their vehicles toward earth at about the same time." (Sullivan. *NYT*, 7/13/65)

July 13: In spite of recent reports from Portugal, the Azores, and the Antarctic of "strange objects moving through space," USAF's Project Blue Book—in charge of investigating Ufo reports—had not yet found any evidence to support the view that flying saucers or anything like them had entered the earth's atmosphere from outer space. 9,118 reports had been investigated since 1947. (Balt. *Sun*, 7/13/65)

* Rep. Donald Rumsfeld (R-Ill.) urged that more effort be devoted to development of U.S. inner space capability, because of its military significance: "If the United States is not striving for military control of inner space, it should, because the safety of the nation and of the non-Communist world depends upon it.

  "The United States can afford to lose the moon race to Russia, which would be a great scientific first. But it cannot afford to lose the race for control of the inner space belt, because it will have lost all." (*CR*, 7/13/65, A3714)

* Sen. William Proxmire (D-Wis.) spoke on the floor of the Senate on potential dangers to universities from Federal research programs and advocated giving more decision-making responsibility on allocation of funds to the universities. (*CR*, 7/13/65, 16075-77)

* U.S. Space Park at the New York World's Fair had had an attendance of 500,000 persons thus far, a 15% increase over last year's attendance, NASA announced. (NASA Hq. *Bull.*, 1-65-14)

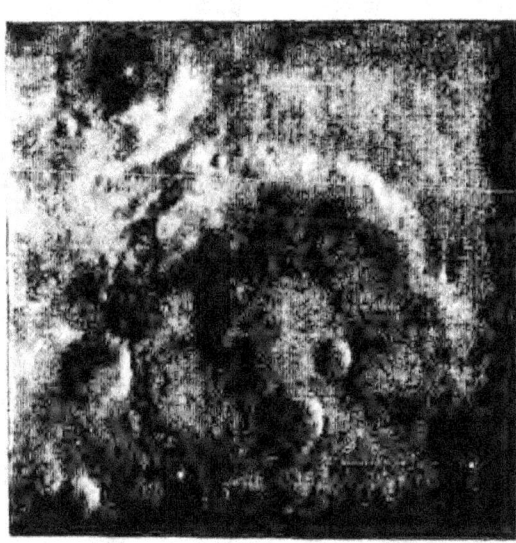

July 14: MARINER IV photograph No. 11 of Mars, taken at a slant range of 7,800 miles.

July 14: MARINER IV approached within 5,500 mi. of Mars and took the first close-up pictures in history of that planet.

At 10:28 a.m. EDT, a signal from the tracking station at Johannesburg, South Africa, had commanded MARINER IV to turn on the encounter equipment. Obeying the command 12 min. later—the time it took the signal to reach the spacecraft across 134,000,000 mi.— MARINER IV's scan platform with TV cameras and two Mars sensors began searching for the planet; the tape recorder began a 10-hr. warm-up.

By 1:10 p.m., the spacecraft's sensors had found the proper angle from which to photograph Mars. On orders from JPL, the Johannesburg tracking station ordered the camera and its light sensor to stop the scan at 7/10 of a degree of the optimum aiming point.

Throughout the day, the spacecraft continued transmitting scientific and engineering information via the telemetry system.

At 7:50 p.m., the Mars probe made its first "sighting" contact with the planet. The telemetry system then shifted and began sending only information from the scientific experiments.

At 8:20 p.m. EDT, the first light from Mars struck the light sensor aligned with the camera and the 25-min. picture-taking sequence began. 21 frames were exposed and recorded on magnetic tape to be telemetered to earth over a 10-day period. Transmission of each picture would require 8 hrs. 25 min.

Near the end of the fly-by, signals received at Goldstone tracking station had indicated a malfunction in the tape system. Telemetry received after the pass, however, showed no indication that trouble had occurred.

At 10:12, MARINER IV flew behind Mars, remained obscured for 53 min., and re-emerged beyond the planet, its transmitter beaming radio signals through the Martian atmosphere. From this occultation experiment, scientists might be able to gauge the depth, thickness, and component gases of Mars' atmosphere by measuring how much these signals were bent and their speeds changed.

Dr. William H. Pickering, JPL Director, told a news conference that signals indicated all instruments on the spacecraft had performed properly during the flyby. (NASA Release 65–227; Sullivan, *NYT*, 7/15/65; Hoffman, *N.Y. Her. Trib.*, 7/15/65; AP, Balt. *Sun*, 7/15/65)

July 14: A 44-day test of NASA Goddard Space Flight Center's Space Environment Simulator (Ses) had just ended, GSFC announced. In this project, the early performance of a scientific spacecraft in orbit around the earth was compared to the performance of a sister craft "orbited" inside the Ses: the UK–D scientific spacecraft—backup model for ARIEL II (UK–II) launched successfully into earth orbit by U.K. March 27, 1964—was "flown" inside the Ses; flight plan for UK–D followed general flight plan of initial flight days of ARIEL II.

Preliminary data indicated operation of the Ses was good throughout the test period. The spacecraft inside the simulator even experienced the same minor malfunction its sister spacecraft had experienced in orbit. (GSFC Release G–17–65)

- Speaking before the Senate Commerce Committee in support of his bill calling for a three-year feasibility study of this country's adopting the metric system, Sen. Claiborne Pell (D-R.I.) said: "Should Canada follow Great Britain's example and convert, the United States would be virtually the only remaining country that uses a system of weights and measures based on other than the metric system. Actually, 90 percent of the people in the world use metric measures right now."

The proposed three-year Dept. of Commerce study, Senator Pell said, would include not only a detailed estimate of the cost of conversion, but also a statement of the possible side benefits in foreign trade. (AP, *NYT*, 7/15/65, 35)

July 14: U.S.S.R.'s ZOND II—launched Nov. 30, 1964—was believed still racing toward Mars, but with its radio power dead. Although ZOND II's specific mission was not disclosed, Tass had announced it was intended to "test the systems of the spacecraft in practical conditions of a prolonged space flight and to accumulate experience." Tass had also said "unprecedented" plasma-jet engines were powering the Mars probe. (UPI, *Houston Chron.*, 7/14/65)

- A full-scale model of NASA's MARINER IV Mars probe went on display at the Smithsonian Institution's National Air Museum. Panels explaining the Mariner program would also be displayed; photographs of the Martian surface would be added to the exhibit as they were received. (NASA Release 65–231)

- 12 delta-winged Mirage IV jet bombers flew across the Paris sky for the first time to open France's annual Bastille Day parade. Two days earlier, the French Atomic Energy Commission had announced production of a smaller, more powerful replacement for the bulky 60-kt. atomic bomb now carried by the Mirage IV. (Breastrup, Balt. *Sun*, 7/15/65)

- Commenting on the U.S. need for a Manned Orbiting Laboratory (Mol), Lt. Gen. Ira C. Eaker (USAF, Ret.) said in the *San Antonio Express:* "The nearly simultaneous launch of NASA's Gemini and Air Force's Titan III–C caused some commentators to suggest that there is competition and rivalry in the space effort. Investigation convinces me that there is complete collaboration and accord between NASA and the military in the space program.

"There is now little reason to doubt that NASA will succeed with Apollo, the Moon mission. Our continued leadership in the exploration of the peaceful uses of space seems assured.

"It is now time to survey where we stand in the military implications and dangers from space. . . .

"The first step in a needed military space program, which the Air Force earnestly and urgently proposes, is a manned orbiting laboratory (MOL). It is visualized as an experimental space station in low orbit to determine what man can profitably do in space.

"The anti-MOL advocates have suggested that the fall-out from the NASA space program will provide all the building blocks to produce military space vehicles quickly when the need is more clearly evident. Fall-out from the automobile industry did not provide the Patton tank. It was necessary to cut and try, build and test many experimental tanks over a period of 25 years.

"Building blocks from civil air transports did not produce the B–52 bomber. We had to build and test 51 earlier bomber models. Effective space weapons will never come solely from the fall-out from the NASA program.

"MOL is said to be held up until the Air Force can precisely define and defend a military mission which MOL can be certain to accomplish. To require anyone to visualize and indicate definitely and in detail now, all that man may ultimately do in space, would be like having asked the Wright brothers in 1910 to lay down the specifications for a supersonic transport." (Eaker, *San Antonio Express*, 7/14/65, 18)

July 15: The first close-up picture of Mars transmitted by NASA's MARINER IV in an eight-hour broadcast over a distance of 134 million mi. clearly showed the edge of the planet. Transmitted to earth as a series of 5 million radio signals representing zeroes and ones, the picture was received at the Madrid and Johannesburg tracking stations and relayed to JPL. A JPL computer reconstructed the digits to produce images consisting of a series of dots of varying darkness. With transmission at 8.3 dots a second, 8½ hrs. were required to receive the photograph.

The photo sequence began when the bright edge, or limb, of Mars was seen by a light sensor which triggered MARINER IV's magnetic tape recorder. Initial information came from an experimental device called a "cluge" which built up a rough pattern of the photo on a monitor screen as it was relayed from the tracking stations to JPL. Taken at a low angle with an oblique view, the first photograph did not show the detail expected in later photographs.

At a news conference at JPL, Dr. J. A. Van Allen, Univ. of Iowa, said that during its Mars flyby, MARINER IV had discovered with a variety of detectors that Mars had little or no magnetic field and, therefore, no radiation belts. This would indicate that the planet lacks a liquid core and thus differs basically from earth. Lack of a metallic core, liquid or solid, would be evidence that the planet never went through the churning internal processes that gave the earth its layered structure. Mars would not have continents formed of lightweight rocks and ocean basins underlain with basaltic rock as found on earth.

W. M. Alexander, NASA Goddard Space Flight Center, explained that there was no evidence of a belt or unusual concentration of cosmic dust around the planet, nor, pointed out Dr. H. S. Bridge, MIT, was there any evidence of a shock wave caused by solar wind flowing across the planet. The solar wind was the steady outflow of thin, hot, high-velocity gas from the sun.

Dr. William H. Pickering, Director of JPL, when questioned by newsmen on the posssibility of higher life existing on Mars, pointed out that the absence of a magnetic field indicated the planet's atmosphere was hit with all types of radiation and the existence of life would depend on how deep the atmosphere was and the extent of radiation that reached the surface. Asked if he were discouraged about the possibility of finding life on the planet, Dr. Pickering replied: "No, I have always felt we will find some sort of life on Mars."

Dr. Pickering pointed out that one explanation for the reddish hue of Mars might be the presence of limonite, an iron oxide. This would suggest that iron was uniformly spread through the planet rather than being largely concentrated in the core, as on earth. (Transcript; Sullivan, *NYT*, 1/16/65, 1, 10; Miles, *Wash. Post*, 7/16/65, A1,A3; Sullivan, *NYT*, 7/17/65, 1, 6)

July 15: Astronauts James A. McDivitt (Maj., USAF) and Edward H. White, II (Maj., USAF) of the June 3-7 GEMINI IV flight received USAF astronaut wings from Gen. John P. McConnell, USAF Chief of Staff, in an Air Force ceremony at the Pentagon. X-15 test pilot Joseph H. Engle (Capt., USAF) also received USAF astronaut wings, thus becoming the 12th American to receive the award and the fifth to wear the insignia for suborbital flight.

Meanwhile the Senate Armed Services Committee approved President Johnson's promotion of Astronauts Walter M. Schirra from commander to captain and John W. Young from lt. commander to commander. (DOD Release 458-65, 7/14/65)

- Dr. Kurt H. Debus, Director of Kennedy Space Center, NASA, said in an address at the First World Exhibition of Transport and Communications in Munich: "It is my personal opinion, shared by some colleagues, that space flight will logically follow the pattern of historical development which has characterized the transportation industry. That is, the exploitation of the system is an outgrowth of its invention.

"This pattern is vividly demonstrated by the popular American legend that Henry Ford invented the automobile, when in fact the automobile is not really an American invention at all. For one thing, it was not invented by any one man, but by a host of inventors—most of them Europeans.

"However, Henry Ford did initiate the exploitation of this invention as a means of transportation for almost every family in America. Ford was a man of great inventive genius who strove to mass produce a highly reliable automobile at low unit cost. So Ford took advantage of this once-novel invention, the automobile, by developing to a high degree the arts of mass production with interchangeable parts, line assembly, and finally conveyor assembly; and thus he ushered in a new age of public transportation.

"The automobile also serves as an excellent example that even the farsighted and visionary inventor often cannot visualize the ultimate utilization of his invention by the public. For the modern automobile is a combination of many inventions—the wheel, the pneumatic tire which in turn depended upon the discovery of vulcanized rubber, the internal combustion engine and gasoline—to name a few. To illustrate an extreme case, I seriously doubt if the inventor of the wheel could visualize its use in such a supernatural machine as the automobile! In the more recent past, Benjamin Silliman, Jr., a brilliant American petrochemist of the late-nineteenth century, considered gasoline a useless and dangerous byproduct of his process for refining lamp kerosene.

"The thing to remember is that the automotive industry did not spring from public demand. It came about because there were inventors who braved ridicule and worse and because others were quick to seize upon their creations. There was no expressed requirement for the airplane, nor for the trans-Atlantic air travel opened up by Charles Lindbergh. The novel products of inventive minds become the everyday products of our society. The full impact of these products is unpredictable at the time of their invention.

"The American scientist, Robert Goddard, who created the first liquid-fueled rockets in the United States, remarked that 'Every vision is a joke until the first man accomplishes it.'

"The men who dreamed up and worked out methods of efficient transportation unwittingly opened up broad new vistas, and touched off the growth of tremendous production and service industries. They showed the way to expedite the movement of people and goods, and they also accelerated the exchange of ideas and customs and thus promoted understanding and cooperation.

"Robert Goddard in America, like Hermann Oberth in Europe, were scorned and laughed at in their time. But while we are meeting here, men are preparing more journeys into the limitless areas beyond Earth's atmosphere, working with much the same theories and techniques proposed by Oberth and Goddard." (Text)

July 15: "One absolute certainty is that if the atomic bomb had not gone off at Alamogordo 20 years ago tomorrow, the spacecraft Mariner 4 would not have flown past Mars yesterday," wrote William Hines in the Washington *Evening Star.* He continued: "Space and the atom are more closely interrelated than most people realize. We are in a space race today because—and only because—big rockets were developed by the East and West starting about 15 years ago. . . . Solely to carry atomic (and later hydrogen) warheads. There would have been no space race without big rockets because the staggering development costs for space applications alone could never have been justified.

"The 'aerospace business' of the non-atomic '60's, then, would have been 'airplane business' pure and simple—and airplanes would be much different than they are today. There would be no 'atomic energy industry,' of course—and these two differences would have a measurable economic impact. Probably a million jobs nationwide depend either directly or indirectly on space and the atom.

"Equally obviously, two vast bureaucracies dedicated to 'running' space and the atom would not exist. The Atomic Energy Commission has about 7,300 employes and the National Aeronautics and Space Administration about 33,000. Together they 'own' something like $18 billion worth of property." (Hines, Wash. *Eve. Star,* 7/15/65)

- A contract for about $60 million would be signed by NASA Goddard Space Flight Center with Republic Aviation Corp. for Phase II development of the Advanced Orbiting Solar Observatory (Aoso), NASA announced. Republic would furnish two flight spacecraft and a prototype; work would also include final development and design, checkout, experiment integration, and launch support services. The Phase I portion of the project was completed in July 1965.

    Launching of the 1,250-lb. Aoso, planned for the late 1960's, would be from Vandenberg AFB by a thrust-augmented Thor-Agena rocket; the observatory would provide a constant search for isolated or unusual solar activity. (NASA Release 65–234)

- *Northwind,* a U.S. Coast Guard icebreaker, left Copenhagen for a three-month scientific voyage of the Soviet Arctic providing a "floating platform" for civilian scientists aboard to conduct oceanographic and meteorological studies in the Barents and Kara Seas—areas for which the U.S. had little scientific data. (*NYT,* 7/16/65)

July 15: Dr. Kurt H. Debus, Director, Kennedy Space Center, NASA, received the Pioneer of the Wind-Rose Award, Order of the Diamond, in Munich, Germany. Dr. Debus, first recipient of the award, was cited for his historical contribution to rocket launch technology and for his contributions to science through his work in rocketry. Award was made at the First International Transport Exhibit. (*Brevard Sent.*, 8/1/65)

- USAF would begin retiring more than 300 of its B-52 strategic bombers next year, predicted Howard Margolis in the *Washington Post*. Margolis theorized that USAF would then seek more money for its follow-on bomber, the Amsa (Advanced Manned Strategic Aircraft), and money for several wings of F-111's for the Strategic Air Command (SAC) bomber force as the B-52's were mothballed. (Margolis, *Wash. Post*, 7/15/65)

July 16: PROTON I, a scientific space station, was orbited by the U.S.S.R. with a "powerful new booster," Tass announced. The 26,880-lb. payload—claimed to be heaviest ever launched—was placed into an orbit with the following initial parameters: apogee, 627 km. (389.4 mi); perigee, 190 km. (118 mi.); period 92 min.; and inclination, 63.5°. In addition to scientific and measuring instruments, the space station was equipped with special equipment for the study of cosmic particles of super-high energies and a radio transmitter. Tass said: "In order to insure realization of the planned space research program, a powerful new booster rocket has been developed in the Soviet Union."

Commenting on the Soviet launching, Dr. Edward C. Welsh, Executive Secretary of the National Aeronautics and Space Council, said he was "impressed but not astonished" by the Russian accomplishment.

"We have been predicting for some time that they would develop into the space station field," he said. "We have anticipated that they would practice rendezvous and docking before they developed what they said was a space station, but they haven't followed that particular line." (Tass, 7/16/65; *NYT*, 7/17/65, 6; *Wash. Post*, 7/17/65, A1)

- U.S.S.R. launched COSMOS LXXI, LXXII, LXXIII, LXXIV, and LXXV into orbit with a single booster. Initial orbital data: COSMOS LXXI, apogee, 542 km. (337 mi.), perigee, 521 km. (324 mi.); COSMOS LXXII, apogee, 588 km. (365 mi.), perigee, 538 km. (334 mi.); COSMOS LXXIII, apogee, 564 km. (350 mi.), perigee, 531 km. (330 mi.); COSMOS LXXIV, apogee, 619 km. (384 mi.), perigee, 537 km. (334 mi.); COSMOS LXXV, apogee. 644 km. (400 mi.), perigee, 539 km. (335 mi.). Period for all five satellites was 95 min.; inclination, 56°. Satellites contained scientific equipment for outer space research as well as "Mayak" radio transmitters. All instruments were said to be functioning normally. (*Izvestia*, 7/17/65, ATSS-T Trans.)

- "Mariner-4's performance . . . constitutes one of man's greatest triumphs to date in the field of science and technology," said a Washington *Evening Star* editorial: "It has blazed the way for the landing of instruments on the Martian terrain within the next decade, and after that, on some day between 1980 and the end of the century the landing of Americans there.

"The wonder and wizardry of it all are nothing less than awesome. Everybody involved with MARINER's success merits the praise of the entire nation." (Wash. Eve. Star, 7/16/65)

July 16: A 707 jet transport, owned and fitted by the Boeing Co. with extensive instrumentation, including an analog computer and a 15-ft. needle-like nose extension to make it an in-flight dynamic simulator, had arrived at NASA Langley Research Center for a four-month program of supersonic transport experiments. The research program would assess simulated landing approaches with both the fixed-wing and variable-sweep concepts, including conditions such as emergency use of cruise sweep-back during a landing. (NASA Release 65–233)

- Gemini V backup crew members, civilians Neil Armstrong and Elliot M. See, Jr., began a training exercise in Gulf of Mexico waters designed to teach the astronauts the proper methods to leave the spacecraft after touchdown. Primary crew members Maj. L. Gordon Cooper (USAF) and Cdr. Charles Conrad (USN) would take the same training July 21. (AP, *Houston Chron.*, 7/16/65)

- FCC rescinded an earlier order and granted a group of communications companies temporary authority to provide television service from EARLY BIRD I comsat. The order authorized AT&T, ITT, RCA Communications, Inc., and Western Union International, Inc., to pool their efforts and resources in a consortium which would buy the television service from ComSatCorp and then sell the service to American television networks. In its earlier order, FCC had granted temporary authority to ComSatCorp to provide the television service directly to the television networks. (Weekley, *Wash. Post*, 7/16/65)

- The successful use in modern surgery of plasma probes developed for the U.S. program by High Temperature Instruments Corp. under contract to NASA Lewis Research Center was reported by Dr. Walter T. Olson, Assistant Director for Public Affairs at LRC. The temperature-sensing elements of these tiny probes were used in a cryogenic cannula for treating Parkinson's Disease, a malady involving control centers in the brain. The surgical probe, through which low-temperature liquid nitrogen flows, is used to destroy parts of the brain by freezing without affecting adjacent tissue. It also permits the neurosurgeon to probe at an intermediate temperature; once located, the selected brain tissues can be frozen by using the thermocouple to monitor the temperature. (LRC Release 65–50)

- NASA awarded $75,000 cost-plus-fixed fee contract to Control Data Corp. for a breadboard model of Scads (Scanning Celestial Attitude Determination System), a simple star-mapping system for use onboard spacecraft to determine the pitch, roll, and yaw attitude errors. Initial tests of the system would be conducted at GSFC. (GSFC Release G–18–65)

- NASA entered competitive negotiations with Documentation, Inc., and Tech/Courier Corp. for a contract valued at approximately $4.5 million to operate NASA's Scientific and Information Facility—containing the world's largest collection of aerospace literature—in a new government-furnished building in College Park, Md. The contract would run through June 1966 and would be monitored by NASA Hq. Scientific and Technical Information Div. (NASA Release 65–236)

July 16: Dr. Edward C. Welsh, Executive Secretary of the National Aeronautics and Space Council, remarked in an interview with *Missile Space Daily:* "I do not doubt that with good fortune, the trip [to the moon] will be made during this decade, which to me means during the 1960's. Replying to a question about the future use of solid fuel rockets, Dr. Welsh said: "I certainly see no reason to doubt that the use of solids will continue to increase. Without attempting to predict specific dates or missions, I believe large solid motor technology will be a very valuable element in our growing national space competence. . . ." On post-Apollo missions: ". . . Under present circumstances, it would seem that our program will include extended exploration of the moon after the first trip, provided conditions are found to be favorable to such exploration. Also, we should be expected to develop a growing capability in earth-orbiting space stations and make an increasing effort in planetary exploration. I do not rate these major objectives one against the other because I believe the national space program will in time include all three of the broad missions I have mentioned." Dr. Welsh said that he hoped the U.S. would be able to "maintain space expenditures at a figure above 1 percent of our Gross National Product." (*M/S Daily,* 7/26/65)

- "We saw it and we were amazed," wrote Pope Paul VI in Latin on a copy of the first photograph returned by MARINER IV. He signed his name underneath with the date. (UPI, *NYT,* 7/17/65)

- 20th anniversary of the first atomic device, at Los Alamos, N. Mex., the beginning of the so-called "atomic age." Since then, atomic devices had been exploded by the U.S.S.R. beginning in 1949, Britain in 1952, France in 1960, and Red China in 1964. Thermonuclear devices had been demonstrated by the U.S. in 1952, U.S.S.R. in 1953, and Britain in 1957. Now Federal spending for the peaceful uses of atomic energy had drawn abreast of spending for nuclear weapons for the first time since the first atomic bomb. Commenting on this fact, Evert Clark wrote in the *New York Times:* "Even in the Atomic Energy Commission there is little tendency to view this as a triumph for peace over preparations for the possibility of war.

 "It is rather that military needs have been largely taken care of, so that they demand less money." (EPH; Clark, *NYT,* 7/17/65, 6)

- A commercial model of the military C-5A subsonic jet was being considered by Pan American World Airways to provide cheap air transportation, reported the *Wall Street Journal.* The aircraft would measure 230 ft. or more from nose to tail, weigh over 700,000 lbs., and carry a payload of over 250,000 lbs.; four jet engines would propel it at 550 mph. The passenger version would hold between 700 and 1,000 people. The jet bus could cut cargo rates 50% or more and cut fares 20 to 25%. Problems would arise, however, in that present airport runways might not be able to sustain the great weight and new procedures would have to be developed to handle the increased traffic. (Cooke, *WSJ,* 7/16/65, 1, 8)

- NASA Administrator James E. Webb was one of four persons named honorary chairmen of the Robert Hutchings Goddard Library Program at Clark Univ. The other three were Mrs. Robert H. Goddard; Dr. Charles G. Abbot, former fifth Secretary of the Smithsonian Institution; and Mr. John Jeppson, executive vice president of the Norton Co. and Chairman of the Board of Trustees of Clark Univ. Mr. J. L. Atwood, president of North American Aviation, was named general chairman of the Library program. (*Goddard*, 7/65)

July 16: Physicist Serafim Nikolayevich Zhurkov, corresponding member of the Soviet Academy of Sciences, was awarded the Order of the Red Banner of labor on his 60th birthday for work in his field. (*Pravda*, 7/16/65)

July 17: Two more MARINER IV close-up shots of Mars were released by NASA. The first three photographs showed an almost unbroken strip of terrain more than 600 mi. long. They revealed features down to two miles in length, including several crater-like objects, a kidney-shaped depression 20–30 mi. in width, and ridges and depressions similar to those on the moon. None of them showed straight-line features that might have been taken by earth-based observers to be canals. Except for the suggestion of an arid, wind-swept, desert-like terrain, the pictures did not bear directly on the question of life on the planet.

During a press conference held at JPL, Dr. William H. Pickering, Director of JPL, said that MARINER IV had been a "magnificent success." The computer which had reconstructed the numbers transmitted by MARINER IV into photographs had proved capable of cleaning up and intensifying the image to a remarkable degree. The pictures were of a high quality, and, in addition, the lighting conditions were expected to improve the detail in later photographs. Comparison of overlapping areas—one picture in a pair exposed through a green filter and one through a red filter—might reveal some things about color on Mars that would not otherwise be obtainable. (Transcript; Sullivan, *NYT*, 7/18/65, 1, 50; Hines, *Wash. Sun. Star*, 7/18/65, A1, A6; *WSJ*, 7/19/65, 1)

- USAF launched Thor-Agena D launch vehicle with unidentified satellite payload from Western Test Range. (*U.S. Aeron & Space Act.*, 1965, 146)

- The Bochum Observatory in West Germany reported it had received signals from Soviet satellite PROTON I which indicated it would "not remain for a long period in a stable orbit." (AP, *Wash. Sun. Star*, 7/18/65, A2)

- First flight of North American Aviation's XB-70A No. 2 from Palmdale to Edwards AFB was almost 100 per cent successful; the drogue chute failed to deploy on landing, but the aircraft braked to a stop in the normal distance. The No. 2 XB-70 reached a speed of mach 1.4 and an altitude of 40,000 ft. Wing tips were folded to the full 65° during the flight, the movable windshield ramp was operated, and some stability and control tests were carried out. Aircraft was piloted by Al White, chief NAA test pilot, and Col. Joseph Cotton, USAF XB-70A test director. (*Av. Daily*, 7/20/65)

- "The President has taken the moderate and, in this case, wise course in deciding to speed up the development of an American supersonic commercial airliner with an additional $140 million in research over the next 1½ years," said a *Washington Post* editorial. It continued: "He could have called a halt to the program and abdicated America's aviation industry leadership to France and Britain, who already are developing their Concorde, or to the Soviets with their TU-144. Or he could have followed the recommendation of some of his advisers and spent even greater sums of money to build two competing types of planes which both might end up as impractical white elephants.

". . . the President's moderate speed-up decision certainly is justified on scientific grounds, for research invariably turns up new ideas for progress. And it is justified on political grounds as well, for this country must maintain its pioneering leadership in all forms of transportation." (*Wash. Post*, 7/17/65)

July 17: Transmissions from MARINER IV were received by the nine-month-old Tidbinbilla tracking station at Canberra, Australia—one of three primary stations in the worldwide network of space tracking stations built for the Mars probe. Since February, a movable reflector antenna had maintained daily contact with MARINER IV, picking up signals from Goldstone, Calif., station in the morning and relaying them to the Johannesburg station nine hours later. As the signals were received from MARINER IV, they were retransmitted over the Pacific cable teletype to the Jet Propulsion Laboratory.

After completion of MARINER IV mission, Tidbinbilla would be used to track Surveyor vehicles. (*NYT*, 7/18/65)

• Dr. Philip Abelson suggested in *Science* that effects on man of a new revolution in genetics would be more profound than the Atomic Age. He said that although controlled laboratory change of human genetic constitution seemed some distance off, he believed it would be done or attempted: "Geneticists will create new knowledge and will have high ideals for its proper application. In practice, power to apply that knowledge, as was the case in atomic energy, will come to rest in other hands." (Abelson, *Science*, 7/12/65)

• West German intelligence had reported that Russia's Antonov An-22 transport plane was designed to carry large missiles undetected, said Omar Anderson in the *Philadelphia Bulletin*. The plane, exhibited June 15 at the International Air Show in Le Bourget, France, could carry up to 720 passengers or 80 tons of cargo. (NASA, Phil. *Sun. Bull.*, 7/18/65, 19)

July 18: ZOND III automatic space station was launched into a heliocentric orbit by U.S.S.R. from a heavy artificial satellite placed in a parking orbit around the earth by a multistage carrier rocket, Tass announced. Tass said that the "trajectory of the automatic station is close to the expected one," that communications were stable, and that all systems were functioning normally. The purpose of the launch was to "check the station's systems in conditions of prolonged space flight and the holding of scientific studies in interplanetary space," Tass said. Pictures of the farside of the moon taken by ZOND III were later released on Aug. 20. (Tass, 7/18/65; Grose, *NYT*, 7/19/65, 1, 31; Loory, *Wash. Post*, 7/19/65)

• A new computer technique had been designed by NASA Goddard Space Flight Center to make repeated rapid checks of effects of a man-made radiation belt of high-energy "Starfish" electrons formed from a U.S. hydrogen bomb test over the Pacific Ocean in 1962, and subsequent tests conducted for Project Fish Bowl, a high-altitude weapon test. Starfish electrons were estimated to have a possible life of up to 20 yrs., during which time electrons in the belt would slowly decay to the energy level of ordinary electrons. E. G. Stassinopoulos, designer of the program, warned that increased solar activity in years ahead would greatly affect the lifetimes of the Starfish electrons, making the computer relatively useless after 1966. (Sci. Serv., *NYT*, 7/18/65, 27)

July 18: MARINER IV's experimental solar-vane aiming system was expected to operate successfully for three and one half years in space, wrote Walter Sullivan in the *New York Times*. The vanes should keep the spacecraft aimed at the sun whose light shining on the vehicle's four-wing solar panels would generate electric power. Acting on pressure of sunlight and the high-velocity solar wind, the vanes would correct tendencies to drift off course.

The basic attitude control system of MARINER IV consisted of 12 jets, at the tips of the solar panels, which squirted cold compressed nitrogen. The purpose of the solar vanes was to conserve the nitrogen and thereby prolong the vehicle's serviceable lifetime. (Sullivan, *NYT*, 7/19/65, 31)

- Editorial comment on the successful transmission to earth of pictures of Mars taken by MARINER IV: "There is something absolutely staggering about the idea of a piece of machinery from this country's workshops finding its way to Mars and then pausing on schedule to make picture signals which bounce back here 12 minutes later.

    "Such a feat cheers up the most confirmed pessimist. After all, if man has the genius to reach back and forth into the universe, surely he will discover a way for peoples, nations and ideologies to live and survive together back here on this insignificant little planet." (*Wash. Post*, 7/18/65)

    "Though all its findings are not yet in, it is already clear that Mariner 4's historic journey to Mars is the most successful and most important experiment man has yet conducted in space, as well as one of the most brilliant engineering and scientific achievements of all time.

    "The triumph scored by MARINER 4 in this first successful attempt at the exploration of Mars emphasizes a point suggested by the earlier achievements of instrumented probes, notably the Ranger photographic voyages to the moon. That point is that a great deal of scientific information about earth's nearest neighbors in the solar system can be obtained relatively cheaply and without risk to human life by utilizing fully the potentialities of existing instruments. Cameras and other research devices can undoubtedly be placed in orbit about or landed softly on the surfaces of the moon and nearby planets. And intensive exploitation of the capabilities of unmanned rockets can make far safer the ultimate dispatch of man into far distant space." (*NYT*, 7/16/65)

- "Soviet propaganda needs in the wake of Mariner 4's flight to Mars appear to have been influential in determining the timing of the latest Soviet space shots," wrote Harry Schwartz in the *New York Times*. He continued: "Moscow's decision to send up last Friday two rockets—one putting five small satellites into orbit and the second orbiting the heaviest man-made object yet put into space—seems to Western observers to be a transparent Soviet effort to counteract the propaganda defeat Moscow suffered as a result of Mariner 4's historic voyage. That defeat was all the greater because of the failure of the Soviet Mars probe, Zond, sent on the same journey at roughly the same time Mariner 4 was launched." (Schwartz, *NYT*, 7/18/65)

July 18: Observers in Moscow believed that the Soviet Union would shortly launch a gigantic spacecraft with as many as four to six people aboard, said an article in Poland's *Trybura Ludu*. The spacecraft would most likely be commanded by one of the experienced cosmonauts and would have a weight of over 12 tons. It would be placed in orbit by the booster which launched PROTON I. The latter, according to the correspondent, probably used solid fuel. The correspondent speculated that the spacecraft would remain in orbit one week and that several cosmonauts would take a "walk in space" simultaneously. (*Trybura Ludu*, 7/18/65)

July 19: W. Averell Harriman, U.S. Ambassador at Large, pressed for more Soviet-American exchanges in scientific and technical fields during a two and a half hour talk with Soviet Deputy Premier Konstantin N. Rudnev. Mr. Harriman reported that the Russians had agreed that such exchanges "had been useful in the past and should be expanded." (Grose, *NYT*, 7/20/65)

• NASA and the Federal German Ministry for Scientific Research (BMwF) signed a Memorandum of Understanding for cooperation in a program of space research on the earth's radiation belts. First phase of the program would consist of sounding rocket launchings and balloon flights to test instrumentation for German experiments. Following successful completion, the second phase would attempt to place a German scientific satellite in polar orbit by 1968. The satellite, to be designed and constructed in Germany, would be launched from the Western Test Range on a Scout vehicle provided by NASA. No exchange of funds between the two organizations was contemplated. Results of the experiments would be made available to the world scientific community. (NASA Release 65–238)

• An unidentified satellite with a Thor-Agena D booster was launched by USAF from Vandenberg AFB. (AP, Wash. *Eve. Star*, 7/20/65)

• U.S.S.R. reported that ZOND III interplanetary spacecraft, launched July 18, had passed the 226,000-km. (140,120-mi.) mark in its flight towards unspecified destination. ZOND III was said to be functioning normally and radioing scientific data back to earth. (*Pravda*, 7/20/65, 1)

• Gemini 5 Astronauts L. Gordon Cooper (Lt. Col., USAF) and Charles Conrad (LCdr., USN) and their backup pilots Neil Armstrong and Elliot See met a news briefing at NASA Manned Spacecraft Center on the Gemini V mission, scheduled for August 19. Cooper explained that during the flight the astronauts would try to sight a USN Polaris missile to be launched from the East Coast and an unidentified missile to be launched from the West Coast. It was not disclosed at what point during the mission either of the launches would occur. In addition, Cooper said he and Conrad would attempt to sight rocket engines ignited on a rocket sled at Holloman AFB, N. Mex.

Also on the program would be an attempt to rendezvous with a Rendezvous Evaluation Pod (Rep) which would be ejected from the spacecraft and would drift to a distance of about 60 mi. from the spacecraft. The radar equipment onboard would then assist in performing maneuvers to achieve rendezvous with the Pod, although actual contact was not planned. Cooper said that backup system for reentry would not be used again because the thrust fuel would be used in the rendezvous experiment. No extravehicular activity would be performed during this mission.

The astronauts announced they would attempt to communicate with fellow astronaut Scott Carpenter (Cdr., USN) participating in the USN Sealab II experiment beneath the Pacific. Flight plans called for a landing in the West Atlantic.

Both astronauts felt that personal hygiene would be one of the main problems of the flight. (Transcript)

July 19: ITT World Communications, Inc., asked FCC for permission to use EARLY BIRD I satellite for live television coverage of part of the Gemini V mission scheduled for Aug. 19. ITT would like to set up a portable ground station in the Atlantic aboard carrier U.S.S. *Lake Champlain*, expected to be the primary recovery ship. (AP, *NYT*, 7/20/65)

- NASA announced it would negotiate an approximate $12-million contract with Douglas Aircraft Corp. for Delta launch support services that would cover an anticipated 15 launchings from Kennedy Space Center, NASA, and the Eastern Test Range for a 12-month period beginning Jan. 1, 1966. The contract would provide for inspection and checkout in addition to actual launching operation. (NASA Release 65–237)

- NASA was acquiring eight KC-135 jet transports and three ships to help maintain communications during Apollo moon flights. In addition, two ships of the existing DOD instrumentation fleet were being remodeled for support of the Apollo lunar mission's reentry phase. The KC-135's would be used during reentry to combat the effects of the plasma sheath blackout which had drowned out communications on previous manned launchings. In addition, three primary ground stations were being prepared at Goldstone, Calif.; Canberra; and Madrid. (NASA Apollo Proj. Off.; AP, *CSM*, 7/19/65)

- Weapons and Ammunition Div. of the Italian Air Force had formed a Missiles and Space Research Center at an airport outside Rome, *Missiles and Rockets* reported. Its establishment marked the separation of Air Force missiles and space R&D from that of the Italian National Committee on Space Research, headed by Prof. Luigi Broglio. (*M&R*, 7/19/65, 9)

- Tactical Air Command's 464th Troop Carrier Wing, Pope AFB, N.C., had been named recipient of the 1964 MacKay Trophy for the most meritorious flight of the year, being cited for its participation in the air lift of 1,500 hostages and refugees from rebel-held territory in the Congo November 1964. (*Av. Wk.*, 7/19/65, 96)

- A newly-created Manpower Utilization and Administration Office, headed by Keith Wible, became operational at NASA Marshall Space Flight Center. Wible had been assistant manager of the Michoud Assembly Facility from November 1961 to February 1965. (MSFC Release 65–181)

July 19: New York Gov. Nelson Rockefeller announced that the New York State Atomic and Space Development Authority had selected American Machine and Foundry Co. for a $2.75-million fixed-price contract to design and build a nuclear-powered plant which would produce a million gallons of fresh water daily from seawater. Named Surfside (Small Unified Reactor Facility with Systems for Isotopes, Desalting and Electricity), the plant would be constructed on Long Island and would also be able to generate 2,500 kw. of electricity and produce high-energy radioactive isotopes. (Sibley, *NYT*, 7/20/65, 1, 14)

- 100 university, research laboratory, and government scientists met in Falmouth, Mass., to recommend modifications or additions to NASA's Apollo Program and Apollo Extension Systems lunar science program. Participants included three astronauts—R. Walter Cunningham and two of the six scientist-astronauts selected on June 28, 1965: Dr. Harrison H. Schmitt of the U.S. Geological Survey and Dr. Frank C. Michel of Rice Univ. This meeting resulted in recommendation of a 10-year post-Apollo lunar exploration program of mainly manned missions, with wide variety of scientific experiments in geology, geochemistry, geophysics, and bioscience. Top priority recommended for early Apollo landing missions was collection of greatest possible number and variety of lunar material samples.

  Conference was sponsored by the Manned Space Science Program of NASA's office of Space Science and Applications. Report of the conference (NASA SP-88) was later published in January 1966. (NASA Release 66-4; NASA SP-88; NASA Release 65-239)

July 20: Saturn IB booster stage (S-IB) was successfully static-fired for 145 sec. at NASA Marshall Space Flight Center—its second captive firing. In early August, the 1.6-million-lb.-thrust, 80-ft.-long stage would be taken by barge to New Orleans and transported from there to Kennedy Space Center, NASA, for launching. (MSFC Release 65-187; *Marshall Star*, 7/28/65, 1)

- Pilot Maj. Robert Rushworth (USAF) flew X-15 No. 3 to maximum altitude of 105,400 ft. at maximum speed of 3,750 mph (mach 5.5) to obtain data with the infrared scanner. (NASA Proj. Off.)

- In a triple launch, USAF launched two 524-lb. Vela Hotel (Sentry) satellites and ORS III-1 (Octahedron research satellite) with an Atlas-Agena D booster from Eastern Test Range. Orbital data: VELA 6577-11: apogee, 72,014 mi. (115,942.5 km.); perigee, 66,583 mi. (107,198.6 km.); period, 6,713 min.; inclination, 34.6°; VELA 6564-10: apogee, 75,761 mi. (121,975.2 km.); perigee, 63,224 mi. (101,790.6 km.); period, 6,716 min.; inclination, 34.8°; ORS III-1: apogee, 69,640 mi. (112,120.4 km.); perigee, 123 mi. (198 km.); period, 6,715 min.; inclination, 34.3°.

  Expected to operate six months, the Vela Hotel satellites were part of DOD's Vela program to monitor space for violations of the nuclear test-ban treaty. ORS III-1 was monitoring natural radiation above the earth and relaying information to ground stations. (USAF Proj. Off.; UPI, *NYT*, 7/21/65, 43)

- Management of the Voyager landing capsule system had been assigned to the Jet Propulsion Laboratory, NASA announced. Harris M. Schurmeier, former Ranger Project Manager at JPL, had been named Voyager System Manager. Currently in a design study phase,

Voyager was planned for unmanned planetary explorations beginning with a Mars mission in 1971. (NASA Release 65–242; JPL Release 346)

July 20: "A manned expedition to Mars within 15 years seems entirely feasible," North American Aviation, Inc., Space and Information Systems Div. President Harrison Storms told William Hines of the Washington *Evening Star*. In charge of building the spacecraft for the Apollo moon landings, Storms offered an approximate timetable for a Mars expedition: "Start planning for it in 1970, start cutting metal in 1975 and go in 1980." A recent report from the National Academy of Sciences had suggested 1985 as a feasible target date. (Hines, Wash. *Eve. Star*, 7/20/65)

- Soviet astronomer Alexander Mikhailov said in Moscow that the MARINER IV photographic mission was a "magnificent feat . . . a staggering achievement." (*Wash. Post*, 7/25/65, A7)
- Soviet astronomer Sofia Kozlovskaya reported at the All-Union Conference on Planetary Cosmogony in Moscow that the density of matter on Mars and Venus was greater than that on earth: matter of Venus had approximately two per cent more iron; Mars had approximately six to eight per cent more iron than earth. In making these calculations, Sofia Kozlovskaya used a new "more exact model of the earth" which she had built with the data from recent seismic observations. (Tass, 7/20/65)
- Prof. S. N. Vernov, corresponding member of U.S.S.R. Academy of Sciences, noted in Tass interview the unusual weight of PROTON I and its special equipment for studying cosmic particles of super-high energy. He said the method for counting the particles, worked out eight years ago by Prof. N. L. Grigorov, had made it possible to determine the total energy of each particle separately. Scientists had hesitated to use the method in space because of the weight of the equipment. This problem had been solved with the creation of a more powerful booster, he said. (*Bakinskoy Rabochiy*, 7/20/65, 2)

July 21: A small roving vehicle as a payload for the Surveyor soft-landing lunar spacecraft would not be developed, NASA announced. Bendix Systems Div. and General Motors Defense Research Laboratories had studied the feasibility and possible scientific value of a rover. NASA's decision was based on a desire to concentrate on the development of the spacecraft itself and on scientific instruments to conduct experiments near the landing area. (NASA Release 65–245)

- "Hopper," a versatile rocket-propelled Lunar Flying Vehicle (Lfv) to transport Apollo astronauts on exploration flights of the moon's surface, had been designed by Bell Aerosystems Co. as a result of a 12-month study conducted for NASA Marshall Space Flight Center, Bell announced. The 400-lb., four-legged, rectangular Lfv would be desk size and propelled by a cluster of five 100-lb-thrust rocket engines; it could fly 50 mi. nonstop.

  Bell also disclosed receipt of a $489,898 follow-on NASA contract to design a Manned Flying System (Mfs) capable of carrying one astronaut-scientist and 300 lbs. of equipment or two astronauts. Intended primarily as an exploratory device, the Mfs would be able to fly 15 mi. round trip without fueling. (Bell Aerosystems Co. Release)

July 21: Tariff amendments for use of EARLY BIRD I satellite for television transmission were filed by ComSatCorp with the FCC. Under the new regulations, which would supersede the May 28 tariff and become effective July 26, circuits between the Andover, Me., ground station and EARLY BIRD I could be leased for a minimum period of 10 min. rather than 30 min. as previously proposed. The European communications agencies were expected to offer comparable 10-min. services from the satellite to their respective ground stations. The rate for one-way TV service during non-peak period would be $1,800 for the first 10 min. and $32 for each additional minute. During peak period, it would be $3,000 for the first 10 min. and $48 for each additional minute. For two-way transmission and reception of black-and-white television signals or one-way color TV, the change would be 150 per cent of the charge for a one-way TV channel. Weekday peak period would be between 7 a.m. and 4 p.m., New York City time; non-peak period would be between 5 a.m. and 7 a.m.; between 4 p.m. and 9 p.m.; and all day Saturday and Sunday. (ComSatCorp Release; *WSJ*, 7/22/65, 2)

- Dr. Albert C. Hall, Deputy Director of Defense Research and Engineering, told the National Space Club meeting in Washington, D.C., that a significant announcement on the MoL program would be made in the "near future," according to *Missile/Space Daily*. When asked if MoL would match the capability demonstrated by the U.S.S.R. in July 16 launch of PROTON I, Dr. Hall replied: "We feel a capability in the order of placing 25,000 lbs. in orbit would enable us to meet DOD needs." Dr. Hall said that DOD was interested in the development of recoverable, reusable boosters and he felt that the reduction of problems in mating payloads to boosters had been a "significant event during the past year." (*M/S Daily*, 7/21/65; NSC Release)

- NASA Marshall Space Flight Center had issued 11 new contracts and modified one other to provide for further studies and component development on the Apollo Extension Systems (Aes). (MSFC Release 65-186)

- Astronomers of the Pulkovo Observatory, Leningrad, had completed installation of a more powerful telescope, Tass announced. Diameter of the instrument's main mirror was 440 mm. (17.3 in.); its focal length, 17.5 m. (57.33 ft.). The new telescope would be able to produce spectrograms from any sector of the sun and simultaneously measure the magnetic field. (Tass, 7/21/65)

July 22: Results of a recent Gallup Poll indicated that 47% of the American public believed that the United States was ahead of Russia in the "space race." A 1961 poll had shown that most Americans viewed the space race as an even contest. Asked if they, themselves, would like to go to the moon, 87% of the people approached said no. (*Wash. Post*, 7/23/65, A2)

- NASA announced four major personnel changes: Willis H. Shapley, Deputy Chief of the Military Div. of the Bureau of the Budget, would become Associate Deputy Administrator effective September 1, 1965; Dr. Harry J. Goett, Director of Goddard Space Flight Center, would become Special Assistant to Administrator James E. Webb, effective immediately; Dr. John F. Clark, Director of Sciences in the NASA Office of Space Science and Applications, would become Acting Director, Goddard Space Flight Center; and Dr. John W. Townsend, Assistant Director, Goddard Office of Space Science and Applications, would become Deputy Director of GSFC —a new post.

When Shapley's appointment became effective, the present Policy Planning Board would be discontinued; the office of Policy Planning would aid him in preparing policy matters for the Administrator, Deputy Administrator, and Associate Administrator. (NASA Release 65-243)

July 22: President Johnson told a group of young foreign journalists at the White House that he envisioned a day when communications satellites would be able to relay telecasts of United Nations' sessions to the homes of "men everywhere." He added that "from better communications must surely come better understanding." (UPI, *Houston Chron.*, 7/23/65)

- Dr. Homer E. Newell, NASA Associate Administrator for Space Science and Applications, spoke at the dedication ceremonies for Martin Co.'s new Research Institute for Advanced Studies in Relay, Md., about some of the progress that had been made in space science: "Because of the space program, geophysics is experiencing a tremendous broadening of its horizons. . . . Moreover, geophysics is being carried forward to new domains, as instruments reach the moon and the planets, giving to the discipline a perspective it could never achieve as long as geophysics was confined to a single body of the solar system.

    "Similarly, the space program is giving a new dimension to astronomy. The ability to observe above the filtering distorting atmosphere in wavelengths not hitherto observable promises exciting new discoveries. . . .

    "The field of physics finds in the regions of outer space a laboratory of challenging opportunity. In interplanetary space, matter and fields exist under conditions unobtainable in the laboratory on the ground. . . .

    "It is, indeed, interesting to observe that one of the impacts of space efforts on physics, geophysics, and astronomy has been to draw the three disciplines together more closely than they have been drawn together in the past. In the investigation of sun-earth relationships, a most complex and challenging area of investigation, all three of these disciplines find themselves in partnership on problems of common interest. . . .

    "The impact of the space program on bioscience is still developing . . . of particular importance is the area of exobiology. . . .

    "The satellite, space probe, and manned spacecraft give the scientist a new approach to the solution of many important problems. They serve to strengthen his hand—if used effectively. . . ." (Text)

- ComSatCorp, in a letter to the FCC, outlined its position regarding ITT World Communications, Inc.'s request to use EARLY BIRD I satellite for live television coverage of the recovery of the Gemini 5 astronauts. George P. Sampson, ComSatCorp's Vice President of Operations, said that although ITT and ComSatCorp had agreed that technical and operational aspects of the shipboard station would be subject to ComSatCorp's control, ComSatCorp took "firm exception" to ITT's insistence that the FCC designate ITT as the responsible operating entity. Sampson noted the FCC's May 12 decision giving ComSat-

Corp sole responsibility for operation of the initial terminal stations for the proposed global satellite system and said: "Despite our fundamental concern with the proposal, ComSat Corp is willing to give consideration to a joint application for the operation of the station." (AP, *Wash. Post*, 7/23/65)

July 22: Roy William Johnson, first director of DOD's Advanced Research Project Agency (ARPA) in 1958 died at the age of 59. Johnson said at the time of his appointment: "Space will tax the imagination of the whole world for the next 100 years." He retired in 1959. (*Wash. Post*, 7/23/65, B8)

- Former astronaut Col. John Glenn (USMC, Ret.) was suffering no effects from last year's ear injury, reported the Wilford Hall Air Force Hospital in Houston. (*Houston Chron.*, 7/23/65)
- Soviet professor V. V. Fedynskiy, in an interview with *Uchitel'skaya Gazeta*, said that PROTON I had demonstrated the possibility of getting more powerful spaceships into orbit, which, if necessary, could lift into space a crew and a large quantity of research equipment. He said use of the new rocket booster would permit more intensive studies of space, including those that would allow a more thorough research of the physical fields and internal structure of the earth directly from space. Such studies would be of primary interest for geophysics and geology, he said. (*Uchitel'skaya Gazeta*, 7/22/65, 4)

July 23: U.S.S.R. launched COSMOS LXXVI with scientific instruments aboard for investigation of outer space, Tass announced. Orbital data: apogee, 530 km. (261 mi.); perigee, 261 km. (161.8 mi.); period, 92.2 min.; inclination, 49°. All systems were reported to be functioning normally. (*Krasnaya Zvezda*, 7/24/65, 1, ATSS–T Trans.)

- Picketing over a contract dispute by Las Cruces, N. Mex., Carpenters Union Local 1962 of White Sands Missile Range construction project resulted in a partial temporary shutdown of the $133,900 project to build an addition to NASA Warehouse No. 1. Activities on the site included testing Apollo service module engines and preparation of test equipment for the Lunar Excursion Module. (Las Cruces *Sun-News*, 7/23/65)

July 24: FAA awarded $268,635 contract to the Flight Safety Foundation, Inc., to launch Project Gape (General Aviation Pilot Education), a safety campaign to reduce general aviation aircraft accidents. Directed primarily at combating the lack of pilot proficiency and knowledge of safety flight procedures involved in most general aviation accidents, Gape would employ a vigorous publicity campaign, displays, meetings, seminars, special conferences, personal contacts, and similar educational activities. (FAA Release 65–60)

- The four largest closed die forgings ever produced from maraging steel were delivered by the Wyman-Gordon Co. to Launch Complex 34 at Kennedy Space Center, NASA. Weighing 15,500 lbs. each and measuring 105 in. by 36 in., they would be used to anchor the mobile service tower for the Saturn IB rocket. (*NYT*, 7/4/65, 25)
- How to achieve the best possible profit from air transportation had been determined at the U.S.S.R.'s Kiev Institute of Civil Aviation Engineers by the computing technology laboratory, *Pravda Ukrainy* reported. The laboratory's chief engineer, V. V. Buryy, said an elec-

tronic computer would suggest the best use of the Ukrainian air fleet. (*Pravda Ukrainy*, 7/24/65, 4)

July 25: "Should the MOL blueprint as envisioned by the Air Force's Systems Command be approved by Defense Secretary Robert S. McNamara, the Navy conceivably could be the first to use this earth-orbiting station as a new technique for charting ship movements on all the oceans of the world," wrote Frank Macomber in the *San Diego Union*. Macomber envisioned "two Navy astronauts . . . spinning around the earth as early as 1968 in a 10-foot diameter, 25-foot long orbital laboratory—about the size of a small house trailer filled with electronic detection gear . . ." (Macomber, CNS, *San Diego Union*, 7/25/65, 13)

- The Mohole project to drill deep in the ocean floor was described during a seminar at the Institute on Man and Science, Rensselaerville, N.Y., by Dr. Columbus Iselin, former director of the Oceanographic Institute at Woods Hole, Mass. In reply to a question about future humans possibly living in the ocean, he said: "I don't see what you'd accomplish down there. It's cold, dark, and nasty. It's a popular idea but an impractical one." (*NYT*, 7/26/65, 13)
- MARINER IV finished transmitting its 21 photographs of Mars and sent about 10 percent of a 22nd picture before its tape ran out. The later pictures were eagerly awaited by JPL scientists because they should show the dark regions of Mars that some people believed harbor life. The photos were scheduled to be released later this week. (*Wash. Post*, 7/25/65, A7)
- Pan American World Airways had ordered four Boeing 727–QC (quick change) jet aircraft that could be converted in less than half an hour from an all-cargo plane to a complete passenger aircraft or a passenger-cargo plane. Aircraft would have a psssenger capacity of 119; in an all-cargo operation the plane could carry a payload of 41,000 lbs. more than 1,400 mi. Delivery was scheduled for summer 1966. (*NYT*, 7/26/65, 39M)
- During a test of the first of two crawler-transporter vehicles at the Apollo-Saturn V Launch Complex 39 at Cape Kennedy, some failures occurred in the roller bearings which support the tracks. At the time of failure, the crawler was carrying a mobile launch stand. (*Miami Herald*, 9/30/65)
- Caption under a photograph in East German newspaper *Berliner Zeitung* read: "A model of a future Soviet space station envisages six hermetically grouped sections around a central core. These are: the control station, a laboratory and a garden, an orientation system, the radar equipment, and a heliostation. In addition, the station has facilities for voice communications with space ships." (*Berliner Zeitung*, 7/25/65, 4)

July 26: Dr. Hugh L. Dryden, NASA Deputy Administrator, said during a recent interview that the next immediate step beyond the initial Apollo lunar landings was to extend the usefulness of both spacecraft and launch vehicle to permit longer stays in earth orbit and on the moon. "This fall we will have to make a definite recommendation to fund one or both," he said. (*M/S Daily*, 7/26/65)

- A Gemini circumlunar flight had not been approved but the idea was receiving serious study, Kenneth S. Kleinknecht, Deputy Manager of the Gemini Project Office at NASA Manned Spacecraft Center, told reporters. The Gemini capsule could circle the far side of the moon—240,000 mi. from earth—and then return to its home planet, a round trip of about six days.

Kleinknecht said if the proposal were accepted, a Titan III–C rocket equipped with two upper stages called "transtages" would place one of the transtages in orbit. A Titan II would send the Gemini craft after the transtage and the two would be docked in orbit. Gemini then would use the transtage propulsion system and its maneuverability for the lunar voyage.

The feasibility was being studied by NASA; the Martin Co., systems manager for Titan III–C; and Aerojet-General, which builds the transtage. "We are always studying possible future missions for Gemini," Kleinknecht added. (Benedict, N.O. *Times-Picayune*, 7/26/65)

July 26: NASA announced the addition of two new pilots to the joint NASA-USAF X–15 research program: Capt. William J. Knight (USAF) and William H. Dana, civilian. They were expected to make their first flights this fall. (NASA Release 65–244; FRC Release 16–65)

• The Agena target vehicle for Gemini 6 was delivered to Cape Kennedy by the Pregnant Guppy aircraft. (*Orl. Sent.*, 7/27/65)

July 26–29: Second annual meeting of the AIAA was held in San Francisco. Dr. William H. Pickering, Director of the Jet Propulsion Laboratory, opened the meeting with a signal from MARINER IV soaring through space 144 million mi. from earth. Received at the Johannesburg tracking station, the signal was relayed to San Francisco from JPL in Pasadena. Speeches at AIAA:

President of Lockheed Aircraft Corp., Daniel J. Haughton, speaking on "Your Role in the New Environment," said: "American aircraft products have dominated world markets for decades. Today at least 90 percent of the airline jets flown in the free world are built in the United States. Aircraft exports, both military and civilian, are $1.2 billion a year. Imports, on the other hand, are almost negligible—$90 million last year.

"But now something new has been added—and that is more thrust toward cooperative programs. . . .

"Cooperative programs are increasing. Only last May Secretary of Defense McNamara proposed a common market for military hardware. If successful, it will help integrate the total NATO market, including the U.S., even more closely, and will also permit foreign firms to sell selectively in the American defense market.

"The total effect of these developments means, for U.S. manufacturers, more competition, more cooperative programs, and a more flexible approach . . . but I do not believe this means we must abdicate our technological leadership. I think, on the contrary, that we must strengthen it. One of my colleagues suggests we should add a new item to our national goals—the goal of overwhelming technical strength. . . ." (Text)

Brig. Gen. Edmund F. O'Connor (USAF), Director of the NASA Marshall Space Flight Center's Industrial Operations, told the AIAA meeting: "The progress of the Saturn IB has been excellent. We are on schedule to begin flight tests next year, to be followed by manned flights in 1967.

"In the Saturn V/Apollo program we are also on schedule. . . ." (Text)

July 26: USAF's Titan III–C rocket could match the Soviet Union's most powerful booster pound for pound in launching heavy-weight satellites into orbit, according to William G. Purdy, General Manager, Launch Vehicles, Martin Co. "Engineering analysis of Titan III–C's performance on its maiden flight of June [18] shows that a payload of nearly 27,000 pounds could have been orbited. In the past, Titan III–C's maximum payload has been computed at 25,000 lbs." Mr. Purdy said that the Soviet Union's best was the 26,840-lb. PROTON I satellite launched July 16 by powerful new rocket. (Martin Co. Release)

NASA Science Advisory Committee was considering a proposal by radio astronomers to create a 10-mi.-dia. antenna array to permit a "look" into the past with radio energy, disclosed Bernard M. Oliver, Vice President of a Palo Alto electronics firm. He estimated that 1,000 antennas, each perhaps 10-ft.-dia., spread out over the area of a 10-mi. circle, would provide the resolution of signal intake necessary for the kind of radio observation he was suggesting. Intake of all the antennas would be focused at a single laboratory. There the radio energy would be converted to sound energy and then into light to provide photographic images equivalent to optical images now taken of the moon. Oliver said: "We quite strongly believe that such an exploration tool can provide the maximum amount of information relating to the origins of the universe, to the life cycle, and to the destiny of the universe, and, in a sense, to the destiny of the human race as a whole." (*Chic. Trib.*, 7/28/65)

Missions to Jupiter could be conducted by 1969, suggested Eugene Lally, Space-General Corp. engineer, in a paper presented to the AIAA outlining a program of six missions to the planet Jupiter, beginning with a fly-by in 1969 and culminating with a Jovian orbit in the mid-1980's. A Jupiter mission would take about two years. Lally postulated that payload weights ranging from a 650-lb. payload to a 12,400-lb. orbiter payload could be easily handled by boosters ranging from Atlas-Centaur with an added kick stage, through the Saturn V which would be used to place a man on the moon. Lally speculated that an instrumented probe would be able to obtain information on the constitution of the core and surface layers, the atmosphere, nature of the largest of the planet's spots, topography, constitution of meteoroids in its vicinity and the presence of small satellites not yet discovered with telescopes. Lally's probe would contain experiments to measure the atmosphere, magnetic fields, and gravitational fields; to conduct infrared and microwave examinations of the surface; and to provide pictures similar to those taken of Mars by MARINER IV. (Space-General Corp. Release)

An Electro-Optical Systems (EOS) bombardment ion engine had successfully operated for more than 2,610 hrs. in vacuum chamber conditions and, as a result of that extended run, "lifetimes in excess of 10,000 hours can now be calculated for the tiny engine under conditions of space flight," Gordon Sohl, Electro-Optical Systems, Inc., told the AIAA meeting in San Francisco. Fueled with cesium, the EOS engine was less than two feet long and weighed 10 lbs. fully loaded with a five-pound fuel supply. It provided a power-to-thrust ratio of 182 kw. per pound. Financed by NASA Lewis Research Center, the EOS research program recently received a follow-on funding from LRC to determine "if the ion engine with 100 pounds of cesium fuel is equivalent in thrust to a conventional chemical rocket carrying a ton of propellant." (EOS Release)

Dr. Raymond L. Bisplinghoff, NASA Associate Administrator for Advanced Research and Technology, presented the third annual Theodore von Kármán lecture. Dr. Bisplinghoff discussed advances in air transport, such as V/Stol aircraft and the hypersonic transport. He pointed out that within 20 yrs. an estimated 130 million persons (about 50% of the U.S. population) would be living in three main metropolitan areas and that there would be an increasing demand for "air buses." Dr. Bisplinghoff suggested that the hydrogen-fueled hypersonic transport could be used as a cheap transport, or as an earth-to-orbit aircraft. He said that the hypersonic transport would carry passengers halfway around the earth nonstop at speeds up to 8,000 mph, but there was one problem area: . . . "Where the airplane threatens to overpower the pilot with characteristics which make the airplane unflyable by human systems." Dr. Bisplinghoff called for increased research for operational experience with the scramjet (supersonic combustion ramjet) engine as a prelude to developing the hypersonic transports.

In a press conference preceding his lecture, Dr. Bisplinghoff predicted "another revolution" for personal aircraft in the form of highly simplified controls similar to those in automobiles. He also urged quick Government action to begin the acquisition of a follow-on hypersonic research aircraft to succeed the X-15 and run the flight profile out to mach 10 to 12. (NASA Release 65-247; Text, *M/S Daily*, 7/28/65; EPH)

NASA Electronics Research Center Director Dr. Winston E. Kock discussed ERC in address. During FY 1965, he said, ERC "awarded 33 contracts totaling almost $2 million. The organizations to whom these contracts were awarded are spread widely throughout the country. . . .

"The average value of the 33 contracts was $59,000, the largest being $285,000. These contracts were almost all in fairly basic research fields, ranging from integrated circuits and thin film space-charge limited triodes, through research in millimeter and submillimeter waves, optical wave-guides and optical components, to space-borne memory organizations, laser gyros and fluid storage and control devices. . . ."

He discussed ERC personnel: "We have grown from a group of 65 at our inception last year [Sept. 1] to a total of 244. As we are still in the formative stages, our scientific and engineering personnel now total only 92 out of the 244, but when we have reached our full strength of 2100, we expect that our staff will be about equally divided between scientific and engineering, technical support and administrative. . . ." (Text)

Thomas Bilhorn, Manager, Mechanics Section, Scientific Balloon Facility at the National Center for Atmospheric Research (NCAR), revealed that facilities for launching balloons capable of carrying 16,000-lb. payloads were expected to be completed by next summer at NCAR's Scientific Balloon Flight Facility in Texas. The device under construction, called an inflation shelter, would have a ceiling height of 140 ft. and interior diameter of 120 ft. It would reduce handling problems during inflation and would give the capability of periods of hold, recall, and post-inflation inspection and repair. (Text)

"There is nothing hostile or aggressive in the military space program we foresee. It is entirely within the context of a national program expressly devoted to peaceful purposes," said Dr. Albert C. Hall, Deputy Director (Space) of Defense Research and Engineering. He continued: "It is likely that military interest will remain focused primarily on near-earth missions, out to synchronous orbit, certainly through 1975. We expect to continue our very large and vigorous unmanned military space program which is performing very important functions. The need for these programs will not diminish since they are by far the most efficient and cheapest way of performing specified tasks. With a steadily increasing experience and know-how in manned space flight, we may expect that spacecraft will acquire characteristics permitting rendezvous, station-keeping, docking, and transfer of man and material. We will likely acquire the means of sustaining military men in space for the periods of time we require. Booster capacities are not likely to limit the applications, but the booster and payload costs will continue to do so." (Text)

In AIAA session on the "History of Rocket Research Airplanes," John Stack, former NACA-NASA designer, pointed out how little was known about transonic and supersonic flight in 1943, when X-1 was conceived. Walter C. Williams, Dr. Raymond Bisplinghoff, and Walter T. Bonney, among other speakers, also stressed the key role of engineer test pilots in the success of the X-1, D-558, X-2, and X-15 programs. Session was chaired by Robert Perry of RAND Corporation, who traced the history of rocket-powered aircraft. (EPH; RAND P-3154)

DeMarquis D. Wyatt, NASA Deputy Associate Administrator for Programing, told the meeting: "Any discussion of the NASA space program for the next 10 years must be given and received with a grain of salt. Reconstruct in your minds the situation 10 years ago and evaluate the validity of any discussion at that time of the NACA program for the period 1955–1965. Such a discussion would not have even mentioned space in any serious fashion. A paper delivered in 1960 that attempted to outline the NASA program for 1960–1970 would have widely missed the mark in the prediction of today's realities. One can, therefore, conclude that one of the major management problems of the national space program is the lack of an adequate crystal ball for forecasting the future.

"In the $6\frac{1}{2}$ years of its existence NASA has carried out a vigorous program of space activities designed to yield:

"(a) a description and scientific understanding of the space environment; (b) the development of a broadly based national capability and capacity for manned and unmanned operations in space, and (c) the development of practical uses of space.

"Decisions will have to be made with an appreciation of, and indeed the shaping of, our whole national attitude toward space in relation to our other national requirements and interests. Far too many future studies within and without the government are predicated on the assumption that the national investment in space research and technology will at least grow at the rate of the Gross National Product. This has not been true for the past several years and does not afford a necessarily sound planning assumption for the future. Our total federal budget has leveled off in spite of the great increase in the national product in recent years. If this trend holds, then marked increases in space expenditures can only come about through decreases in other federal spending. Such an assumption would indeed be a slender reed upon which to prognosticate the future." (Text)

Dr. Vincent P. Rock, Director of George Washington Univ.'s Program of Policy Studies in Science and Technology, said: "People are shaped by their environment. Technology is creating a new environment. In these circumstances power flows to those with access to technology—all technology, not simply military weapons. The exercise of power brings with it responsibility. The ultimate responsibility of those who exercise technological power is the achievement of a hospitable environment for mankind." "Text)

July 27: USAF XB-70A research bomber reached a speed of 1,850 mph at 66,000 ft. in a test flight at Edwards AFB. (AFFTC Release 65-7-18)

- The Civilian-Military Liaison Committee created by Section 204 of the National Aeronautics and Space Act of 1958 was abolished by Reorganization Plan No. 4 of 1965, effective this date, which President Johnson had submitted to Congress May 27, 1965. CMLC functions were transferred to the President of the United States. (*F.R.*, 7/28/65, 9353; NASA Hq. Memorandum)

- Dr. George E. Mueller, NASA Associate Administrator for Manned Space Flight, speaking at the Annual Convention of the American Trial Lawyers Association in Miami Beach, said: "I would like first to comment on what appears to be a general misconception about the overall purposes of the Apollo program. Many people seem to believe that a landing on the moon, ahead of the Soviets, is the paramount objective. This is not so. The principal goal is to make the United States first in space by the end of this decade, and to make this pre-eminence unmistakably clear to the world.

  "Why . . . is it so vital that the United States be pre-eminent in space? There are many reasons that can be cited, and they fall generally into two major categories—the imperative reasons and the ancillary, or spin-off benefits. It is imperative, in the Cold War arena, that the United States be first for reasons of national security, national pride of achievement, and the international prestige. The ancillary . . . reasons include the benefits of scientific discovery; the stimulation of economic and social progress; technological advancement, including the civilian application and utilization of the products of space-oriented research; and the compelling urge of man to explore and to discover." (Text)

- Edward Z. Gray, Director of NASA's Advanced Manned Missions Program, in an interview with *Missile Space Daily*, said that the logical step toward manned flights to Mars "would be a nine-man space station, about the same size as a crew module on a planetary flight which would be continuously deployed and periodically supplied." He added, however, that such a station was "beyond the FY '68 budget."

Gray said that meanwhile NASA would continue its study program on planetary flight. Propulsion was the limiting factor, he explained; other problems such as power, life support, and communications were expected to evolve satisfactorily by the late 1970's. He foresaw the possibility of a manned flyby mission around Mars by the mid-1970's.

"Development of a new launch vehicle is probable in another four or five years," he said, "and a reusable system is a 'good bet.' Large solids will also be a candidate and could be competitive with the best of the others until the reusable vehicle comes along." Even then, Gray saw a division of labor in which solids would handle payloads in the 1-to-5-million-lb. class and reusable vehicles would concentrate on 20,000-to-50,000-lb. payloads. Reusable solids were also under consideration. (M/S Daily, 7/27/65)

July 27: NASA Ames Research Center was spending over $20 million in an expansion program designed to provide the necessary research tools to stay ahead of the industry's hardware designers, reported Robert Lindsey in *Missile Space Daily*. Major facets included a new advanced space flight guidance simulator costing more than $12 million, a $4.2 million life sciences building, and a $1.4 million supersonic transport flight simulator.

Space-related research would account for about 75% of the center's projects, but ARC was continuing a broad-based program in atmospheric flight, ranging from V/Stol research to a current major study of hypersonic transports. (Lindsey, *M/S Daily*, 7/27/65, 1, 2)

- From July 28 to October 15, 1965, U.S.S.R. would conduct launchings in the Pacific Ocean basin of new types of rockets carrying space objects, Tass announced. Test area would be 80-n.mi.-dia. circle centering on 37°39′ N and 173°25′ E. Governments were requested to warn their nationals not to be in the ocean and air space from 12 noon to 12 midnight local time daily. (Tass, *Komsomolskaya Pravda*, 7/27/65, 3, ATSS-T Trans.)

- Gen. Omar N. Bradley (USA, Ret.), Chairman of the Board of the Bulova Watch Co., disclosed at the annual stockholders meeting that Accutron-type electronic clocks were being designed for use on the control panel of NASA's Project Apollo Lunar Excursion Module. Star-tracking devices incorporating Accutron would be in the moon vehicle's navigation system. A special "moon van" containing components and examples of hardware for use in Project Apollo was on display at the meeting. (*NYT*, 7/28/65, 43C)

- Construction of ESRO's first launch site, located 100 mi. north of the Arctic Circle at Kiruna, Sweden, was nearing completion, reported John Herbert in *Missile Space Daily*. A $10 million investment, the range was scheduled to open in May 1966 with the launch of a French-made Centaure sounding rocket to probe the ionosphere. At least 50 launchings of high-altitude probes were programed annually for the next seven years with English Skylarks slated for use after the Centaure series.

The rocket experiments would aid European researchers in their studies of magnetic storms, the Northern lights, temperatures, air currents, and communications—disturbing phenomena in an area ranging between 20 and 150 mi. above the earth. (Herbert, *M/S Daily*, 7/27/65)

July 27: Describing the growing space role of Woomera, Australia's equivalent to Kennedy Space Center, NASA, R. N. Hughes-Jones said is *Missile Space Daily:* "Cooperation between Australia and the United States in the field of space research began [in 1957] when installations were set up at Woomera for the International Geophysical Year. They occupied a couple of caravans.

"In 1960, in a formal exchange of notes, the governments of the two countries affirmed their intention to extend the cooperative program to space flight operations.

"Under it, the Australian Department of Supply establishes and operates stations on behalf of NASA, for tracking, communicating with and obtaining telemetered information from U.S. space vehicles.

"Establishment of the station at Carnarvon . . . consolidated at one site NASA's ground support facilities for the Gemini project.

"Tidbinbilla was officially opened on March 19 of this year, and is the first of three NASA facilities programed for the Australian Capital Territory.

"The second of the three is at Orroral Valley and is currently under construction. It will track the larger and the more complex of the U.S. scientific satellites, while the third, at Honeysuckle Creek, will support the U.S. Lunar Manned Space-flight Project." (Hughes-Jones, *M/S Daily*, 7/27/65)

- American Airlines would directly support San Francisco Oakland Helicopter Airlines under the terms of an agreement recently submitted to the CAB for approval. (*NYT*, 7/28/65, 54M)
- FAA awarded IBM $1,761,470 contract for two semi-automatic air traffic control systems. Scheduled for operation in 1967, one system would be located at FAA's National Aviation Facilities Experimental Center in Atlantic City, N.J., and the other at FAA's field site in Jacksonville, Fla. (FAA Release T65–37)

July 28: A 28-day-old wage strike by 60 guards at NASA Goddard Space Flight Center ended with agreement on a two-year contract by the United Plant Guard Workers of America and Wackenhut Services, Inc., which employed the GSFC guards. (*Wash. Post*, 7/29/65, A18)

- NASA awarded $1,190,000 facilities grant to the Univ. of Florida for the construction of a Space Science Building on campus. James E. Webb, NASA Administrator, said the new grant "will permit the expansion of theoretical and experimental research in aeronautical and space sciences and will enable the university to train an increased number of highly qualified young researchers." (NASA Release 65–248)
- NASA Administrator James E. Webb, speaking on statistical standards at the National Governors' Conference in Minneapolis, said: ". . . the timely availability of accurate, comprehensive data—based on valid and accepted concepts and definitions—will become increasingly important to the effective conduct of state government. The need for

compatibility between federal and state data systems is recognized at the top levels of government. . . .

"It is clear that the goals of complete uniformity, perfect comparability and total integration of statistical and management information systems will not soon be accomplished. But it is equally clear that urgent efforts toward these goals are being made and are necessary if state and local governments are to keep pace with the needs of modern society." Webb recommended that each state: "1. *Establish a state statistical standards unit* . . . 2. *Sponsor a National Conference on the Comparability of Statistics Among the States* . . . and 3. *Examine applicability of modern information technology at the state and local levels.* . . ." (Text)

July 28: Brig. Gen. Joseph S. Bleymaier (USAF), Deputy Commander for Manned Systems, AFSC Space Systems Div., was cited by President Johnson for cost reduction efforts on the Titan III program. The citation said his efforts had enabled the Air Force to achieve on time all major program objectives with fewer test flights than programed, "thus reducing defense costs $33 million in fiscal year 1965." (DOD Release 487–65)

- Rules for assigning specific emergency evacuation duties to crew members of helicopters operating in scheduled air carrier operations, controls to cover drinking aboard helicopters, and measures to prevent intoxicated persons from boarding helicopters were proposed by FAA in an effort to bring operations governing scheduled helicopter operations in line with rules governing air-carrier fixed-wing operations. (FAA Release T65–38)

- Almost every major aspect of European space programs, both international and national, would be intensively reviewed within the next four months, wrote *Missile Space Daily*. ESRO would meet in Paris to review early progress of its eight-year $310 million program to place 17 satellites of varying size and complexity into orbit. ELDO planned to meet in early fall to discuss continuation of the Eldo-A booster program and possible speed-up in the Eldo-B follow-on vehicle. In Great Britain and Germany, key government reviews and position papers about roles in future space activities were expected to occur within the next several weeks. (Getler, *M/S Daily*, 7/28/65)

July 29: Dr. Robert B. Leighton, Cal Tech professor, summarized the results of the MARINER IV mission to President Johnson in a White House ceremony during which the remaining photos transmitted by the spacecraft were presented to the President and the Nation. Dr. Leighton said: "Man's first close-up look at Mars has revealed the scientifically startling fact that at least part of its surface is covered with large craters. . . .

"The existence of Martian craters is demonstrated beyond question; their meaning and significance is, of course, a matter of interpretation. The seventy craters clearly distinguishable on Mariner photos Nos. 5 through 15, range in diameter from three to 75 miles. It seems likely that smaller craters exist, and there also may be still larger ones than those photographed, since the Mariner photographs, in total, sampled only about one percent of the Martian surface.

"The observed craters have rims rising a few hundred feet above the surrounding surface and depths of a few thousand feet below the rims. Crater walls so far measured seem to slope at angles up to about 10 degrees.

"The number of large craters per unit area on the Martian surface is closely comparable to the densely cratered upland areas of the Moon."

Dr. Leighton said that no earth-like features were recognized and that clouds "were not identified and the flight path did not cross either polar cap."

Some of the fundamental inferences drawn from the MARINER IV photos were:

"1. In terms of its evolutionary history, Mars is more Moon-like than Earth-like. Nonetheless, because it has an atmosphere, Mars may shed much light on early phases of Earth's history.

"2. Reasoning by analogy with the Moon, much of the heavily cratered surface of Mars must be very ancient—perhaps two to five billion years old.

"3. The remarkable state of preservation of such an ancient surface leads us to the inference that no atmosphere significantly denser than the present very thin one has characterized the planet since that surface was formed. Similarly, it is difficult to believe that free water in quantities sufficient to form streams or to fill oceans could have existed anywhere on Mars since that time. The presence of such amounts of water (and consequent atmosphere) would have caused severe erosion over the entire surface.

"4. The principal topographic features of Mars photographed by Mariner have not been produced by stress and deformation originating within the planet, in distinction to the case of the Earth. Earth is internally dynamic giving rise to mountains, continents, and other such features, while evidently Mars has long been inactive. The lack of internal activity is also consistent with the absence of a significant magnetic field on Mars as was determined by the Mariner magnetometer experiment.

"5. As we had anticipated, Mariner photos neither demonstrate nor preclude the possible existence of life on Mars. The search for a fossil record does appear less promising if Martian oceans never existed. On the other hand, if the Martian surface is truly in its primitive form, the surface may prove to be the best—perhaps the only —place in the solar system still preserving clues to original organic development, traces of which have long since disappeared from Earth."

Dr. Leighton noted that "one of the most difficult problems associated with the Mariner photographic mission to Mars was the wide illumination range" that was encountered. Assisting Dr. Leighton in his presentation were: Prof. Bruce C. Murray, Cal Tech; Prof. Robert P. Sharp, Cal Tech; Richard K. Sloan, JPL; and J. Denton Allen, JPL. President Johnson said he was a bit relieved that MARINER's photographs "didn't show more signs of life out there." He described the Mars pictures as "awe-inspiring" and said that "the flight of Mariner 4 will stand as one of the great advances of man's quest to extend the horizons of human knowledge."

President Johnson presented the following awards: to Dr. William H. Pickering, Director of JPL, the NASA Distinguished Service Medal;

to Jack N. James, Assistant Director of JPL for Lunar and Planetary Projects, the NASA Medal for Exceptional Scientific Achievement; and to Oran W. Nicks, Director of Lunar and Planetary Programs, the NASA Medal of Outstanding Leadership. (Transcript; Sullivan, *NYT*, 7/30/65; Simons, *Wash. Post*, 7/30/65, A1, A3)

July 29: During MARINER IV press conference at NASA Hq., NASA Administrator James E. Webb, in response to a question about the possibility of "another Mariner mission in the relatively near future," replied: "We had decided some time ago not to fly another one of these missions with the equipment we used on MARINER IV but to concentrate on much more important work that we can do with advanced equipment.

"So I should say that it's highly unlikely that we would revive consideration of another flight."

Mr. Webb said that the MARINER IV flight, including the MARINER III attempt that failed, "cost over $100 million." He continued: "Dr. Pickering and his group with the American industrial companies proved that we could move out from the earth and get to the planet and do what we intended to do. . . .

"Second, the scientific experimenters worked in close harmony, and the relationship between the experimenters and the people responsible for making the flight get to its destination and bring the data back was I think a very outstanding achievement and also is a part of the learning process in the space program.

"Lastly, it certainly is very important, as we have emphasized in the manned spaceflight program, to gain some knowledge, even though it is not full and complete, at as early a stage as possible, because we have planned a broad-based program over a ten-year period. But we also have the capability of change and modification in the program. And this gives us a good deal of information that in my view will have a strong bearing on the decisions to be made in the 1967 budget."

Mr. Webb said that he did not see "very much difference between our capability and their [Soviet] capability at this time but that we are both moving into a period when we will be able to select certain options for further emphasis and development."

It was revealed that the results of the MARINER IV mission would be published in *Science*.

Participating in the press conference with Mr. Webb were: Dr. William H. Pickering, Director, JPL; Edgar M. Cortright, Deputy Associate Administrator for Space Science and Applications, NASA; Prof. Robert B. Leighton, Cal Tech, Principal Investigator; Dan Schneiderman, Mariner Project Manager, JPL; Prof. Bruce C. Murray, Cal Tech; Julian Scheer, Assistant Administrator for Public Affairs, NASA. (Transcript)

- PROTON I, unmanned space station launched by U.S.S.R. July 16 with instrumentation for studying high-speed cosmic particles, was functioning normally, Tass reported. (Tass, 7/29/65)
- At the Honors Convocation of the AIAA meeting in San Francisco the following presentations were made: Rodney C. Wingrove, Research Scientist at NASA Ames Research Center, received the Lawrence Sperry

Award for his contribution to controlled reentry and precise landings of U.S. manned spacecraft; Dinsmore Alter, Director Emeritus of Griffith Observatory, received the G. Edward Pendray Award for an "outstanding contribution to aeronautical and astronautical literature"; Lloyd L. Kelly, President, Link Group, General Precision, Inc., received the DeFlorez Training Award for "an outstanding improvement in aerospace training"; and Dr. Wernher von Braun, Director of NASA Marshall Space Flight Center, received the premier Louis W. Hill Space Transportation Award "for significant contributions indicative of American enterprise and ingenuity in the art and science of space flight." This award included a $5,000 honorarium. (ARC Release 65–15; MSFC Release 65–151; AIAA Honors Convocation Program)

July 29: Ball and roller bearings assembled according to the Lewis Hardness Differential Guide could be expected to have four to five times greater fatigue life, NASA announced. The Guide was developed by a NASA Lewis Research Center team of engineers whose tests showed that bearing load capacity and fatigue life were greatest when the rolling elements of the bearing were between one and two points harder (measured on the Rockwell C Scale) than the races. Manufacturers usually made bearings with balls and races of the same hardness. (NASA Release 65–246; LRC Release 65–51)

- Rep. Charles McC. Mathias (R–Md.), advocating increased research in weather modification, introduced in the House a bill (H.R. 10173) requiring a study of current public and private efforts and a Presidential report thereon: "The science of weather modification is still in its infancy, but it is a very active youth. Public and private efforts in this field have expanded greatly in the past decade. Several Federal agencies, primarily the Department of Commerce, the National Science Foundation, NASA, and the Departments of Interior and Defense, have increased their support of atmospheric research to a total Federal investment of $3,529,683 in fiscal 1964. . . .

  "Before committing this country to such a massive and sustained effort, we should know where we stand now. A comprehensive Presidential report such as the one required by my bill would give the Congress and the Nation the fundamental information which we need before attempting to evaluate expert recommendations on methods and goals. For, like nuclear physics, the science of weather modification has an infinite capacity for mischief or for good. We must be sure that man's efforts to tame the elements proceed along paths beneficial to mankind." (CR, 7/29/65, 18071)

- BOAC announced in its annual report that it had asked British and American aircraft manufacturers to submit plans for a subsonic airliner accommodating up to 250 passengers. Sud Aviation of France and Hawker-Siddeley of Great Britain told the Corporation they were not interested in the project. (Reuters, *NYT*, 7/29/65, 48)

July 30: PEGASUS III meteoroid detection satellite was launched into orbit from Eastern Test Range, with a Saturn I booster—last (SA–10) in a series of ten launch vehicle test flights. Initial orbital data: apogee, 336 mi. (541.9 km.); perigee, 324 mi. (522.6 km.); period, 95.5 min.; inclination, 28.9°. Main assignment of the 3,200-lb. spacecraft with wing-like panels was to add information on the frequency of meteoroids in near-earth environment, for use in the design of future manned and unmanned spacecraft. Eight of its detachable panels carried 352 thermal surface samples collected from the aerospace industry. If NASA should program an astronaut-PEGASUS III rendezvous, the astronaut would detach as many panels as possible and return them to earth for study. Also orbited was Apollo command and service module boilerplate (BP-9), which served as shroud for PEGASUS III. Apollo launch escape system was jettisoned during launch vehicle's ascent.

PEGASUS III—expected to return meteoroid data to ground stations for at least one year—was identical to PEGASUS II, in orbit since May 25, 1965. PEGASUS I was sent into orbit Feb. 16, 1965. (NASA Release 65–232; MSFC Release 65–185; MSFC Release 65–190)

July 30: NASA's seven-year Saturn I program was concluded with the successful launch of PEGASUS III. It marked the tenth success in as many attempts for the Saturn I booster. A significant development of the program was the clustering of several large rocket engines: the power plant in the first stage of the Saturn I was a cluster of eight H-1 engines each with 188,000 lbs. of thrust to give this stage 1,504,000 lbs. of thrust. The first four flight tests (SA-1 through SA-4) were with dummy upper stages; beginning with SA-5, both stages (S-I and S-IV) were "live"; and the last three Saturn I's each orbited a Pegasus. Other significant developments growing out of the Saturn I program included:

"1. First extensive use of multi-engines (six RL-10-A3's) and liquid hydrogen in the upper stages.

"2. Advancement of guidance and instrumentation technology.

"3. Facility expansion, and development of new transportation modes for large rockets.

"4. Orbiting meteoroid technology satellites, the largest instrumented satellites launched to date.

"5. Developing the capability of placing into earth orbit payloads of more than 37,000 pounds.

"6. Developing guidance and instrumentation technology which could be used in other programs." (NASA Release 65–253)

- Third earth landing (twelfth test drop) of steerable parachute-retrorocket landing system for Gemini-type spacecraft was termed "100% successful" by engineers at NASA Manned Spacecraft Center. Dropped from a C-119 aircraft at 10,000-ft. altitude, the vehicle was turned into the wind, downwind, and fully around several times by remote control before it was brought to a landing 40 ft. from the target. (*Houston Post*, 7/31/65; Maloney, *Houston Post*, 8/2/65)

- NASA Marshall Space Flight Center awarded two parallel one-year contracts to two firms to study feasibility of developing a drill for probing some 100 ft. below the moon's surface: Northrop Space Laboratories received $509,992; Westinghouse Electric Corp. Defense and Space Center, $570,624. Astronauts on post-Apollo lunar missions would use the drill as a geological research tool to bore holes for geophysical measurements. (MSFC Release 65–194)

- ComSatCorp announced it would request proposals from several U.S. manufacturers for a new space exploration communications system that would be associated with "certain space exploration activities, particularly the Apollo program." Proposals would be for five synchronous-orbit satellites each capable of being launched with a thrust-augmented Delta booster, with options for additional satellites. First delivery would be within eight months. ComSatCorp also established requirements for four transportable satellite earth stations with options for additional stations—first two stations to be delivered in eight months. (ComSatCorp Release)

July 30: Technical Systems Office had been newly established by MSFC to handle launch vehicle technical systems problems, MSFC announced. Dr. J. C. McCall, deputy director of Research and Development Operations, would serve as acting director of the office until Oct. 1, when L. G. Richard, assistant director of the Astrionics Laboratory, would become director. (MSFC Release 65-193)

- A prototype experimental life-support system enclosed in a simulated space cabin was placed on a ship in San Diego for delivery to NASA Langley Research Center. Designed and constructed by the Convair Div. of General Dynamics Corp., the system was intended for use in a research program to enlarge the scope of life-support technology. Within the device were subsystems to: (1) extract oxygen from the carbon dioxide exhaled by the occupants; (2) convert waste liquids and humidity condensates to drinking water; (3) provide control of internal temperature and humidity; (4) allow storage and handling of a freeze-dried food supply for four test subjects for 90 days at a time; (5) remove from the cabin atmosphere all contaminating vapors which might be generated; and (6) provide personal hygiene facilities. (LaRC Release)

- Commenting on the success of MARINER IV, a *New York Times* editorial said: "A whole host of new sciences is being born—extraterrestrial geology most obviously among them. By learning more about Mars—even a lifeless Mars—men will understand better the origin of the solar system. And, by being able to compare the red planet in greater detail with this earth, new understanding will evolve of why there is life here and, apparently, none there. The exploration of the planets has begun and more than one generation will be required to finish that task. But, so long as men stand on this puny globe and gaze wonderingly at the lights in the sky, they will remember that the first successful pioneer was named Mariner 4." (*NYT*, 7/30/65, 24C)

- Robert N. Allnutt was appointed Assistant General Counsel for patent matters in the NASA General Counsel's office, effective Sept. 13. Mr. Allnutt, who had been with the NASA office of General Counsel since 1961, would succeed Gerald D. O'Brien who had been appointed an assistant commissioner of patents by President Johnson. (NASA Release 65-251)

- U.S. Justice Dept. opposed "at this time" the enactment of measures involving the Government's rights to inventions discovered by private research and development contractors working with Government funds. The Department set forth its view in a letter from Deputy Attorney General Ramsey Clark to Sen. John L. McClellan (D-Ark.), Chairman of the Senate Judiciary Patents Subcommittee. (Mintz, *Wash. Post*, 7/30/65, A12)

July 31: DOD refused request by television networks to present live coverage of the recovery portion of the Gemini V flight scheduled for Aug. 19. Equipment necessary to set up a portable ground station on the aircraft carrier U.S.S. *Lake Champlain* "might interfere with operational requirements and shipboard communications," DOD said. The networks had intended to send pictures to EARLY BIRD I satellite for relay to the Andover, Me., ground station. (*NYT*, 7/31/65, 35)

- First anniversary of historic mission of taking and relaying to earth the first closeup pictures of the lunar surface, by NASA's RANGER VII. (EPH)
- Pictures of Mars taken by MARINER IV during the July 14 flyby received editorial comment in the *Washington Post:*

    ". . . If some people are disappointed because Mariner 4 did not produce any conclusive documentation on the existence of the long-suspected life on Mars, it should be remembered that this was not one of Mariner 4's objectives. Even the Tiros, circling earth at far less distance from its surface than Mariner 4 was from Mars, has indicated only once in the thousands of pictures taken that the life we know exists here actually can be spotted from far out in space." (*Wash. Post*, 7/31/65)

During July: The support of science in the U.S. was discussed by Dael Wolfle in the *Scientific American:*

"From 1953 until 1960 about 8 percent of the Nation's research and development budget was devoted to basic research. The percentage has been rising since 1960, reaching almost 12 percent in 1965. As for the Federal Government's funds, in 1953, less than 7 percent went for basic research. The figure has been rising since 1960, to about 11 percent in 1965. The universities are relatively much more prominent in basic research than in the total research and development effort, being responsible for almost half of all basic research. In contrast the industrial laboratories, which dominate in development activity, conduct only about a fourth of the basic research." (*CR*, 7/14/65, A3760-61)

- Inventors of Lunar Landing Research Vehicle, lunar landing simulator referred to as "the belching spider," in use at NASA Flight Research Center, were granted a patent. Built by Bell Aerosystems Co. under contract to NASA, the research craft had a jet engine that supported five-sixths of its weight; the pilot would manipulate lift rockets that would support the remaining one-sixth. The craft's attitude would be controlled with jets of hydrogen peroxide. Inventors were Kenneth L. Levin and John G. Allen, Jr., of the Bell staff. (Jones, *NYT*, 7/3/65, 23)
- U.S.S.R. satellites ELECTRON I and ELECTRON II were among the new exhibits in the "Kosmos" (Space) Pavillion at the Soviet exposition on achievements of the U.S.S.R. national economy, reported *Kryl'ya rodiny*. These satellites, launched to study the near-earth radiation belts, were said to have made it possible for Soviet scientists to safeguard the cosmonauts during their flights in this region. (*Kryl'ya rodiny*, 7/65, 1)
- In answer to the query, "What is an orbital analyst?" an article in *The Airman* said: "The complex tasks of the analyst involve

the use of a high order of mathematics, laws of celestial mechanics, and adaptation of orbital analysis problems to high-speed scientific computer solutions. These highly specialized activities are absolutely essential to carrying out the vast mission of the Air Force SPACETRACK system: detection, tracking, and identification of all man-made objects orbiting the earth." (*Airman*, 7/65, 24)

During July: Fortune magazine recounted the story of the development of the Lear jet—a small jet aircraft for the corporate market. William Lear, said the article, became the first man in history to design, build, and win certification for a jet airplane—all with his own money. (*Fortune*, 7/65, 137–140, 185)

- Prospects for U.S.-European industrial cooperation in space were discussed in *Air Force and Space Digest* by Elmer P. Wheaton, Vice President of Lockheed Missiles & Space Co. Wheaton offered four "guiding principles" for any program of U.S.-European industrial effort:

    "The program to be jointly undertaken should avoid unnecessary duplication of an existing program. . . .

    "The project or program should contribute to a better understanding of the space environment. . . .

    "The program must provide a logical extension of our current space technology. . . .

    "The program must not require such a substantial increase of knowledge in either the space environment or the space technology that it involves a high risk of failure. . . ."

    Wheaton suggested: scientific areas that could be usefully investigated by Europe; advanced programs for "more distant European exploration"; and satellites that could be profitably developed by Europe—notably, applications satellites. In such company-to-company cooperation, contribution of the U.S. companies should be chiefly in space systems management. (*AF* Mag., 7/65, 53–57)

- European contributions to international space communications were recommended by Dalimil Kybal, Senior Consulting Scientist, Lockheed Missiles & Space Co. Summary of his article in *Air Force and Space Digest*:

    "Contrary to the views of those Europeans who tend to think of European and American space communications systems as separate entities, existing international agreements clearly call for a *global* system. Europe, as it develops space technology skills, ought to concentrate on developing next-generation satellites as replacements for existing hardware, in keeping with the competitive approaches contemplated in the international agreement. . . ." (*AF* Mag., 7/65, 60–61)

- Soviet aircraft designer Oleg K. Antonov provided technical and performance specifications of the large An–22 aircraft during an interview conducted by J. Marmain for *Wehr und Wirtschaft* (W. Germany): first flight was made in February 1965; aircraft shown at International Air Show, Paris, was transport model; considerable rebuilding, including 15-m. extension of the tail, would be required for the craft to accommodate 720 passengers. Antonov said his design office was not concerned with development of a passenger version of the An–22, a new version of the An–24 for 64 passengers, and a smaller special purpose aircraft. (Marmain, *Wehr und Wirtschaft*, 7/65, 390–91)

During July: Paul Tillich, Protestant theologian, commenting upon the "Pacem in Terris" of Pope John XXIII, wrote: "A . . . genuine hope for peace is the technical union of mankind by the conquest of space. Of course, nearness can intensify hostility; and the fact that the first manifestations of the technical oneness of the world were two world wars proves this possibility. But nearness can also have the opposite effect. It can change the image of the other as strange and dangerous; it can reduce self-affirmation and effect openness for other possibilities of human existence and—particularly as in the encounter of religions—of other possibilities of genuine faith." (Tillich, "The Limits of Peace," *Chicago Today*, Summary 1965, 2–5)

- More than 10,000 natural scientists, social scientists, and engineers were admitted to the U.S. as immigrants during fiscal years 1962 and 1963, according to National Science Foundation. (NSF *Reviews* . . ., 7/65)

August 1965

August 1: S–IB–2, the first stage of the second Saturn IB booster, left NASA Marshall Space Flight Center aboard NASA's barge *Palaemon* for Michoud Assembly Facility, where it would undergo post-static-firing checkout. The 80-ft.-long launch vehicle had been fired for 30 sec. at MSFC on July 8, and for 2½ min. on July 21. (MSFC Release 65–195)

August 2: MARINER IV's tape recorder was turned off at the end of its second playback of the 21 pictures it took of Mars on July 14. A spokesman for the Jet Propulsion Laboratory said the second run of pictures would be compared with the first as a check against possible errors in transmission and reception. No significant differences had been reported yet by scientists studying the photographs. (UPI, *Chic. Trib.*, 8/3/65; NASA Proj. Off.)

- NASA announced plans to install Unified S-Band System equipment at Corpus Christi communications station for use with Apollo spaceflights. With the system, the station would be able to combine in a single two-way transmission all types of communications with the three Apollo astronauts.

    Seven kinds of communications would be conducted simultaneously, including tracking the spacecraft; commanding its operations and confirming execution of commands; two-way voice conversation; continuous checks on the astronauts' health; continuous check on the spacecraft and its functions; continuous information from onboard experiments; and television pictures of the astronauts and their exploration of the moon. All communications would be conducted with one 30-ft.-dia. parabolic ground antenna to be constructed at Corpus Christi. (NASA Release 65–250)

- Astronaut Edward H. White II (L/Col., USAF) and David S. Lewis, president of McDonnell Aircraft Corp., launched Operation Zero Defect at the McDonnell plant in St. Louis, Mo., county. They addressed an outdoor gathering of 3,400 employees, asking them to continue doing a good job of producing spacecraft and Phantom F–4 jet fighters for USAF, USN, and the Marines. (*St. Louis Post-Dispatch*, 8/2/65)

- Hamilton Standard delivered to NASA Manned Spacecraft Center a prototype portable life support system (Plss) to be used by Project Apollo astronauts. Weighing about 60 lbs., unit was designed for use with water-cooled undergarment astronauts would wear beneath a spacesuit during lunar surface exploration. Water-cooled undergarment would cool the astronaut by conducting the metabolic heat generated by his motions into water which would circulate through a web of plastic tubing in contact with the skin. Water would carry the heat into the portable life support system which would recool and recirculate it. Contract called for delivery of 52 packs.

Testing of the undergarment and the Plss would be conducted at MSC. (*Houston Post*, 8/3/65)

August 2: ComSatCorp received a check from AT&T for use of 60 channels on EARLY BIRD I comsat during its first month of operation. It was the first operating revenue ComSatCorp had had; previous income had been interest on the $200 million received from sale of stock to the public. The AT&T check was only half as much as had been expected originally; ComSatCorp's initial estimates had been based on expectation that AT&T would use 100 channels on EARLY BIRD I. (*Wash. Post*, 8/3/65)

- Alternate methods for re-establishing communications with MARINER IV on its next closest approach to earth around Sept. 4, 1967, were being considered by Mariner project planners at Jet Propulsion Laboratory, *Aviation Week and Space Technology* reported: (1) attempted reacquisition beginning in early February 1967, making use of the spacecraft's high-gain directional antenna to obtain data from the spacecraft for up to 10 mo.; (2) reliance solely on MARINER IV's low-gain antenna for transmission, permitting two-way communications for four to six weeks around September 1967. In the 10-mo. plan, reacquisition would be initiated when the spacecraft was about 135 million mi. from earth. Due to relative sun-earth positions in February 1967, the angle at which the high-gain directional antenna would be permanently fixed would enable it to be aimed at the earth by having MARINER IV roll about its longitudinal axis, which would be pointed at the sun. The 100-kw. transmitter at Goldstone tracking station would send the necessary commands to MARINER IV. In the latter plan, use of the 210-ft. antenna, expected to become operational at Goldstone in January 1966, could extend reception of intelligible telemetry signals to as long as six weeks, compared with four using the standard 85-ft. dishes of the Deep Space Network.

W. A. Collier, assistant Mariner project manager at JPL, told *Aviation Week* that MARINER IV would be of particular scientific interest in 1967. First, there were no other interplanetary probes being sent away from the sun at that period. Second, when MARINER IV passed within 6,000 mi. of Mars July 14, the gravitational pull of the planet had tilted the plane of the spacecraft out of the plane of the ecliptic. MARINER IV, 5.3 million mi. above the ecliptic in September 1967, would give scientists their first chance to compare interplanetary findings outside this plane with those obtained in it.

Preliminary estimates were that the 10-mo. plan would cost between $5 million and $15 million, while the four-to-six-week project would cost less than $1 million.

MARINER IV, launched Nov. 28, 1964, was in solar orbit with a period of 567.11 days, perihelion of 103.1 million mi., and aphelion of 146.2 million mi. (Watkins, *Av. Wk.*, 8/2/65, 32)

August 3: U.S.S.R. launched COSMOS LXXVII, 13th Soviet spacecraft orbited in the last two months. Initial orbital data: apogee, 300 km. (187 mi.); perigee, 184 km. (114 mi.); period, 89.3 min.; inclination to the equator, 51.8°. Tass said the unmanned satellite would gather data to prepare equipment for manned flights; instruments were functioning normally. (*Pravda*, 8/4/65, 1; UPI, *NYT*, 8/4/65, 4)

August 3: USAF launched two unnamed satellites from the Western Test Range with an Atlas-Agena D booster. (UPI, *Wash. Post*, 8/4/65; *U.S. Aeron. & Space Act., 1965*, 149)

- X-15 No. 2, piloted by Maj. Robert Rushworth (USAF), attained a maximum altitude of 208,700 ft. and a maximum speed of 3,602 mph (mach 5.14) in a flight to obtain data for the reaction augmentation system, and to check out ultraviolet photographic experiment, advanced landing dynamics, and to continue pilot altitude build-up. (NASA X-15 Proj. Off.; *X-15 Flight Log*)

- Brazilian Space Commission (CNAE) would cooperate with U.S. and Argentine scientists in a study of hemispheric weather patterns, NASA announced. NASA and CNAE had signed an agreement July 1 providing for the cooperative project, which would be part of the inter-American Experimental Meteorological Sounding Rocket Network (EXAMETNET).

  First launchings were scheduled before the end of 1965 from Natal, Brazil, and Wallops Station. (NASA Release 65–258)

- First static test firing of European Launcher Development Organization's (ELDO) Europa I booster was conducted at Spadeadam Rocket Establishment in Cumberland, England. The launch vehicle was composed of Blue Streak first stage, French Coralie second stage, West German third stage, and Italian satellite and nose casing. For this test, all but Blue Streak were dummy stages. (*Av. Wk.*, 8/2/65, 35)

- NASA still planned to launch an eight-day, 121-orbit, two-man Gemini V flight August 19, but fuel cells were causing some problems. The cells' oxygen and hydrogen tended to evaporate too quickly to keep the Gemini spacecraft electrically "alive" for such a long trip, NASA spokesmen said. (UPI, *NYT*, 8/3/65, 15; *WSJ*, 8/3/65, 1)

- Reviewing the results of the successfully completed Saturn I rocket program, Bob Ward listed in an article in the *Huntsville Times* eight major contributions which the work had made to launch-vehicle and other aerospace technology: (1) clustering of large rocket engines—a cluster of eight H-1 engines forming the power plant of the booster's 1,504,000-lb.-thrust first stage; (2) first extensive use of multi-engine power plants and liquid hydrogen fuel for upper stages; (3) advances in the fields of rocket guidance and instrumentation; (4) expansion of facilities and development of new methods of transporting large rocket stages, including special-purpose barges and aircraft; (5) development of fabrication techniques needed for large rockets; (6) orbiting meteoroid technology satellites—the three Pegasus orbiters were the largest instrumented satellites yet launched; (7) proving the aerodynamics of the Apollo spacecraft by orbiting five boilerplate versions of the command and service modules; and (8) developing sufficient launching power to place payloads of almost 20 tons into orbit around the earth—the seventh Saturn launch September 18, 1964, orbiting a 39,200-lb. payload. (Ward, *Huntsville Times*, 8/3/65)

- House Committee on Science and Astronautics began hearings on H.R. 2626, a bill to provide that the National Bureau of Standards conduct investigations to determine the practicability of U.S. adoption of the metric system of weights and measures.

J. Herbert Hollomon, Assistant Secretary of Commerce for Science and Technology, told the Committee that if the U.S. did not adopt the metric system it would stand alone in this regard in 10 to 15 yrs. (*CR*, 8/3/65, D740; Transcript)

August 3: GEMINI IV spacecraft arrived at the New York World's Fair where it would formally be placed on view in the U.S. Space Park starting August 4. (Dougherty, *NYT*, 8/4/65, 41)

* A bolt of lightning struck and killed Albert J. Treib, a construction superintendent at Kennedy Space Center, NASA, as he and his crew poured concrete during a misty rain. The bolt slightly injured five other workmen. (Schreiber, *Miami Her.*, 8/4/65)

* North American Air Defense Command said reported Ufo sightings from six states were probably the planet Jupiter or one of the stars Rigel, Capella, Betelgeuse, or Aldebaran.

    A USAF weather observer in Norman, Okla., taking issue with the Air Force's stand, said: "What we saw was not an aircraft . . . nor was it a planet or star . . . It was about 22,000 feet high and pitched at about a 45 degree angle." He said he and a friend had observed the object, which was "moving quite rapidly," for a little less than five minutes through a 40-power telescope. (*Wash. Post*, 8/3/65)

August 4: The Senate-House conference committee reported the Independent Offices Appropriation to the House and Senate. The report (#727) provided for NASA $4,531,000,000 for research and development instead of $4,521,000,000 proposed by the House and $4,536,971,000 proposed by the Senate; $60,000,000 for construction of facilities as proposed by the House instead of $62,376,350 proposed by the Senate; $584,000,000 for administrative operations instead of $579,000,000 proposed by the House and $590,957,850 proposed by the Senate. Senate language authorizing appropriation reimbursement was retained. House provision on payment of indirect costs of research grants was retained. (*CR*, 8/4/65, D746; Conf. Rpt. 727)

* TIROS X meteorological satellite photographed an area of unusual cloudiness in the Atlantic about 2,400 mi. east-southeast of Miami, the Miami Weather Bureau said. A hurricane-hunter aircraft would be sent to check. (*Miami Her.*, 8/5/65)

* There exists a serious misunderstanding about the U.S. space program, Dr. Edward C. Welsh, Executive Secretary of the National Aeronautics and Space Council, told the Tenth Symposium on Space and Ballistic Missile Technology in San Diego: "All too many people seem to have the impression that part of our program is peaceful in intent while the other part is something different, presumably non-peaceful. This misconception goes further by attempting to identify the non-peaceful and the non-scientific with the military and to credit the peaceful and scientific to the civilian. . . . The fact is—in both policy and practice —that all of our space activities are peaceful. . . . Just in case it may have been forgotten, let me quote from our highest policy level. In 1962, President Johnson, then Vice President and the Chairman of the National Aeronautics and Space Council, stated: 'The United States does not have a division between peaceful and non-peaceful objectives for space but rather has space missions to help keep the peace and space missions to improve our ability to live well in peace. . . .'

"In 1964, as President, he said: 'Our space program, in both its civil and military aspects, is peaceful in purpose and practice. . . .'

"I . . . am not saying that space cannot be used for purposes of aggression . . . no nation should bury itself in sands of complacency and thereby neglect to develop the technological and military strength so necessary for deterring potential aggressors. The maintenance of such strength in no respect conflicts with the policy of peace. In fact, the more competent we are to prevent surprise, to discover aggressive maneuvers, and to intercept hostile weapons in any medium, the better chance we have of living in peace. . . .

"When I state, therefore, that our entire national space program is peaceful, I mean that we have no aggressive intent, that we seek no domination over other peoples, and that we are eager to share the benefits of space exploration with all mankind." (Text)

August 4: NASA selected three firms to design the Apollo Lunar Surface Experiments Packages (Alsep) under separate and concurrent $500,000, six-month, fixed-price contracts. The firms were Bendix Systems Div., Bendix Corp.; Space-General Corp.; and TRW Systems Group, Thompson-Ramo-Wooldridge, Inc. Packages would contain scientific instruments to measure the moon's structure and surface characteristics, atmosphere, heat flow, solar wind, radiation, and micrometeorite impacts. They would be carried to the moon on the initial Apollo spaceflights and placed on the surface by astronauts. Instruments would transmit data back to earth for six months to one year. (NASA Release 65-260)

- Personnel of NASA Manned Spacecraft Center would be augmented to meet the increasing tempo of Gemini and Apollo manned space flight operations, NASA Marshall Space Flight Center announced. Over the next ten months, approximately 200 persons would be transferred from MSFC to MSC. Total number of personnel to be provided from other NASA activities had not been determined. (MSFC Release 65-199)

- In an article in the Orlando *Evening Star*, Barry Goldwater said: "Defense Secretary Robert McNamara has indicated that space weapons are too costly, as though any dollar cost is too high for the security of 190 million Americans and a billion allies and friends. The only major space-military program McNamara has permitted to stay alive is the Manned Orbital Laboratory, and he has so slowed and limited this that its orbit is apparently toward nothing but bureaucratic extinction.

"We have deployed one or two so-called 'satellite killers,' but they are a pitiful particle of what really is needed.

"Pictures of Mars are fine. So is a trip to the moon. But the first job of any administration is to secure the nation against its enemies.

"We will not remain the most powerful nation on earth for long if we do not reverse the suicidal Johnson-McNamara refusal to let us arm ourselves in space." (Orl. *Eve. Star*, 8/4/65)

- "Absence of proper space laws may lead to dangerous conflicts and complications not only in this sphere of man's activities but also in purely terrestrial affairs," said Genadii Zhukov, scientific secretary of the Soviet Space Law Commission in an interview with *Krasnaya Zvezda*. The resolution adopted by the U.N. General Assembly on Dec. 13, 1963, had confirmed that outer space was open to all states. However, he stressed, the resolution made special reservations about the impermissibility of using sputniks for war propaganda and for inciting enmity between peoples: "All states must refrain from potentially harmful experiments in outer space or any other steps liable to interfere with the peaceful use of such space by other countries."

Zhukov noted that in connection with the prospects of creating permanent orbital stations, the need would arise to determine their legal status, as well as conditions of their use by other countries. As human beings made further inroads into outer space, the problem of determining the legal status of stations and settlements on the moon and on other celestial bodies, the conditions for tapping their natural resources, would become important. He added: "Space law must also guarantee protection of other living worlds if such are discovered on distant planets." (Tass, 8/4/65)

August 4–5: Some 300 representatives of industry, NASA, and other agencies attended a conference on design of leak-tight fluid connectors at NASA Marshall Space Flight Center. Sponsored by MSFC and the Society of Automotive Engineers, the conference was planned to promote direct exchange of technical information on separable, semipermanent, and permanent fluid connectors. (MSFC Release 65-196)

August 5: S–IC–T, 138-ft.-tall test version of Saturn V's first stage, was static-fired for 2½ min. at NASA Marshall Space Flight Center in first full-duration test-firing. The five F-1 engines, each consuming liquid oxygen and kerosene at the rate of three tons a second, generated 7.5 million lbs. thrust. Ability of the engines to steer the rocket was also successfully demonstrated. The five-engine cluster was mounted so that only the one in the middle of the cross-shaped array was stationary. The others could gimbal slightly in pairs. (MSFC Release 65-197; *Marshall Star*, 8/11/65, 1)

- NASA announced selection of Documentation, Inc., Bethesda, Md., to operate its Scientific and Technical Information Ficility, the world's largest collection of aerospace literature, in a Government-provided building in College Park, Md. Contract negotiations were expected to result in a cost-plus-award fee contract for approximately $3.6 million. In Fiscal Year 1965, the contract figure was $4.9 million.

  Mission of the facility was to acquire and organize worldwide technical reports in the aerospace sciences, indexes, abstracts, and items on space exploration; prepare announcement journals; process selected items on microfilm; and provide a central reference service to NASA and its contractors. (NASA Release 65-263)

- James C. Elms and L/Gen. Frank A. Bogart (USAF-Ret.) had been appointed NASA Deputy Associate Administrators for Manned Space Flight effective Sept. 1, Dr. George E. Mueller, NASA Associate Administrator for Manned Space Flight, announced. Elms had served as Deputy Director of NASA Manned Spacecraft Center before assuming his present position as vice president of Raytheon Co. General Bogart, former USAF comptroller, had served as Director for Management Operations, OMSF, since February.

  Paul E. Cotton, who had been assistant to Dr. Mueller since November 1963, would become Director of Manned Space Flight Management Operations, succeeding General Bogart. B/Gen. Julian H. Bowman (USAF, Ret.) would succeed Cotton. General Bowman had been a special assistant to Dr. Mueller. (NASA Release 65-264)

August 5: First full-duration static test of Saturn V first stage at NASA Marshall Space Flight Center.

August 5: Gifford K. Johnson, former president of the Graduate Research Center of the Southwest, Dallas, was sworn in by NASA Administrator James E. Webb as a NASA consultant. He would provide advice and guidance in the areas of technology utilization and technology reporting programs. (NASA Release 65-268)

- Harro Zimmer, chief of a West Berlin satellite tracking station, said the U.S.S.R. had secretly launched a second spacecraft with PROTON I July 16. The unannounced spacecraft, said Zimmer, had been brought back to earth, landing near Magnitogorsk, 300 mi. north of the Aral Sea, between 3 a.m. and 3:25 a.m. (EDT) July 31. (UPI, *Wash. Post,* 8/6/65)

- An editorial discussing space weapons appeared in the Washington *Evening Star:* "Secretary of State Rusk . . . is on record as having warned that the ocean of space might support 'huge nuclear-propelled dreadnaughts armed with thermonuclear weapons. The moon might be turned into a military base. Ways might be found to cascade radioactive waves upon an enemy' and there might be other equally deadly spatial advances. The same opinion is held by highly placed military officers.

  "As General Ferguson of the Air Force put it a couple of years ago, in urging the swiftest possible development of an American 'military patrol' in space, no one even dimly foresaw the nuclear bomber when the airplane began to operate a half-century ago.

  ". . . if we ignore General Ferguson, we could lose everything. As a matter of prudence, our country should at least maintain a program of research and development designed to insure it against the danger of becoming second best in the military uses of space." (Wash. *Eve. Star*, 8/5/65)

August 5: Saline water would be distilled and made potable under a $185-million study program adopted by both the House and Senate. Shortly before the measure passed, President Johnson had told his science advisers to push desalinization "as if you knew you were going to run out of drinking water in the next six months." (*CR*, 8/5/65), 18756–57)

August 5–6: NASA held an international meeting at Wallops Station to discuss overall objectives and conduct of the Inter-American Experimental Meteorological Rocket Network (EXAMETNET). Representatives from Argentina, Brazil, Canada, Mexico, Peru, the Weather Bureau, and NASA took part in the meeting. Preliminary plans called for network stations to be located at Wallops Station; Natal, Brazil; and Chamical, Argentina, with others to be added in both hemispheres later. Personnel from all participating countries would receive training at Wallops Station with NASA providing the training and the launch vehicles for sounding rocket launches from stations throughout the northern and southern hemispheres.

General purpose of the network was to contribute to studies of atmospheric structure and behavior in the southern hemisphere and to help explain atmospheric differences and similarities between the northern and southern hemispheres. (NASA Release 65–45; *SBD*, 8/5/65, 172)

August 6: Milton O. Thompson (NASA) flew X–15 No. 1 to maximum altitude of 103,200 ft. and maximum speed of 3,534 mph (mach 5.15) to obtain data for the infrared scanner program and the stability and control system. (NASA X–15 Proj. Off.; *X–15 Flight Log*)

- NASA had awarded a contract to Rice Univ. for a new type of Explorer satellite designed to extend studies of near-earth atmosphere phenomena, especially auroral phenomena. Under a cost-reimbursement contract, the university would provide two Owl Explorer spacecraft and a flyable prototype at a total estimated cost of $3,676,100. Earliest launch would be in 1967 from the Western Test Range. Launch vehicle would be a four-stage Scout. (NASA Release 65–266; Wallops Release 65–46)

- A Saturn/Apollo Applications Directorate had been established in the Office of Manned Space Flight, NASA announced. The new directorate would plan and direct programs utilizing technology developed in Project Apollo.

  M/Gen. David M. Jones (USAF) would be Acting Director of Saturn/Apollo Applications in addition to his duties as Deputy Assistant Administrator for Manned Space Flight (Programs). Deputy Director would be John H. Disher, formerly Test Director in the Apollo Program Office. Melvyn Savage, who had served under Disher as Chief of Test Planning, would become Apollo Test Director. (NASA Release 65–265)

- Venus' surface was dry, radio astronomers at Cal Tech reported, probably consisted of sand or porous rock, and was much too hot for any known form of life—up to 675° F. The planet's cloudy atmosphere was at least 40-mi. thick consisting mostly of carbon dioxide with some nitrogen and a trace of water vapor. Observations were made at Cal Tech's Owens Valley Radio Observatory by Dr. Barry Clark of the National Radio Astronomy Observatory at Green Bank, W. Va., and Dr. Arkady Kuzmin of the Lebedev Institute of Physics in Moscow. (AP, *NYT*, 8/7/65, 10)

August 6: Tass announced that U.S.S.R.'s 12.2-ton PROTON I satellite was orbited by a booster whose main engines were rated at thrust of more than 60-million horsepower. PROTON I was orbited July 16. PROTON I marked "the beginning of a new phase in the exploration and domestication of space." It would study solar cosmic rays, spectrum and composition of cosmic ray particles, nuclear interactions of galactic origin, and galactic gamma rays. (Tass, 8/6/65)

- A new helicopter world record was set by Soviet aviatrixes T. Russiyan and L. Isayeva, who flew 1,040 km. (645 mi.) in six hrs. 58 min. in an Mi-4 at an altitude of 1,000 m. (3,280 ft.). (*Pravda*, 8/6/65, 4)
- Hiroshima was devastated by world's first operational atomic bomb 20 years ago. (*WSJ*, 8/6/65, 1)

August 7: Operation Firefly ended as some 3,000 fireflies snared by Rockville, Md., children were turned over to NASA Goddard Space Flight Center. An extract of the firefly's lamp would be used in a life detection instrument under development at GSFC. Goddard's "Firefly" instrument would contain a mixture of all the contents of the insect's glow system except ATP (adenosine triphosphate), a high-energy compound essential to all life as it is known on earth. Thus, when the instrument encountered a live microorganism in space, the ATP contained therein would complete the circuit and a glow would be produced. (GSFC Release G-19-65; *Wash. Post*, 8/7/65)

- M/Gen. Samuel C. Phillips (USAF), director of the Apollo program in the NASA Office of Manned Space Flight, addressed the American Bar Association's Seminar of the Committee on the Law of Outer Space in Miami Beach: "In many respects, the Gemini IV flight of Jim McDivitt and Ed White could well be viewed as a turning point in the American manned space program. The success of this mission has given us greatly increased confidence that we will be able to carry out our national goal of a manned landing on the moon in this decade.

    "It is a pleasure to report that the Apollo Program is also moving ahead very rapidly, and we are meeting our key milestones on schedule. The progress on Apollo is especially rewarding since it is the largest and most complex single research and development project ever undertaken. This is a crucial year for Apollo, but our prospects look good and we are becoming increasingly confident that the lunar landings will take place before the end of 1969." (Text)

- Paul Haney, Chief of Public Affairs for NASA Manned Spacecraft Center, addressed Oklahoma members of the American Legion to open their state convention: "Although we are not pursuing any military objectives as such, every once in a while somebody turns over a rock and finds something which could be of immense military value.

    "As Detroit learned to produce millions of cars in a hurry by production line development and thus was able to convert without a bobble to tank production in World War II, so could our entire manned space flight program be converted." (AP, *Houston Post*, 8/8/65)

August 7: Soviet aircraft designer Andrei N. Tupolev wrote in the magazine *Aviatsiya i Kosmonavtika* that one of his assistants, Aleksey Cheryomukhin, had built and flown "the world's first helicopter capable of flying and not just jumping into the air for several seconds." Cheryomukhin's first flight took place in November 1930, Tupolev said, and by 1932 he was flying his aircraft, designated Ea-1, to 2,000-ft. altitude. "I am very sorry that we did not publish Cheryomukhin's records at the time," Tupolev added.

It had been generally believed that Russian-born Igor Sikorsky had developed the first successful helicopter in the U.S. during the mid-nineteen-thirties. (UPI, *NYT*, 8/7/65, 34)

- Gleb Chebotarev, head of Leningrad's Theoretical Astronomy Institute, said the solar system extended far more than 21 trillion miles from the sun—nearly 6,000 times the distance from the sun to Pluto, the most distant planet now known, UPI reported. Estimate of the solar system's size was based on mathematical calculations of the gravitational interaction of the sun and various stars. (UPI, *Wash. Post*, 8/7/65, A2)

August 7–9: 13 Nike-Cajun sounding rockets were launched from Wallops Station, Va.; Point Barrow, Alaska; and Churchill Range, Canada, to implement studies of atmospheric phenomena and conditions at about 100-mi. altitude. Onboard each was a grenade experiment (an ejected explosive that detonated as the rocket climbed) and/or a sphere experiment (an ejected mylar ball that inflated and drifted for tracking by radar).

The 26-in.-dia. metalized mylar spheres were developed by the Univ. of Michigan. (Wallops Release 65–48; *SBD*, 8/11/65, 199)

August 8: Photographs of Mars by MARINER IV did not contradict his theory that life may exist on that planet, according to Dr. Joshua Lederberg, prof. of genetics at Stanford Univ. School of Medicine. His views were supported by Dr. Carl Sagan of Harvard Univ. and the Smithsonian Astrophysical Observatory. Dr. Lederberg said important point was that "we still do not know the abundance and distribution of water on Mars. However much there is, almost all of it must be frozen." Presence of water on Mars could be confined to frost covering the polar regions or could even be locked within the Martian crust. Pockets of water could be warmed by volcanic activity, forming pools where life could exist. (Sullivan, *NYT*, 8/8/65)

- Nikolay P. Dubinin, biologist and corresponding member of the U.S.S.R. Academy of Sciences, wrote in *Bakinskiy rabochiy* that not only cosmic radiation but also weightlessness and vibration would have harmful effects on the human organism during space travel. He said exposure of fruit flies to weightlessness and vibration produced genetic changes which became apparent in the F_2 generation. Genetic effects were most apparent in the offspring of flies exposed to actual spaceflight conditions. (Dubinin, *Bakinskiy rabochiy*, 8/8/65, 4)

- NASA Manned Spacecraft Center Director Dr. Robert Gilruth received the China-Burma-India World War II service group's annual Americanism award at the national convention in Houston as "the man who has contributed the most during the past year to the American way of life." (*Houston Chron.*, 8/8/65)

- Indonesia test-fired a research rocket from a site "somewhere in

Indonesia," Djakarta radio reported. The rocket reached 210-mi. (338-km.) altitude. (UPI, *Houston Chron.*, 8/8/65)

August 9: The first flight-model S–IVB second stage for the Saturn IB booster was test-fired at Douglas Aircraft Co.'s Sacramento Test Center to demonstrate its flight readiness before formal acquisition by NASA. The stage's 200,000-lb.-thrust J–2 engine was fired for 452 sec. (7½ min.) without mishap. Test was computer-controlled throughout, marking "the first time a fully automatic system has been used to perform a complete checkout, propellant loading, and static firing test on a space vehicle." (NASA Release 65–267; MSFC Release 65–202; *Marshall Star*, 8/11/65, 1, 10)

- A six-and-one-half-minute captive test firing of Saturn V booster's second (S–II) stage was conducted at the Santa Susana, Calif., static test laboratory of North American Aviation, Inc., NASA announced. The test—first full-duration firing—continued until the stage's propellants were depleted, slightly longer than normal flight duration. (MSFC Release 65–203; *Marshall Star*, 8/11/65, 1, 7)
- A small number of Redstone missiles formerly deployed with the U.S. Seventh Army in Europe would be recalled to duty in connection with an Advanced Research Projects Agency (ARPA) classified program, *Missiles and Rockets* reported. DOD officials had reportedly indicated that the total number of Redstones that could be refurbished and made ready for firing was less than 30. (*M&R*, 8/9/65, 11)
- In an address to employees at NASA Marshall Space Flight Center, Dr. Wernher von Braun expressed his appreciation for their efforts in the highly successful Saturn I program: "What you have done reflects the ability of our nation to move forward quickly. The Saturn I proved that we were correct in our heavy duty launch vehicle theories. One of the most important products of our Saturn I program is that we have built up confidence—both self confidence and the trust of the American people who are depending upon us for American progress in space exploration." (Text)
- An explosion triggered a flash fire in a 170-ft.-deep Titan II missile silo, killing 53 men. The silo was part of Complex 4 near Searcy, Ark., about 50 mi. from Little Rock AFB. Two construction workers survived the explosion. (*Wash. Post*, 8/10/65, A1)

August 9–30: NASA Wallops Station was host for the joint NASA-Univ. of Virginia Bio-Space Technology Training Program to provide on-site experience for 32 professional biologists in the operational and engineering aspects of the national space program. The biologists attended seminars, training, and laboratory exercises emphasizing biotechnical aspects of spaceflight and spacecraft design. Four sounding rockets were launched with white rat biological payloads to provide bio-space experiments. (NASA Release 65–44)

August 10: A Scout Evaluation Vehicle (SEV) was successfully launched by NASA from Wallops Station. Primary purpose of the mission was to demonstrate in flight the operation of improved vehicle features: (1) use of new second- and fourth-stage rocket motors with improved thrust characteristics; (2) test of Scout's capability to fly a "dog-leg" course from Wallops Station by yaw torquing, performed during the third-stage coasting period; (3) test of in-flight performance of improved spin motors to stabilize the fourth stage of the vehicle; (4)

demonstration of the Scout air transportability concept by launching a vehicle which, after complete assembly at Wallops, had been airlifted from and returned to the launch site in simulation of a transcontinental trip.

All flight objectives were met, including injection of Army Corps of Engineers SECOR V (Sequential Collation of Range) geodetic satellite into an orbit with apogee, 1,504 mi. (2,421 km.); perigee, 702 mi. (1,130 km.); period, 122 min.; inclination, 69.23°. Elliptical orbit should enable SECOR V to transmit measurements of distances up to 2,000 mi.—twice the distance possible with earlier Secor versions which had been launched into 600-mi. circular orbits. The satellite would map the surface of the earth, pinpointing the location of land bodies separated by large expanses of ocean. (Wallops Release 65-49)

August 10: X-15 No. 3, piloted by Capt. Joseph Engle (USAF), reached maximum speed of 3,550 mph (mach 5.20) and a maximum altitude of 271,000 ft. The purpose of the flight was to obtain data on the boundary layer noise, and reentry maneuvering techniques. (NASA X-15 Proj. Off.; *X-15 Flight Log*)

- NASA Goddard Space Flight Center had selected the Apparatus Div. of Texas Instruments, Inc., for contract negotiations expected to exceed $1 million for development of a weather-measuring device to be carried on the Nimbus B weather satellite: an experimental sensor, Iris (Infrared Interferometer Spectrometer), would be designed to collect information on the atmosphere's vertical temperature, ozone, and water vapor distribution on a worldwide basis.

  Nimbus B was scheduled for launching by a Thorad Agena D booster in 1967. (GFSC Release G-21-65)

- Explanation of rash of Ufo sightings was given by Howard Margolis in the *Washington Post:* "The latest flurry of saucer reports, according to the Air Force, seems to be based on such things as a conjunction of the planet Jupiter and some bright stars, compounded by atmospheric conditions that produce an enhanced twinkling effect, further compounded by the annual summer meteorite showers; still further compounded by some other atmospheric effects that tend to produce bogus radar reflection, and finally compounded by the well-marked tendency of any good Ufo sighting reported in the newspapers to lead many more in the same area—in this case the Midwest—for a time thereafter." (Margolis, *Wash. Post,* 8/10/65)

- First regularly scheduled air cushion service in the U.S. was initiated between Oakland and San Francisco over San Francisco Bay in a year-long test, authorized by the Civil Aeronautics Board, to determine feasibility of using Hovercraft in ferrying passengers in metropolitan areas. Eight round trips a day would be provided by a seven-ton, $300,000 craft by Bell Aerosystems. (*NYT,* 8/8/65, 74)

August 11: NASA Atlas-Centaur 6 launched a dummy Surveyor spacecraft along a simulated lunar trajectory in the fourth successful vehicle flight of six made to date. The Atlas booster operated as planned; Centaur ignited, burning for 7 min. 12 sec., and injected the dummy Surveyor spacecraft into a highly elliptical simulated lunar transfer orbit with apogee, 509,829 mi. (820,824 km.); perigee, 105 mi. (169.15 km.); period, 31 days; inclination, 28.55°. Speed of injection into orbit was 23,700 mph. The 2,084-lb. metal payload continued on a path toward a point 240,000 mi. from earth which would simulate a lunar transfer orbit. The target zone was on the path the moon follows as it orbits the earth.

The 113-ft., 303,000-lb. launch vehicle produced 389,000 lbs. thrust at liftoff. The Centaur 2nd stage was powered by two RL-10 liquid hydrogen-liquid oxygen engines with a thrust of 15,000 lbs. each.

The AC-6 mission, sixth in a series of eight scheduled Centaur development flights, was a rehearsal for the AC-7 flight scheduled for later 1965 in which an engineering model of Surveyor was to make a soft landing on the lunar surface. Flight was designed to obtain data on several new Atlas-Centaur features and to continue evaluation of other components and systems tested during previous missions.

Atlas-Centaur vehicle development for direct ascent missions was complete, and the vehicle was now capable of supporting fully operational Surveyor missions. (NASA Release 65-235; KSC *Spaceport News*, 8/12/65, 1; AP, *NYT*, 8/12/65, 11; NASA Proj. Off.)

August 11: A sounding rocket for study of ultraviolet radiation was fired from Salto di Quirra AFB in Sardinia, the Italian Defense Ministry announced. The rocket was one of 400 to be fired by the Italian Air Force in collaboration with the European Space Research Organization (ESRO). (AP, *Wash. Post*, 8/11/65, 15)

- Indonesia successfully launched the second (of ten) Japanese Kappa 8L two-stage meteorological sounding rocket from a site near Bandung, West Java. The 62-kg. (136-lb.) rocket reached an altitude of 90 km. (56 mi.). First firing had been Aug. 7. (*Interavia Air Letter*, 8/16/65, 5)

- NASA Goddard Space Flight Center announced it had selected four companies for feasibility studies for experiments in applications satellite technology. Contracts awarded were: Control Data Corp., $45,000 to examine a technique for determining a satellite's orbit by using only spacecraft observation; Philco Corp., $50,000 to study a device capable of determining attitude of a spin-stabilized spacecraft from star measurements; Bell Aerospace Corp., $47,000 to study an electrostatic accelerometer which could provide information about the relative motion of a gravity gradient stabilized spacecraft; and Electro Optical Systems, Inc., $37,000 to study measuring of degradation of optical characteristics of materials in space. Studies should be completed in eight weeks. (GSFC Release G-20-65)

- A blotting material to absorb excess moisture, which might have caused the eye and nose irritation of Astronauts Edward White (L/Col., USAF) and James McDivitt (L/Col., USAF) during the June 3 GEMINI IV flight had been eliminated from the GEMINI V spacecraft, MSC spokesmen said. (*Houston Chron.*, 8/11/65)

- GSFC announced management changes "designed to meet the increasing demands of advanced space programs by strengthening lines of authority and responsibility at the Center." Dr. John F. Clark was Acting Director and John W. Townsend, Deputy Director.

Three additional Assistant Directorships and a Chief of Advanced Plans Staff were created. GSFC personnel were appointed to fill these posts: Herman E. LaGow, Assistant Director for Systems Reliability; Daniel G. Mazur, Assistant Director for Technology; George F. Pieper, Jr., Assistant Director for Space Sciences; Robert E. Bourdeau, Acting

Assistant Director for Projects; and William G. Stroud, Chief, Advanced Plans Staff.

Dr. Michael J. Vaccaro and John T. Mengel continued as Assistant Directors. Dr. Vaccaro's responsibilities were expanded to include Technical Services as well as Administration. Mengel's position as Assistant Director for Tracking and Data Systems was unchanged. (GSFC Release G–22–65)

August 12: ECHO I, launched by NASA five years ago, had traveled more than 659 million miles and circled the earth more than 22,600 times. Orbital data: apogee, 1,165 mi. (1,875.7 km.); perigee, 560 mi. (901.6 km.); period, 113 min. The satellite had demonstrated that large inflatable spheres could be used as passive communications reflectors in space. (GSFC Release G–19–65)

- Meteoroids probably would not be unduly hazardous to spacecraft flying for short periods in the near-earth environment, a NASA report indicated. Based on data from EXPLORER's XVI and XXIII and PEGASUS I and II, report was presented by Charles T. D'Auitolo, NASA Hq. Office of Advanced Research and Technology; William H. Kinard, Langley Research Center; and Robert J. Naumann, Marshall Space Flight Center, at the Symposium on Meteor Orbits and Dust conducted in Cambridge, Mass., by the Smithsonian Astrophysical Observatory.

  EXPLORER XVI, launched Dec. 16, 1962, had registered 62 meteoroid penerations during its lifetime. EXPLORER XXIII, launched Nov. 6, 1964, had reported 103 penetrations. PEGASUS I, launched Feb. 16, 1965, carried three thicknesses of panels but mechanical malfunctions had destroyed the usefulness of data on the .008- and .016-in.-thick panels; 104 penetrations had been reported from the .0015-in.-thick panels. PEGASUS II, launched May 25, 1965, was returning useful data from all three thicknesses of panels and had registered 61 penetrations. The higher frequency of penetrations recorded by PEGASUS II was thought to have been caused by a meteoroid shower. PEGASUS III, launched July 30, had not had time to return significant data. (NASA Release 65–205)

- First S–IB flight model stage for the Saturn IB booster which left NASA Marshall Space Flight Center's Michoud Assembly Facility for Kennedy Space Center, NASA, aboard the barge *Promise,* was scheduled to arrive at KSC August 14. The 80-ft.-long S–IB stage would be joined with S–IVB second stage to make up the first Saturn IB launch vehicle. NASA planned to launch the 225-ft.-long booster in 1966. (MSFC Release 65–206)

- Facility grants in the amount of $2,226,000 to the Case Institute of Technology and $1 million to the Univ. of Rochester, had been approved by NASA. The grant to Case, together with funds from Institute sources, would permit construction, on land owned by Case, of the Case Laboratory for Space Engineering Research. The Univ. of Rochester grant would be used to build a five-story addition to be used for space-related research in optics, geology, exobiology, physiology, and cosmic ray physics.

  NASA Administrator James E. Webb said: "The new facilities will permit expansion of NASA-supported research directly related to the national space effort and will enable both institutions to train greater numbers of highly qualified young researchers." (NASA Release 65–270)

August 12: In a report to President Johnson on the Titan II missile silo disaster of Aug. 9 which had claimed 53 lives, Air Force Secretary Eugene Zuckert said: "The cause of death was almost exclusively asphyxiation. A number of men had attempted to escape by the emergency ladder which apparently was blocked by two men who became jammed together in trying to pass simultaneously through a restricted area on the ladder, thus denying access to those on the ladder below them." (Text)

- Aerospace Corp. was accused of practices that were "uneconomical, unnecessary, unreasonable, or unjustified" by the Special Investigations Subcommittee of the House Armed Forces Committee, in a report on an examination of business management and fiscal controls at Aerospace that had culminated in hearings in May 1965. The report recommended a reappraisal of the USAF concept that had led to the creation of Aerospace, the abolishment of the fee-funding system, a review and reform of Aerospace's security procedures, and a study of personnel policies and salaries. It also said USAF exercised improper and inadequate control of its contracts with Aerospace, often resulting in fees being provided "for purposes for which they were never used," and Aerospace using "fees for purposes never intended by the Air Force . . ." (Committee on Armed Services Report)

August 13: Five SURCAL satellites and one unidentified satellite were orbited by single USAF Thor-Able-Star launch vehicle from Western Test Range. The satellites all were in similar orbits of about 735-mi. (1,183-km.) apogee, 677-mi. (1,090-km.) perigee, 108-min. period, 90° inclination. (*U.S. Aeron & Space Act., 1965*, 150)

- Gemini V Astronauts L. Gordon Cooper (L/Col., USAF) and Charles Conrad, Jr. (LCdr., USN) would undergo intensive debriefing and medical tests for 11 days following their spaceflight, NASA announced. Dr. George E. Mueller, NASA Associate Administrator for Manned Space Flight, said: "Producing scientific and technical information is the purpose of the flight. This information is vital in determining the effects of long-duration flight on the human system and in proving out flight systems for future flights." (NASA Release 65–273)

August 14: COSMOS LXXVIII was launched by the Soviet Union, Tass announced. Initial orbital parameters: apogee, 329 km. (204.3 mi.); perigee, 209 km. (129.8 mi.); period, 89.8 min.; inclination, 60°. Equipment "for continuing the exploration of outer space" was functioning normally. (Tass, 8/14/65)

- NASA Lewis Research Center project officials said tracking data from JPL's Deep Space Network on Aug. 11's successful Atlas-Centaur launch from Kennedy Space Center, NASA, had indicated precise guidance system accuracy for lunar and planetary trajectories. Less than one-tenth of the midcourse correction capability in the Surveyor model payload would have been needed to put the spacecraft on the final trajectory for a soft landing at a preselected site on the moon. With the success of this mission, Centaur was first U.S. launch vehicle to qualify operationally an all-inertial guidance system for deep space application. (NASA Release 65–271)

- Potential of the Saturn IB was noted in the New Orleans *Times-Picayune*: "The payload potential of the Saturn I–B boosters now under production at Michoud seems to have escaped general attention in discussions about space missions and the rocket assemblies assigned or assignable to carry them out. These and other factors are significant in connection with efforts under way, just coming to light, to intrude on, supplant or degrade the si–b in the Apollo or other programs for which it is fitted.

"The first of the 'new Saturns,' or intermediate Saturns, of the Chrysler Corporation's Space Division has reached Cape Kennedy for the initial lift-off in tests and flights that will lead to human exploration of the moon.

"Payload requirements vary of course for different objectives of this 12-shot series. Regardless of what is first put aloft, the i–b assembly is designed to send into desired orbit 36,000 pounds of functional vehicle—not to mention the 30,000 pounds of a burned-out second-stage, should that lagniappe be added.

"This rating compares with the 26,000-pound payload Soviet spacemen boosted into orbit last July 16; with 21,000 pounds sent into orbit by the Titan 3-C June 18; with the rated 25,000-pound payload capacity of the Titan 3-C; with the 22,000-pound payload rating of the 'old Saturn'; with the 18,000-pound payload dispatched by an 'old Saturn' a year or two ago; and with 4,000 pounds orbited by the Atlas." (*CR*, 8/19/65, A4674)

August 14: The first stage of the first Saturn ib launch vehicle to be flown arrived at Kennedy Space Center, nasa, aboard the barge *Promise*. This was the first transit of the new Port Canaveral locks, due for formal dedication Aug. 21. (*Brevard Sentinel*, 8/15/65)

• Photographs of Mars returned by nasa mariner iv showed surface features which could be interpreted as possible Martian canals, wrote Eric Burgess, a fellow of the Royal Astronomical Society in London in a letter to the Society. Burgess interpreted a dark, 30-mi.-wide streak shown on photograph No. 11 as a rift valley. This surface feature occurred at the same location on Mars where some astronomers claimed to have seen canals. Burgess said the photograph revealed that the escarpments passed through the rim of a large 100-mi.-wide Martian crater, indicating that this particular rift valley appeared after the formation of the crater. (*L.A. Times*, 8/16/65)

August 15: zond iii's lunar photography mission was announced by Tass: "Automatic station Zond 3 is continuing its flight along a heliocentric orbit. . . . Photography of the moon began on July 20, 36 hrs. after launch, at 0424 hrs., Moscow time, when the automatic station was at a distance of 11,600 km. from the lunar surface, and was concluded at 0532 hrs. at a distance of about 10,000 km. After the lunar flyby the station is continuing its motion in a heliocentric orbit, becoming increasingly more distant from the earth and the sun. Image transmission began in accordance with the program of July 29 at a distance of 2.2 million km., when the angular size of the earth was small enough for the precise tuning of the on-board parabolic antenna towards the earth. Zond 3 for the first time photographed a part of the moon that cannot be seen from the earth and which was not included in the photos made by the Soviet interplanetary station in October 1959. At the moment photography began the phase of the moon visible from Zond 3 was close to full moon, while when the photography ended almost half of the lunar disc was in shadow. The first frames show a considerable part of the moon visible from the earth. Later frames show the surface of the moon when there was considerable solar side illumination, when the relief formations cast distinct shadows. The images from Zond 3 are sent with a clarity of 1,100 lines. When photographing from a distance of the order of 10,000 km., images of a lunar surface area of about 5 million square km. are possible. The quality of the photographs obtained make it possible to see numerous details of the lunar relief, which are of considerable interest. The photo obtained from Zond 3 will be published in leading newspapers and scientific journals. To analyze the result obtained and to name the craters, ranges, and other formations on the newly photographed sections of the moon, the Academy of Sciences of the U.S.S.R. has set up a special commission. Scientific investigations are continuing on Zond 3. For further testing of the radio line the transmission of photographs of the far side of the moon from the station will continue during subsequent communications sessions, right up to the greatest distance from the earth. Thirty-eight communications sessions have been conducted from Zond 3. All on-board systems are functioning normally." (*Izvestia*, 8/15/65, 1)

August 15: ZOND III's photographs of the hidden side of the moon received editorial comment in the *New York Times:* "Scientists and laymen from all nations will look forward to the pictures of the dark side of the moon that Moscow reports its Zond 3 satellite took last month. Unitl then the only photographic intelligence on the area had come from the pictures an earlier Soviet satellite took in 1959, a time when the available equipment—and consequently the quality of the result— was primitive by present standards. Six years ago the marvel was that this feat could be accomplished at all. Now, in the wake of the photographic knowledge of the moon and of Mars obtained by this country's Ranger and Mariner vehicles, world judgment of the importance of the Soviet accomplishment will depend upon the quality and the quantity of the new knowledge it provides." (*NYT*, 8/16/65, E9)

- Comparing LUNA III [LUNIK III] and ZOND III, H. Pfaffe said in an article in East Germany's *Berliner Zeitung:* "Owing to its trajectory, Zond 3 was able to photograph the moon under favorable illumination conditions for one hour and eight minutes. Luna 3 was able to photograph only 40 minutes. Zond 3's trajectory was so chosen that the probe assumed a planetary orbit around the sun. Luna 3, on the other hand, assumed an elliptical orbit with the earth as one focal point. While Luna 3 transmitted its photos of the moon from distances of 400,000 km. and 40,000 km., Zond 3 began to transmit from a distance of 2,200,-000 km. This and the fact that the new photos were on a 1,100-line basis indicated the advances made in the Soviet radio and photographic systems." (Pfaffe, *Berliner Zeitung*, 8/15/65, 3)

- Sonic boom damage claims during three months of supersonic training jet flights over Chicago had totaled $52,434 paid to 707 claimants, UPI reported. Some 1,434 claims had been filed. (AP, Wash. *Eve. Star*, 8/15/65, 17)

August 16: Dr. Raymond L. Bisplinghoff would not terminate his service with NASA September 1, 1965, to assume the presidency of Case Institute of Technology as NASA had previously announced, but would continue as Associate Administrator for Advanced Research and Technology until a successor was selected and had assumed these duties. Dr. Bisplinghoff would then become Special Assistant to NASA Administrator James E. Webb in the advanced research and technology field. (NASA Release 65-274)

August 16: Robert Hotz commented in an *Aviation Week and Space Technology* editorial: "The summer of 1965 seems destined to appear in space history as a period of formidable success in extending the horizons of man's efforts to explore his universe. The two spectacular achievements of Mariner 4 and Gemini 4 have been interspersed with several other milestones in space technology as the summer unfolded.

"Even after all the millions of words that have been poured out in describing the amazing feats of the Mariner 4 spacecraft, it is difficult to comprehend the full magnitude of its achievements. The 228-day voyage of Mariner 4 across 135 million mi. to the vicinity of Mars, and its transmission of the first relatively close-up pictures of that planet back across the void to earth, probably will rank as the most spectacular feat of space exploration in the first decade since Sputnik 1.

"More than any other space flight since the first manned space mission by Vostok 1, this successful reconnaissance of Mars has demonstrated the infinite possibilities that lie ahead in obtaining a truly scientific grasp of the universe around us. The evidence that was provided by the Mariner 4 photographs was a scientific discovery of the first magnitude and certainly showed the limitations of earth-based astronomy. But perhaps its real significance lies in the demonstration that a complex spacecraft can voyage, communicate, and function over the vast reaches of space under human control and command from earth. . . .

"There is little doubt that the U.S. space program would not be operating at its present pace or scope without the early, humiliating prod of the Soviet Union's pioneering in this field. However, the results now being produced from the U.S. program are proving so useful in such a variety of applications that there should be little doubt that it is worth pursuing for its own goals, even without the stimulus of international competition. It may well be that when historians write of the space age they will point to the summer of 1965 as the time when the U.S. effort finally reached maturity and began breaking its own new ground, in contrast to the earlier era of stern-chasing the Soviets." (Hotz, *Av. Wk.*, 8/16/65, 21)

- NASA was seeking buyers for rocket launches, and the improved Delta, incorporated into the launch package, might prove most economical, William S. Beller reported in *Missiles and Rockets*. Cost of launching the Delta, either to low-earth orbit or into interplanetary space, would be the same. Cost to send a package to the moon would be $16,000/lb.; on a Mars flyby, $21,000/lb.

Beller's article continued: "It is ironic that the fabulously successful series of Delta rockets is based on much of the hardware and concepts used in the presumed failure called Vanguard. In 1959, the Air Force took the upper two Vanguard stages, mounted them on Thor, and called the combination Thor-Able. NASA then made minor modifications in the vehicle's coast attitude-control system and put in a new auto-pilot control system; but the basic Thor-Able shell was used. This was the Delta that failed in its first firing in 1960 with an Echo satellite payload. The subsequent Deltas orbited their payloads in 22 successive launchings.

"The Vanguard program, too, besides giving birth to Delta was probably one of the most economical buys of the U.S. space program. Originally budgeted for a total cost of $20 million, Vanguard ended up costing $105 million, which was paid not only for development of the Vanguard hardware and several launchings but also for setting up the worldwide Minitrack network used in Project Mercury.

"Improvements in Delta from 1960 to the present involved using more energetic propellants, lengthening the second-stage tanks, replacing the third-stage motor for a more active one, adding solid strap-ons to the first stage, and again enlarging the second stage.

"The result is an inexpensive and reliable bus whose launch can be bought for less than $3.5 million." (Beller, *M&R*, 8/16/65, 24)

August 16: Ten U.S.-made Skyraider bombers were turned over to Cambodia by France. (UPI, *Wash. Daily News*, 8/16/65, 17)

- AFSC announced development of an ultrasonic corrosion detector which could inspect aircraft fuel tanks quickly and accurately. The corrosion detector, mounted on a trailer, would be guided under the wing of a parked aircraft and raised until it touched the underside of the wing; ultrasonic sound waves would be bounced against the skin of the aircraft. Facsimile recordings of corrosion with good definition and sensitivity had been obtained through metal surfaces one and one half inches thick. (AFSC Release 89.65)

August 17: The large "crawler-transporter" scheduled to move the 500,000-lb. Saturn V booster and an 11-million-lb. launching tower from the assembly area at Kennedy Space Center, NASA, to the launching site three and a half miles away, had been crippled by repeated failure of roller bearings. Eight tractor trucks equipped with treads and rollers like those of a bulldozer would propel the crawler. It was the bearings in 88 rollers that had caused the problem. Tests of the $6-million vehicle, manufactured by Marion Power Shovel Co., had been suspended until a solution could be found.

A NASA spokesman stated that neither tests nor launchings of the Saturn V boosters should be delayed. (Clark, *NYT*, 8/18/65, C13)

- ComSatCorp requested industry proposals for an advanced satellite for a worldwide commercial communications system. The Corporation requested a satellite for use in a phased system at altitudes between 6,000 and 12,000 mi., or in a synchronous system at an altitude of 22,300 mi.; that had 1,000 two-way voice channels; that weighed approximately 240 lbs.; that could be launched alone or in groups of up to six satellites; that measured not more than 56-in.-dia. and 40-in. high; and that had a five-year lifetime. ComSatCorp stipulated the proposal should cover construction of six to 24 satellites, the number depending on the type of system, with the first six to be delivered within 24 months after the contract award. Proposals would be opened on October 25. (ComSatCorp Release)

- The Gold Hodgkins Medal for pioneer work in space age science, plus a cash award, was presented to three scientists during the second week of the Smithsonian Astrophysical Observatory's 75th anniversary observance in Cambridge, Mass.: Prof. Marcel Nicolet, National Center for Space Research, Brussels; Dr. Joseph Kaplan, UCLA; and Dr. Sydney Chapman, Geophysical Institute, Univ. of Alaska, (AP, *Houston, Chron.*, 8/18/65)

August 17: Dr. Winston E. Kock, Director of NASA Electronics Research Center, announced that NASA had authorized architects to proceed with detailed designs for a 26-story tower building, a three-story microwave laboratory, and an auditorium-cafeteria building at ERC. Construction would begin next spring, provided approval were given for urban renewal proceedings in the Kendall Square-Cambridge site. The buildings would be built with $15 million already approved by Congress. (ERC Release 65-31)

- Final approval for the Aug. 19 launch of Gemini V on an eight-day orbital mission was given by Mission Director E. E. Christensen. Astronauts L. Gordon Cooper (L/Col., USAF) and Charles Conrad (LCdr., USN) passed a final physical examination, weather conditions were favorable, and the tracking network was in good operating condition. The astronauts and their backup pilots, Neil Armstrong and Elliot See, Jr., went through a complete review of the mission. (AP, *NYT*, 8/18/65, 12C)

- JPL Director Dr. William H. Pickering, speaking to the National Space Club in Washington, D.C., on "Exploring the Planets," said: "Scientists interested in the solar system point out that all of the planets of the solar system present interesting challenges to the explorers. A study group at Woods Hole in Massachusetts this summer has set priorities on the exploration of the planets in the order of Mars, Venus, Mercury, Jupiter. We hope to make exploratory flights to all of these planets." (Text)

- The *New York Times* commented on the success of the U.S.S.R. ZOND III mission: "Like the Ranger and Mariner feats before it, Zond 3 has again shown what enormous gains in man's knowledge can be obtained by using instruments alone, a technique much cheaper and less hazardous than sending men to the moon or the planets. There is every reason to suppose that a United States satellite could have gotten similar photographs of the other side of the moon even earlier than Zond 3 had a sustained effort to this end been made.

    "The restraining element has been the limitations imposed on this country's exploration of space with instrument-carrying rockets by the overriding priority given to the enormously expensive Apollo project for landing a man on the moon by 1970. The Soviet propaganda and prestige gains that will result from the current accomplishment represent one of the costs of the decision to put so much emphasis upon a manned voyage to the moon." (*NYT*, 8/17/65, 30)

- Aviaexport Chairman B. I. Kharchenko said in an interview with *Sovetskaya Rossiya* that Soviet aircraft were greatly valued in the world market. At present, more than 1,500 Soviet airplanes and helicopters were being exported to about 40 countries. This year, Soviet aviation equipment was being sent to 16 countries. Last year, 400 pilots and technicians from various countries were trained. (*Sovetskaya Rossiya*, 8/17/65, 4)

August 18: USAF launched unidentified satellite from Vandenberg AFB, with Thor-Agena D launch vehicle. (UPI, *Wash. Daily News*, 8/18/65, 19)

August 18: The launch of Oso-C was postponed from Aug. 24 to Aug. 25 because NASA's Fort Myers tracking station might be acquiring data from EXPLORER XXVI satellite about the same time it was scheduled to conduct Oso-C's critical first-orbit interrogation. Recent calculations at NASA Goddard Space Flight Center indicated the conflict was expected to last for only two minutes, but project officials decided to avoid jeopardizing Oso-C's first orbital data. (NASA Release 65-227)

- The Aug. 19 launch of Gemini V and most of the day's coverage would be televised in color by all three major networks—the American Broadcasting Co., the Columbia Broadcasting System, and the National Broadcasting Co. The only previous color broadcast of a space flight had been NBC's telecast of the GEMINI IV launch. (Adams, *NYT*, 8/18/65, 55M)

- On U.S.S.R.'s Aviation Day, Soviet Defense Minister, Commander-in-Chief of Antiaircraft Units, Air Marshal V. A. Studets said in an interview with *Nedelya* that Soviet antiaircraft units could destroy any flying object with the first rocket: ". . . we have means that can in good time detect and destroy any flying apparatus of any size, at all altitudes, even the lowest. If the country is threatened by a ballistic missile, it will be met far away by an antimissile."

    Soviet TV showed a film of a new aircraft reported to fly to nearly 95 km. (59 mi.) altitude at a speed of 6,500 km/h (4,500 mph)—mach 6. This aircraft would be used for cosmonaut training (as with the U.S.'s X-15A or NF-104A). No further details were available. (*Nedelya*, no. 33, 1965, 4–5; Loory, *N.Y. Her. Trib.*, 8/18/65; *Interavia Air Letter*, 8/20/65, 5)

- Dr. Duane Graveline resigned as one of the nation's six new scientist-astronauts "for personal reasons," NASA Manned Spacecraft Center announced. Dr. Graveline would remain with MSC as a flight surgeon. (AP, *NYT*, 8/19/65, 3; AP, *Chic. Trib.*, 8/18/65)

- General Dynamics/Convair received from USAF a $1,000,000 increment to a $7,385,383 cost-plus-incentive-fee contract for the production of Atlas-Agena boosters. (DOD Release 536–65)

- USAF awarded General Electric Co. a $1,494,000 increment to a $33,552,-224 cost-plus-fixed-fee contract for flight testing of the Maneuvering Ballistic Re-entry Vehicle. (DOD Release 536–65)

August 18–20: Vice President Hubert H. Humphrey gave the keynote address at the opening of the AAS meeting on "The Impact of Space Exploration on Society" in San Francisco: "Although the accomplishments of the last seven years are impressive, they will be dwarfed by the events of the coming 20 years:

"Before the end of this decade, man will have set foot on the Moon's surface. This will be an event of historical magnitude.

"By 1970, manned satellite stations will be circling the Earth for a month or more. A crew of three or four will be on board, performing many kinds of scientific experiments, ranging from astronomy to zoology. Much larger manned space stations could become a reality in future years.

"Before the end of the next 5 years, world-wide communication by means of satellites will be a routine accomplished fact, as will detailed observation of weather patterns and collection of meteorological data around the globe.

"Radio and TV broadcasting to large areas of the Earth from satellites could become a reality in the next 10 years.

"Before 1975 unmanned vehicles from Earth will be circling Mars and Venus and sending exploratory probes to the planets' surfaces.

"To look into the future beyond 1975 is much more uncertain, but experts judge that if Man has the desire and will, he can have gone, by 1985, a long way toward the goal of sending a manned expedition to Mars, perhaps with the aid of some kind of nuclear propulsion. . . ." (Text)

Leonard Jaffe, Director of NASA Communication and Navigation Programs, told the Conference that recent announcement that DOD's navigation satellite system was operational demonstrated the great potential of artificial earth satellites for providing position determinations to the maritime industry. He said NASA had under study a number of techniques, different from those employed in the DOD system, by which satellites located in medium or synchronous orbits could provide position fixes to ships and aircraft. Techniques whereby the craft's position would be determined at ground computer stations might make the satellite system attractive to U.S. industry, he said.

For the future Jaffe foresaw the combination of a position determination method and a communications system integrated into a single satellite to provide capability of aiding in air traffic control. He said: "Accurate position data and reliable communications readily available to air traffic controllers by use of satellites may provide the answer to the acute problem of safely reducing the present separation standard of aircraft flying across the crowded Atlantic Ocean region. Similar data supplied to a maritime regulatory agency may provide a means for preventing marine collisions." (Text)

August 18-20: Gen. Bernard A. Schriever, AFSC Commander, participating in a panel discussion on the "Social Implications of Space Exploration," said: ". . . there are two ways of looking at space. The first way is simply as a place—as the vast expanse which is an extension of the earth's atmosphere in every direction, and which can serve as an operational environment. There is much we need to discover about this region, but it is already clear that space adds a new dimension to our thinking about national security.

"Space has at least four unique characteristics which offer advantages for military operations. These are: first, extremely high altitudes which offer line of sight to large areas of the earth's surface; second, very high speeds; third, almost infinite flight duration; and fourth, highly predictable flight paths.

"These recognized advantages have led to the development of a number of space systems which provide support to military land, sea, and air operations. These include satellite systems for communications, weather, and navigation. The capabilities they provide are either unique or represent a distinct improvement over other means of accomplishing the same task . . .

"On the other hand, the advantages of space are frequently offset by several disadvantages. The first of these is the cost of launching payloads into space. Presently this amounts to about $1000 a pound in low earth orbit, although it is estimated that the Titan III will reduce this figure to about $500 a pound, and we foresee methods of reducing the cost to the neighborhood of $100 a pound. The second is the requirement for extremely high reliability—the need to build satellites that will operate for long periods without maintenance and repair. Third is the absence of a demonstrated method of returning from space to a controlled landing at a specified point. And fourth is the environment of space itself—the problems associated with hard vacuum, temperature extremes, and intense radiation.

"The need to overcome these disadvantages leads to consideration of the second way in which space may be regarded—not merely as a place, but as a primary focus for man's scientific and technological efforts in the years ahead. The success of these efforts will determine the practical use that can be made of space for a variety of purposes, including military purposes." (Text)

August 19: Gemini V launch counted down to T–10 min. before it was canceled and reset for Aug. 21. Astronauts L. Gordon Cooper (L/Col., USAF) and Charles Conrad, Jr., (LCdr., USN) were disappointed, but took the delay in stride. Cooper commented: ". . . you promised a launch today and not a wet mock."

Five hours before the scheduled launch, computer studies suggested that the liquid hydrogen in the spacecraft's fuel cells was boiling off too quickly. Engineers' attempts to force additional liquid hydrogen into the fuel cells succeeded only in hastening the escape of hydrogen gas. Finally, a colder liquid hydrogen was used with success. Ten minutes before lift-off, apparent troubles with a telemetry programer caused another hold in the countdown. During a thunderstorm, a lightning-induced power surge affected the electrical instruments.

At 5:00 p.m., EDT, as NASA officials were opening a press conference, they were informed that a fire had broken out in a communications cable linking the Gemini V launch complex with mission control center. The officials said that there was a "very good possibility" that Gemini 5 could be launched on Aug. 21, but that no definite date could be set until a partial countdown was held the next day to determine the damage caused by the power surge and the cable fire. (Clark, *NYT*, 8/20/65, 1; Simons, *Wash. Post*, 8/20/65; A1; *WSJ*, 8/20/65, 1)

- Hughes Aircraft Co. was the only concern to respond to ComSatCorp's request for proposals to build four advanced Early Bird-type satellites with options for additional satellites, ComSatCorp announced. Five companies had bid for construction of four transportable ground stations with options for additional stations: Hughes Aircraft Co.; International Telephone and Telegraph Corp.; Page Communications Engineers; and Radio Corp. of America. Delivery of the first satellite and the first two stations would be within eight months. (ComSatCorp Release)
- Two groups of youths were apprehended by a security patrol at Kennedy Space Center, NASA, for trespassing near Gemini 5 on Launch Complex 37. Unauthorized entry to the site is a Federal offense. (*Wash. Post*, 8/20/65)
- ComSatCorp asked 24 firms to propose earth station construction plans for antenna systems in Brewster, Wash., and Paumalu, Hawaii, by Sept. 17. The RFP specified four 85-ft. parabolic antennas capable of

working with satellites at altitudes from 6,000 mi. to 22,300 mi. Requirements were for one operational antenna system at each of the earth stations within a year after the contract award, and the second antenna system at each station two months later. The stations would provide high quality two-way commercial communications, track spacecraft, and send commands and receive data from satellite both during launch and while satellites were in orbit. (ComSatCorp Release; UPI, *NYT*, 8/20/65, 41)

August 19: Environmental Science Services Administration (ESSA) asked that NASA temporarily assign Dr. Joachim P. Kuettner, deputy director of the Technical Staff (Saturn/Apollo Systems Office) of the Research and Development Operations, NASA Marshall Space Flight Center, as Chief Space Scientist of the National Weather Satellite Center. Dr. Kuettner, expected to leave for Washington, D.C., about Sept. 1, would administer ESSA's space program, specifically, manned and unmanned advanced systems. (MFSC Release 65–207)

August 20: MARINER IV spacecraft, which relayed to earth the first closeup pictures of Mars July 14, was still transmitting engineering and scientific data. On its 265th day of travel, the spacecraft was 163,162,460 mi. from earth, 8,622,011 mi. from Mars, and had traveled 365,000,000 mi. since its November 28 launch from Kennedy Space Center, NASA. (NASA Release 65–278)

- NASA announced that Gemini V would be launched from Kennedy Space Center, NASA, Aug. 21, at 10:00 a.m., EDT. Problems which had collectively postponed the launch on August 19, had been resolved: (1) the malfunctioning telemetry programer had been replaced; (2) the underground communications cable had been repaired; (3) the technique for loading the fuel cells with liquid hydrogen had been perfected; (4) security police would take extra precautions to prevent trespassing onto the launch area; and (5) the Gemini 5 computer, which officials feared had been damaged by the lightning-induced power surge, was functioning normally. (Clark, *NYT*, 8/21/65; *Wash. Post.*, 8/21/65, A1; *WSJ*, 8/16/65, 1)

- S–IVB stage for the Saturn V booster was successfully test-fired for full duration at Douglas Aircraft Co.'s Sacramento Test Center. In a test simulating a lunar flight, the stage ran for three minutes, was cut off for 30-min. orbital coast period, and then re-ignited for 355-sec. run. (*Marshall Star*, 8/25/65, 1)

- Two USAF Athena missiles were fired from Green River, Utah, to White Sands Missile Range, N. Mex. Although one of the flights was terminated because of a malfunctioning computer, both missiles landed within WSMR. (AP, *NYT*, 8/22/65, 72)

- Dr. Mac C. Adams, vice president and assistant general manager for space systems at Avco Corp., would succeed Dr. Raymond L. Bisplinghoff as NASA's Associate Administrator for Advanced Research and Technology. Dr. Adams, a consultant to NASA and NACA, was expected to assume his new duties in October. Dr. Bisplinghoff would become Special Assistant to NASA Administrator James E. Webb. (NASA Release 65–276)

- FAA announced that Gordon Bain, Deputy Administrator for Supersonic Transport Development, had resigned for personal reasons effective

Sept. 15, and that M/Gen. M.S. White, Federal Air Surgeon, was returning to the Air Force. B/Gen. Jewell C. Maxwell, commander of the USAF Western Test Range, had been assigned to FAA on active duty to replace Bain and Dr. Peter Siegel, Chief, Aeromedical Certification Div., Office of Aviation Medicine, would replace Dr. White. (FAA Release 65–65)

August 20: Several U.S. and foreign scientists would be guests on the 11th Antarctic expedition of the Soviet Union, Tass announced. Leonid Balakshin of the Arctic and Antarctic Institute in Leningrad said that 50 foreign scientists had already spent winters at the Mirny Observatory and other Russian stations near the South Pole. "The exchange of polar scientists is striking evidence of the constructive cooperation between Soviet and foreign explorers on the sixth continent," Balakshin said. (Reuters, *NYT*, 8/21/65, 21)

- Fred W. Friendly, president of the CBS news div., criticized the three national television networks for devoting seven consecutive hours Aug. 19 to the scheduled Gemini V launch which was finally postponed. Friendly told reporters: "We abdicated journalistic responsibility and participated in competitive inanities that are the very opposite of the service that the public expects and demands of us." In the future, Friendly said, CBS would abridge its coverage of space missions. (Adams, *NYT*, 8/21/65)

- A system for the docking and sealing together of two orbiting spacecraft, designed to allow men and tools to be moved back and forth between the two vehicles while they were joined, was granted a patent. Invented by John P. Dunn, Martin Co. engineer, the equipment was being built in full scale for tests in Martin's rendezvous and docking simulator (Jones, *NYT*, 8/21/65, 25)

- USAF named four contractors to participate in a six-month design study effort on an advanced vertical short take-off and landing (V/Stol) tactical fighter program. The firms, which would be funded at approximately $1 million each, were: Lockheed Aircraft Corp.; McDonnell Aircraft Corp.; Republic Aviation Corp.; and the Boeing Co. Studies would cover aircraft configurations for tactical missions. (DOD Release 541–65)

- United Airlines announced that fog-dispersal plans involving aerial seeding of clouds were being discussed in nine cities in addition to those where the technique had already been successful. United had conducted successful seeding operations last winter using light planes to drop ice pellets into the top layers of super-cooled fog. The airline estimated that the fog-abatement program had permitted 135 of its flights, carrying 3,200 passengers, to operate at airports that otherwise would have been closed. (AP, *NYT*, 8/21/65, 48)

August 21–29: NASA's GEMINI V spacecraft, piloted by Astronauts L. Gordon Cooper (L/Col., USAF) and Charles Conrad, Jr., (LCdr., USN), was successfully launched with two-stage Titan II booster from Eastern Test Range on an eight-day mission (190 hr. 55 min.) comprising 120 revolutions of the earth. Flight objectives were: (1) demonstrate and evaluate performance of the Gemini spacecraft; (2) evaluate performance of the rendezvous guidance and navigation system using the Radar Evaluation Pod (Rep); and (3) evaluate effects of prolonged exposure to the space environment of the two-man crew.

August 21–29: GEMINI V launch (*left*), and Astronauts Charles Conrad, Jr., and L. Gordon Cooper, Jr., onboard U.S.S. *Lake Champlain* after eight-day mission and recovery (*above*).

First stage of the booster burned for 156.8 sec. Second stage then separated, burned for 179.6 sec. Traveling at 17,605 mph, 7,947-lb. GEMINI V was inserted into orbit: apogee, 215.4 mi. (347.4 km.); perigee, 100 mi. (161.3 km.); period, 89.58 min.; inclination to the equator, 32.6°.

Fifty-six minutes after liftoff, astronauts fired thruster rockets to raise the perigee to 106 mi. (170.9 km.) in preparation for rendezvous with Rep, a practice maneuver to prepare for Gemini 6's rendezvous in orbit with an Agena stage. At beginning of 2nd orbit, astronauts ejected Rep from the adapter section behind the spacecraft. Rep moved away at five miles per hour.

About two hours after insertion of GEMINI V into orbit, oxygen pressure for the new fuel cell dropped from 800 to 60 lbs. psi because of a failure in the heater element circuitry. This pressure-drop severely reduced the fuel cell's output of electrical power. The Rep experiment was abandoned and there was some apprehension as to whether the flight could continue.

At 27 hrs. into the flight, Cooper and Conrad successfully locked their spacecraft radar on a spare Rep transponder mounted on a tower at NASA Merritt Island Launch Area during a six-minute pass. Oxygen pressure began building up in the fuel-cell system and hopes of completing the mission were raised.

On Aug. 23, as GEMINI V made its 32nd revolution around the earth, the spacecraft began a series of maneuvers to rendezvous with a simulated Agena upper stage in lieu of the Rep experiment. By the 34th orbit GEMINI V was only 16 mi. behind the simulated Agena, very close to the planned 15-mi. separation NASA hoped to have on GEMINI VI prior to final closing and docking maneuver. Last maneuver was not attempted because of the need to conserve fuel on GEMINI V.

On Aug. 24, Astronauts Cooper and Conrad twice spotted objects on the ground; they observed a Minuteman ICBM launch from Vandenberg AFB, and tracked and photographed it; they sighted a huge checkerboard design laid out on the ground near Laredo, Tex., and reported direction in the which arrows in the center of the blocks were pointed.

August 21-29: On Aug. 25, the astronauts saw, tracked, and photographed another Minuteman ICBM fired from Vandenberg AFB and identified a rocket sled test on Holloman AFB. At request of U.S. Weather Bureau, they attempted to photograph the eye of tropical storm Doreen about 200 mi. south of Hawaii.

During the flight, the astronauts both remained in good spirits. Sixteen of the planned seventeen experiments were conducted. Despite the heavy work schedule, Cooper and Conrad each slept about five hours a day. The astronauts ate about 2,000 calories of food a day, less than the 2,900 calories prescribed for the mission, and each drank six pounds of water daily.

Aug. 27 saw problems with the steady accumulation of water in the fuel cell, which threatened to exceed water-storage capacity; faulty attitude and maneuvering thrusters; and the boiling off of hydrogen gas which caused the spacecraft to roll.

On Aug. 28, the fuel cell and the left thruster were functioning well. Conrad reported seeing "a carrier and a destroyer steaming right into Jacksonville"; it was actually a tug pulling a large barge.

On Aug. 29 Cooper and Conrad talked via radio with Astronaut M. Scott Carpenter (Cdr., USN), in Sealab II on the floor of the Pacific off La Jolla, Calif. Approach of Hurricane Betsy from the coast of South America forced NASA officials to order GEMINI V to land one revolution early. Two more of GEMINI V's thrusters froze, but there was still sufficient steering power to correct minor variations.

To begin reentry, the astronauts fired four braking rockets as the spacecraft passed 700 mi. north of Hawaii. Reentry was normal. GEMINI V splashed down into the Atlantic at 8:56 a.m. EDT, 600 mi. east of Jacksonville—90 mi. off target. A human error in computing the elapsed time of flight caused the target error. Assisted from the capsule by Navy frogmen 45 min. after splashdown, the astronauts were taken by helicopter to U.S.S. *Lake Champlain* for medical debriefing. (NASA Release 65-262; NASA Proj. Off.; *NYT*, 8/22/65, 1, 72, 73, E1; 8/23/65, 1, 16; 8/24/65, 1, 16; 8/25/65, 1, 24; 8/26/65, 1, 15; 8/27/65, 1, 12; 8/29/65, 1, 68, 69; 8/30/65, 1, 16, 17; *Wash. Post*, 8/22/65, 1, 34; 8/23/65, 1, 3; 8/24/65, 1, 6; 8/25/65, 1, 10, 24; 8/26/65, 11; 8/27/65, 1, 6, 7; 8/28/65, 1, 7; 8/29/65, 1, 8, 10; 8/30/65, 1, 9, 11; *WSJ*, 8/23/65, 1; 8/24/65, 1; 8/27/65, 1; 8/30/65, 1; *Av. Wk.*, 8/30/65, 24-28)

August 21: NASA Administrator James E. Webb told newsmen at Manned Spacecraft Center that the U.S. might land a man on the moon as early as late 1968. He expressed great confidence in the space program, but cautioned: "I'm not going to say we're ahead of schedule. We've had a remarkable set of flights, and we're now entering a period of intensive testing." (Freelander, *Houston Post*, 8/22/65)

August 21: The first stage of the Titan II booster which orbited GEMINI V was recovered intact in the Atlantic about 450 mi. northeast of Cape Kennedy. An Air Rescue Service plane reported that the 71-ft. stage was in the water 25% submerged. (AP, *Wash. Post,* 8/22/65, A34; UPI, *NYT,* 8/22/65, 13)

- President Johnson watched the GEMINI V launching on color television for seven minutes at the White House in a direct broadcast from Kennedy Space Center, NASA. (AP, *NYT,* 8/22/65, 72)

- PROTON I space station, launched July 16 by the U.S.S.R., had, for the first time, trapped particles with energy of 100 trillion electron volts and a measured energy spectrum of cosmic rays containing such particles, Nikolai I. Grigorov, head of Cosmic Ray Laboratory at Moscow Univ. Institute of Nuclear Physics, told *Pravda.* He said that precise measurements of high-energy particles required the use of a heavy ionizing calorimeter which could not have been orbited without the lifting capacity of Russia's new large rockets. Prof. Grigorov explained that high-energy particles never reached earthbound registering devices because they were absorbed and transformed as they collided with atoms of earth's atmosphere. (*Pravda,* 8/20/65, ATSS–T Trans.; Shabad, *NYT,* 8/22/65, 72)

- The Guggenheim International Astronautics Award for 1965 had been conferred upon Professor Mstislav V. Keldysh, President of the Soviet Academy of Sciences, announced Dr. C. Stark Draper, President of the International Academy of Astronautics. The award, which carried with it a $1,000 honorarium, would be presented during the XVIth International Astronautical Congress to be held in Athens September 12–18, 1965. (IAA Release 26)

- An article in *Human Events* expressed fear that Russian military space capability surpassed that of the U.S.: ". . . Four times in the past 2 months, *Red Star* (a newspaper published by the Defense Ministry) has trumpeted the claim that Russia now has orbital space rockets with the capacity of wiping any aggressor off the face of the earth.

    "There is no doubt at all about the intended identity of the aggressor and among American scientists back home there is little doubt about the weapon.

    "There is little doubt because the United States spent 7 years and $10 million between 1958–65 to plan the nuclear-powered space orbiting system called Project Orion. We have high-ranking scientists who believe the Russians have copied or adapted Orion, which has been declassified since October, 1964. The Russians might easily have cracked the secrecy before that.

    "It is a first generation vehicular system that is capable of carrying 8 men and 100 tons of payload to Mars and back. It is a jet-propelled engine or engines powered by a nuclear reactor which, in effect, bombs a shield at the rear of the vehicle and drives it forward. The usual limitations of temperature do not affect this vehicle because the hot debris of the explosions makes only a momentary contact with the shield. The limitations of using full nuclear power are also absent because the reactor is outside the vehicle.

    "Guidance, mobility, range, the exchange of crew-members, and the restocking of the payload are all within this space cruiser's capability. Its potentials as a raider of earth targets, and as an interceptor of our space satellites, are almost limitless. It may not be the never-never ultimate weapon, but it will do until another comes along." (CR, 8/25/65, A4807–08)

August 22: New York Times commented on the malfunction of the fuel cell in the GEMINI V spacecraft launched August 21: "Past manned space flights have gone so well in most respects that yesterday's tension-filled hours were a shock and a surprise. It has all seemed too easy to the millions of television viewers who have observed previous man-in-space experiments. It was natural to assume that once Gemini 5 was so successfully launched, all would once again go more or less as planned. But those intimately involved with these efforts have known all along of the numerous possibilities of trouble. They were not greatly surprised that one of those possibilities for major difficulty—in this case a malfunction of the fuel cell on Gemini 5—turned into reality.

"There will undoubtedly be future emergencies in space. Yesterday's events are a sobering reminder that the effort to reach the moon during this decade is straining the limits of available technology, and that serious risks are being taken. These hazards are also faced, of course, by the Soviet manned space program. The brave men both countries are sending into space would be the prime beneficiaries of full Soviet-American cooperation that would pool both countries' knowledge and experience to minimize the dangers inherent in these pioneering efforts." (NYT, 8/22/65, 10E)

- Officials at the Milwaukee Public Museum complained that sonic booms from Strategic Air Command supersonic jets had dismembered a 3,000-yr.-old Egyptian mummy. (UPI, NYT, 8/22/65, 9)
- Dispatch to Moscow from Tass' New York correspondent, Leonid Ponomarev: "The flight of the U.S. spaceship Gemini-5 . . . proceeded normally for only two hours, although the flight program was calculated for eight days.

"The main reason [was] that a new system for providing electric power. . . was being used in the Gemini-5 spacecraft for the first time. . . .

"It is a noteworthy fact that the new system for providing electrical power has never once been tested in spaceflight conditions. From the test stands on earth, the installation was put directly into working conditions in space.

"There are several reasons that explain this haste and a certain degree of risk. The program chiefs were given the task of 'catching up with the Soviet Union' at any price regarding the length of flight in space. However, the present power of U.S. rockets and the size of the astronauts' capsule are utterly inadequate for lifting and storing reliably tested battery-powered electrical power installations, which would weigh more than the new system. U.S. scientists were compelled to rapidly create a system, new in principle, in order to provide power for the Gemini-5 spacecraft for eight days so that it could fulfill its tasks." (Tass, 8/22/65

August 23: Astronauts L. Gordon Cooper (L/Col., USAF) and Charles Conrad, Jr. (LCdr., USN), carried two one-dollar bills during the GEMINI V mission to prove on landing that they were the same spacemen who were launched, UPI reported. A representative of the Fédération Aéronautique Internationale recorded the serial numbers on the bills before the launch and would check them after the landing. The Fédération Aéronautique Internationale, with 54 member nations, is the agency that certifies world aviation and space records. (UPI, *NYT*, 8/25/65, 1)

August 23: It was too early to set a realistic date for man's landing on the moon, Mstislav V. Keldysh, President of the Soviet Academy of Sciences, said during a Moscow press conference devoted to the ZOND III mission. He explained: ". . . we have . . . learned quite a lot about the surface of the moon. Yet this does not seem to be a great deal . . . where the choice of a landing spot is concerned. . . . Our knowledge of the soil of the moon is still somewhat contradictory, and I think it is much too early to start looking for a future landing site . . ."

Yuri Khodarev, engineer in the Soviet space program, revealed that ZOND III had used photographic film instead of magnetic film to record pictures for later playback on earth command. The mission had been essentially a test in which the spacecraft would continue to play back the moon pictures repeatedly at increasingly greater distances up to "hundreds of millions of kilometers," Khodarev said, adding, "we have created a new, compact photo television system intended for photographing and transmission of images of planets under conditions of prolonged space flight." The wide-angle camera lens had a 106.4-mm. focal length at a relative aperture of 1:8. A 25-mm. film was used with exposures of 1/100th and 1/300th of a second. Describing the ZOND III television system, Khodarev said ZOND III had taken 25 photos of the far side of the moon in little more than an hour. Only one command had been given from earth—to start taking the photographs. All remaining operations occurred automatically.

The ZOND III photographs confirmed earlier conclusions concerning the scarcity of dark depressions on the moon's reverse side, but also revealed sea-like formations up to 500 km. (310.5 mi.) across and about as long as the maria. These formations differed structurally from the maria and lacked their characteristic dark color. The northern part of the side of the moon was occupied by a bright elevation pitted with craters. Concentration of craters throughout the reverse side of the moon was high.

Soviet astronomer Alla Masevich said the moon photographs were as good as those taken of the visible side from earth-based observatories: "They show beautiful half-tone transitions, reflecting the character of detailed surface relief."

Alexander A. Mikhailov, director of the Pulkovo Observatory, attributed the observed difference between the moon's visible side and its reverse side to geological rather than astronomical factors. He noted the appearance on Zond photographs of chains of craters along elongated fracture lines which he attributed to volcanic activity.

Prof. Keldysh noted that the trajectory of ZOND III had passed far from the planets but said that "the experience we gained in launching this probe into remote space will be used in subsequent launchings to photograph the planets." He said the U.S.S.R. had already begun preparations to achieve a soft lunar landing. Experiments aboard the spacecraft were for study of magnetic fields, solar wind, low-frequency radio emission of the galaxy, micrometeors, cosmic rays, and the infrared and ultraviolet spectra of the lunar surface. (Tass, 8/23/65; Shabad, *NYT*, 8/24/65, 1)

August 23: ComSatCorp had selected Holmes and Narver, Inc., Los Angeles architectural and engineering firm, for design work on two new earth station sites in Hawaii and Washington—the U.S. ground links in a worldwide commercial communications satellite system. The contract provided for $186,000, plus additional costs for optional on-the-site construction supervision and related services. Detailed drawings and specifications would be supplied to ComSatCorp no later than December 1. (ComSatCorp Release)

- West Germany and the U.S. would jointly develop a "jump-jet" fighter and reconnaissance aircraft to be battle-ready in the 1970's, West German Defense Military announced. Two West German and five U.S. aircraft companies had each been awarded $1-million contracts to produce the studies for the Vtol aircraft. A joint American-German group of companies would build the prototypes.

  Boeing Co., Lockheed Aircraft Corp., McDonnell Aircraft Corp., Republic Aviation Corp., and United Aircraft were the U.S. firms involved. (Reuters, *NYT*, 8/24/54, 3)

- Seven Soviet An-24B airliners had been sold to the United Arab Republic, the *New York Times* reported. Details of the sale had not been officially disclosed, but European sources said that the cost of each plane was about $375,000—less than half the true value. (*NYT*, 8/24/65, 50M)

- Dr. John Strong of Johns Hopkins Univ. told the Fifth Annual Space Conference at VPI that Venus was almost as hot on the side facing away from the sun as on the side facing the sun because the entire planet was kept warm by snowfalls. An instrumented balloon launched in 1964 had identified ice crystals in the Venusian atmosphere. Strong speculated that a 120-mph wind on the surface of the planet dragged the ice crystals around to the back side where they presumably became a "warming snowfall." Dr. Strong, who intended to launch another instrumented balloon during 1965, said that an automated, unmanned balloon was preferable to a manned vehicle. (AP, *Wash. Post*, 8/24/65)

August 24: First sounding rocket launching undertaken jointly by the U.S. and Brazil under a memorandum of understanding signed April 21, 1965, was conducted successfully at NASA Wallops Station. The Nike-Apache rocket carried a 60-lb. payload to peak altitude of about 101 mi. (162.6 km.) during the seven-minute flight.

Instrumentation for payload and telemetry ground support equipment were constructed jointly by Brazilian Space Commission (CNAE) and NASA engineers at NASA Goddard Space Flight Center. NASA furnished the rocket and use of Wallops Station facilities. The Brazilian launch team, presently in training at Wallops Station, directed the launch operations and acquired telemetry data during the flight. (Wallops Release 65–51)

- House Committee on Science and Astronautics favorably reported out H.R. 10329, which would authorize the Secretary of Commerce to conduct a feasibility study on adoption of the metric system. (*CR*, 8/24/65, 19007)

August 24: Dr. Donald F. Hornig, science adviser to President Johnson, told the Senate Committee on Aeronautical and Space Sciences that a manned Mars expedition would take from four hundred to six hundred days. He continued: "The longest manned mission we are currently planning is some 14 days. Whether such a long mission is physiologically or psychologically feasible is almost impossible to judge before we have more experience with Apollo, AES, and other manned systems. In any case, we would have to build systems of greater complexity and reliability than we have yet dreamed of.

". . . if we compare the probable scale and technical difficulties of a manned Mars expedition with Apollo it is hard to conclude that its probable cost could be much less than perhaps five times that of Apollo—that is, of the order of one hundred billion dollars." (Text)

- Rusted metal remains of a boilerplate Mercury spacecraft with parachute attached was found in Galveston Bay, Tex., by the shrimp boat "Nancy" and returned to NASA Manned Spacecraft Center. The spacecraft was lost May 31, 1962, when MSC conducted a drop test from a C-119 aircraft at 1,500 ft. altitude; a lanyard broke and the parachute failed to open. (MSC *Roundup*, 9/17/65, 1, 3)

- Eighty per cent of the $4 million needed to construct new bridges, modernizing the access road complex to Cape Kennedy, would be contributed by the Federal Government. The announcement was made jointly by Vice President Hubert Humphrey, Florida Governor Haydon Burns, and Florida Senators George Smathers and Spessard Holland. NASA and USAF would divide the $3.2 million Federal share and the state of Florida would provide $.8 million. (*Cocoa Tribune*, 8/24/65; KSC *Spaceport News*, 8/26/65, 1)

- Soviet claim that the GEMINI V spaceflight was undertaken with "haste and definite risk" received editorial comment in the Washington *Evening Star*: "Tass, the official Soviet news agency, should feel a little bit embarrassed at the moment. It has given a big play to a story by Leonid Ponomariov [sic], its New York correspondent. The story has charged that American space officials have been ordered by the Johnson administration to 'beat the Soviet Union at any price with regard to the duration of orbital flight.'

". . . our country may indeed be forging ahead. But nothing could be more ridiculous than Ponomariov's [sic] studied suggestion that Gemini 5 has been put into orbit in a slapdash manner, with callous disregard for the safety of the astronauts aboard, just to steal a space scene from the Soviets. The fuel system and everything else aboard the Cooper-Conrad vehicle are not whimsical improvisations. That has been made dramatically clear by the way in which Gemini's initial power failure has been eliminated with help from ground control, by the craft's built-in means of self-correction.

"As a result, it seems fairly certain right now that Astronauts Cooper and Conrad will be able to complete their eight-day mission. . . .

"God willing, the Gemini flight will succeed to a degree that may persuade the men of the Kremlin to pay serious attention to our country's standing proposal for a joint Soviet-American effort to promote a manned landing on the moon." (Wash. *Eve. Star*, 8/24/65)

August 24: Gen. Nikolai Kamanin, commander of Soviet cosmonauts, said the premiere showing in Moscow of "A Man Walks in Space"—film describing the March 18 VOSKHOD II flight—that Soviet data had helped advance the date of L/Col. Edward White II's "walk" in space June 4. Kamanin said: "The Americans had not planned to let a man out until the end of this year. But after our flight they became bolder. A small group of American specialists, with the permission of our government, came to the Soviet Union and talked with cosmonauts [Col. Pavel I.] Belyayev and [L/Col. Aleksey] Leonov about their flight and we didn't hide anything."

Questioned by newsman about the visit by "American specialists," Kamanin said either three or five persons had interviewed the Russian cosmonauts for several days, but he could not recall their names. Kamanin was asked if the visitors were officials of the U.S. National Aeronautics and Space Administration. "I don't know," he replied. "Officially they were here with a television company—allegedly."

NASA categorically denied General Kamanin's allegation, UPI reported. A spokesman said: "We have been unable to uncover any United States delegation of scientists addressed by the Soviet cosmonauts." Reuters, *NYT*, 8/26/65, 40; UPI, *NYT*, 8/26/65, 40; Loory, *N.Y. Her. Trib.*, 8/25/65, 1)

- The *New York Times* editorialized: ". . . The care that has marked every phase of the [Gemini V] flight thus far provides assurance that the decision on whether to call the capsule back to earth before that time will be based solely on the best judgment of the scientific team as to whether any element of needless risk would be involved in keeping the astronauts aloft. Their safety must, of course, take precedence over the fact that the eight-day goal is the most important single objective of the flight.

    "This is no matter of oneupmanship over the Russians, whose Valery Bykovsky stayed up for five days in 1963. The eight days it is hoped to keep Gemini 5 in space are roughly the time that woud be required for a round trip to the moon. If Colonel Cooper and Commander Conrad can function effectively in a state of weightlessness for that long and then return without injury, they will have given strong support to the hypothesis on which the whole program for a manned moonshot is based.

    "Whenever that shot is made—and remain convinced that it is foolhardy to chain it to any fixed deadline—it will have to be based on the fullest possible advance knowledge of man's ability to cope in another environment with challenges his evolution on earth never required him to encounter." (*NYT*, 8/26/65, 29N)

- The *Washington Post* editorialized: "Decision of the Gemini-5 space flight commanders to try for the scheduled eight-day voyage, despite the power shortages that threatened the exploit during the first two days is one that the whole country will greet with delight. The astronauts have accomplished no mean feat already, but the execution of the intended plan will prove the sophistication of the American space engineers. . . .

    "The power difficulties on Gemini-5 have, in a curious way, reminded us that these incredible penetrations of space are not yet so commonplace as to be devoid of risk and misadvanture. A succession of triumphant experiments by Americans and Russians have made it look easy. The powerpack troubles, like the dips and bobs and pauses of gifted high-wire performers, have succeeded in reminding the 'audience' that the whole act is indeed very difficult and dangerous. We may have forgotten it for a moment, but now we are convinced." (*Wash. Post*, 8/24/65, A12)

August 25: President Johnson announced he had approved DOD development of a Manned Orbiting Laboratory (Mol) at a cost of $1.5 billion. At a White House news conference, the President said: "This program will bring us new knowledge about what man is able to do in space. It will enable us to relate that ability to the defense of America. It will develop technology and equipment which will help advance manned and unmanned space flight and it will make it possible to perform very new and rewarding experiments with that technology and equipment. . . .

"Unmanned flights to test launching, recovery and other basic parts of the system will begin late next year or early 1967. The initial unmanned launch of a fully equipped laboratory is scheduled for 1968. This will be followed later that year by the first of five flights with two–man crews.

"The Air Force has selected the Douglas Aircraft Company to design and to build the spacecraft in which the crew of the laboratory will live and operate. The General Electric Company will plan and develop the space experiments.

"The Titan 3C booster will launch the laboratory into space and a modified version of the NASA Gemini capsule will be the vehicle in which the astronauts return to earth."

President Johnson emphasized that the U.S. would "live up to our agreement not to orbit weapons of mass destruction and we will continue to hold to all nations, including the Soviet Union, the hand of cooperation in the exciting years of space exploration which lie ahead for all of us. . . ." He directed NASA Administrator James E. Webb "to invite the Soviet Academy of Sciences to send a very high level representative next month here to observe the launching of Gemini 6." (Transcript, *Pres. Doc.*, 8/30/65, 142; DOD Release 551–65)

- OSC–C was launched from Eastern Test Range, with a Delta booster, fell into the south Atlantic Ocean after failing to achieve orbit. First two stages of the launch vehicle performed perfectly: following second stage cutoff and an approximate 7-min. coast period, small rockets mounted on a table between the second and third stage ignited and spun the table up to 120 rpm; second stage separated, but the third stage ignited about 5½ sec. prematurely, was not properly aimed, and did not attain orbital speed.

OSO–C was third in a series of eight orbiting solar observatories planned by NASA. First two satellites were launched from Kennedy Space Center, NASA, March 7, 1962, and February 3, 1965. OSO–C I had provided more than 2,000 hrs. of scientific information during its lifetime; OSO II was still operating. The Orbiting Solar Observatory program was designed to advance understanding of the sun's structure and behavior and to determine the physical processes by which the sun influences the earth. Next Oso launch would occur in mid-1966.

(NASA Release 65-261; *Wash. Post*, 8/26/65, A 4; *N.Y. Her. Trib.*, 8/26/65)

August 25: COSMOS LXXIX was orbited by U.S.S.R., Tass announced. The satellite contained instrumentation "to continue space research under the program announced by Tass on March 16, 1962." Orbital data: apogee, 359 km. (223 mi.); perigee, 211 km. (131 mi.); period, 89.7 min.; inclination, 64.9°. (AP, *St. Louis Post-Dispatch*, 8/26/65; GSFC *SSR*, 9/1/65; *U.N. Registry*/INF. 117)

- X-15 No. 1, piloted by Milton O. Thompson (NASA), attained maximum velocity of 3,511 mph (mach 5.11) and maximum altitude of 214,100 ft. The purpose of the flight was to obtain data for the MIT horizon scanner program, basic stability and control, and the Pace transducer. (NASA X-15 Proj. Off.; *X-15 Flight Log*)

- Argo D-4 Javelin sounding rocket was successfully launched with 80-lb. instrumented payload from NASA Wallops Station to peak altitude of 549 mi. (883.9 km.). Furnished by the Univ. of Pittsburg under contract to NASA, the experiment measured the quantity of helium and hydrogen gases and the ionization of helium in the exosphere. Impact occurred 653 mi. (1,051.3 km.) downrange in the Atlantic. (Wallops Release 65-52; NASA Rpt, SRL)

- Rep. Roman C. Pucinski (D-Ill.), speaking on the floor of the House, commented on President Johnson's decision to develop a manned orbital laboratory: "I believe . . . the most significant and important aspect of this latest development is the peaceful aspect of this program. This indeed, can provide mankind with the open sky policies that we have been working for in order to let people know that we have no intentions for any agressive moves and to let them know that we certainly know what is going on in the rest of the world. . . ." (*CR*, 8/25/65, 21024)

- GEMINI V mission director Christopher Kraft, asked at a press conference at NASA Manned Spacecraft Center if he would welcome a Soviet observer at the Gemini 6 launching, said "yes." Kraft said he felt he might learn something about the way the Russians handled their flights. "If you're going to ask if I'd like to go over there [to observe a Russian flight], you bet your sweet life I would." (Sehlstedt, Jr., *Balt. Sun.*, 8/26/65, 2; *NYT*, 8/27/65, 2)

- Three members of the NAS-NRC Space Science Board, Dr. Lloyd V. Berkner, Dr. Harry H. Hess, and Dr. Gordon J. F. MacDonald, appeared before the Senate Committee on Aeronautical and Space Sciences in its hearings on post-Apollo goals of the U.S. space program. They repeated the recommendation of NAS in its 1964 *National Goals in Space 1971-1985* that emphasis be placed on unmanned exploration of the planets, especially Mars, and added that supplementary data would be forthcoming with the release of reports on the Space Research Summer Study—1965 (June 21-July 16). Dr. Hess noted that the 1964 study recommended highest priority be given to search for life on a neighboring planet. He said the only difference in the 1965 recommendation was to "give a somewhat higher priority to Venus but leaving Mars the number one objective." (NAS-NRC *News Report*, 9/65, 5)

- NASA had awarded Douglas Aircraft Co. $16,200,000 fixed-price, incentive-fee contract for 15 Improved Delta launch vehicle upper stages and associated equipment. The Improved Deltas would have larger fuel tanks which would extend the burning time to 400 sec. from 160 sec. for the standard 2nd stage of the Delta. (NASA Release 65–280)

August 25: Preparation for landing a man on the moon by 1970 was not the main purpose of the Gemini program as the U.S. contended, according to *Krasnaya Zvezda*, official newspaper of the Soviet Defense Ministry: "The main purpose is testing the capability of intercepting artificial satellites and conducting reconnaissance from space." The newspaper said long-range cameras aboard could provide detailed photographs of cities, railroads, posts, and ships, and that the astronauts were also equipped to "carry out visual intelligence." It noted that GEMINI V was scheduled to pass above Cuba 11 times, North Vietnam 16 times, and China 40 times. (Shabad, *NYT*, 8/26/65, 15)

- DOD would initiate a demonstration program of a high-performance cryogenic engine in FY 1966, which could provide up to 50% payload increase in the future, Dr. Harold Brown, Director of DOD Research and Engineering, told a closed session of the Senate Aeronautical and Space Sciences Committee. Brown said that the program, which would be closely coordinated with NASA, would provide the design data necessary to initiate a high-performance engine development program in the future.

  Dr. Brown suggested that the 1975–1985 time period might include operational reusable spacecraft, possibly growing out of an Mol program requirement for data return or logistic resupply. He foresaw an evolution of "reentry spacecraft shapes which will provide greater flexibility of operations, enjoy higher reusability, and possess close to the same volumetric efficiency of current reentry spacecraft. These high performance maneuverable reentry spacecraft should be capable of performing missions calling for fast tactical response and greater option in the selection of landing sites. . . ." (Text)

- Distinction between "orbit" and "revolution" was discussed by John A. Osmundsen in a *New York Times* article: "An orbit is simply the completion of a circuit in space.

  "A revolution, on the other hand, is two consecutive passages of a satellite over a particular meridian on earth. . . .

  "The switch in terminology from orbits to revolutions was desirable for practical reasons involving ground tracking, ground-to-satellite communications and the execution of satellite experiments geared to terrestrial features, such as volcanoes and clouds.

  "It is important to make the distinction between orbits and revolutions . . . because a satellite moving east completes an orbit in less time than it takes to complete a revolution. Hence, it will make fewer revolutions than orbits." (Osmundsen, *NYT*, 8/26/65, C17)

- Plans for the exploration of Mars called for an engineering test shot in 1969 and an unmanned landing in 1971, Robert F. Fellows, NASA program chief for planetary atmospheres, said at Fifth Annual Space Conference at VPI. Additional unmanned Mars landings were scheduled for 1973 and 1975; larger capsules—up to three tons—were expected to land in the early 1980's. Fellows said that the 1971 mission would require two pieces of hardware: an orbiter that would circle Mars for up to 50 yrs. and a capsule that would be lowered to the surface of the planet. The orbiter would have a scientific payload up to 300 lbs. to relay information to earth for about a year. The capsule would penetrate the thin Martian atmosphere protected by a heat shield and, after landing, deploy its instruments, take soil samples, and relay the data to earth. Fellows confirmed that the interplanetary mission after Mars would be to Venus, with Jupiter next. (AP, *Wash. Post*, 8/26/65, A10)

August 25: Representatives of 27 firms interested in providing computer services to support NASA Michoud Assembly Facility and NASA Marshall Space Flight Center's Mississippi Test Facility attended a pre-proposal conference at the New Orleans installation. The computer services contract, which would cover a one-year period with three one-year renewal options, would include operation or maintenance of about 20 digital and analog computers, a data transmission system, a data reduction system, and related electronic equipment. (MSFC Release 65–211)

- x–19 experimental V/Stol aircraft crashed and burned during its initial flight test at National Aviation Experimental Facilities Center, Pomona, N.J. Both pilots parachuted safely from the aircraft, which had been designed by Curtiss-Wright Corp. for DOD. (*NYT*, 8/26/65, 24; *WSJ*, 8/26/65, 1)

- Wendell F. Moore of Textron's Bell Aerosystems Co. was recipient of The Franklin Institute's John Price Wetherill Medal for his invention of small rocket lift device, Franklin Institute announced. (*Av. Wk.*, 9/13/65, 23; Franklin Institute)

- A speech by Rep. John Brademas (D–Ind.) that had been made at Purdue Univ. on the role science and technology would play in economic development of the Midwest was inserted in the *Congressional Record:* "Based on . . . studies one might put forth the following hypothesis: The Midwest is exceedingly successful in obtaining nonmission-oriented basic research funds; holds its own in general university research, basic and applied; does very poorly in industrial development related to Federal research and development problems; and does very well as a supplier of production items in support of Federal research and development projects. . . .

  "A recent study by the National Academy of Sciences . . . shows that the great universities of the Midwest—I still refer to the five states of the east northcentral region—were the source of well over one-fourth of all the doctorates produced in the United States during the 1920's, a figure above that of any other statistical region of the country. In the 1960-1961 period, our region is still producing over one-fourth of the Nation's doctorates and continues to surpass any other region." (*CR*, 8/25/65, 2083–84)

August 26: Maj. Robert Rushworth (USAF) piloted X–15 No. 3 to maximum speed of 3,409 mph (mach 4.79) and maximum altitude of 239,600 ft. to obtain data on the NSL scanner and to measure boundary layer noise. (NASA X–15 Proj. Off.; *X–15 Flight Log*)

- USAF would train about 15 astronauts to operate manned orbiting laboratories, the *Washington Post* reported. Trainees would be chosen from graduates of the Aerospace Flight School, Edwards AFB. Gen. Joseph Bleymaier (USAF), Commander of Western Test Range, suggested that in later Gemini flights the two-man crew might include one USAF astronaut. (AP, *Wash. Post*, 8/27/65, A7)

August 26: Washington *Evening Star* editorial: ". . . there is no point in trying to argue that the projected 'manned orbiting laboratory' is purely scientific in character. It is not purely scientific. It has military potentials of considerable significance, and our country should not apologize to anybody for that fact. The Russians are busy in this field. Why should we ignore it? The simple reality is, as Secretary of State Rusk has warned, that the ocean of space might become a fearsome theater of war in the not distant future. In the circumstances, our country would be guilty of a kind of suicidal passiveness if it failed to develop the 'manned orbiting laboratory.'" (Wash. Eve. Star, 8/26/65)

- Approval by President Johnson of USAF's Manned Orbiting Laboratory received editorial comment in the *New York Times:* "The United States Air Force has been demanding orbiting laboratories for years. The reason is essentially defensive. Such space laboratories could have the capacity to intercept and destroy enemy satellites. Their value for surveillance is obvious. As President Johnson said, they will provide 'new knowledge about what man is able to do in space' and 'will enable us to relate that ability to the defense of the nation.'

  "The MOL's however, will naturally be capable of offense as well as defense. It was therefore significant that President Johnson should have taken the opportunity to couple his announcement with a new peaceful gesture toward Moscow for cooperation in the space program. . . ." (*NYT*, 8/26/65, 30M)

- The *Wall Street Journal*, commenting on President Johnson's decision to develop Manned Orbital Laboratory, wrote: "The President didn't enumerate the MOL experiments, but there is little doubt they would involve such possible missions as:
    —High-altitude reconnaissance over the Soviet Union and China;
    —Inspection and possible destruction of non-U.S. space satellites;
    —Antisubmarine surveillance of the oceans . . ." (Beecher, *WSJ*, 8/26/65, 2)

- President Johnson's order for the Air Force to proceed with development of a Manned Orbital Laboratory produced a quick reaction in Wall Street: stock of the companies involved registered rapid and sometimes sizable gains. (*NYT*, 8/26/65, 46)

- In a *New York Times* letter to the editor, Daniel Murphy commented on President Johnson's decision to develop Manned Orbital Laboratory: "Once again we are asked to silently acquiesce to a questionable military dictum—'if we don't, they will.' Because Russia has the capacity to send up such armaments, why does it so logically follow that they will do so? It is we, not they, who have a three or four to one superiority in ICBM's. . . .

  "By our action will we allow the Soviet Union any alternative but to respond in kind? Would not President Johnson's invitation to the Soviet scientists stand a much greater likelihood of acceptance if coupled with an offer of diplomatically exploring the possibility of not constructing such armaments?

  "The Administration's proposal can only increase the tensions of a world troubled enough." (*NYT*, 9/1/65, 32)

August 26: United Press International (UPI) reported that NASA had postponed until next week an attempt to have MARINER IV photographic probe take pictures of the blackness of space. The pictures would provide scientists with a means of editing MARINER IV's photos of Mars. Communication problems forced the delay. (UPI, *NYT*, 8/27/65, 19)

: Policy shift reflected in Administration approval of the Manned Orbiting Laboratory was discussed by Howard Simons in the *Washington Post*: "Presidents Eisenhower and Kennedy virtually ignored Air Force arguments that it had a manned role in space.

". . . President [Johnson] has not committed the Nation to a long-term Air Force manned space flight effort. But what he has done is to protect himself politically against potential political repercussions should the Russians send a six- to eight-man orbiting space station aloft, as they now appear to have the wherewithal to do. . . .

"Meanwhile the Administration is very wary about destroying the image of a peaceful American manned space flight program, which it has so carefully and lovingly constructed." (Simons, *Wash. Post*, 8/26/65)

August 27: At the Seventh International Conference on Phenomena in Ionized Gases in Belgrade, NASA Lewis Research Center scientist Charles M. Goldstein outlined the effect of collisions on a monoenergetic electronbeam current in the presence of a low-density neutral scattering gas. Research had been conducted in connection with electromagnetic propulsion studies. (LRC Release 65-56)

: *Washington Post* editorial comment on President Johnson's assignment of the Manned Orbiting Laboratory project to the Air Force:

". . . Anxiety arises out of the Air Force commitment to total secrecy in its space operations. . . . [NASA] has pursued a brilliantly successful open public information policy. It has made every American a participant in its exciting conquest of space, aroused the national interest in the whole world of science, stirred the youth of the country to enthusiasm and stimulated national pride. The Vandenberg Air Force Base which will become a major site for MOL launching has operated under a veil of secrecy. . . .

"The Pentagon has not yet devised a policy for informing the public on the man-in-space MOL project. If it continues the Vandenberg policies of the past, the country is going to know very little about MOL. Such secrecy is bound to arouse international suspicions and alarms, particularly since the flights will be over Soviet territory. Either this is primarily a project in the peaceful penetration of space that requires little secrecy; or it is a secret military project that cannot be reconciled with our previous professions." (*Wash. Post*, 8/27/65, A23)

: Geologist G. J. H. McCall of Univ. of Western Australia said he believed the lunar surface to resemble plastic, and he recommended another Ranger spacecraft be sent to the moon to test his theory. McCall was in Bend, Ore., attending the International Lunar Geological Conference. (AP, San Diego *Eve. Trib.*, 8/27/65)

August 28: Prof. Robert H. Dicke of Princeton Univ. suggested in the *Astronomical Journal* that a way to map the sun's shape as a test of Einstein's theory of relativity would be to use measurements of the orbit of the asteroid Icarus as it passed within 4 million mi. of earth on June 15, 1968. Two peculiarities would make Icarus, discovered in 1949, valuable for such an experiment: (1) Icarus was the only body in the solar system known to pass inside the orbit of Mercury; and (2) Icarus' orbit was highly inclined to that of the other planets. Dr. Dicke proposed that measurements be taken when the planet cut across the plane of the earth's orbit. If the two measured points changed with each close passage to the sun, then the sun would be known to be less than a perfect sphere. And, the amount by which it was not truly spherical would be known precisely.

If the sun's radius at the equator exceeded its polar radius by as little as 0.005%, the distortion of the sun's gravitational field that this would cause would be enough to account for 10% of the predicted effect on Mercury's orbit. But the general theory of relativity had already accounted for the peculiarities in Mercury's orbit. Therefore, if the distortion were found, the theory must be incorrect. In other words, said Dr. Dicke, gravitational attraction of one body for another must be accounted for, not by relativity, but by some other mechanism. (*AIP News*, 8/28/65)

August 28: Three gold religious symbols which Astronaut Edward H. White, II (L/Col., USAF) carried with him on his walk in space on June 4 during the GEMINI IV mission were displayed at the New York World's Fair: a cross, a Star of David, and a St. Christopher medal. (*Wash. Post*, 8/28/65, C6)

• Astronaut M. Scott Carpenter (Cdr., USN) and four aquanauts began a 45-day, USN-sponsored experiment in 12 x 58-ft. Sealab II to test how well man can function at 205-ft. depth in the Pacific under pressure six times that at the surface. Carpenter would try to stay down 30 days to set a record; the other aquanauts would remain below for varying periods. The group would collect and study marine specimens, map the ocean floor, and explore a deep marine canyon. (*Wash. Post*, 8/29/65)

August 29: President Johnson commented on the success of the GEMINI V flight during a news conference at his Texas ranch: "This is a moment of great achievement, not only for astronauts Gordon Cooper and Charles Conrad, but for those whose hopes have ridden with Gemini 5.

"I am so happy that Mr. Webb and Mr. Seamans, who had so much to do directing this very successful venture, are here to share with us the pride we all feel today. And I deeply regret that our late, beloved President Kennedy, under whose leadership all of this work was so carefully planned and thought out, can't be here to enjoy the fruits and success of his planning and his forethought.

"The successful completion of the eight-day, 3-million mile flight of the Gemini 5 proves, I think, not only man's capacity for endurance in space, but it proves that man is in space to stay.

"We can be, and we are enormously proud of every member of our space team. That means all the scientists, and the technicians, and the controllers, and the trackers—to everyone who contributed in any way. As President of this country, I want, this afternoon, to extend the thanks of the entire nation for a job well done. To Gordon Cooper and Charles Conrad, and to their wonderful families, I want to simply repeat again: we are all very deeply proud of you.

"The difficulties and disappointments of this flight have served to increase our appreciation, respect and trust for the skill and ability of all the men involved—at all their posts of duty. We can face the challenges and opportunities with far greater confidence and certainty —and this is an unmistakable gain.

"Only seven years ago we were neither first nor second in space— we were not in space at all. Today the capacity of this country for leadership in this realm is no longer in valid question or dispute. Openly, proudly, we are proceeding on our course willing always to share our knowledge, our gains with all mankind. I would repeat and renew this country's invitation to all nations to join together to make this adventure a joint adventure.

"This globe seems smaller today than ever before.

"Somehow the problems which yesterday seemed large and ominous and insoluble today appear less foreboding. As man increases his knowledge of the heavens, why should he fear the unknown on earth? As man draws nearer to the stars, why should he not also draw nearer to his neighbor?

"As we push ever more deeply into the universe—probing its secrets and discovering its way—we must also learn to cooperate across the frontiers that divide earth's surface.

"No national sovereignty rules in outer space. Those who venture there go as envoys of the human race. Their quest must be for all mankind—and what they find belongs to all mankind. That is the basis of the program of which astronauts Cooper and Conrad are a part. . . .

"This flight of Gemini 5 was a journey of peace by men of peace. Its successful conclusion is a noble moment for mankind—and a fitting opportunity for us to renew our pledge to continue our search for a world in which peace reigns and justice prevails.

"To demonstrate the earnestness of that pledge, and to express our commitment to the peaceful uses of space exploration, I intend to ask as many of our astronauts as possible—when their schedule and program permit—to visit various capitals of the world. Some, I hope, will be able to journey abroad soon.

"Gemini is but the beginning. We resolve to have many more such journeys—in space and on earth—until man at last is at peace with himself." (Transcript, *Pres. Doc.*, 9/6/65, 170–72)

August 29: Within an hour after landing on the *Lake Champlain*, Astronauts L. Gordon Cooper (L/Col., USAF) and Charles Conrad, Jr. (LCdr., USN) received a telephone call from President Johnson congratulating them on their courage in the face of disappointments and discouragement: "You have certainly proved once and for all that man has a place in the exploration of the great frontier of space." (Clark, *NYT*, 8/30/65, 1)

- The U.S. GEMINI V mission established eight new world records: (1) longest manned space flight—190 hrs., 55 min.; former record, 119 hrs., six min. held by U.S.S.R.; (2) national man hours in space—641 hrs., 24 min.; former record, 507 hrs., 16 min. held by U.S.S.R.; (3) longest multi-manned space flight—190 hrs., 56 min.; former record, 97 hrs., 48 min. set by U.S.; (4) most revolutions for a manned space flight—120; former record, 81, set by U.S.S.R.; (5) most manned flights—9; former record held by U.S.S.R., 8; (6) first man to make a second orbital flight—L. Gordon Cooper (L/Col., USAF); (7) individual with most space flight time—Col. Cooper; (8) individuals making the longest single space flight—Cooper and Conrad (LCdr., USN); former record held by U.S.S.R. (AP, *NYT*, 8/30/65, 17)

August 29: Rep. Olin E. Teague (D–Tex.), chairman of Subcommittee on Manned Space Flight of the House Committee on Science and Astronautics, said at MSC that NASA's FY 1967 budget would be limited to about the same level as the FY 1966 budget ($5.3 billion) because of military expenses in Viet Nam. He said NASA would like "$200 million or $300 million more" for FY 1967. "With the war in Viet Nam, I doubt there'll be more money for NASA. And NASA needs more money than it is getting now to use its team with top efficiency." (*Houston Post*, 8/30/65)

August 30: GEMINI V astronauts Cooper and Conrad flew to Kennedy Space Center, NASA, following a night onboard the aircraft carrier *Lake Champlain*. Upon their arrival, Florida Governor Haydon Burns presented them with plaques in recognition of the record-setting, eight-day mission. Intensive medical examinations begun on the carrier were continued on Merritt Island. (Toth, *Wash. Post*, 8/31/65; Clark, *NYT*, 8/31/65)

- The appropriateness of President Johnson's comments on the achievements of Astronauts Cooper and Conrad during the GEMINI V spaceflight was discussed editorially in the *New York Times:* "He did not gloat over the evident Soviet discomfiture at the United States forging ahead in the space race. Nor did he follow the repeated Soviet practice of claiming that a particular achievement in the cosmos somehow 'proved' the superiority of one political system over another. Instead, he appealed to all nations to join together for the conquest of space, an arena in which cooperation could help ease world tensions.

  ". . . the President's statement . . . should help to dispel . . . fears. He has made clearer than ever that this country sincerely wants full cooperation in space. If Gemini 5's flight helps to bring that objective closer, its immediate political contribution will rival its enormous gains for science and for technology." (*NYT*, 8/30/65, 24C)

- MARINER IV spacecraft took and recorded 10½ photographs of black space on command from the Goldstone Tracking Station in California. Playback of five of the black-space pictures began on command. MARINER IV now was more than 171 million miles from earth, 11 million miles from Mars. (JPL Release)

- NASA Manned Spacecraft Center had developed a 175-lb. scoop attachment for helicopters' rescue and recovery operations. The device consisted of a retractable rigid boom of tubular aluminum and a plastic-covered wire net, both attached to the forward section of the helicopter. Use of the boom would enable the pilot to keep the subject in view at all times and would eliminate the need for a crew member to leave the helicopter to help secure the object being recovered—the net would scoop it up. (NASA Release 65–283)

- NASA had already received 170 suggested experiments for Voyager—the largest response to any scientific satellite program NASA had conducted, reported *Aviation Week and Space Technology*. NASA would refine the experiments in consultation with the originators by November 19; preliminary selection would be made by March, and final one by July. (*Av. Wk.*, 8/30/65, 21)

August 30: NASA would negotiate with Ball Brothers Research Corp. an approximate $12-million contract for procurement of three additional Orbiting Solar Observatory (Oso) spacecraft, bringing to eight the number of satellites contracted for in the Oso series. (NASA Release 65-282)

- Aerospace engineers hired by California to study the state's crime problem recommended a five-year effort to make crime fighting more efficient. Report was prepared by Space-General Corp. under a $100,000 contract. California had ordered four such reports on the assumption that analytical methods used in setting up aerospace and defense systems could be applied to social problems. (*WSJ*, 8/30/65, 1; Davies, *NYT*, 8/30/65, 35)

August 31: "I certainly have seen nothing in the eight-day data that would lead me to be worried about a 14-day flight," Dr. Charles A. Berry, chief physician for the astronauts, told reporters at a news briefing in Cocoa Beach. He said that the GEMINI V astronauts were in good physical condition and were returning to normal as quickly as had the pilots of the four-day Gemini GT-4 flight in June. Changes in such indicators as the heart rate and blood pressure characteristics seemed to reflect adaptation to the weightless space environment and, later, to the stress of returning to normal gravity of earth, Dr. Berry explained. Heart rates became lower in space and rose above normal on returning to earth. At first, on return to earth, blood pressure did not respond as promptly as normal to tilt-table tests designed to show how well the circulatory system copes with sudden changes from horizontal to nearly-vertical posture.

Cooper and Conrad lost about 7½ and 8½ lbs. respectively, but had gained most of it back. This loss might have been a response to the space environment but was believed to be partly water loss after reentry while waiting in the sun to be rescued, Berry said. Both seemed to be less tired than their predecessors. (Schmeck, *NYT*, 9/1/65, 1, 15)

- The world applauded the success of the GEMINI V mission: many leaders in Western Europe sent congratulatory cables to President Johnson, and newspapers in Great Britain, Malaysia, India, and Japan were among those giving the flight front-page coverage. The Soviet Union televised pictures of the astronauts and broadcast news reports of their landing. Only a few dissenting voices were heard—among them the Cuban newspaper *Revolucion* which described the flight as "cosmic espionage." Communist China remained silent. (*Wash. Post*, 8/31/65, A3)

- Soviet President Anastas Mikoyan cabled President Johnson: "Please accept, Mr. President, our congratulations to the American astronauts Cooper and Conrad on the safe conclusion of their Gemini 5 space flight." At the same time, Mstislav Keldysh, President of the Soviet Academy of Sciences, told Johnson that the Academy would reply promptly to a U.S. invitation to send a Russian scientist to observe the next Gemini launching. (*Wash. Post*, 9/1/65, A28)

August 31: First flight model of the S–IVB, Saturn IB second stage and Saturn V third stage, was formally presented to NASA by California Governor Edmund Brown in ceremonies at Douglas Aircraft Co.'s Sacramento Test Center. A hydrogen-fueled vehicle powered by a 200,000-lb.-thrust Rocketdyne engine, the 58-ft.-long 21½-ft.-dia. S–IVB would be shipped to Kennedy Space Center, NASA, aboard the *Steel Executive*. The stage had successfully undergone a 2½ min., full-power, preflight test-firing August 8. (MSFC Release 65–215)

- A Centaur rocket similar to the vehicle launched successfully into a simulated lunar-transfer trajectory Aug. 11 from Kennedy Space Center, NASA, and a model of MARINER IV photographic Mars probe were being exhibited in the NASA display at Cleveland's Natural Science Museum. NASA had contributed displays to the museum for about 10 yrs. (LRC Release 65–60)
- Secretary of the Air Force Eugene M. Zuckert named Gen. Bernard A. Schriever (USAF) as Director of the Manned Orbiting Laboratory Program. This assignment was in addition to his duties as Commander, Air Force Systems Command. B/Gen. Harry L. Evans was named Vice Director of the MOL Program and B/Gen. Russell A. Berg was named Deputy Director. (DOD Release 560–65)
- Gen. Bernard A. Schriever, AFSC Commander, named B/Gen. Joseph S. Bleymaier as commander of the USAF Western Test Range with headquarters at Vandenberg AFB. Presently deputy commander for manned systems at AFSC Space Systems Div., General Bleymaier replaced B/Gen. Jewell C. Maxwell, recently named head of the FAA Supersonic Transport Program. In his new command, General Bleymaier would be responsible for maintaining, operating, and modifying as needed, the western portion of the national range in support of DOD, NASA, and other agency programs. (AFSC Release 130.65)
- First flight of Super Guppy, developed by Aero Spacelines, Inc., for NASA and the world's largest aircraft in terms of cubic capacity. The aircraft was made up of sections from four Boeing 377 Stratocruisers as well as extensive new manufacture. Specifically designed to airlift outsized cargoes such as the S–IVB stage of the Saturn V launch vehicle and the Lunar Excursion Module Adapter for the Apollo program, the aircraft made its first flight from Van Nuys Airport to Mojave, Calif. (*Huntsville Times,* 9/9/65)
- President Johnson approved the following policy on the promotion and decoration of astronauts:

    "1. Each military astronaut will receive a one grade promotion as a direct result of the first successful space flight, but not beyond the grade of colonel in the Air Force and Marine Corps or captain in the Navy. Promotions to general officer rank will be accomplished through usual military selection board process.

    "2. Each Gemini astronaut will be awarded the NASA Medal for Exceptional Service (or Cluster) after completion of a successful space flight. The NASA Medal for Distinguished Service, the highest award which can be given by that agency, will be awarded for exceptional accomplishments in the Gemini program, including but not limited to accomplishments in actual flight.

    "3. Military decorations associated with space flights, such as awards for exceptional heroism or other distinguished service, will be determined on an individual basis consistent with general policy governing the award of traditional military decorations." (Text, *Pres. Doc.*, 9/6/65, 183-84)

During August: Karl G. Harr, Jr., president of Aerospace Industries Assn., wrote in *Aerospace:* "Almost every aspect of our national life has begun to show a positive and beneficial impact of this nation's decision to be a major participant in man's greatest adventure.

"There has been the direct economic impact on the communities and regions where the major parts of this large national event are centered. There has been introduced into our national economy a wide variety of radically new industrial techniques. There have evolved new standards of excellence in engineering, testing, design, reliability, environment, control, and the skill and dedication of personnel . . .

"In space research, development and exploration, the identity becomes . . . more imperative. Aerospace companies go many routes attempting to achieve that goal. Leading aerospace companies, for example, developed a 'Zero Defects' program aimed at reminding, rewinding, and emphasizing to the worker that rejects are expensive and, if undetected, would contribute to tragedy. Other companies have instituted variations of the zero defects promotional program.

"A second motivation campaign called 'The Critical Parts Program,' to insure that all 98 subsystem assemblies that go into a major space booster system will work, precisely in sequence, has been sold as an industry wide concept.

"Another company reports that employes were amused and, more importantly, impressed, when a sign was tacked above the door of its main assembly room that reads: 'In NASA we trust. Everything else we check . . .'" (*Aerospace*, Summer 1965, 25)

- Experts had observed that some plants had an unusual ability to adapt to the most severe physical conditions, reported *Aviatsiya i Kosmonavtika*. Plants had been deprived of oxygen or supplied with very small amounts of it, had been exposed to argon or nitrogen atmospheres, and had been exposed to extreme cold in experiments which tested the effects of a simulated Martian climate on plant growth. It had been found that the cucumber, a heat-sensitive variety of plant, could withstand an argon atmosphere and frost; lettuce, tomatoes, beans, turnips, and other plants had been grown in compartments supplied with extremely small amounts of oxygen. This led to the speculation that not only lichens but small shrubs might be indigenous to Mars and that there may in fact be a rich and varied Martian flora. (*Aviatsiya i Kosmonavtika*, 8/65, 96)

- A pressurized suit identical to that worn by Astronaut Edward H. White II (L/Col., USAF) during his walk in space June 4, was donned by George C. Wiswell, Jr., founder and head of Marine Contracting, Inc., for a repair job 200 ft. below water at the American Electric Power Co.'s Smith Mountain Dam near Roanoke, Va. Wiswell headed a team of divers who lived in a pressurized tank for a week at the bottom of the dam, using a pressurized diving bell while making the necessary repairs. "The job could have been done by conventional methods," Wiswell asserted, "but the divers would each have been limited at that depth to 20 min. of work a day." With the David

Clark Co.'s "aquanaut" suit, four hours work each day for each man was possible, he said. (Smith, *NYT*, 8/9/65, 35)

During August: Gen. Bernard A. Schriever, AFSC Commander, announced the assignment of B/Gen. Daniel E. Riley as commander of the USAF Contract Management Div. (AFCMD) at Inglewood, Calif. General Riley replaced Col. Fred L. Rennels, who had been commander of AFCMD since its activation in January 1965 and who was named vice commander to General Riley. (AFSC Release 132.65)

- Report to industrial and defense management on the technical and economic status of magnesium-lithium alloys had been published by NASA's Technology Utilization Div. The report was prepared for NASA by the Battelle Memorial Institute. A NASA technology utilization report based on cryogenic research directed toward evaluation of the stress distribution near abrupt changes in wall thickness of pressure vessels also became available. (NASA Releases T-65-12 and T-65-10)

- Data were summarized from NIMBUS I meteorological satellite, which had provided first high-resolution, nighttime, cloud-cover pictures and cloud-top temperatures taken from a satellite. Circular 575-mi.-altitude orbit had been planned for the 830-lb. spacecraft; but short burn of the Agena stage—resulting when about 100 lbs. of fuel which should have been loaded on board the Agena had escaped through a faulty valve prior to launch—had injected NIMBUS I into an elliptical orbit. Launch was made Aug. 28, 1964.

    Results from Hrir (High Resolution Infrared Radiometer) data had demonstrated: (1) feasibility of complete nighttime surveillance of surface and land features on a global scale; (2) detailed vertical structure of intertropical convergence zone and formation of tropical storms and of frontal zones; (3) capability to detect temperature gradients over earth's surface under clear skies; (4) applicability of high-resolution radiometry for glaciology, geology, and oceanography.

    Apt (Automatic Picture Transmission) system experiments provided almost instantaneous data on clouds for thousands of square miles around Apt ground stations, demonstrating that the system could provide cloud-cover data for almost all local forecast requirements; thus, it would be a basic element in the Tiros Operational Satellite program.

    Avcs (Advanced Vidicon Camera System) experiment provided first near-global, relatively high-resolution cloud pictures ever assembled. Proved capabilities of camera assembly and confirmed decision to use it as basis for first operational meteorological satellite system.

    On Sept. 3, 1964, during 381st orbit, NIMBUS I stopped operating. Deterioration of the bearing grease at high temperatures had caused the paddles in the solar array drive system to lock.

    Final contact with NIMBUS I occurred Nov. 20, 1964. Data revealed that (1) all batteries were in trickle charge; (2) voltage regulation had failed; (3) both Pcm and command clock subsystems were operable; (4) spacecraft tape recorders were not operating; and (5) controls power supply had failed. (NASA Proj. Off.)

- Lowell Thomas, news commentator, explorer, and author, was named the fifth Honorary Chairman of the Robert Hutchings Goddard Library Program (Clark Univ.). The four other honorary chairmen were previously announced [see July 16]. Appointments to the Goddard Library's International Sponsors Committee were announced: Dr. Hugh L. Dryden, NASA Deputy Administrator; Dr. Robert R. Gilruth, MSC Director; and Dr. William H. Pickering, JPL Director. (*Goddard, 8/65*)

September 1965

September 1: Nike-Apache launched from NASA Wallops Station carried 80-lb. instrumented payload to 94-mi. peak altitude and impacted 74 mi. downrange in the Atlantic. Conducted for the Southwest Center for Advanced Studies and the Central Radio Propagation Laboratory, the experiment telemetered measurements of electron and ion densities and temperatures, to test and compare the operation and performance of five different types of ionospheric plasma probes. (Wallops Release 65–53)

- GEMINI V Astronauts L. Gordon Cooper (L/Col., USAF) and Charles Conrad, Jr. (LCdr., USN) finished the last day of their general debriefing at Kennedy Space Center, NASA. (UPI, *NYT*, 9/2/65, 15; *Wash. Post*, 9/2/65, A14)
- Fuel cells used in the GEMINI V mission August 21–29 were being developed by General Electric Co. for commercial use, Dr. Arthur M. Bueche, GE vice president for research and development, told a news conference in New York City. The first models, expected to be ready next year, would provide only 12 watts [GEMINI V cells provided 2,000 watts] but could power remote television cameras and other communications equipment. Larger units might soon provide emergency power to homes, Dr. Bueche said. (*NYT*, 9/2/65, 38C)
- Man can withstand spaceflights as long as 30 days without suffering serious biological damage, L/Col. Edward C. Knoblock, director of the Walter Reed Army Institute of Research and member of the medical debriefing team for Project Mercury flights, told the annual convention of the American Association of Clinical Chemists in Chicago. Although GEMINI V relayed signals on the rate of breathing, body temperature, heart beat, and perspiration, the more sophisticated equipment needed to reflect the astronauts' body chemistry would be available on the proposed 14-day flight of a larger manned orbital laboratory. (Powers, *Chic. Trib.*, 9/2/65)
- "The Pentagon has surprised almost everyone with its promptness in applying the first squeeze of censorship and news management to its new manned orbiting laboratory (MOL) program," wrote William Hines in the Washington *Evening Star*. "Most people assumed that soon after the military got a manned role in space, it would start classifying it, but few could have foreseen the rapidity with which restrictions came. The elapsed time from President Johnson's announcement of the start of MOL at his press conference last week to the Pentagon's first fumbling bit on news management was exactly 2 hours. . . .

    "Reporters trooping to an MOL briefing at the Pentagon were instructed that they would not be allowed to make tape recordings or to mention the name of the official (Dr. Albert C. Hall, Deputy Director of Research and Engineering), who was briefing them. . . .

"The Defense Department is not the only traducer of a free news flow. The space agency gives news management the old college try every time a manned spacecraft goes up.

"Of all the significant news locations in a Gemini flight, the only one not covered by the combined news media . . . is the most important one of all, the mission control center at Houston.

"It is not a secret place, not one in which unnecessary traffic is discouraged. . . .

"But neither camera nor tape recorder nor pen-and-paper reporter is allowed in the nonsecret room at any time during a flight. . . . This is not to suggest that there has been any coverup to date. In the course of missions, [Christopher] Kraft gives regular, full, and apparently frank accounts of flight activities and opens himself to detailed questioning. So do his associates. A mission commentary of less consistent accuracy and authenticity is broadcast.

"But whether or not there has been suppression to date is not the point. All flights so far have ended happily, and nothing succeeds like success. There has been no reason for a coverup.

"The point is that the opportunity for news management definitely exists in mission control—and it is an axiom of political science that where opportunity exists, there are always people waiting to seize it." (Hines, Wash. *Eve. Star*, 9/1/65)

September 1: Backup Interceptor Control (Buic), first site of USAF 13-site computer controlled radar system for detection of enemy aircraft, became operational at North Truro, Mass. The 13-site system, which would eventually cost $100 million, was designed by AFSC to assume air defense of the U.S. should the Semi-Automated Ground Environment (Sage) system fail or be destroyed by enemy attack. (UPI, *NYT*, 9/2/65, 18)

September 2: X-15 No. 2, piloted by John B. McKay (NASA), reached 239,000-ft. altitude and 3,511-mph speed (mach 5.16) in a flight to obtain information on the reaction augmentation system, to check out star-tracking ultra-violet stellar photography experiment, and to gain information on advanced X-15 landing dynamics and pilot altitude buildup. (NASA X-15 Proj. Off.; *X-15 Flight Log*)

- GEMINI V Astronauts L. Gordon Cooper (L/Col., USAF) and Charles Conrad, Jr. (LCdr., USN) were briefly united with their families at Ellington AFB near Houston. The astronauts had flown to Ellington from Kennedy Space Center, NASA, where they had been undergoing medical debriefing. After a short visit at Manned Spacecraft Center, the astronauts left their families for more debriefing sessions and tests. (UPI, *NYT*, 9/3/65, 10)

- NASA Marshall Space Flight Center Director, Dr. Wernher von Braun, in Munich for the International Transport Fair, said at a press conference: "The fact is that never to date has a manned American satellite flown over even one square centimeter of Soviet soil." Von Braun denied rumors that he might accept a position with a European aeronautics and space organization. He was presented with the international aviation decoration, Pioneer Chain of the Compass Card with Diamonds, while at the Fair. (Reuters, Balt. *Sun*, 9/3/65)

September 2: Commenting on American contributions to space science, N. M. Sissakian, Soviet biochemist, wrote in *Pravda:* "Soviet scientists note with satisfaction the considerable contribution of American science to the study and opening up of cosmic space, which specifically manifested itself in a series of flights of cosmonauts in one-seater and two-seater ships. Most essential results were obtained by the American scientists from the flights of the ships Gemini 4 and Gemini 5. . . .

". . . The flight of G. Cooper and C. Conrad was additional proof that the state of weightlessness during the minimum time required for a flight to the moon and back does not cause essential physiological changes in the human organisms and has no noticeable influence on his ability to work.

"Of great interest in the cosmic flight of man from the biological viewpoint is the study of such problems as the functional state of the cardiovascular system, the water-salt exchange, and the vestibular apparatus.

"Aboard the ship Gemini 5 a number of physiological experiments . . . were performed. The method tested on the Cosmonaut C. Conrad of inflated cuffs around the extremities, which were intended to maintain the necessary functional vigor of the cardiovascular system, merits attention. . . .

"The flight of Gemini 5 is a great success for U.S. scientists. The Soviet scientists sincerely congratulate Cosmonauts G. Cooper and C. Conrad on the successful conclusion of their flight, giving due credit to their courage and endurance. They transmit to the scientists and the entire personnel which prepared and carried out the flight their congratulations and wishes for further successes in the peaceful opening up of the cosmos." (*Pravda*, 9/2/65)

- Astronauts L. Gordon Cooper and Charles Conrad, Jr., had demonstrated during their August 21–29 GEMINI V flight that a manned weather satellite could become a valuable part of Weather Bureau forecasting, Howard Simons reported in the *Washington Post.* U.S. weather forecasters needed information on tropical storm Doreen moving through the Pacific. Cooper and Conrad observed the storm and fixed its position, acquiring the information earlier than TIROS X meteorological satellite and more precisely than a ship 150 mi. from the storm. The San Francisco Weather Bureau Office had been able to combine GEMINI V data with that from TIROS X to prepare an advisory on Doreen. (Simons, *Wash. Post*, 9/2/65, A3)

- ComSatCorp invited 29 manufacturers to submit by September 30 proposals for communications and control equipment to be installed at two ground stations in Hawaii and Washington state—U.S. links in the proposed worldwide communications satellite system. Two sets of ground communications equipment with one control unit would be integrated with two antenna systems in both earth stations. Each station was expected to cost about $6 million. Work was to begin in the fall and to be completed in about a year. (ComSatCorp Release)

- *Washington Post* concluded that it might be embarrassing for the Russians to refuse President Johnson's invitation to send a representative from the Soviet Academy to view the next Gemini launching. An editorial elaborated: "It just so happens that a high-level Russian space delegation will be in New York City in mid-October to attend a meeting of the United Nations Committee on the Peaceful Uses of Outer Space.

"It just so happens, too, that the Gemini 6 launching is scheduled for late October.

"And, it just so happens that Mr. Johnson knew the dates of both scheduled happenings. . . .

"There is no guarantee that the Russians will accept the President's invitation—the first formal invitation of its kind extended to them. Hitherto, the Russians have not accepted informal invitations to witness an American space launching. . . .

"Experts have suggested that one reason for this Soviet shyness was the tacit implication that if the Russians accept an American invitation to Cape Kennedy, Russia would have to respond with a like invitation. . . ." (*Wash. Post*, 9/2/65, A3)

September 2: U.K. should build small supersonic aircraft capable of operating on short runways and huge submarine merchant ships capable of operating beneath polar ice, Dr. Barnes Wallis, head of British Aircraft Corp.'s research dept., told annual meeting of the British Association for the Advancement of Science in Cambridge. He said the supersonic design would be Britain's solution to possible U.S. dominance in the construction of supersonic aircraft. (*NYT*, 9/3/65, 44M)

September 3: Nike-Apache launched from Wallops Station to peak altitude of 95.3 mi. (153.1 km.). Purpose of flight, second of a series of two, was to compare five ionospheric plasma probes, using stable ionosphere as a laboratory, to evaluate the probes and check relevant theories of probe operation. Little scientific data were obtained; the nosecone did not eject and only one antenna deployed. (NASA Rpt. SRL)

• U.S.S.R. launched COSMOS LXXX, LXXXI, LXXXII, LXXXIII, and LXXXIV into orbit with a single booster. Initial orbital data: COSMOS LXXX, apogee, 1,552 km. (963.3 mi.); perigee, 1,356 km. (842 mi.); period, 115 min.; COSMOS LXXXI, apogee, 1,556 km. (967 mi.); perigee, 1,385 km. (860.2 mi.); period, 115 min.; COSMOS LXXXII, apogee, 1,563 km. (971 mi.); perigee, 1,410 km. (876 mi.); period, 115 min.; COSMOS LXXXIII, apogee, 1,569 km. (975 mi.); perigee, 1,438 km. (893.1 mi.); period, 116 min.; COSMOS LXXXIV, apogee, 1,574 km. (978 mi.); perigee, 1,467 km. (911.1 mi.); period, 116 min. Inclination for all five satellites was 56°. One of the satellites was powered by a radioactive isotope, but Tass announced that full measures had been taken "to preclude the possibility of the radioactive isotope contaminating the atmosphere or the surface of the earth." All onboard equipment was functioning normally. (Tass, 9/3/65)

• Second two-stage Dragon sounding rocket launched by France from Skogasandur, Iceland, to 248-mi. peak altitude. Payload was instrumented to study protons and electrons in the Van Allen radiation belt. (AP, *NYT*, 9/5/65, 48)

• In a congratulatory wire to President Johnson, Yugoslavia's President Tito called the GEMINI V spaceflight a "major achievement of American science" and expressed hope that it would "serve peace in the world." (UPI, *NYT*, 9/5/65, 28)

September 3: U.S. and Soviet astronauts had accumulated 19,610,000 accident-free miles—greatest total in the history of transportation—reported the National Geographic Society. A compilation released by the society showed that U.S. astronauts had traveled 10,692,000 mi. and Soviet astronauts, 8,918,000 mi. (UPI, *Houston Chron.*, 9/3/65)

- Refined data on Martian atmosphere's density, ionization, and composition, based on analysis of radio signals from MARINER IV during its occultation experiment, were reported by scientists at American Geophysical Union meeting held at Southern Methodist Univ., Dallas, Tex. Martian atmosphere was now thought to be much thinner than previously believed: it now seemed that air pressure at Martian surface was only about one-5,000th that of earth. Theory that Martian atmosphere was more dense at heights of 25 mi. or more above the surface than earth's at comparable elevations was not upheld; observations reported today indicated Martian atmosphere was thinner than earth's at all elevations. Temperature was about $-250°$ at location of maximum ionization in Martian atmosphere (78-mi. altitude, about one-fourth the expected height). In earth's ionosphere, temperature is higher than $2,200°$ F. Atmospheric composition that would account for the MARINER IV observations, the scientists said, would be largely carbon dioxide. (Earth-based observations of Mars had shown the presence of carbon dioxide in its atmosphere.) Only scant amount of carbon dioxide is present in earth's nitrogen-oxygen atmosphere. The scientists making this report were Dr. A. J. Kliore of JPL and Dr. Von R. Eshleman of Stanford Univ. (Sullivan, *NYT*, 9/4/65, 1, 12)

- Paraglider, in its first successful manned free flight, demonstrated it could guide a spacecraft to preselected landing site. Test pilot Donald F. McCusker landed a Gemini-type test craft at Edwards AFB, following a four-and-one-half-minute flight suspended beneath the wing. He put the previously-inflated paraglider and simulated spacecraft through a series of turn and pitch maneuvers enroute to his landing, after being towed aloft by helicopter and dropped at approximate 8,000-ft. altitude.

    In actual spacecraft recovery, the paraglider would be stored in a ten-cubic-foot canister onboard the spacecraft. After reentry when the vehicle would be slowed by the earth's atmosphere and parachute drogue, the paraglider would be deployed to its full size—31 ft. long with a 31-ft. span—using an onboard nitrogen supply. The astronaut would control his glide through an onboard flight-control system.

    North American Aviation, Inc., was conducting the paraglider program for NASA Manned Spacecraft Center, studying the feasibility of a controlled earth landing system. (NAA *S&ID Skywriter*, 9/3/65, 1)

- Patent granted to Barnes Engineering Co. on an instrument that could distinguish a rocket flash from a sunset—an important distinction in the automatic tracking of missiles. The invention was already in use at a Government launch site for research in missile development, reported the *New York Times*.

    Lockheed Aircraft Co. received a patent for a reentry vehicle designed for more accurate landings than had been made with the Mercury and Gemini spacecraft. Invented by Charles H. Christenson, the craft had wings that could be extended for flight in the atmosphere or folded for launching and space travel. The vehicle's tail section would exert considerable drag after reentry; ring-shaped control surfaces could reduce the drag when extended rearward. (Jones, *NYT*, 9/4/65, 27)

September 3: Speculation that the earth might briefly have had another moon, one of pure iron which disintegrated and fell to earth in a series of fireball showers, was reported in *Science*. Discovery of a 47-mi.-long string of craters and meteorites extending northeast to southwest in northcentral Argentina led scientists to propose that the rare meteorites, consisting of almost pure iron, might be remnants of a single object which had orbited inside the Roche Limit, been torn apart by the gravity of the earth, and fallen from orbit. Perigee of the hypothetical orbit would have been over the latitude of Argentina and inclination would have been about 40° to account for the alignment of the nine craters. (*Science*, 9/3/65, 1055–1064)

- Soviet scientists, writing in the bulletin of the Soviet Academy of Sciences, asserted that a flight to the moon can now be regarded as fully reliable from the medical and biological viewpoint. This conclusion was reached in summing up the results of the medical-biological research carried on in the two Voskhod spaceflights. Careful training and selection of visual reference points in advance helped to offset the anticipated disorientation of spaceflight. (Tass, 9/3/65)

- FAA announced allocation of $84.5 million in Federal matching funds for the construction and improvement of 445 civil airports under the Federal-aid Airport Program for FY 1966. The program provided for the improvement of 371 existing airports ($74.7 million) and the construction of 74 new airports ($9.8 million). Major emphasis was placed on airports used by the airlines, airports used by general aviation which would relieve congestion at crowded metropolitan airports, and airports used by commercial air taxis. (FAA Release 65–67)

- All bids submitted July 21 for surplus Titan/Atlas missile silos were rejected by General Services Administration because they were too low. The silos had cost the Government between $12 million and $25 million each; the highest offer was $26,110 each. GSA said disposal plans were now being reviewed. (AP, *NYT*, 9/4/65, 7)

September 4: NASA astronauts gathered at NASA's Manned Spacecraft Center for a debriefing by GEMINI V Astronauts L. Gordon Cooper and Charles Conrad, Jr. The meeting was held primarily for the Gemini VI and VII crews, but other astronauts attended as observers. (AP, *NYT*, 9/6/65, 6; *Wash. Post*, 9/5/65, A8)

- Aerospace industry would probably increase its employment of scientists and engineers by more than seven percent this year, announced Karl G. Harr, Jr., president of Aerospace Industries Association. He made public a study which indicated that employment of scientists and engineers in aircraft, missile, and space fields would increase by about 13,000 from March to December, reaching a total of 190,000 by the end of the year. He said total aerospace employment would increase by 9,000 from 1,123,000 in March to 1,132,000 in December. This forecast contradicted other recent surveys which had predicted a decline. (Text; AP, *NYT*, 9/5/65, 5)

September 5: The first photographs taken at the Yale-Columbia Observatory in the Andes in western Argentina had fulfilled the "best optimistic expectations," reported Professor Dirk Brouwer, director of the Yale Observatory. Brouwer said three test photographs of Orion, nebulae, and Omega Centauri had indicated the lenses apparently were the best of their kind. Principal mission of the Ford Foundation-financed observatory was to make two-hour exposures of the Southern Hemisphere sky to obtain positions of the stars in our galaxy against a background of distant galaxies. (UPI, *NYT*, 9/6/65, 31C)

- Immanuel Velikovsky, author of the controversial *Worlds in Collision* (1950), submitted that many of his rejected cosmological hypotheses had been confirmed by space science events: surface temperature of Venus, moon-like surface of Mars, the earth's magnetosphere, the interplanetary magnetic field, the radio noises of Jupiter, and other predictions. "A Tempest in the Cosmos," (*Book Week, Wash. Post,* 9/5/65, 2, 8, 9)

- *Washington Post* discussed the role of a manned orbiting laboratory (MOL) in arms control: "The primary mission of MOL, a canvass of Washington officials makes clear, is without a doubt to have man supplement the machine as a shutterbug spy in the sky. Hence man will advance the sensationally successful camera work of the unmanned Samos series of photo reconnaissance satellites now producing thousands of pictures of the Soviet Union and Communist China.

    "Human judgment is the critical new factor being added by use of the MOL . . .

    "Here is what the Air Force believes . . . men will be able to do:

    "They can use their judgment on what to photograph. They can be selective on when and where to aim not only one camera but a series of cameras including new and experimental photographic equipment. They can shoot accurately through a hole in clouds. And they can maintain and repair equipment that may have become inoperative. . . .

    "How . . . will MOL contribute to arms control?

    "The thesis is that the two major nuclear powers are deterred not only by the nuclear weapons they possess but by what each knows—and how fast it knows—about what goes on, in a military sense, in the other's country. It is hoped that MOL will immensely add to that American capability in relation not only to the Soviet Union, but to China and, indeed, the whole world.

    "Indeed, it is not inconceivable that these developments could lead to manned synchronous satellites, able to hover over Russia and the United States ready to flash instant word on missile firings, rocket tests, nuclear explosions, mass troop movements or other important military activities.

    "The logic of this, from the arms controllers' viewpoint, is that all this could lead to a formal worldwide inspection agreement—peacekeeping from space. But short of such agreement, such space surveillance would vastly add to the deterrence on which today's nuclear peace is founded." (Simons and Roberts, *Wash. Post,* 9/5/65, A1, A5)

September 5: Significance of supersonic transport development was discussed in *Nashville Tennessean:* "Unless the United States goes ahead with the SST, domestic airlines will probably be forced to buy the supersonic Concorde being developed by Britain and France.

"The 1,450-mile-per-hour Concorde is due to be ready for service in 1971 while the present schedule would have a U.S. SST flying by 1974.

"Because the U.S. plane will be faster and carry 220 to 250 passengers as opposed to 118 to 132 for the Concorde, informed sources believe the 3-year lag will not be too damaging.

"Some visionaries say the SST will be out-dated before it ever realizes the potential forecast by its proponents.

"This school of thought suggests that 20 years from now, rocket-boosted passenger vehicles will hurl travelers across the seas at near orbital velocities, making cities on opposite sides of the earth less than an hour apart.

"But each new generation of commercial air transports has descended from military parentage. And today the military has no active program which would logically produce this speedy type of civilian travel." (*CR,* 9/21/65, A5530–31)

- *New York Times* editorialized: ". . . Until recently the idea of men returning to the seas to live and work seemed fantasy. But increasing population and the strain it has put on the resources of the land are forcing close attention to the enormous food reserves in the oceans and the incalculable mineral wealth under them.

  "It is these practical considerations that inspire the Sealab 2 project now begun 205 feet below sea level off the California shore. . . .

  "The work of the aquanauts may not be as spectacular as the exploits of astronauts in outer space. But the chances are good that the experience they gain in what many now call inner space will help bring humanity rich material returns in the decades immediately ahead." (*NYT,* 9/5/65, 8E)

September 6: NASA awarded $100,000 in initial funds to General Dynamics Corp.'s Convair Div. to determine vehicle design modifications and ground support requirements for combining the Centaur as a third stage with the two-stage Saturn IB launch vehicle. The combination was being proposed to launch unmanned Voyager spacecraft to Mars in 1971. Study would be under direction of LRC. (NASA Centaur Proj. Off.; Gen. Dyn. Corp. Release)

- Capt. Joseph Engle (USAF), youngest holder of Air Force astronaut wings, was honored by his hometown of Chapman, Kansas, with a parade. Engle flew the X-15 to peak altitude of 282,000 ft. on June 29, qualifying for the astronaut wings. (AP, Wash. *Eve. Star,* 9/7/65, A-2)

- German Space Research Commission (DKFW) recommended $460-million West German space program between 1966 and 1970: $250–$280 million would be spent on broad-based "vital-point" program; the balance would be used for West German participation in the

European Space Research Organization (ESRO) and the European Launcher Development Organization (ELDO). The Commission's report emphasized three programs from its list of "vital points": development of satellites, development of a recoverable sounding rocket, and construction of research installations. The report deplored the previous neglect of space research in Germany, emphasized that the funds recommended were minimal and that Germany's late start in the space field would entail a high rate of annual increases in cost. At the end of the building stage, costs could "be adjusted to fairly fixed share of the gross national product." (Wetmore, *Av. Wk.*, 9/6/65, 51)

September 6: "The world's first international school for training specialists in satellite tracking," would be opened in Tashkent, capital of Uzbekistan in Soviet Central Asia, Tass announced. Fifteen young scientists from Soviet satellite countries and 20 Soviet scientists would pursue each nine-day course. The school was set up within the framework of the U.N. International Commission for Space Research. (AP, *NYT*, 9/8/65, 5)

- ". . . The last two Gemini missions have shown clearly that the U.S. manned space flight program has come of age technically. They also demonstrated the operational flexibility that can be achieved in long-duration manned missions," wrote Robert Hotz in *Aviation Week*.

  "With each new manned mission, the Gemini program is forging a record that insures its place in space history as the great leap forward in space flight that transformed the role of man himself from that of a surviving passenger to a useful performer of vital functions in the space environment. . . .

  "It was really the conclusive demonstration that adding a long-term manned capability would improve satellite reconnaissance by orders of magnitude that finally forced the decision to proceed with MOL after nearly two years of delay under the guise of 'further study.' " (Hotz, *Av. Wk.*, 9/6/65, 17)

- "Adaptation of Apollo technology to sustain the pace of the U.S. space program between initial manned lunar landings and still-undefined exploration of the future is accelerating toward selection of contractors to integrate a wide variety of potential applications," wrote William Normyle in *Aviation Week and Space Technology*. "Creation of an Apollo applications directorate in the Office of Manned Space Flight is the culmination of plans first generated months ago . . . and formed the basis of pledges to Congress that the National Aeronautics and Space Administration intends to exploit every available use of Apollo hardware and techniques. . . .

  "Apollo applications is considered by NASA to be the logical stepping-stone linking Gemini and Apollo to advanced earth-orbital stations, lunar-orbital surveying vehicles and extended lunar surface exploration.

  "It is also intended to define advanced logistics for larger orbiting space stations—based on Apollo and using several lunar excursion modules—and these more specific missions which NASA planners now consider of particular importance:

"Orbiting launch complex involving several Apollo spacecraft joined into a multi-module space station that would be launching platform for an unmanned fly-by of Mars and possibly other planets. . . .

"Recovery and repair of unmanned satellites. . . .

"Extensive extra-vehicular activity permitting such innovations as remote manipulative procedures for maintenance and repair. . . ."

Normyle felt that the "largest single decision this fall will be to select two contractors to integrate the experiments which have been proposed." (Normyle, *Av. Wk.*, 9/6/65, 25)

September 6: Two Soviet scientists retained their full capacity to work after spending 30 days in a closed test chamber—said to be the period necessary on a spaceflight to circle the moon. Psychologists did not favor a two-man crew for long space flights, however, since the experiment showed that "even trifles psychologically are blown out of all proportion." Results and data of the experiments were published in *Problems of Space Biology*. (Tass, 9/6/65)

- Karl G. Harr, president of Aerospace Industries Assn., made public a study indicating that employment of scientists and engineers in aircraft, missile, and space fields would increase by about 13,000 from March to December 1965, reaching a total of 190,000 by the end of 1965. He said total aerospace employment would increase by 9,000—from 1,123,000 in March to 1,132,000 in December 1965. (AP, *L.A. Times*, 9/6/65)

September 7: Soviet Cosmonaut Aleksey Leonov was quoted by *Zolnierz Wolnosci* on the future Soviet space program: spacecraft would remain in orbit much longer and their crews would be larger; highly trained teams would assemble permanent orbital space stations, orbital laboratories, and interplanetary spacecraft while the remainder of the crews would work inside the craft. Spacecraft participating in these flights would rendezvous and dock, using a flexible concertina-like passageway. Leonov said problems still to be solved included transfer of crews for interplanetary flights, accomplishment of engineering work in space, and return of spacecraft to earth. (Badowski, *Zolnierz Wolnosci*, 9/7/65, 1)

- Members of the House Committee on Science and Astronautics, meeting to consider H.J. Res. 597 providing for a monument to Dr. Robert H. Goddard at Clark Univ., agreed to limit cost of the memorial to $150,000. (*CR*, 9/7/65, 23315)

- *New York Times* editorialized: "The Aerospace Corporation, a multi-million–dollar nonprofit organization sponsored and financed by the Air Force, has been subject to a searching scrutiny by a House Armed Services subcommittee. Its report is a devastating indictment of Aerospace's financial management. . . .

"The abuses at Aerospace are mainly the fault of the Air Force, which indulged its prodigal offspring by failing to exercise any real control. Aerospace is only one of a growing number of Government-owned non-profit companies working on important defense assignments; there is no evidence that all are as expensive in their spending practices, but we agree with the subcommittee that it would be a good idea to examine the function and the costs of the rest and bring their fees under strict regulation. There is no reason for maintaining such operations if their work can be done cheaply and effectively either by the Government itself or by private enterprise." (*NYT*, 9/7/65, 34M)

September 8: Four NASA Tiros meteorological satellites observing Hurricane Betsy were providing the most extensive picture coverage of a storm ever obtained in space. TIROS X had photographed Betsy Aug. 27 before she reached hurricane strength; since then TIROS VII, VIII, and IX had been photographing her about eight times daily. (GSFC Release G–26–65)

- Stanford Univ. received a $2,080,000 facilities grant from NASA to assist in the construction of a space-engineering research building. After signing the NASA/Stanford agreement, NASA Administrator James E. Webb said: "Construction of these facilities will help Stanford provide the laboratory areas so urgently needed by scientists engaged in NASA-supported research and training activities. The national space effort will further benefit by thus increasing the capability of Stanford to train greater numbers of highly qualified young scientists, engineers and researchers." (NASA Release 65–285)
- 250 photos taken by GEMINI V Astronauts L. Gordon Cooper and Charles Conrad, Jr., during the GEMINI V spaceflight were released by NASA Manned Spacecraft Center. (AP, *NYT*, 9/10/65, 17)
- U.S.S.R. declined President Johnson's invitation to send an observer to the launching of Gemini VI in October. M. V. Keldysh, president of the Soviet Academy of Sciences, in a letter to NASA Administrator James E. Webb, said that the Academy was "grateful" for the invitation, but that "at the present time, our representative cannot avail himself of your invitation.

  "Availing myself of the opportunity, I would like on behalf of the Academy of Sciences of the U.S.S.R. to congratulate you, the cosmonauts Charles Conrad and Gordon Cooper and also the scientists and engineers who took part in the preparation and execution of the flight of the space ship Gemini 5."

  Presidential press secretary Bill Moyers said that President Johnson regretted the refusal, but hoped that the Soviet Union would send observers in the future. (*NYT*, 9/9/65, 11; *Wash. Post*, 9/9/65, A9)
- The appointments of Dr. George P. Cressman as director of the Weather Bureau and R/Adm. James C. Tison (USN) as director of the Coast and Geodetic Survey were announced by Dr. Robert M. White, head of the Environmental Science Services Administration (ESSA). The men had been serving as acting directors since the formation of ESSA July 13. (ESSA Release 65–1)
- FAA's failure to inform the airlines that navigational radar at O'Hare International Airport, Chicago, would be closed for repairs Sept. 8, reportedly caused a delay of more than 300 flights and cost the airlines more than $1 million. An FAA spokesman said steps were being taken to prevent future delays during radar repairs including installation of standby units at airports with heavy traffic. (UPI, *NYT*, 9/15/65, 77M)

September 8-10: Fifteen Arcas and Hasp meteorological rockets were launched from NASA Wallops Station in the first series of firings for the Experimental Inter-American Meteorological Rocket Network

(EXAMETNET). Instrumented rockets achieve altitudes of about 180,000 ft.; wind and temperature measurements were obtained as payloads, descended by parachute. Primary purpose was to determine daily variations of winds and temperatures and estimate solar radiation correction for meteorological payloads planned for use by EXAMETNET. Participants included Argentine and Brazilian trainees, U.S. Weather Bureau, and Naval Ordnance Laboratory. (Wallops Release 65–57)

September 9: Two-stage Nike-Apache launched from NASA Wallops Station carried 55-lb. payload to peak altitude of 106 mi. (170.9 km.) and impacted 82 mi. downrange in the Atlantic. Conducted by GSFC, the flight measured antenna impedance characteristics in the ionosphere. (Wallops Release 65–56)

- U.S.S.R. launched COSMOS LXXXV containing scientific equipment for investigation of outer space. Orbital parameters: apogee, 319 km. (198.1 mi.); perigee, 212 km. (131.6 mi.); period, 89.6 min.; inclination, 65°. All onboard equipment was functioning normally. (Tass, *Krasnaya Zvezda*, 9/10/65, 1, ATSS-T Trans.)

- GEMINI V Astronauts L. Gordon Cooper and Charles Conrad, Jr., together with Dr. Robert R. Gilruth, Director, NASA Manned Spacecraft Center, Dr. Robert C. Seamans, Jr., NASA Associate Administrator, and Paul P. Haney, MSC Public Affairs Officer, held a news conference at MSC. Both astronauts recommended that in future flights both pilots sleep at the same time; more rehydratable foods be included in the diet; daily inflight exercises to compensate for muscular inactivity of the legs be included; the crew travel without spacesuits for greater comfort and flexibility; and that the training period be less compressed. Cooper and Conrad agreed that stowage had been the main problem throughout the flight. "Any small little item that was misplaced or wasn't placed in the proper location seemed to multiply," Cooper said. They had seen "a great many micrometeorites, and one meteorite when it was reentering the atmosphere down below us. And on any night side, during a very short interval of time, you could see micrometeorites reentering."

    Dr. Gilruth said he considered both astronauts "very potential candidates" for the first trip to the moon. (Transcript)

- NASA would contract a total of about $50 million with Convair Div. of General Dynamics Corp. and Honeywell, Inc., for five Centaur stages and guidance systems for use in Surveyor lunar-landing missions. Convair would manufacture, test, erect, and launch Atlas-Centaur boosters; Honeywell was developing the all-inertial guidance system under separate contract with LRC. (NASA Release 65–286)

- Pilot Robert A. Rushworth (L/Col., USAF) flew X–15 No. 1 to maximum altitude of 97,200 ft. at maximum speed of 3,545 mph (mach 5.22) to obtain data on the infrared scanner program and information on ablative coatings under consideration for use on future mach-8 flights. (NASA X–15 Proj. Off.; *X–15 Flight Log*)

- Nine of the ten aquanauts in USN Sealab II had developed ear infections, but the ailments would not curtail the aquanauts' activities, a spokesman said. Physicians attributed the infections to microorganisms in the sea. (*Wash. Post*, 9/9/65, A4)

September 9: House Rules Committee deferred bill to study conversion to the metric system, proposed by Rep. George P. Miller (D–Calif.), Chairman of the House Committee on Science and Astronautics. (AP, *NYT*, 9/10/65, 41)

- Rep. Joseph Karth (D–Minn.) told the House that the flight of GEMINI V had successfully tested not only the high qualities of Astronauts Gordon Cooper, Jr., and Charles Conrad, but the reliability and suitability of many items involved in America's space program, including food. He inserted into the *Congressional Record:*

    "The Pillsbury items scheduled by NASA for the Cooper-Conrad mission—each ready to eat and requiring no reconstitution—were brownies, gingerbread, pineapple fruitcake, date fruitcake, apricot cereal bars, strawberry cereal bars, toasted bread cubes and cinnamon toast.

    "Cooper and Conrad were programed to eat 4 meals each day, providing a daily average of approximately 2,500 calories. The selected menu . . . was designed to maintain body weight at zero gravity under light physical activity. The 8 Pillsbury items were designed to provide a high energy source that supplemented the rehydratable foods in the menu for each meal. They required no preparation time and provided solid food consistency. . . ." (*CR*, 9/9/65, A5095)

- William Hines commented on the economy of low bidding in an article in the Washington *Evening Star:* "The lowest bidder still produces hardware for Uncle Sam's space effort, and this arrangement does not always prove the most economical in either time or money. A glaring example has developed at the Kennedy Space Center in Florida, where America's moonport is rapidly nearing completion.

    "The lowest bidder has built two huge tank-like vehicles called 'crawler-transporters' there, at a saving of almost $3 million below a competitive bidder's price. The only trouble is that the crawlers will not do the job they were designed for, which is to carry a Saturn V moon rocket on a mobile launching tower some three miles from an assembly building to where the rocket will actually blast.

    "Trouble has been traced to some tapered roller bearings. . . .

    "There seems to be nothing wrong with the bearings, which are produced by the biggest and best known manufacturer of such items. The fault apparently lies in incorrect design assumptions which result in overloads sometimes being imposed on the bearings, causing them to fail.

    "Since the design—offered by the contractor, Marion Power Shovel Co.—was approved by NASA technical experts, the government will probably have to pay for the redesign necessary to make the crawler work. These redesign costs could well wipe out the difference between Marion's bid and the higher proposal of Bucyrus-Erie, another large builder of heavy mobile equipment. . . .

    "As it turns out, the false economy of low bidding in this instance may cause the United States to miss its stated goal in space: men on the moon in the '60s. Entirely aside from whether this is a worthwhile goal, it is a national objective and the national prestige is riding on it. . . ." (Hines, Wash. *Eve. Star*, 9/9/65)

- *Wall Street Journal* space reporter wrote a "Memo on Space Semantics" to his editor: "I think you should be aware of some problems in semantics and etymology that are becoming acute as a result of the activities at Cape Kennedy. Unless we resolve some of these problems in the near future, I fear that a new space age neurosis may appear among our reporters and copy editors.

"If you recall, since late 1957 there have been a number of hints that the advent of space exploration would create some new difficulties with our editorial style. First, there was the problem of deciding whether to italicize the Russian word 'sputnik.' Then we had to decide on the spelling and capitalization of 'a-okay' (or is it 'A–OK'?).

"Unfortunately, we are now encountering some difficulties which would tax the abilities of Clifton Fadiman, himself. . . .

"Perhaps the most serious problem . . . stems apparently from an oversight on the part of our ancestors. In writing out the names of the nine planets of our solar system it became evident that we have neglected to formally name our own planet. At least this is the only conclusion one can come to when it's noted that we capitalize the names of all the planets except our own. Webster, himself, insists on using a lower case 'e' for earth while dignifying tiny Mercury with a capital. Even the minor planets such as Eros carry formal names.

"The neglect is even more conspicuous in the instance of our own natural satellite. While the moons of other planets have exotic names such as Phobos we refer to ours in the generic and, consequently, with a lower case 'm'.

"I suggest we either start capitalizing Earth and Moon or else ask the American Society of Newspaper Editors to convene an international conference to pick a name for our planet. While they're at it, they might decide whether 'marsography' is a word and straighten out the planetary adjectives." (Bishop, *WSJ*, 9/9/65, 12)

September 10: NASA began recruiting to select additional pilot/astronauts for manned space flight missions. There were presently 28 pilot/astronauts and five scientist/astronauts participating in the NASA manned space flight program. To be eligible for selection, applicants had to: (1) be a U.S. citizen, no taller than six feet, born on or after Dec. 1, 1929; (2) have a bachelor's degree in engineering, physical or biological science; and (3) have acquired 1,000 hrs. jet pilot time or have been graduated from an armed forces test pilot school. (NASA Release 65-288)

- USAF launched unidentified satellite payload from WTR using a Thor FW4s booster. (*U.S. Aeron. & Space Act.*, 1965, 151)
- The U.S. manned orbiting laboratory was aimed at "the practical testing of orbital nuclear weapons, not scientific space laboratories" in violation of a U.N. agreement not to orbit weapons of mass destruction, wrote Col. Gen. Vladimir Tolubko, deputy commander of Soviet rocket troops in *Za Rubezhom*. His remarks were Russia's first comment on President Johnson's Aug. 25 announcement that U.S. would develop a manned orbiting laboratory. (*Wash. Post*, 9/10/65, A25; Reuters, *NYT*, 9/10/65, 12C)
- President Johnson ordered two-year extension of active duty for V/Adm. Hyman G. Rickover (USN), developer of the atomic submarine. Without Presidential intervention, Rickover, who would be 66 yrs. old in January, would have been forced to retire from active duty early next

year. Rickover attributed the success and relatively low cost of his expanding atomic submarine program to the fact that he had directed the program for 18 consecutive years. (Finney, *NYT*, 9/11/65, 54C)

September 11: The White House announced that Astronauts L. Gordon Cooper and Charles Conrad, Jr., would leave Sept. 15 on a six-nation goodwill tour to Greece, Turkey, Ethiopia, Madagascar, Kenya, and Nigeria. The astronauts would be accompanied by their wives, Cooper's daughters, Dr. Charles A. Berry, chief surgeon for the astronauts, and President Johnson's Army aide, Maj. Hugh Robinson. (Pomfret, *NYT*, 9/12/65, 1, 33)

- NASA awarded $900,000 facilities grant to Univ. of Denver for construction of Space Sciences Laboratories building. (NASA Release 65-289)
- The possible implications of President Johnson's decision to develop a manned orbiting laboratory were discussed by Raymond Senter in *The New Republic:*

  "President Johnson's decision to allow the Air Force to build and launch five manned orbital laboratories (MOL), at a cost of $1.5 billion or more, is likely to increase tensions between the US and the USSR and to spark a similar military space program by the Russians. If so, the Air Force will certainly urge further escalation of its own military space program, raising the specter that space will become a fantastically expensive battlefield of the future. . . .

  "It is, however, possible that MOL will demonstrate the feasibility of a few American and Soviet spacemen in their respective spacecraft operating a continuous space watch. If it does, and if both nations exercise restraint, it could have a stabilizing effect, as have our mutual unmanned reconnaissance satellites. If man can be an efficient observer in orbit for extended periods, the time may come when the U.S. should invite the United Nations to maintain a continuous space patrol, with a multi-national crew, to warn of any impending or surprise attack." (Senter, *The New Republic*, 9/11/65, 9)

September 12: Weathermen detected Hurricane Betsy early, tracked her accurately, and warned islands and coastal areas of her approach well in advance, reported Evert Clark in the *New York Times*. He explained: "Two major advances have made this possible in recent years. One is in technology—the satellites, reconnaissance planes, radars, computers and communication networks that have become the forecaster's new set of tools.

  "The other is in the use of the new knowledge provided by these tools to help the public deal more readily with the violent natural forces unleased by a hurricane. . . .

  "A great improvement in advisories, warnings, community planning and evacuation—sponsored by the Weather Bureau—has dramatically cut deaths and destruction from hurricanes in recent years. . . ."

  Hurricane Betsy was first detected by NASA's TIROS X meteorological satellite Aug. 27, and was followed by TIROS VII, VIII, IX, and X. (Clark, *NYT*, 9/12/65, E7)

- President Johnson nominated John S. Foster as Director of Defense Research and Engineering. Dr. Foster, formerly director of the Lawrence Radiation Laboratory, would replace Dr. Harold Brown, who would become Secretary of the Air Force on Oct. 1. (White House Release)

September 12: Dr. Cyril Ponnamperuma, chief of the chemical evolution branch at NASA Ames Research Center, reported to the American Chemical Society meeting in Atlantic City that he had joined together in a laboratory two subunits (nucleotide molecules) of RNA (ribonucleic acid) under conditions simulating earth as it might have been 3 billion yrs. ago. His successful experiments offered further scientific support that life on earth might have evolved chemically. (Haseltine, *Wash. Post*, 9/13/65, A6)

- Activities of the Clearinghouse for Federal Scientific and Technical Information, operated by the Commerce Dept. at N. Springfield, Va., to distribute reports on non-classified Government projects to industry, scientists, and engineers, were summarized by Charles Covell in the Washington *Sunday Star*. Reports acquired, principally the result of research performed in Government laboratories or by contractors for the various defense agencies, NASA, and AEC, were expected to number about 65,000 this fiscal year. (Covell, Wash. *Sun. Star*, 9/12/65)

- A scorpion fish stung aquanaut M. Scott Carpenter as nine of his fellow-aquanauts were preparing to leave USN Sealab II on the bottom of the Pacific. Their departure was delayed until a Navy physician had determined that Carpenter could remain in the Sealab 15 additional days as planned. Nine new crewmen descended to join him. (AP, *NYT*, 9/14/65, 61M; AP, Balt. *Sun.*, 9/13/65)

- High winds from Hurricane Betsy inflicted extensive damage to glass windows, roofing, and metal sheeting on almost all major structures at NASA Michoud Assembly Facility. Winds and rising tides forced the Saturn barge *Promise* onto the west levy of the Michoud slip. Damage was not expected to delay the production schedules of the Saturn IB and Saturn V boosters, however. At Mississippi Test Facility, only minor damage was incurred and area residents took shelter in MTF's main administration building. (*Marshall Star*, 9/15/65, 1, 4; *Marshall Star*, 9/22/65, 7; AP, *Wash. Post*, 9/7/65, A1)

- *Kansas City Star* editorial: "In the long book of history, the 20th century is likely to have a place alongside the 15th and 16th centuries that saw historic explorations of the surface of the earth. Today, man is on the edge of the conquest of another world—the world of space. He now possesses the capability of leaving this planet and setting foot on the moon or another body of the solar system.

    "In fact, it now begins to appear that American astronauts will settle gently onto the surface of the moon in 1969. It will be an exploit that knows no counterpart in more than 4 billion years of earth's history. . . ." (*Kansas City Star*, 9/12/65)

- *Washington Post* contended in an editorial that long-range achievements of oceanographic research might have a greater impact than the space program on the lives of Americans: The reactions of the aquanauts "to prolonged periods in Sealab's artificial environment and in the ocean depths will provide vital information on man's ability to explore and work on the world's continental shelves. Difficulties that have already occurred indicate that undersea exploration, heretofore granted only limited funds for research and development, is not likely to advance as spectacularly as the space program . . ." (*Wash. Post*, 9/12/65)

September 12: U.S.S.R. made public a statement issued by Soviet Foreign Ministry accusing Britain of using its territory to train West German troops in "handling rocket weapons that can bear nuclear warheads." The British Embassy in Moscow, recipient of the statement, made no comment. (AP, *NYT*, 9/13/65)

September 13: In a Cabinet report for President Johnson, NASA Deputy Administrator Hugh L. Dryden said: "The primary objective of the Gemini V mission to demonstrate man's ability to function in the space environment for 8 days and to qualify the spacecraft systems under these conditions was met. This milestone duplicated the period required for the manned lunar exploration mission.

"Gemini V also demonstrated the capability of man to withstand prolonged periods of weightlessness. The adaptability of the human body was indicated by the performance of the astronauts. For example, their heartbeat rates gradually dropped to a level significantly lower than their preflight normal rates, but by the fourth day, adapted to the weightless condition and leveled off. Upon return to Earth, the heartbeat rates were slightly higher than normal, as expected, but returned to normal rates during the second day. This has assured us of man's capability to travel to the Moon and return." (*Pres. Doc.*, 9/20/65, 258)

- President Johnson, in a statement to the Cabinet on strengthening academic capability for science, said: "Almost all of the Federal research money is provided to produce results that are needed now and in the future to achieve our many national goals in health, in defense, in space, in agriculture, and so on. Of the total provided to universities, 34 percent comes from the National Institutes of Health, 23 percent from the Department of Defense, 9 percent from NASA, 6 percent from the AEC, and 4 percent from Agriculture. Only 13 percent is provided by the National Science Foundation—the only agency which supports science and science education as such." (*Pres. Doc.*, 9/20/65, 268)

- First flight model of S-IVB, second stage of NASA's Saturn IB booster, was en route via water from Douglas Aircraft Co., Sacramento, to Kennedy Space Center, NASA, where it might be flown as early as December in the first launch of an Apollo flight-type spacecraft. (*Marshall Star*, 9/22/65, 10)

- Two USAF Athena missiles were fired within hours of each other from Green River, Utah, into the White Sands Missile Range. Vehicle performance was said to have been good. (DOD Press Office; *M&R*, 9/27/65, 15)

- Philip E. Culbertson became Director of Lunar Mission Studies, Advanced Manned Missions Program Office, Office of Manned Space Flight, NASA Hq. Culbertson would direct studies concerning manned exploration of the moon in the time period following Project Apollo. Before joining NASA, he had been associated with General Dynamics/Astronautics since 1958, where he was Chief Project Engineer for Atlas launch vehicle systems, manager of the Atlas launch vehicle systems engineering and reliability program, manager and director of manned space studies, and director of advanced development. (NASA Release 65-308)

September 13: Dr. Charles C. Price, president of the American Chemical Society, told the annual meeting of the Society in Atlantic City that the synthesis of life should be made a national goal: "We have been making fantastic strides in uncovering the basic chemistry of the life process and the structure of many of the key components of living systems.

". . . The political, social, biological and economic consequences of such a breakthrough would dwarf those of either atomic energy or the space program. Success could lead to modified plants and algae for synthesis of foods, fibers, and antibiotics, to improved growth or properties of plants and animals, or even to improved characteristics for man himself. . . ." (Text)

- Robert Hotz wrote in *Aviation Week and Space Technology:* "A strong smell of sour grapes is beginning to emanate from the Soviet propaganda machine with its waspish comments on recent U.S. space achievements. . . .

"This . . . was evident in the Soviet propaganda barrage aimed at the Gemini 5 flight when it became apparent that Astronauts Cooper and Conrad would surpass the world space flight endurance record set by Soviet Cosmonaut Valery Bykovsky in June 1963. Tass . . . charged that U.S. space officials were endangering the lives of the two Gemini Astronauts by keeping them in flight after the initial fuel cell problems in a supposedly rash effort to surpass the Soviet mark. *Red Star,* the official Defense Ministry newspaper, joined in by 'exposing' the Gemini 5 flight as a Pentagon plot to develop better space reconnaissance techniques and charged the peaceful scientific aims of space exploration were being subverted in Gemini by its military experiments. . . .

"What is really significant about these Soviet sour grapes is that they reflect a realization that the USSR has been overtaken and is being passed in the space race by the U.S. program. The U.S. superiority in unmanned space probes has become clearly recognized by the world with the Ranger lunar pictures and the Mars data from Mariner 4. Gemini 4 and 5 finally erased the Soviet lead in the manned space flight. The solid operational foundation of the Gemini program has made possible the swift and sure pace which promises to eclipse significantly the Voskhod flights in the immediate future. . . ." (Hotz, *Av. Wk.,* 9/13/65, 21)

- Harry Schwartz discussed the "New Arms Race in Space" in the *New York Times:* "The President's M.O.L. announcement came at a time when the Soviet Union was publicly attacking Gemini 5 as primarily a venture in military espionage. Just why the President chose precisely that time to supply the Russian propagandists with supporting ammunition is still a puzzle.

"Moreover, by underlining the potential military significance of space, the President may have finally ended any last lingering hope that there might be a joint Soviet-American program for sending a man to the moon or to some more distant objective in the solar system.

"Most serious is the indication that the Soviet Union and its allies have interpreted the M.O.L. decision as the American signal for an arms race in space.

". . . Would it not have been better for the National Aeronautics and Space Administration to have been given the M.O.L. assignment and thus avoid the provocation and propaganda setback represented by the decision the President actually took? And once a NASA M.O.L. were developed, it could be used by the military if the need arose. That need would presumably be evidence that the Russians or others were actively exploiting space for military purposes other than intelligence collection." (Schwartz, *NYT*, 9/13/65)

September 13: ComSatCorp was seeking establishment of a world-wide satellite communications network to be leased jointly by NASA and commercial telecommunications carriers, reported Katherine Johnsen in *Aviation Week and Space Technology*. Miss Johnsen reported that last week the National Communications System (the U.S. Government's centralized communications agency) and ComSatCorp appeared to be approaching "agreement on an arrangement under which it would pay about $80 million over a 10-year period for use by NASA of the global network, primarily for the Apollo manned lunar landing program." Several variations of the proposal were also presented and ComSatCorp executives were optimistic that one would be approved. (Johnsen, *Av. Wk.*, 9/13/65, 35)

- NASA Michoud Assembly Facility resumed partial operations after suffering extensive damage from Hurricane Betsy. The Tri-State Roofing Co. and J. A. Jones Construction Co. were selected to begin emergency repairs. (MSFC Release 65–237)

- *U.S. News and World Report* commented on President Johnson's decision to develop a manned orbiting laboratory: "President Johnson has now committed this country to a vital and far-reaching race against Russia for military supremacy in space.

  "It is a strategic decision regarded as important as any since World War II.

  "The race actually has started. At this point it is largely silent and secret. But both powers are known to be conducting military missions with unmanned satellites, and results are described as striking and enormous.

  "Next, both countries will make major strategic moves by putting up space stations with military crews. Beyond that, plans are being readied for armadas of orbiting platforms and fleets of space planes— and, in case of hostilities, possible confrontation in space.

  "The U.S. goal: to make certain Russia is denied control of space that might enable it to tip the balance of power here on earth. . . ." (*U.S. News*, 9/13/65, 10)

- In a Washington *Evening Star* article, "Costs Soar in Lagging TFX Project," Richard Wilson wrote: "Here's the latest word on that $8 billion miracle airplane program, the TFX. It is costing about $5 million per plane instead of the estimated $2.2 million. It is too heavy for the Navy to use. It could have been ready now but it isn't because the Kennedy administration lagged a year while evaluating and reevaluating the aircraft so that General Dynamics Corp. could get the contract. . . .

  "Nobody knows for sure that the Boeing plane would have worked out any better but the top military men all thought so. . . . So far as realizing any billion dollars savings as was at first advertised, this doesn't appear very likely if the Navy has to have a lighter version. In view of the nearly doubling of cost per plane, it does not seem likely in any case. . . ." (Wilson, Wash. *Eve. Star*, 9/13/65)

September 13–18: International Astronautical Federation's 16th Congress was held in Athens. Participants were greeted by King Constantine of Greece. Introductory speeches were made by the president of the IAF, Dr. William H. Pickering; chairman of the International Academy of Astronautics, Dr. Charles S. Draper; and president of the International Institute of Space Law, Dr. I. Pepin. (Hines, Wash. *Sun. Star*, 9/12/65; *Izvestia*, 9/14/65, 5, ATSS–T Trans.)

Dr. Wernher von Braun, Director of NASA Marshall Space Flight Center, reported at a plenary meeting of the congress that the Apollo program had now passed the half-way point and was moving forward at full momentum. He listed three primary commitments:

"—the first Saturn V carrying an unmanned Apollo spacecraft is scheduled to be launched in 1967.

"—the first Saturn V carrying a manned Apollo spacecraft is scheduled to be launched in 1968.

"—the manned lunar landing is scheduled to be accomplished prior to 1970."

Following his presentation, von Braun told a news conference that "many Soviet and American scientists will participate to lay the foundation for extensive cooperation on the moon." He said there were indications that the moon's surface was solid enough to bear the weight of the spacecraft. Soft spots had been identified and would be avoided. He estimated that a landing on Mars would occur 15 yrs. after a successful lunar landing. (Text; *NYT*, 9/15/65, 24)

Dr. George E. Mueller, NASA Associate Administrator for Manned Space Flight, outlined broad objectives of the Saturn-Apollo applications missions, earth-orbital and lunar:

"(1) Evaluate and extend man's capabilities to operate in space effectively as an astronaut and as a scientist.

"(2) Conduct observations of the Earth, extra-terrestrial phenomena, and experiments dependent on the space environment.

"(3) Qualify systems and crews for subsequent long-duration space missions.

"(4) To explore, map and survey the Moon.

Throughout these extended Apollo missions, the primary emphasis will be on science and applications experiments.

"The foundation for any long-range and broad national space program is a strong manned Earth-orbital program. For example, the approach to future manned planetary missions will depend not only on what is learned from unmanned missions such as Mariner and Voyager, but also what we can learn in Earth orbit about the effects on crew members of long-term confinement and reduced and/or intermittent artificial-gravity fields, as well as on the Earth-orbital qualification of extended-duration life support and power systems.

"Man's greatest contributions in space will come when he can bring his intelligence to bear on the spot. . . ." (Text)

Michael Stoiko, technical director of Martin Co.'s Gemini advance design section, suggested to the IAF congress that an international space rescue service to aid astronauts stranded in orbit be devised. The U.S. and Russia would organize the service with available equipment, and eventually rescue would be conducted on an international basis. Stoiko said the Martin Co. estimated 280 manned spaceflights would be flown by various nations in the next 20 yrs., with crews totaling 800 men. There was a "reasonable probability" that one or more U.S. crews would be stranded in orbit and a 62% probability that at least seven space ships with crews totaling 22 astronauts and cosmonauts would have to be rescued. (AP, *N.Y. Her. Trib.* 9/15/65)

Paper on "Research and Systems Requirements for a Lunar Scientific Laboratory" was presented at the IAF congress by C. William Henderson, NASA Office of Manned Space Flight, and Grady L. Mitcham of the Boeing Co.: "It is our obligation to define a meaningful lunar laboratory program which will extract the maximum scientific return for the investment of resources. Our scientific methods may have to change radically in order to achieve this goal of efficiency; perhaps by using the mantime on the moon only for the extraction of data, leaving the analysis of this information to the far less costly man power on earth. Most probably, leading scientists will not be, themselves, on the lunar surface, but rather will be at earth-based television consoles directing and guiding their counter-parts on the moon. . . ." (Text)

Soviet Cosmonaut Pavel Belyayev reported at the IAF congress that the exit of cosmonauts into space from orbital stations to relieve the crews and perform work in space would be made only through air locks, making it possible to avoid dehermetization of the spacecraft. Belyayev said visual acuity of astronauts was lessened by from 20–30% during spaceflight. Sensitivity of the eye to color was also affected. Drop of 50% occurred in the case of green and purple.

Cosmonaut Aleksey Leonov described in his report the functioning of the self-contained life-support system he wore in open space. He was able to regulate the atmosphere in his spacesuit, maintaining within it an excess pressure of 0.4 atmospheres or 0.27 atmospheres. When he stepped out of VOSKHOD II, Leonov said, the excess pressure in the spacesuit was 0.4 atmospheres and before reentry into the spacecraft, he said: "I once more set the pressure back to 0.27 atmospheres." Data obtained, he summed up, permitted the conclusion that stepping out into open space does not cause serious specific changes in psychophysiological functions.

Of the future of the Soviet space program, he said: "In time there will be many space laboratories, with crews being periodically exchanged. Then there will be a spaceship for the moon, and a landing on the moon, followed by shots to other planets." (AP, *Wash. Post* 9/17/65; AP, *Balt. Sun,* 9/17/65; *National Zeitung,* 9/16/65, 3)

A voyage in a seven-man spacecraft, past Venus to land on Mars, was proposed by Robert L. Sohn of the Space Technology Laboratories in a paper at the 16th Congress of the International Astronautical Federation in Athens. Sohn said the flight could be conducted in the 1980's using launching equipment and spacecraft already in development, including Saturn launch vehicles and modified Apollo spacecraft. He further claimed that the billion-mile, 400-day trip would cost about $20 billion. A 400,000-lb. craft would be assembled in earth orbit after component parts were placed into orbit by four or five separate launches. (*NYT*, 9/14/65)

Soviet Professor Savenko reported to the congress about the study of primary cosmic rays by PROTON I, launched July 16, 1965. Volume of information radioed back to earth was 60 million units daily. Equipment was sensitive to particles of energy of "billions and more of electron volts which makes it possible for the station to explore the depth of matter." Design of the ionization calorimeter in PROTON I permitted improved methods of research, Professor Savenko said. Information relayed from PROTON I processed so far showed that distribution of cosmic ray particles of very high energies differed from picture obtained by indirect measurements. (Tass, 9/17/65)

Mstislav Keldysh, president of the Soviet Academy of Sciences, received the International Academy of Astronautics' 1965 Daniel and Florence Guggenheim International Astronautics Award. The congress elected the leadership of IAF and reelected Dr. William H. Pickering as president. (Hines, Wash. *Sunday Star*, 9/12/65; Tass, 9/18/65)

September 13: Reporting on the Lunar International Laboratory, under consideration by the International Astronautical Federation meeting in Athens, William Hines, in the Washington *Evening Star*, said the case for Lil had been outlined this way:

"1. A spaceship is horribly expensive; it costs $15,000 now to boost one pound of instruments to the moon, and will cost $180,000 a man hour to sustain scientists in space in the 1970s.

"2. The precedent exists for international scientific efforts: nuclear physicists already are discussing a trillion-volt atom smasher of which the world needs only one.

"3. The concept of international science has been accepted by the United Nations since 1946. . . .

"4. While the Russians have not taken an official stand, Sedov and others last year openly discussed Soviet studies on moon construction.

"[C. Stark] Draper [Director of MIT Instrumentation Laboratory] . . . has called for a computer center on the moon to collect data there and pass it on to earthly campuses and laboratories with information literally unavailable here. The Lil program has not yet reached the cost effectiveness and design stages yet, but at the rate the Apollo program to land men on the moon by 1970 is going, backers say only 10 to 20 years is needed. A curbstone opinion is that $20 billion to $30 billion . . . comparable to the Apollo costs—would handle the construction and start of Lil operations." (Hines, Wash. *Eve. Star*, 9/13/65, A8)

September 14: In a White House ceremony, President Johnson conferred Exceptional Service Medals on GEMINI V Astronauts L. Gordon Cooper and Charles Conrad, Jr., and Dr. Charles Berry, chief surgeon for the astronauts. The President announced that military promotions for the astronauts would be submitted to the Senate immediately. (Richard, *Wash. Post*, 9/15/65, A3)

Following the ceremony, Vice President Hubert Humphrey accompanied the astronauts, their families, and Dr. Berry to the National Academy of Sciences. Humphrey told the Academy: "The achievements of our astronauts are not just the achievements of two men. They are the achievements of family, country and of a society. . . .

"This is the best example of cooperation between Government and industry, between public and private, that I know in this nation, and might I add, that it is an example that should be emulated by many others in many other endeavors in this country.

"But our space program . . . is more than national. It would be well enough if it were only national, but it is more than that. It is truly international. And I want . . . to stress our space program's potential as an arm of our foreign policy, of our international relations, and particularly of our peace policy, because this Government has but one objective in the world: a just, honorable and enduring peace. . . ."

Astronaut Cooper summarized some of the things that man could do in space:

". . . we again showed that man can do a great deal of geological observations while in space. . . . I think it is very significant that although you pass across some [geological formations] rather rapidly you still are able to . . . observe them and you are very capable of photographing them in great detail.

". . . man is very capable of doing oceanographic observations. We have noted that you can detect a great deal of detail as to the depths of water, as to the currents of running water, and even to wind patterns that run on the waters, the wave patterns caused by the wind.

". . . man could do a great deal of weather observations, both in real time and photographically. And we . . . were able to pinpoint one tropical storm in between some of the planned observations, and apparently that was very accurate pinpointing of its position.

"We determined in our discussions later with the Weather Bureau people, that for future use . . . we could get much greater accuracy even on pinpointing these by the use of our platform combination and taking angles to a great deal more accuracy, rather than just estimations of the angular distance and locations.

"We found that you could do a great deal of visual observations of the phenomena around the earth . . . we picked up ship wakes, several hundred miles away of . . . the contrast of the white wake against the blue background in very clear weather areas. . . ."

Conrad revealed that photographs taken during the GEMINI V mission at the request of the Mexican government, had located volcanoes in Mexico that "the Mexicans didn't know they had. . . ." He also said: "I think a system that has been left out and nobody has talked about is our environmental control system, which performed in a most outstanding manner for eight days. We had probably the nicest, freshest air that anybody could want to breathe, and we had an unusual first in the space business: most environmental control systems had been loaded down and pilot after pilot came back and said they were too warm, and we fought and fought it, and at McDonnell, to have a little change made because of some of the problems of heat exchange between the inlet and outlet pipes. It turned out that we were too cold.

"With two tubes running, we were probably the first ones to shut the cabin heat exchanger down and run everything as hot as we could get it in there. We were just right. And the thing that most surprised us on opening the hatch on landing, we couldn't tell the difference between inside and out, and that is a monumental job for that small system."

Dr. Berry concluded the program by presenting a review of the medical data received from the GEMINI V mission.

The astronauts' party then proceeded to the Capitol where Vice President Humphrey presented them to the Senate and then to the House. Cooper told the House: "We set out with our prime objective on Gemini 5 to fly 8 days. We were to fly this 8 days primarily to show that man, with all the various equipment on board, and the machines, could safely and very functionally do this 8-day mission with no adverse effects. We were entirely successful in this. . . ." Conrad explained: "Gemini 5 was a significant advance in the whole program in that we flew the first all-out spacecraft. We would not have been able to go without the fuel cells for 8 days. This was the first time they were flown. We had the first radars from which we got some very excellent data to enable the GT-6 to continue with their present rendezvous plans. It was the first time that the cryogenic storage of liquid hydrogen and liquid oxygen had been run that long in space. . . ."

Dr. Berry told the Senate: "We are conducting a program such as this [Gemini program] . . . to make man a vital part of such a research effort and to show the facility that man has to gain scientific information using vehicles such as we are able to build in this country. . . .

"I think I can report, so far as information has been obtained in this country to date and at the moment, that we are the only ones who have that sort of information for the duration of which we are speaking. We can confidently say that man has been able to perform very well up to 4 days in a weightless state earlier, and on this mission 8 days in a weightless state. He has then been able to re-adapt back to a 1-G environment. We have living proof of that." (Texts; *CR*, 9/14/65, 22924–22927; NAS-NRC *News Report*, 9/65, 1-2)

September 14: Commenting on the visit to Washington, D.C., by GEMINI V Astronauts L. Gordon Cooper and Charles Conrad, Jr., the *Washington Post* wrote: "Colonel Cooper has been through all this before following his 22-orbit flight in May, 1963. The reception this time will be less spectacular. But this does not mean that the astronauts have become passé or that the American people are indifferent to their achievements. Rather it is a reflection of the public's trust in the National Aeronautics and Space Administration's space program. For if the people no longer gawk and shout, they still applaud vigorously, with their hearts as well as their hands, each monumental step in the great venture into space." (*Wash. Post*, 9/14/65)

• Pilot John B. McKay (NASA) flew X-15 No. 3 to maximum altitude of 239,000 ft. at maximum speed of 3,545 mph (mach 5.03) to obtain data on pilot altitude buildup. (NASA X-15 Proj. Off.; *X-15 Flight Log*)

September 14: Nine aquanauts who emerged from a decompression chamber after spending 15 days in USN Sealab II reported it took longer than expected to accomplish tasks; the helium atmosphere in the lab made their voices sound squeaky; the men's efficiency began to fall off after two weeks. (UPI, *NYT*, 9/15/65, C20; *Wash. Post*, 9/15/65, A9)

- The House Ways and Means Committee considered H.R. 8210 to exempt the European Space Research Organization from U.S. taxation. ESRO would operate a tracking station in Alaska. (NASA LAR IV/167–68)

- Soviet scientist Dr. V. S. Troitsky, of the Scientific Research Radiophysics Institute in Gorky, said that radio emissions from the moon indicated that its interior was so hot that it must be four times as radioactive as the inside of the earth. He also said that radio observations indicated a surface layer of light material about 20-ft. thick, overlying denser rock. Dr. Troitsky made these statements at an international conference on the solar system, being held at Cal Tech. (*NYT*, 9/15/65)

- Participants in the World Peace through Law Conference in Washington, D.C., agreed that advances in international communications, including EARLY BIRD I communications satellite, were important milestones on the road to world peace. John A. Johnson, vice president of ComSatCorp, told the international communications panel that EARLY BIRD I had provided a case study of progress in "the development of legal principles and institutions which further international cooperation." He said global coverage in the satellite communications system was planned for 1967. Prof. Ivan A. Vlasic, McGill Univ., warned we may be facing "a proliferation of competing, possibly even antagonistic communications systems. Given the present state of international relations the chances of conflict are considerable." Jean Evensen, Norway's government broadcasting system adviser, felt "a pressing need for ways of enforcing international communications rules. One rebel, one nonconformist . . . can create havoc . . . to the damage of the rest of the world," he said. (Bullen, *Wash. Eve. Star*, 9/15/65)

- NASA had selected Perkin-Elmer Corp. and Chrysler Corp. for contracts to study feasibility of including optical-technology experiments—particularly lasers and large telescopes—in future extended Apollo flights. NASA was also interested in optical communication in deep space, the effects of space environment on optical systems, and related secondary experiments. The program would be directed by MSFC. (MSFC Release 65–223)

- Secretary of Commerce John T. Connor told a White House Cabinet meeting that Government scientists wanted to undertake a national weather-control program to determine whether fog could be dispelled, hail suppressed, hurricanes diverted, and rain made, reported Howard Simons in the *Washington Post*. Connor's report represented a change of view in the scientific community, heretofore reticent about weather modification. (Simons, *Wash. Post*, 9/14/65, A9)

- Former FAA Administrator Najeeb E. Halaby was appointed senior vice president of Pan American World Airways and elected to its board of directors. Halaby's initial responsibilities would embrace the airline's defense activities, its services for USAF at the Eastern Test Range, and its business-jets division. (*NYT*, 9/15/65, 77M; *Wash. Post*, 9/15/65, B9)

September 14: Dr. William W. Hagerty, president of Drexel Institute of Technology, was sworn in as consultant to NASA Administrator James E. Webb, in university/industry relations. (NASA Release 65–291)

• At a meeting of the American Chemical Society in Atlantic City, Dr. Emmanuel Roth of the Lovelace Foundation proposed neon as a gas that astronauts might breathe during spaceflights: "Its permeation coefficient is lowest. That is, neon does not go into solution or diffuse through tissue as readily as other gases do. As a result, the deadly bubbles form at a much slower rate." No tests of neon with human subjects had been run. (*Newsweek*, 9/27/65)

September 15: GEMINI V Astronauts L. Gordon Cooper and Charles Conrad, Jr., left with their wives on a six-nation goodwill tour arranged by President Johnson to demonstrate the U.S.' peaceful intentions in space. They would visit Greece, Turkey, Ethiopia, Madagascar, Kenya, Nigeria, and the Canary Islands. At a State Department luncheon prior to departure, Cooper told the audience, which included ambassadors from the six countries he and Conrad would visit, that from the GEMINI V cockpit "you don't see any of the combat, you don't see any of the fighting and bickering, the world looks like a very peaceful place." (UPI, *NYT*, 9/16/65, 27)

• MARINER IV, after 291 days in space, had exceeded its design-mission lifetime by nearly 500 hrs. Having traveled 400 million mi., MARINER IV had been reporting on cosmic dust, magnetic fields, and interplanetary levels of cosmic rays and radiation for ten months—about 7,000 hrs. In addition to scientific information, the spacecraft was reporting engineering data on its own condition as it orbited the sun. Data from MARINER IV were being transmitted to earth by radio over a straight-line distance of nearly 182 million mi. (NASA Release 65–293)

• NASA launched Nike-Apache sounding rocket, with 51-lb. payload instrumented to measure electron and ion density and solar radiation in the D and E regions of the ionosphere to peak altitude of 110 mi. Experiment, conducted for the Univ. of Illinois and the GCA Corp. from Wallops Station, was part of International Quiet Sun Year 1964–65. (Wallops Release 65–58; NASA Rpt. SRL)

• USAF awarded $6.5-million contract to Boeing Co. to build one ground-test and three flight-test models of a highly-reliable, low-cost upper stage for orbiting small and medium unmanned satellites. Called "Burner II," the stage would be used with Thor standard launch vehicles and be adaptable for use with Atlas and Titan boosters. Burner II, which include a spherical solid-propellant rocket motor, inertial guidance system, and attitude-stabilization system, would bridge the payload gap between the DOD/NASA Scout launch vehicle and the more expensive USAF Agena and Able-Star upper stages. (AFSC Release 95.65)

• Spokesmen for some 43,000 United Aerospace Workers Union members employed by North American Aviation, Inc., announced plans for possible strike action Oct. 10 unless negotiators reached agreement on a new wages and hours contract. UAW members involved worked in plants in Los Angeles, Calif.; Tulsa, Okla.; Columbus, Ohio; and Neosho, Mo. (UPI, *Houston Chron.*, 9/16/65)

September 15: Soviet Cosmonaut Vladimir Komarov said at a press conference at the International Transport Exhibition in Munich that the Soviet Union planned to land a man on the moon in 1970. Komarov, in charge of the first spacecraft with a three-men crew, was awarded the Pioneer Chain of the Compass Card—a top international aviation decoration. (Reuters, *Houston Post*, 9/16/65)

- Western Union International, Inc., asked FCC for authority to conduct a series of satellite communications tests should NASA and DOD permit live coverage of the Gemini 6 recovery in October. The proposal included testing a portable ground station at Taylor's Island, Md., and then on a ship at sea. Trials would include live television transmission from the portable ground station to EARLY BIRD I communications satellite, which would relay the transmission to U.S. and Europe. ComSatCorp approval would also be necessary before tests could begin. (*Wash. Post*, 9/15/65)

- France hoped to orbit its first earth satellite in November—nearly two months ahead of schedule, reported the *Houston Post*. An Armed Forces Ministry satellite would be fitted to the first Diamant booster and test-fired from Hammaguir, Algerian Sahara. (Reuters, *Houston Post*, 9/15/65)

- Suggestion that synthesis of life should be a national goal by Dr. Charles C. Price, president of the American Chemical Society, received comment from the *New York Times:* "Could such an effort be added to the total national scientific enterprise or could it be fitted in only by cutting back on other sectors, particularly the very expensive programs in space or particle physics? Would mankind benefit more if the funds needed for this project were devoted instead to less exotic but perhaps more vital needs of ending the pollution of the air we breathe and the water we hope to drink? The verdict on Dr. Price's proposal is by no means immediately obvious.

    "The most unfortunate result that could come from Dr. Price's suggestion would be the mounting of a new international competition, a 'life race' that would produce the same tension and needless duplication that the space race has produced. Men of many nations have contributed to the progress that makes it possible now to consider the goal Dr. Price has put upon the public agenda. If any such effort is undertaken it should be as international as the common humanity that makes all men brothers." (*NYT*, 9/15/65, 43)

- On the occasion of the Smithsonian Institution Bicentennial Celebration, President Johnson said:

    ". . . the Institution financed by Smithson breathed life in the idea that the growth and the spread of learning must be the first work of a nation that seeks to be free.

    "These ideas have not always gained easy acceptance among those employed in my line of work. The government official must cope with the daily disorder that he finds in the world around him.

    "But today, the official, the scholar, and the scientist cannot settle for limited objectives. We must pursue knowledge no matter what the consequences. We must value the tried less than the true.

"To split the atom, to launch the rocket, to explore the innermost mysteries and the outermost reaches of the universe—these are your god-given chores. And even when you risk bringing fresh disorder to the politics of men and nations, these explorations still must go on." (*Pres. Doc.*, 9/20/65, 276)

September 15: GEMINI V Astronauts L. Gordon Cooper and Charles Conrad, Jr., arrived in Athens with their wives and children at the start of a six-nation goodwill tour and were greeted with cheers from the crowd. The astronauts would attend the International Astronautical Federation Congress which opened Sept. 13. (UPI, Phil. *Eve. Bull.*, 9/16/65)

September 16: NASA announced selection of Aero Spacelines, Inc., for negotiation of a $1.5-million one-year contract, with provision for three one-year renewal options, to provide air transportation service on the Super Guppy for boosters and spacecraft from factories to test sites and launch centers. The Super Guppy was the only aircraft in existence that could fulfill size and weight requirements for cargoes such as the S–IVB stage of the Saturn IB and V launch vehicles, the Saturn IB and Saturn V instrument units, and the Lunar Excursion Module adapter. (NASA Release 65–296; MSFC Release 65–229)

- France's first satellite, FR–1, arrived from Paris at Dulles International Airport and was taken to NASA Goddard Space Flight Center for testing. The satellite, to be launched for France by NASA in a joint project to study very low frequency radio waves and the distribution of ionization in the earth's magnetosphere, would be sent to Ling-Temco-Vought, Texas, where it would be modified to fit a Scout-type rocket. Launching was scheduled for late 1965.

  Memorandum of understanding for the FR–1 program was signed Feb. 18, 1963, between NASA and the French Space Agency, Centre National d'Etudes Spatiales (CNES). (NASA Release 65–59; *Wash. Post*, 9/16/65)

- Third Saturn S–IB stage (S–IB–3) arrived at MSFC aboard the NASA barge *Palaemon*. First stage for the third Saturn IB launch vehicle, it would be captive-fired at least twice before being returned to Michoud Assembly Facility for post-static-test checkout. (MSFC Release 65–228; *Marshall Star*, 9/22/65, 1)

- Modest turnout of spectators at the Sept. 14 motorcade in Washington, D.C., for Astronauts Cooper and Conrad received editorial comment in the Washington *Evening Star:* ". . . We think it is an encouraging trend. Just possibly the lack of hoopla and gapers suggests that America has accepted the space program as serious business instead of a circus of stunt men." (Wash. *Eve. Star*, 9/16/65)

- Picket lines were established at all entrances to Kennedy Space Center, NASA as the International Assn. of Machinists and Aerospace Workers began a nationwide strike against the Boeing Co. Key issue was union demand that Boeing abandon performance analysis system of rating employees for promotion, demotion, and discharge. Union contended that seniority should be the sole factor in job security; Boeing claimed its system was necessary to keep the company competitive in the aerospace industry. (AP, Balt. *Sun*, 9/16/65; UPI, *Houston Chron.*, 9/16/65)

September 16: Senate Commerce Committee reported favorably S.774, which would authorize a study of the practicality of adoption of the metric system. (CR, 9/16/65, 15856)

- A turbocompressor, first flight-weight hardware for an eight-kilowatt space power generating system, arrived at NASA Lewis Research Center for testing. It would use a 30-ft. solar mirror to gather the sun's rays and operate on a Brayton cycle with an unreactive gas as its working fuel. In space, heat source could be a reactor, a radio-isotope, or a solar heat receiver supplied by concentrated sunlight reflected from a mirror. Turbocompressor was built for NASA by Air Research Manufacturing Co. as part of NASA's Advanced Technology Program. (LRC Release 65–64)
- South African police and scientists investigated report that a flaming "saucer" about 30 ft. in diameter had landed on a main highway near Pretoria. Scientists who examined the alleged landing site were reported to have found a six-ft.-wide section of the tarred road badly burned. (AP, Balt. *Sun*, 9/17/65)
- Deploring the "belated decision" to finance the space defense of the U.S., syndicated columnist David Lawrence wrote: "It is interesting to note that on October 7, 1963, Louis C. Wyman, Republican, who was then a member of Congress and formerly was attorney general of New Hampshire, submitted alone a minority report to the House appropriations committee, in which he criticized at length the administration's indifference to the military significance of space developments. He said: 'As a first priority, rather than racing to the moon, the United States should establish and maintain an integrated weapons system in inner space within manned space capsules that have a capability to observe, intercept and, if necessary, destroy other objects in space.'

    "Mr. Wyman's extensive report will go down in history as a remarkable prophecy as well as an interesting example of how often minority reports become majority opinion." (Lawrence, *Kansas City Times*, 9/16/65)
- Secretary of Defense Robert S. McNamara announced that USAF would send 12 F–5A jet fighter aircraft to Vietnam this fall to evaluate the performance of this new multipurpose aircraft under combat conditions. Built by Northrop Corp., the F–5A was a lightweight supersonic aircraft that could carry 6,200 lbs. of bombs, rockets, air-to-air missiles, or other ordnance. With two 20-mm. cannons mounted in its nose, it could take off fully loaded from unpaved runways. (DOD Release 614–65)

September 16-17: Rendezvous of four spacemen—Soviet Cosmonauts Belyayev and Leonov and American Astronauts Cooper and Conrad—at the IAF Congress was described in *Life* by Jim Hicks: "First man to try his hand at arranging a rendezvous . . . was Jules Bergman, ABC's space reporter. The Russians agreed. Bergman thought the Americans had agreed . . . and on the afternoon of the day Cooper and Conrad arrived in Athens . . . [they] waited for Cooper and Conrad to appear . . . Finally Bergman phoned Julian Scheer, the NASA publicity man who was traveling with the astronauts. When would the Americans show?

"Why they would not appear at all, said Scheer. He said no one had cleared such a meeting with him . . . The Russians were angry and left the scene in a huff, and at a news conference later, Sedov made no secret of the fact that they considered the incident a snub and an affront.

"An air of mistrust now prevailed. There was, as Pete Conrad observed later, 'some real bad feeling around this place.'

"Scheer, trying to right things, sent a letter to Dr. William Pickering, president of the International Astronautical Federation, urging that the astronauts be brought together. . . .

"Friday afternoon Cooper and Conrad spoke before a Congress session. . . .

"As soon as that meeting was over, the two Americans jumped from the conference stage and headed for the white-uniformed Belyayev . . . Conrad [said] through the Russian's interpreter 'Tell him we would like to show them all our pictures. . . . Tell him we've got 250 slides and they can see them all. When would they like to see them?'"

Hicks said Belyayev exchanged his lapel pin with Cooper and Conrad gave his to Belyayev to be delivered to Leonov. Hicks continued: "That night, considerably more progress was made at the Congress's large banquet. Cooper and Belyayev, strategically seated within talking distance, even went so far as to exchange watches. . . . Finally the astronauts warmed up the cosmonauts from the earlier chill. They all posed for a four-way handshake photograph. No more fooling around through official channels, drawled Cooper. Why don't you fellows come up to our place for breakfast tomorrow morning?

"The Russians agreed. . . . The talk, through an interpreter, was almost entirely personal and non-technical. . . . Conrad and Leonov exchanged writing pens. All traded autographs and Leonov . . . drew a picture of himself walking in space. . . .

"As they departed, the spacemen locked in Russian-style bearhugs. . . . Said one of the Soviet pilots, 'We are colleagues and we have a full understanding. Gordon Cooper and Charles Conrad are good boys.'" (Hines, Wash. *Eve. Star*, 9/17/65, A10; Hicks, *Life*, 10/1/65, 113-116)

September 16-20: Four NASA Nike-Apache sounding rockets with Univ. of California (Berkeley) experiments to measure fluxes and spectra of precipitating electrons responsible for auroral x-ray activity were launched from Ft. Churchill, Canada. The first rocket reached only 17-mi. altitude because Apache motor did not ignite; so, although instrumentation functioned properly, no data were obtained. Second Nike-Apache reached 100-mi. altitude, all instruments functioned as planned, and good data were expected. The third also performed well, although peak altitude (91 mi.) was somewhat below predicted, and good data were expected. No data were obtained from the fourth flight, although instrumentation functioned well: peak altitude was only 14.8 mi. because of a burnthrough at the Apache headcap and separation of the payload from the rocket. (NASA Rpts. SRL)

September 17: NASA launched an Aerobee 150A sounding rocket from Wallops Station, Va., with 150-lb. payload containing French-built radio propagation experiments designed for later flight on the FR-1 satellite. It reached 114-mi. (183-km.) peak altitude during an eight-minute flight. (Wallops Release 65–59)

September 17: OGO I (Orbiting Geophysical Observatory) spacecraft had completed its first year of operation in space Officially classified as a failure when a major objective—three axis stabilization—was not achieved, the 1,200-lb. satellite launched Sept. 4, 1964, had exceeded its one-year design lifetime and was still transmitting valuable data from 16 experiments. OGO I's scientific objectives were to conduct time-correlated measurements of space phenomena to help in the understanding of earth-sun relationships. Although its scientific usefulness was lessened when the earth-run stabilized orientation was not achieved, 16 scientific papers had been presented by experimenters on findings of their instruments aboard OGO I. (NASA Release 65–294)

- NASA requested that final proposals for scientific experiments for the 1971 Voyager missions to Mars be submitted by Nov. 19. Selection would be made by July 1, 1966. All 1971 Voyager experiments on the landing craft would be subject to strict sterilization requirements. (NASA Release 65–297)

- King Constantine of Greece invited Astronauts L. Gordon Cooper (L/Col., USAF) and Charles Conrad, Jr. (LCdr., USN), their wives, and other U.S. officials in Athens for the 16th International Astronautical Congress, to a royal palace banquet. NASA Marshall Space Flight Center Director Dr. Wernher von Braun and JPL Director Dr. William H. Pickering were among the guests. (AP, Phil. *Eve. Bull.*, 9/17/65)

- The task of effecting a soft landing on the moon had been placed on the agenda of space projects in the U.S.S.R., Academician Leonid Sedov, leader of the Soviet delegation to the 16th IAF Congress, said at a press conference. He added: "Such a landing has already been worked out on the ground." (*Tass*, 9/17/65)

- A simple, two-door pneumatic tube device for garbage disposal during the two-week Gemini VI manned space flight scheduled for December was being considered by NASA Manned Spacecraft Center officials, the *Houston Post* reported. The astronauts would open the first door, stuff in the material to be disposed of, close the first door, and open the second door. Space being a nearly perfect vacuum, the material would be forced out of the tube. (*Houston Post*, 9/17/65)

- USAF Chief of Staff Gen. J. P. McConnell told a meeting of the Air Force Association in Washington, D.C., that "any reports of the impending demise" of the flying Air Force were "slightly exaggerated." In fact, he added, a wide variety of new and better aircraft was needed to meet "the ever-changing nature and scope of the threat to our national interests." (Text)

- David Sarnoff, Chairman of the Board of RCA, warned the Conference on World Peace Through Law that communications progress would beget serious problems: "By the end of the decade there will be not only one communications satellite but many; not a single global satellite system but possibly several in competition with one another; not a sole operating agency dealing with many nations, but many nations with their own operating agencies pursuing different satellite communications plans and objectives.

"As the number of satellites multiplies in space, a corresponding number of problems will multiply on earth. . . . When we can communicate instantly to everybody, everywhere, we will set in motion a force whose ultimate political, social and economic impact upon mankind cannot be calculated today." (Text)

September 17: Col. John H. Glenn, Jr. (USMC, Ret.), would make a three-week goodwill trip to Western Europe in October, the White House announced. (UPI, *Wash. Post*, 9/18/65)

- Commenting on how swiftly man had moved into the space age, an article in the Baltimore *Sun* said: "Travel in space . . . remains and will remain dangerous—a work for none but the brave, the skilled and the dedicated. There will be moments of breathless drama, many of them, as when the first man arrives on the moon, and of tragedy. But short of those moments, travel in space near the earth is beginning to be routine." (Balt. *Sun*, 9/17/65)

- Decision to proceed with the Mol received comment in *Science:* "Some proponents of MOL believe that, as insurance against 'technological surprise' and as a test of improved methods of intelligence gathering, the project will lead to greater stability in relations between the United States and the Communist world. But skeptics fear that MOL will carry the arms race into space. Despite a long hunger, the Air Force has never before been permitted a role in manned space flight, a function heretofore reserved exclusively for the National Aeronautics and Space Administration." (*Science*, 9/17/65, 1357)

September 18: U.S.S.R. launched COSMOS LXXXVI, LXXXVII, LXXXVIII, LXXXIX, and XC artificial earth satellites with a single booster. Orbital parameters: apogee, 1,609 km. (999.3 mi.); perigee, 1,380 km. (857 mi.); period, 116.7 min.; inclination, 57°. An electrical pack operating on energy released by a radioactive isotope was onboard power system on one of the satellites. Instruments were functioning normally. (*Krasnaya Zvezda*, 9/21/65, 1, ATSS-T Trans.)

- First cooperative sounding rocket experiment sponsored by NASA and the Netherlands Organization for the Advancement of Pure Research (ZWO) was successfully conducted by Dutch scientists at Coronie, Surinam. Object of the experiment was to measure winds in the equatorial upper atmosphere by releasing a cloud of sodium vapor to be illuminated by the sun and tracked. Launching, using a Nike-Apache rocket, was first of four to be conducted under a memorandum of understanding signed in June 1964. The three subsequent launches were successfully conducted Sept. 21, Sept. 24, and Sept. 27. Dutch launch team had trained at Wallops Station, Va., earlier this year. (NASA Release 65–299; NASA Rpts. SRL)

- Astronauts L. Gordon Cooper and Charles Conrad, Jr., arrived in Izmir, Turkey. Only about 5,000 of the city's 750,000 residents turned out to watch their motorcade drive the 15 mi. into Izmir from the NATO air base at Cigli. Spokesman for the Turkish Foreign Office said the visit was an "unofficial culture visit," and the governors of three cities on the astronauts' itinerary were instructed not to welcome the party. Coolness was attributed to U.S. stand on Cyprus in 1964. (*Wash. Post*, 9/19/65)

- Defense Communications Agency and the U.S. Army had announced that site preparation had begun at Helemano, 20 mi. north of Honolulu on

Oahu Island, for installation of the first earth terminal of a planned worldwide military experimental satellite communication system. Southwest Constructors had been awarded the contract for construction expected to begin in mid-October. (*J/Armed Forces*, 9/18/65, 16)

September 18: In a farewell speech at a dinner meeting of the Air Force Association in Washington, D.C., Secretary of the Air Force Eugene M. Zuckert said that in contrast to five years ago, the Air Force today was better. He predicted it would continue to improve: "We have not reached a plateau in technology, strategy, concepts, or doctrine. All kinds of changes will continue—in both hardware and ideas." (AP, Balt. *Sun*, 9/18/65, 5)

September 19: Progress report in *Pravda* on ZOND III: "Soviet automatic space station Zond 3 launched July 18, 1965, continued its orbital flight gradually moving farther away from the sun. On September 15, the space station was 12.5 million kilometers away from the earth. During this period 75 radio-communication sessions were held with the station. During these sessions, photos of the far side of the moon, comprehensive telemetric data on the physical processes in interplanetary space, and data on the station's instruments and systems were transmitted to earth. To check the systems that make it possible to change flight direction, the flight trajectory of Zond 3 was successfully corrected on September 16 for experimental purposes. The astroorientation system, which automatically orients on the sun and the star Canopus, turned the station into the specific position commanded from the earth and has maintained this orientation with great accuracy. Then the correcting engine was switched on and changed the flight speed of the station 50 meters per second at an angle of 45° with respect to the direction to the earth. Radio control sessions with the station confirmed that the correction maneuver had been executed correctly. They also showed that the station's systems continue to function normally.

"The space station relayed to the earth comprehensive data on the interplanetary magnetic field, cosmic radio emission, interplanetary ionized plasma, longwave cosmic radio emission, and micrometeorite particles. . . . During the flight around the moon, data were also obtained on the spectrum of its infrared and ultraviolet radiation." (Tass, 9/18/65; *Pravda*, 9/19/65, 2)

- Astronauts L. Gordon Cooper and Charles Conrad, Jr., received a warm welcome in Ankara, Turkey, in contrast to cool receptions in Izmir and Istanbul. President Cemal Gursel told the astronauts their flight of nearly eight days had demonstrated "great courage . . . to your nation and to the whole world." (AP, Balt. *Sun*, 9/20/65)

- Announcement by Soviet Cosmonaut Aleksey Leonov at the 16th International Astronautical Congress in Athens that the U.S.S.R. planned to orbit a permanent manned space station as the next major project in its space program received editorial comment in the *Philadelphia Inquirer*: "To be sure, Cosmonaut Leonov adds that after 'many space laboratories' have been established, 'with crews being periodically changed,' his country expects to give attention to 'a spaceship for the moon, and a landing on the moon.' It has been understood all along, of course, that this was the method Russia would use in sending men to the moon. The emphasis on many permanent manned space stations orbiting close to earth, however, shows all too plainly that the real concern of the generals running the Communist space program is in the military advantages that lie in control of the space immediately beyond the atmosphere and over the heads of free nations around the world." (*Phil. Inq.*, 9/19/65)

September 19: According to Dr. Colin Pittendrigh, Dean of the Graduate School and biology professor at Princeton Univ., MARINER IV photographs of Mars did not provide any new evidence that life could not exist on that planet. Even though Mars appeared virtually waterless in the photographs, he pointed out, the area photographed was only a small fraction of the planet's total area. "A scan across our Atlantic Ocean might lead some to say there is no land on earth." (AP, *Miami Her.*, 9/19/65)

- S-IVB second stage for the first Saturn IB launch vehicle to be launched arrived at Kennedy Space Center, NASA, aboard the cargo vessel *Steel Executive.* (*Orl. Sent.*, 9/20/65)
- Parallel problems facing astronauts and aquanauts were noted by Assistant Manager of Lockheed Missiles & Space Co. Bioastronautics Div., Dr. J. A. Kraft. "These problems are both biomedical and mechanical. Because of their similarity, we have in the ocean a readily available laboratory environment. In it we can investigate the more significant problems common to both. . . ." One parallel is the changed pressure of the environment: for the aquanaut, pressure problems involve return from extremely heavy pressure to normal pressure at earth's surface; for the astronaut, pressure problems involve functioning in lower pressure of spacecraft and vacuum of space itself. Other shared biomedical problems: dysbarism, oxygen toxicity, trace contamination, and anoxia. (Macomber, CNS, *San Diego Union,* 9/19/65)

During week of September 19: Medical checkup indicated that Astronaut Alan B. Shepard (Capt., USN) still suffered from labyrinthitis—an inner ear infection. Because of the illness, Shepard had not been selected for any of six Gemini flight crews. He also could not fly aircraft alone. (AP, *Wash. Eve. Star,* 9/28/65, A10)

September 20: Dr. Smith J. DeFrance, Director of NASA Ames Research Center, would retire Oct. 15 after 45 yrs. of service, announced NASA Administrator James E. Webb. "Dr. DeFrance's leadership at Ames has brought about many engineering and scientific achievements in our country's aviation and space programs," Webb said, "and we all owe him a great debt of gratitude." Following distinguished service as a combat pilot in France during World War I, DeFrance served for 18 yrs. at Langley Research Center. He became director of ARC, when it was created in 1940. In 1947 he received the Presidential Medal of Merit for designing and building the Center.

H. Julian Allen, present Assistant Director at ARC, would succeed Dr. DeFrance. Allen was recognized as an international authority on reentry physics, having conceived a solution to the reentry heating problem. In 1957, he received NACA's Distinguished Service Medal for this work. Allen received the Sylvanus Albert Reed Award in

1955 from the Institute of Aerospace Sciences "for contributions and leadership in solving problems in the design of supersonic airplanes and missiles, especially thermal problems at hypersonic speeds."

John F. Parsons, Associate Director of Ames, would remain in this post. (NASA Release 65-298)

September 20: Paul Haney, NASA Manned Spacecraft Center Public Affairs Officer, announced at a news conference the crew selected for the Gemini VIII spaceflight: Neil A. Armstrong, a civilian, would be command pilot; David R. Scott (Maj., USAF) would be copilot. Backup crewmen named were Charles Conrad, Jr. (LCdr., USN), and Richard F. Gordon, Jr. (LCdr., USN). Gemini VIII was scheduled to include practice on rendezvous and docking maneuvers and a space walk that could last as long as one orbit of the earth—about 95 min. (AP, Wash. *Eve. Star*, 9/21/65; AP, Balt. *Sun*, 9/21/65, 6)

- CBS interviewed Dr. Edward Welsh, Executive Secretary of the National Aeronautics and Space Council, on Cosmonaut Leonov's affirmation at the IAF Congress in Athens that the Soviet Union would first rendezvous, dock, and assemble systems in orbit before proceeding to a lunar flight. "They have a lot of things they have to do before they can really put a so-called permanent platform up there. They have to rendezvous. They have to engage in docking. They haven't had enough time experience yet of men in space to really say what they can do on a permanent platform . . ." (*SBD*, 9/20/65, 88)

- U.S. policy decision to conduct space operations in an atmosphere of maximum public exposure received editorial comment from Robert Hotz in *Aviation Week and Space Technology:* "The course of space technology has proved the U.S. policy to be far more effective than that of the Soviets. It has projected an international aura of leadership and achievement that has permitted the whole world to share in U.S. space projects, both scientifically and emotionally. It has also applied a steady and increasing pressure on the Soviets to abandon their super-secrecy. . . .

  "There is little doubt that the leadership in space technology passed to the U.S. during the course of last summer. But who, outside a small internal bloc of techno-politicians, would have realized this under a blanket of supersecrecy?" (Hotz, *Av. Wk.*, 9/20/65, 21)

- Three basic capsule designs for the Project Voyager Mars lander vehicle were being studied by a special planetary missions technology steering committee at NASA Langley Research Center, reported *Aviation Week and Space Technology*. Low atmosphere density values for Mars determined by the MARINER IV occultation experiment would have to be taken into consideration before final capsule design specifications could be prepared. (*Av. Wk.*, 9/20/65, 28)

- Move begun within USAF to offer all scientific experiments previously proposed for Mol to NASA was reported by Donald E. Fink in *Aviation Week and Space Technology*. Proposal was that NASA fly these experiments in its Apollo Extension System (Aes) program. (Fink, *Av. Wk.* 9/20/65, 26)

- Hundreds of construction workers returned to their jobs, following removal of Boeing pickets from four of the five entrances at Kennedy Space Center, NASA, ending the eighth major construction work stoppage at KSC within 20 mo. Confinement of the striking IAM members to the one gate used by Boeing personnel had been ordered September 18 by NASA and USAF. (AP, Balt. *Sun*, 9/21/65, 10)

September 21: In advance excerpts of his book, *Waging Peace, 1956–1961*, President Dwight D. Eisenhower related his "reactions to Sputnik and the Gaither Report" in October and November 1957. President Eisenhower reviewed a meeting on October 8, 1957, with his principal military and scientific advisers. The late Deputy Secretary of Defense Donald A. Quarles reported: "There is no doubt that the Army Redstone could have orbited a satellite a year or more earlier. . . ." Quarles reviewed the reasons for the separation of the IGY Vanguard satellite from the ballistic missile programs, saying: "The Russians have done us a good turn, unintentionally, in establishing the concept of international space," a principle confirmed by the orbiting of Sputnik over the air space of country after country. "Late that same morning," President Eisenhower wrote, "I directed Secretary of Defense Charles Wilson to have the Army prepare its Redstone at once as a backup for the Navy Vanguard."

Of SPUTNIK II, which orbited "Laika" on November 3, 1957, President Eisenhower said it did not arouse American public opinion because people "seemed to resent the sending a dog to certain death—a resentment that the Soviet propagandists tried to assuage a few days later by announcing that they had put poison in the last of Laika's rations. . . .

"The public, however, became bewildered and upset when word got out that a far from optimistic secret report had been made to me in the National Security Council . . . called the Gaither Report.

"This was a period of anxiety. Sputnik had revealed the psychological vulnerability of our people. The Communists were steadily fomenting trouble and rattling sabers; our economy was sputtering somewhat, and the ceaseless and usually healthy self-criticism in which we of the United States indulge had brought a measure of self-doubt. Added to this was the failure of our first satellite launching attempt [Vanguard] in the full glare of publicity and the alleged missile 'gaps' which political observers claimed they had detected . . .

"The Soviet satellites," President Eisenhower said, "were a genuine technological triumph, and this was exceeded by their propaganda value. To uninformed persons in the world, Soviet success in one area led to the belief that Soviet communism was surging ahead in all types of activity." (*Wash. Post*, 9/21/65, A1, A19)

- Computer guidance system, Mod 1, that had piloted the Nation's first Atlas ICBM, was presented to the Smithsonian Institution by USAF at a ceremony in the Institution's Museum of History and Technology. (*Phil. Inq.*, 9/20/65, 23)
- NASA announced it would not extend the contract for supply of liquid hydrogen with Union Carbide Corp. Needs for liquid hydrogen on the West Coast would be met by continuing the contracts with Linde and with Air Products and Chemicals, Inc. (NASA Release 65–300)
- American Broadcasting Co. requested permission from FCC to own and operate a communications satellite system linking an estimated 200

affiliated television stations. The proposed $21.5-million system would use a synchronous orbit satellite similar to EARLY BIRD I.

ABC's request challenged the right of ComSatCorp to be sole operator of U.S.-launched commercial communications satellites. (ABC Release)

September 21: Rep. William S. Moorhead (D–Pa.) spoke on the floor of the House on the problem of collecting and using information for decision-making in government and industry: "Experts say that the human mind has difficulty in considering more than 10 or 20 factors at the same time in making decisions. Yet, the unsolved problems of our society may require thousands or hundreds of thousands of factors or subfactors to be considered. Industry has learned to simulate mathematically a given environment. By varying the input assumptions or by varying subdecisions the decision maker can be given rational basis on which to make alternative decisions." (*CR*, 9/21/65, 23755)

- Statement by President Johnson at a meeting with representatives of larger research universities: "Creative research through free inquiry is the working way to new greatness in our society. It can open roads to

    "—man's mastery of his environment,

    "—sufficient food, water, and energy to sustain the massed population that is making ours a crowded planet,

    "—the building of corridors linking the earth to the stars,

    "—ultimate victory over the tragedy of mental and physical afflictions, and

    "—progress in helping man live in peace with his neighbor." (*Pres. Doc.*, 9/27/65, 294)

- Charles R. Able, group vice president, Missile and Space Systems, Douglas Aircraft Co., defined for the National Space Club in Washington, D.C., the reasons the U.S. must have an on-going space program. He concluded: "The most important single factor in deciding what space programs we will implement in the future . . . is going to be, simply, how well we perform in the programs now going and those just getting started.

    "The extent to which this Nation will be willing to commit itself to new ventures in space in the years ahead depends entirely on how well we stay on schedule and within the budget set for us.

    "The most important task we have in preparing for the future is to do a good job now." (Text, *CR*, 9/22/65, A5370–71)

- Astronauts L. Gordon Cooper and Charles Conrad, Jr., visited Ethiopian Emperor Haile Selassie in the royal palace in Addis Ababa, then made the fourth stop of their goodwill tour—after Greece, Turkey, and Ethiopia—the Malagasy Republic. They were received cordially. (AP, *Wash. Eve. Star*, 9/21/65; UPI, *Miami Her.*, 9/22/65)

- Technical Systems Office, Technical Staff Office, Operations Management Office, and Missions Operation Office had been established at MSFC "to centralize future projects and advanced systems operations and to streamline MSFC's effort . . . with manned and unmanned launch vehicles." The move abolished as entities the Saturn/Apollo Systems Office and Advanced Systems Concepts Office. (MSFC Release 65–232)

September 21: International Association of Machinists had accused NASA and USAF of using pressure to help Boeing as the IAM strike went into its sixth day, according to *Aviation Daily*. IAM said NASA and USAF had demanded that Boeing employees at Cape Kennedy use only one gate and then followed this with an order restricting IAM pickets to that gate. The union complied and removed picket lines at other gates. Hundreds of building trades workers who had been respecting the picket lines had returned to work. W. J. Usery, IAM representative at the Cape, called the Government move illegal support of Boeing and said that "if pressure is to be applied it should be equal pressure," referring to the absence of Government pressure to end the strike. (*Av. Daily*, 9/21/65)

September 22: Pilot Robert A. Rushworth (Lt. Col., USAF) flew X–15 No. 1 to maximum altitude of 100,300 ft., at maximum speed of 3,545 mph (mach 5.18). (X–15 Proj. Off.; *X–15 Flight Log*)

- USAF launched Thor-Agena D launch vehicle with unidentified satellite from Western Test Range. (*U.S. Aeron. & Space Act.*, 1965, 152)
- NASA selected Ling-Temco-Vought for negotiation of a one-year cost-plus-award-fee contract to provide engineering support services at White Sands Missile Range. It would contain provisions for two additional one-year renewals. Estimated cost for the three year period was more than $5 million. (NASA Release 65–301)
- NASA Marshall Space Flight Center had awarded a $4,514,295 modification to its existing contract with Boeing Co. to provide services in Saturn V ground support. Tasks included analysis of ground support equipment, monitoring equipment qualification testing, acceptance testing, conducting design verification, and interface and installation control documentation. (MSFC Release 65–234)
- Soviet Union had asked the Fédération Aéronautique Internationale to officially confirm two spacecraft world records established by the crew of VOSKHOD II on March 18, 1965, reported *Interavia Air Letter*. Applications were for first extravehicular activity of a man from a spacecraft in orbit and a new altitude record of 497.7 km. (309 mi.). (*Interavia Air Letter*, 9/22/65, 6)
- NASA had awarded a six-month, $117,175 cost-plus-fixed-fee study contract to Ball Brothers Research Corp. to define and determine engineering requirements for a solar telescope mount to be used on proposed Apollo applications manned missions. System would provide a capability of observing the sun with relatively large astronomical-type telescopes and in having an astronaut available to correct errors in alignment and make other adjustments. (NASA Release 65–302)
- U.S. Army's Nike-X Project Office had signed two contracts with Western Electric for FY 1966 work on the antimissile missile program. First, at $221,216,696, was for "continued development and testing." Second, at $7,283,304, was for "production engineering and planning." (DOD Release 646–65)
- Officials at the Smithsonian Astrophysical Observatory confirmed discovery by two Japanese astronomers of a new comet, Ikeya-Seki, named after its discoverers. (UPI, *Wash. Daily News*, 9/22/65)

September 23: Four-stage Javelin (Argo D–4) sounding rocket with an ionosphere experiment was launched by NASA from Wallops Station,

Va. The 138-lb. payload reached 495-mi. (797-km.) altitude and impacted 625 mi. downrange in the Atlantic Ocean. Primary objectives of the flight were to measure ion and electron densities and temperatures in the upper atmosphere. Launch was timed to coincide with a pass of the Canadian ALOUETTE satellite; measurements would be compared with those of the satellite and with ground-based ionosonde measurements. (Wallops Release 65-60)

September 23: Soviet Union launched COSMOS XCI unmanned earth satellite, carrying instrumentation designed to continue the space exploration program begun March 16, 1962. Orbital parameters: apogee, 341 km. (212 mi.); perigee, 211 km. (131 mi.); period, 89.8 min.; inclination, 65°. Apparatus was functioning normally. (*Komsomolskaya Pravda*, 9/24/65, 1)

- NASA would negotiate a contract with the Bendix Field Engineering Corp. to provide operations and maintenance support for portions of the Space Tracking and Data Acquisition Network (STADAN) facilities over a two-year period from Oct. 1, 1965, through Sept. 30, 1967. Total cost was estimated at $25 million. (NASA Release 65-305)

- Contrasting the "image" of U.S. and U.S.S.R. at Sept. 13-18 IAF Congress, William Hines said: "Whether through the Russians' skill at brainwashing or our own ineptitude, we emerged . . . as the rocket rattlers and they as the peace lovers. . . ." (Hines, Wash. *Eve. Star*, 9/23/65)

- GEMINI V Astronauts L. Gordon Cooper, Jr., and Charles Conrad, Jr., were lost briefly over Kenya while flying a small aircraft to Keekerok Game Lodge to visit President Jomo Kenyatta. With Cooper at the controls, they landed at Keekerok only 20 min. late. (AP, Balt. *Sun*, 9/24/65)

September 24: Dan Schneiderman, Mariner project manager at JPL, had invented "Conrad," a self-contained radio navigation aid for amateur sailors. The device would sell for less than $100 and function as an entertainment radio when not in use. (Hines, Wash. *Eve. Star*, 9/24/65)

- Water could have been present on Mars during the first 3.5 billion years of the planet's history and then disappeared before its craters were created by meteorites, said researchers Edward Anders of the Univ. of Chicago and James R. Arnold of the Univ. of California at San Diego in a report in *Science*. The scientists said they believed the Martian craters were only 300 to 800 million years old, compared with up to 4.5 billion years for those on the moon. They added: ". . . The crater density of Mars no longer precludes the possibility that liquid water and a denser atmosphere were present on Mars during the first 3.5 billion years of its history." (*Science*, 9/24/65)

- Ecumenical Council delegates studied theological problems involved in space travel and the possibility of life on other planets. Some bishops believed the question should have a place in the Council document entitled "On the Church in the Modern World" now under debate by the 2,500 prelates assembled in St. Peter's Basilica. (UPI, *Wash. Daily News*, 9/24/65)

- Columbus Association of Genoa, Italy, announced that Astronauts Charles

Conrad, Jr., and L. Gordon Cooper, Jr., had been selected to receive the 1965 Columbus Prize. (AP, *Wash. Post*, 9/25/65)

September 25: A half-length 260-in.-dia. solid rocket motor generated 3.5 million pounds thrust during a successful two-minute test firing at Aerojet General's Dade County, Fla., facilities. Firing tested strength of the maraging steel motor case, structural integrity of the cast propellant, the insulation, and the ablative nozzle; it demonstrated that massive quantities of solid fuel could be controlled under firing conditions. Test was part of the national large solid motor technology program initiated by USAF in 1963 and transferred to NASA in 1965. (NASA Release 65-295, 65-311; LRC Release 65-65)

• NASA launched a second Aerobee 150A sounding rocket from Wallops Station containing French-built radio propagation experiments designed for later flight on France's FR-1 satellite. The 197-lb. payload reached peak altitude of 120 mi. (192 km.) during the eight-minute flight. First launch had carried an identical payload Sept. 17. (Wallops Release 65-61)

• Super Guppy, a five-story-high aircraft designed to carry large rocket sections, made a safe emergency landing at Edwards AFB after part of its aluminum skin peeled off during a high-speed dive. No one was injured. Aircraft was undergoing flight tests for certification by FAA. (AP, *Wash. Post*, 9/26/65)

• Launch of Gemini VI from Eastern Test Range on a two-day mission no later than Oct. 25 was announced by NASA. It would be man's first attempt to rendezvous and dock with an orbiting spacecraft. Pilots for Gemini VI were Astronauts Walter M. Schirra (Capt., USN) and Thomas P. Stafford (Maj., USAF). Backup pilots were Astronauts Virgil I. Grissom (Maj., USAF) and John W. Young (Cdr., USN). This would be Schirra's second space flight. His first was Oct. 3, 1962, a six-orbit flight in Mercury spacecraft SIGMA VI (NASA Release 65-307)

• Dr. Donald Young and Dr. Ralph Pelligra of NASA Ames Research Center had been studying "the use of high caloric diets for prolonged space flights on the theory that it may be possible for astronauts to use their own fatty deposits as a source of energy." It might be possible that future astronauts would prepare for spaceflight by eating "certain types of fats that would build up in their bodies," thus providing them with "storage depots of their own fat" that could serve "as a backup if food supplies were limited." (UPI, St. Louis *Post-Dispatch*, 9/26/65)

• Apollo Extension Systems (Aes) schedule was summarized by James J. Haggerty, Jr., in the *Journal of the Armed Forces:* "NASA has worked up a tentative Aes schedule. It calls, first, for perhaps three or four 14-day earth-orbital missions, using a basic Apollo spacecraft only slightly modified, starting in 1968. This phase would be followed by extended earth-orbital (45-day) missions at the rate of five or six a year in 1970-71 and 14-day lunar exploration missions at the rate of one or two a year in 1970-71." (Haggerty, *J/Armed Forces*, 9/25/65, 23)

September 26: NASA ceased operation of RELAY II communications satellite after a final demonstration broadcast during which Sen. B. Everett Jordon (D-N.C.) spoke via the satellite from Exposition Hall in Atlantic City in opening the week-long International Exposition of the American Textile Machinery Assn. there. After the demonstration, the Mojave Desert Ground Station, the only one in the world equipped to communicate with RELAY II, was closed to begin modifications for use in the Ats (Applications Technology Satellite) program. Since its launch Jan. 21, 1964, RELAY II had performed experiments demonstrating the feasibility of a worldwide system of communications by spacecraft and the technology for such a system. (NASA Release 65–306)

September 26: Astronaut M. Scott Carpenter and nine fellow aquanauts rose 205 ft. from Sealab II in a pressurized personnel transfer capsule. They would decompress on the support ship *Berkone* for about 30 hrs. before being exposed to atmospheric pressure. Carpenter had completed a record 29 days and 14 hrs. in the ocean-bottom laboratory; his associates had been below 15 days as had another team before them. A third team of ten men would complete the Navy's 45-day experiment in underwater living and working. President Johnson phoned congratulations to Carpenter. (*Wash. Post,* 9/27/65)

September 27: Six rhesus monkeys exposed to a total radiation of 500 rads during a 10-day simulated space flight experienced no performance decrement, reported Heather M. Davis in *Missiles and Rockets.* Radiation was administered at rate of 2 rads per hour from an 80-curie gamma radiation source at the Los Alamos Scientific Laboratory. Thirty-day physical examinations just conducted showed the animals were still in good health. (David, *M&R,* 9/27/65, 38)

- U.S. Army's Nixe-X Project Office at Redstone Arsenal announced appointment of Dr. Oswald H. Lange as Chief Scientist and Charles E. Richardson as Chief Engineer. Lange was returning to the Army after more than five years with NASA Marshall Space Flight Center, where he directed the Saturn vehicle project and for the past two years was assistant MSFC director for scientific and technical analysis. Richardson was formerly chief of the Nike-X Project's Test and Range Operations Div. (*Marshall Star,* 9/29/65, 1)

- GSFC was pushing Aerobee 350 sounding rocket toward operational status to get second- and third-generation upper altitude data, wrote William S. Beller in *Missiles and Rockets:* "With much of the first-generation work already performed in the altitudes from about 50 mi. to 100 mi., a more sophisticated rocket is needed. . . .

    "The Aerobee 350 shows more than twice the performance of the 150: for heavy payloads, the factor tends toward 2.5, showing the most significant advantage; for lighter payloads, the factor approaches 2.0.

    "The Aerobee 350 launched from Wallops Island will carry a payload of 150 lbs. to 290 mi. altitude; or 500 lbs. to 210 mi. If the launching takes place at White Sands, N.M., which is at 4,000 ft., the rocket will put out 15–20 mi. more altitude performance for the same payload weight. An Aerobee 350 launch tower is to be built at White Sands." (Beller, *M&R,* 9/27/65, 26)

- Philco Corp. had been awarded a $3 million increment to a contract with USAF for work on a satellite tracking network. (DOD Release 651–65)

- S–IC–1, first stage of the Saturn V booster, rolled out of the Manufacturing Engineering Lab at MSFC exactly on schedule. The stage was moved to the Quality and Reliability Assurance Lab where it would be checked out automatically in preparation for static firings. (MSFC Release 65-239)

September 27: USAF named Robert G. Loewy chairman of the U.S. Air Force Scientific Advisory Board. (Hussie, *Phil. Inq.*, 9/26/65)

- ComSatCorp asked FCC for permission to make available service via EARLY BIRD I outside of normal hours to permit transatlantic televising of Pope Paul VI's visit to the U.S. October 4. The Corporation normally could offer its services between 5 a.m. and 9 p.m. daily; it was seeking to extend them from 9 p.m. to 5 a.m. during the papal visit. (ComSatCorp Release)

- Potential of a manned orbiting vehicle to obtain advanced photographic and electromagnetic intelligence on Soviet ICBM defenses, on Russian deployment of new smaller solid propellant ICBMs, and on Chinese progress toward a strategic missile force was suggested by *Aviation Week and Space Technology* as being significant factor behind Defense Secretary Robert S. McNamara's decision to proceed with manned military space missions. It was photographic evidence provided by Samos, shortly after McNamara had come to the Pentagon, that had enabled the U.S. to determine that the Russians were not building and deploying ICBMs as fast as had been believed and which had allowed McNamara to establish more modest missile production goals. (*Av. Wk.*, 9/27/65, 26)

September 28: In its 150th flight, the X-15 (No. 3) was flown to 295,600-ft. altitude and top speed of 3,682 mph (mach 5.33) by NASA research pilot John B. McKay. Altitude marked the fourth highest reached by the X-15. Purpose of flight was to measure boundary layer noise; test a horizon scanner; and measure aerodynamic and structural loads on the horizontal tail surfaces. (NASA Release 65-310; *X-15 Flight Log*)

- Astronaut M. Scott Carpenter (LCdr., USN), after a record 30 days 205 ft. below the surface of the Pacific, said he was convinced "men can

September 28: John B. McKay, pilot of 150th flight, in front of X-15 No. 3.

live forever—any length of time they wish—beneath the surface of the ocean." He told a news conference of the beauty, pain, and hard manual labor and added: "The ocean is a much more hostile environment than space. I worked much harder in Sealab 2 than in the Mercury capsule. More energy is required just to stay warm, because of the 50-degree cold. But the real key is the isolation. I think men can live as long in underwater habitats as they can in dewline stations or any place else where they are isolated." (AP, Balt. *Sun*, 9/29/65)

September 28: Astronaut Charles Conrad, Jr., (LCdr., USN) discussed the GEMINI V spaceflight with the Oba (King) of Benin in his palace in Nigeria. In his luncheon speech, the Oba said: "In this age, astronauts are physically making visits to some sphere in the universe our ancient people thought was possible to visit only in spirit form. . . . Such journeys may prove or disprove such theories that the space is limitless, according to scientists, or that over and above the outer space there is a canopy, according to the saying of our ancient people." (Louchheim, *Wash. Post*, 9/29/65, A1)

- U.S. Army had awarded a contract for $21,580,464 to Western Electric Co. for development of an advanced version of the Zeus antimissile missile. The long-range Zeus was one of the two interceptor missiles in the Nike-X system; it would employ two solid propellant motors and carry a nuclear warhead. (DOD Release 653–65)

- Smithsonian Astrophysical Observatory announced the discovery in England of a new comet. Discovered Sept. 26 by astronomer G. E. D. Alcock (and named after him), the 10th-magnitude comet appeared as part of the constellation Hercules. (AP, *Wash. Post*, 9/28/65, A1)

- Gen. Bernard A. Schriever outlined examples of the current USAF effort to upgrade its data management activities in both Government and industry: (1) new manuals would be forthcoming on contract definition, systems engineering, and cost estimating; (2) the USAF Seed (Supply of Essential Engineering Data) concept was being considered for application to new types of data besides engineering items; and (3) a cadre of Data Management Officers and supporting personnel would work with their counterparts in industry to iron out data problems. General Schriever delivered the keynote address to the Air Force/Industry Data Management Symposium in Los Angeles. (Text)

- Dr. Donald Hornig, President Johnson's science adviser, told the Associated Press in Washington, D.C., that the Government planned to proceed with Project Mohole despite mounting costs. The project called for sinking a metal drill six miles through Pacific Ocean and floor about 100 mi. northeast of Maui in the Hawaiian Islands. (*Wash. Post*, 9/28/65, A5)

- Former Astronaut John H. Glenn, Jr., and his wife left for Europe on a goodwill tour ordered by President Johnson. Cities on the schedule included London, Berlin, Bremen, Hamburg, Amsterdam, Rotterdam, Genoa, Rome, Naples, Florence or Venice, Madrid, and Lisbon. (Wash. *Eve. Star*, 9/29/65)

September 29: ALOUETTE I, Canadian scientific satellite launched by NASA on September 29, 1962, had completed its third year in space and was still operating normally. Satellite was using the swept-frequency topside sounding technique to gather information about ionospheric electron density and distribution during the four-to-six hours daily it was activated by command from the ground. (NASA Release 65-312)

September 29: Bureau of Naval Weapons announced it had placed two more navigation satellites in orbit. The two 135-lb. spacecraft, launched June 24 and Aug 13, were intended "to augment the now operational all-weather satellite navigation system and to allow for more frequent position fixes by ships at sea." (AP, Balt. *Sun.* 9/29/65)

- XB-70A research aircraft No. 2 made its sixth flight, reaching a speed of mach 2.23 (about 1,460 mph) and altitude of 54,000 ft. During the one-hour-and 44-minute flight (32 min. of which were at supersonic speeds), several studies were conducted: runway noise-levels were studied, sonic-boom tests conducted, experiments related to supersonic transport program made, and aircraft's air-inlet control system was operated. Pilots were Al White, chief test pilot, and Van Shepard, co-pilot. (NAA *S&ID Skywriter,* 10/1/65, 1)

- Thiokol Chemical Corp. would receive from USAF a $1,562,000 cost-plus-fixed-fee contract for work on a solid-fuel rocket program. (DOD Release 659-65)

- U.S. Army Corps of Engineers had awarded Baxter Construction Co. a $1,224,271 firm fixed-price contract for construction of the Atmospheric Reentry Materials and Structures Evaluation Facility at MSC. (DOD Release 659-65)

- A rocket-driven sled that would reach a speed of about 2,300 mph was being built for USN, Thomas Henry reported in the Washington *Evening Star.* Designed to carry a simulated nose cone, the sled would run into a high-energy blast about halfway along its course. Measurements of the shockwaves would help determine hardening required to protect a missile nose cone. (Henry, Wash. *Eve. Star,* 9/29/65, 30)

- Aviatrix Jerrie Mock broke the speed record for single-engine aircraft over a 500-km. (304-mi.) course. She established an average speed of 203.858 mph on her flight, which lasted one hour, 31 min., and 27 sec. Former world record of 178 mph was set in 1956 by Czech pilot Lubos Stastny. (UPI, *Wash. Post,* 9/30/65, D6)

- Tribute to Secretary of the Air Force Eugene M. Zuckert by Sen. Howard W. Cannon (D-Nev.) was read on the floor of the Senate: "It is no surprise that Eugene Zuckert served in the office of the Secretary for longer than any other man. His leadership spans nearly 20 years, dating almost from the time that the Air Force became an independent service while Mr. Zuckert served as special assistant to the Assistant Secretary of War for Air. He served for a time on the Atomic Energy Commission and brought a high degree of competence and ability to this important operation.

    "After more than $4\frac{1}{2}$ years as Secretary, Eugene Zuckert has established an enviable record, and I suggest that his devotion to the service and his unique skills will make his absence from Government of very short duration. No man who has done what he has for the Air Force in the critical years when that service entered the space age can be forgotten or easily replaced." (*CR,* 9/29/65, 24571)

- H. Z. Hopkins, Jr., chief of North American Aviation, Inc., flight test operations at Edwards AFB, Calif., was named president of the Society of Experimental Test Pilots (SETP). Also honored by SETP during its annual meeting in Los Angeles was Al White, NAA's chief test pilot, who was recipient of the Ivan C. Kincheloe Award. White was cited for his role in development and testing of XB-70A research aircraft. (NAA Release NL-19; NAA *S&ID Skywriter*, 10/1/65,2)

September 30: X-15 No. 1 was flown by Capt. William J. Knight (USAF) to an altitude of 76,000 ft., at a speed of 2,761 mph (mach 4.06). (NASA X-15 Proj. Off.; *X-15 Flight Log*)

- USAF launched Atlas-Agena D launch vehicle from Western Test Range with unidentified satellite payload. (*U.S. Aeron. & Space Act., 1965*, 152)

- President Johnson signed into law the High-Speed Ground Transportation Act of 1965 (PL 89-220) in White House ceremony. He said:

    ". . . In recent decades, we have achieved technological miracles in our transportation. But there is one great exception.

    "We have airplanes which fly three times faster than sound. We have television cameras that are orbiting Mars. But we have the same tired and inadequate mass transportation between our towns and cities that we had 30 years ago.

    "Today . . . an astronaut can orbit the earth faster than a man on the ground can get from New York to Washington. Yet, the same science and technology which gave us our airplanes and our space probes, I believe, could also give us better and faster and more economical transportation on the ground. And a lot of us need it more on the ground than we need it orbiting the earth. . . .

    "The High-Speed Ground Transportation Act of 1965 really gives us, for the first time in history, a coordinated program for improving the transportation system that we have today, and making it a better servant to our people. . . ." (Text, *Pres. Doc.*, 10/4/65, 329-30)

- Secretary of Defense Robert S. McNamara directed USAF to proceed immediately to develop and produce the C-5A transport plane. Gross weight of aircraft would be nearly 350 tons—twice that of present largest military cargo plane. It would be able to carry loads of a quarter million pounds 3,200 mi., and loads of 100,000 lbs. nonstop across the Pacific Ocean. It would be more than 230 ft. long, 63 ft. high at the tail, have a 220-ft. wing span, and be able to land on unprepared airfields of 4,000 ft. Development cost and initial production order for 58 planes, including engines, would be about $2 billion. Aircraft would be bought under new contracting concept under which both the airframe and engine manufacturers would receive contracts covering not only development but also production. USAF had selected Lockheed Aircraft Corp. to develop and produce the C-5A. Four new fan-jet engines for the aircraft, each capable of 40,000 lbs. thrust, were being developed and would be produced by General Electric Co. (DOD Release 663-65)

- General Dynamics/Convair was being awarded a $2,198,000 firm fixed-price contract by U.S. Army for designing, furnishing, and installing an operational TV system for Launch Complex 39 at Kennedy Space Center, NASA. (DOD Release 664-65)

September 30: ComSatCorp asked FCC for authority to build four new satellites to provide communication services for the Apollo project as well as for other commercial uses. (ComSatCorp Release)

- Highlights of report by Civil Service Commission Chairman John W. Macy, Jr., to President Johnson on Government savings during FY 1965 resulting from employee suggestions showed that one of the top suggestion awards went to NASA personnel. The $1,400 award to MSFC aerospace technologists Emmett L. Martz, John L. Burch, and William L. Kimmons was for their design which reduced by $133,438 the cost of certain bearings in Saturn launch vehicle's guidance system. NASA was among the five nondefense Government agencies that exceeded the million-dollar mark in savings from adopted suggestions. (*Pres. Doc.*, 10/4/65, 335)

- NASA Deputy Associate Administrator for Advanced Research and Technology Alfred Eggers, speaking to the Aviation Space Writers Association in Washington, D.C., said one of the basic programs now underway in OART was to establish a mission capability flexible enough to satisfy most mission requirements for the balance of the century. Describing the capability as a "platform" that could support whatever mission the Nation wanted, he cited the present indecision over the mission that should be undertaken after the lunar landing. Given the difficulty of obtaining a consensus from the American public and government in the matter, the only useful alternative was to develop such a platform. He said the present OART program would enable this country to go to Mars, Venus, the asteroids, or the moons of Jupiter before the end of the 20th century. (Text)

- Jet airline pilots would no longer be required to wear oxygen masks when flying above 35,000 ft. according to a FAA rule effective as of this date. Under the new rule, the requirement for one pilot at the controls to use oxygen would apply now only above 41,000 ft., where the time element would be much more critical in case of sudden decompression. (FAA Release 65–84)

During September: First useful photograph of the nightglow was recorded aboard a NASA Aerobee 150 sounding rocket launched from White Sands Missile Range. Rocket was at 173-km. (106.8-mi.) altitude; distance from the earth's horizon to the center of the nightglow was 90 km. (55.9 mi.). Tri-X film was used. (*M&R*, 9/27/65, 15)

- 44 of the 107 pages of "Opportunities for Participation in Space Flight Investigations"—NASA's semi-annual publication inviting scientists to propose spaceflight experiments—were devoted to Apollo manned missions, some of which were under study. Detailed descriptions and timetables covering a wide range of NASA flight projects, manned and unmanned, were provided in the publication. Flight dates generally covered the period from 1966 through 1972. Proposals would be reviewed and evaluated for scientific merit and technological feasibility, the competence and experience of the investigator, assurance of institutional support, and the scientific adequacy of proposed apparatus. (NASA Release 65–284)

- A "caloric water and protein balance study" was in progress at Wright-Patterson AFB, Ohio, under guidance of NASA Manned Spacecraft Center's Dr. Paul LaChance, to verify hardware and procedures for waste management proposed for the 14-day Gemini VII mission in late December. According to *Missiles and Rockets*, four volunteers would be sustained on a controlled diet of bite-sized rehydratable food over a six-week period. Midway through the study, they would enter an altitude chamber for two weeks. This would be followed by two weeks of "post-flight" checks. Careful measurement of calcium intake and loss through body fluids—one of the Gemini VII medical experiments—would be made on each subject. (*M&R*, 9/27/65, 9)

During September: A light metal sphere found in a remote part of Australia was described in a paper by Peter M. Twiss in *Journal of Spacecraft and Rockets*. The paper analyzed the likely reentry history of the sphere during decay from circular orbit, and concluded the sphere came from a U.S. spacecraft in orbit. (Twiss, *J/Spacecraft & Rockets*, 9/10/65, 660–663)

- USAF had developed and proved a new method enabling recovery and consequent reuse of expensive research balloons. Developed by James Payne of AFCRL, the recovery system employed two parachutes in tandem—the lower one to return the scientific payload and the upper one to return the balloon. Following controlled deflation of the balloon, the upper parachute—with an opening in its center through which the neck of the balloon would be fitted—would ride up the balloon's neck as it was deflated. The center hole would have a nylon sleeve attached to it. The sleeve would envelop the balloon material as it would deflate and literally wrap it in a protective nylon package for its return to earth and later reuse. (AOAR Release 9–65–1)

- Paris newspaper *Le Monde* reported that first launch of a satellite by the Diamant booster might come at the end of November, before the December 5 presidential election, and might include an attempt to put the A-1 satellite—an 80-lb. experimental spacecraft with an orbital life of several weeks—in orbit. In case of an A-1 failure, *Le Monde* added, a second launch might take place before Dec. 5. Mid-January was quoted as the deadline for orbiting the D-1 satellite, and D-1B might be launched at an earlier date. (*Av. Wk.*, 9/20/65, 29; *M&R*, 9/27/65, 9)

- According to a report in a West German publication, *Luftfahrttechnik Raumfahrttechnik*, Soviet communications satellite MOLNIYA I transmitted color television programs for nine hours in May 1965. Quality of the transmissions was good. (*Luftfahrttechnik Raumfahrttechnik*, 9/65, 237)

- Evaluating the controversy on manned versus unmanned exploration of space, Robert Colburn said in *International Science and Technology*: "I see no meaningful issue between (unmanned) space science and (manned) space exploration. The real question is what sort of space science contributes best to the success of space exploration. More concretely, the question is not whether to investigate the moon with an unmanned Surveyor shot or a manned Apollo; it's whether an extra Surveyor mission now would significantly reduce the cost or danger of the eventual Apollo mission and, if not, whether it may be wasteful to send a Surveyor now when the same information can be had more easily later, once there is a man on the moon to help." Rep. Olin Teague (D–Tex.), member of the House Committee on Science and Astronautics, inserted Colburn's editorial in the *Congressional Record*. (*CR*, 9/14/65)

During September: NASA's Technology Utilization Div. published a 66-page illustrated book outlining potential uses of space telemetry techniques in the biological and medical fields. (NASA Release 65-309)

- U.S. Army Engineer Geodesy, Intelligence and Mapping Research and Development Agency (GIMRADA) awarded $557,000 contract to Cubic Corp. for four additional Secor (Sequential Collation of Range) satellites for use in gathering information on the exact location of land bodies. Although similar to the Secor satellites presently in orbit, the four new satellites would have a transponder capable of operating at higher altitudes. Delivery of the first of the four was scheduled for next February. (GIMRADA Release)

- Commenting on "race" between the U.S. and the U.S.S.R. to be first on the moon, Soviet Cosmonaut Pavel Belyayev said: "I simply don't understand such appeals. As far as I know, we have no intention of competing in this area." Lt. Gen. Nikolai Kamanin said: "In a space research program, haste is out of place. Each space flight, especially with a cosmonaut aboard, must be the product of a sustained effort that may take months. Improvisation is impermissable in space research; we don't believe in crash programming." Cosmonaut Aleksey Lenov said: "I also think that all this talk of getting to the moon first is nonsense. What we must do is work toward that goal calmly and thoroughly and without any rush. I think that those who work that way will be first on the moon." (*Soviet Life*, 9/65, 28-29)

- North American Aviation, Inc., transportation study which warned that within 50 yrs. the demand in California for transportation of people and commodities might increase 500% and 700% respectively, received comment from California Governor Edmund (Pat) Brown: "North American Aviation's report on transportation makes the answer unanimous: systems engineering is not only a sound approach to social problems, it may well be the only approach.

    ". . . The idea of transferring talented systems engineers from the field of space hardware to the broader field of human need is, in itself, a breakthrough of significant proportion. The talent has been there for some time. It just took us awhile to see its tremendous proportion. . . ." (NAA *S&ID Skywriter*, 10/1/65, 1, 2)

- Dr. Robert G. Loewy, Associate Professor of Mechanical and Aerospace Sciences, Univ. of Rochester, was named Chief Scientist of USAF. (*Av. Wk.*, 10/11/65)

October 1965

October 1: NASA's MARINER IV Mars probe stopped transmitting continuous data reports when a JPL command switched its transmitter from a high-gain directional antenna to a low-gain all-direction antenna to permit periodic tracking by NASA's Deep Space Network as spacecraft orbited the sun. MARINER IV, 19,359,086 mi. from Mars, traveling 90,499 mph relative to earth, had completed a total of 418,749,386 mi. in its solar orbit. Although it had achieved its mission objectives, additional scientific and engineering data might be obtained if the spacecraft were still operating in 1967 when the earth-spacecraft distance would narrow to about 30 million mi. (NASA Release 65–316; NASA Proj. Off.)

- NASA consolidated its unmanned launch activities at both the Eastern and Western Test Ranges under Kennedy Space Center, NASA. At Cape Kennedy, the Launch Operations Div., GSFC, would become an integral element of KSC; at Western Test Range, Goddard personnel permanently assigned there and NASA Pacific Launch Operations Office which logistically supported them would also be placed under KSC. Robert Gray would be Assistant Director for Unmanned Launch Operations. Checkout and launch of all NASA launch vehicles, except the solid propellant Scout rockets developed and launched by LaRC at Wallops Station and Western Test Range, would be supervised by KSC. (NASA Release 65–313)

- Secretary of Defense Robert S. McNamara administered the oath of office to four new defense officials in a Pentagon ceremony: Dr. Harold Brown, Secretary of the Air Force; Dr. John S. Foster, Jr., Director of Defense Research and Engineering; Norman S. Paul, Undersecretary of the Air Force; and Thomas D. Morris, Assistant Secretary of Defense for Manpower. (DOD Release 666–65)

- First complete test model of the S–II stage of the Saturn V launch vehicle left North American Aviation's Space and Information Systems Div., Seal Beach, aboard the U.S.S. *Point Barrow* for NASA Marshall Space Flight Center. The 4,000-mi. trip via the Panama Canal would take about two weeks. (MSFC Release 65–246)

October 1–2: Contributions of NASA's space program to the future of the Nation were outlined by NASA Deputy Administrator Dr. Hugh L. Dryden at the Governor's Conference on Oceanography and Astronautics in Hawaii:

"We are building toward pre-eminence in every phase of space activity. . . .

"We are building a network of large-scale engineering facilities, spaceyards, proving grounds, and spaceports to assemble, test, and launch the space vehicles we need now and in the future.

"We are creating new national resources of lasting value in these facilities; in the industrial and managerial capabilities we are developing; and in the growing number of scientists and engineers who are learning about space and space technology.

"We are filling the pipelines of hardware and knowledge, and, as measured by the financial resources required, we're about halfway toward our first manned lunar mission in mid-1965.

"We are accumulating, in space, the basic scientific knowledge about the earth, the solar system, the universe, and about man himself.

"We are bringing benefits not only to the United States but to all the world through the use of space and space technology. . . .

"We are providing a much-needed stimulus to the energies and creativity of people everywhere. . . .

"We are bringing about increased economic activity. . . .

"And we are making certain . . . that the realm of space now opening up to us shall be a domain of freedom.

"It is for these reasons that we have mounted the greatest peacetime undertaking in the history of mankind. . . ." (Text)

"The Military Implications of Space" were discussed for the Governor's Conference by Gen. Bernard A. Schriever, AFSC Commander: "Our military efforts in space can be placed into three broad categories: first, the development of space systems to support military missions on earth; second, the development of defensive measures against possible enemy actions in space; and third, the conduct of experimentation and of programs aimed at pushing technology forward. . . ."

Schriever noted two examples of unmanned satellite systems: (1) the Nuclear Detection Program consisting of six Vela satellites gathering information on radiation backgrounds in far space and defining an operational nuclear detection system; and (2) communications satellites. He revealed that an initial R&D system to satisfy military communications' requirements had been authorized by DOD for launch early next year and would include up to 23 satellites orbiting earth at random spacing at about 21,000-mi. altitude. The satellites would be launched in clusters of eight from three Titan IIIC boosters and then would be positioned along an orbital path. An advanced system was also planned in which expected life of each satellite would be increased from one and a half years to between three and five years.

In the advancement of space technology he listed several major programs: manned orbiting laboratory, space cabin experiments, development of cryogenic rocket engines, development of the scramjet, development of a spacecraft capable of maneuvering during reentry, and the Spacecraft Technology and Advanced Reentry Program (Start). (Text)

October 2: NASA Marshall Space Flight Center shipped the instrument unit for the first Saturn IB launch vehicle to Kennedy Space Center, NASA aboard the barge *Palaemon*. First Saturn IB was scheduled for launch from KSC early next year. (MSFC Release)

- USN recommissioned two rocket launching ships, the *Carronade* and the *White River*, and assigned them to the Pacific Fleet to support amphibious landings. On Sept. 18 the rocket launching ships *Clarion River* and *St. Francis River* had been recommissioned. (DOD Release 652-65)

October 2: Soviet Cosmonauts Pavel Belyayev and Aleksey Leonov arrived in East Berlin to begin a ten-day visit to East Germany. (Reuters, *Wash. Post*, 10/3/65, A31)

October 3: Former Astronaut John H. Glenn, Jr. (Col., USMC, Ret.), in Western Europe on a three-week goodwill tour at the request of President Johnson, saw the Berlin wall during a helicopter sightseeing trip. (Reuters, *Chic. Trib.*, 10/4/65)

- The extent of the economic impact of the space program on a seven-county Central Florida area around Cape Kennedy was disclosed in a report by the Univ. of Florida Bureau of Economic and Business Research. Total personal incomes in the seven counties had increased from $372,779,000 in 1950 to $572,375,000 in 1954, then more than tripled to $1,738,566,000 in 1963. (*Houston Post*, 10/6/65; KSC Historian)

October 4: LUNA VII 3,313-lb. instrumented moon probe was successfully launched by U.S.S.R. Tass said the last stage of the multistage launching rocket was put into parking orbit and then LUNA VII was fired on a trajectory toward the moon. All onboard equipment was said to be functioning normally. Launching occurred on the eighth anniversary of the orbiting by the Soviet Union of the first earth satellite, SPUTNIK I. (AP, *Wash. Eve. Star*, 10/4/65, 1; AP, *Wash. Post*, 10/5/65, A11; Nordlinger, *Balt. Sun*, 10/5/65, 5)

- Dr. Mac C. Adams was sworn in at NASA Hq. as Associate Administrator for Advanced Research and Technology by Dr. Robert C. Seamans, Jr., NASA Associate Administrator. Dr. Adams came to NASA from Avco Corp., Wilmington, Mass., where he was vice president and assistant general manager for space systems. In 1949–51 he was an aeronautical research engineer with NASA Langley Research Center (then NACA Langley Memorial Aeronautical Laboratory). (NASA Release 65–317; NASA Ann.)

- Dr. George E. Mueller, NASA Associate Administrator for Manned Space Flight, discussed the national space program in address before the Hartford Rotary Club, Hartford, Conn. Looking toward the future, he said:

    "The Gemini and Apollo-Saturn programs are providing this country with a broad base of technological, managerial, and resources capability which makes feasible a wide spectrum of space missions beyond the first lunar landings. This capability can be exploited in a wide range of earth orbital, lunar orbital, and lunar surface missions. . . .

    "Over 90 percent of the Apollo program is directly applicable to our earth orbital capability. Many interesting ideas are under consideration on how to use this extended earth orbit capability to yield great benefits to mankind in his daily life on earth, and to conduct much significant scientific research.

    "We are also studying the use of modified Apollo vehicles for missions both in orbit about the moon and for exploration of the moon's surface. In lunar orbit, we can do mapping, surveying, and exploration of the moon. On the moon, we can conduct detailed exploration of the surface environment and the moon's overall properties. We are also looking at the moon as a base for astronomical and for biological studies.

"All of these applications will capitalize on the presence of men in space. There is no substitute for having man's abilities available right on the spot during a mission, with his human intellect to cope with the unexpected and operate in an unplanned or unprogramed manner...." (Text, CR, 4/7/65, A5656–58)

October 4: NASA was negotiating estimated $25-million follow-on contract with Bendix Field Engineering Corp. for operations and maintenance support of the Space Tracking and Data Acquisition Network (STADAN), reported *Missiles and Rockets*. (M&R, 10/4/65, 10)

• Dr. Willard Libby, UCLA, recommended development of a nuclear spaceship in his address before the Aerospace Instrumentation Symposium in Los Angeles: "The only possible way of reaching into the deep reaches of space is by accelerating to incredible speeds, speeds that go far beyond those which we have obtained so far in the space program, and it is completely clear that the only way of reaching these speeds is by using atomic energy." Libby specifically recommended consideration of a reactor complex capable of supplying 10,000 megawatts of power to an engine with a designed thrust of 15,000–20,000 lbs. Such a spaceship would cost billions of dollars and be a decade in the building, but "with it we could expect literally to explore the Solar System," Libby said. He noted that Project Orion—the use of atomic explosions to propel massive space platforms—might be useful for delivering freight to the moon, but the shock of the explosions would pose difficulties for manned spaceflight. He was pleased with the success of the Rover-Phoebus thermally-heated atomic engines and felt they might be used to propel ships and run a ferry service to a moon base. (Text)

In a press conference, Dr. Libby urged a program to put more science into space exploration because "the dominant factor so far in the manned space program has been the drive to establish good engineering technology." He suggested: (1) NASA establish a six-man scientific task force to live and work for a year with training astronauts; (2) close cooperation between scientific research at outstanding universities, engineering efforts at aerospace industries, and NASA facilities; and (3) a mutually-happy "marriage" between space and education to attract and inspire students. (M&R, 10/11/65, 23)

JPL Director Dr. William H. Pickering told the Symposium that adaptive instrumentation was an important requirement for future unmanned interplanetary missions. In view of the increasing number of instruments on interplanetary spacecraft, more effort should be concentrated on the onboard data handling and command system, he suggested. This might transmit data only when meaningful changes were recorded by sensors. Also, the data system should monitor the performance of instruments and take corrective action should a failure occur. (Pay, M&R, 10/11/65, 36)

• FREEDOM 7 Mercury spacecraft, flown by Astronaut Alan B. Shepard, Jr. (Cdr., USN), on May 5, 1961, was put on exhibit at the Science Museum in London. (Reuters, *Boston Globe*, 9/17/65)

October 5: Aerospace Research Satellite was placed in orbit as a pick-a-back satellite on USAF Atlas D missile test from Western Test Range, the satellite entering the first highly retrograde orbit. Orbital data:

apogee, 2,141 mi. (3,447 km.); perigee, 255 mi. (410.6 km.); period, 125.7 min.; inclination, 144.3°. Carrying onboard instruments to map and monitor energetic particles, mass spectrometers, detectors imbedded in radiation shield, and instruments measuring dose rates in tissue-equivalent medium, the satellite successfully returned data. (*U.S. Aeron. & Space Act., 1965*, 152; AP, Wash. *Eve. Star*, 10/6/65)

October 5: USAF launched Thor-Agena D launch vehicle from Western Test Range with unidentified satellite. (*U.S. Aeron. & Space Act., 1965*, 152)

- At NASA Hq. Annual Honors Ceremony, Dr. T. Keith Glennan, first NASA Administrator, said: "Mine was the relatively easy task of putting together an organization. Yours has been the much more difficult task of making good the promises we held out, at times so naively, to the Congress and the people of this Nation and the world. But handling these heavy responsibilities cannot be the duty of just one man, or a small group of men at the top.

  "This Agency's success attests to the skill with which NASA employees have been and are carrying out their jobs. . . . To all of you and to the thousands of NASA people working at their tasks at whatever location throughout this nation and the world, I am proud to say—well done! And I am confident that the hard won leadership position you have attained will be maintained throughout the years ahead." (Text)

  NASA Administrator Webb presented Exceptional Scientific Achievement Awards to H. Julian Allen, Ames Research Center; Leslie H. Meredith and William Nordberg, GSFC; Dan Schneiderman and Eberhardt Rechtin, JPL. Smith J. DeFrance, Ames Research Center, and Bruce T. Lundin, LRC, received Outstanding Leadership Medals. Associate Administrator Dr. Robert C. Seamans, Jr., presented Exceptional Service Medals to John R. Cassani, JPL; Seymour C. Himmel, LRC; and William Lilly, Hq. He presented Group Achievement Awards to the Agena Project, LRC; Florida Operations Team, KSC; Launch Support Equipment Engineering Div., KSC; and Scout Project Office, LARC. Seventeen thirty-year service awards were presented by Earl D. Hilburn, Deputy Associate Administrator. (NASA Release 65–315)

- Tass reported that Crimean Astrophysical Observatory of the Soviet Academy of Sciences had photographed LUNA VII moon probe with 2.6-meter telescope, largest in U.S.S.R. and Europe. Photographic method used made it possible to obtain an image of the object in motion in the form of a chain of three dots. This made it easier to distinguish the object from weak stars. Photographs would help to determine LUNA VII's orbit with greater precision. (Tass, 10/5/65)

- ComSatCorp invited 20 construction companies to bid on site preparation for the Brewster Flat, Wash., earth station. Expected to cost approximately $6 million, the station would be a U.S. link in a worldwide commercial communications satellite system. (ComSatCorp Release)

- Machinist union members at Boeing Co. installations throughout the country voted to accept a new contract and end their 19-day-old strike against Boeing Co., announced Charles F. West, the union's general vice president. Under the contract, IAM would be free to strike again after six months if agreement were not reached on performance analysis—system by which the company graded employees for promotion, demotion, or discharge. (AP, Phil. *Eve. Bull.*, 10/5/65; *Wash. Post*, 10/5/65)

October 5: USAF officials closely associated with the manned orbital laboratory (Mol) program denied a published report that the Central Intelligence Agency had requested control of the program, reported *Missile Space Daily*. They said the report that a USAF–CIA disagreement over control of Mol had delayed President Johnson's August 25 announcement of the program was "without foundation." (*M/S Daily*, 10/5/65)

October 7: Soviet probe LUNA VII was proceeding on course and all onboard systems were functioning normally, Tass announced. Lunar landing was scheduled for October 8 at 6:08 p.m. EDT. (Tass, 10/7/65; AP, Wash. *Eve. Star*, 10/7/65, 1)

- Wright Brothers Medal for 1964 was presented to three NASA LaRC scientists—Marion O. McKinney, Jr., Richard E. Kuhn, and John P. Reeder—at the meeting of the Society of Automotive Engineers in Los Angeles. Cited "for meritorious contribution to aeronautic engineering," the scientists were selected for their paper "Aerodynamics and Flying Qualities of Jet/Vertical Takeoff-Landing and Short Takeoff-Landing Airplanes," presented in 1964 at the SAE-American Society of Mechanical Engineers meeting in New York. (LaRC Release)

- At the United Nations, Astronauts James A. McDivitt (L/Col., USAF) and Edward H. White II (L/Col., USAF) presented Secretary General U Thant a small U.N. flag carried by White on his June 4 walk in space. (AP, Wash. *Eve. Star*, 10/5/65)

- FCC approved ComSatCorp's proposal to build a ground station at Brewster Flat, Wash.—a U.S. link in their proposed worldwide communications system. The approval applied only to transmission and receiving facilities; other portions of the application such as ownership of the station and establishment of its links with San Francisco were still under consideration. (Wash. *Eve. Star*, 10/8/65, A17; *WSJ*, 10/14/65)

- U.S. suspended practice firing of Little John missiles in the foothills of Mt. Fuji at the request of the Japanese government to avoid clashes with about 2,000 farmers demonstrating near the target area. The farmers claimed that practice firing of the Little John, capable of carrying nuclear warheads, would lead to nuclear armament of Japan and would involve Japan in the Vietnam war. Little John was a short range, Army battlefield support missile. (AP, Wash. *Eve. Star*, 10/6/65, A21; AP, Wash. *Eve. Star*, 10/7/65, A13)

- A projection of the U.S. space program was outlined by Dr. Edward C. Welsh, Executive Secretary of the National Aeronautics and Space Council, at the Building Products Executives Conference in Washington, D.C.:

    "1. After we have made initial landings on the moon, we will, if conditions warrant, make many other trips to explore the various parts of the lunar surface and possibly to establish one or more bases there.

    "2. Not only will unmanned probes be sent throughout the solar system, but manned expeditions will visit the planets whenever that becomes scientifically promising and practicable.

"3. We will develop a family of useful manned earth-orbiting stations, growing from relatively small orbiting laboratories to large multi-manned permanent stations. Regular ferry service will transfer personnel and supplies to and from such spacecraft.

"4. Global communications via satellites will become a fact in the very near future and will be followed by direct broadcast of both voice and TV by satellite to home receivers throughout large sections of the world.

"5. Orbiting spacecraft will increase annually in numbers, in size, and in sophistication. Through such activity we will greatly increase our knowledge about the earth as well as about the heavens.

"6. We can expect a marriage of the major features of both aeronautics and astronautics . . . lifting bodies and winged spacecraft with maneuverable reentry ability will be launched into near and distant space by means of recoverable and re-usable launch vehicles. Drastic reduction in the mileage cost of space travel will result.

"7. Spaceports for the coming and going of spacecraft will be built in a number of parts of this country as well as elsewhere.

"8. Improved propulsion—faster, more powerful rockets using nuclear as well as chemical energy—will characterize space transportation. Planetary trips which today would take many months will be done in perhaps a week's time or less, while carrying substantial payloads of passengers, equipment, and supplies. And,

"9. As competence in space increases . . . we can expect to see greater economic progress and international cooperation. Then we can truly say that our space efforts have made major contributions to world peace. . . ." (Text)

October 7–8: About 700 engineers and scientists attended the Fourth X–15 Technical Conference at NASA Flight Research Center for a technical review of the NASA–USAF X–15 flight research program. Paul F. Bikle, director of FRC, and John S. McCollum, director of New Programs and Research Projects, USAF Aeronautical Systems Div., said future plans for the X–15, not yet approved, included modification to carry experimental hypersonic ramjet engines and study of a new delta-wing configuration. Such changes would provide vital information for future hypersonic aircraft design. The X–15, which had completed more than 150 successful flights since June 1959, had more than doubled the limits of manned flight in winged vehicles. Speed had been increased from about 2,000 mph to more than 4,100 mph. Built for altitudes up to 250,000 ft., the plane had surpassed 300,000-ft. altitudes four times and had reached a maximum of 354,200 ft. (NASA Release 65–318; FRC Release 20–65)

October 8: LUNA VII moon probe probably was destroyed in impact on the moon because the firing of the retrorockets was "only partially successful," speculated Sir Bernard Lovell, director of Jodrell Bank Observatory. Sir Bernard said variations in the signals heard at Jodrell Bank showed LUNA VII's retrorockets were fired at 4:58 p.m. EDT for six minutes. He estimated that LUNA VII, traveling about 2 mps., was slowed down by five-eighths of a mile a second during this firing. Signals began again at 5:20 p.m. and then stopped at 6:08 p.m.—the precise time when Russia had said LUNA VII would reach the moon. There was no official word on the fate of the ton-and-a-half space probe, but Western experts believed the Soviets had failed in their third attempt to softland on the moon. (UPI, *Wash. Post*, 10/9/65, A6; AP, Wash. *Eve. Star*, 10/8/65, A5)

October 8: Test model of the Lunar Landing Research Vehicle designed to simulated lunar landing was flown by former NASA X-15 pilot Joseph Walker to 300-ft. altitude. Built by Bell Aerosystems Co. under contract to NASA, the research craft had a jet engine that supported five-sixths of its weight; the pilot manipulated solid-fuel lift rockets that supported the remaining one-sixth. The craft's attitude was controlled with jets of hydrogen peroxide. (*Wash. Post*, 10/9/65, A4; AP, *Houston Chron.*, 10/9/65)

- NASA had asked ComSatCorp to provide communications satellite services in support of tracking and data acquisition needs for Project Apollo. The facilities—six highly-reliable telecommunications channels to be made available to three Apollo tracking ships and three remote land stations—were to be in operation by the fall of 1966. Negotiation of a definitive contract would begin shortly. (NASA Release 65-320)

- NASA and Soviet Academy of Sciences representatives reached two satisfactory understandings in their New York discussions on space cooperation. The first, which reaffirmed the existing agreement for exchange of weather satellite data, provided that satellite data would be available on a continuing basis from both sides within a few months. The second, an agreement for the preparation and publication of a joint review of research in space biology and medicine, provided for a joint editorial board to receive full cooperation from both sides in the preparation of materials available in the two countries, selection of authors, and publication of their work. Barring an objection from either side within two months, the agreement would become effective automatically. (NASA Release 65-325)

- House Committee on Science and Astronautics' Subcommittee on NASA Oversight issued a report on investigation of Project Surveyor

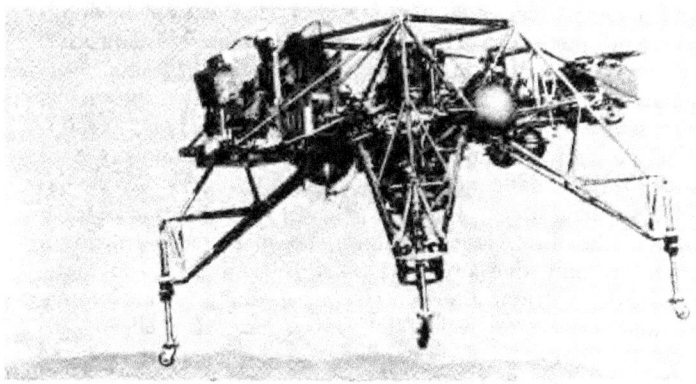

October 8: Lunar Landing Research Vehicle flight by Joseph Walker at Edwards.

by Rep. Joseph E. Karth's (D–Minn.) Space Science and Applications Subcommittee. Karth committee summarized the most serious reasons for Surveyor's delay: 1) the failure to do adequate preliminary work; 2) a series of major modifications; 3) inadequate staffing by JPL in the first three years; 4) too rapid expansion of personnel at a later date; 5) poor management on the part of Hughes; 6) a resistance on the part of Hughes to making management changes; 7) an improperly conducted testing program by Hughes; 8) management weaknesses in NASA; and 9) a lack of vigorous centralized control by NASA.

To describe the magnitude of the Surveyor problems, the Congressional investigators highlighted a NASA press release of January 1961, which outlined Project Surveyor as costing about $50 million, consisting of seven spacecraft, each weighing 2,500 lbs., capable of placing over 200 lbs. of scientific payload on the lunar surface, with launches scheduled during the 1963–66 time period. The Subcommittee then noted that the first launch had been delayed about two and a half years, payload capability had been reduced to only a TV camera on the first four spacecraft and only 114 lbs. on the next three operational models, with total costs so far of nearly $350 million for the spacecraft R&D alone. The investigation concluded that the responsibility for the Surveyor situation could be found in the inadequate preparation for the project, the performance of JPL, the performance of Hughes Aircraft, and the performance of NASA Headquarters. "A serious mistake" was made at the very beginning of the Surveyor project when insufficient preliminary work was done to identify serious technical problems, judge feasibility, consider alternatives, and more accurately estimate costs. The investigation concluded further that this was "the inevitable result of a poorly defined project." (House Rpt.)

October 8: Sweden, Norway, and Denmark were planning to establish a Nordic chain of detector stations to register underground atomic blasts. Sweden and Norway would each construct at least one detector station and join their stations; Denmark would build a large station in Greenland. Swedish experts believed that the new U.S. detector station in Montana together with the Nordic chain would make detections 100 per cent certain. (Fleisher, *Wash. Post*, 10/9/65, A18)

• Until man can duplicate responsibility—society's basic ingredient—in test tubes and mathematical equations, he should not attempt to duplicate himself in a laboratory, asserted Dr. Johannes M. Burgers. Speaking at the Symposium on Fluid Dynamics and Plasma, Burgers recommended a 50-yr. moratorium of artificial-life experiments because "our society is not ready for it. . . . Too many groups would take it out of the hands of science and make unholy use of it. . . . Wait until the educational level of man is higher. Wait until you know more about life." (Homan, *Wash. Post*, 10/8/65, B10)

• U.S. would make the first lunar landing because the "Saturn 5 booster . . . is the only one anywhere powerful enough to do that," predicted NASA Administrator James E. Webb, speaking to the International Radio and Television Society in New York. He believed the U.S. was ahead of Russia in cumulative knowledge of manned spaceflight, but conceded that Russia was probably 18 mo. ahead in other areas—specifically, multimanned spacecraft and the building and use of big boosters. Webb refused to classify Russia's LUNA VII mission as unsuccessful, explaining: "They're practicing how to do these things. They must have gained a good deal of information from it." (*San Diego Union*, 10/10/65)

October 8: $12 million, 210-ft.-dia. antenna at JPL's Goldstone Tracking Station, Calif., was being readied to permit the tracking of very distant space vehicles sending faint signals to earth, including MARINER IV, now in solar orbit. Called the Mars station, the antenna would be able to track a signal with a strength of only one-thousandth of one-billionth of one watt. (*Wash. Post*, 10/8/65, A3)

October 9: Two-stage Saphir rocket successfully fired by France from Hammaguir Range, Algeria, reached 720-mi. altitude, French Armed Forces Minister announced. Saphir, combined with a third stage already tested successfully, would be used to orbit the first French satellite toward the end of November 1965. (Reuters, *Wash. Post*, 10/12/65, A3)

- Gemini 7 spacecraft, destined for the long-duration 14-day mission, arrived at KSC from the McDonnell Aircraft Corp. factory in St. Louis. The spacecraft was moved to the Pyrotechnic Installation Building for receiving inspection and fuel cell installation. (KSC Historical Office)

- Crown Prince Harald of Norway toured NASA Manned Spacecraft Center. (MSC *Roundup*, 10/15/65, 8)

- *Pravda* announced completion of flight of LUNA VII: "The automatic station Luna 7 reached the lunar surface on October 8 at 1:08 p.m. Moscow time in the region of the Sea of Storms west of the crater of Kepler. After the correction made on October 5, during the lunar approach, most of the operations were conducted that were necessary for accomplishing a soft landing. Certain operations were not conducted in conformity with the program and require further development. . . . During the flight of the Luna 7 spacecraft, much practical material was collected for further studies." (*Pravda*, 10/9/65, ATSS-T Trans.)

- Soviet Professor V. Ivanchenko, writing in *Izvestia*, implied that LUNA VII spacecraft had crashed against the moon's surface because its braking rockets had failed to fire in time. He described the difficulties of activating the rockets at "precisely" the right moment to cut the spacecraft's speed of almost two mps to zero. Ivanchenko said only the trial-and-error system would give a successful result and predicted many more Soviet launchings. (*Wash. Post*, 10/10/65, A25)

- British Astronomer Professor Fred Hoyle, long-time adherent to the steady-state universe theory, wrote in *Nature* that he now believed that the universe or parts of the universe is in a state of flux, expanding for billions of years, then contracting to a dense ball of matter, and expanding again. In view of recent observations, especially those of a quasi-stellar radio source which indicated that the universe had been much denser, "the steady-state theory is out," he said. (Malkin, *Sun-News*, 10/11/65)

October 10: Final team of aquanauts surfaced from Sealab II, ending the U.S. Navy's successful 45-day, $1.8-million project. The Sealab capsule, sealed by the aquanauts before they surfaced, would be raised Oct. 13 and towed to the Long Beach, Calif., Naval Shipyard. (UPI, *NYT*, 10/11/65, 48L; AP, *Wash. Post*, 10/11/65, A21)

October 10: Urban Renewal Div. of Massachusetts Department of Commerce and Development approved Cambridge's Kendall Square urban renewal project, thereby opening the way for construction of NASA Electronics Research Center there. (*Boston Sun. Globe*, 10/10/65)

- North American Aviation, Inc., and United Aerospace Workers Union announced approval of a new three-year contract covering 35,000 workers in four states. (AP, *NYT*, 10/11/65, 28; AP, *Wash. Post*, 10/11/65, A19)

October 11: Formation of a Voyager Landing-Capsule Task Team at Ames and Langley Research Centers to assist JPL during early phases of the design was announced by NASA. This would include research and analytical support. (ARC Release)

- NASA's attempt to launch 99-lb. "mother-daughter" payload from Wallops Station for Pennsylvania State Univ. was unsuccessful because of a launch vehicle malfunction. Launched with four-stage Javelin (Argo D-4), the two-part payload, which only reached 20-mi. altitude, had been programed to separate at a 170-mi. altitude to obtain a profile of electron density in the upper atmosphere as the two parts ascended. In a similar launch Oct. 5, the mother-daughter sections of the payload had failed to separate. (Wallops Release 65–64)

- Dr. Hugh L. Dryden, NASA Deputy Administrator, addressed symposium on universities and Federal science policies held during meeting of National Academy of Sciences, in Seattle. He discussed the role of universities in space exploration and specified NASA-university relationships: (1) direct project support by means of research contracts to universities; (2) provision of new facilities, including those to encourage interdisciplinary groups for research in the interest of both the university and NASA; and (3) training support, including program to turn out 1,000 Ph.D.'s annually. He then turned to evolution of the university's future role:

  "It is not expected that the university will become the primary action agency for solving the economic and social problems of our society or for setting basic social objectives, but it must have a prominent voice and influence. The university should, I think, aspire to intellectual leadership, to observe and analyze, to apply the power of the human mind—that faculty which distinguishes man from animal, to establish a rational basis for policy and action. The direct attack must be undertaken by other institutions, mainly by political agencies. We are now witnessing the evolution of government agencies, particularly those of the Federal government, to meet social needs by direct action in all fields from economics and public health to science and technology. A beginning is being made in the mobilization and integration of the total available resources to attack the most complex social problems. ". . . I believe that . . . the traditional values of pursuit of excellence, freedom of inquiry, preservation of creativity, and the support of the unconventional new ideas, particularly those of young scientists, . . . may not only be preserved but may even be strengthened within the larger framework of increased scope which can provide a sounder human motivation, bring allocation of increased resources, provide a satisfaction of great accomplishment in terms of human goals, and a sense of a community of interest with all mankind. If you will take the time to discover what is now going on in the exploration of space, you will find a free association of individuals, each with complete freedom of individual choice, in the largest and most challenging venture of man, the search for knowledge of his environment, and the application of this knowledge to his benefit. In this group of several hundred thousand people you will find innumerable examples of the pursuit of excellence from the renewed pride of the workman in his handicraft, to the intellectual effort of the scientist to analyze and observe. You will find mainly young men, unafraid of large problems or rapid change. You will find the unconventional idea being given attention. The university and the university scientist must and will adapt to the age of space exploration. . . ." (Text; *Science*, 11/26/65, 1129–33)

October 11: This has been the best year in history for airline-equipment orders, Stuart G. Tipton, president of the Air Transport Association, told a news conference in a summary marking the tenth anniversary of the first order for jet airplanes. Tipton said that U.S. airlines would make delivery this year on 161 jet and propjet airplanes costing $880.6 million; 239 jets and propjets costing $1,201,600,000 would be delivered to the U.S. airlines next year, on the basis of orders already placed; and 187, costing $1,033,900,000, in 1967. (AP, *NYT*, 10/12/65, 69)

October 11–13: The AIAA's Fourth Manned Space Flight Meeting was held in St. Louis.

William B. Taylor, Director of NASA Apollo Applications Office, outlined some long-range goals of Saturn/Apollo: "By 1980, it seems likely there will be a vigorous program for manned exploration of the near planets, probably based on a new generation of space vehicles capable of flights for durations of 2 to 3 years with crews consisting of 6 to 8 men. Although by 1980 it is not likely that men will have landed on Mars, it is possible that a Mars flyby or manned orbiting mission will be imminent if not already accomplished.

"On the moon by 1980, there may be in operation one or more multi-manned research observatories, conducting astronomical and geophysical operations on a year-round basis. These observatories may also be home base for mobile lunar explorations, using self-sufficient surface vehicles.

"In earth orbit by 1980 there may be one or more multi-manned, orbiting space stations, functioning as research laboratories for the advancement of human knowledge and increase of living standards. In addition to these permanent space stations, there could be a variety of smaller earth orbiting spacecraft, operated by private industry or government agencies in much the way aircraft operate today, to carry out such applications as long term weather forecasting and control; surveys of water resources, crops, forests and ocean currents; continuous, wide-band worldwide communications; air, sea and space navigation, and traffic control.

"These, then are some of the longer range objectives of the 1970's and 1980's which we feel can be achieved with the wide base of space technology which can be developed in alternate missions using the Saturn/Apollo systems. . . ." (Text)

Undiscovered biological rhythms in man could prevent his visiting other planets, reported biologist Eugene Spangler, TRW Systems. "The consequences of the removal [of man from earth] are in many respects unknown. The solution may not come except by means of gradual accretions in the total time that man spends in space together with careful monitoring of his reactions as that time is increased. But it seems necessary to come to further grips with the problem insofar as possible before prolonged space flight is attempted. The goals must be first to define the problem in two steps: (1) what specific correlations exist between the life processes of man and the variations in his geophysical environment? and (2) what are the results of substantial alteration of that environment on these correlations?"

Spangler noted that all of the known human rhythmic cycles could be interrupted without apparent harm, yet even the adjustment to a rapid change of several time zones after a jet flight could involve unpleasant recycling of the body's "biological clock." (Text)

Suits of armor might be worn by astronauts on long lunar explorations, reported Richard S. Johnston, James V. Correale, and Matthew I. Radnofsky, NASA Manned Spacecraft Center. The suits must be reliable, mobile, leak-resistant, and have lightweight construction and high durability for repeated wear. Two models of a metal spacesuit had met these requirements and been delivered to MSC for further testing, they said.

Spacesuits for the Apollo mission, which would involve relatively brief lunar walks, would be of fabric and consist of several layers, starting with water-cooled underwear, followed by pressure layers to hold the suit's atmosphere, and finally an "overcoat" of fabric. (Text)

Greater space cooperation between the U.S. and U.S.S.R. was predicted by Astronaut L. Gordon Cooper (L/Col., USAF) in a news conference. Cooper based his prediction on discussions he and fellow GEMINI V Astronaut Charles Conrad, Jr. (LCdr., USN), had had with Cosmonauts Col. Pavel I. Belyayev and L/Col. Aleksey A. Leonov at the International Astronautical Congress in Athens. Cooper said the cosmonauts were "mighty nice fellows" who "appeared to be very happy that we could get together and talk." (Clark, *NYT*, 10/12/65, 9)

Dr. Alfred J. Eggers, NASA Deputy Associate Administrator for Advanced Research and Technology, told the meeting that the nation as a whole should be ready to decide whether to send men to other planets. He expressed faith that the public would want to continue space exploration: "It is all part of man's wanting to know what the hell he's doing in his environment.

"The American people are just about what they claim they are—a people very fascinated with their environment, and when they can find another way to explore their environment they are very willing within reason to put their resources where their interest is.

"They will do that so long as they are getting return on their investment. But there has to be a payoff here. . . ." The "payoff" would not necessarily come before the mid-nineteen seventies' planetary decision, so the public would have to continue to have faith that the space program had such potential, Eggers said. (Clark, *NYT*, 10/14/65, 8)

October 12: X–15 No. 3 was flown by Capt. William J. Knight (USAF) 3,136 mph (mach 4.62) to 94,400-ft. altitude for pilot checkout. (NASA X-15 Proj. Off.; *X-15 Flight Log*)

- Four were honored with AIAA awards for achievements in manned flight at the AIAA 4th Manned Space Flight Meeting in St. Louis: the 1965 Astronautics Award to B/G Joseph S. Bleymaier (USAF), Commander of the Western Test Range, for "exceptional direction to Titan III space booster R&D program"; the Octave Chanute Award to Alvin S. White, Chief Engineering Test Pilot/XB–70 Project Pilot, North American Aviation, for "continued significant contributions to the aerospace sciences in the development of flying skills and techniques necessary to flight of an advanced aircraft at unprecedented speeds, altitudes, and weights"; the John Jeffries Award to Col. William K. Douglas (USAF), Director of Bioastronautics, USAF National Range Div., for "outstanding contributions to the advancement of aeronautics and astronautics, his writings, lectures, and research, and his continued dedication to the study of medical problems relating to the aeronaut and astronaut"; and the Robert M. Losey Award to George P. Cressman, Director of the Office of National Meteorological Services, Weather Bureau, for "his rapid and effective application of numerical weather analysis and forecasting techniques to aeronautical operations." (*A&A*, 11/65)

- "The United States . . . has gone into space for many compelling reasons," NASA Administrator James E. Webb told the National Association of Retail Druggists meeting in Washington, D.C. "It is . . . imperative that the U.S. be first in space for reasons of national achievement, for the pride that goes with such achievement, and for the international prestige that accompanies both.

  "We must go into space to reap the benefits of scientific discovery, to stimulate our economic and social progress, to advance our technological advancement, including the civilian application and use of the products of space-oriented research, and to fulfill what has been called the compelling urge to explore and discover. . . ." (Text)

- Within five years aquanauts could conduct systematic explorations of the ocean's floor at depths up to 1,700 ft., Capt. George Bond (USN), chief scientific investigator for the Sealab II project, predicted at a news briefing in La Jolla, Calif. Bond said that the Sealab II aquanauts had performed their tasks so satisfactorily that he could envisage a time when men could live indefinitely in underwater capsules while performing extended work on the ocean floor. Bond said that within a year, another capsule would be placed at a depth of between 400 ft. and 450 ft. for further experimentation. (Bart, *NYT*, 10/13/65, 16)

- Discussing the problems of organizing and consolidating the vast amounts of technical and scientific data at the Congress of the International Federation for Documentation in Washington, D.C., Dr. Eugene B. Konecci, National Aeronautics and Space Council, said: "It is felt by many that the most highly advanced information system in the Federal Government has been and remains the program of the National Aeronautics and Space Administration. This program has been a pace-setter for information advances, both on the part of the Government and private sectors of our economy. The NASA contractors and grantees get the benefit of the most widely-diversified and promptest information service rendered by any government agency. The NASA attempts to cover world space literature. STAR [Scientific and Technical Aerospace Reports] is reproduced semi-monthly and through arrangement with the American Institute of Aeronautics and Astronautics, the world's open literature is covered in the international aerospace abstracts. A total of more than 50,000 items a year are indexed in depth on magnetic tape for use in the NASA computer base literature research system. Tapes are updated monthly and distributed to the NASA Research Centers, a dozen major NASA contractors, and three university research centers that use them in dissemination of technological information to industry. . . ." (Text)

October 12: At a background briefing at NASA Hq. on Advanced Chemical Propulsion, Adelbert O. Tischler, Director, Chemical Propulsion Div., NASA Office of Advanced Research and Technology, revealed that NASA would probably be "confined to the Saturn IB and V for the next decade." A maneuvering vehicle would be developed sometime around 1970 but would not be available for manned use until 1975. Tischler said that an ion propulsion system would never be applicable for deep manned space flight because of "its abhorrent thrust-to-weight ratio . . . it is going to take a long time to accelerate the vehicle out of Earth orbit to arrive then later at the planetary orbit, and to decelerate in the planetary orbit, and vice versa. . . ."

Ion propulsion systems would be suitable for unmanned missions, however, because unmanned systems generally were small and ion systems were small; also, time was not a factor. In addition, they might be practicable as auxiliary devices for midcourse corrections on manned interplanetary trips.

After agreeing that present planning limited the application of nuclear systems to interplanetary trips, Tischler said: "There has been, during the last two years, a real marked decline of interest . . . about exploring Venus. I don't think that is a correct viewpoint at all. I think there is a great deal . . . to learn about Venus.

"We shouldn't eliminate the possibility of some other planets. We could easily . . . visit the asteroid belt. There are some good-sized asteroids. Ceres is one of them. We could ultimately . . . land on the Moon of Jupiter with such a system. I doubt if we will undertake a landing on Jupiter until we know how to fight gravity. . . .

"All of these possibilities exist. In fact, I would even mention Mars has a couple of moons that we might use to land on in preference to Mars itself." (Transcript)

- Philadelphia's Mayor James H. J. Tate, presenting a bronze medal to Astronaut Charles Conrad, Jr. (Cdr., USN), said: "I cannot imagine a more appropriate time to honor a space explorer than on the anniversary of the day back in 1492 when an Italian explorer named Christopher Columbus first set eyes on the New World and made a similar enormous contribution to history." (*NYT*, 10/13/65, 47)

October 12: USA–USAF project to test plastic covering for assault airlift airstrips had been completed at Ft. Campbell, Ky. Armed Forces News Service spokesman said the plastic covering, held in place by special anchoring pins, might prove a practical, all-weather landing-zone surfacing. (*NYT*, 10/12/65)

- USAF Athena missile fired from Green River, Utah, impacted on target at White Sands Missile Range, N. Mex. It was the 25th successful firing in 30 attempts of a 78-shot series to study how warheads behaved during reentry. (AP, *Wash. Post*, 10/13/65, A19)

- Tass announced conclusion of U.S.S.R. launchings in the Pacific Ocean basin of new types of rockets carrying space objects. "The flights of the rockets and the functioning of all their stages proceeded normally. The mockups of the last but one stage of the carrier rockets reached the water surface in the present area with high precision," the announcement said. The test area—80-n.mi.-dia. circle centering on 37°39′ N and 173°25′ E—was now free for navigation and aircraft flights. (Tass, 10/12/65)

- $10-million seismometer post was dedicated in Billings, Mont., at ceremonies attended by more than 60 scientists from 30 nations who would tour the 525 seismometer sites in Montana. President Johnson, in Washington, D.C., said: "This new Department of Defense installation, the first of its kind, will help detect and identify earthquakes. It brings us closer to the day when science can distinguish with certainty between underground nuclear explosions and other forms of seismic disturbances." Scientific data would be made available to the world, Johnson added. (AP, *NYT*, 10/14/65, 3)

- Soviet Cosmonauts Pavel I. Belyayev and Aleksey Leonov fled West Berlin to East Berlin under police escort after demonstrators splattered their limousine with red paint. The cosmonauts had visited West Berlin to attend a rally at the invitation of the West Berlin German-Soviet Friendship Society. (AP, *NYT*, 10/13/65, 1, 3; *Wash. Post*, 10/13/65, A27)

October 13: NASA Administrator James E. Webb, speaking at the dedication of the Space Science Laboratory at Univ. of Chicago, said: "The space effort is really a research and development competition—a competition for technological pre-eminence, which demands and creates the quest for excellence. . . . Should we fall behind in the area of space technology, we would jeopardize our ability to progress, on earth as well as in space. At stake is the technological balance of power in the world. . . .

"The influence of our technological progress and prowess is and has been the deciding factor in keeping the peace; technological and scientific capabilities constitute a basic source of national power. Pre-eminence in the field is an instrument in international relations, and influences our dealings with other nations involving peace and freedom in the world. . . .

"By mastering the restrained but decisive use of the power that modern science and technology give us, we are providing the Nation with the capability to achieve greater cooperation toward a world consistent with the ideals we seek for others as well as ourselves. . . ." (Text)

NASA still assumed that life on Mars was possible, Dr. Norman H. Horowitz, JPL, said at the dedication. Although photographs relayed by MARINER IV had indicated Mars was extremely cold and lacked oxygen and water vapor, some terrestrial organisms might live comfortably in this environment, he said. Horowitz felt that the unexplained seasonal color changes in the dark regions of Mars might indicate plant growth. (Kotulak, *Chic. Trib.*, 10/14/65)

October 13: Aerobee 150 sounding rocket, launched by NASA from White Sands Missile Range to study the ultraviolet radiation of Orion and a group of stars, reached 114-mi. (183.5-km.) altitude and performed successfully. (NASA Rpt. SRL)

- NASA had awarded Thiokol Chemical Corp. approximate $16-million, multiple-incentive-development contract to develop a 100-lb.-thrust rocket engine for use in spacecraft attitude control and maneuvering systems and for launch vehicle ullage and attitude control systems. Designated C-1, the engine would be powered by storable hypergolic liquid propellants. (NASA Release 65–322; MSFC Release 65–262)

- NASA selected Lockheed Electronics Co. to negotiate cost-plus-award contract to provide computer, programing, and operational support services to Manned Spacecraft Center. The approximate $3,000,000 contract would be for one year with provisions for two one-year renewals. (NASA Release 65–323)

- Telecomputing Services, Inc., was selected by NASA for negotiations leading to two-year, cost-plus-award-fee contract for computer operations and data processing services at NASA Goddard Space Flight Center. The contract, expected to exceed $4 million, would include an option for an additional year increasing the value by $2 million. (NASA Release 65–324)

- Former Astronaut John H. Glenn, Jr. (Col., USMC, Ret.), visited Pope Paul VI at Vatican City and accepted a medal commemorating the Pontiff's October 4 visit to the United Nations. (Reuters, *NYT*, 10/14/65, 32; AP, Wash. *Eve. Star*, 10/14/65, A–2)

- The problem of the moon's origin should be solved when Surveyor spacecraft had sent samples of the lunar surface to earth, predicted Dr. Harold C. Urey, Nobel prize-winning chemist, at a news conference during the meeting of the National Academy of Sciences in Seattle. Absence of terrestrial matter in lunar soil would indicate that the moon originated elsewhere in the solar system, Urey said. (AP, Wash. *Eve. Star*, 10/13/65)

- Secretary of Defense Robert S. McNamara, addressing 4,000 employees and officials of Grumman Aircraft Engineering Corp., warned Grumman to reduce contract proposals sharply if it expected to gain a share of future defense contracts. McNamara said Grumman's proposals for construction of a portion of the F–111 fighter's tail assembly were "higher than other competitive sources.

  "There is no question in my mind that you can reduce costs on this $1 billion contract. They've got to come down," he said. The contract he was referring to had not yet been awarded. McNamara, accompanied by Senator Robert F. Kennedy (D–N.Y.) and Asst. Secretary of Defense Arthur Sylvester, was making a brief tour of defense plants on Long Island. (Maiorana, *NYT*, 10/14/65, 41)

October 13: Pre-Continent III, 30-ton sphere in which six men lived and worked 330 ft. beneath the sea off the French Riviera for three weeks, surfaced after its inhabitants had erected and maintained a mock sea-bottom petroleum drilling rig without ever surfacing for air. The men would remain in the sphere until Oct. 16 when the atmospheric pressure in the capsule was expected to be low enough to permit emergence without danger. Object of this experiment, conceived by Jacques-Yves Cousteau, was to avoid need for frequent decompressions by placing workers' dormitories on the ocean's floor to allow more working time and less transition time per diver. Such a technique would permit man to mine minerals and cultivate underwater plants to ease the world's food shortage problem, Cousteau said. Sealab II, a similar experiment by U.S. Navy, was completed Oct. 10. (Reuters, *NYT*, 10/15/65, 12; UPI, *Wash. Post*, 10/14/65, A27)

- $13-million, 140-ft.-wide radiotelescope, one of the largest in the world, was dedicated at Green Bank, W. Va., by the National Radio Astronomy Observatory. It would be able to amplify radio emissions from deep space. (*Wash. Post*, 10/14/65, A15)

- Increase in private flying was reported by FAA, with prediction that pilot certificates issued during 1965 would exceed the previous high of 117,902 certificates issued in 1957. During the first six months of 1965, FAA issued 61,744 pilot certificates, an increase of 22% over the same period in 1964. (FAA Release 65–92)

- In answer to the query "Is a permanent role waiting in space for military astronauts," Robert Brunn in the *Christian Science Monitor* quoted an Air Force official as saying: "No strategy is known for space as yet. We can't use space the way the cowboys and Indians used high ground for advantage, nor the way the Air Force in World War II climbed to altitude and dove out of the sun.

    "Something with military importance may turn up in the next five years. The Air Force doesn't know what the usefulness of space will be. . . ." (Brunn, *CSM*, 10/13/65, 14)

October 14: OGO II Orbiting Geophysical Observatory (Ogo–C) was launched by NASA from Vandenberg AFB with Thrust-Augmented Thor-Agena D booster. Because of a slight malfunction in the launch vehicle's primary guidance system, orbit achieved had the following parameters: apogee, 939 mi. (1,511 km.); perigee, 250 mi. (403.1 km.); period, one hour 45 min.; inclination, 87°. Planned orbit had been somewhat lower: 575-mi. (925-km.) apogee, 207-mi. (333.3-km.) perigee, one-hour 37-min. period, and 86° inclination. Most of the 20 onboard experiments were not expected to be affected by the malfunction; OGO II's solar panels and instrumented booms deployed as planned and the spacecraft achieved three-axis stabilization.

Second of the seven spacecraft programed for NASA's Ogo program, OGO II weighed 1,150 lb. and contained more than 100,000 parts, including a communications system capable of transmitting 64,000 bits per second in real time and 128,000 bits of information per second from tape playback. Its mission was to concentrate on near-earth space phenomena with emphasis on global mapping of the geomagnetic field, measurements of the neutral, ionic, and electronic composition of the earth's atmosphere. Data would be correlated with solar ultraviolet and x-ray emissions, and such events as particle dumping in the auroral zones and airglow. Although it returned good experiment data, ogo II was to be considered a failure (see Oct. 24). (NASA Proj. Off.; NASA Releases 65-314, 65-368; AP; *NYT*, 10/15/65, 14; AP, *Wash. Eve. Star*, 10/15/65, A8; AP, *Wash. Post*, 10/15/65, A4)

October 14: Capt. Joseph Engle (USAF) flew X-15 No. 1 to 266,500-ft. altitude at 3,580 mph (mach 5.08). The flight carried a slightly modified horizon scanner and device for measuring microscopic atmospheric pressure. (NASA X-15 Proj. Off.; *X-15 Flight Log*)

- A second MOLNIYA I communications satellite was launched by U.S.S.R. into a high elliptical orbit with the following parameters: apogee, 40,000 km. (24,800 mi.); perigee, 500 km. (310 mi.); period, 11 hrs., 59 min.; inclination, 65°. Tass said: "With the launching of the second MOLNIYA I satellite, verification will begin of the possibility of organizing a communications system with the simultaneous use of several Sputniks. . . .

  "The main purpose of the launching . . . is to further develop the system of two-way long-distance television and telephone-telegraph radio communication, and its experimental operation. The entire equipment . . . is functioning normally.

  "The second MOLNIYA–I . . . has already been used for telephone calls and an exchange of television programs between Moscow and Vladivostok." (Tass, 10/14/65; *Wash. Post*, 10/15/65, A4)

- Italy's Order of Merit was awarded former Astronaut John H. Glenn, Jr. (Col., USMC, Ret.), during his visit to Rome. He and his wife were also received by Prime Minister Aldo Moro. (Reuters, *NYT*, 10/16/65, 54)

- Saturn V launch vehicle was under consideration by NASA for its Voyager program of unmanned planetary exploration, beginning with a 1971 Mars mission. NASA had considered developing a Saturn IB/Centaur launch vehicle, but the greater thrust of the three-stage 7.5-million-lb. thrust Saturn V would give more flexibility to planning early Voyager missions and would also provide the launch vehicle capability at the beginning of the Voyager program necessary for future missions. Saturn V, being developed by MSFC, would undergo its first flight test in 1967. (NASA Release 65-328)

- NASA would negotiate $60,000 contracts with Consolidated Systems Corp. and Perkin-Elmer Corp. for design studies of an instrument to analyze and control a two-gas atmosphere system for use on manned spacecraft conducting missions of 45 days or more. Manned spacecraft flown to date had used a single gas system—oxygen. After a comparative evaluation of concepts derived from the two four-month studies, NASA might award a second-phase contract leading to construction of one or more prototype atmosphere sensor systems. (NASA Release 65-326)

- NASA was negotiating with General Electric Co. to provide 50-w. isotopic power generator for Apollo Lunar Surface Experiment Packages (Alsep) program. The thermoelectric nuclear power unit, designated Snap-27, would use plutonium-238 as the heat source. The packages would transmit selected measurements back to earth. AEC would manage detailed design and development of the unit based on studies of prototypes by NASA Manned Spacecraft Center. (AEC–NASA Release H-226)

October 14: NASA Wallops Station conducted flag-raising ceremony to dedicate its new international court which would fly the flags of foreign countries with personnel visiting Wallops on any given day. The court was constructed in recognition of the important role that Wallops Station plays in NASA's international program for cooperation in space research. (NASA Release 65-329)

- Soviet astronomers at the Institute of Theoretical Astronomy predicted that the Ikeya-Seki comet, discovered Sept. 18 by two Japanese astronomers, would collide with the sun Oct. 21. U.S. astronomers predicted that the comet would pass about 300,000 mi. from the sun on Oct. 20. (Bishop, *WSJ*, 10/14/65, 1)
- Writing about Surveyor lunar probe in the Washington *Evening Star*, William Hines said: "Surveyor will give science a view of the moon that will be different not just in degree but in kind. A 'soft' landing, with the spacecraft braked by automatic rockets, should enable scientists to learn—rather than infer—much about the lunar surface.

    "Even the first Surveyor . . . should tell a lot and this one will lack most of the cleverly designed scientific instruments of later probes. Equipped with television, the initial Surveyor may settle many points of contention over surface conditions . . .

    "Assuming success with Surveyor 1, scientists are hoping that its findings won't evoke a repetition of the short-sighted clamor for 'economy' that followed the successful flight of Ranger 7 in July 1964 . . . owing to the mistaken impression that when you've seen one moon picture you've seen them all.

    "To suggest that photographic and physical evidence obtained by one Surveyor in one isolated spot could yield results characteristic of the whole moon is senseless. . . .

    "Since the safety of astronauts reaching the moon a few years from now may well depend on the wise selection of their landing site, it would be advisable to get as much first-hand Surveyor-type data as possible and let the nation's best qualified scientists study this information as thoroughly as they know how." (Hines, Wash. *Eve. Star*, 10/14/65, A12)
- The Duke of Windsor toured NASA Manned Spacecraft Center. (NASA Off. Int. Aff.)

October 15: USAF launched Titan III-C from ETR with triple-satellite payload: LCS II radar calibration sphere, OV2-1 radiation sensor satellite, and a metal-ballasted dummy. About the time of the second burn of the transtage engines, the total assemblage exploded into hundreds of fragments and therefore the satellites were not useful beyond the earlier steps of launch vehicle testing. The Titan III-C transtage was to have ignited its engines 10 times, changed course 4 times, and ejected the satellites. Primary mission of the flight was launch vehicle testing, and USAF officials noted most major objectives were met. (*U.S. Aeron. & Space Act., 1965*, 153; AP, Newport News *Daily Press*, 10/17/65)

- Gen. Mark E. Bradley (USAF, Ret.), former commanding general of USAF Logistics Command, was appointed consultant to NASA Administrator James E. Webb on management matters and evaluation of NASA's contributions to supersonic transport development and aeronautical research programs. (NASA Release 65-330)

October 15: NASA awarded Tri-State Roofing Co. $534,817 contract and J. A. Jones Construction Co. $1,130,531 contract for repair of buildings at Michoud Assembly Facility damaged by Hurricane Betsy. Both contractors had begun emergency repairs September 13. (MSFC Release 65-263)

- North American Aviation's XB-70A supersonic bomber reached 2,000 mph and 60,000-ft. in a one-hr. 47-min. experimental flight from Edwards AFB across Calif., Ida., Nev., Utah, and Ariz. The aircraft was powered by six 1,500-lb. turbojet engines, each with more than 30,000-lb. thrust. XB-70A flights were being used to obtain data valuable for design of the supersonic transport (Sst) now in the design study phase. (*NYT*, 10/15/65, 45)

- Vesta, a new French meteorological sounding rocket, was successfully fired for the first time from the Hammaguir Range, Algeria. The single-stage rocket, propelled by liquid fuel, carried an instrumented payload to 118-mi. (190-km.) altitude. (*NYT*, 10/28/65, 74M)

- The quality of radio communications and telecasts between Moscow and Vladivostok via the second MOLNIYA I communications satellite was better than via the first MOLNIYA I, Soviet Deputy Minister of Communications Ivan Klokov wrote in *Izvestia:* "Due to its elongated orbit, during the greater part of the 11 hours 59 minutes while the satellite makes one orbit, it is over the northern hemisphere and insures a reliable link between the most remote points of the U.S.S.R.

    "The orbiting . . . will offer a chance to check the possibility of establishing a communications system based on several satellites. This opens the prospect of a sharp increase of communications time, to the point of making it round the clock. . . ." (Tass, 10/15/65)

- M/Gen. Don R. Ostrander, Commander of USAF's Office of Aerospace Research, retired after 28 yrs. of service. He received the Distinguished Service Medal. (OAR Release 10-1-65)

- Senate passed joint resolution (H.J. Res. 597) authorizing $150,000 appropriation for a memorial to Dr. Robert H. Goddard at Clark Univ. (*CR*, 10/15/65, 26107)

- In a report on the pacing systems of the Apollo Program, the Subcommittee on NASA Oversight of the House Committee on Science and Astronautics concluded that the "general progress of the Apollo program is good at this time with a reasonable expectation of a lunar landing in this decade. . . ." It noted two major pacing items requiring maximum effort: lunar excursion module and command and service module. The report pointed out that NASA had been maintaining the Apollo program schedule within funds authorized and appropriated; "however, additional funds of the order of $200 million over and above the funds requested by NASA for fiscal year 1966 would have provided:

    "1. Increased assurance of meeting the Apollo goal of a lunar landing in this decade;

    "2. Broader latitude in providing system alternatives when engineering difficulties are identified; and

"3. Additional management latitude in utilizing all of NASA resources of personnel, facilities, and equipment in the Apollo effort. . . ." (Text)

October 15: James McCormack (M/Gen., USAF, Ret.), vice-president of MIT and chairman of the Massachusetts Bay Transportation Authority, was elected Chairman and Chief Executive Officer and Chairman of the Board of ComSatCorp by the Corporation's Board of Directors. McCormack would take office December 1, succeeding the retiring Leo D. Welch. (ComSatCorp Release)

- A shift in emphasis in the U.S. missile program "from production to testing, updating, and improving firing, guidance and targeting systems" was reported by Richard Rutter in the *New York Times.* Speculating on the future of the missile program, Rutter quoted B/Gen. H. J. Sands, USAF Space Systems Div.: " 'Now we are in a period of apparent tapering off of missile program activity deploying the final units of the last of our presently operational ballistic systems. On the surface it would appear that having put our instant defenses in place we are coasting to a gradual stop on a plateau of preparedness where we can relax and take a breather.

  " 'The appearance is deceptive. Actually, we are in a critically important germination period of research and development of future missile systems.' " Rutter noted that the shift in the missile program "does not mean . . . that its wings have permanently been clipped or that the scores of companies with a stake in the missile field face a bleak future. The missile era, in fact is still aborning." (Rutter, *NYT*, 10/15/65, 65, 69)

- In an interview with Tass, Soviet rocket designer Prof. Petrovich said that Soviet scientists intended one day to put scientific expeditions on the planets of the solar system to domesticate them. Designers were concentrating their efforts on developing highly efficient new engines which could lift extremely heavy payloads to the prescribed orbits, he added. (Tass, 11/15/65)

- Concorde, British-French supersonic passenger aircraft, would be test-flown in 1968 and ready for service by 1970, reported a spokesman for France's Sud Aviation Co. in Genoa. (AP, *Wash. Post,* 10/16/65, A5)

- NASA Hq. announced appointment of William C. Schneider as Deputy Director, Mission Operations and Gemini Mission Director, effective immediately. Schneider, formerly Deputy Director of the Gemini Program, was replaced by LeRoy E. Day, who became acting Deputy Director of the Gemini Program. (NASA Release 65-331)

October 16: U.S.S.R. launched COSMOS XCII unmanned satellite, equipped to continue studies of the infrared and ultraviolet radiation of the earth's atmosphere. Orbital parameters: apogee, 329 km. (204 mi.); perigee, 199 km. (123 mi.); period, 89.8 min.; inclination, 65.02°. All systems were functioning normally.

Soviet Prof. Aleksandr Lebedinskiy told Tass that two spectrophotometers would point vertically part of the time and at the horizon part of the time. One instrument would cover the spectrum from seven to 20 microns and the other from 14 to 38 microns; it would take 20 sec. to obtain one spectrogram. (Tass, 10/16/65)

October 16: A secret weapon, about which former Soviet Premier Nikita S. Khrushchev had boasted, blew up in 1960 killing the commander-in-chief of the Soviet missile forces, Marshal Mitrofan I. Nedelin, and 300 others, the *Manchester Guardian* reported. The weapon was a missile with a nuclear propellant. The *Guardian* quoted secret reports of Oleg V. Penkovsky, a Russian executed in 1963 for spying for the West, which would be published in November 1965. (*NYT*, 10/17/65, 25; *Wash. Post*, 10/17/65, 1)

October 17: In an original report on communications satellites, Soviet Prof. M. Kaplanov said: "The transmitter on the Molniya 1 Comsat had a power of 40 watts, i.e. 10 times that of the Early Bird. In this regard, the Soviet scientists and engineers have solved the most complex technical problems in the development of a sputnik equipped with onboard antennas oriented on the Earth with a strong power system assuring a supply to the transmitter throughout the entire communication session.

"The development of a wideband space transmitter of 40 watts power operating continuously is an innovation in world technology. This distinguishing feature of the Molniya-1 Comsat permits a considerable simplification of the onground equipment and a great reduction in the cost of its construction and operation. Evidently the great complexity of servicing and the inadequate reliability of the [ComSatCorp] equipment compelled France, England and West Germany to organize the alternate operating of their ground receiving points of space communication. Each of these points operates only one week out of three, the second week it is in reserve, and the third is spent in repair work. We note that the Soviet onground facilities for space communication in Moscow and Vladivostok have been operating daily since the launching of the Molniya-1. This is quite important for a satellite communication system, since upon the operating cost there depends greatly its ability to compete with other means of communication." (*Izvestia*, 10/17/16, 5, ATSS-T Trans.)

- Boris Yegorov, Soviet physician-cosmonaut, said in interview in *Izvestia's* Sunday supplement *Nedelya* that it was quite possible that in prolonged spaceflight the human organism would develop a new complex of protective-adaptive mechanisms which would safely protect it under conditions of weightlessness. At the same time, Yegorov voiced the fear that these new changes might persist and not be easily removed after return to earth. To aid in re-adaptation to earth's gravity, he said future Soviet cosmonauts would use exercises and drugs. (Tass, 10/23/65; Shabad, *NYT*, 10/24/65, 59)

October 17–18: "Galileo," a Convair 990 jet equipped with $3 million of sensitive equipment, took off from Hickam AFB, Hawaii, and raced with the sun to keep comet Ikeya-Seki in view for the longest possible period. A team of 30 scientists aboard the aircraft, under NASA sponsorship, would (1) make spectral observations of Ikeya-Seki's coma and tail in ultraviolet and infrared, and (2) photograph comet's tail in the vicinity of the sun. The former observations would pertain to the composition and structure of the comet; the latter would permit studying the solar corona and solar wind at high inclination angles and would assist research on comet disintegration. (NASA Release 65–332; *NYT*, 10/18/65, 10)

October 18: U.S. should build a supersonic transport regardless of the fate of the Concorde or any other country's effort because "this is the next technological step in air transportation," FAA Administrator William F. McKee told the House Appropriations' Subcommittee on Independent Offices requesting $140 million for the project for FY 1966, according to *Aviation Week and Space Technology.* The House subcommittee responded by approving the money request. Acting Chairman Joe L. Evins (D–Tenn.) said: "Once we have made a basic decision to go forward with this program, I think we ought to go forward with it. This continual question of reservations as to whether we are going to build a supersonic transport is disruptive to continued progress." (*Av. Wk.*, 10/18/65, 28)

- Ikeya-Seki comet might produce a brighter-than-expected spectacle when it swept close to the sun Oct. 20. Cal Tech astronomers said observation through infrared telescopes showed the comet was hot; previously, comets had been believed to be made of space dust and frozen gases that became luminous only when they neared the sun. Spokesman for the scientists said "an explosion of some sort is a possibility." (*WSJ*, 10/18/65, 1)

- Test results were received by NASA which promised a solution to the crawler-transporter bearing failure without jeopardizing the key milestones in the Apollo program. Roller bearings in a crawler-transporter had failed during a July 25 test at KSC and subsequent investigation had defined a new bearing system. (*Cocoa Tribune*, 10/18/65)

- Evidence that the bright areas of the Martian surface contained iron oxides in a loose combination with water was reported by Dr. Carl Sagan of Harvard Univ. to the Biophysical Society's annual meeting in San Francisco. He said reflected light of hydrated iron oxides in an earth-bound laboratory matched the sunlight reflected from the Martian surface when viewed in the ultraviolet and infrared. Dr. Sagan believed that confirmation of the iron-water compounds on Martian deserts, which showed up as bright areas to earth astronomers, suggested conditions were once right for life to originate on Mars. He said he was convinced that when life of a primitive form had started, the chances were it had survived. (*Balt. Sun*, 10/18/65)

- USAF had assured the House Armed Services Committee Military Airlift Subcommittee that the Lockheed C-5A transport would be developed with commercial as well as military requirements in mind. Subcommittee Chairman Melvin Price (D–Ill.) told *Aviation Week and Space Technology* that military airlift had improved dramatically over the last few years but that requirements kept rising with no immediate leveling off in prospect. One area which still needed improvement, he said, was modernizing commercial cargo fleets. He said the subcommittee was hopeful the C-5A would be a big step in this direction. (*Av. Wk.*, 10/18/65, 28)

- NASA's budget for FY 1967 received comment in *Aviation Week and Space Technology:* "NASA is nervously asking the White House for $5.76 billion for Fiscal 1967 and contends it must have at least $5.25

billion compared with the $5.176 billion Congress appropriated for this fiscal year.

"The Apollo program to land a man on the moon appears—thanks to Russian competition—pretty safe from the budget cutters. But every other program is a potential target, including the $1.2 billion Voyager program to explore Mars and all kinds of proposals to exploit the Apollo hardware. . . .

"The Johnson Administration has prepared a pat answer to complaints that its tight Fiscal 1967 budget will thwart new initiatives. With so much accomplished in this session of the 89th Congress, the argument goes, next year must be 'a management year.'" (*Av. Wk.*, 10/18/65, 15)

October 18: British scientist Dr. J. E. Lovelock told the Third International Symposium on Advances in Gas Chromatography in Houston that although pictures taken from space probes indicated no sign of life on Mars as we know it, life may be there. Dr. Lovelock said it was urgent that the unmanned spacecraft that would land on Mars be designed "so they would not rule out the possibility that life may take different forms and be of some entirely different nature on planets beyond our own." (*Houston Post*, 10/19/65)

- DOD's Project Hindsight, intended to isolate improvements in military hardware resulting from DOD's fundamental research programs, was discussed in *Aviation Week and Space Technology*. Hindsight would also identify which in-house laboratories, profit or non-profit firms, or individuals made the most useful technological contributions to military systems. End object was improved management of the research and exploratory development programs. (Johnsen, *Av. Wk.*, 10/18/65, 47)

- Incorporation of a severely swept delta wing on X-15 No. 3 had been proposed by NASA Flight Research Center for hypersonic (mach 5+) flight tests beginning in late 1968, *Aviation Week and Space Technology* reported. Decision on the proposal, under study at NASA Hq., was not expected for six months. X-15 program is a joint USAF-NASA venture, with the three aircraft belonging to the Air Force and NASA operating the research flights program. (*Av. Wk.*, 10/18/65, 22)

- During May 7 flight of USAF-NAA XB-70A No. 1, Michael L. Yaffee reported in *Aviation Week and Space Technology*, five of the experimental bomber's six General Electric YJ93 engines were damaged when the wing apex, a steel honeycomb structure, tore apart and scattered pieces of steel into both sides of the aircraft's divided, rectangular air intake beneath the wing. The No. 5 engine was most heavily damaged, a post-flight examination had shown, but it was the only engine on the right side of the aircraft operating during the half-hour flight back to Edwards AFB. When the apex broke, the XB-70A was flying at mach 2.56. About 30 min. later, NAA test pilot Alvin S. White landed the aircraft with all left-hand engines (1, 2, and 3) operating at full power and engine No. 5 at 90% power. (Yaffee, *Av. Wk.*, 10/18/65, 64–65, 67)

- A gas turbine-powered tugboat, the first ever built, was delivered to NASA's Mississippi Test Facility for use in the canal system there. The

69-ft.-long *Clermont* would perform a variety of chores in connection with future space rocket static firings at the Hancock County, Miss., facility, an element of Marshall Space Flight Center. Main duties would be to berth and tow space vehicle and cryogenic tanker barges in the inland canal system. (MSFC Release 65-264; *Marshall Star*, 10/20/65, 1)

October 18: ComSatCorp had filed a contract with FCC for purchase of four satellites from Hughes Aircraft Co. to supply communication services for NASA's Project Apollo and for other commercial services. Two spacecraft would be placed in synchronous orbits at about 22,300 mi.: one would be located over the Pacific between Hawaii and Midway Island; the other over the Atlantic near the west coast of Africa. Of the other two, one would be used as a reserve for replacement and one as a backup in case of a launch failure. Contract also provided an option for purchase of up to 15 of the satellites. Delivery date of the first satellite would be eight months after ComSatCorp would give Hughes notice to proceed, with one additional spacecraft of the four to be delivered every 21 days thereafter. (ComSatCorp Release)

October 18-19: Scientific results of 15 scientific and technological experiments carried on GEMINI III and IV were presented by the principal investigators during a symposium at the Museum of Natural History in Washington, D.C. Experiments were in weather and terrain photography, bone demineralization, astronomy, communications, and radiation in the spacecraft.

Dr. George E. Mueller, Associate Administrator of NASA for Manned Space Flight, introducing the symposium, noted three important general conclusions to be drawn from the results. First, data on man himself indicates that astronauts can adapt to the space environment, giving great confidence that manned space flight effort can move according to plan without the necessity of introducing special requirements such as artificial gravity. Second, further substantiation of the Mercury program evidences that man can see things from space which are not programmed on instruments to see. Third, the outstanding quality of photographs of earth from space demonstrates the feasibility of a number of significant applications of manned space flight for improvement of life here on earth. (NASA Release 65-321; NASA Rpt. of Symposium)

October 19: NASA launched an Aerobee 150A sounding rocket from Wallops Station, Va., to investigate characteristics of the Ikeya-Seki comet. Reaching peak altitude of 111 mi., the 245-lb. payload, designed by scientists at the Univ. of Colorado and JPL, contained a scanning spectrometer, a filter wheel photometer, and related equipment to obtain spectra of the head and tail of the comet and to measure radiation. Data would assist in analyzing the materials in the comet and would be coordinated with observations conducted by other scientists. (Wallops Release 65-67)

- COSMOS XCIII unmanned satellite was launched by U.S.S.R. Onboard was scientific apparatus for continuing studies of outer space in accordance with the program announced by Tass on March 16, 1962. Orbital data: apogee, 522 km. (324.2 mi.); perigee, 220 km. (136.6 mi.); inclination, 48°24'. All instrumentation was functioning normally. (Tass, 10/19/65)

October 19: Gen. Bernard A. Schriever, AFSC Commander, told a National Space Club luncheon that he had met with NASA Associate Administrator for Manned Space Flight Dr. George E. Mueller and had reached a formal agreement on cooperation in R&D, including Mol experiments. Discussions were also underway, he said, for joint use of NASA and DOD global range facilities—tracking station, communications, and other supporting units—in the Apollo and Mol programs.

"Now we are in the beginning stages of two new ventures, the Manned Orbiting Laboratory (MOL) and the C-5A large transport aircraft. . . .

"Both the MOL and the C-5A have great potential significance for our continuing national security. They also promise to yield a number of 'fallout benefits' of a scientific or economic nature. The knowledge gained from the MOL program, in addition to its possible defense applications, should also prove very useful for future manned scientific ventures into space. The C-5A transport not only will greatly increase our ability to respond to military crisis situations and other emergencies such as earthquakes or floods in remote areas, but also may prove quite attractive to the commercial airlines. . . ."

General Schriever said he felt a follow-on to the B-52 was needed; that an Advanced Manned Strategic Aircraft (Amsa) would have considerably longer range than a contemplated B-111 and would be superior; and that a maneuverable ferry vehicle for supplying Mol could very well come out of the Start program. (Text; AP, *Wash. Eve. Star*, 10/20/65, E8; *M/S Daily*, 10/20/65)

- Dr. P. H. Fang, a researcher at NASA Goddard Space Flight Center, had discovered that silicon solar cells damaged by electron radiation could be completely recovered any number of times when exposed to high temperatures. Report was given at a Photovoltaic Specialists Conference at GSFC Oct. 18-20. (GSFC Release G-28-65)

October 20: NASA Aerobee 150 sounding rocket with four NRL solar-study experiments was launched from WSMR to 115-mi. altitude. Rocket performed well and three experiments functioned as planned, providing good data: ultraviolet spectroheliograms were obtained, spatial detail in Lyman-Alpha light on the solar disk was photoelectrically recorded, and solar Lyman-Alpha flux measurements were made by the ion chamber. The fourth experiment was to have recorded the comet Ikeya-Seki against the corona, but because of a mechanical failure in the white-light coronagraph experiment these data were not obtained. (NASA Rpt. SRL)

- Federal Communications Commission approved live television coverage from the aircraft carrier *Wasp* of Gemini VI recovery in the Altantic Ocean. This would be the first live transmission from a recovery area since the U.S. man-in-space program began. Television signal from the *Wasp* would be sent to EARLY BIRD I which would relay it to the satellite ground station in Andover, Me., for transmission to the three television networks. (*NYT*, 10/22/65, 3)

- NASA announced it would negotiate a contract with the Bunker-Ramo Corp. for installation and operation of a small-scale prototype document information retrieval system. Negotiations would be based on a cost-plus-fixed-fee proposal of $86,000. (NASA Release 65-334)

October 20: USAF awarded Philco Corp., Aeronutronic Div., a $1,200,000 increment to a previously awarded contract for fuzing and arming tests and evaluation of reentry vehicles. (DOD Release 730–65)

- Apollo Spacecraft 009, first Apollo flight spacecraft of the type that would carry three U.S. astronauts to the moon and back, was accepted by NASA Manned Spacecraft Center during informal ceremonies at North American Aviation, Inc.'s, Downey, Calif., facility. Largest U.S. manned spacecraft built to date, Spacecraft 009 included a command module, service module, launch escape system, and adapter. (*Marshall Star*, 10/27/65, 1, 10; NAA *S&ID Skywriter*, 10/22/65, 1)
- Ikeya-Seki comet reached perihelion (closest approach to the sun) with the comet only about 300,000 mi. from the visible solar disk and within a solar radius of the sun's surface. Traveling along an elliptical path that would carry it around the sun and deep into the solar corona, Ikeya-Seki had a visual magnitude of −10, nearly as bright as the moon. It was the brightest comet since the one in 1882 which reached an intensity of −7. (NASA Release 65–332; Osmundsen, *NYT*, 10/20/65, 39; Sullivan, *NYT*, 10/21/65, C23)
- Third annual Albert A. Michelson Award of Case Institute of Technology was presented to Prof. Luis Alvarez, physicist at the Berkeley campus of the Univ. of California. Dr. Alvarez was honored "for the discovery of significant properties of cosmic rays, neutrons, isotopes and nuclear transformations; for leading in the development of quantitative tools for nuclear physics and for pioneering in radar and aircraft landing systems." (*NYT*, 10/21/65, 53)
- Patent for a flying belt capable of propelling its passenger to 350-ft. altitude was granted Robert F. Courter, Jr., flight test engineer for Bell Aerosystems Co. The 155-lb. machine would have three tanks strapped to the passenger's back: two for fuel and one for nitrogen to push the fuel into the fuel tanks. Two handles—one in each hand— would control the steering. Courter envisaged the flying belt of the future as a solution to the commuting problem. (Lardner, *Wash. Post*, 10/20/65, A24)

October 21: Second Aerobee 150 sounding rocket with experiment to obtain measurements of comet Ikeya-Seki was launched by NASA from Wallops Station, Va. The 205-lb. payload, consisting primarily of an Ebert-Fastie scanning spectrometer and an attitude control system for pointing the instrumentation at the comet, was boosted to peak altitude of 117 mi. (Wallops Release 65–69)

- NASA announced it would negotiate a contract extension with IBM for continued support of the Real-Time Computer Complex at MSC. The extension was expected to cost about $80 million and would contain provisions for converting to incentive arrangements. (NASA Release 65–336)
- NASA had awarded a contract to the Univ. of Iowa for preparation of an Injun spacecraft to be used in a dual satellite launch in 1967. The Injun Explorer would be teamed with a 12-foot inflatable Air Density Explorer and flown on a single Scout launch vehicle in the same way EXPLORERS XXIV and XXV were orbited Nov. 21, 1964. The contract, valued at $1,070,488, covered construction of the Injun spacecraft,

preparation and integration of the onboard experiments, and integration with the Air Density Explorer spacecraft as a single Scout payload. Dr. James A. Van Allen of the Univ. of Iowa would be the principal scientific investigator. (NASA Release 65-338; LaRC Release)

October 21: U.K. Minister of Aviation Roy Jenkins and U.S. Secretary of Defense Robert S. McNamara had signed a memorandum of understanding approving a joint project for development of an advanced lift jet engine, DOD announced. Development work would be performed by Rolls Royce, Ltd., and a U.S. contractor yet to be chosen. Engines of this kind would be used for takeoff and landing of V/Stol aircraft. (DOD Release 732-65)

- French and Soviet scientists ended a one-week conference on possible cooperation in space programs, but did not issue a communique. Howard Simons in the *Washington Post* quoted "informed sources" as saying the subject of a French-Russian communications satellite had been raised by the Soviet Union and that the French were lukewarm to the idea but had not rejected it. (Wash. *Daily News,* 10/21/65, 22; Simons, *Wash. Post*, 10/29/65, A6)

- Gen. Curtis E. LeMay, former Air Force Chief of Staff, was awarded the 1965 Collier Trophy, U.S. aviation's highest honor, in a ceremony at the Executive Office Building, Washington, D.C. General LeMay was cited for "development of high performance aircraft, missiles, and space systems which in 1964 significantly expanded the frontiers of American aeronautics and astronautics." (NAA *News*, 10/18/65; Raymond, *NYT*, 10/22/65, 1)

- Distinguished Lecture Series sponsored by the Metropolitan Washington Board of Trade was opened by Dr. James A. Van Allen, head of the Univ. of Iowa's physics and astronomy department. Dr. Van Allen discussed "Space Science: Past, Present, and Future" at the Univ. of Maryland. Series of seven lectures was designed "to foster better understanding between industry and local universities in an effort to create a research community in the Washington [D.C.] area equivalent to the Harvard-MIT complex in the Boston area." (*Wash. Post,* 10/3/65, B2)

- Evaluation of the Soviet space effort should be done in terms of Russian rather than U.S. requirements and considerations, suggested William Hines in the Washington *Evening Star:* "To look at things from the Russian angle for a change may provide a new view of the whole space race." Hines suggested that the orbits of the Molniya satellites were egg-shaped because with 12-hr. periodicity "the Molniya-type satellite stays over the homeland about two-thirds of the time and comes back at exactly the same time each day." He said that the U.S.S.R. probably had no weather satellites "because it does not need this branch of the weatherman's art." A possible reason that there were no polar-orbiting Soviet photographic reconnaissance satellites like the U.S.'s Samos was that they were not needed. No Saturn V-type launch vehicle development was evident, Hines concluded, possibly because "Russia's approach to [lunar landing] is not necessarily . . . the same as America's. . . . A revealing hint along this line is contained in a recent issue of Moscow News. . . . The author puts forward the suggestion that a lunar landing and return mission might be organized using two spacecraft rather than one, each of which could be launched by a rocket smaller than Saturn V." (Hines, Wash. *Eve. Star*, 10/21/65, A14)

October 21: 1965 Nobel Prize for physics was awarded jointly to Julian Schwinger of Harvard Univ., Richard Feynman of Cal Tech, and Shinichero Tomonaga of Japan. The three scientists were cited for "their fundamental work in quantum electrodynamics, with deep-ploughing consequence for the physics of elementary particles." The Nobel chemistry prize went to Prof. Robert Burns Woodward of Harvard for "his meritorious contribution to the art of organic synthesis." (*Wash. Eve. Star*, 10/21/65, A4)

- Technical program of the first autumn meeting of the National Academy of Engineering, held in New York, was a Symposium on Ocean Engineering. (NAE Release)

October 22: Ten areas on the moon had been selected by NASA for planning photography by the Lunar Orbiter spacecraft next year. Areas included examples of all major types of moon surface to permit assessment of their suitability for spacecraft landings. Nine of the sites were within the area proposed for Apollo manned landings; eight were potential sites for Surveyor softlanding spacecraft. The 10 areas were located along the moon's equator beginning at about 43° east longitude and stretching to 65° west longitude. (NASA Release 65-335)

- In a letter transmitting to Congress NASA's 12th *Semiannual Report*, covering the period July 1–Dec. 31, 1964, President Johnson said: "In 1958, it was my privilege to introduce the legislation to create the National Aeronautics and Space Administration. I stated then: 'I confidently believe that the developments of the Space Age will bring the beginning of the longest and greatest boom of abundance and prosperity in the history of man.'

  "Time is bearing out that belief." (*Pres. Doc.*, 11/1/65, 430)

- Sen. Karl E. Mundt (R–S.D.) inserted in the *Congressional Record* the letter of transmittal accompanying a report on trends in appropriations for Federal departments and agencies for Fiscal Years 1967–70, prepared by the Library of Congress. The table accompanying the letter indicated the NASA appropriation for 1967 would be $5.4 billion; for 1968, $5.7 billion; for 1969, $6.1 billion; for 1970, $6.4 billion. (*CR*, 10/22/65, 27360–61)

- Sen. Philip A. Hart (D–Mich.) introduced S. 2715, a bill to establish a Government patent policy. He said: "The evidence vividly demonstrates that the Government's research undertakings yield a great many inventions. Between 1945 and 1962 Government-financed R&D produced more than 40,000 patentable inventions. Nearly one-third were patented in the 4-year period ending in 1962. The great bulk of these inventions were made by private contractors whose research efforts were supported by the Federal Government.

  "Each day the problem—judged in quantitative terms—is becoming more serious. Let me give an illustration. In January 1963 . . . NASA . . . reported that its work, conducted both in Government laboratories and private facilities, had led to 786 inventions. By August 1964 that number had increased to 2,500. And by May 1965

—in less than 9 months—the number had doubled to nearly 5,000." (*CR*, 10/22/65, 27127)

October 22: Jack G. Webb was named director of the Federal Aviation Agency's National Aviation Facilities Experimental Center near Atlantic City, N.J., headquarters for FAA's research and development activities. (FAA Release 65-97)

- Concern regarding the secrecy shrouding the Manned Orbiting Laboratory (MOL) program received editorial comment in the *Washington Post*: "The President has given assurances that although information gained through MOL will relate to America's defense capability, the thrust of the project is peaceful. On the other hand MOL's director seems quite willing to provoke a military race in space. There is one easy, proven method by which the doubts and misgivings about MOL can be allayed. The Air Force should adopt NASA's open public information policies at once and apply them to every stage of MOL." (*Wash. Post*, 10/22/65, 24)

- Address by James G. Allen of the Univ. of Colorado before the American Astronautical Society on the impact of space exploration was inserted in the *Congressional Record* by Sen. Peter H. Dominick (R–Colo.): "The space revolution of the mid-20th century must be regarded as the most significant of all those great revolutions of history which have affected the fate of man. By definition, a revolution does produce a series of pyramiding effects, one building up on another. But, in the last analysis, these effects focus upon, and culminate in, truly significant changes in the social and economic relationships which shape one society.

  "The space revolution of the mid-twentieth century thus is unique, as its course and direction will affect every individual on each of the five continents. Moreover, its effects will lie in an entirely new dimension—beyond the physical realm of man's earthly existence and into the boundless areas of space itself. The space revolution of our day has resulted from the theories, hypotheses and observations of the scientists as their conclusions were tested and applied by the eingineers and technicians." (*CR*, 10/22/65, 27386–88)

- Editorial by Henry Eyring, Univ. of Utah, in *Science:* ". . . The crash program on the atomic bomb grew out of groundless fears that our antagonists would get the atomic bomb first. The vast sums being spent at present on a crash program for an early landing on the moon have their own somewhat obscure, psychological basis. If the moon program is really the most effective means of staving off all-out war, expensive as it is, it is still a bargain. On the other hand, the attempts which are sometimes made to sell the moon program on its scientific merits alone, in competition with other scientific uses of the money, are less convincing. The charitable conclusion is that in public affairs it is deemed necessary to oversimplify actual objectives so that the general public will best serve its own interests for the wrong reasons. This oversolicitude is probably neither necessary nor desirable. . . ." (*Science*, 10/22/65, 439)

October 23: Dr. Charles A. Berry, chief physician for the astronauts, said in an interview with the *New York Times* that spaceflights had caused subtle, temporary changes in the human body that must be watched with care for possible importance to longer future flights. Measurements of the apparent density of heel and finger bones of the GEMINI IV astronauts, who flew for four days, and the GEMINI V astronauts, after an eight-day flight, had indicated that calcium was being lost, or at least shifted to some other part of the body. "We have really found no gross changes," he said. "In my opinion, we are going to see the body adapt to space flight. You are never going to have the guy's bones turn to jelly so that when he stands on his leg it just goes 'pfft.'" (Clark, *NYT*, 10/23/65, 58)

October 24: OGO II, NASA geophysical observatory satellite launched Oct. 14 at Western Test Range, ceased normal operation about 24 hrs. after its attitude control system gas supply was depleted. Gas depletion was associated with automatic maneuvers to restabilize OGO II because one or more of its horizon scanners tracked temperature gradients (possibly cold clouds) near the earth's surface rather than the horizon itself. Although NASA considered the mission a failure, OGO II had provided valuable data from 19 of its 20 onboard experiments. (NASA Release 65–368; NASA Proj. Off.)

- Lt. Gen. Nikolai Kamanin, commander of Soviet cosmonauts, said during a meeting with Moscow writers: "After the flight of GEMINI V, Johnson said officially that the United States was inviting representatives of the Soviet Union to the next flight so that they could see U.S. rockets and rocket devices. We told him: Thank you but we do not intend to visit your launching site. We do not intend to go because we know all too well that all space devices are as a rule launched into space with the aid of military rockets. You do it this way and we do it that way. While military rockets are used for these purposes we are unable to show you our Soviet rockets because we know all too well that our rockets were, are, and we are convinced, will be the most powerful, with the greater distance and load capacity. This is of decisive importance for the defense of the country." (Tass, 10/24/65)

- In an article in the *New York Times* discussing the "marriage" of medicine and electronics, William D. Smith said the monitoring of the astronauts' health while in flight had produced several electronic medical systems. He added that NASA was expected to spend about $50 million during 1965 on electronic medical equipment. (Smith, *NYT*, 10/24/65, F1)

October 25: Gemini VI, scheduled to be launched from Eastern Test Range, with Astronauts Walter M. Schirra (Capt., USN) and Thomas P. Stafford (Maj., USAF), was canceled by failure of the Agena rocket, vehicle with which Gemini spacecraft was to rendezvous and dock. The mission was to have been the fourth manned flight and the first rendezvous and docking mission in the Gemini program. Simultaneous countdown of both the Titan II-Gemini 6 and the Atlas-Agena had been developed to maximize launch reliability for a "same day" launch.

Atlas-Agena liftoff was at 10:00:04 EST. The Agena Target Vehicle separated from the Atlas at 10:05:12 with all sequences and parameters nominal at separation. Following a nominal Agena coast period, loss of Agena telemetry and radar beacon track occurred at 10:06:20 after initiation of primary propulsion system burn. Attempt to establish radar or telemetry track by the Canary Islands tracking station and subsequently by the Carnarvon station was unsuccessful. A hold was called at T-42 in the Gemini 6 countdown to determine the status of the Agena Target Vehicle. Mission was terminated at 10:54 a.m. because the Agena Target Vehicle had failed to achieve orbit. Formal Gemini VI mission failure investigation was immediately initiated. (NASA Release 65-237; *Marshall Star*, 11/3/65, 1; Hines, Wash. *Eve. Star*, 10/25/65, 1; Clark, *NYT*, 10/26/65, 1)

October 25: Dr. Michel Bader, NASA Ames Research Center scientist, reported that comet Ikeya-Seki had apparently suffered no structural deformation as a result of its close passage to the sun October 20. The tail of the comet now appeared to be about 12° (25 million miles) in length. (ARC *Astrogram*, 10/28/65, 1)

- ComSatCorp announced agreement on a $4,512,772 contract with Page Communications Engineers, Inc., for three transportable earth stations to be used as links in providing communication services for Project Apollo. Contract was filed with the Federal Communications Commission in conformity with FCC rules requiring 10 days notice for such awards. ComSatCorp also had filed with the Commission a related application seeking authority to construct the stations and asking FCC approval of their technical characteristics. (ComSatCorp Release)

- Command Module of Apollo spacecraft 009 was delivered to KSC from North American Aviation's Downey, Calif., plant aboard a C-133B aircraft. (KSC Historical Office)

- Sir William Hildred, Director General of the International Air Transport Association, said in an address at the Association's 21st annual meeting that proposed supersonic jet airliners would face stiff competition from giant subsonic jets expected on international airways by 1967. Sir William said the airlines favored supersonic airliners, such as the planned British-French Concorde, if they could provide faster transportation "without insuperable financial, technical or operational problems." He said the Concorde was likely to be available in 1971–72, and its American counterpart in 1973–74. But before that, Sir William continued, giant air buses, "stretched" subsonic jets seating up to 250 passengers, would be available. (*NYT*, 10/25/65, 66)

During week of October 25: First two H-1 rocket engines uprated to 205,000 lbs. thrust were delivered from Rocketdyne's Neosho, Mo., plant to Chrysler Corp.'s Space Division in New Orleans. Later, at the Michoud Assembly Facility in New Orleans, H-1 engines would be installed in Saturn IB vehicles to be used in the early stages of the Apollo program. The uprated engines would add a total of about 40,000 lbs. thrust to NASA's Saturn IB booster, increasing the total thrust of the eight-engine cluster to 1,640,000 lbs. (MSFC Release 65-269)

October 26: A "catastrophic failure" had occurred some 10 min. after launch of the Atlas-Agena for the Oct. 25 Gemini VI mission, M/Gen. Ben I. Funk, commander of the Air Force Space Systems Div., revealed. M/Gen. Vincent Huston, commander of the Eastern Test Range, said radar at Patrick AFB had "picked up five pieces" of debris at the point in space where the Agena was supposed to be. G. Merritt Preston, NASA, remarked that just before ground stations lost contact with the Agena, telemetry records from the spacecraft showed a marked rise in pressure in both propellant tanks. Despite the inferred explosion of Agena, no fragments were recovered from the Atlantic Ocean where tracking radar screens had indicated the rocket should have fallen. None was seen by planes alerted to watch for their reentry.

Lockheed Missiles and Space Co., manufacturer of the Agena, had scheduled a news conference to try to explain the problem, the Washington *Evening Star* reported. Conference was canceled, reportedly on instructions from the Air Force, and a secrecy lid imposed. (Wash. Eve. Star, 10/26/65, A6)

October 26: S–IB–3, Saturn IB launch vehicle's first stage, was successfully captive fired for 2½ min., its full flight duration, at MSFC. The stage was powered by eight Rocketdyne H–1 engines, developing a total of 1.6-million pounds thrust. (MSFC Release 65–267)

- NASA reported that efforts to regain control of OGO II—the nation's second orbiting geophysical observatory—had failed: "OGO II is assumed to be tumbling in orbit. The prospect of ever achieving useful operation of spacecraft under present conditions is slim." The electrical power supply of the 1,150-lb. spacecraft had been practically depleted. (AP, *NYT*, 10/27/65, 19)

- U.S.S.R. believed it was possible to land men on the moon and bring them back to earth with the same rocket, Soviet cosmonaut Lt. Col. Andrian G. Nikolayev said in Tokyo. Colonel Nikolayev and his wife, Valentina Tereshkova, also a cosmonaut, were on a two-week visit to Japan. (Reuters, *NYT*, 10/27/65, 19)

- Commenting on cancellation of the Gemini VI mission, an editorial in the *New York Times* said: "The Atlas-Agena rocket has performed so well on so many different missions in the past that its failure to achieve orbit—and the consequent impossibility of the planned rendezvous and docking experiment—comes as a sharp disappointment. But it will be worth while if it destroys the tendency toward complacency that has been increasingly visible recently. The difficulties that still lie ahead of the effort to land a man on the moon are far greater than those that have been surmounted to date." (*NYT*, 10/26/65, 42M)

- U.S. would put a supersonic airliner into the skies no later than 1974, William F. McKee, Federal Aviation Administrator, told the Aero Club of Washington. He said FAA had a target date of no later than 1974 for finishing all tests before regular passenger flights. (UPI, *NYT*, 10/27/65, 95)

- Describing the size, capacity, and productivity of the C–5A aircraft, recently approved and funded by DOD and ordered by the Air Force from Lockheed Aircraft Co., Ira C. Eaker said in the *San Diego Express:* "A fleet of 130 C–54 aircraft, the best then available, were required for the Berlin airlift. Five C–5As could have done that job more efficiently and with a saving of 5,000 crew and support personnel." (Eaker, *San Diego Express*, 10/26/65, 29)

October 27: NASA named an Agena Review Board to try to identify the causes of the failure which prevented the Agena stage from fulfilling its mission in the Gemini VI flight Oct. 25. Board would be headed by co-chairmen Dr. Robert R. Gilruth, Director of NASA Manned Spacecraft Center; and Maj. Gen. O. J. Ritland (USAF), Deputy Commander for Space, Air Force Systems Command. (NASA Release 65-342)

October 27: X-15 No. 3 was flown to 236,900-ft. altitude and top speed of 3,477 mph (mach 5.06) by NASA research pilot John B. McKay to obtain data on the NSL scanner, boundary layer noise, and horizontal stabilizer loads. (NASA X-15 Proj. Off.; *X-15 Flight Log*)

- Service module of Apollo spacecraft 009 was delivered to KSC from North American Aviation's Downey, Calif., plant aboard the Pregnant Guppy aircraft. (KSC Historical Office)

- U.S.S.R.'s photographic moon probe, ZOND III, was 33.4 million miles (52.8 million km.) from earth. Transmission of photos of the moon's hidden side had been resumed according to the flight program. Pictures were of good quality. (Tass, 10/28/65)

- Two workmen were killed and three were injured in an explosion and fire in the propellant preparation building at Thiokol Chemical Corp.'s Minuteman missile facility. The accident occurred at Minuteman Plant 78, an Air Force facility operated by Thiokol. An investigation was underway to determine the cause. (AP, *NYT*, 10/28/65, 43M)

- USAF had awarded Radiation Inc. a $1,044,120 increment to an existing contract for modification of automatic tracking telemetry antennas used is space tracking at ETR. (DOD Release 759-65)

- Soviet deputy minister of communications I. V. Klokov, commenting for *Izvestia* on orbiting of the second MOLNIYA I comsat, said: "The orbiting of the second Soviet Comsat 'Molniya-1' will permit us to verify the feasibility of organizing a communication system envisaging the combined use of several sputniks. This will offer the possibility of a marked increase in the duration of contact to the point of becoming a round-the-clock system. . . .

    "With the aid of the new sputnik, the first telephone communication sessions have already been conducted; Moscow and Vladivostok have exchanged the first TV programs. These tests have shown that the quality of the radio communications and telecommunications is superior to that found in the experiments with the first Soviet Comsat, Molniya-1." (*Izvestia*, 10/16/65, 3; ATSS-T Trans.)

- Aerospace and defense planning concepts could be applied to many of the Nation's major social problems, Dr. Ruben F. Mettler, executive vice president of TRW, Inc., told a luncheon meeting of AIAA in New York. Dr. Mettler listed transportation, medical services, and housing as areas where the systems approach to problems would prove beneficial. He said that he thought the application of aerospace systems concepts had a bright and profitable future in civilian endeavors. (*NYT*, 10/27/65, 74M)

- Vice Adm. Hyman G. Rickover (USN), in a London lecture sponsored by the British Association for the Advancement of Science, warned against uncontrolled exploitation of science that "may become a Frankenstein monster, destroying its creator," man. He listed misuses of new technology, including one that may be irreparable: "We may be damaging the atmosphere permanently by changing its chemical composition." (UPI, *NYT*, 10/28/65, 39)

October 28: Soviet Union launched COSMOS XCIV artificial earth satellite carrying scientific instrumentation to continue the space exploration program. Orbital parameters: apogee, 293 km. (181.9 mi.); perigee, 211 km. (131 mi.); period, 89.3 min.; inclination, 65°. Equipment was operating normally. (Tass, 10/28/65)

October 28: USAF launched Thor-Agena D launch vehicle from WTR with an unidentified satellite. (*U.S. Aeron. & Space Act.*, 1965, 153)

- In a memorandum report to President Johnson on Gemini VI NASA Administrator James E. Webb said: "This is to report to you that the Titan II booster which we expected to use on October 25 to launch Gemini 6, carrying astronauts Schirra and Stafford, is now being removed from the launching pad. We have examined carefully the question of whether this booster could be used for the launching of Gemini 7 into a 14-day orbit, and our studies show that the Titan booster which we have especially prepared for Gemini 7 is more suitable. This is the reason for the change from the booster now on the pad to the one especially designed for the Gemini 7 launch.

  "Also, we have examined a number of ways to speed up the accumulation of the information which the Gemini 6 rendezvous flight was designed to give us. We find that it may be possible to take advantage of the work we have already done in mating the Gemini 6 to its booster and to the launching facility and thus save considerable time in its re-erection. If we can launch Gemini 7 without serious damage to the launching pad, there is some possibility that we could immediately re-erect the Gemini 6 spacecraft and booster and launch it in time to rendezvous with Gemini 7 before the 14-day flight comes to an end." (Text; *Pres. Doc.*, 11/1/65, 734)

- Presidential Press Secretary William D. Moyers announced from the Texas White House that the U.S. would launch Gemini 6 and Gemini 7 about 10 days apart and have them rendezvous in space. The double launching would probably take place in January with the two spacecraft scheduled to maneuver within a few feet of each other but without touching. Astronauts Walter M. Schirra (Capt., USN) and Thomas P. Stafford (Maj., USAF) would be in Gemini 6, whose Oct. 25 mission was canceled after an Agena vehicle with which they were to dock exploded in space. In Gemini 7, set for a 14-day orbital journey, would be Astronauts Frank Borman (Maj., USAF) and James Lovell (LCdr., USN). (Pomfret, *NYT*, 10/29/65, 1; Chapman, *Wash. Post*, 10/29/65, A1)

- Dr. Robert R. Gilruth, Director of NASA Manned Spacecraft Center, said in a telephone interview with the *New York Times* that the proposal to fly the Gemini 6 and Gemini 7 spacecraft on a dual flight originally had been made by Walter F. Burke and John F. Yardley of McDonnell Aircraft Corp., the spacecraft's manufacturer. This might permit rendezvous, but no docking, earlier than if NASA waited for the Agena malfunction to be found and corrected. If the launching pad, the Titan booster, and Gemini 6 could not be made ready in time to catch the Gemini 7 in orbit, "we would not have lost anything but the trying," Dr. Gilruth said. As for the risk of having two manned spacecraft flying a few feet apart, "That's no more dangerous than two fighters up there flying with each other," he added. (Clark, *NYT*, 10/29/65, 13)

October 28: Dr. Edward C. Welsh, Executive Secretary of the National Aeronautics and Space Council, said in an address to the American Ordnance Association: ". . . In the case of the most recently announced space project, the Manned Orbiting Laboratory, we again have an example of a highly valuable exchange of technology and experience by two operating agencies of the Government. In this instance, NASA's considerable success in manned space flight and in the development of spacecraft will assist the Air Force substantially in carrying out the Manned Orbiting Laboratory project. Such interagency cooperation will tend to improve rather than impair the peaceful image which this country has established.

"Since I have mentioned the Manned Orbiting Laboratory, it is worth pausing right now to challenge forthrightly those who have asserted or intimated that it has something to do with a weapons race. We expect misinterpretations of that sort to come from unfriendly countries and sometimes from ignorant domestic critics. However, I was disappointed to find that a few otherwise well informed publications and invididuals have asserted that the MOL is a weapons carrier and a project contrary to our peaceful progress in space.

"I assert as positively as I can that the MOL is *not* a weapons system, is *not* a means by which aggressive actions can be perpetrated, and is in *no* way in conflict with the established peaceful policies, objectives, or methods of the United States. *Rather*, it is a program that will increase our knowledge of man's usefulness in space and will relate that ability to our national defense." (Text)

- Months-long breathing of pure oxygen at the pressure used in Gemini spacecraft might damage the lungs and interfere with blood cell manufacture in the body, reported Col. Harold V. Ellingson (USAF) at a meeting of the American College of Preventive Medicine in Chicago. For that reason, he said, pure oxygen would not be used in Air Force manned orbiting laboratories in which astronauts would remain in space 30–90 days. Instead a mixture of oxygen and helium was being considered. Colonel Ellingson, Commander of the Air Force School of Aerospace Medicine, Brooks AFB, emphasized he was not referring to brief orbital trips such as the Gemini flights, but to missions of one to three months duration. (Lewis, *Wash. Post*, 10/28/65, G3)
- A Group Achievement Award was presented at NASA Langley Research Center's annual awards ceremony to Eugene Schult, Head of the Scout Project Office, in recognition of "the outstanding Scout vehicle success record during the past eighteen months." (LaRC Release)
- Recording of powerful radio waves by U.S.S.R.'s instrumented space probe ZOND II had been reported to a conference of Soviet astronomers by Vyacheslev Slish, according to Tass, which said the astronomer had no "plausible theory" to account for the radio beam, which was said to have been 100 times stronger than anything expected from man's present knowledge of space. (*NYT*, 10/29/65, 13)

October 29: NASA had selected Philco Corp. for negotiation of an estimated $1 million contract to build magnetometers to fly on interplantetary Pioneer spacecraft. The instruments would survey the interplanetary magnetic field during 1967 and 1968. (ARC Release 65–24)

- A new telescope on Mt. Wilson, after surveying more than half the sky visible from that site, had detected from 400–1,000 celestial objects cooler than 1,500° F, reported Dr. Bruce T. Ulrich of Cal Tech at a meeting of the American Physical Society in Chicago. Since between 70% and 80% of them lay along the Milky Way, it was theorized that they were "very large—probably thousands of times larger than the sun, very distant, and quite rare."

The new telescope was novel in several ways: its concave mirror was of an epoxy plastic coated with aluminum; the mirror rocked back and forth at 20 cps, so that light from an infrared source in the sky would sweep across detectors of lead sulphide and silicon; the detectors, kept at $-320°$ F, were sensitive enough to detect the slight increase in the infrared as the arm of the telescope swept past one of these objects. (Sullivan, *NYT*, 10/30/65, 28C)

October 29: American Newspaper Publishers Association asked the FCC to permit wide use of communications satellites by newspapers and wire services. In a brief filed with the Commission the publishers' group said: "Specifically, we propose that a basic policy determination be made at this time, assuring access by the press to Comsat facilities for news media determination." (*NYT*, 10/31/65, 22)

- Soviet Union announced new rocket tests in the Pacific Ocean west of Hawaii and asked all countries to leave the area during the testing period. An official announcement by Tass, the Soviet press agency, said "a new modification of a space vehicle booster" would be launched. Tass said the test area covered "a circle with a radius of 40 n. mi., with a center of 19°8′ north and 178°46′ west." The test firings would take place between October 31 and December 30. (*Pravda*, 10/30/65, 1, ATSS–T Trans.)

- U.S. exploded an 80-kiloton hydrogen bomb 2,300 ft. below the surface of Amchitka Island. The experiment was expected to produce important data for the monitoring of future bans against nuclear testing. The experiment, called Project Longshot, had required two years of preparation at a cost of $10 million. The readings of shock waves at 211 stations throughout the world were expected to enable scientists to distinguish between manmade explosions and natural seismic disturbances. (UPI, *NYT*, 10/30/65, 1)

- William F. R. Ballard, chairman of the New York City Planning Commission, had proposed in a letter to Robert C. Weaver, administrator of the Federal Housing and Home Financing Agency, that a major study be made of land use problems around New York metropolitan airports. Ballard sought Federal aid for the proposed study, which called for an inquiry into methods for reducing disturbances to home owners rising from the noise of aircraft, under the Housing and Urban Redevelopment Act. (*NYT*, 10/30/65, 14)

- Lt. Frank K. Ellis (USN), who had lost his lower legs in an aircraft crash, was advised that the Navy was considering his application to become an astronaut. "The only difference between me and any normal man," he said, "is running and jumping ability. There is no change in my flying ability . . . I've wanted to be an astronaut ever since I first heard the word. That field is moving more and more into actual controlled flying. I'm a fly boy. Count me in." (AP, *NYT*, 10/30/65, 28C)

- First full-scale testing of a Coralie engine took place at Vernon, France,

15 days ahead of schedule. Firing of the engine, second stage of the European Launcher Development Organization's ELDO Europa booster, lasted 96 sec. Compatibility of the four-nozzle liquid engine with the flight-type structure was checked out, and the low level of vibration was called highly encouraging. (*M&R*, 12/6/65)

October 29: Dr. Philip H. Abelson, editor of *Science*, editorialized:

"To date the purely scientific results from our manned space program have not been impressive. With good reason, the engineering and medical aspects have been given overriding priority. In effect, our manned space program has consisted of a series of great technological stunts. . . .

"Will the Space Agency be able to devise a continuing series of spectaculars of ascending dramatic quality? I think not. The first successful landing on the moon will be a climax. Just as succeeding climbs of Mt. Everest, after the first ascent, have drawn diminishing attention, later lunar travel will lose its novelty.

"As for Mars, how many popular songs have been written about it? . . . More fundamental is the question, 'How many people know where Mars is, or even care?' Perhaps man will one day go to the planet, but the psychological and emotional impact of the trip will be pale in comparison with that of the first successful landing on the moon." (*Science*, 10/29/65, 557)

October 30: U.S. Geological Survey, on behalf of NASA, had prepared geologic interpretative maps of approximately 3,000,000 sq. mi. of the moon's surface. These maps, prepared by astrogeologists at Flagstaff, Ariz., were part of a program to aid manned exploration of the lunar surface. Dr. Eugene Shoemaker, chief of the Astrogeology Branch, had been designated by NASA as the principal scientist for Project Surveyor spacecraft. He would make a geologic analysis of photographs of the lunar surface which would be made during a "soft" landing of instruments. (*NYT*, 10/30/65, 14)

- NASA Administrator James E. Webb, in dedicating Boeing's new Space Simulation Facility at Kent, Wash., said NASA planned to increase its aeronautical research. Webb pointed out that while the industry-Government relationship sometimes appeared to be that of vender and buyer, it was actually a partnership with all of its problems. Significant influences on this partnership in recent years had included: the demand on industry for faster rates of technical advance; the increased complexity and technical difficulty of major programs with consequent delays and cost overruns; the decreasing volume of production work and increasing volume of research and development contracts; the steady increase in the requirements for technical and program management personnel; the requirement for Government to better define its objectives and requirements; the emphasis in the procuring agencies on increasing competition at all stages, including research and development; changes in contracting methods which offered more incentives but imposed more risk on contractors; and necessary increases in Government controls on configuration, quality, and on financial data in multiple contract, large and long-lead-time projects. (Text)

- Application to "Spudnik I" — a potato in orbit — of his theory of biological rhythms was explained by Prof. Frank A. Brown, Jr., biologist at Northwestern Univ., in an interview with the *New York Times*. About two thirds of the way to the moon, the earth's magnetic field and other earth-related forces that might help run the biological clock, including gravity, electrostatic fields, and barometric pressure, would be absent, he said. Suggesting that NASA orbit a potato, he added, "If the potato dies, it [the space program] better be checked before a man is sent out there." (Wehrein, *NYT*, 10/31/65, 73)

October 30: U.S. would install new equipment costing an estimated $5 million at one of its three satellite tracking stations in the Pretoria area of South Africa. Arrival of the expensive new equipment was taken by observers as a sign that NASA officials were determined to hold on to the stations. Possibility that they might be closed had been raised last summer when Prime Minister Hendrik F. Verwoerd had said American Negroes could not be assigned to the stations. American officials insisted they had accepted no racial restriction on their personnel in South Africa. (*NYT*, 10/31/65, 12)

• Telephone calls between two Moscow city exchanges would be routed soon over a three-mile-long laser beam, the newspaper *Trud* reported. The laser link, developed by the Soviet Central Scientific Research Institute of Communications, was already in experimental use. (*NYT*, 10/31/65, 26)

October 31: A new theory stating that molecules adsorb on the surfaces of ionic (electrically-charged) surfaces by unbalanced electrical fields of force had been developed at the Jet Propulsion Laboratory under sponsorship of the NASA Office of Advanced Research and Technology, NASA announced. Knowledge of the exact mechanics of gas adsorption would be vitally important in innumerable scientific, industrial, and medical fields. (NASA Release 65-340)

During October: JPL Director William H. Pickering, describing MARINER IV's Mars mission in *Astronautics and Aeronautics*, wrote: "This mission has proven to be of immense scientific and engineering importance. New scientific information is now available on regions of the solar system never before penetrated with instruments. Observations from the vicinity of Mars suggest entirely new concepts about the nature of that planet. Spacecraft performance has shown our ability to design and construct a remotely operated device of astonishing complexity. Its continued operation establishes a standard of reliability seldom, if ever, attained. Even maintaining two-way communications over distances exceeding 100-million mi. remarkably demonstrates advances in communication technology which were not thought possible a decade ago.

"The design concepts underlying Mariner date back to 1959 when the Jet Propulsion Laboratory began the Vega program. In 1960, Vega was modified to become Ranger. The Mariner series took its present form in 1961, when a mission to Venus was planned for the planetary opportunity in August 1962. . . ." (Pickering, *A&A*, 10/1965, 20-21)

• Dr. Charles S. Sheldon, National Aeronautics and Space Council Staff, in NASC compilation of international space programs, predicted that future Soviet activities in space would include: manned circumlunar flight; development of a Soviet version of Mol; large manned stations maintained for indefinite periods by supply ferries; manned lunar landing; major new version of an interplanetary probe, notably on using Venus as a target this fall. (Normyle, *Av. Wk.*, 10/11/65, 32)

During October: USAF had formed a seven-member, top-level policy committee to function as chief advisory group of manned orbital laboratory (Mol) objectives and problems, reported *Aviation Week and Space Technology.*

Chaired by Air Force Secretary Harold Brown, the committee would establish program objectives, plans, schedules, milestones, and development and test goals; make program and system changes; define major technical developments; and identify management and fiscal problems, as well as problems that affect other military departments and Government agencies. (*Av. Wk.*, 10/4/65, 25)

- "It seems clear that preeminence in aeronautics and space in the future will certainly demand *a continuous evolvement of new technology*," Dr. Raymond L. Bisplinghoff, Special Assistant to the NASA Administrator, said in *Air Force and Space Digest.*

  "The consequences of not having done our homework prior to undertaking a system development to meet an explicit requirement are overruns in cost and time. These costs in resources are often so high that the means must be found to evolve new technology in advance of requirements. One would conclude from this that the formulation of a *responsible* requirement demands an underlying body of technology.

  "One of the most important purposes served by the creation of technologies is to provide options in the selection of new requirements or missions. The existence of several technical options is fundamental to sound planning. Because of the many important implications of space activities, policy planners will require that they be given options. It therefore seems desirable that the development of a new branch of technology be directed toward a related class of requirements or missions rather than a single requirement. . . ." (*AF* Mag., 10/65, 61–64)

- NASA Office of Technology Utilization issued a report on metal-forming techniques currently in use in the aerospace industry including those in the experimental stage; a survey, "Handling Hazardous Materials," dealing with such oxidizers as liquid fluorine, chlorine trifluoride, nitrogen tetroxide, and ozone; a book containing 71 ideas for shop techniques and applications used successfully in space-related research at NASA centers; and a technical survey tracing significant recent progress in plasma jet technology. (NASA Releases)

- USAF had ordered an additional six Titan IIIC launch vehicles for unmanned payload applications, bringing to 19 the number of firm orders, reported *Aviation Week and Space Technology.* (*Av. Wk.*, 10/11/65, 23)

- Dr. Kraft A. Ehricke, Director of Advanced Studies of General Dynamics Corp., in a series of talks at Evanston College, predicted several "realistic possibilities" for the 1980's and 1990's: (1) a manned space station to handle the world's communications needs, including global telephone calls complete with televised views of the speakers; (2) space vehicles to keep man informed of all weather developments, detect forest fires, and inform firefighters; (3) orbiting manned information centers to supply doctors and other scientists with data an any subject regardless of how distant they were from the source of material on earth; (4) orbiting hospitals or lunar hospitals to relieve persons suffering from certain ailments by providing gravityless or very low gravity conditions. (*Chic. Trib.*, 10/10/65)

During October: Dr. Walter Dornberger, Vice President for Research of Bell Aerosystems Co., former chief of German v–2 missile program, and father of the Dyna-Soar concept, was interviewed on the eve of his retirement from Bell by Claude Witze in *Air Force and Space Digest*. Witze said:". . . [Although Dornberger is] the outstanding pioneer in the development of ballistic rockets, he feels strongly that we have erred in relying on this single vehicle, with its inherent limitations." He quoted Dr. Dornberger:

" '. . . our whole approach to space is no good, if we really want to think of space as an operational area.

" 'We must use a completely different approach. We must get away from this launching from pads, which costs millions and billions of dollars, to the more conventional way of taking off from a runway.' "

Witze continued: "He predicts that after we have gone to the moon we will start over again. Project Apollo cannot be turned back, he says, but the next time 'we must create an environment in space that can be used by men, not only for research but for commercial and military purposes.'

"This environment he is talking about is a permanent space station, one that will never come back into the atmosphere. This will require a logistics system—a recoverable, reusable space transporter to carry people and supplies back and forth to space. The Dyna-Soar was a crude but necessary step in this direction, replaced today by the concept of the aerospace plane. . . .

"The Dornberger thesis is that a manned station in near space is as practical as a military base on Okinawa, a laboratory at the South Pole, or an aircraft parts and maintenance depot in Chateauroux. Once established, all that is needed is a logistics system to keep it going. From such a space station, expeditions can be sent to the moon, or many other places in space, with engines that give a thrust of only 20,000 pounds, instead of the millions of pounds required on one-way booster trips. . . ."

Quoting Dr. Dornberger: " 'It took mankind hundreds of years to learn to use the sea, but only fifty years to use the air. Space can be used in twenty-five years, if we get started. I doubt that the use of space vehicles will be more hazardous than the use of submarines.' " (*AF* Mag., 10/65, 80–88)

November 1965

November 1: The rendezvous of Geminis 6 and 7 had been scheduled tentatively for the eighth day of Gemini 7's 14-day endurance flight, revealed Command Pilot Frank Borman (L/Col., USAF) at a news conference held by Gemini VII's prime and backup astronaut crews at MSC. Gemini 7 would act as a passive target vehicle, but would expend fuel to circularize its 108-mi.-perigee orbit to 161-mi.-perigee orbit and to maneuver into range of Gemini 6's radar, if necessary. Borman said there would be no Eva on Gemini VII because the crew would be wearing new lightweight spacesuits unsuitable for work in the vacuum of space; he was unaware of any plans for Eva on Gemini VI. During ten days of the mission, the Gemini VII crew would fly in USAF-issue longjohn underwear. Pilot James Lovell (Cdr., USN) said that training for Gemini VII had emphasized stowage reviews because of the serious stowage problems encountered by Astronauts Gordon Cooper and Charles Conrad, Jr., on their eight-day GEMINI V flight August 19. (Transcript; Hines, *Wash. Eve. Star*, 11/2/65, A3)

- FAA became the first civilian Government agency to recover the entire cost of developing a device produced on Government contract by a private manufacturer and sold to the public. Wilcox Electric Co. paid FAA $142,540—total cost of developing a general aviation transponder that would identify airplanes for air traffic control purposes—in conformance with FAA policy that "where the national interest requires government action in the form of Federal expenditures, those expenditures which do not accrue to the benefit of the public at large should be recovered to the maximum extent possible." (FAA Release 65–104)

- DOD had warned that contracts for the USN version of the F–111 aircraft would be canceled if contractors and involved parties did not resolve arguments and furnish sufficient reliable performance data to warrant production, reported George C. Wilson in *Aviation Week and Space Technology*. General Dynamics was the prime contractor; Grumman Aircraft Engineering Corp. was building and refining most of the USN version; and Hughes Aircraft Co. was developing the Phoenix air-to-air missile for the F–111B. (Wilson, *Av. Wk.*, 11/1/65, 16)

- In a brief filed with the FCC, ComSatCorp commented on the problem of direct purchase of communications services: ". . . Comsat believes that as a general rule it should afford the other carriers the first opportunity to provide satellite services desired by users other than the U.S. Government and foreign communications entities. However, in the event that the other carriers, owning and operating non-satellite facilities in which they have substantial investments, do not provide a satellite service to any customer who desires such service, Comsat should be able to provide such service directly to that customer. . . ." (Text)

November 1: By a 45%-to-42% margin, the American people believed the space program was worth an annual $4-billion expenditure, according to a Harris poll. By a 50%-to-38% margin, the public would oppose continuing the program at the present rate of expenditure if it were not for Russian exploits in space. (Harris, *Wash. Post,* 11/1/65, A2)

- U.S. aviation's trunkline scheduled passenger mileage was expected to increase two to three times during the next ten years whether fares increased, decreased, or remained at 1964 levels, according to the Civil Aeronautics Board's traffic forecast for the 1965–1975 period. (Text)
- U.S. experts believed that Russia might attempt to launch a multimanned spacecraft on a 15-day mission as her next manned space effort, reported William J. Normyle in *Aviation Week and Space Technology.* Based on recent conversation with Soviet scientists, U.S. specialists felt that there was complete confidence in the Voskhod environmental-control system which had been qualified for 30-day missions, Normyle explained. They also believed that the Soviets were prepared to attempt a rendezvous mission more ambitious than Gemini 6. (Normyle, *Av. Wk.,* 11/1/65)
- President Johnson issued the annual proclamation inviting Americans to observe "Wright Brothers Day, December 17, 1965, with appropriate ceremonies and activities, both to recall the accomplishments of the Wright brothers and to provide a stimulus to aviation in this country and throughout the world." *(Pres. Doc.,* 11/8/65, 448)
- Commenting on the Project Surveyor report of Rep. Joseph E. Karth's (D-Minn.) Subcommittee on NASA Oversight [see Oct. 8], William J. Coughlin said: ". . . the really amazing thing is that what the Karth report calls the 'stormy 4½-year history' of the project has not been stormier. Given a limited budget to accomplish the impossible, anyone will have difficulties. . . .

  "The Karth report, in fact, acknowledges this when it states: 'In essence, insufficient preliminary work was done prior to the decision to go ahead with the project, and the award of a contract for development of the Surveyor spacecraft.'" (Coughlin, *M&R,* 11/1/65, 46)
- Reaction to the cancellation of the NASA Gemini VI mission October 25 indicated that the space program and its public acceptance had matured during the first eight years of the space age, wrote Robert Hotz in *Aviation Week and Space Technology.* He recalled the "abuse heaped on the Vanguard program, the demands for Ranger's cancellation, and the timorous whimpering that tried to suppress the Mercury program and bury the Apollo plans," and suggested that "the lesson we should learn as a nation from our space program is that no task is too difficult to achieve and no challenge too great . . . to surmount if its top-level leadership points out the goals and sounds the charge." (Hotz, *Av. Wk.,* 11/1/65, 11)

During week of November 1: Two 16-in. guns welded together end to end would serve as first stage of a three-stage launch device believed capable by its developers—USA Ballistic Research Laboratories and McGill Univ. Space Research Institute, Montreal—of orbiting a 50-lb. satellite.

The gun had launched 185-lb. payloads to approximate 84-mi. altitudes last year. Components of the rocket assembly had been successfully tested separately, but not as a unit; test firing of the first stage would take place in Barbados within four months. Dr. Charles H. Murphy, Ballistic Research Laboratories, and Dr. Gerald V. Bull, McGill Univ., reported on the project at a meeting of the New York Academy of Sciences. (Sullivan, *NYT*, 11/9/65, 3)

November 2: PROTON II 26,900-lb. unmanned scientific space station was launched by U.S.S.R. into orbit: apogee, 638.7 km. (396 mi.); perigee, 191.9 km. (119 mi.); period, 92.6 min.; inclination, 63.5°. Tass said that instrumentation on PROTON II would study cosmic particles of superhigh energies, cosmic rays and their radiation danger, and the nuclear interaction of cosmic particles with super-high energies up to 1,000 billion electron volts. All onboard equipment was said to be functioning normally. (*NYT*, 11/3/65, 5; AP, Balt. *Sun*, 11/3/65, 4; UPI, *Wash. Post*, 11/3/65, A21)

- The Interim Communications Satellite Committee (ICSC), on behalf of the International Telecommunications Satellite Consortium (Intelsat), approved a communications satellite system to provide services for NASA's Project Apollo and for other commercial users. The system, which would provide the first commercial comsat service to the Pacific area and supplement services across the Atlantic as part of the Intelsat system, would include four satellites: two in synchronous orbit at about 22,300-mi. altitudes, one over the Pacific, the other over the Atlantic; and two in reserve. Larger and more versatile than EARLY BIRD I comsat providing commercial service over the Atlantic, these satellites would weigh 150 lbs. compared to EARLY BIRD I's 85 lbs. ICSC also approved ComSatCorp's contract with Hughes Aircraft Co. to buy four satellites for the system. (ComSatCorp Release)

- NASA would negotiate with Ling-Temco-Vought, Inc., for 12 Scout launch vehicles under a 27-month, firm-fixed-price contract, valued at more than $8 million exclusive of option provisions for several additional vehicles. (NASA Release 65–343)

- Rep. John W. Wydler (R–N.Y.) urged the Subcommittee on NASA Oversight of the House Committee on Science and Astronautics to consider priorities in the U.S. space program. He noted that the Agena Target Vehicle failure which postponed the Oct. 25 Gemini VI mission might have been avoided if the modified Agena had undergone flight-testing before the Gemini mission attempt. Reiterating his view that "the last two Pegasus [satellites] shots were not clearly necessary," he suggested: ". . . this modified Agena rocket could have been test fired by means of utilizing the last two Saturn I rockets, which carried the additional Pegasus satellites instead. Such a test might have avoided the failure. . . ." (*CR*, 11/2/65, A6259)

- Soviet-French communique was issued after six days of talks between Soviet Foreign Minister Andrei Gromyko and French Foreign Minister Maurice Couve de Murville in Moscow expressing the desire to sign "an appropriate agreement" on space cooperation. This apparently referred to a Soviet invitation that France join in a communications satellite system which would compete with the U.S.-led ComSatCorp, to which France already belonged. (Rosenfeld, *Wash. Post*, 11/3/65, A9)

November 3: X–15 No. 2, piloted by L/Col. Robert Rushworth (USAF), attained 70,600-ft. altitude at 1,432 mph (mach 2.31) in the first of two flights to test the inflight tank ejection procedure and tank-recovery systems of two external propellant tanks. The external tanks would carry added propellant on future flights that could increase the X–15's speed to more than 5,000 mph. (NASA X–15 Proj. Off.; *X–15 Flight Log*)

- Launch of Geos A Geodetic Explorer satellite, originally scheduled for Nov. 2 and later postponed to Nov. 5 because of an electric circuit break, had been rescheduled for Nov. 6. Geos A would contain five geodetic instrumentation systems to more accurately map the earth and to serve as a precise space reference point for ground surveyors. (AP, *NYT*, 11/4/65, 24; *Wash. Post*, 11/5/65, A16)
- Development of three-dimensional television which could be viewed in color or black and white without wearing special glasses might be possible utilizing a technique developed by Spaco, Inc., while performing research for NASA MSFC, Marshall's Office of Technology Utilization announced. (MSFC Release 65–276)
- ComSatCorp requested bids from 12 companies by Nov. 22 for site preparation and construction of buildings and facilities for a fixed earth station at Paumalu, Oahu, Hawaii—a U.S. link for worldwide commercial satellite communications. The Hawaiian station and one at Brewster Flat, Wash., were the only fixed stations planned by ComSatCorp. (ComSatCorp Release)
- The 1965 International Space Electronics Symposium was held in Miami Beach. Dr. Edward C. Welsh, Executive Secretary, National Aeronautics and Space Council, told the Symposium that electronics would continue to play a prime role in the U.S. space program. Emphasizing the difficulty and complexity of the electronic problems which would be encountered in manned planetary exploration, Dr. Welsh noted that for manned Mars travel "we'll need reliability for at least 400 days contrasted with more immediate goals of 14 to 30 days; we'll need electrical 'on board' power measured in kilowatts instead of today's tens of watts; we'll need the data rates of 5 million bits per second instead of the current capability of 8 bits per second; and there is a still debatable possibility that we will need laser beam pointing accuracy several magnitudes better than what we now have. In components, we shall require new or improved concepts such as self-healing, self-checking, and accurate failure prediction to give the greatly increased reliability required. . . ." (Text)

    Dr. Homer E. Newell, NASA Associate Administrator for Space Science and Applications, told the Future Space Programs panel that the use of satellites for scientific research and the development of practical applications in meteorology, communications, navigation, and geodesy were important components of the total progress of the space program. He said science had been the prime beneficiary of skills developed for using satellites and space probes as scientific tools: "The vitality of space science has been its close association with various disciplines of science, and its ability to offer those disciplines new ways of solving old problems, while at the same time opening up new horizons." (Text)

The military space program had established the feasibility of using space as a medium for military support missions and now had the initial techniques to exploit the military usefulness of space, M/G Jerry D. Page, AFSC, told the panel on Future Space Programs. Application of these capabilities involved: (1) programs to provide near-term applications such as the Nuclear Detection Satellite program, a research-oriented effort which, at the same time, provided a global nuclear detection capability; and (2) consideration of multi-mission satellites for the future. (AFSC Release)

November 3: Evidence of the impact on education of achievements in space technology was cited by James V. Bernardo, Director of NASA Educational Programs Div., at the Michigan Education Association Meeting in Detroit: (1) greater demand for scientists, engineers, and technicians; (2) examination and revision of the science and mathematics courses in high schools and colleges to include new concepts and to provide enrichment of basic principles through space-related materials; (3) recognition of the need for better teacher training to meet the increasing needs of science and technology in our society; (4) recognition that science should be a basic part of general education for all students and that we must develop a science-literate public; (5) effort to develop a well-balanced national effort in science, technology, the social sciences, and the humanities; (6) need for advancing scientific research activities related to space exploration problems, including the development of training programs for scientists, engineers, and technicians. (Text)

- U.S. Army announced selection of Lockheed Aircraft Co. to develop "on an expedited basis" ten prototype high-speed helicopters, the first conceived and designed exclusively as weapon platforms. Called the Advanced Aerial Fire Support System (Aafss), the new compound helicopter would be capable of firing a variety of weapons and flying 50% faster than any other operational Army helicopter. (DOD Release 781–65)
- Air-to-surface Hound Dog missile was successfully fired from USAF B–52 aircraft over Green River, Utah, to White Sands Missile Range. (AP, *NYT*, 11/4/65, 35)
- Wrecks of ten U.S. U–2 photoreconnaissance planes downed over Communist China during the last three years were on exhibit at the Museum of the Revolution in Peking. (Reuters, Balt. *Sun*, 11/4/65, 7)

November 4: U.S.S.R. launched COSMOS XCV artificial earth satellite carrying scientific instrumentation to continue the space exploration program. Orbital parameters: apogee, 521 km. (323 mi.); perigee, 207 km. (128.3 mi.); period, 91.7 min.; inclination, 48°. Equipment was functioning normally. (*Pravda*, 11/5/65, 2)

- NASA issued a phased-planning policy directive, effective immediately, prescribing sequential steps for each major project: Phase A—Advanced studies; Phase B—Project Definition; Phase C—Design; Phase D—Development/operations. Each phase would be a specifically approved activity undertaken after management review of the preceding phase. (NASA Release 65–345)
- NASA pilot William Dana flew X–15 No. 1 to 80,200-ft. altitude at 2,795 mph (mach 4.22) in a pilot-checkout flight which also carried a

slightly modified horizon scanner and device for measuring microscopic atmospheric pressure. (NASA X-15 Proj. Off.; *X-15 Flight Log*)

November 4: Data obtained by the Pegasus meteoroid-detection satellite program and estimates by Dr. Fred L. Whipple, Smithsonian Astrophysical and Harvard College Observatories, prompted Dr. Ernst Stuhlinger, Director of MSFC Research Projects Laboratory and one of the scientists who had conceived the Pegasus program, to estimate that a Pegasus-sized sensor with a 1-mm. (40-mil) aluminum sheet would be perforated in earth orbit about 40 times annually and a sensor with a 2-mm. aluminum sheet only one to three times annually.

Dr. Stuhlinger told the Northeast Electronics Research and Engineering Meeting in Boston of initial evaluation of the Pegasus data: in the smallest size meteoric particles, fewer particles were encountered than expected; in the largest, more were encountered than planned; and in the mid-range, actual results agreed closely with theoretical predictions. Dr. Stuhlinger reported on the Pegasus project from inception in 1962 through reduction of data as late as October 8, 1965. (MSFC Release 65-275; *Marshall Star*, 11/17/65)

- Six key appointments at NASA's Mississippi Test Facility were announced by MTF Manager Jackson M. Balch: Henry F. Auter, Deputy Manager and Chief of Projects Control Office; L/Col. Frederic C. French (USA), Assistant Manager for Construction and Installation; Waldo H. Dearing, Chief of Management Support Office; Myron L. Myers and Robert A. Bush, project managers for S-IC and S-II operations, respectively; and Myrl E. Sanders, project manager for support activities. (MSFC Release 65-278)

- U.S. had no knowledge that any Soviet cosmonaut had ever died in spaceflight, Dr. Charles S. Sheldon II, of the National Aeronautics and Space Council, told an AIAA luncheon meeting in Washington, D.C. Posthumously published notes of convicted Soviet spy Col. Oleg V. Penkovsky had contended that "several cosmonauts had lost their lives in spaceflight." Sheldon said that statistically the U.S. was leading the U.S.S.R. in the number of successful satellite launches by a 2.3-to-1 ratio, but that the Russians were leading in total weight of payloads launched and their lead had been increasing each year for the past five years. He noted, however, that "the true value of scientific findings made by each country is harder to measure statistically, and neither country has been wholly capable of objective judgment in this regard." (Text)

- NASA's increased use of real-time planning in the Gemini program, in contrast to the "cast-in-concrete" planning of the Mercury program and the first Gemini flight, was praised by William Hines in the Washington *Evening Star*. He said this new elasticity had made possible the "revolutionary reprogramming" of the Gemini VI and Gemini VII missions where "quick and drastic re-thinking of almost every aspect of space flight operations was required. Also required was the junking of a great many shibboleths, not the least important being the fixed opinion long weeks of launching-pad checkout must precede liftoff of a man-carrying rocket...." (Hines, *Wash. Eve. Star*, 11/4/65, A12)

November 4: The impact of a meteorite might have begun life on earth, Dr. Brian Mason, U.S. National Museum, told the annual meeting of the Geochemical Society of America in Kansas City. Scientists had speculated that the presence of organic compounds—substances containing carbon and nitrogen—in meteorites indicated that life exists, or had once existed, elsewhere in the universe. Mason noted that such compounds could be made without life, but added: "I believe . . . they may be a forerunner of life. A falling meteorite may be the way life got started." Mason said that most meteorites appeared to come from the asteroid belt which orbits the sun between Mars and Jupiter. This belt might be a planet that never collected to become a single body, he suggested. Dr. Mason was conducting research on the chemical differences a man-earth environment had imposed on meteorites. (*Houston Chron.*, 12/5/65)

November 5: NASA would negotiate with International Latex Corp. and Hamilton Standard Div. of United Aircraft Corp. for development and production of Project Apollo flight suits and a portable life support system for extravehicular activities during earth-orbital flights and on the lunar surface. International Latex would receive about $10 million to produce the flight suits, consisting of a liquid-cooled undergarment, constant-wear garment, pressure garment assembly, and thermal-micrometeoroid protective over-garment. Hamilton Standard, under separate contract, would receive about $20 million to produce the life-support system: a backpack weighing about 65 lbs. containing an oxygen system, thermal control system, and communications equipment. Present plans called for the pressure suit to be worn during the latter phase of the Apollo/Saturn IB earth orbital mission series and during Apollo/Saturn V missions. Gemini pressure suits would be used on initial Saturn IB missions. (NASA Release 65–346)

- Final test in North American Aviation Space and Information Systems Div.'s seven-month paraglider operational test program was successfully completed at Edwards AFB, when Gemini boilerplate suspended beneath an inflated paraglider was towed to 9,000- to 10,000-ft. altitudes and released for free flights that averaged four and one half to five minutes. The test program, in which 12 consecutive successful flights and landings in tow-test vehicles were executed by company pilots, was not related directly to the NASA Gemini program but was an investigation in general operational aspects of manned landing using deployable maneuverable landing systems and emphasizing pilot problems. No further NASA funding was expected but NAA was performing some company-funded work on adapting the paraglider for controlled delivery of air-dropped cargo and had submitted a proposal to the U.S. Army for further work on this technique. (NAA *S&ID Skywriter*, 11/12/65, 1)

- Thirty-six MSFC employees received a variety of awards, including a Presidential Citation and six inventions awards, in a local ceremony observing NASA's seventh anniversary. Dr. Wernher von Braun, Director of MSFC, addressed the gathering and William Rieke, NASA Deputy Associate Administrator for Industry Affairs, presented the invention awards. (MSFC Release 65–273)

- Dr. William R. Lucas, chief of the Materials Div. of the Propulsion and Vehicle Engineering Laboratory, MSFC, received the Hermann Oberth

Award from the Alabama Section of the AIAA for his "outstanding individual scientific achievements in the field of astronautics and for the promotion and advancement of the aeronautical sciences." (MSFC Release 65-279)

November 5: Man might be able to change the orbit of asteroid Icarus and make it an orbiting earth station, suggested Soviet scientist K. Stanyukovich in an interview published by *Economichesky Gazetta.* Icarus has almost a one-mile diameter and weighs over six billion tons; every 19 yrs. it passes within 4 million mi. of earth. "There is a great demand for a natural moon of Icarus' size," Stanyukovich said, where man could build scientific observatories and warehouses to store fuel for spaceships departing on interplanetary trips. He doubted that it could be captured on its next approach in 1968, but foresaw a possibility for 1987. (Burke, *L. A. Times,* 11/6/65)

- Five lunar and Martian experts applied geological methods to interpreting Mars and the moon at the Geological Society of America Symposium in Kansas City. They found similarities between Mars and the moon: both bodies were subjected to slow erosion from showers of micrometeoroids from space; there was no evidence of surface water on either. Dr. Robert P. Sharp, Caltech, said that Martian craters were three-and-one-half times more numerous than on the lunar maria, but not as numerous as on the moon's uplands. He believed the planet had an extremely thin atmosphere of carbon dioxide, temperatures ranging from $35°$ C to $-100°$ C and no liquid water. Dr. Eugene Shoemaker, U.S. Geological Survey, related experiments on earth with features on the moon in effort to explain why the moon was being eroded. A nuclear explosion in Nevada, for example, created a meteorcrater-like hole, resulting in hundreds of secondary impact craters. "I think the surface is a fluffy material, about one meter thick, resting on the floor of small craters," he said. "The bulk of matter is moved by the impact of small micrometeoroids. Probably a good deal of the material is melted and the melt may fly off into space." Bruce C. Murray, Caltech, said that physical observations by studying emissions and radiation showed the moon emits light and other wave lengths in the same way that "fine, loose powder does." He believed the outer one-half millimeter is covered with dust. Infrared heat samplings showed variations indicating a heterogenous type of rock near the surface. Dr. E. C. T. Chao, U.S. Geological Survey, discussed tektites—small, dark, glass, button-shaped objects, which he believed had been formed by meteoric impact, presumably on the moon, and had splashed off to fall on earth. Harold Masursky, U.S. Geological Survey, showed a geological map of the moon with the ages of various areas in different colors. He said the Survey had mapped three million square miles in the potential landing areas on the equatorial belt. He found five different "episodes of movement" on the surface. (McCoy, *Kansas City Times,* 11/6/65)

November 6: EXPLORER XXIX Geodetic Explorer satellite (Geos A) was successfully launched by NASA from ETR by a Improved Thrust-Augmented Delta with a new enlarged second-stage fuel tank to provide longer engine burn. Because the guidance system did not shut down the second stage at the desired time, the achieved orbit had the following parameters: apogee, 1,412.4 mi. (2,274 km.); perigee, 695.6 mi.

November 6: EXPLORER XXIX launch with first NASA Improved Thrust-Augmented Delta from Cape Kennedy.

(1,118.1 km.); period, 120 min.; inclination, 59°. Planned orbit had called for apogee of 920 mi.

The 385-lb. satellite, designed and built by Johns Hopkins Univ. Applied Physics Laboratory, contained five geodetic instrumentation systems to provide simultaneous measurements necessary for a more precise model of the earth's gravitational field and to map a world coordinate system relating points on or near the surface to the common center of the mass: (1) four flashing light beacons to be photographed against the background of stars to define the arc of orbit and angular data; (2) corner cube quartz reflectors to pinpoint the satellite's position by reflecting laser beams; (3) three radio transmitters for Doppler-shift determination of the precise orbit; (4) radio range transponder to fix the positions of the satellite and interrogating ground stations; and (5) range and range-rate transponder to determine the changing range and radial velocity of the satellite. Simultaneous operation of the five independent and diverse geodetic-tracking systems would permit cross-checking and evaluation of the different techniques and was expected to enhance the accuracy of each system. Other objectives of the Geos program were: (1) to map with a high degree of mathematical exactness the structure of the earth's irregular gravitational field; and (2) to compare and correlate results from different instrumented techniques employed simultaneously so as to assure greater accuracy and reliability.

Critical to optimum use of the radio and optical beacons on EXPLORER XXIX was the gravity-gradient attitude stabilization system to keep the satellite antennas, laser reflectors, and optical beacons pointing earthward at all times. In two to three days the attitude control would be initiated. (NASA Release 65-333; NASA Proj. Off.; AP, *NYT*, 11/7/65, 10; UPI, *Wash. Post*, 11/7/65, A16)

November 6: Comet Ikeya-Seki was leaving beaded trails of nuclear condensation as if it were breaking up, reported Howard Pohn, lunar geologist for U.S. Geological Survey's astronomy branch in Flagstaff, Ariz. Pohn had discovered one trail Nov. 4 and photographed what appeared to be another Nov. 6. He noted that this was similar to what had happened to the great comet of 1882, adding: "That one was visible for some five months after its perihelion passage—closest to the sun— and the way the Ikeya-Seki comet is acting indicates it may also be visible for that long a time." Ikeya-Seki's perihelion passage had occurred Oct. 21. (AP, Wash. *Sun. Star,* 11/7/65, A25)

- The suggestion that the Gemini VII/VI endurance-rendezvous mission scheduled by NASA for Dec. 4–13 might be a "space spectacular for spectators sake," was firmly rejected by Rep. George P. Miller (D–Calif.), Chairman of the House Committee on Science and Astronautics, at an MSC press conference. Miller said he believed the U.S. would land a man on the moon by 1970 and emphasized that information from the Gemini VII/VI mission would further the orderly exploration of outer space "for this country and the world." (Maloney, *Houston Post,* 11/17/65)

November 7: Brandeis Univ. awarded honorary degrees to 12 persons, including five Nobel laureates, at the dedication of the university's new science center. James E. Webb, recipient of one of the degrees and main speaker at the convocation, described the events leading to his appointment as NASA Administrator: "Near the end of January 1961, my good friend, Jerry Wiesner, [Dr. Jerome B. Wiesner, MIT] whom you are honoring here today, telephoned to me in Oklahoma to say that Vice President Johnson and President Kennedy wanted me to come to Washington immediately to talk about serving as Administrator of the National Aeronautics and Space Administration . . . the following Monday I found myself in President Kennedy's office between him and the Vice President and saying that I thought they needed a scientist or any engineer. 'No,' President Kennedy said, 'the issues involved in the development of space are policy issues—of great national and international significance. You . . . have some familiarity with how policies are established and how they are carried out.' At the time, that seemed reasonable. . . ." (Text; AP, *NYT,* 11/8/65, 52)

- Three new rockets described by Tass as "elusive to the enemy's air and space reconnaissance and . . . constantly ready to strike a crushing nuclear blow at an aggressor" were displayed by the Soviet Union in a parade in Moscow commemorating the 48th anniversary of the Bolshevik Revolution. Tass said one of the missiles had warheads that could "deliver their surprise blow on the first or any other orbit around the earth." (Grose, *NYT,* 11/8/65, 1, 6; Nordlinger, Balt. *Sun,* 11/8/65, 1)

November 8: Failure of the Gemini Agena Target Vehicle (GATV) which aborted the Gemini VI mission October 25 was probably caused by a hard start of the primary propulsion system which either shook the rocket apart or caused an explosion that destroyed it, reported a USAF Flight Safety Review Board which had met with the NASA Design Certification Review Board. Series of continuing tests was being scheduled to assure reliability of future Gatv flights. (Text; Clark, *NYT*, 11/9/65, 3)

November 8: USAF launched two unidentified satellites with Atlas-Agena D booster from Vandenberg AFB. (UPI, *Chic. Trib.*, 11/9/65; *U.S. Aeron. & Space Act, 1965*, 154)

- The Gemini IX crew was announced by MSC Public Affairs Officer Paul Haney at a press conference in Houston: Elliot M. See, Jr. (civilian), command pilot; Charles A. Bassett II (Capt., USAF), pilot; Thomas P. Stafford (Maj., USAF) and Eugene M. Cernan (LCdr., USN), backup crew. Scheduled for the third quarter of 1966, the mission would probably last two or three days, and would include rendezvous and docking and extravehicular activity. Bassett, who would remain outside the spacecraft for at least one revolution, would wear the USAF-designed manned maneuvering unit (Mmu) backpack, a self-propelled hydrogen peroxide system with gyro stabilization. (Transcript)

- Production of Gemini spacecraft and F-4 Phantom jet fighter aircraft was halted when 16,000 machinists struck McDonnell Aircraft Corp. in a wage dispute after rejecting the company's contract proposals and the ten-day contract extension recommended by IAM leaders. The unauthorized strike was not expected to interfere with the scheduled December flights of Geminis VI and VII from KSC. (AP, *Balt. Sun.*, 11/9/65)

- An 11-ton, 60-ft.-dia. radio antenna was erected on a mountain near San Diego, Calif., by USN Electronics Laboratory to bounce signals off the moon and certain artificial satellites. R. U. F. Hopkins, director of the Microwave Space Relay Project, said the antenna could improve the tracking and monitoring of active satellites such as the NASA Tiros weather-forecasting series, aid in studying refraction of radio waves caused by atmosphere, and beam powerful signals into space. (AP, *NYT*, 11/17/65, 33)

November 8-9: The NASA-Western University Conference at JPL presented to over 200 educators from 13 western states a comprehensive view of NASA's current programs of interest to universities and described ways for faculty members to participate in these programs. JPL Director Dr. William H. Pickering was the conference chairman. (JPL Release)

November 9: Four flashing light beacons on NASA EXPLORER XXIX satellite had been turned on and were functioning normally, NASA officials announced. The beacons, each emitting a light of 1,580 candleseconds per flash, would permit the satellite to be photographed against the stellar background. (AP, *Wash. Eve. Star*, 11/10/65, A6)

- NASA Nike-Apache sounding rocket with Univ. of Michigan pitot-static probe experiment to measure pressure, temperature, and density from 15-km. to 115-km. altitude was launched from Ft. Churchill, Canada. Rocket and experiments functioned properly and good data were obtained. (NASA Rpt. SRL)

- Snap 10-A, first nuclear reactor to operate in space, was probably shut down prematurely May 16, 1965, because of a spurious command from a decoding device triggered by a voltage regulator failure, AEC reported. Launched April 3 from Vandenberg AFB, with an Atlas-Agena booster, Snap 10-A achieved an 806-mi. circular orbit and produced over 500,000 watt-hrs. of electricity before it shut down May 16. It would orbit for more than 4,000 yrs. (AEC Release H-247)

November 9: USAF fired Minuteman ICBM from Vandenberg AFB, to target area in the Eniwetok Lagoon, about 4,500 mi. across the Pacific. (AP, NYT, 11/10/65, 6)

• USAF launched an Honest John-Nike Hydac high-altitude research rocket with 250-lb. payload from Eglin AFB, Fla. (Eglin AFB Release 65-449)

• Gemini VII would be launched Dec. 4 and Gemini VI Dec. 13 in a combination long-duration, rendezvous mission, NASA announced. Astronaut Frank Borman (L/C, USAF) would be command pilot and Astronaut James A. Lovell, Jr. (Cdr., USN), pilot, on a 14-day mission to determine the effects of long-duration spaceflight on man; 20 scientific, medical, and technological experiments would be performed. Astronauts Edward H. White II (L/C, USAF), and Michael Collins (Maj., USAF) would be Gemini VII's backup crew. Gemini VI mission would be nearly identical to the original rendezvous flight postponed October 25 when the Agena Target Vehicle failed to achieve orbit, and would demonstrate rendezvous of two vehicles in space. Command Pilot Walter M. Schirra (Capt., USN) and Pilot Thomas P. Stafford (Maj., USAF) would maneuver the Gemini 6 spacecraft within close proximity of Gemini 7 during the fourth revolution and station-keep for a period of time, but would not dock. Astronauts Virgil I. Grissom (L/C, USAF) and John W. Young (Cdr., USN) would be Gemini VI's backup crew. (NASA Release 65-347; UPI, *NYT*, 11/10/65)

• Britain's Princess Margaret and the Earl of Snowdon tourned JPL. (NASA Off. Int. Aff., 11/8/65)

November 10: NASA EXPLORER XXIII meteoroid-detection satellite had successfully completed its one-year expected lifetime, NASA announced. The 295-lb. satellite, launched November 6, 1964, from Wallops Station to measure the rate of meteoroid punctures at 300-mi. to 600-mi. altitude, had recorded 122 punctures as of September 30. These results indicated that an exposed area of 10 sq. ft. made of metal one-thousandth of an inch thick would experience penetration by a meteoroid about once a week. (NASA Release 65-351)

• NASA Nike-Apache sounding rocket launched from Ft. Churchill, Canada, with Univ. of Michigan pitot-static probe to measure pressure, temperature, and density from 15-km. to 115-km. altitudes. Experiment was not successful because of undetermined malfunction in the rocket during Apache-stage propulsion. (NASA Rpt. SRL)

• Two simultaneous IQSY launches for high-altitude research were made by USAF from Eglin AFB, Fla., in support of each other, one with a Nike-Cajun rocket and the other with a Sparrow-Arcas. (Eglin AFB Release 65-449)

• Four Nike-Apache sounding rockets would be launched from Chamical, Argentina, for the study of an ionospheric phenomenon called "Sporadic E" under an extension to a cooperative U.S.-Argentine agreement. The Argentine Comision Nacional de Investigaciones

Espaciales (CNIE) would provide the personnel for payload fabrication at NASA's Goddard Space Flight Center, procure the rockets, operate the range at Chamical, and be responsible for the reduction and analysis of data obtained. NASA would provide the equipment and facilities for construction of the payloads and a Nike-Apache launcher on loan. No exchange of funds between CNIE and NASA was contemplated. (NASA Release 65-350)

November 10: Dr. Hugh L. Dryden, NASA Deputy Administrator, delivering the annual Robert Thurston Lecture of the American Society of Mechanical Engineers in Chicago, said that man woud probably never explore the stars since a round trip to the nearest would take 160,000 yrs. Dr. Dryden spoke on the impact of man-in-space on engineering. This was to be his last public appearance. (Manly, *Chic. Trib.*, 11/11/65)

- NASA selected Telecomputing Services, Inc., and Ling-Temco-Vought's Range System Div. for competitive negotiations of a cost-plus-award-fee contract to provide computer support services for the Michoud Assembly Facility and the Mississippi Test Facility. The $1.5-million contract would be negotiated for a one-year period with provisions for three consecutive one-year renewals. (NASA Release 65-349)

- Laboratory research prompted by weight losses of American and Russian spacement on orbital flights had indicated a relationship between a person's water-drinking habits, working ability, and real or imagined stresses and strains, reported William J. Perkinson in the Washington *Evening Star*. American astronauts had lost between three and five percent of their body weight in orbital flights; Russian cosmonauts had lost less—between one and three percent—partly because they perspired less in the shirt-sleeve environments of Soviet spacecraft than Americans did in spacesuits. One NASA report on hypohydration—condition when a person drinks too little water—noted a 5% weight loss due to water imbalance was tolerable, but a 10% loss could cause gross mental and physical deterioration. (Perkinson, *Wash. Eve. Star*, 11/10/65, B10)

- Christopher Kraft's hometown of Hampton, Va., honored him with a full day of ceremonies on Christopher Kraft Day. (*Langley Researcher*, 11/5/65)

- U.S. authorities had considered and rejected the idea of building orbiting nuclear missiles because it would be a clumsy, inaccurate method of waging atomic war, reported the Associated Press. U.S. experts had calculated that a warhead launched from orbit would not come within 50 mi. of its target on earth whereas U.S. ICBMs and submarine-launched Polaris missiles were accurate within one mile. In addition, U.S. had developed antisatellite rockets that could intercept enemy satellites in orbit. Disclosure was made because of a November 7 Tass announcment that one of the missiles paraded through Moscow for the Anniversary of the Bolshevik Revolution had warheads that could "deliver their surprise blow on the first or any orbit around the earth." (AP, *Wash. Eve. Star*, 11/10/65, H2)

November 11: Production of Gemini spacecraft and F-4 Phantom jet fighter aircraft resumed when 16,000 machinists ended their four-day wildcat strike at McDonnell Aircraft Corp. IAM, which had refused to sanction the strike, announced it would authorize a strike beginning

Nov. 18 if a new contract were not accepted by midnight Nov. 17. Strikers were seeking a 20-cent-an-hour wage increase each of the next three years. (AP, Balt. *Sun*, 11/12/65)

November 11: A guidance control system for the Little Joe II booster was released from a Las Cruces, N. Mex., hospital in satisfactory condition following emergency x-ray treatment. White Sands Missile Range technicians preparing for a Dec. 1 Apollo escape system test took the unit to the hospital after they failed to determine the cause of a malfunction. Industrial x-ray facilities at White Sands were closed for Veterans Day. (AP, Wash. *Sun. Star*, 11/15/65, A15)

* Tass reported that astronomers at the Vladivostok Observatory had taken clear photographs of the Ikeya-Seki comet. (UPI, *Wash. Post*, 11/12/65, C4)

* Thirty years ago a helium-filled balloon, Explorer II, carried Capt. Orvil A. Anderson (USA) and Capt. Albert W. Stevens (USA) to a 14-mi. altitude—highest ever attained by man at that time—in a 230-mi., eight-hour 13-min. flight which proved that man could survive at great heights. Sen. Stuart Symington (D-Mo.) paid tribute to the late Anderson and Stevens at an anniversary ceremony near Rapid City, S. Dak., attended by officials of the National Geographic Society and the Army Air Corps, cosponsors of the flight. (Schaden, Wash. *Eve. Star*, 11/11/65, A8)

* Christopher Kraft, Assistant Director for Flight Operations, MSC, received Virginia Polytechnic Institute's distinguished alumnus citation, the university's highest award, in a ceremony in Blacksburg. (VPI Release)

* Former German missile experts who helped foster American rocketry after World War II met at NASA MSFC to celebrate their 20th year in the U.S. Dr. Wernher von Braun, Director of MSFC, attended. (UPI, *Cocoa Trib.*, 11/4/65; *Marshall Star*, 11/10/65)

November 12: VENUS II 2,123-lb. unmanned space probe was successfully launched by U.S.S.R. into a heliocentric orbit on a three-and-one-half month journey toward Venus. Tass announced that the trajectory was "close to the prescribed one" and that all onboard equipment was functioning normally. During its flight, VENUS II "would carry out an extensive space research program using onboard scientific instruments." (Tass, 11/12/65; *NYT*, 11/13/65, 10)

* FAA awarded $2.6-million contract to Texas Instruments, Inc., for 20 airport surveillance radar (Asr) systems. Seventeen of the systems would replace obsolete equipment at USN and Marine air stations and would be paid for by USN; two systems would be installed in and paid for by Brazil; and one system would be used and paid for by USA. (FAA Release 65-107)

* Fourth and final flight of the Stellar Acquisition Feasibility Flight (Staff) program from Eastern Test Range aboard a USN-supplied Polaris A-1 was termed an unqualified success by USAF and General Precision, Inc. GPI made the Staff guidance system, intended as forerunner of a Stellar Inertial Guidance System (Stings). During the flight, guidance equipment sequentially acquired both the star Polaris and second star of less magnitude. Ability to move through a two-star sequence would be necessary for guidance systems used in con-

junction with any future mobile missile system to accurately determine coordinates of the initial launch position, predict trajectories, measure deviation, and make corrections. (USAF Staff Proj. Off.)

November 12: The vast accumulation of knowledge generated by the space program required a "new wave" of science reporters to challenge and stimulate "a new wave of readers and viewers who are seeking answers," Julian Scheer, NASA Assistant Administrator for Public Affairs, told the Sigma Delta Chi National Convention in Los Angeles. He referred to a need to look "beyond the obvious and the glamorous to what is happening in space and science and what impact this will have socially, diplomatically, politically and economically." (Text)

- DOD selected the first eight of a planned 20 astronauts for USAF's Manned Orbiting Laboratory program: Maj. Michael J. Adams (USAF); Maj. Albert H. Crews (USAF); Lt. John L. Finley (USN); Capt. Richard E. Lawyer (USAF); Capt. Lachlan Macleay (USAF); Capt. F. Gregory Neubeck (USAF); Capt. James M. Taylor (USAF); and Lt. Richard H. Truly (USN). Formal training would begin at Edwards AFB, early next year. (USAF Release)

- A new kind of scientist—an interdisciplinarian to transcend specialization— was needed to solve many of the problems created by specialists, said Dr. John H. Heller, director of the New England Institute for Medical Research, in a lecture at American Univ., sponsored by the Metropolitan Washington Board of Trade. He said the interdisciplinarian would receive as much training in depth as a scientific specialist but would be instructed "far more broadly in depth." (*Wash. Sun. Star*, 11/14/65, A3)

- USAF Hound Dog missile fired from B-52 bomber crashed near Ft. Wingate, N. Mex., and ignited several fires in Cibola National Forest. Missiles were programed to land at White Sands Missile Range, 175 mi. southeast of impact point. (UPI, *NYT*, 11/13/65, 12)

- Soviet Cosmonaut Aleksey Leonov, first man to walk in space, displayed seven paintings at Moscow's Palace of Pioneers, headquarters of the National Children's Organization. Explaining one painting of the cosmos, Leonov told reporters: "Previous cosmonauts tried to photograph three vivid belts of light—red, orange and blue—that ring the earth, but they never showed up in photographs. When I got back I painted the belts of light, each fading into the next, just as I had seen them. The sun glowed through the colors looking like a strange object with little wings coming directly at me." (Grose, *NYT*, 11/13/65, 10)

- North American Aviation President J. L. Atwood discussed "dramatic strides" made in civil aviation, at the dedication of the Albuquerque Sunport, N. Mex.: "In just 35 years—from 1929 to 1964—the number of passenger-miles traveled annually on the world's civil airlines increased from 105 million to 105 billion—almost a thousandfold. . . . In only 15 years, between 1949 and 1964, the annual number of passengers on world civil airlines increased from 27 million to 154 million —nearly sixfold . . . A Stanford Research Institute study predicts that passenger-miles on the free world's scheduled airlines—which were 105 billion in 1964—will reach approximately 229 billion in 1975. . . ." (Text)

November 13: Russia was developing long-range intercontinental missiles that could maneuver in flight, Col. Gen. Vladimir Tolubko, First Deputy Commander of the Soviet Strategic Rocket Forces, told Tass. (UPI, *NYT,* 11/14/65, 74)

November 15: EXPLORER XXIX, first NASA satellite to use a gravity-gradient system for stabilization, had pointed its instrumentation toward earth and was expected to be programed for operation within several days. Achievement of the desired attitude was confirmed by magnetic and solar sensors and by marked increase in the strength of radio signals. Project officials at NASA GSFC and Johns Hopkins Univ. Applied Physics Laboratory reported four EXPLORER XXIX geodetic measurements systems had been checked out and were performing as expected; the fifth, using laser beam reflectors, would be tested when the spacecraft had completed its stabilization phase. (NASA Release 65–354)

- Tass reported all equipment onboard Soviet probe VENUS II, launched November 12 on a three-and-one-half month journey toward Venus, was functioning normally. Radio communications were excellent and commands were being followed. VENUS II was 1,149,000 km. (712,380 mi.) from earth. (Tass, 11/15/65)

- Agreement was signed for a cooperative Brazilian-U.S. sounding rocket project to obtain meteorological information, NASA announced. Project provided for cooperation in obtaining wind, temperature, and other meteorological information between 40 km. and 100 km. by rocket soundings using the acoustic grenade technique. Experiments would be conducted from the Brazilian launch range at Natal. The agreement, in form of a memorandum of understanding, was signed by representative of the Momissão Nacional de Atividades Espaciais (CNAE) of Brazil and NASA. It was the third to be concluded between CNAE and NASA during 1965. (NASA Release 65–372)

- A group of amateur astronomers working on a lunar research project for NASA reported observations of unusual color glows on the moon and photographed this phenomenon. The group told NASA it saw the color in the crater Aristarchus during a four-hour period before daylight through a 16-in. telescope at Port Tobacco, Md. Although such events had been observed several times since the Russian astronomer N. Kozyrev first recorded observations or red glows on spectrograms, this was the first time photographic equipment was used successfully to record the sightings in the crater Aristarchus. Observation was culmination of a 16-mo. vigil by members of a "Moon-Blink" team from Annapolis, Md. The team had made two previous confirmed sightings, including one in the crater Alphonsus last year, but they were much shorter and were not photographed. (NASA Release 65–370)

- A five-week technical assessment of supersonic transport air-frame designs proposed by the Boeing Co. and Lockheed Aircraft Corp. was begun by an 82-member Government technical team with representatives from NASA, FAA, USAF, and USN. Formed under the FAA Director of Supersonic Transport Development B/G J. G. Maxwell (USAF), the group would emphasize operational performance of the proposed configurations in terms of program objectives for a safe, economical aircraft compatible with present airports, airline operating requirements, and engine noise and sonic boom criteria. Performance characteristics of each design would be examined in detail, both analytically and through wind tunnel tests at NASA Langley Research Center and NASA Ames Research Center. Results of the assessment would provide guidance for further contractor programing in the present detailed design and hardware test phase of the Sst development program which called for prototype construction to begin by the end of 1966. (FAA Release 65-110)

November 15: Nucleus of comet Ikeya-Seki had split into three pieces, all of which were traveling together in parallel courses into deep space. The three-piece break, observed Nov. 4 and 5 by Howard Pohn at U.S. Geological Survey Observatory in Flagstaff, had been confirmed by Mt. Palomar Observatory and Boyden Station, South Africa. (Sci. Serv., *NYT*, 11/15/65, 74)

- Modified Boeing 707-349C jet carrying 40 scientists left Honolulu to begin a 26,263-mi. around-the-world flight over the North and South Poles. The jet would cruise at between 30,000 ft. and 40,000 ft. to allow study of high-altitude meteorology, clear-air turbulence, the jet stream, and cosmic radiation; time-lapse cameras would photograph the weather at five-minute intervals. Environmental fatigue problem of flight's participants resulting from the cramped quarters and long duration of the 52-hr. to 56-hr. flight would also be studied. Dr. Serge A. Korff, New York Univ., headed the scientific team; Rockwell Standard Corp. was sponsor; NASA and U. S. Weather Bureau were two participating agencies. (*NYT*, 11/14/65, 23; AP, Wash. *Eve. Star*, 11/15/65, A1)

- Vice President Hubert H. Humphrey said in address before American Nuclear Society-Atomic Industrial Forum in Washington, D.C.:

"If anyone doubts the impact that science and technology have had on society, in just the past 20 years, he need look only at the new industries employing many thousands of people which have come into existence during that time—industries based solely on the evolution of new products and services. These include—and I mention only a few —television, the computer, the jet engine and nuclear energy.

"Some of these industries have sprung from the application of a single invention or chain of scientific thoughts.

"During these past 20 years, too, we have seen in the United States the creation of a constructive partnership unknown in our previous history—a partnership of government, university, management, labor, science and citizen—a partnership devoted to maximum development of science and technology not for the narrow interest of any single group, but for the common good.

"To witness the space launchings at Cape Kennedy, as I have, is to see this partnership in action. . . ."

Turning to the peaceful uses of atomic energy, he said:

"As chairman of the National Aeronautics and Space Council, I am particularly concerned with the atom's work in space.

"In 1961 our first operating space radioisotopic power source was orbited. And in 1965 our first space reactor was operated in orbit.

"The atom will soon become a major power source for our space program—an auxiliary source for spacecraft and life support systems and a necessary source of propulsion for extended space exploration.

"I can, in fact, foresee the time when our space efforts will be able to continue *only* through use of rocket and nuclear power units transported and assembled in space. . . ." (Text)

November 15: Reports on the Saturn V/Apollo crawler-transporter bearings had just begun and might not be completed until next year, Don Buchanan, project engineer for prime contractor Marion Power Shovel Co. told *Missiles and Rockets*. (*M&R*, 11/15/65)

- Death-ray use of the laser was impractical, the 72nd annual meeting of the Association of Military Surgeons of the U.S. was told by Dr. Donald H. Glew, Jr., George Washington Univ. surgeon who had conducted laser research for the Armed Forces. Dr. Glew said that, with present sources of laser beams at least, such rays would be impractical as weapons because the necessary equipment would be "far too massive for field use." (AP, *Wash. Post*, 11/16/65, C5)
- In a special report on Project Apollo, Edward Kolcum wrote in *Aviation Week and Space Technology* that NASA was now driving to simplify Apollo spacecraft systems, subsystems, and components, and had embarked on an educational program with one goal: to stop the mushrooming tendency to build elements that were more complex and did more than was necessary. (Kolcum, *Av. Wk.*, 11/15/65, 55)
- U.S. would withdraw in six months from the Warsaw Convention limiting the liability of international airlines to $8,300 for each passenger killed or injured in air accidents, unless a new agreement was reached substantially increasing carrier liability, the State Department reported. U.S. had refused to ratify a 1955 amendment to the Convention raising the limit to $16,000, was arguing for a $100,000-limit liability. (U.S. State Dept. Release 268)

November 15–17: Dr. Edward C. Welsh, Executive Secretary of the National Aeronautics and Space Council, told a panel at the American Nuclear Society-Atomic Industrial Forum in Washington, D.C.:

". . . advancing our competence in nuclear technology is of critical importance to the future of the national space program. Nuclear power and nuclear propulsion are musts for the more complex and long duration missions of the future. . . .

"We all know that the nuclear going in space won't be easy. . . . Yet I am confident that the technical problems are resolvable and the investments warranted. . . ." Dr. Welsh delivered a warning to "space planners" to stop sitting on nuclear propulsion concepts and to expedite development programs. "We must not wait for clear-cut space requirements for these nuclear systems before moving forthrightly with their development. If we wait for precise mission definitions the technology will not be available when it's needed. Moreover, we can be sure that our Soviet competitors will take positive actions, whether we do or not. For this country to sit back on its technological haunches and let someone else, bolder and more imaginative, show us the way is unthinkable to me." (Text)

November 15–18: The International Congress on Air Technology was held in Hot Springs, Ark. Dr. Robert M. White, Administrator, Environmental Science Services Administration (ESSA), U.S. Dept. of Commerce, referred to the developing World Weather Watch in his address: "The work . . . comprises two broad, continuous, and parallel streams of action. The first will be a process of introducing into the present international weather system already proven equipment, techniques, and procedures. We hope that by 1971 . . . we will have made three significant improvements in the international system. First, we hope to improve the ability of the system to observe the global atmosphere and to provide fuller data for weather forecasting. In part this improvement will rest on the Tiros Operational Satellite System . . . which the United States will launch into orbit next year. Second, we hope to extend the benefits of modern computer technology throughout the world—by utilizing the computer to prepare weather analyses and forecasts routinely for the entire globe. And third, we hope to develop an international communications network for the timely exchange of raw data and for the rapid dissemination of analyses and forecasts." (Text)

Dr. Floyd L. Thompson, Director of NASA LaRC, delivered a keynote address on "Advances in Aircraft Technology." He discussed the importance of scientific research to the continued leadership of the U.S. in the field of air transportation with particular reference to the next decade.

Representing NASA Flight Research Center at the technical sessions was Joseph Walker, Chief of Research Pilots, who spoke on the X-15 research airplane as a tool for progress in hypersonic flight. (LaRC Release)

November 16: Project Luster, managed by NASA Ames Research Center, successfully recovered samples of matter from interplanetary space with an Aerobee 150 sounding rocket launched from White Sands Missile Range, N. Mex., during the annual Leonid meteor shower. The sampling instrument comprising the payload consisted of three deployable arms, each containing four flat pans holding special collection surfaces. The arms deployed as programed at 47-mi. altitude on the way up, closing again at that altitude on the way down. Payload reached 89-mi. altitude, traveled 44 mi. downrange, and landed by parachute in soft sand. Vacuum seal of the 12 collection pans was found to be perfect on recovery. Unopened collectors would be distributed by ARC to 12 guest experimenters in the U.S. and Europe for study of the nature of comets and of extraterrestrial material in general. (ARC Release 65-26; NASA Rpt. SRL)

- U.S.S.R. launched VENUS III unmanned space probe towards the planet Venus. Tass said the technique for placing the 2,112-lb. spacecraft into heliocentric orbit was similar to that used in orbiting VENUS II on Nov. 12. Trajectory was close to the one calculated. Purpose of both launches was "to augment the volume of scientific information and to obtain additional scientific data regarding Venus and outer space," but VENUS III contained equipment to carry out different scientific investigations from VENUS II. All systems were functioning normally. VENUS II and VENUS III were expected to reach the vicinity of Venus about March 1. (Tass, 11/17/65; *Krasnaya Zvezda, Pravda,* and *Komsomolskaya Pravda,* 11/17/65, ATSS-T Trans.)

- Astronaut M. Scott Carpenter (Cdr., USN) told the National Press Club in Washington, D.C., that while both space and underseas programs were potentially important from the standpoint of material benefits, "I feel that whatever material gains come from the two programs, the ones we might get from the assault of the ocean are much more immediate than those we might get from an assault on the moon, for example." Later Carpenter received the Legion of Merit in a Pentagon ceremony for his work as one of the two team captains in Project Sealab II Aug. 28-Sept. 26. (AP, Balt. *Sun,* 11/17/65; UPI, *NYT,* 11/17/65, 21)

November 16: G. Mervin Ault, NASA LRC engineer, discussed the status of development of high-temperature materials for advanced turbojet engines at the International Conference on Aircraft Design and Technology in Los Angeles. Meeting was jointly sponsored by the AIAA, the Japan Society for Aeronautical and Space Sciences, and the Royal Aeronautical Society. He said although the past decade had seen dramatic progress in such high-temperature or refractory materials, research had indicated that further improvements would be possible, especially in the structural properties of these materials. (LRC Release 65–81)

- NASA disclosed plans for the new Applications Technology Satellites (Ats) that would appear to hover over a given spot on earth to check out communications, weather, scientific and engineering ideas. There would be five in all, with four in synchronous orbit at 22,300-mi. altitude. First Ats was scheduled for launch in 1966; two more would be launched in 1967; the last two in 1968. Primary aim of the Ats program would be to find out (1) what happens to satellites in synchronous orbit, and (2) what can be done by a satellite continuously above one area of the earth. Among experiments planned were color television transmission, the first attempt to "talk" among unmanned satellites and aircraft, and photographing cloud formation changes in one place over a long period of time. (NASA Release)

- "We'll be lucky if we have nuclear propulsion around 1980," and electric propulsion is still further down the line for sometime after 1985, according to Dr. John C. Evvard, Deputy Associate Director for research at NASA Lewis Research Center. Evvard gave his views on advanced propulsion systems at the sixth annual Space Technology Series sponsored by the Canaveral Section, Institute of Electrical and Electronics Engineers, at Brevard Engineering College, Patrick AFB, Fla. A pioneer in electric propulsion and advanced nuclear rocket studies, Evvard believed such concepts should not be considered for Mars missions of under 100 flights per year. Only frequent flights would increase the need for higher specific impulse and would justify the expense, he said.

  Meanwhile, he predicted, chemical rockets of less than 500 sec. specific impulse would lift men to Mars, not by direct ascent but through a succession of parking orbits and assembly of hardware in space. Evvard told the meeting that since May 13, 1964, at least nine tests on Kiwi, Nerva, and Phoebus nuclear reactors had proved out a thrust greater than 50,000 lbs., at a specific impulse in excess of 750 sec. (Text)

- U.S.S.R. had offered to launch a French satellite with a Soviet rocket during space cooperation talks in October, a Centre National d'Etudes Spatiales (CNES) official told *Space Business Daily.* CNES had proposed launching an Imp-type payload into a 125,000-mi.-apogee orbit, but the Soviets preferred a 25,000-mi.-apogee orbit. (*SBD,* 11/16/65, 1)

November 16–18: The First Annual National Conference on Spacecraft Sterilization Technology, sponsored by NASA and hosted by Cal Tech, was held at Cal Tech to brief the space industry and the academic world on NASA's needs for spacecraft sterilization. (NASA Release 65–290)

November 17: International Association of Machinists union rejected McDonnell Aircraft Corp.'s latest contract proposal and prepared for a strike at midnight tomorrow that would halt the production of Gemini space capsules and F-4 Phantom jet fighter planes. McDonnell had offered a nine-cent-an-hour wage increase in each of the next three years and various improvements in fringe benefits and working conditions. Union members said, however, that the improvements still would not put them on a par with other workers in the aerospace and aircraft industries. (AP, *NYT*, 11/18/65, 27; AP, Wash. *Eve. Star*, 11/18/65, A3)

- Tracking and telemetry station to support Applications Technology Satellites (Ats) would be established by NASA at Toowoomba in eastern Australia near Brisbane at an approximate cost of $6 million. Plans for the new station were jointly announced by NASA Administrator James E. Webb and Australian Minister for Supply Allen Fairhall. (NASA Release 65–357)

- Balloon experiment with a collection system to study the Leonid meteor shower was flown from Palestine, Tex., the National Center for Atmospheric Control announced. It cruised at about 95,000-ft. altitude for about 10 hrs. and ended its flight near Concord, N.C. (AP, *NYT*, 11/20/65, 11)

- NASA Associate Administrator Dr. Robert C. Seamans, Jr., told the National Space Club in Washington, D.C., that the time had come for the Administration to decide on what goals it wanted to set for the period after landing astronauts on the moon. The U.S. must use the tremendous space capability it had carefully built up since SPUTNIK I, he said, "or see its value erode. If we do not use what we have created, continued expansion of the Soviet program will likely lead to future Soviet missions that will have the impact of Sputnik I." (Text)

- Discussing long-term plans of the nuclear rocket program, Dr. Harold B. Finger, Manager, AEC–NASA Space Nuclear Propulsion Office, said he foresaw development of a nuclear rocket engine having a thrust of 200,000 lbs. to 250,000 lbs., using reactors designed for 4,000 to 5,000 megawatts, and capable of direct-flight lunar landing missions; deep space unmanned space missions; and manned planetary missions. He said development of such an engine would utilize the technology already available and being developed through the Kiwi, Nerva, and Phoebus reactor programs. (AEC Release, 11/17/65)

- Christopher Kraft, flight director for the Gemini program, escaped death or serious injury when the gun pointed at him by a teenage boy wanting to go to Cuba to aid anti-Castro prisoners misfired. Incident occurred aboard a jet airliner enroute to Miami from New Orleans. The young gunman, identified as Thomas Robinson, 16, of Brownsville, Tex., was disarmed by one of the passengers and subdued by Kraft and Paul Haney, chief of public information at MSC. (AP, Balt. *Sun*, 11/18/65)

November 18: EXPLORER XXX (IQSY Solar Explorer) was successfully launched by NASA from Wallops Station, Va., by a four-stage Scout booster. Orbital data: apogee, 548 mi. (883 km.); perigee, 440 mi. (709 km.); period, 100.8 min.; inclination, 59.7°. Ninth Solrad Satellite developed by Naval Research Laboratory, EXPLORER XXX would monitor and measure x-ray emissions from the sun as part of the U.S. contribution to the International Quiet Sun Year project. (NASA Release 65–352; GSFC)

- All five geodesy experiment systems on the new NASA EXPLORER XXIX satellite had been tested successfully and operational programing of the spacecraft was expected within a few days, NASA announced. EXPLORER XXIX, popularly referred to as GEOS I, was launched from Eastern Test Range Nov. 6. (NASA Release 65–359)

- NASA Nike-Apache sounding rocket was launched in the second apparently successful attempt this week of Project Luster to capture particles of interplanetary matter during the Leonid meteor (shooting star) shower. The 105-lb. payload, containing special collection surfaces, was launched from White Sands Missile Range to peak altitude of about 100 mi. Instrument package separated from the rocket and descended into the range recovery area. Initial radar and telemetry data indicated that inflight performance of the rocket and its payload was as planned. Similar flight conducted Nov. 16 with an Aerobee 150 sounding rocket from White Sands also performed flawlessly. (NASA Release 65–358)

- Trouble with power-producing fuel cells threatened to delay the December 4 launching date of Gemini VII. During a test, pressure apparently built up and forced liquid hydrogen into the two fuel cells which combine hydrogen and oxygen to produce electrical power. Although the cells might not have been harmed, the decision was made to replace them. (AP, Wash. *Eve. Star*, 11/18/65, A3)

- NASA Associate Administrator for Space Science and Applications Dr. Homer E. Newell said at the Univ. of Pittsburgh dedication of its Space Research Coordination Center: ". . . one basic principle has governed our policy regarding relationships with educational institutions. That is: NASA intends to work within the structure of the colleges and universities in a manner that will strengthen them and at the same time make possible the accomplishment of our mission." (Text)

- Hughes Tool Co. announced it had obtained $297,000 Army contract to provide a helicopter that could convert in flight to a 400-mph airplane. It would lift by means of a single helicopter rotor with a large triangular hub. After reaching a horizontal speed of about 150 mph, the rotor would be stopped in such a position that the hub could then serve as the wing of a craft that would assume a delta-wing shape. The rotor would also be used in landing. When operating as a helicopter, the turbine engine would drive hot gases through the rotor hub to the rotor tips, and the rotor would be powered like a fireworks pinwheel. This would eliminate gearboxes and shafting. Hughes had tested the principle in flight with the experimental XV-9A research aircraft. (UPI, *NYT*, 11/18/65, 95)

- Display of what the Soviets called an orbital missile at the military parade in Moscow Nov. 7 and "related Russian statements do seem to place upon the Russian government an obligation to make clear its future intentions with respect to the [U.N.] resolution" against weapons in space, State Department press officer Robert J. McCloskey told a news conference. Noting that the U.S. could easily put such a weapon into orbit, McCloskey said that the military value of this would be negligible. (AP, *Wash. Post*, 11/19/65, A10)

November 18: Bernard J. Vierling, Director of FAA's Systems Maintenance Service since mid-1962, was named Deputy Director of FAA's Office of Supersonic Transport Development. (FAA Release 65–111)

November 19: NASA successfully conducted the first tethered test firing of the Apollo 009 Service Module propulsion system at Kennedy Space Center. Test was marred by failure in the master countdown clock, which—a split second after ignition—recycled to T minus 99 sec., throwing the count out of real time. The ground computer, which operated in plus time only, was thus unable to send the signal to shut off the engine. Instead of the two planned 15-sec. firings, the system was fired only once for a total of 20 sec. and shut off with a signal from the blockhouse. For the same reason, the ground computer was unable to send the signal to gimbal the engine and this apparatus was not tested. (UPI, *NYT*, 11/20/65, 21; *M&R*, 11/29/65, 28)

- First successful firing of the Sprint anti-missile missile in guided flight from an underground cell of the type conceived for operational use was conducted by U.S. Army at White Sands Missile Range. The first Army missile to use "pop-up" launch technique, Sprint was under development as a companion to the Zeus missile for the Nike-X missile defense project: Zeus was designed to intercept attacking warheads outside the earth's atmosphere and Sprint, targets that had penetrated the atmosphere. (DOD Release 832–65)

- Extravehicular equipment for Gemini VIII underwent two qualification tests in the 20-ft. vacuum chamber at NASA Manned Spacecraft Center. The Extravehicular Life Support System was combined with the Extravehicular Support Pack (Esp), a 92-lb. backpack in which the astronaut would carry his oxygen and maneuvering gas supply. First test evaluated capability of the oxygen bottle in the pack to supply the extravehicular astronaut with air at a high rate of flow. Second test was a propellant blowdown qualification in which the space gun was fired in 30-sec. bursts to exhaust the freon supply. A vacuum equal to 150,000-ft. altitudes existed in the chamber, and the walls of the thermal box had been cooled to $-300°$ F to simulate the conditions of orbital night for both tests. (MSC Release 65–105)

- Institute of Strategic Studies, a private organization in Britain, reported that the U.S. margin over the Soviet Union in numbers of missiles had dropped from 4–1 in early 1965 to 3–1 now. The Institute said the margin had been cut when the U.S. scrapped such obsolete missiles as the Atlas, while the Soviets boosted by 40% the number of their operational intercontinental ballistic missiles. (AP, Wash. *Eve. Star*, 11/19/65, A3; Myer, *Wash. Post*, 11/19/65)

- State Dept. official said at a press conference that radioactivity released by a Soviet underground atomic test in January 1965 was the result of a technical "miscalculation" and not a violation of the 1963 nuclear test ban treaty. (*Wash. Post*, 11/20/65, A4)

November 20: Plans for the side-by-side orbiting of Canadian Alouette and U.S. Explorer satellites to study the ionosphere were announced. Previously scheduled for Nov. 23 from Western Test Range, the launch had been postponed until at least Nov. 26 because of telemetry problems with the U.S. satellite. Double-launch project was called Isis-X (International Satellites for Ionospheric Studies). (AP, *NYT,* 11/21/65, 11; *Wash. Post,* 11/21/65, A14)

- William E. Simkin, director of the Federal Mediation and Conciliation Service, reported that "some progress" had been made in efforts to settle the IAM strike against McDonnell Aircraft Corp. which began Nov. 18. Union had demanded changes in job specifications and was dissatisfied with machinery for filing grievances against management. McDonnell was prime contractor for the Gemini spacecraft at Cape Kennedy. (Jones, *NYT,* 11/21/65, 43)
- Gemini 7 fuel cell system that would power the spacecraft during its 14-day mission was successfully turned on and was reportedly "working very well." Original fuel cell system had suffered possible damage because of a testing error Nov. 15 and had been replaced Nov. 17. Checks of the new unit were delayed by the strike against McDonnell Aircraft Corp. by IAM machinists at Kennedy Space Center. (*NYT,* 11/22/65, 40)
- For more than six years, U.S. nuclear warheads had been mounted secretly on planes and missiles of West Germany and other North Atlantic Treaty Organization allies, the *New York Times* reported. Through a combination of physical and electronic controls, the warheads had remained under American custody, and the allies could not use the weapons without specific approval of the U.S. (Finney, *NYT,* 11/21/65, 1)
- Japanese Navy icebreaker *Fuji,* a floating laboratory with a battering ram for a bow, left Tokyo on a voyage of scientific exploration in frozen Antarctic seas. Observations planned by the 18 scientists aboard were linked to worldwide scientific activity during the International Quiet Sun Year. (Trumbull, *NYT,* 11/21/65, 30)

November 21: Creation of an Institute for Earth Sciences to conduct advanced and applied research in seismology, geomagnetism, and geodesy, was announced by Dr. Robert M. White, Administrator of the Environmental Science Services Administration (ESSA). Under the direction of Dr. Leroy R. Alldredge, the Institute would actively seek new knowledge of the properties of the earth's interior and develop methods of using this knowledge to predict earthquakes. (ESSA Release 65-7)

Week of November 21: MSC remote site flight controller teams for the Gemini VII/VI mission began deploying to the seven locations around the world where they would exercise detailed real-time mission control during the upcoming flights of the two Gemini spacecraft. (MSC Release 65-109)

November 22: Gemini 7 spacecraft was mechanically mated with its launch vehicle following activation of two replaced fuel cell sections. The cells originally installed in the spacecraft were thought to have been damaged when one of hte cryogenic reactant tanks was inadvertently overpressurized. Preparations at Launch Complex 19 and flight controller simulations in Mission Control Center, Houston, were proceeding smoothly toward supporting the scheduled December 4 launch date. (MSC *Roundup*, 11/26/65, 1)

November 22: NASA Flight Research Center had received a flying laboratory that would be used to provide airborne simulation of advanced aircraft with particular emphasis on the proposed supersonic transport. Called the General Purpose Airborne Simulator (Gpas), the new system was a Lockheed JetStar capable of speeds greater than 550 mph and altitudes up to 40,000 ft. that had been modified by the Cornell Aeronautical Lab. under $1.3-million contract with NASA. It would enable NASA engineers and pilots to evaluate specific future aircraft designs in a wide variety of actual flight conditions. (FRC Release 24-65)

- At the request of Labor Secretary W. Willard Wirtz and chief Federal mediator William Simkin, the International Association of Machinists authorized 200 strikers to resume work on the Gemini spacecraft at Kennedy Space Center. The machinists struck McDonnell Aircraft Corp., Gemini's builder, on Nov. 18 in a dispute over wages and working conditions. Not only had the strike threatened to ground Gemini VII, scheduled for launch December 4, it had also halted, and continued to halt, work on the F-4 Phantom jet fighter at the McDonnell St. Louis plant. (AP, *Wash. Post*, 11/23/65, A8; Hoffman, *N.Y. Her. Trib.*, 11/23/65)

- NASA selected four companies to perform four-month design studies on an experiments pallet to fly aboard Project Apollo missions: Lockheed Missiles and Space Co.; McDonnell Aircraft Corp.; Martin Co.; and Northrop Space Labs. The firms, under separate and concurrent fixed-price contracts valued at approximately $375,000, would design, develop detailed specifications, and produce mock-ups of a pallet to house scientific, technological, and engineering experiments to be carried on Apollo missions of up to two-weeks duration beginning in 1968. After review and evaluation of the design studies, NASA planned to select one of the firms to develop the experiments pallet flight hardware under a cost-plus-incentive-fee contract. (NASA Release 65-361)

- AFSC Space Systems Div. would like to run a four-month series of wind-tunnel tests to requalify the Gemini Agena Target Vehicle, reported *Missiles and Rockets*. Test series would be part of USAF's and NASA's continuing effort to determine what caused the Agena failure on Oct. 25 and what modifications would be needed to prevent a recurrence. (*M&R*, 11/22/65, 13)

- In an editorial in *Missiles and Rockets*, William J. Coughlin said: "A start on a correct MOL public information program should be made immediately by taking SAMOS off the dirty-word list and bringing it back out in the open. To do otherwise is to keep the U.S. in the position of accepting reconnaissance as something offensive, in the most literal sense. It is not. The U.S. conducts reconnaissance over the Soviet Union for its own protection against a closed society. It should not be afraid to acknowledge that fact." (*M&R*, 11/22/65, 46)

- Surveyor mission to softland a television camera on the moon had been postponed until May 1966, JPL announced. Unspecified technical problems in the spacecraft and testing gear were blamed. (AP, *NYT*, 11/23/65, 11)

November 22: France postponed her first attempt to orbit a satellite. No reason was given. (UPI, *NYT*, 11/23/65, 52)

November 23: COSMOS XCVI was launched by the Soviet Union carrying scientific instruments for continued space research under the program announced by Tass Mar. 16, 1962. Orbital data: apogee, 310 km. (192.5 mi.); perigee, 227 km. (140.9 mi.); period, 89.6 min.; inclination, 51°54'. Equipment was functioning normally. (Tass, 11/23/65)

- Last in series of three NASA Nike-Apache sounding rockets with Univ. of Alaska instrumented payload was launched from Ft. Churchill, Canada, to obtain data on the visible aurora. First in the series of nighttime experiments was launched Nov. 16 and the second, Nov. 20. All three rockets performed satisfactorily. On first two flights experiment instrumentation performed normally, but on the third flight portions of the instrumentation did not function normally because nose cone failed to eject. The complete series resulted in an excellent collection of scientific data. (NASA Rpts. SRL)

- A team of scientists from USAF Office of Aerospace Research (OAR) and the Lockheed-California Co., under OAR sponsorship, recorded the annular eclipse at an observing station established in the Burma-Thailand area. Purpose of the expedition was to verify deviations in the moon's shape which appeared in photographs taken by the Lockheed-California Co. during two similar eclipses—one in West Africa in 1962, and the other in South Africa in 1963. (OAR Release)

- Sea bottom between New Guinea and the Solomon Islands, which was affecting the orbits of artificial satellites, had become subject of a survey by British and American geophysicists on the British survey vessel, *Dampier*. Gravitational pull of exceptionally dense rock about 24,000 ft. below the surface was so strong it tended to drag the satellites out of their intended orbits. (*NYT*, 11/23/65, 23)

- Preliminary results of ARPA-sponsored Project Longshot—detonation October 29 of an 80 kiloton nuclear device buried 2,300-ft. deep on Amchitka Island, the Aleutians—indicated that seismic wave arrival times from the region were earlier than predicted at all locations monitoring the event. This indicated a pronounced seismic velocity anomaly in the crust and mantle of the Amchitka region. If further analysis confirmed this result, it could lead to revisions in curves of energy loss for seismic wave propagation to long range. Results of Longshot also provided data of use in research on distinguishing earthquakes from nuclear events. (DOD Release 846–65)

November 24: Ground test version of the Saturn V launch vehicle's first stage (S–IC–T stage) was static fired for its full flight duration of about 2½ min. at NASA Marshall Space Flight Center. S–IC–T, designed by MSFC and the Boeing Co., generated its full thrust of 7.5-million lbs., equivalent, at maximum flight velocity, to about 160 million hp. Firing was the second conducted by the Boeing Co., MSFC's prime contractor for the S–IC. All early firings were conducted by the MSFC Test Laboratory, which directed today's test. (MSFC Release 65–287)

- In a 6,087-to-2,841 vote, machinists at the McDonnell Aircraft Corp. approved a new three-year contract with McDonnell Aircraft which had been worked out by McDonnell, IAM, and the Federal Mediation

Board to end the six-day walkout of 17,000 workers. The strike had only lasted four days for the 200 IAM members at Kennedy Space Center who had been allowed to return to work Nov. 22 on preparations for the Gemini 7–6 launch. (AP, *NYT*, 11/25/65, 56)

November 24: NASA had requested industry to submit proposals for a study of the feasibility of a satellite capable of broadcasting directly to conventional home FM radios and/or shortwave radios. Potential contractors were expected to have their proposals back to NASA in 45 days. Following a NASA evaluation, one or more contracts would be awarded for a detailed six-month mission study. (NASA Release 65-363)

- Tass announced that "in connection with extension of the program of scientific research aimed at further studying outer space, the Soviet Union will fire rocket boosters from November 25 to December 25, 1965, into a circular target area in the central Pacific having a radius of 40 n.m. and a center of 0°5 min. south and 163°45 min. west." Governments of countries using sea and air routes in the Pacific were requested not to enter this area from noon to midnight local time each day. (Tass, 11/24/65)

November 26: France successfully launched A–1, her first satellite, with the Diamant booster from Hammaguir Range, Algeria. Countdown had been interrupted for seven hours when a faulty diode was discovered in the boosters' third stage. Initial orbital data: apogee, 1,768 km. (1,098 mi.); perigee, 525 km. (326 mi.); period, 108 min.; inclination, 34.65° (compared with planned 2,154-km. [1,562-mi.] apogee, 553-km. [331-mi.] perigee, 53° inclination). The 88-lb. satellite, comparable to the U.S. Vanguard, carried a radio and radar transmitters but no scientific equipment; it was designed for a two-week lifetime. Radio was functioning feebly.

Primary purpose of launch was to test the Diamant booster, whose three stages had been tested individually but not as a three-stage launch vehicle. Second and third stages used solid fuel; the first used liquid. (Root, *Wash. Post*, 11/27/65, A1; AP, *NYT*, 11/27/65, 1, 4; *SBD*, 12/11/65, 172)

- U.S.S.R. launched COSMOS XCVII artificial earth satellite "containing scientific equipment for continuing outer space research," Tass reported. Orbital parameters were close to the calculated ones: apogee, 2,100 km. (1,304 mi.); perigee, 220 km. (136.6 mi.); period, 108.3 min.; inclination to equator, 49°. Onboard equipment was operating normally. (*Pravda*, 11/27/65, 1, ATSS–T Trans.)

- British Black Knight research rocket reentered earth's atmosphere at 10,800 mph after a successful firing at the Woomera range, Australia. Rocket had reached 390-mi. altitude. (AP, *Wash. Post*, 11/27/65, A2)

- National Aeronautic Assn. named Jerome Lederer, director of the Flight Safety Foundation of New York, as winner of the Wright Brothers Memorial Trophy for 1965. Trophy is awarded annually for significant public service of enduring value to aviation. (AP, *NYT*, 11/27/65, 65)

- NASA Marshall Space Flight Center announced that 48 additional J–2 liquid-hydrogen rocket engines would be purchased from Rocketdyne Div. of North American Aviation, Inc., under an amendment which converted the engine production contract to a cost-plus-incentive-award-fee agreement. The initial contract was a cost-plus-fixed-fee agreement. Cost of amendment was $75.8 million which brought the total value of the contract to approximately $206 million. A total of 103 J-2 engines was now on order for the Saturn IB and Saturn V launch vehicle program. (MSFC Release 65-289)

November 26: ComSatCorp announced the selection of Vern W. Johnson & Sons, Inc., Spokane, Wash., for site preparation and construction of buildings and other facilities for ComSatCorp's fixed earth station at Brewster Flat, Wash. The construction contract, totaling $909,382, was filed with FCC. The Brewster Flat station, when completed, would serve as a U.S. link in a worldwide commercial satellite communications system. (ComSatCorp Release)

- U.S. officials said this country lacked enough intelligence information to know with certainty if the Russians were developing a solid-fuel intercontinental ballistic missile, reported the Baltimore *Sun*. Weapons described by the Soviets as solid-fuel were displayed in a Nov. 7 Moscow parade observing the 48th anniversary of the Bolshevik Revolution. (Sehlstedt, Balt. *Sun*, 11/27/65, 5)

- Dr. Homi J. Bhabha, chairman of India's Atomic Energy Commission, denied reports that India was secretly preparing to explode a nuclear device. In an interview, he said: "We are still eighteen months away from exploding either a bomb or a device for peaceful purposes and we are doing nothing to reduce that period." (*NYT*, 11/29/65, 8)

November 27: Soviet Union launched COSMOS XCVIII unmanned satellite with "scientific apparatus to continue space investigations," Tass announced. Orbital data: apogee, 570 km. (354 mi.); perigee, 216 km. (134 mi.); period, 92 min.; inclination, 65°. Equipment was functioning normally. (*Pravda*, 11/28/65, 4)

- A-I, France's 88-lb. first satellite, continued orbiting, but its radio signals had become weaker. Telemetry analysis indicated part of the difficulty was damage to the antennas at launch Nov. 26 from Hammaguir Range, Algeria. (AP, Wash. *Eve. Star*, 11/27/65, A2; Reuters, Balt. *Sun*, 12/28/65, 6)

- With a direct $40,000 grant from NASA, the Staten Island Public Health Hospital was using convicts for a research program entitled "The Effects of Acute Heat Stress and Simulated Weightlessness." According to Warden Frank Kenton of the Federal Correctional Institution, Danbury, Conn., which supplied the men, those who had volunteered for the project were minimum risks and were carefully screened before selection was made. "It's a good break for these fellows. They get a $25 honorarium, three days off their sentences, an opportunity to get out of prison for a while, and a change in routine and surroundings." (*NYT*, 11/27/65, 32)

- Figures compiled by NASA indicated that since Jan. 1, 1964, 75 work stoppages had cost more than 92,000 man days of work, mostly at Kennedy Space Center, Associated Press reported. (AP, *NYT*, 11/28/65, 4)

November 28: Canadian ALOUETTE II and American EXPLORER XXXI (Direct Measurement Explorer) were launched in a pick-a-back configuration by NASA from the Western Test Range with a single Thor-Agena B booster. Their orbital parameters were nearly identical: apogee, 1,837 mi. (2,958 km.); perigee, 329 mi. (516.8 km.); period, 121 min.; inclinations, 80°. The two satellites would make related studies of the earth's ionosphere as they orbited in close proximity. Both the 323-lb. ALOUETTE II and the 218-lb. EXPLORER XXXI were performing well. Called Isis-X, the double-launch project was first in a new NASA–DRB (Canadian Defense Research Board) program for International Satellites for Ionospheric Studies (Isis). (NASA Release 65–355; GSFC)

November 28: Lights flashing earthward from 1,000 mi. in space were giving scientists their first operational workout with geodetic satellite EXPLORER XXIX launched by NASA Nov. 6 from Eastern Test Range. From a site about 100 mi. south of Cape Kennedy, five different types of camera systems were set to record on each clear night the one millisecond bursts of light from EXPLORER XXIX's four lamps. In this first programed exercise, the Jupiter Calibration Experiment, the cameras were positioned side-by-side at the Jupiter, Fla., Baker-Nunn Camera Station of the Smithsonian Astrophysical Observatory. Simultaneous use of the different camera systems would permit validation of data and procedures required for later phases of the NASA EXPLORER XXIX project that called for mutal visibility by ground stations equipped with optical and electronic tracking equipment. Calibration of the camera systems would be an important step in realizing maximum utilization of the NASA satellite and participating ground stations in the National Geodetic Satellite Program. (NASA Release 65–365)

- *Jane's All the World's Aircraft* carried a description of how Soviet cosmonauts were ejected from their space capsules at 23,000 ft. after reentry into earth's atmosphere. The yearbook also published detailed descriptions of the Soviet spaceship VOSTOK I, the world's first manned earth satellite. It showed a picture of the 120-ft. ballistic missile that was displayed last May in a Moscow parade and reported that the missile was a sister vehicle to the "mighty booster" that launched the Vostok in 1961. *Jane's* also obtained information on a large Soviet hovercraft still under construction and published a photograph taken at the shipyard. (AP, *NYT*, 11/29/65)

- The growing debate about what the next step in manned spaceflight should be was roughly divided into three schools, according to Howard Simons in the *Washington Post:* (1) a negative "Let's do nothing at all after landing men on the moon" school; (2) a "wild-space-yonder" school wanting to try everything; and (3) a "school of thought that would make do with what we are now developing through modification and imagination before taking expensive new steps into the heavens." He said the latter view would mean using Saturn IBs and Vs to launch modified Apollo spacecraft on largely scientific space ventures, possibly as early as 1968. (Simons, *Wash. Post,* 11/28/65)

November 28–December 1: The Committee on Space of the National Citizens' Commission on International Cooperation recommended nine possible areas of international space cooperation at the White House Conference on International Cooperation: (1) new launching sites for satellites; (2) multipurpose navigation satellite systems; (3) experimental data-gathering satellite system; (4) synoptic sounding rocket investigations; (5) applications of communications satellites; (6) mutual assistance between national and international tracking and data acquisition networks; (7) exploration of the distant planets; (8) remote sensing; and (9) international convention to govern human activity on the moon. (Text)

Vice President Hubert H. Humphrey, in opening remarks at the session on international space cooperation, said: "I want to support what I understand to be the central argument of the report.

"We have done well so far in pursuing our objective of international cooperation.

"Our cooperative projects have borne witness to our peaceful aims in space . . . involved foreign intellectual and material resources in our programs . . . and established patterns of cooperation that further our goal of a stable world.

"But the future demands a new level of effort." (Text)

The Honorable Earl Warren, Chief Justice of the U.S. Supreme Court, speaking at the White House Conference, said: "We live at a time when the whole world is being made over socially, economically, scientifically and even intellectually. Our era has witnessed such dramatic achievements as flight faster than sound, the splitting of the atom, miracle drugs, and manned satellites hurling through space. Rapid and turbulent changes in the scientific, economic, and social fields, almost too numerous to name, daily defy evaluation on the basis of prior standards and experience. Vistas of endless space have opened as man's horizons have widened to encompass the universe. At a pace beyond dreaming the whole pattern of our existence is being reshaped. To ensure that these changes work for the benefit of mankind, law must be developed rapidly enough in the world community to cope with the problems they raise and to harness their potential for peaceful and productive ends." (Text)

The Committee on Science and Technology of the National Citizens' Commission on International Cooperation said at the Conference that cooperation in science already "is an extensive and integral part of life" but that the same thing was not true in the development and application of technology. The Committee urged an international earthquake prediction program, a world oceanography organization, and a "greatly accelerated" program to disseminate technology "for improving the basic needs of man—such as nutrition, sanitation, health, shelter, and communications." (Text)

An Inter-American Skyway, linking cities in Latin America with each other and to North American centers, was suggested by the Committee on Aviation. The Skyway was needed, the Committee said, because airports, controlled airways, navigation aids, safety rules, and other services "lag behind the speed, number and variety" of aircraft on Latin American routes, as they do throughout the world. (Text)

November 29: Second Orbiting Solar Observatory (OSO II), after exceeding its operational life expectancy by 50 per cent, had been placed in coasting mode by NASA. Gas supply for the pitch-control in the stabilization system had been depleted, allowing the spacecraft to exceed acceptable tolerances in pointing at the sun. Data would be collected from time to time when the spacecraft did properly observe the sun. OSO II, launched Feb. 3, 1965, had completed more than 4,100 orbits and returned some 2,200,000 bits of scientific data each orbit. The nine advanced OSO II experiments were designed to further the work of OSO I as well as to extend measurements of the study of solar x-rays, gamma rays, and ultraviolet radiation. (NASA Release 65–367; GSFC Historian)

November 29: Gemini VII astronauts Frank Borman and James A. Lovell, Jr., reviewed for several hours where they would put food wrappings and other waste materials during their 14-day spaceflight scheduled to begin Dec. 4. Study of the problem was part of the final phase of training for the mission. (AP, *NYT*, 11/29/65, 55)

- M/G O. J. Ritland (USAF), retiring Nov. 30, was awarded the NASA Exceptional Service Medal by NASA Administrator James E. Webb. General Ritland who, as Deputy Commander for Space of the Air Force Systems Command, had been closely associated since 1962 with NASA's Office of Manned Space Flight, was cited for his contributions to the Mercury and Gemini manned space flight programs. (NASA Release 65-369)

- U.S.S.R. transmitted a color television program by satellite to France for the first time. Tass said the transmission, using the French Secam color system, followed a series of technical tryouts through MOLNIYA I, first of the Soviet Union's two communications satellites now in orbit. (Reuters, *NYT*, 11/30/65, 49)

- A moving model of a proposed lunar exploration vehicle under development by the Grumman Aircraft Engineering Corp. was demonstrated on two acres of simulated moonscape. The two-section wheeled craft, known as a mobile base simulator, lumbered at five mph around moonlike craters made of cinders and coal dust at the Grumman test field, Calverton, L. I. The Grumman engineer in charge of the project, Edward G. Markow, explained that the aluminum working model, valued at $250,000, was being developed in the hope that NASA officials would buy the project for use possibly by 1975. The exploration craft would have a lunar range of 250 mi. and accommodate two astronauts with 50 lbs. of equipment for perhaps two weeks. It was designed for use with the Saturn V booster. (*NYT*, 11/30/65, 22)

- Dr. Warren Weaver, former president of the AAAS, said in an interview with *U.S. News & World Report* that it was wrong to spend $30 billion on getting an American to the moon by 1970 when so much else to benefit the human race could be done with the money. He suggested: "We could give every teacher in the U.S. a ten percent raise a year for ten years; endow 200 small colleges with $10 million each; finance the education through graduate school of 50,000 scientists at $4,000-a-year; build ten new medical schools at $200 million each; build and endow complete universities for more than 50 developing countries; create three new Rockefeller Foundations worth $50 million each." (*U.S. News*, 11/29/65)

- Rep. Adam Clayton Powell (D-N.Y.), Chairman of the House Committee on Education and Labor, personally wrote the presidents of the Nation's 116 predominantly Negro colleges and universities urging them to apply for NASA grants and research contracts. In his letter he noted that of 187 colleges and universities receiving NASA grants, only one—Howard Univ.—was predominantly Negro and that 20 universities (10%) received $61,451,000 (50%) of the $121,115,000 total expenditure. (House Comm. on Education and Labor Release)

November 30: A suggestion made at the White House Conference on International Cooperation that the U.S. seek international agreement on a legal code for human activity on the moon received editorial comment in the *New York Times:* "An important precedent for a code of lunar law exists in the Antarctic Treaty of Dec. 1, 1959. That pact, to which both the United States and the Soviet Union adhere, provides that Antarctica shall be used only for peaceful purposes and shall contain no military bases. It suspends all territorial claims to the Antarctic and establishes the principle that all settlements and activities in the area are open to inspection by observers designated by the nations ratifying the treaty.

"The contracting nations agree to make available to each other all scientific information gathered there, and to exchange personnel among their expeditions and stations.

"Adoption of an analogous code for the moon would be a major triumph of international cooperation . . ." (*NYT*, 11/30/65, 40)

November 30: Science interest among U.S. high school students was declining, Dr. Robert Jastrow, director of the New York branch, NASA Goddard Institute for Space Studies, told the New York City Youth Board, meeting to inaugurate Youth Week. Citing figures from a study covering 1958 to 1963, Jastrow reported a 30% decline in the number of semi-finalists competing for National Merit Scholarship Awards who whose physics as a career preference. He suggested that the image of science as "a dehumanized field of work open only to a gifted few" could be counteracted only if scientists "are willing to step down from this pedestal and interact with young people." (Whitehouse, *NYT*, 12/5/65, 71)

• U.S. Army modified Redstone missile was successfully test-fired from WTR as part of Advanced Research Projects Agency's Project Defender. Some 24 Army-stockpiled Redstones had been ordered by ARPA for use in reentry measurements. This seven-year-old Redstone, among the first Redstones reactivated last June for Project Defender, was the first Redstone to be launched since November 1963. (*SBD*, 1/13/66, 72; *M&R*, 12/13/65, 17)

During November: Wendover Air Force Auxiliary Field, Utah, an empty facility in "caretaker" status, was reported to be leading candidate for site of the first American inland landing station for spacecraft. A study made by a DOD committee had described Wendover as "the only site known to satisfy the recovery requirements of polar orbiting vehicles." (*NYT*, 11/28/65, 50)

• The Daniel and Florence Guggenheim Foundation would award nine or more graduate fellowships to train men for engineering and scientific leadership in spaceflight, rocket-propulsion, and flight-structures research in 1966–67. The fellowships were for residents of the U.S. and Canada. (*NYT*, 11/29/65, 41)

• General Electric researchers at the Philadelphia Aquarama studying weightlessness problems of the Manned Orbiting Laboratory (Mol) in underwater tests substituting neutral buoyancy for weightlessness concluded that in weightlessness the time required to perform a task would be substantially increased in some instances by as much as 50%. The company-funded program had also succeeded in developing: (1) a tether line that could be made rigid in any position or left slack for use as a restraining device, a ladder, or means of moving; (2) a better understanding of procedures for movement between the Gemini spacecraft and Mol and out of the Gemini spacecraft for extravehicular experiments in the weightless state; (3) a conceptual design of restraining devices for use inside the cabins of both vehicles to aid in performance. It had been verified that neutral buoyancy appeared to simulate reasonably the parameters of weightlessness that have a significant effect on human motor performance, but could not simulate effects of weightlessness on any activity in an important role. (David, *M&R*, 11/8/65, 34)

During November: Compact, remote transmitters similar to those designed for NASA by USAF School of Aerospace Medicine, Brooks AFB, Tex., were being worn by cardiac patients as part of the Central Cardiac Monitoring System (Ccms) at the hospital at Andrews AFB, Md., reported *The Airman*. Use of Ccms, a centrally-located computer capable of receiving telemetered data from several ambulatory patients simultaneously, improved patient care and provided physicians advance warning of deterioration in a patient's condition. (*The Airman*, 11/65, 9)

- Policy statement on participation by NASA employees in "widely-attended dinners honoring aerospace pioneers" was issued by Deputy Administrator Dr. Hugh L. Dryden: "NASA policy does not prevent acceptance of an invitation from a company to attend such affairs, provided the expense borne by the company is limited to the cost of tickets. Lodging or travel expenses may not be accepted." (NASA Ann.)

- FCC received 27 statements on the question of direct purchase of communications services from ComSatCorp. A brief submitted by the American Newspaper Publishers Assn. argued that "a basic policy determination [should] be made at this time, assuring access by the press to ComSat facilities for news media dissemination. . . ." A brief by AT&T argued that "authorized users" of ComSat services should be common carrier companies only not any communications organization wishing to rent satellite channels. The "only interpretation . . . consistent with the specific language of the Communications Satellite Act of 1962," argued AT&T, is "the view that ComSat is to function as a 'carriers' carrier,' except possibly in the case of the U.S. Government. . . ." Weekley, *Wash. Post*, 11/4/65, C9)

- Study of approximately 300 photographs taken over the Antarctic by NASA NIMBUS I meteorological satellite had caused the U.S. Geological Survey to plan three major changes on its relief maps: (1) Mount Siple, 10,000-ft.-high Antarctic mountain used by pilots as a navigational aid, would be repositioned 45 mi. further west; (2) a mountain group in the Kohler range, positioned by two expeditions in two different locations, would appear as one group rather than two; and (3) ice front information on the Filshner Ice Shelf, Weddell Sea, and Princess Martha Coast areas would be updated with photographs producing better definition of the ice shelf's shape. (GSFC Release G-29-65)

- A Soviet booster capable of generating more than 60-million hp. had launched the instrumented spacecraft of the Proton series, Prof. Georgi V. Petrovich disclosed in *Aviatsiya i Cosmonavtika*. Each engine had produced about 3-million-lbs. thrust at liftoff. USAF Titan III-C had produced 2.4-million-lbs. thrust when it was launched with an instrumented payload June 18. (Shabad, *NYT*, 11/14/65, 74)

December 1965

December 1: Full-duration test firing of the second flight S–IVB stage of the Saturn IB launch vehicle was conducted at Sacramento by Douglas Aircraft Co. The 7½-min. acceptance test was run to check engine performance, propellant tank pressurization systems, data acquisition systems, power and control systems, and structural reliability of the rocket stage. Following detailed post-test evaluation, the S–IVB would be shipped to Kennedy Space Center for launch as part of a complete Saturn IB vehicle in 1966. (*Marshall Star,* 12/8/65, 4)

- Gemini VII Astronauts Frank Borman (L/Col., USAF) and James A. Lovell, Jr. (Cdr., USN), passed their 5-hr. preflight physical examination at KSC. Dr. Charles Berry, Gemini medical director, said he could find no medical problem that might interfere with the scheduled launch of Gemini VII Dec. 4. (Wash. *Eve. Star,* 12/1/65, A3; Wilford, *NYT,* 12/2/65, 22)
- Dr. Thomas L. K. Smull, Director of NASA Office of Grants and Research Contracts, was temporarily assigned as Special Assistant to Administrator James E. Webb. He would be concerned with examining, in depth, the progress made in the development and conduct of university activities. Dr. John T. Holloway would serve as the Acting Director of the Office of Grants and Research Contracts. (NASA Ann.)
- Dr. Donald P. Burcham, Voyager project manager at JPL, told the *Christian Science Monitor* in an interview that NASA had decided to launch the Voyager spacecraft with the Saturn V booster instead of the Saturn IB. "This," Dr. Burcham said, "means we can put a heavier retrorocket pack on the spacecraft. It will enable us to release the lander after the spacecraft has gone into orbit around Mars instead of 10 days before reaching the planet, as planned earlier, and that will give us a better chance of putting a lander on the surface." (Cowen, *CSM,* 12/1/65)
- Col. John H. Glenn (USMC, Ret.) was guest of honor and elected to membership at the 148th annual dinner of the New York Academy of Sciences. He said that 100 yrs. from now "space travel will not be a form of international competition, nor will it have political implications; it will be the beginning of the greatest exploration period in world history." (*NYT,* 12/3/65, 42)
- NASA Marshall Space Flight Center had awarded a $358,808 contract to Bryson Construction Co., Inc., Ala., for building a non-destructive facility for testing rocket materials in simulated space environments. Four shielded "radiographic laboratories" would provide areas where rocket materials would be exposed to radiation and would have equipment for searchray, motion radiography, radioisotope, and radiography testing. (MSFC Release 65–290)

December 1: ComSatCorp announced selection of J. P. Finan General Contractor, Inc., for site preparation and construction of buildings and other facilities for ComSatCorp's earth station at Paumalu, Oahu, Hawaii, under a contract totaling $948,362. The Paumalu station, when completed, would serve as part of a worldwide commercial satellite communications system. (ComSatCorp Release)

- Unpublished study of the Organization of Economic Cooperation and Development (OECD)—international, intergovernmental agency—to determine importance of military and space research for civilian technological advance was reported by the *Washington Post* to contain three key conclusions: (1) military and space research is a costly and inefficient way of spurring technological advance for the civilian economy; (2) although there is some direct civilian application of new products and techniques developed for the military, the amount is limited and is shrinking; (3) while a massive military-space program like that of the U.S. may indirectly stimulate innovation by civilians, some of the same results could probably be obtained more cheaply by direct government aid to nondefense research and development. The study implied there were better ways for Europe's industry to catch up. The report also stated the U.S. and the Soviet Union were putting about the same effort into research, and Western Europe was not far behind. In terms of manpower involved, U.S. total was over 1 million; European, about 500,000; Soviet, 1.5 million. In international earnings of patents and licenses, the U.S. was ahead. (Nossiter, *Wash. Post*, 12/2/65, A24; *NYT*, 12/19/65, 20)

December 2: NASA launched an Aerobee 150 sounding rocket from White Sands Missile Range to peak altitude of 115 mi. (185 km.). Objectives were to photograph the solar disc and to measure solar flux. Parachuted payload was recovered without observable damage; development of film was in process. (NASA Rpt. SRL)

- Dr. Hugh L. Dryden, NASA Deputy Administrator since 1958, died of cancer at age 67. He was a former Director of the National Advisory Committee for Aeronautics, a former Associate Director of the National Bureau of Standards, and—for forty-five years—an ordained minister in the Methodist Church.

    An internationally renowned scientist-engineer who once said he and the airplane grew up together, he was recognized for his leadership in the development of aeronautics and astronautics. He generally was regarded as the man who guided the United States into the space age. Dr. Dryden served the U.S. Government with distinction since 1918 in science and technology, as an administrator, and more recently as a skilled diplomat in negotiating international agreements for cooperative efforts in the peaceful exploration of space. The recipient of many awards and honors, Dr. Dryden was especially known for his scientific contributions to fluid mechanics and boundary-layer phenomena.

    At Johnson City, Tex., President Johnson said that the death of Dr. Dryden "is a deep personal loss and a reason for national sorrow. . . . No soldier ever performed his duty with more bravery and no statesman ever charted new courses with more dedication than Hugh Dryden. Whenever the first American space man sets foot on the moon or finds a new trail to a new star, he will know that Hugh Dryden was one of those who gave him knowledge and illumination."

From Great Falls, Mont., Vice President Humphrey, Chairman of the NASC, said: "The death of Dr. Hugh Dryden is a sad loss to all of us, and especially to those of us who have been guided by his wisdom, experience, and great common sense in planning the nation's space program.

"We shall miss him sorely as we plot our course for the decade ahead. So much of what this nation has been able to do in aeronautics and space over the past 40 years we owe to the creative science and the confident, skillful leadership of this great public servant. I know of no finer example of modern man in all his versatility than Dr. Hugh Dryden, whose vision, courage, and lifetime of service have helped to lead the way into the Space Age. . . ."

NASA Administrator James E. Webb said: "Dr. Hugh L. Dryden was a man of his time—of the air age, of the nuclear age, of the space age, with all the implications of modern science and technology for the accomplishments of his time. He was also a man for all times because his courage in undertaking the conquest of air and space, his scientific, engineering, and administrative competence and his qualities of humanity and leadership mark him as one of the truly great men who have contributed over the years to these fields. He will be sorely missed in NASA, in the nation and, indeed, in many nations."

Dr. Edward C. Welsh, Executive Secretary of the National Aeronautics and Space Council, said: "The country has lost one of its greatest scientists and one of its most dedicated citizens. The country will have to continue with its space and aeronautics programs without him —but will not be able to carry on as well without his inspiring leadership and ability." (Texts)

December 2: Tass reported that U.S.S.R.'s VENUS II had covered 6,540,000 km. (4,062,111.8 mi.); VENUS III, 5,310,000 km. (3,298,136 mi.). Both spacecraft were in trajectories close to those calculated; all onboard equipment was functioning normally. (Tass, 12/2/65)

- President Johnson named Dr. Finn J. Larsen, Honeywell vice president, as Principal Director of Defense Research and Engineering for DOD, succeeding Dr. E. G. Fubini, who left the job July 15. (AP, *NYT*, 12/3/65, 22; *Pres. Doc.*, 12/6/65, 546)
- Lt. Frank K. Ellis (USN), double-amputee pilot whose name was among astronaut candidates submitted to NASA by Chief of Naval Operations Adm. David L. McDonald, had been nominated for "special capacity" work in the space program, unidentified NASA sources revealed. While not nominated to become an actual astronaut, Ellis would be considered for work that would use his aeronautical knowledge. He lost both legs in a jet crash in July 1962. (CNS, *San Diego Eve. Trib.*, 12/3/65)
- 65-ton Saturn S–IVB "battleship" tankage and associated vehicle equipment were shipped from San Francisco by MSFC to USAF Arnold Engineering Development Center, Tullahoma, Tenn., for use in high-altitude tests of the stage's J–2 engine. (MSFC Release 65–291)
- A new alloy of hafnium and tantalum had been developed which could withstand temperatures as high as 4000° F. It would raise tempera-

ture limits formerly placed on such engine components as nozzles and throat inserts by the lack of reliable protective coatings. Application of the alloy was developed on a NASA contract by the IIT Research Institute. (NASA Release 65-365)

December 2: Physicist and author Dr. Ralph E. Lapp said at Central Connecticut State College that the space budget would decline from $5 billion to about $2 billion by 1970 unless new space projects were proposed and approved soon. He was pessimistic about the possibilities of finding worthwhile proposals among those now being considered by NASA and suggested that (1) Presidential science adviser Donald Hornig issue a report which would be "an accounting and a forecast" of the possibilities in space, and (2) the National Academy of Sciences poll its 700 members for their views on how Federal funds for research and development should be allocated for space and other projects. Lapp said "near-earth" projects such as manned earth stations offered some posibilities, but not enough to fill the gap which would soon appear in the aeronautics and space industry. (Garwood, *Wash. Post,* 12/3/65, A8)

- Need for creation of a "Buck Rogers rescue squad" with trained men and equipment to give aid on short notice to spacemen stranded or stricken in orbit was discussed by William Hines in the Washington *Evening Star.* Hines said a persuasive case had been made by Michael Stoiko of the Martin Co.: "Stoiko estimated that in the next 20 years there will be about 280 manned space flights involving about 800 men. Half of each—men and flights—will be American, half Russian. Total flight time will be about 2.4 million man hours, compared with the world-wide total of about 1,200 man hours to date."

  Stoiko projected "a 62 percent probability of at least seven emergency situations involving 22 men in the next 20 years; a 58 percent probability of two or three emergencies in the coming decade."

  Hines said that now "astronauts stranded in orbit would have no hope of rescue and only a melancholy choice between two fatal alternatives: to sweat it out until the oxygen finally was gone, or to make a quick end of it by explosively decompressing both spacecraft and suit." (Hines, Wash. *Eve. Star,* 12/2/65)

- Sir Francis Vallat, acting director of McGill Univ.'s Institute of Air and Space Law, told a news conference that if a piece of a satellite were to hit an individual on the head, he could sue the government that launched it and probably win the case. (AP, Wash. *Eve. Star,* 12/3/65, A8)

- Avco Corp. was issued a $1,600,000 increment to a previously awarded contract for design, development, fabrication, test, and evaluation of Minuteman Mark IIA reentry vehicles for AFSC. (DOD Release 871-65)

December 3: U.S.S.R. launched LUNA VIII unmanned spacecraft toward the moon. Objectives were testing of soft lunar landing system and scientific research. Weighing 1,552 kg. (3,422 lbs.), the spacecraft was following a trajectory close to the calculated one. Equipment was functioning normally. (*Komsomolskaya Pravda,* 12/4/65, 1, ATSS-T Trans.)

- AEC announced the U.S. had conducted a weapons-related nuclear test in Nevada with a yield equivalent to an explosive force of 200,000 to one

December 4–18: Photograph of GEMINI VII taken from GEMINI VI during rendezvous December 15 (left); and, GEMINI VII Astronauts Frank Borman (leading) and James A. Lovell, Jr., step onboard U.S.S. *Wasp* after recovery by helicopter from 14-day mission.

million tons of TNT. It was the 22nd test announced in 1965 and reportedly the year's biggest. (*NYT*, 12/4/65, 3)

December 3: ComSatCorp announced a $4,650,000 contract with Sylvania Electric Products, Inc., for two large antenna systems to be installed at earth stations in Paumalu, Hawaii, and Brewster Flat, Wash. Movable portions of the 85-ft. dish antennas would weigh more than 135 tons and, when in place atop concrete foundations, be up to 110 ft. high. When completed, the stations would serve as links in a worldwide commercial satellite communications system. (ComSatCorp Release)

- Dr. John A. O'Keefe, NASA Goddard Space Flight Center, said previous Russian efforts to land a vehicle on the moon had failed because they could not slow it up enough at the point of contact. Dr. O'Keefe also advanced the view that the moon originally broke off from the earth and eventually assumed its present position in the universe. He spoke at the monthly meeting of the Catholic Laymen's First Friday Club in Washington, D.C. (*Wash. Eve. Star*, 12/4/65, A3)

- Secretary of Defense Robert S. McNamara directed USAF to proceed with development of a reconnaissance version of the F–111. New version would be called the RF–111A and would be developed by General Dynamics Corp., prime contractor for the USAF F–111A and USN's F–111B tactical aircraft. More than $12,000,000 had been authorized for the initial development program, which would be directed by AFSC. (DOD Release 873–65)

December 4–18: NASA's GEMINI VII spacecraft, piloted by Astronauts Frank Borman (Maj., USAF), command pilot, and James A. Lovell, Jr. LCdr., USN), pilot, was successfully launched from Eastern Test Range at 2:30 p.m. EST on 14-day mission—longest U.S. flight to date. It would be target vehicle in rendezvous with Gemini 6 spacecraft, scheduled for launch December 13. Titan II booster's first stage burned 155 sec.; second stage separated and burned 182 sec. Traveling at 17,586 mph, GEMINI VII was inserted into elliptical orbit with 203-mi. (327.4-km.) apogee; 100-mi. (161.2-km.) perigee; 89-min. period; and 28.9° inclination. Immediately after spacecraft separation, GEMINI VII turned blunt end forward and began stationkeeping on Titan II's second stage at distances from 20 ft. to 50 mi. for 17 min. Several hours later Borman made the first of a series of course corrections to position GEMINI VII for the scheduled rendezvous with Gemini VI by firing the thruster rockets 1¼ min. to raise the apogee from 100 mi. to 138 mi. Apparent loss of pressure in spacecraft's fuel cells during flight's early stages was later determined to be faulty indicator light.

On Dec. 5, the astronauts encountered difficulties at first in sighting the rectangular panels displayed near Laredo, Tex., but later they successfully identified the patterns of panels and provided evidence that their visual acuity was not degraded over the 14-day period.

On Dec. 6, Lovell removed his 16-lb. spacesuit, becoming first U.S. astronaut to fly in undergarment. The astronauts visually tracked Polaris A-3 missile fired from USS *Benjamin Franklin,* submerged nuclear submarine off Cape Kennedy, and successfully tested onboard radar receiving unit needed for rendezvous.

On Dec. 7, Borman fired thruster rockets to achieve a higher orbit for rendezvous attempt: apogee, 197 mi. (301.7 km.); perigee, 145 mi. (233.8 km.).

On Dec. 8, clouds over New Mexico caused astronauts to cancel scheduled experiment to communicate with ground by laser beam.

On Dec. 9, Borman executed a posigrade maneuver to circularize GEMINI VII's orbit, firing thruster rockets one minute 18 sec. to raise perigee to 185.8 mi. (299.7 km.); 43 min. later, he fired braking thrusters 15 sec. to lower apogee to 188.3 mi. (103.7 km.) and provide a proper target orbit for Gemini 6. Dr. Berry recommended that Borman apply ointment to ease the nasal dryness caused by the 100%-oxygen atmosphere of his spacesuit.

On Dec. 10, when temperatures in GEMINI VII's cabin rose to 85°, Flight Director Christopher Kraft ordered Lovell to put on his spacesuit to permit Borman to remove his. One astronaut was to wear his spacesuit at all times during the mission. Dr. Berry recommended that Lovell apply antihistamine ointment to ease nasal dryness.

On Dec. 11, laser experiment was conducted with partial success: the Hawaiian ground station received signals from the spacecraft when Lovell locked his hand-held laser onto the ground-based one, but contact was not precise or strong enough to carry a human voice.

On Dec. 12, the astronauts tried again to establish ground laser communications over White Sands Missile Range, but were able to pick up only two quick flashes from the ground beacon.

On Dec. 13, when GEMINI VII's fuel cell warning light flashed on, the astronauts flushed excess water from the system by forcing in additional oxygen borrowed from the cabin's oxygen supply, thereby correcting the difficulty. A taperecorder malfunction prevented the spacecraft from supplying automatic data on performance of its systems between tracking stations.

On Dec. 14, Borman observed reentry of a Minuteman missile fired from Vandenberg AFB to Eniwetok—first time a missile reentry had been sighted by an astronaut. Both GEMINI VII astronauts were beginning to feel the strain of their ten days in space: "Jim and I are beginning to notice the days seem to be lengthening a little," Borman said. "We're getting a little crummy."

On Dec. 15, GEMINI VII and GEMINI VI achieved their historic rendezvous in orbit (see Dec. 15–16).

On Dec. 16, a report from GEMINI VII Astronauts Borman and Lovell that three of the six fuel cell stacks were not operating aroused apprehension that the flight might end prematurely.

Next day GEMINI VII's fuel cells began to function satisfactorily, thereby assuring a full-duration mission. Astronauts Borman and Lovell checked out all their spacecraft's systems and received reentry instructions.

On Dec. 18, GEMINI VII began a normal reentry when its four retrorockets fired automatically in correct sequence above the Pacific southeast of the Philippines. After a controlled reentry to the predetermined landing point, the spacecraft splashed down in the Atlantic at 9:06 a.m. EST, 700 mi. southwest of Bermuda—only 7.6 mi. from target. The astronauts, after their record-breaking 330-hr. 35-min. spaceflight, were assisted from the capsule by Navy frogmen and taken to the carrier *Wasp* for medical debriefing. (NASA Release 65–362; NASA Proj. Off.; Transcript; *NYT*, 12/5/65, 1, 72; 12/6/65, 1, 43; 12/7/65, 24; 12/13/65, 1, 46; 12/15/65, 23; *WSJ*, 12/9/65, 1; 12/13/65, 1; 12/14/65, 1; *Wash. Post*, 12/6/65, A3; 12/7/65, A1, A3; 12/9/65, A9; 12/10/65, A1, A3; 12/12/65, A1, A8; 12/13/65, A1; 12/14/65, A3; Wash. Sun. *Star*, 12/5/65, A1, A8; Wash. Eve. *Star*, 12/6/65, A1, A12; 12/7/65, A3; 12/8/65, A3; 12/12/65, A7; 12/15/65, A1, A2, A6; *Time*, 12/24/65, 32–36; MSC *Gemini VII/Gemini VI Fact Sheet*)

December 4: President Johnson, after viewing on television the launching of GEMINI VII, said: "Once again, two brave Americans have carried the quest for knowledge to the threshold of space. They also take with them our prayers, and our pride. As they orbit the earth in the days and weeks ahead, Astronauts Frank Borman and James Lovell will broaden our knowledge of space. But they will do more. Their voyage will be a continuous reminder that the peaceful conquest of space is the only form of conquest in which modern man can proudly and profitably engage. In this struggle, all men are allies, and the only enemy is a hostile environment. The victory over the final enemy will belong, not just to Americans, but to all the world. We are proud that these fine young Americans have brought us one step closer to that goal." (*Pres. Doc.*, 12/4/65)

- KSC technicians began to repair Pad 19 for the Gemini VI launch immediately after GEMINI VII had lifted off. Damage was minimal, NASA officials reported, and they were optimistic about being able to launch Gemini VI Dec. 13. (Simons, *Wash. Post*, 12/5/65, 1)

December 5: LUNA VIII would land on the moon Dec. 6 at about 4:50 p.m. EST, Tass reported. Telemetry data indicated the spacecraft was on a correct trajectory and functioning normally. (Reuters, *NYT*, 12/6/65, 43)

- In an editorial titled "Loss of a Leader," the Washington *Sunday Star* concluded a review of Dr. Hugh L. Dryden's contributions to American aeronautics and astronautics: "It is in the tradition of NASA, and of NACA before it, to name its research centers and after important figures in the development of aeronautics and astronautics: Langley, Lewis, Ames, Marshall, Kennedy. One major center—the one at Houston—remains unnamed. It would be a fitting tribute to a dedicated American if that center were henceforth to bear the name of Hugh L. Dryden." (Wash. Sun. Star, 12/5/65, B2)

December 5: Hamilton Standard Div. of United Aircraft Corp. successfully tested a life-support back pack designed to meet requirements of the lunar surface suit for the Apollo lunar-landing mission. System functioned as planned for over three hours inside a vacuum chamber while the test subject walked on a treadmill to simulate the metabolic load of an astronaut on the lurain. The 65-lb. portable life support system supplied oxygen, pressurized to a minimum 3.7 lbs. psi, controlled its temperature and relative humidity, and circulated it through the suit and helmet. The pack pumped cooled water through the tubing of the undergarment for cooling inside the pressure suit. A canister of lithium hydroxide trapped carbon dioxide and other air contaminants to purify the oxygen for reuse. (UPI, *NYT*, 12/6/65, 42)

December 6: FR–I (FR–1A), second French satellite to be launched within two weeks, was successfully injected into near-polar orbit from the Western Test Range by a NASA Scout booster. Orbital parameters: apogee, 480 mi. (773 km.); perigee, 462 mi. (743.8 km.); period, 100 min.; inclination, 75.9°. NASA provided the four-stage launch vehicle and launch service as part of its cooperative agreement with the French Centre National d'Etudes Spatiales (CNES), which designed, built, and tested the 135-lb. satellite. FR–I would study propagation of very low frequency (Vlf) radio waves and measure electron densities. Telemetry reports indicated all systems were operating nominally and were returning useful data. (NASA Release 65–366; *Wash. Post*, 12/7/65, A6; UPI, *NYT*, 12/8/65, 13)

- Funeral in Washington for late Deputy Administrator of NASA, Dr. Hugh L. Dryden, attended by hundreds of friends and high Federal officials. Pallbearers were Gen. James H. Doolittle (Chairman of the NACA, 1956–58), Dr. T. Keith Glennan (NASA Administrator, 1958–1961), and Astronauts Scott Carpenter, Gordon Cooper, James McDivitt, and John Glenn. (EPH)

- Results of an experiment to be performed on GEMINI VII to measure amount of bone demineralization brought on by 14 days of weightlessness in space could lead to a preventive that would protect aged persons suffering from brittle bones, said Dr. George P. Vose, professor of radiographic research at Texas Women's Univ. (Lee, *Houston Post*, 12/6/65)

- Course of Soviet spacecraft LUNA VIII was successfully corrected and it was now on a trajectory toward the moon close to the calculated one, Tass announced. (Tass, 12/6/65)

- Test of the Apollo command module and launch escape system, scheduled for Dec. 8 at White Sands Missile Range, was postponed due to a malfunction in the Little Joe II booster rocket guidance system. Malfunction was caused by bad soldering connections. (MSC Info. Proj. Off.; Las Cruces *Sun News*, 12/6/65)

December 6: The $2-billion Rover nuclear-propelled rocket program was approaching a decision whether to move forward and prepare for early post-Apollo missions or let the technology dwindle, Harold B. Finger, Director of NASA-AEC Space Nuclear Propulsion Office, told *Aviation Week & Space Technology*. The two-part decision was (1) whether to move forward with a flight-type engine of 200,000–300,000-lb. thrust— an advanced Nerva (Nuclear Engine for Rocket Vehicle Application), and (2) whether to start a $50-million construction program in Nevada for testing in the vicinity of the Nuclear Rocket Development Station near Las Vegas. Unless these decisions were made soon, "there will be a tail-off of technology," Finger said. (*Av. Wk.*, 12/6/65, 57)

- Vice President Hubert H. Humphrey urged establishment of research and experimental institutions in underdeveloped nations that would enable young scientists and technologists there to develop their own countries. "To train people without giving them the opportunity to put their training to full use can only result in frustration at home or migration abroad," he said.

  The Vice President, speaking at a dinner given in New York by the American Committee for the Weizman Institute of Science, emphasized that measures must be taken "to make these talented people actually want to work in their own countries, by opening up to them careers which are genuinely and deeply rewarding in professional advancement and service to their people." (*NYT*, 12/7/65, 10)

- Scientists and broadcasting executives from 20 countries began a four-day conference in Paris, sponsored by UNESCO, on communications satellites and their probable effect on press and radio. (Reuters, *NYT*, 12/6/65, 3)

- NASA would sponsor six 10-week summer programs of study and advanced research for almost 150 young university engineering and science faculty members in 1966. Recipients of the summer fellowships could participate in an ongoing space research project at a NASA center and at the same time carry on related seminars at nearby universities. 11 schools were participating in the program with the NASA centers. Primary objectives of the program were to stimulate and enrich university activities and to encourage exchange of ideas between NASA and university personnel. (NASA Release 65–374; Proj. Off.)

- USAF announced that first firing of an operationally configured Minuteman II ICBM from an operational silo was "100 percent successful with all research and development objectives achieved" and that Minuteman II performed with "accuracy and reliability," impacting more than 5,000 mi. downrange from the launch site at Vandenberg AFB. (AFSC Release)

- USAF was considering using obsolete Boeing Minuteman missiles as space launch vehicles with the United Technology Center FW–4s solid-propellant motor as a fourth stage, *Aviation Week and Space Technology* reported. Studies by Boeing and DOD had indicated such a vehicle would be cheaper than a new Ling-Temco-Vought Scout. (*Av. Wk.*, 12/6/65, 23)

- The Soviet Union could upset the U.S.'s military superiority by scientific breakthroughs in any one of four military areas, reported Ray Cromley in the *New York World Telegram:* (1) ability to disrupt U.S. communications completely just before or during a war; (2) effective defense against Polaris-type submarines; (3) adequate system of defense against ICBMs; (4) ICBMs so accurate that hardened silos would not be adequate protection for unfired Minutemen. (Cromley, *N.Y. World Telegram,* 12/6/65, 7)

December 7: Soviet spacecraft LUNA VIII impacted on the moon. Indications were that it was destroyed instead of making a soft landing. Tass reported that "the systems of the station were functioning normally at all stages of the landing except the final touchdown."

Sir Bernard Lovell, director of the Jodrell Bank radiotelescope tracking station, said: "The recordings of the final stages of Luna 8 made at Jodrell Bank indicate that a minor fault probably developed approximately four minutes before touchdown. The retrorockets did not have full effect and the probe probably made a hard rather than a crash landing, but nevertheless sufficient to put the instruments out of action. It seems clear the Russians narrowly missed complete success. . . . They have probably obtained a great deal of new information which will enable them to correct the remaining minor faults. . . ." (Tass, 12/7/65; Balt. *Sun,* 12/8/65)

- Discovery of a malfunction in the computer on Gemini 6 spacecraft might prevent a December 13 launch, NASA officials speculated. The computer, whose "memory" had been altered, was removed from the spacecraft to have new information fed into it. (Wilford, *NYT,* 12/8/65, 10C)

- Soviet Premier Aleksey Kosygin was asked in an interview at the Kremlin by *New York Times* associate editor James Reston if expenditures for space exploration were questioned by Soviet government officials. He replied: "I would say that man will always go on seeking a solution to the problems of the universe. There will always be funds that will be set aside to resolve the problems relating to the world and the universe; this is all to the good, if it's purely scientific. . . . We don't have any contradiction in the Soviet Union between appropriations for space research and for the needs of the population. The funds appropriated are relatively small compared with the funds serving the needs of the population on education and such. They are negligible on the over-all expenditure. Space expenses do not detract from the needs of the population."

  Asked if the U.S.S.R. wanted to go forward with peaceful space arrangements, Kosygin said: "We have expressed our view on this. We are in favor of peaceful, not military, uses of outer space. Now, to insure that peaceful research in space does not place too much of a burden on us, states might reduce military expenditures and channel the funds into peaceful space explorations."

  Reston queried: "You have made great progress in space with the use of the most sophisticated and complicated computers and other modern science, yet your speech on Sept. 27 was highly critical of the lag in the use of modern science in Soviet industry. Why this discrepancy?" Kosygin replied: "I don't understand your distinction between space and economic and industrial development. We judge our economy as a whole—we include all spheres. It is true that not everything is going satisfactorily in science and technology. Not everywhere are the most advanced methods being used. We are now taking all possible measures and we shall remedy this situation. For this it is necessary to utilize all the achievements of engineering thought. We are doing our utmost to insure the further rise of our economy and our technology. In four months I will make my report to the party congress on our forthcoming five-year development plan. In this plan our most advanced scientific and technical achievements will be mobilized." (Text, *NYT*, 12/8/65, C20)

December 7: Australian National Development Minister Allen Fairhall said that contracts were being arranged for new spacetracking stations at Cooby Creek, in Queensland, and Honeysuckle Creek, near Canberra, in support of NASA's programs. Stations were expected to be operational by 1966. (*NYT*, 12/27/65, 23)

- Designs for an aircraft with speed potential of 17,500 mph were being studied by both USAF and NASA, reported the *Washington Daily News*. Called the Scramjet (Supersonic Combustion Ramjet) by USAF and the hypersonic Ramjet by NASA, the aircraft could reach about 180,000-ft. altitude with the Ramjet engine and then be boosted into space by a small auxiliary hydrogen rocket motor. Above 180,000 ft. there would not be enough oxygen to support the Ramjet engine's combustion. The aircraft would fire retrorockets to reenter the atmosphere and return to earth. One of three competing companies would be selected to begin work on test engine hardware by summer 1966. (*Wash. Daily News*, 12/7/65, 3)

- Survey issued by the North American Air Defense Command revealed 879 objects orbiting in space: American, 672 (164 payloads, 508 pieces of debris); Soviet, 198 (53 payloads, 145 pieces of debris); Canadian, 4 (2 payloads, 2 debris); British, 3 (2 payloads, 1 debris); French, 2 (1 payload, 1 debris). Except for the payload and debris from France's recent launching, all the objects in space were launched on American or Soviet rockets. (*Wash. Eve. Star*, 12/7/65, A11)

- NASA Electronics Research Center had accomplished a promising advance in microwave research by generating higher frequency microwaves by interaction of hot electrons in solid-state component material. Technique applied low voltage across a bulk gallium arsenide semi-conductor crystal less than one-thousandth of an inch thick and might improve microwave signal transmission efficiency in space. (NASA Release 65–373)

- Fourth Chrysler-assembled Saturn IB first stage left NASA Michoud Assembly Facility aboard the NASA barge *Palaemon* for static-firing tests at Marshall Space Flight Center. Trip would take about seven days. (MSFC Release 65–295)

- Italy launched a French Centaure rocket in collaboration with the European Space Research Organization (ESRO), the Italian Defense Ministry announced. The rocket was launched from the Salto di Quirra range in Sardinia to gather information on atmospheric particles at between 60- and 100-mi. altitude. (Reuters, *Wash. Post*, 12/8/65, 3)

December 8: 200,000-lb.-thrust J–2 engine was captive fired for 388 sec. on a new test stand at NASA MSFC. The J–2 engine would be used to power the Saturn S–IVB stage, second stage for the Saturn V. Ten tests of the liquid hydrogen-liquid oxygen powered rocket engine had been conducted at MSFC since the J–2 engine test facility was put into use in August 1965. (MSFC Release 65–300)

December 8: Full-scale Saturn V booster (s–ic stage) weight simulator was shipped from NASA Marshall Space Flight Center to Kennedy Space Center on NASA barge *Poseidon.* Trip would require five days. Simulator would be used in checking equipment and handling procedures at KSC's Launch Complex 39 in preparation for arrival of the Saturn V facility vehicle early in 1966. (MSFC Release 65–295)

- Soviet Union gave assurances to U.S. through Ambassador Anatoly F. Dobrynin that it was abiding, and would continue to abide, by the 1963 U.N. resolution calling on all states not to orbit weapons of mass destruction. U.S. had raised the question whether the resolution was being violated after the Nov. 7 display during a military anniversary parade of "an orbital missile" capable of delivering a surprise blow from space.

  Comment in *Pravda* on American press reaction to display of orbital rockets during the Nov. 7 parade: "By raising a racket about the Soviet orbital rocket, somebody in the USA evidently calculated to divert the attention of the world public from the American military preparations in the cosmos. The activity of the USA . . . is . . . subordinated to the idea of using space for military purposes . . . Program MOL. The military equipment reviewed on Red Square on November 7 demonstrated . . . the power of our rocket weapons not in order to threaten anyone. Nuclear rocket weapons, which the Soviet Government has at its disposal, are the powerful means of guaranteeing the peace." (*Pravda,* 12/8/65, DOD Trans.; Finney, *NYT,* 12/11/65, 1)

- A temporary injunction was issued banning a steel fabricating firm from running internal combustion engines while GEMINI VII orbited over the Corpus Christi, Tex., area. The Government, in asking for the injunction, said machinery at Safety Steel Services, Inc., interfered with radio signals to the spacecraft and "threatened the safety of the astronauts." (AP, *Wash. Post,* 12/9/65)

- Secretary of Defense Robert McNamara announced plans to phase out all 80 U.S. B–58 bombers and 345–350 of the Nation's 600 B–52 bombers by 1971. Action was part of his program to consolidate or eliminate 149 of military installations in the U.S. and abroad at a yearly estimated saving of $410 million. In ordering the bomber phase-out, McNamara said certain bases from which B–52 operations were being removed were being retained "for a new mission which will be disclosed subsequently." Observers believed DOD might order production of the bomber version of the F–111 fighter. (DOD Release 887–65; Raymond, *NYT,* 12/9/65, 1; Corddry, *Wash. Post,* 12/9/65, A2)

- French Ambassador-designate Charles Lucet awarded gold medals and citations to French and American scientists who had worked on the FR–1 project. The French satellite was launched from the Western Test Range Dec. 6 with a NASA Scout booster. NASA officials honored at the French Embassy ceremony in Washington included Arnold W. Frutkin, Assistant Administrator for International Affairs; Robert C. Baumann, Chief of Spacecraft Integration and Sounding Rocket Div., GSFC; and Dr. Robert W. Rochelle, Chief of Flight Data Systems Branch, Spacecraft Technology Div., GSFC. (Ross, *Wash. Post,* 12/9/65, K1)

December 8: Australia was negotiating with the U.S. to use American Redstone rockets instead of Britain's Blue Streak for research at Woomera on rocket reentry into the atmosphere. National Development Minister Allen Fairhall told the Australian Parliament that the Redstone, though obsolete for orbital purposes, would be more suitable because it achieved greater altitude. (Reuters, *NYT*, 12/9/65)

• Thomas Carroll, pioneer NACA test pilot in the 1920's and chief of safety design for the old Washington Airport, died after a long illness. As first and chief test pilot for NACA, he tested planes at Langley Laboratory in Virginia from 1920 to 1930. (*Wash. Eve. Star*, 12/10/65)

• R. Gordon Gould asked the U.S. Court of Customs and Patent Appeals to declare that he was first to conceive a particularly promising version of the laser and to grant him patent rights. A patent application on the device had already been filed by Nobel prize winner from MIT Dr. Charles H. Townes and Dr. Arthur L. Schawlow of Stanford Univ. Gould's lawyers contended he had conceived the idea first, and despite the other application, had preserved his rights to the patent by working to perfect the device. (*NYT*, 12/9/65, 96L)

• W. A. Patterson, Chairman of the Board of United Air Lines, told the Aviation-Space Writers Assn. in Washington, D.C., that his company had refrained from ordering a supersonic transport "because it's a phony deal. You don't put a deposit on a plane that may cost $40 million, that you have never seen, and that you don't know anything about." Patterson said the deposit plan was designed to create in Congress an "atmosphere of enthusiasm" to obtain appropriations for Sst research and development. (AP, *NYT*, 12/9/65, 93)

• Soviet communications expert Dr. N. I. Chistiakov, speaking at the UNESCO-sponsored space communications conference in Paris, called for an international convention to govern the use of satellites for broadcasting. He said a draft agreement should be drawn up by the International Telecommunications Union and should make satellite communications available to all countries on a non-discriminatory basis. (*Wash. Post*, 12/8/65)

• Commenting on Dr. Warren Weaver's Nov. 29 statement to *U.S. News and World Report* questioning the wisdom of spending $30 billion to get an American to the moon by 1970, the *Wall Street Journal* said: "Now Dr. Weaver is not against going to the moon . . . what he is against is the hell-for-leather way the moon program is being whipped along . . . where the emphasis rightly should fall is on Dr. Weaver's assertion that 'the great ideas that develop within the body of science —strange and improbable as this sounds—arise from curiosity and not from urgency,' and that the moon program has caused a massive diversion of scientists and engineers from possibly more productive fields. . . ." (*WSJ*, 12/8/65, 16)

December 8-9: NASA Lewis Research Center hosted in Cleveland a Conference on Selected Technology for the Petroleum Industry. Session was sponsored by the NASA Technology Utilization Program whose objective was to make available to industries the results of knowledge gathered from space research and development.

NASA Administrator James E. Webb, addressing the Conference, warned of the possible consequences of reduction in the NASA budget request for FY 1967: ". . . history should have taught us that new space capabilities, in which we have made a considerable investment, must be used or their benefits will be lost." Webb said history should have taught us also "that plans for the future should not be drawn by a timid hand." He quoted the late Dr. Hugh Dryden, NASA Deputy Administrator, as saying "the present gap in manned flight activity [between the U.S. and the U.S.S.R.] is a direct consequence of a postponement of the decision to proceed beyond Project Mercury from September 1960 until May 1961." (Text; LRC Release 65-88)

December 9: Ground test version of Saturn V's first stage, S-IC-T, was captive fired for the 14th time at NASA Marshall Space Flight Center. This was a full-duration firing in which the unit developed 7.5-million lbs. thrust for 150 sec.; it was conducted by Boeing Co., S-IC prime contractor. (MSFC Release 65-301)

- USAF launched an unidentified satellite from Vandenberg AFB with a Thor-Agena D booster. (*Wash. Post*, 12/10/65, A29)
- An exhibit showing the contributions to space travel of scientists buried or commemorated in Westminster Abbey would be part of the 900th anniversary of the Abbey's founding, Westminster officials announced. (AP, *NYT*, 12/9/65, 70)

December 10: At Austin, Tex., Administrator James E. Webb announced that Dr. Robert C. Seamans would become the Deputy Administrator of NASA, filling the position held by the late Dr. Hugh L. Dryden. Dr. Seamans joined NASA as Associate Administrator of NASA on September 1, 1960. In this post he was general manager of NASA's operations, including field laboratories, research centers, rocket testing and launching facilities, and a world-wide network of tracking stations. (EPH)

- COSMOS XCIX unmanned satellite, carrying scientific instruments "for continued space research," was launched into orbit by the Soviet Union, Tass announced. Initial orbital data: apogee, 320 km. (198.7 mi.); perigee, 199 km. (123.6 mi.); period, 89.6 min.; inclination, 65°. Equipment was functioning normally. (Tass, 12/10/65)
- The Gemini 6 spacecraft underwent final checkout of all systems; its crew, Astronauts Walter M. Schirra, Jr., and Thomas Stafford, received final briefings. NASA announced that Gemini VI would be launched December 12, one day ahead of schedule. (*WSJ*, 12/10/65, 1; Wilford, *NYT*, 12/11/65, 1, C54)
- NASA announced EXPLORER XXXI and Canadian ALOUETTE II, launched together on the same booster Nov. 28, were functioning as planned. EXPLORER XXXI's apogee was less than a mile higher than ALOUETTE II's and its perigee less than a mile lower. Orbits were some 1,850 mi. (2,978.5 km.) at apogee and 310 (499 km.) at perigee. The 13 experiments on the ISIS satellites were working well with excellent data on the ionosphere being obtained. (NASA Release 65-377)
- Possibility of an atomic-powered artificial human heart being developed sometime in the future was mentioned by AEC Chairman Glenn T. Seaborg in discussion with President Johnson at the LBJ Ranch: "There is a possibility of using isotopes produced in this way as a source of energy for pacemakers for the human heart, and possibly

even as a source of energy for a completely artificial heart implanted in man. The isotope which would be used in this case happens to be the isotope plutonium 238. That is the isotope that has the best qualifications for this purpose." (Wash. *Eve. Star*, 12/11/65, A5)

December 11: XB-70 research aircraft, flown from Edwards AFB by North American Aviation pilots Alvin S. White and Van H. Shepard, reached 1,920 mph (mach 2.9) for five minutes. Purpose of 123-min. flight was to test the XB-70's stability and control at nearly triple sonic speed and the effect of 556°C heat from air friction on the aircraft's surface. (AP, *Virginian-Pilot*, 12/15/65)

- Oak Ridge National Laboratory was investigating possibility of building compact, low-weight, nuclear-electric power plants for space systems, the *New York Times* quoted AEC officials as saying. One critical component of a space reactor system—a potassium vapor turbine—had completed more than 2,000 hrs. of test operation. Reactor system would use boiling potassium to cool the reactor's enriched uranium fuel elements. Potassium vapor produced in the reactor core would drive a turbine generator to produce electric power. (*NYT*, 12/12/65, 78)

- Dr. Hugh L. Dryden, late NASA Deputy Administrator, was among the 11 recipients of the 1965 National Medal of Science announced by President Johnson. Established by Congress in 1959 for scientists who had made outstanding contributions in their fields, the award never before had been made posthumously.

    President Johnson said: "Hugh Dryden's recent death ended nearly 50 years of singleminded devotion and effort by one of the most distinguished civil servants this country has ever known. Beloved by all his associates and respected throughout the world, Dr. Dryden more than any other man led us into the age of jet aircraft and space exploration." (*Pres. Doc.*, 12/20/65, 585; *Wash. Post*, 12/12/65, A5; Pomfret, *NYT*, 12/12/65, 78)

- In an article discussing preparations for the Gemini VI flight, William Hines noted in the Washington *Evening Star:* "Total propellant load of a ready-to-go Titan 2 consists of 13,700 gallons of a fuel called aerozine-50 and 15,900 gallons of an oxidizer, nitrogen tetroxide. Their principal combustion product is nitric acid." (Wash. *Eve. Star*, 12/11/65, A2)

December 12: Attempt to launch Gemini VI from ETR was unsuccessful when an electric plug connecting the Titan II booster to the launch pad fell loose 2¼ sec. early, causing the automatic sequencer to shut down the engine booster 1.2 sec. after ignition. Inside the Titan II a wire extended from the tail plug to tap a circuit which carried current from the batteries to an intervalometer—electric clock—which controls the first 2½ min. of powered flight. When the plug shook loose prematurely and started the clock, signals that liftoff had occurred were sent to the Titan II's automatic pilot, local guidance stations, and an automatic sequencing device at KSC Launch Control Center. The sequencer registered the mishap and shut down Titan II's engines. Astronauts Stafford and Schirra remained calm throughout the misfire with Command Pilot Schirra rejecting the option to actuate the ejection seats. The astronauts were removed 99 min. later from the spacecraft and the launch rescheduled for Dec. 15. (Hines, *Wash. Eve. Star*, 12/13/65, A1; Simons, *Wash. Post*, 12/13/65, A1; MSC *GEMINI VII/GEMINI VI Fact Sheet*)

December 12: Statement by President Johnson on the delay of the Gemini VI flight: "We are all disappointed that Gemini 6 did not go off as expected. But our disappointment is exceeded by our pride in Astroauts Walter Schirra and Thomas Stafford and the flight directors of NASA. With the world watching, they acted with remarkable courage in the face of danger and potential disaster. Their eager desire and determination to try again proves once more that men are the real heroes—and the essential factor—in space exploration." (*Pres. Doc.*, 12/20/65, 587)

- Soviet Cosmonauts Konstantin P. Feoktistov and Lt. Boris B. Yegorov, two of three crew members in the 24-hr. spaceflight of VOSKHOD I, launched Oct. 12, 1964, had experienced space sickness during the flight, Tass reported. They felt nausea while in the state of weightlessness and imagined themselves to be suspended in strange positions. The disorders, Tass said, stemmed from specific irritations to the vestibular organs and were related to the duration of training and what Tass called the cosmonauts' "sensitivity to imponderability." (Tass, 12/12/65; Grose, *NYT*, 12/13/65, 47)

- Radio Prague disclosed that at a November meeting in Moscow of Soviet-bloc countries the U.S.S.R. had agreed to launch Communist nations' artificial satellites, sounding rockets, and probes for scientific research. Countries involved were preparing research programs which would be announced in 1966. (*NYT*, 12/12/65, 141)

- Sigvard Eklund, general director of the International Atomic Energy Agency, told Tass that Soviet Union had a new type of nuclear reactor that could be used as a power station in space. He said the Romashka (Daisy) reactor provided electricity "on the basis of direct conversion of heat given out by a chain reaction from nuclear fission," and that such a source could "feed scientific instruments on sputniks or satellites." (*Wash. Post*, 12/12/65, A28)

- Analysts of the 1966 budget of the U.S.S.R. had suggested that the 9.9% increase in expenditures for scientific research might be aimed at intensification of the space race, Harry Schwartz reported in the *New York Times*. (Schwartz, *NYT*, 12/12/65, F1)

December 13: A plastic dust cover carelessly left in a fuel line would have blocked the Gemini VI launch even if an electrical plug had not dropped out of the tail and shut down the Titan II engines, NASA officials revealed at a press conference. The device apparently had been installed at the Baltimore Martin Co. plant and was not removed due to "human error." The Martin Co., makers of the Titan II booster for Gemini spacecraft, could lose a $15,000 bonus because of the launch failure. (AP, *Balt. Sun*, 12/14/65; Simons, *Wash. Post*, 12/14/65, A1)

- Abort Dec. 12 of the scheduled Gemini VI flight received editorial comment in the *New York Times:* "Disappointing as this was, the event had its brighter aspects as well. It gave a vivid demonstration of the effectiveness of the failsafe arrangements. These permitted survival of the rocket and its crew despite the fact that ignition and a buildup of power had taken place before the safety mechanism sensed trouble and shut down the powerful motor." (*NYT*, 12/13/65, 38)

December 13: Missions currently performed at Vandenberg AFB, reportedly now busier than Cape Kennedy, were listed by *U.S. News and World Report:* launching "spy" satellites into polar orbits to take detailed photos of military bases in Communist lands; testing Titan and Minuteman ICBMS over the Pacific; launching weather, mapping, navigation, and other scientific satellites; tracking space vehicles—U.S. and Soviet. Programs projected included: launching of the first Manned Orbiting Laboratory (Mol) in 1969 or 1970, following unmanned test shots in 1968; rapid buildup of manned military flights that could reach the level of 50 military man-in-space launches a year by the early 1970s; establishment of regular shuttle service in small rockets to and from large orbiting military spacecraft. Vandenberg was described as "a nest of pads and silos, some underground and some above, from which every U.S. missile and satellite can be fired." In the past year, there were 117 launchings at Vandenberg, compared with 93 at Cape Kennedy. (*U.S. News*, 12/13/65, 5)

- Dr. W. Randolph Lovelace II, NASA Director of Space Medicine, his wife, and a pilot were reported missing on a private plane flight from Aspen, Colo., enroute to Albuquerque. (AP, *NYT*, 12/14/65, 51; AP, *Wash. Post*, 12/14/65, A1; AP, Wash. *Eve. Star*, 12/14/65, A9)
- NASA had awarded a new $47,655,103 cost-plus-incentive/award fee contract to TRW Inc., for work on the Gemini/Apollo mission trajectory control program and the Apollo spacecraft systems analysis program. (NASA Release 65–378)
- British Defense Secretary Denis Healey told Parliament the U.S. had agreed to a two-month extension to the original Jan. 1 option deadline for decision on purchase of the F–111. Britain took the option to buy the F–111 after scrapping the British TSR–2 low-level strike-bomber early in 1965 because of rising costs. (Reuters, *Wash. Post*, 12/14/65, A20)
- ComSatCorp had awarded three six-month study contracts totaling $240,000 to the U.K. General Post Office (GPO) for research into certain aspects of satellite communications. Studies would be carried out in cooperation with U.K. industry—Marconi, Plessey, General Electric Co., Mullard, and Standard Telephone Labs.—and would involve research into three major areas of technology involved in commercial communications of all types via satellite. (ComSatCorp Release)
- USN had completed tests at Kirkland AFB, N. Mex., of a revolutionary new airport concept employing a completely circular runway. Under the proposed design, the main runway would be in form of a banked track, and would form a perfectly circular perimeter around the airport. At the very center would be the control tower, with an unobstructed view of every portion of the runway. The passenger terminal would also be circular, ringing the control tower.

  Airport would provide a number of special advantages: (1) aircraft would have an infinitely long runway; (2) airport would require only ⅔ the acreage needed for an equivalent conventional airport; (3) because crosswind effect would be minimal, all incoming aircraft could be brought in through a single air corridor; (4) single approach air corridor would need only one set of electronic landing guides. Construction costs would be slightly higher than for conventional runways, due to requirement for precise banking of the runway. (Appel, *NYT*, 12/13/65, 33)

December 13: Earth's third largest meteorite, weighing 30 tons, had been discovered in Communist China's Gobi Desert, the *Washington Daily News* reported. It was said to have been found in northern Sinkiang and was on display in Urumchi, capital of Sinkiang. (Wash. *Daily News*, 12/13/65)

- Group headed by Junkers Flugzeug- und Motorenwerke AG, and including Lockheed Missiles and Space Co., had been selected by the European Space Research Organization (ESRO) to develop the Highly Eccentric Orbit Satellite (Heos). Junkers team was selected from eight international groups competing to build Heos, which would gather interplanetary particle data. Heos would be the first satellite developed in West Germany, but it would include the combined technology of several nations. Value of the over-all contract for satellite development was about $6.5 million. (*Av. Wk.*, 12/13/65, 36)
- A new planetoid had been discovered by East German astronomer Cuno Hoffmeister, East German news agency ADN reported. Planetoid orbited the sun once every 3.67 yrs., was 56 million mi. from the earth at its nearest point of orbit, and could be seen only every 11 years. (Reuters, *Houston Post*, 12/14/65)

December 14: Lockheed's 156-in.-dia., solid-fueled rocket motor fired for approximately 58 sec., developed over 3-million-lbs. thrust, and produced about 715 psi of pressure. Vector control system went through its three programed cycles without flaw. Fifth firing of a 156, test was conducted at Lockheed's Potrero, Calif., facilities. (*Av. Wk.*, 12/27/65, 61)

- Launch Dec. 15 of Pioneer 6 sun-orbiting satellite was postponed to avoid conflict with launch of Gemini VI. (AP, *NYT*, 12/15/65, 22)
- Four USAF officers emerged from a simulated space capsule, where they spent 56 days in an oxygen-helium atmosphere that would have sustained them for 900 revolutions of the earth, longest simulated space voyage to date in an oxygen-helium atmosphere. They existed on bite-sized dehydrated foods supplemented by a liquid similar to a milkshake, had television, radio, exercise, and were able to take sponge baths and change clothing. They said the thing they missed the most was a "home-cooked" meal. (UPI, Phil. *Eve. Bull.*, 12/14/65)
- NASA added a $13,121,252 renewal contract to the Mason-Rust Co. for continued provision of support services at NASA Michoud Assembly Facility. (MSFC Release 65–303)
- NASA Goddard Space Flight Center awarded a $500,000 contract to ITT to develop a lower-altitude space weather camera for NASA's Applications Technology Satellite (ATS). (AP, *NYT*, 12/15/65, 2)
- Between December 16, 1965, and June 11, 1966, the Soviet Union would conduct tests of a space vehicle landing system, Tass announced. Some elements of the booster-rockets would fall in the area of the Pacific with the following coordinates: 43°44 min. north latitude, 179°7 min. west longitude; 44°17 min. north latitude, 177°49 min. west longitude; and 41°33 min. north latitude, 177°22 min. west longi-

tude. U.S.S.R. asked government of countries using sea and air lanes in the Pacific to instruct ships and planes not to enter this area daily from midnight to 1200 hrs. local time. (Tass, 12/14/65)

December 14: FAA awarded two research contracts to obtain jet operations data on atmospheric turbulence: General Dynamics/Convair Corp. was awarded a $30,100 contract to develop a test program for obtaining highly precise data on the responses of pilot and aircraft to turbulence encountered in regular jet airline service; a $23,700 contract was awarded Eastern Air Lines to conduct a meteorological study of clear air turbulence (Cat). (FAA Release 65-118)

December 15-16: GEMINI VI, piloted by Walter Schirra, Jr. (Capt., USN), command pilot, and Thomas P. Stafford (Maj., USAF), pilot, was successfully launched from ETR with two-stage Titan II booster on NASA's fifth manned spaceflight in the Gemini series and first rendezvous mission. It achieved an elliptical orbit: apogee, 161 mi. (259.7 km.); perigee, 100 mi. (161.1 km.); inclination, 28.9°; period, 88.7 min. Toward the end of the first revolution, GEMINI VI was trailing GEMINI VII by 1,200 mi. (1,935 km.). Schirra began series of posigrade maneuvers in preparation for rendezvous in the fourth revolution by firing thruster rockets 18 sec. to shorten the distance between the two spacecraft to 730 mi. (1,177 km.). Schirra fired rear thruster rockets 77 sec. during the second revolution and GEMINI VI drew within 300 mi. (483 km.) of GEMINI VII; 30 min. later he fired another brief burn to move GEMINI VI 7/1000ths of a degree south to the same plane as GEMINI VII. With the spacecraft 431 mi. (695 km.) apart, Schirra fired the thrusters one second to shift GEMINI VI into a somewhat higher orbit—the only maneuver not prearranged by the flight plan. To shorten the 230-mi. (371-km.) distance between the two spacecraft as GEMINI VI entered its third revolution, Schirra fired his thrusters 53 sec. to shift into near-circular orbit: apogee, 170 mi. (274 km.); perigee, 165.5 mi. (267 km.); inclination, 89.9°. Schirra initiated terminal phase of the rendezvous maneuver with a 32 fps posigrade burn as GEMINI VI entered its fourth revolution. GEMINI VI and VII were then flying nose to nose about 25,000 ft. apart. Schirra applied a posigrade velocity of 43 fps and GEMINI VI approached within six feet—later determined to be one foot—of GEMINI VII. VI then performed an in-plane fly-around maneuver around VII, and later VII maneuvered beside VI. Flight plans had specified four hours of station keeping, but the two Geminis flew in formation, keeping within 20 to 100 ft. of each other, for 5 hrs., 19 min., during which time the astronauts photographed each others' crafts, sighted a fire in Madagascar, and conversed. All crewmen took turns in the formation flying activities, to obtain rendezvous maneuvering experience. After 15 hrs., 19 min., Schirra fired GEMINI VI's thruster rockets to separate the two spacecraft about 15 mi. for drifting flight during the sleep period.

On Dec. 16, GEMINI VI Astronaut Schirra reported to Mission Control that an unidentified satellite in a low trajectory in polar orbit was trying to contact him. Before ground officials could respond, he and Stafford played Jingle Bells with a harmonica and bells.

GEMINI VI began reentry northeast of Canton Island in the Pacific during its 15th revolution; four retrorockets fired automatically in the correct sequence, each providing 2,500 lbs. thrust. Reentry was normal and was the first controlled reentry to a predetermined landing point in the U.S. manned spaceflight program.

GEMINI VI splashed down in the Atlantic at 10:29 a.m. EST, 700 mi. south of Bermuda—only 14 mi. off target—after a 26-hr., 1-min. flight. Assisted from the spacecraft by Navy frogmen, the astronauts arrived by helicopter at the carrier *Wasp* at 11:20 a.m. for medical debriefing.

Recovery and rendezvous section of the GEMINI VI spacecraft splashed down near the spacecraft itself and was retrieved by another team of swimmers. This was first time the service section—complete with rendezvous radar equipment—had been recovered. Main parachute also was retrieved. (NASA Release 65–362; NASA Proj. Off.; Transcript; *NYT*, 12/16/65, 1, 28, 29, 30; 12/17/65, 1, 28; 12/18/65, 1, 16; 12/19/65, 1, 68, 69; *WSJ*, 12/17/65, 1; 12/20/65, 1; *Wash. Post*, 12/16/65, A1, A14, A15; 12/17/65, A1, A8, A17; 12/18/65, A1, A7; 12/19/65, A1, A16, A17; 12/20/65, A1, A3; Wash. *Eve. Star*, 12/16/65, A1, A14; 12/18/65, A1, A3; 12/19/65, A1, A8, A9; *Time*, 12/24/65, 32–36; MSC *GEMINI VII/GEMINI VI Fact Sheet*)

December 15: Success of the GEMINI VII–VI mission received worldwide acclaim: newspapers in France, Great Britain, Switzerland, Italy, and Cuba were among those which gave the flight front-page coverage and radio stations throughout the world interrupted regular programs to broadcast minute-by-minute accounts of the historic rendezvous. The Soviet newspaper *Izvestia* printed a detailed description of the rendezvous which it described as "a great success." Communist China remained silent. (Reuters, *Wash. Post*, 12/16/65, A14; UPI, *NYT*, 12/16/65, 29; Reuters, *Wash. Post*, 12/19/65, A17; AP, *NYT*, 12/17/65, 29)

- Dr. Edward C. Welsh, Executive Secretary of the NASC, after watching the successful GEMINI VII–VI rendezvous on television, told reporters that the accomplishment would lead to: (1) operation of manned space stations with crews replaced and supplies renewed by rendezvous methods; (2) assembly of large observatories and spacecraft hundreds of miles above earth; (3) ability to inspect foreign spacecraft; and (4) techniques for visiting and rescuing astronauts stalled in orbit. (UPI, *Wash. Post*, 12/16/65, A14)

- In a telegram of congratulations to NASA Administrator James E. Webb on the rendezvous of GEMINI VI and GEMINI VII, President Johnson said: "You have all moved us one step higher on the stairway to the moon. By conducting this adventure for all the world to see, you have reaffirmed our faith in a free and open society. We invite those throughout the world who have shared our suspense and suffered with us during our temporary failures to share with us this triumph, for it belongs not just to the United States but to all mankind." (*Pres. Doc.*, 12/20/65, 593)

- Brazil successfully launched the first Nike-Apache sounding rocket from its Natal Range in a joint program with NASA to investigate the lower regions of the ionosphere with emphasis on the effects of cosmic rays. Launching was conducted by the Brazilian Space Activities Commission (CNAE). Instrumentation for the rocket payload and the telemetry ground support equipment was constructed by Brazilian technicians at NASA Goddard Space Flight Center. (Wallops Release 65–80; NASA Release 65–328)

December 15: Nike-Apache sounding rocket launched by NASA from Wallops Station, boosted 51-lb. payload with ionosphere experiment for the Univ. of Illinois and the GCA Corp. to 113-mi. (182-km.) peak altitude. Payload carried instrumentation to measure electron and ion density and solar radiation in D and E layers of the ionosphere. No recovery was required since data from the experiments were radioed to ground stations during the seven-minute flight. Experiment was last NASA 1964–65, IQSY project. (Wallops Release 65–79; AP, *Wash. Post,* 12/17/65, A8)

- Bodies of NASA Director of Space Medicine Dr. W. Randolph Lovelace II, his wife, and pilot were found near the wreckage of their two-engine aircraft about 40 mi. southeast of Aspen, Colo. Aircraft had been missing since Dec. 12. Coroner's report later said that the cold and not injuries had apparently caused the deaths.

  A space medicine pioneer, Dr. Lovelace had parachuted from record 44,000-ft. altitude in 1944, had helped determine criteria for selection of Mercury astronauts, and had founded the famous Lovelace Clinic. (Wash. *Eve. Star,* 12/16/65, A18; *Wash. Post,* 12/16/65, A3; *NYT,* 12/16/65, 50; AP, *NYT,* 12/18/65, 16)

- AP quoted informed sources as saying that negotiations were in the final stages to allow the U.S. to fire Redstone rockets in Australia as part of the development of an antimissile missile. Agreement would call for about 12 of the rockets to be fired at the joint British-Australian Woomera Weapons Research Range. (AP, *NYT,* 12/16/65, 12)

- Ralph E. Cushman had been named Director of the new NASA Facilities Management Office, reporting to the Deputy Associate Administrator for Industry Affairs. Cushman had been Director of Management Coordination in the Hq. Office of Administration. Appointment was effective immediately. (NASA Ann.)

- New NASA Space Radiation Effects Laboratory, Oyster Point, Va., was dedicated in ceremonies attended by Gov. Albertis S. Harrison, Jr. LaRC facility would provide LaRC the means to test and study the effect on spacecraft and their systems of particle radiation from the sun or in the earth's magnetic field; scientists would be able to simulate space radiation and conduct studies to increase reliability and safety of spacecraft and space missions. (LaRC Release)

December 16: NASA's PIONEER VI interplanetary probe was successfully launched into heliocentric orbit from Eastern Test Range with an Improved Thrust Augmented Delta booster. Main Delta engine and three solid strap-on motors fired together and burned for 43 sec.; burned-out casings were jettisoned 70 sec. after launch; main engine burned out after two minutes 45 sec. Delta second stage ignited after two minutes 59 sec.; first stage separated and fell away; shroud was jettisoned at two minutes 59 sec. Second stage burned six and two-thirds minutes, then stage coasted for some 16 min. During coast phase, spacecraft was pointed in direction for injection into solar orbit and, with third stage, was spin-stabilized. Nine seconds after spin-up third stage separated from second, then third stage ignited and burned for 23 sec. Two seconds after third stage burnout, PIONEER VI separated from the burned-out stage 346 mi. above Africa and was now in solar orbit. Two seconds after third stage separation (about 25 min. after launch), spacecraft booms automatically deployed; automatic changes began to orient the spacecraft perpendicular to the sun. Sun orientation maneuver took about five minutes.

The 175-lb., drum-shaped PIONEER VI, first of four Pioneer spacecraft to be launched at six-month intervals, would study the solar windstream; investigate the sun's magnetic field and chart it from several locations in the plane of earth's orbit; attempt to differentiate between cosmic rays coming from the sun and galactic cosmic rays originating from beyond the solar system. Closest approach to the sun would be about 76 million miles and would be reached after 155 days of flight. The six scientific experiments, provided by four universities and ARC and GSFC, were functioning normally. (NASA Release 65–375; AP, Wash. *Eve. Star*, 12/16/65, A14; *Wash. Post*, 12/17/65, A8; AP, *NYT*, 12/17/65, 29; *WSJ*, 12/17/65, 1)

December 16: NASA canceled for budgetary reasons further development of the Advanced Orbiting Solar Observatory (Aoso). From its 1963 start through the fiscal year ending June 30, 1966, a total of $39 million had been budgeted for Aoso development; NASA said some of the $24.9 million appropriated for FY 1966 would be recoverable. Aoso was to have been a 1,250-lb. satellite that would accurately point 250 lbs. of scientific instruments at the sun to measure solar radiation. A prototype spacecraft was being developed and the first Aoso flight had been scheduled for sometime in 1969. (NASA Release 65–380)

- Successful 41-sec. test firing of the Saturn V booster (S–IC–T) at NASA Marshall Space Flight Center concluded a test series underway since April. 15 tests of the booster were held, 14 of them at full thrust of 7.5-million lbs. The 14 full-power tests accumulated a running time of 862 sec. First test was of only one engine for 17 sec. (MSFC Release 65–306; *Marshall Star*, 12/15/65, 2)

- Statement made by President Johnson on the death of the NASA Director of Space Medicine: "A day of great achievement in space was marred by news of the death of Dr. William R. Lovelace, II. His life was too short, although his legacy to space medicine will endure and will be a resource of assurance to future astronauts whose names and deeds are yet unknown." (*Pres. Doc.*, 12/20/65, 593)

- Editorial commentary in *Washington Post* on GEMINI VII–VI flight: "On the crowded globe beneath the soaring astronauts men were still fighting each other, cursing each other, starving each other and maltreating each other. Not yet have mortals risen above man's inhumanity to man. Not yet have they surmounted the emotions that set at naught the generations of light and learning. Not yet are they wise enough to govern their passions. But hope soars aloft with the astronauts. The creatures who can do this, the beings who can defy gravity, disregard distance, conquer space, circumnavigate the planet and mingle with the stars, may yet make the larger conquests of mind and spirit that are necessary if human beings are to live together in peace." (*Wash. Post*, 12/16/65, A20)

December 16: ITT-designed portable transmitter onboard the *Wasp* sent pictures of splash-down and recovery of GEMINI VI Astronauts Schirra and Stafford to EARLY BIRD I comsat which relayed them to the Andover, Me., ground station from which they were transmitted to major television networks in U.S. and Canada. It was the first time recovery of astronauts had been shown in real-time on television. (Gould, *NYT*, 12/17/65, 28)

- GEMINI VI Astronauts Schirra and Stafford were "very healthy," reported Dr. Howard Minners at the conclusion of their post-flight physical examination aboard the carrier *Wasp*. (AP, Wash. *Eve. Star*, 12/17/65)

- 152 colleges and universities would participate in the NASA graduate training program during the 1966–67 academic year, NASA announced. In all, 1,335 graduate students would begin work toward doctoral degrees in space-related areas under grants to be received by schools in all 50 states. About 3,100 graduate students already were in training under this program. (NASA Release 65-379)

- NASA and FAA announced a cooperative research project to find out whether passengers on supersonic airliners would be exposed to danger from cosmic radiation. To collect data, USAF would send RB–57 jets to the altitude between 40,000 and 80,000 ft. where the supersonic jetliners were expected to fly. (NASA Release 65–383; UPI, *NYT*, 12/18/65, 16)

- Donald E. Gault and William L. Quaide, NASA Ames Research Center scientists, believed the moon's surface was a loose layer of fine sand-like material at least 10 ft. and possibly "tens of meters" deep, reported Associated Press. Their theory was based on laboratory experiments in which they closely matched the moon crater pictures transmitted to earth by the Ranger spacecraft. Laboratory results were confirmed in field tests. (AP, Wash. *Eve. Star*, 12/16/65)

- ComSatCorp announced it was negotiating with TRW Systems, Inc., for development of at least six and perhaps 24 satellites for use in a global system. ComSatCorp said the global system should be in operation by 1968, relaying telephone, television, and data messages between continents. No estimate of the value of the contract was made. (ComSatCorp Release)

- United Air Lines placed a $39-million order with Univac Div. of the Sperry Rand Corp. for a computer system that could handle 140,000 transactions an hour, including passenger reservations, crew and aircraft scheduling, meal planning, cargo billing, and other bookkeeping items. (O'Toole, *NYT*, 12/16/65, 75)

- Eight pioneers in aviation were enshrined in the Aviation Hall of Fame, Dayton, O.: Eddie Rickenbacker, World War I pilot who later became president of Eastern Air Lines; Alexander Graham Bell, cited for research on principles of aerodynamic lift propulsion and control in the early 1900's; Eugene Burton Ely, whose work led to practical use of aircraft carrier ships; Alfred Austell Cunningham, the Marine Corps' first aviator; Thomas Etholen Selfridge, first American to die testing an experimental aircraft; Charles Edward Taylor, builder of the first successful airplane engine; A. Roy Knabenshue, pioneer in building and flying steerable balloons; and Albert Cushing Read, participant in the first successful trans-Atlantic flight. (AP, *NYT*, 12/17/65, 77; AP, Wash. *Eve. Star*, 12/17/65, A4)

December 16: Ground-breaking ceremonies were held for a $14 million science and mathematics building at the U.S. Naval Academy, its first new academic building in 25 years. (*NYT*, 12/19/65, 58)

December 16–17: Scientific results of the International Quiet Sun Year 1965 NASA solar eclipse expedition were presented by experimenters from several nations at NASA Ames Research Center. Purpose of expedition was to measure and study the structure, composition, and temperature of the chromosphere and corona of the sun. New information was gathered on the atomic process in the chromosphere and corona, including radiation and collision phenomena; findings were made on electron densities and dust concentration. Presence of a number of spectral lines was discovered, and others that had been previously suspected were confirmed. The advantages of an airborne laboratory were demonstrated in that the corona of the sun could be observed out as far as 12 solar radii, whereas ground observations would have been limited to about three radii; not only could more solar phenomena during a total eclipse be observed from an airborne laboratory, but also the eclipse was visible for twice as long. (ARC Release)

December 17: JPL announced that PIONEER VI, launched Dec. 16, had made final vernier adjustment relative to pointing the high-gain, narrow-beam antenna toward the earth. Spacecraft was 230,000 mi. from earth; all systems were functioning normally. (NASA Pioneer Proj. Off.; AP, Wash. *Eve. Star*, 12/18/65, A3; AP, *NYT*, 12/19/65, 67)

- COSMOS C, containing scientific equipment, was launched by the Soviet Union into circular orbit at 650-km. (403.7-mi.) altitude, with 97.7-min. period and 65° inclination. Onboard equipment was functioning normally. (*Komsomolskaya Pravda*, 12/18/65, ATSS–T Trans.)

- NASA announced management assignments in the Apollo Applications area: MSC would be responsible for development and procurement of all standard and modified spacecraft (Command, Service, and Lunar Excursion Modules), astronaut activities, flight operations, and integration of experiments in the command and service modules; MSFC would be responsible for development and procurement of launch vehicles, integration of experiments into the Lunar Excursion Module, Saturn instrument units, and S–IVB stages (top stages of both Saturn IB and Saturn V vehicles); KSC would assemble, check out, and launch Apollo Applications space vehicles and their associated payloads.

  Proposals for possible Apollo Applications experiments were expected to be submitted by the world scientific community, industry, other Government agencies, and the entire NASA organization. (NASA Release 65–381)

- On arrival at KSC, Astronauts Schirra and Stafford personally thanked the 400 men responsible for the successful GEMINI VI launching. The astronauts would undergo medical examinations and debriefings for three days and then fly to MSC for a reunion with their families. (AP, Wash. *Eve. Star*, 12/18/65, A1; UPI, *NYT*, 12/18/65, 16)

December 17: NASA officials released clear, detailed, color and black and white photographs of the GEMINI VII–VI rendezvous taken by the GEMINI VI astronauts. (UPI, *Wash. Post*, 12/18/65, E1)

- Jerome Lederer, director of the Flight Safety Foundation, Inc., received the Wright Brothers Memorial Trophy at the Aero Club's annual Wright Brothers Memorial Dinner in Washington, D.C. He was cited for "35 years of distinguished service and unceasing devotion to increasing the safety of flight throughout the world . . ." (NAA *News*; AP, *L.A. Times*, 11/27/65)
- National Science Foundation announced that a party of U.S. scientists had landed on an icecap 630 mi. from the South Pole to establish a station that would study Antarctic weather; the earth's magnetic field; naturally-produced, very-low-frequency (Vlf) radio waves; and the aurora australia. Project was sponsored by NSF. (AP, *NYT*, 12/18/65, 16)
- NASA's decision to divert funds from its basic science research program to Project Apollo as a budgetary expediency was criticized by the *New York Times:* "[NASA] is sacrificing scientifically important projects whose sole defect is that they lack the spectacular publicity value of Project Apollo, which already consumes most of NASA's huge appropriation.

  "We believe this is an irrational set of priorities, the result of the public-relations approach. There is no compelling scientific reason why a man should be landed on the moon by 1969 rather than in 1971 or 1973. . . ." (*NYT*, 12/17/65, 38)
- A 160-ft.-tall Saturn I rocket had been erected in display of missiles and space vehicles at the NASA Marshall Space Flight Center Orientation Center. (MSFC Release 65–305)

December 18: President Johnson sent letters to Astronauts James A. Lovell, Jr. (Cdr., USN), Walter M. Schirra (Capt., USN), Frank Borman (L/Col., USAF), and Thomas P. Stafford (Maj., USAF), congratulating them on successful completion of the GEMINI VI and GEMINI VII space missions and advising them that promotions for new astronauts would be submitted to Congress for confirmation in January. (*Pres. Doc.*, 12/27/65, 606)

- Nike-Apache sounding rocket launched by Brazil from the Natal Range reached an altitude of 117 mi. The second Brazilian sounding rocket launch in a cooperative meteorological program with NASA, the flight was a nighttime twin-experiment of daytime experiment Dec. 15. (NASA Rpt. SRL; Reuters, *NYT*, 12/20/65, 45)
- World records set by NASA's GEMINI VII/VI: (1) longest manned spaceflight (VII)—330 hrs. 35 min.; (2) first rendezvous of two manned maneuverable spacecraft; (3) total man hours in space for one nation—1,354 hrs. 38 min. compared with 507 hrs. 16 min. for U.S.S.R.; (4) individuals with most spaceflight time—Col. Borman and Cdr. Lovell with 330 hrs., 35 min.; (5) longest multimanned spaceflight; (6) most revolutions for a manned spaceflight—206; (7) most miles traveled on a manned spaceflight—5,129,400; (8) most manned flights—U.S., 11, U.S.S.R., 8; (9) most men sent into space—U.S. 16 (13 astronauts with three making two flights), U.S.S.R., 11; (10) most manned flights in one year by one nation—5; (11) most

men sent into space in one year by one nation—U.S., 10. (MSC Gemini Proj. Off.)

December 17: All major goals of the Gemini program except actual docking of two spacecraft had been achieved with the flights of GEMINI VI and VII, MSC Director Dr. Robert R. Gilruth said at an MSC news conference. Gemini program director Charles Mathews said the remaining five Gemini launchings, all scheduled for 1966, would attempt to demonstrate some complex rendezvous techniques useful for Project Apollo. (AP, *NYT*, 12/19/65, 69; AP, *Wash. Post*, 12/19/65, A17; MSC *Roundup*, 12/23/65, 6)

- Physical condition of GEMINI VII Astronauts Borman and Lovell was "better than expected," reported Dr. Charles Berry, Gemini medical director. (Waldron, *NYT*, 12/19/65, 68)

December 18–22: Dr. Robert Jungk, director of the Institute for Research into Problems of the Future at Vienna, told an international gathering of scholars at the Center for the Study of Democratic Institutions that man cannot allow forms of technology that destroy nature rather than cooperate with it. Jungk said scientists, philosophers, and experimenters in technology must act as intellectual missionaries to the common man and to the young. Intellectual leaders, he said, must try to influence the power structure to harness technological development and divert it toward the needs of man. Dr. Robert Maynard Hutchings, president of the center, in summing up the five-day symposium, said: "Technology, at this moment, in the United States, is not directed toward making a decent habitation for man. . . . It is directed in piecemeal fashion by the wrong people, in the wrong direction, to the wrong ends." (AP, *Wash. Post*, 12/24/65, A2)

December 19: The successful GEMINI VII-VI mission received extensive editorial comment. *New York Times:* "The exploits of astronauts Borman, Lovell, Schirra and Stafford and the extraordinary pictures they brought back from space have thrilled men everywhere. Admiration for their high achievements has—if only for the moment—overcome most of the usual division of this quarrelsome world. . . .

"Borman and Lovell in Gemini 7 took the longest journey in history —more than five million miles. They showed that creatures of the earth's surface can live and work effectively for fourteen days in a weightless environment, in which men must encapsulate themselves as in the womb and bring their own air, food and water to survive . . . they proved that a human organism is up to the task of staying in space long enough to fly to the moon, to do work there for several days, and then to return to this planet.

"The unprecedented precision navigation feat of Schirra and Stafford in Gemini 6 was important primarily because, for the first time, it brought two spaceships close enough together to be joined. Their demonstration opens the way for construction in space and vastly expands the potentialities for human activity in that realm. . . ." (*NYT*, 12/19/65, 8E)

Washington *Sunday Star:* "Now it is all over, but it will be remembered in history as one of man's finer achievements. . . ." (*Wash. Sun. Star*, 12/19/65, C2)

December 19: U.S. Ambassador Arthur J. Goldberg told the U.N. General Assembly's Political Committee that progress had been made on a start toward a basic international law for space and for the assistance and return of astronauts forced down on foreign soil. He said that an additional topic should be brought under study by the U.N. Committee on the Peaceful Uses of Outer Space: "Within a few years, the need for a treaty governing activities on the moon and other celestial bodies will be real. My government plans to present a definite proposal as to the contents of such a treaty." (UPI, *Wash. Post*, 12/20/65, A15)

- France was exploring with the U.S. the possibility of establishing an international space launching center in French Guiana, John Finney reported in the *New York Times*. Center's location near the equator would make it possible to launch satellites directly into either equatorial or polar orbits valuable for both scientific and communications satellites. Preliminary surveys were under way. (Finney, *NYT*, 12/20/65, 8)

December 20: NASA announced it had extended the Scout Reentry Heating Project to include one more ballistic flight and had invited industry to submit proposals for the design of the spacecraft. Sub-orbital reentry experiment was scheduled for 1967 and would be sixth in the Langley Research Center project. It would require launching of a 13-ft.-long, pointed cone from Wallops Island, Va., using a modified three-stage Scout launch vehicle with no heat shield. Reentry would take place near Bermuda at a velocity between 12,000 and 13,000 mph. (NASA Release 65-384)

- In season's greetings to NASA employees, Administrator James E. Webb said: "During 1965 we have continued our effects to place and keep the United States in a position second to none in space and aeronautics. This is significant for many reasons, but the most important of these, perhaps, is the fact that our efforts constitute a very real and significant contribution toward the ultimate realization of the true meaning of Christmas—peace on earth." (NASA Hq.)

- GEMINI VII Astronauts Frank Borman and James Lovell flew from the carrier *Wasp* to KSC where they had a brief reunion with GEMINI VI Astronauts Walter M. Schirra, Jr., and Thomas P. Stafford, who were departing later in the day for MSC for further medical debriefing. (AP, *NYT*, 12/20/65, 10C)

- L/Col. Pavel R. Popovich, who orbited the earth 48 times in VOSTOK IV in August 1962, said in an interview with *Izvestia* that the rendezvous of GEMINI VI and GEMINI VII was "a great achievement of American cosmonautics on the way to exploiting space around the earth." (Reuters, *NYT*, 12/22/65, 19)

- Dr. Caryl P. Haskins, president of Carnegie Institution of Washington, warned in his report for 1964–1965 against confusing technology with science. He said the main job of technology was to turn out socially useful products while the essential goal of science was the search for truth. (*Rpt. of Pres.*, 12/20/65)

December 21: USAF Titan III-C, launched from Eastern Test Range, encountered trouble with its transtage—third stage—and failed to reach near-synchronous equatorial orbit with four pick-a-back satellites. At lift-off, the booster's two 1.2-million-lb.-thrust strap-on engines and the Titan III-C core vehicle functioned as scheduled, propelling the booster to 400,000-ft. altitude at 10,300 mph. Second stage ignited and the stages separated simultaneously with first stage burnout. Ten-second coast period followed second stage shutdown before stage was separated; three seconds after separation, the transtage, powered by twin engines rated at 8,000-lb. thrust each, ran for 17 sec. to drive vehicle into temporary parking orbit with apogee, 194 mi. (311 km.); perigee, 103.6 mi. (167 km.); inclination, 28.6°; orbital insertion velocity, 25,609 fps. About three-quarters through its first orbit (75 min. after lift-off) guidance system commanded transtage engines to restart and burn for 297 sec. to drive transtage and payload upward toward synchronous altitudes; transtage entered transfer orbit with apogee, 20,948 mi. (33,725 km.); perigee, 113.4 mi. (182.6 km.). Vehicle coasted in this deep elliptical orbit almost five hours before transtage was scheduled to restart its engines for third and final time to circularize its orbit. For unknown reasons, the engines did not restart and the vehicle entered a highly elliptical earth orbit with apogee, 20,900 mi. (33,649 km.); perigee, 120.8 mi. (194.4 km.). Telemetry indicated the LES III and IV satellites were released, as was the OSCAR IV satellite, but all three much later than planned and into the wrong orbits. Fourth payload, OV2-3, remained attached to transtage.

Titan III, most powerful rocket currently in use by the U.S., was being developed by USAF for use with the Manned Orbiting Laboratory (Mol). This was the second time in three Titan III tests that the transtage had failed. (AP, *Wash. Post*, 12/22/65, A1; AP, *NYT*, 12/22/65, 14; AP, *Houston Post*, 12/22/65; *Av. Wk.*, 12/27/65, 27; *U.S. Aeron. & Space Act., 1965*, 157–158)

December 21: COSMOS CI unmanned satellite was launched by the U.S.S.R. "to continue space investigations," announced Tass. Orbital data: apogee, 550 km. (341 mi.); perigee, 260 km. (165 mi.); period, 92.4 min.; inclination, 49°. Equipment was functioning normally. (*Pravda*, 12/22/65, 1)

- GEMINI VII Astronauts Lovell and Borman completed the medical phase of their debriefing at KSC. Dr. Charles Berry, Chief of Medical Programs for MSC, said that "a quick look at data available . . . indicates man has fared extremely well in two weeks of space environment." A detailed medical analysis would be available later. NASA released several terrain photographs taken by Astronaut Lovell during GEMINI VII's spaceflight Dec. 4–18. (AP, *NYT*, 12/22/65, 14; AP, *Wash. Post*, 12/22/65, A3)

- In conversations with West German Chancellor Ludwig Erhard, President Johnson announced he would send a commission to Europe early in 1966 to consult with nations wishing to join the U.S. in "a major endeavor" in space exploration. The group would be headed by NASA Administrator James E. Webb. The President said: ". . . we would like to discuss with you—and others—an even more ambitious plan to permit us to do together what we cannot do so well alone. Examples would be two projects which stand high on the space agenda. Both are very demanding and complex. One would be a probe to the sun, and another a probe to Jupiter. To cooperate on such a major endeavor would contribute vastly to our mutual knowledge and our mutual skills." (Marder, *Wash. Post*, 12/21/65, A1)

December 21: Dr. Robert C. Seamans, Jr., was sworn in as NASA Deputy Administrator, succeeding Dr. Hugh L. Dryden who died Dec. 2. Dr. Seamans, who had been Associate Administrator, would also retain that position for an indefinite period. (NASA Release 65-388)

• At NASA Industry Briefing on Computer Procurement, held at Hq., NASA briefed representatives of 14 leading companies in the automatic data processing industry on NASA's present and future plans and policies for the procurement of large-scale computing systems. William Rieke, NASA Deputy Associate Administrator for Industry Affairs, pointed out that NASA's principal purpose in future Adp procurements would be to increase competition throughout the computer industry. Citing the growing concern in recent years over expanding Adp inventories and increasing dollar expenditures for computers throughout Government, Rieke outlined a three-point plan to improve the exchange of information between NASA and the computer industry: (1) NASA would hold annual industry briefings to inform computer firms of its long-range plans, problems, needs for improved system technology, and projected procurements with the first to be held next May; (2) individual computer manufacturers would be given an opportunity to conduct annual briefing for NASA's personnel to inform NASA of their plans, problems, and developments; (3) in connection with large or unusual Adp procurements with restrictive requirements, NASA would hold prespecification briefings for interested companies in advance of issuing requests for proposals.

Edmond C. Buckley, director of NASA's Office of Tracking and Data Acquisition, outlined NASA's methods for long-range planning on computer acquisition and utilization, and emphasized that the agency's future Adp procurements would be for computing systems rather than just hardware.

NASA Administrator James E. Webb traced the cooperative history of the NASA-Industry-University "team" in contributing to NASA's accomplishments since 1958. He emphasized the importance of the computer industry's contribution to the space effort, in particular, and solicited the industry's continued cooperation in carrying out NASA's mission. (NASA Release 65-391)

• L/Col. Frank Borman and Cdr. James A. Lovell, Jr., had brought back proof of the endurance records set in space. Each had a $1 bill with recorded serial number at lift-off. Bills were turned over to a member of the National Aeronautic Association and then to the Fédération Aéronautique Internationale in Paris—world flight recordkeeping organization.

FAI announced NASA had filed spaceflight records set by GEMINI VI and VII. (UPI, *NYT*, 12/19/65, 67; UPI, *NYT*, 12/22/65, 4)

• GEMINI VI Astronauts Schirra and Stafford were made honorary members of the American Federation of Musicians for their rendition of Jingle Bells during the GEMINI VII–VI mission. (AP, Wash. *Eve. Star*, 12/21/65, A2)

• NASA had selected the Range Systems Div. of Ling-Temco-Vought, Inc., to provide computer support services for the Michoud Assembly Facility. A cost-plus-award-fee contract would be negotiated for a one-year period with provisions for three consecutive one-year renewals. Cost for the first year was estimated to exceed $1.5 million. (NASA Release 65–386; MSFC Release 65–310)

December 21: Karl G. Harr, president of the Aerospace Industries Assn., told the Aviation-Space Writers Assn. in Washington, D.C., that the aerospace industry would "jump at the chance" to attack social problems such as urban congestion and water pollution. He said this was so even though space companies had lost money in the first experiment in applying their techniques to urgent public problems. (Clark, *NYT*, 12/26/65, 43)

- EARLY BIRD I comsat relayed pictures for a televised debate between students of Oxford and Harvard on whether the U.S. should carry out its Vietnam commitment. (Adams, *NYT*, 12/10/65, 87)
- New York's Mayor-elect John V. Lindsay announced formation of a Science and Technology Advisory Council composed of representatives of universities, foundations, and corporations. Council would attempt to attract science-oriented industries into New York City and assist those already there. It would meet monthly with the Mayor and conduct studies into the scientific and technological advances that might affect industries in the city. (*NYT*, 12/22/65, 12)
- Second XB-70A research bomber underwent a 10-min. heat-friction test from Edwards AFB withstanding 530°C heat at 1,900 mph at 70,000 ft. The air inlets control system was also tested at twice the speed of sound at 63,000 ft. in the 145-min. flight. (AP, *NYT*, 12/23/65, 43)

December 22: NASA deferred the first unmanned Voyager planetary exploration mission until 1973 and scheduled one 1967 Mariner flight to Venus and two 1969 Mariner flights to Mars. First Voyager mission had been planned to orbit Mars in 1971 and a second to orbit and land instrumented capsules in 1973; no further Mariner missions had been scheduled. Changes in the planetary exploration program were being made to obtain the greatest possible return from funds available for FY 1966, those anticipated for FY 1967, and from information available from previous missions. (NASA Release 65–389)

- USAF launched unidentified satellite payload with Scout booster. (*U.S. Aeron. & Space Act., 1965,* 159)
- NASA had awarded a $67,135 contract to the Univ. of Utah for a one-year study of chemical processes occurring during combustion of solid-propellant rocket motors. Study would attempt to better understand the interaction of gaes in the combustion zone by using recently developed fast-scanning spectroscopic instruments. (NASA Release 65–390)
- NASA Marshall Space Flight Center had awarded a one-year $1.6-million contract to Aero Spacelines, Inc., for flying large rocket cargoes in its Super Guppy aircraft—the only aircraft in existence which could fulfill the size and weight requirement for the S–IVB stage of the Saturn vehicle, the Saturn IB and Saturn V instrument units, and the Lunar Excursion Module adapter. The agreement also provided for an extension of a previous NASA-Aero Spacelines contract for use of the Pregnant Guppy aircraft. (MSFC Release 65–311)
- A German amateur radio station used OSCAR IV, sent into an unplanned orbit Dec. 21 by the Titan III–C rocket, to relay a signal to the Bo-

chum Institute for Satellite and Space Research which had spotted the amateur radio satellite. Signal was reportedly loud and clear. USAF said 42-lb. comsat was expected to be useful to amateur radio operators around the world despite its poor orbit. (*Wash. Post*, 12/23/65, A6)

December 22: Analyzing President Johnson's proposal for Europe to join America in a "major endeavor" in space exploration, Howard Simons said in the *Washington Post* that the offer had three main objectives: (1) to satisfy European industrialists desirous of a share in 20th century technology; (2) to involve Europeans in developing complex spacecraft rather than launch vehicles that could be developed into ballistic missiles; (3) to generate interest in West European scientists as well as industrialists and engineers. As pace goal for a joint U.S.-European effort, President Johnson had cited the sun and the planet Jupiter. (Simons, *Wash. Post*, 12/22/65, A6)

- L/Col. Frank Borman (USAF) and Cdr. James A. Lovell, Jr. (USN), arrived in Houston to see their families for the first time since the start of their 14-day GEMINI VII space mission December 4. They would spend the Christmas holidays at home. (UPI, *NYT*, 12/23/65, 14)

- Gen. Thomas Dresser White, USAF Chief of Staff from 1957 to 1961 and one of the chief proponents of a balanced air-space defense system, died. He had been a member of the NACA 1957–58, and the first U.S. military attache in Russia in 1933. (*Wash. Post*, 12/23/65, D4)

- Spanish scientists reported recovery near Seville of three heavy metal spheres and other metal objects bearing Soviet markings. Debris was believed to be from Soviet spacecraft or rockets which disintegrated as they reentered earth's atmosphere. (AP, Wash. *Eve. Star*, 12/22/65)

- USAF XB–70A research bomber was flown from Edwards AFB by NAA pilot Van A. Shepard for 155 min.—10 min. longer than on any of its previous 28 test flights. (AP, *Wash. Post*, 12/23/65, A3)

December 23: GEMINIs VII and VI apparently approached within one foot of each other during their rendezvous, Robert Aller, NASA Chief of Mission Planning, told the National Press Club in Washington, D.C., when he narrated films of the rendezvous. Aller also reported that GEMINI VII command pilot Borman appeared to have reentered the earth's atmosphere without his helmet on. Mission rules ordinarily required astronauts to wear their suits and helmets during launchings, space maneuvers, and reentry. (Clark, *NYT*, 12/24/65, 1)

- NASA announced appointment of Bernard Moritz, NASA Assistant General Counsel for Procurement, as Assistant Deputy Associate Administrator for Industry Affairs. S. Neil Hosenball, Chief Counsel at NASA Lewis Research Center, would succeed Moritz. (NASA Release 65–387)

- An escalation of the war in Vietnam might consume the military men and equipment currently at NASA's disposal for spaceflight recovery operations, suggested William Hines in the Washington *Evening Star*. He said that L/Gen. Leighton I. Davis (USAF), head of spaceflight recovery operations, had commented on the possibility: "I'm sure if there were (a diversion from Viet Nam to support space flights) the NASA officials would be the first to relieve us of the requirement for support." (Hines, Wash. *Eve. Star*, 12/23/65, 27)

December 24: USAF launched Thor-Agena D booster from WTR with unidentified satellite payload. (*U.S. Aeron. & Space Act., 1965*, 159)

- Article in *Science* discussed "factors pressing France toward greater emphasis on international collaboration" in space exploration, then said:

    "Hence there was warm, if somewhat skeptical, interest in France when, in October, a Soviet scientific delegation brought up the idea of launching French payloads on Soviet rockets. It appeared, according to the one authoritative account of this proposal (*Le Monde*, 11 November), that the Soviet scientists did not envisage French requests for precise knowledge about acceleration and vibration from Soviet rockets so that the payload design could proceed, nor did they expect that French scientists would wish to be present in Soviet launch-bases and tracking stations . . . it was expected that the negotiations would be long and delicate. . . ." (McElheny, *Science*, 12/24/65, 1700–01)

- Commenting on election of political scientist Don K. Price to presidency of American Association of the Advancement of Science, *Science* editor Philip H. Abelson quoted Price himself: " 'The union of the political and scientific estates is not like a partnership, but a marriage; it will not be improved if the two become like each other, but only if they respect each other's quite different needs and purposes. No great harm is done if in the meantime they quarrel a bit.' " (*Science*, 12/24/65, 1669)

- Commenting on budgetary considerations as they might affect the Titan III–C, William Hines said in the *Washington Evening Star:* "If the economy ax falls on the Titan 3C program in the wake of an unsuccessful launching Tuesday, it will be sad indeed. The program is one of the few major space efforts in recent memory to have gone along so far on schedule, close to budget limits, and with a reasonably high degree of success." (Hines, Wash. *Eve. Star*, 12/24/65, A3)

December 26: Planning Research Corp. of Los Angeles received a one-year, $48,229 contract from NASA Goddard Space Flight Center to conduct an independent reliability assessment of the Radio Astronomy Explorer satellite. (Wash. *Eve. Star*, 12/26/65, D6)

- Ocean data station buoy Bravo was being used by USN to study influence of the moon's tidal pull on the Gulf Stream near Hollywood, Fla., according to General Dynamics Corp., developer of the buoy. The round steel hull, 40 ft. in diameter, was equipped with instruments and a 40-ft. mast that collected information on wave heights and velocity of the stream and transmitted it to a data center on shore. (*NYT*, 12/26/65, 10E)

December 27: Life Sciences Research Laboratory dedicated at NASA Ames Research Center. The laboratory was organized into divisions for exobiology, environmental biology, and biotechnology. In the dedication address, Rep. George P. Miller (D–Calif.) said that "the work done at Ames is a reaffirmation of the avowed goals and aspirations of a world that is painfully searching for peace and hope."

Dr. Mac C. Adams, NASA Associate Administrator for Advanced Research and Technology, said during the ceremonies: "This new structure represents many things. It represents efficiency and economy for it brings together a staff and associates numbering almost 250 who formerly were scattered about in 22 separate quarters. It represents a grouping of new research tools which can be used for new advances in science and technology; but most important, I believe, this Life Sciences Laboratory, placed within the complex of physical laboratories, represents the interdisciplinary approach to solving new problems." (NASA Release 65-394; Text; ARC *Astrogram*, 1/6/66, 1)

December 27: Pocomoke City, Md., radio station WDMV proposed that Wallops Island, Va., be renamed Dryden Island in memory of Dr. Hugh Dryden, late NASA Deputy Administrator. (AP, Wash. *Eve. Star*, 12/27/65, B1)

• AEC reported it had decided not to build a new type of breeder nuclear power reactor that had been considered as a source of electricity for pumps for California's water project. The reason given was that technical problems had been encountered in research and development. The reactor would have used thorium as the key fuel. It would have been designed to breed more fuel than it consumed and run about nine years on one fuel charge. (AEC Release H288; AP, *NYT*, 12/28/65, 4)

• William Hines, science writer and columnist for the Washington *Evening Star*, was presented the AAAS's top award for science writing in 1965 at the Association's annual meeting in Berkeley, Calif. His award-winning entry was a series of articles on the journey of MARINER IV to Mars. The articles appeared from Nov. 6, 1964, to Aug. 12, 1965. (Wash. *Eve. Star*, 12/17/65, A2)

December 28: PIONEER VI interplanetary probe, launched by NASA Dec. 16, had completed shakedown operations and was encountering fair space weather as it began its long cruise around the sun, NASA announced. Scientific data telemetered to earth indicated that the solar wind was blowing at relatively slow speeds of about 670,000 mph compared with 2 million an hour registered in periods of high solar activity; magnetic fields were fairly unfluctuating; and comparatively few charged particles were being encountered. Information being received from all six experiments was reported to be of excellent quality. (NASA Release 65-392)

• U.S.S.R. successfully launched COSMOS CII and COSMOS CIII unmanned satellites to continue space investigations, Tass announced. Orbital parameters for COSMOS CII: apogee, 172 mi. (278 km.); perigee, 135 mi. (218 km.); period, 89.24 min.; inclination, 65°. COSMOS CIII had been placed in near-circular orbit at 372-mi. (600-km.) altitude, with a period of 97 min. It was not specified whether the two spacecraft were launched simultaneously or whether there was an interval between launchings. All onboard systems were functioning normally. (*Pravda*, 12/29/65, 1)

• LES IV communications satellite, released into an unplanned orbit Dec. 21 by Titan III-C, had been activated and was operating well, Lincoln Laboratory reported. Tumbling had prevented restart of Titan III-C for intended orbit. (AP, *NYT*, 12/30/65, 11)

• Formation of Institute for Oceanography, part of the Environmental Science Services Administration, was announced by Dept. of Commerce. Interaction between ocean, earth, and atmosphere would be studied in programs designed to gain new knowledge of the ocean. Headed by Dr. Harris B. Stewart, Jr., formerly chief oceanographer of the Coast and Geodetic Survey, the Institute had headquarters in Washington, D.C. (Dept. of Commerce PIO; *NYT*, 12/30/65, 42)

December 29: Dr. John W. Salisbury and Joel Adler of Cambridge Laboratories reported to 132d meeting of the AAAS in Berkeley, Calif., that American astronauts landing on the moon might encounter lunar dust that would stick like wet snow to their windows, faceplates, and camera lenses. Salisbury and Adler said they had reached those conclusions by two separate experiments. (*Wash. Post*, 12/30/65, A4)

Medical and physical data from the two-week GEMINI VII flight indicated that man could withstand a lunar mission without any serious harm, reported Dr. Charles Berry, Chief of Medical Programs for NASA MSC, at the AAAS meeting. Dr. Berry revealed that the astronauts' heart rates and blood pressures during the flight had remained within normal ranges and that within 10 hrs. after splashdown, Borman had regained 4.8 lbs. of the 9.6 lbs. he had lost; Lovell, who had lost 5.9 lbs., had regained 6.6 lbs. Berry, who doubted the astronauts could have endured 14 days in spacesuits, attributed much of the success of the mission to the freedom of flying in underwear. He reported that the electroencephalogram studies recording brain wave activity had shown no abnormalities during the first two flight days, but that the electrodes on the astronauts' scalps had come loose before the end of the planned 4-day experiment. As to the radiation dosage hazards on the 14-day flight, Dr. Berry said the measured amounts were "peanuts." The calcium balance studies which required measurements of calcium loss to sweat, blood, urine, and feces might take "several more months to complete," he added. Astronaut Edward H. White (L/Col., USAF) accompanied Dr. Berry to the meeting to brief scientists on preparations necessary to plan and execute successful space missions. (Haseltine, *Wash. Post*, 12/30/65, A1)

- NASA Hq. reorganization plan, effective Jan. 2, 1966, was disseminated throughout NASA. Two main effects of the changes: establishment of Office of the Administrator, in which the Administrator and Deputy Administrator would be supported by the Associate Deputy Administrator with a strong Secretariat; and, establishment of operating pattern within this office delegating authority and responsibility to the Deputy Administrator, who would serve as general manager as well as Acting Administrator in the Administrator's absence. Dr. Robert C. Seamans, Jr., Deputy Administrator, retained the additional title of Associate Administrator.

    Other changes included: Director of the Office of Tracking and Data Acquisition would be made Associate Administrator for Tracking and Data Acquisition; heads of functional staff offices except General Counsel would have the title Assistant Administrator. Heads of all functional staff offices as well as the four program offices would be responsible to the Deputy Administrator. (NASA Memo)

- Article in Japanese newspaper said U.S. Defense Secretary Robert S. McNamara's estimates of Red Chinese nuclear potential were probably conservative, citing increasing appropriations by Red China for scientific research as indication of rapid progress that might be expected. From $16 million in 1955, the amount rose to $340 million in 1959; it was estimated that $2 billion had been spent since

1960. It was calculated that Red China should have a minimum arsenal of between 150 and 200 atomic bombs by the end of 1967; production of delivery vehicles should by that time be keeping pace. (Elegant, *Wash. Post*, 12/30/65)

December 29: USAF awarded Thiokol Chemical Corp. a $2,000,000 initial increment to a fixed-price contract for production of solid rocket motors. Estimated final amount of contract was $5,000,000.

General Dynamics Corp. received from AFSC a $1,655,299 fixed-price contract for design and fabrication of reentry vehicle instrumentation and range safety systems. (DOD Release 946-65)

- Mounting cost of the Vietnam war was apt to slow the USAF Manned Orbiting Laboratory project, the *New York World Telegram* quoted informed sources as saying. (Troan, *N.Y. World Telegram*, 12/29/65)

December 30: Astronaut Walter Schirra said that he and Astronaut Thomas P. Stafford had maneuvered the GEMINI VI spacecraft to within one foot of GEMINI VII during their Dec. 15 rendezvous and that although they were backed up by an advanced type of computer, they probably could have effected rendezvous without it. Schirra and the three other astronauts involved in the GEMINI VII-VI flights gave reports on their missions at an MSC news conference. (AP, *Wash. Post*, 12/31/65, A1; MSC *Gemini VII/Gemini VI Fact Sheet*)

- Special NASA awards ceremony was held at MSC, with NASA Administrator James E. Webb making the presentations. Rep. Olin Teague, Chairman of Manned Space Flight Subcommittee of the House Committee on Science and Astronautics, also was present for the occasion.

NASA Distinguished Service Medal, NASA's highest honor, was presented to Donald K. Slayton, Assistant Director for Flight Crew Operations, MSC, and to Astronaut Walter M. Schirra, Jr., Command Pilot for GEMINI VI mission. Slayton's award was for "his outstanding performance in directing NASA flight operations and for his leadership of the continuous and rapid adaptation of NASA's astronaut training activities to the experience gained from Mercury and Gemini flights. . . ." Schirra's DSM cited "his courage and judgment in the face of great personal danger, his calm, precise and immediate perception of the situation that confronted him and his accurate and critical decisions that made possible the successful execution of the Gemini VI mission." This ceremony marked the first time the top NASA medal was presented away from Washington. (With these two presentations, all seven original Mercury astronauts had received the Distinguished Service Medal.)

NASA Exceptional Service Medal was presented to the crews of GEMINI VII (Astronauts Borman and Lovell) and VI (Astronauts Schirra and Stafford); William C. Schneider, Deputy Director of the Gemini Program for Mission Operations, OMSF; and John T. Mengel, Assistant Director for Tracking and Data Systems Directorate, GSFC.

Group Achievement Awards were presented to the following groups from the Cape: KSC Launch Operations; KSC Spacecraft Operations; Martin Co. Gemini program group; McDonnell Aircraft Corp. Gemini program group; Gemini Launch Vehicle Div., AFSC 6555th Aerospace

Test Wing: and AFETR Test Operations Div. (MSC *Gemini VII/ Gemini VI Fact Sheet;* Exec. Secy., MSC Awards Committee; NASA Proj. Off.)

December 30: ComSatCorp invited design proposals for a multi-purpose satellite having at least 20 times the communications capacity of EARLY BIRD I. Firms throughout the world were asked to submit proposals. ComSatCorp said more than one study contract could be awarded. Proposed new satellite would make nationwide and international distribution of television feasible, provide a sending and receiving capability between ground stations and ships at sea or aircraft in flight, provide complete interchangeability from one type of service to another within a single satellite, and permit access to these services by any number of earth stations. (ComSatCorp Release)

- NASA and USAF announced agreement for extended XB-70 flight research beginning next spring in joint project to obtain supersonic operational flight information impossible to get in ground facilities. Among the items of interest were skin friction, stability and control, drag, boundary layer flow, air loads, thermal environment, sonic boom, landing, and crew workload. (NASA Release 65-393)

December 31: Soviet space probe VENUS II, launched Nov. 12, would pass "at the prescribed distance" from Venus, Tass announced. VENUS III, launched Nov. 16, had been corrected in flight "in order to bring it closer to Venus," the announcement added. Telemetered data showed that conditions aboard both probes were normal; scientific instrumentation was functioning normally. VENUS II was 15.5-million km. (9.6-million mi.) from earth; VENUS III was 14.3 million km. (8.9 million mi.) away. (Tass, 12/31/65)

- Cosmonaut Yuri Gagarin, the first man to fly in space, said in an interview with *Krasnaya Zvezda* that an increasing number of professions would take part in future spaceflights. He noted that pilots, engineers, a scientist, and a doctor had already flown in Soviet spacecraft. Gagarin described recent American spaceflights as "a major achievement," but said it would be a simplification to interpret U.S. accomplishments as "first steps on an unexplored way." He said Soviet spacecraft VOSTOK III and VOSTOK IV had flown in formation in August 1962. (AP, Wash. *Eve. Star,* 12/31/65, A2)

- U.S. and Yugoslavia would exchange visits and information under a new, non-Governmental agreement reached between the U.S. National Academy of Sciences and the Yugoslav Academy of Sciences. Under the terms of a memorandum of understanding, scientists of both countries would be able to take advantage of short- and long-term visits to lecture, conduct seminars, or carry out laboratory research. Money for the exchange program would come from the U.S. National Science Foundation. (*Wash. Post,* 12/1/65, A3)

- In a report of shareholders for fall 1965, ComSatCorp said it had realized revenues of $966,000 from the operation of EARLY BIRD I covering a period from June 28, when the satellite began commercial service between North America and Europe, through Sept. 30. As of Sept. 30, the Corporation's total cash and temporary cash investments amounted to $187,767,000. (ComSatCorp Release)

During December: NASA Flight Research Center had completed analysis of flight handling characteristics of six representative light private aircraft currently manufactured in the U.S. Aircraft flown in the study were considered a good cross-section of this type of aircraft and included high- and low-wing and single- and twin-engine configurations. Report would be published as a technical note in spring 1966. (*Av. Wk.*, 12/27/65, 13)

• Snap–10A nuclear ground test system, designated Flight System–3 (FS–3) and a flight-qualified copy of the SNAPSHOT I orbital test system, had operated continually since January 22, 1965, exceeding by four months the previous record for continuous power operation of any known reactor. By the time of the SNAPSHOT I launch (April 3), FS–3 had accumulated more than 70 days of operating time. Continuing to operate throughout the remainder of the year, the system operated satisfactorily but with a gradual degradation in power output. (*Atomic Energy Programs, 1965*, 151–152)

• The Council of the NAS announced that a statement had been placed in the Minutes of the Council as a memorial to the late Dr. Hugh L. Dryden:

"Although the service and devotion of Hugh L. Dryden to the National Academy of Sciences, where he was ten years Home Secretary, twelve years our colleague in the Council, and twenty-one years a member of the Section of Engineering, have in our time rarely been equalled and certainly not surpassed, they represent only a portion of his service and devotion to several national institutions closely linked to the welfare of our people. . . . In achieving so much for his country and its institutions, he gave of himself without thought of self. He was deeply admired and loved by all who came in association with him. The sorrow felt at his passing by all members of the Academy is accompanied by an enduring pride in honoring his memory." (NAS–NRC–NAE *News Report*, 12/65)

• In his book *Galaxies, Nuclei, and Quasars* British scientist Fred Hoyle declared that on the basis of new evidence the "steady-state" theory of cosmology—of which he had been a leading proponent since 1948—was now untenable. He discussed his new theory, a variation of the oscillating-universe concept. (*Science*, 12/24/65, 1708)

• NASA Marshall Space Flight Center ended 1965 with 7,522 employees earning in excess of $82.8 million—about 700 of whom were located in contractor plants throughout the U.S. There were some 4,280 contractor employees working at MSFC's Redstone Arsenal complex with estimated earnings of $43 million bringing the Center's direct and indirect payroll for 1965 (combined Civil Service and contractor) to about $125.8 million. An estimated 10,000 other contractor workers were employed by contractors in Huntsville in connection with MSFC programs. (MSFC Release 65–313)

• Principal source of advanced technology in the U.S. had been and would remain the aerospace industry, postulated an article in *Aerospace*. Only in its programs were technical goals high enough and national requirements urgent enough to move forward in major steps. As these steps were completed and technical goals achieved, the entire economy fell heir to the new technology. Key goals in defense and space programs were listed as: (1) lowering costs; (2) improving the efficiency of motors, generators, and all other energy conversion devices and processes; (3) improving design, that is, reducing the weight and increasing the strength of all machines, by either improved knowledge of the machine or by using lighter, stronger materials; (4) improving the accuracies to which machines could be controlled; (5) improving reliability; (6) improving communications between men, between men and machines, and between machines. (*Aerospace*, Winter 1965)

During 1965: In 1965, NASA attempted 28 missions with 23 successes, a score of 82% mission success. Two spacecraft were launched on a single booster. NASA attempted 30 launches of space boosters and had 26 successes, a score of 87% launch vehicle success.

Of the 92 payloads orbited by the U.S., NASA orbited 25—five of which were two-manned spacecraft and four were escape-mission probes. U.S.S.R. orbited 64 payloads, of which one was a manned spacecraft and seven were escape-mission probes. France entered the space age with two satellites, one orbited by France herself (A-1) and one by the U.S. (FR-1).

Most spectacular of NASA's space missions were the real-time reception of close-up lunar photographs by RANGER IX, first U.S. extravehicular activity by GEMINI IV Astronaut White, man's first close look at Mars (including 21 photographs) by MARINER IV, and manned rendezvous to within one foot by GEMINI VI and VII. The Gemini program began 1965 with GT-2 unmanned suborbital flight; this was followed by four two-man orbital flights which logged more than 1,300 manhours in space. GEMINI V and VII, long-duration missions (8 and 14 days respectively) proved man can withstand extended conditions of spaceflight and validated plans for manned Apollo lunar flights. At the year's end Gemini spaceflights had set for the U.S. more than 10 records, among them the record for total manhours in space: 1,354 hrs., 38 min., vs. U.S.S.R.'s 507 hrs., 16 min.

1965 was a year of extensive ground tests of the Apollo spacecraft and the Saturn launch vehicles. There were static firings of engines for the various Saturn IB and Saturn V stages, highlighted by full-duration firing of all three Saturn V stages. Saturn I program ended with a record of 10 successes out of 10 attempts, its three 1965 launches orbiting Pegasus meteoroid detection satellites. Fire II provided valuable data for Apollo on reentry from simulated lunar trajectory.

In lunar and planetary achievement, RANGERS VIII and IX provided more than 13,000 lunar surface photographs and brought that project to a close. Atlas-Centaur AC-6 launched a dummy Surveyor on a simulated lunar transfer orbit and proved itself capable of operational Surveyor missions. NASA stepped up its interplanetary research with PIONEER VI, first of four projected interplanetary satellites.

TIROS IX became the first weather satellite to provide close to 100% coverage of the earth daily, and TIROS X was the first in a series of interim operational satellites for the U.S. Weather Bureau. Five Explorer scientific satellites, OSO II, Canada's ALOUETTE II, and France's

FR–I were orbited. NASA orbited SECOR V for the U.S. Army, SOLRAD IX (EXPLORER XXX) for the U.S. Navy, and EARLY BIRD I communications satellite for the Communications Satellite Corporation. NASA turned over SYNCOMs II and III to DOD for operational use at the completion of their R&D function.

The five NASA mission failures: vehicle test of Atlas-Centaur AC–5; test of Apollo launch escape system on a Little Joe II booster; failure of OGO II, which did return good (but incomplete) experiment data; failure to orbit a third Orbiting Solar Observatory; and failure of Gemini 6 because Agena Target Vehicle did not achieve orbit.

In 39 successful launches, DOD orbited 67 satellites. In addition, there were four unsuccessful DOD space launches, losing five payloads. Decision was made that DOD proceed with the Manned Orbiting Laboratory, and launches were begun of the powerful Titan III–C which would eventually orbit the MoI.

Highlighting the U.S.S.R.'s busy space year was man's first extra-vehicular space activity, by VOSKHOD II's Cosmonaut Leonov. Soviet lunar exploration intensified, with apparent soft-landing attempts by LUNAS V through VIII, as well as photographs of the moon's hidden side by interplanetary probe ZOND III. U.S.S.R. launched VENUS II and III on the long flight toward the planet Venus. (Western experts speculated that COSMOS LX was an unsuccessful lunar soft-landing attempt and COSMOS XCVI an unsuccessful Venus probe.) U.S.S.R. orbited 52 Cosmos satellites, two heavyweight Proton spacecraft, and her first communications satellites—two Molniyas I. (NASA Release 65–368; MSFC Release 65–312; NASA HHR–8; NASC Staff; NASA *A&A 1965*; Simons, *Wash. Post*, 12/12/65)

During 1965: Some NASA research highlights of 1965:

NASA launched 191 scientific sounding rockets to obtain a variety of scientific data and about 10 for advanced research and technology. By the end of the year, the three X–15 research aircraft had made 156 flights, 32 of them in 1965. Tests proceeded of the modified X–15 (No. 2), expected to exceed 5,000-mph flight speed. NASA supported the National supersonic transport program and evaluated wind tunnel models of the two proposed Sst designs. NASA continued using USAF XB–70 aircraft for research in flight problems of Sst and large supersonic aircraft in general. Feasibility studies of adapting V/Stol concepts to commercial transportation and wind-tunnel studies of several V/Stol designs continued. As part of research toward efficient design of hypersonic aircraft, comparative performance and heat transfer measurements were obtained on a variety of aircraft wings, bodies, and wing-body combinations. Ramjet engine research was in the engine design concept phase. NASA developed and tested a new near-field theory useful in predicting sonic boom characteristics of Sst configurations during transonic climb-out. Flight research of M–1 lifting body was advanced, and construction of HL–10 lifting body was being completed.

Solid propulsion technology was advanced with static-firing of the 260-in. solid-fuel rocket motor. In nuclear rocket technology, three Nerva reactor experiments were completed and the new Phoebus advanced graphite reactor test program was begun. The success of a

2,600-hr. ion engine test provided evidence that life-times of 10,000 hrs. for electric rocket thrusters may be within reach. Among highlights in electronics and control was reentry communications experiment on GEMINI III demonstrating feasibility of water injection to overcome communications blackout during reentry. (*U.S. Aeron. & Space Act.*, 1965, 25–38; OSSA & OART S.Rkt. Proj. Off.; NASA Release 65–368)

During 1965: About 90% of NASA's budget was going to contractors to pay for work being done by nearly 400,000 people of about 20,000 prime and subcontractors. In the university program, NASA was doing business with about 200 universities in every state on space-related projects at the year's end. NASA's requirement to transfer its technology to other sectors of the economy had developed into an information system with more than 200,000 technical documents abstracted, indexed, and filed in a computer-based nationwide system. (NASA Release 65–368)

- NASA's FY '65 top 100 contractors, according to the net value of direct awards, were headed by North American-Downey ($1,099,448,000), Boeing-New Orleans ($305,988,000), Grumman-Bethpage ($267,226,000), Douglas-Santa Monica ($251,668,000), GE-Huntsville ($181,472,000), McDonnell-St. Louis ($166,670,000), IBM-Huntsville ($128,312,000), Aerojet-Sacramento ($123,186,000), General Dynamics/Convair ($111,148,000), and RCA-Princeton ($106,552,000). Total NASA procurements for the year totaled $5,187,000,000 with the top 100 firms accounting for $4,141,434,000 of that amount. (NASA *Ann. Procurement Rpt. FY 1965*)

- Some highlights of the year in physics: Planet Mars has no substantial magnetic field, according to scientific experiments by MARINER IV Mars probe. Because of this, Martian life-forms would have to be able to withstand intense radiation, but possibility of Martian life was not ruled out. MARINER IV also reported Mars' atmosphere was too thin to readily support parachute or glider-type descents for soft landings on the planet.

    Comet Ikeya-Seki, the brightest comet to enter the solar system this century, was discovered Sept. 18 and subsequently studied from the ground, airplanes, and rockets. The comet passed within 300,000 mi. of the sun and made a hairpin turn Oct. 21, when it broke into three pieces.

    Results of NASA's testing of public reaction to sonic booms, begun in 1961, showed that no serious psychological or physiological effects result from sonic booms. Commercial use of the supersonic transport, to begin in early 1970's, was expected to create sonic booms that, when Sst's come into full use in the U.S., would be heard by everyone about once a day.

    On the subject of antimatter, two items: Swedish physicist Hannes Alfven theorized such cosmic phenomena as quasi-stellar radio sources ("quasars"), radio stars, and supernovae may be powered by matter-antimatter annihilations; and, antideutrons were produced in the laboratory by scientists of Columbia Univ.

    Unwanted noise from a BTL horn-reflector antenna being used in comsat research may have been radiation from the birth of the universe. This interpretation was based on the "big bang" theory of the universe's formation and was propounded by a group of Princeton Univ. scientists.

Decision was believed near on which of the three principal models of the universe was correct—the steady-state, expanding, or oscillating universe theory. Results in astronomy based on red-shifts of "blue-galaxies" (discovered in 1965) and quasars hinted at an oscillating universe. (AIP *News*, 12/22/65)

During 1965: Among highlights in the U.S. communications satellite program were: initiation of commercial comsat operations, with EARLY BIRD I in synchronous orbit over the Atlantic; initiation of a synchronous-orbit comsat program to furnish communications for Project Apollo and for expanded commercial service; and growth of the International Telecommunications Satellite Consortium to 48 member nations, representing every continent. Communications Satellite Corp. represented the U.S. and served as manager for the consortium. (ComSatCorp Release)

- LRC completed 10 experiments to explore nuclear engine chilldown (cooling by liquid hydrogen propellant) and to determine the range of conditions for which flow oscillations would occur in an engine system. An additional 16 runs were conducted to obtain data on the "bootstrap" starting of a nuclear engine. Results of these tests indicated that nuclear rocket engines should be able to start smoothly and stably over a wide range of startup conditions. (*Atomic Energy Programs, 1965,* 148–49)

- Nearly 390,000 visitors toured NASA Kennedy Space Center and Cape Kennedy. The Space Center's doors were first opened to visitors Jan. 3, 1965. In its first year, the Center was host to visitors from all 50 states and 57 countries. (*NASA Space Sheet,* 3/31/66)

- Visitor attendance at the MSFC Space Orientation Center was 202,445, an increase of about 37% over 1964. (MSFC Historian)

- In the last half of the year, USAF doubled its airlift to Southeast Asia, mainly Vietnam. In the first six months of 1965 the Military Air Transport Service (MATS) airlifted 37,684 tons of cargo and 91,994 passengers to the area. In the last half, it flew in 58,858 tons of cargo and 183,132 passengers.

    USAF flew 10,570 tactical sorties over North Vietnam and 37,940 over South Vietnam (compared to 764 for 1964); VNAF flew an additional 23,700 sorties during 1965, mostly over South Vietnam. (Watson, Balt. *Sun,* 1/6/66; Brownlow, *Av. Wk.,* 1/10/66)

Appendix A

SATELLITES, SPACE PROBES, AND MANNED SPACE FLIGHTS

A CHRONICLE FOR 1965

The following tabulation was compiled from open sources by Dr. Frank W. Anderson, Jr., Deputy NASA Historian. Sources included the United Nations Public Registry, the *Satellite Situation Report* issued by the Space Operations Control Center at Goddard Space Flight Center, public information releases of the Department of Defense, NASA and other agencies, and the *Report to the Congress from the President of the United States: United States Aeronautics and Space Activities, 1965*. Russian data are from the U.N. Public Registry, the *Satellite Situation Report*, translations from Tass News Agency statements in the Soviet press, and international news services' reports.

It might be well to call attention to the terms of reference stated or implied in the title of this tabulation. This is a listing of payloads that have (a) orbited, (b) as probes, ascended to at least the 4,000-mile altitude that traditionally has distinguished probes from sounding rockets, etc., or (c) conveyed one or more human beings in space. Furthermore, only flights that succeeded—or at least are not known to have failed—in doing one of the above are listed. Date of launch is referenced to local time at the launch site. An asterisk by the date marks those dates that are one day earlier in this tabulation than in listings which reference to Greenwich time.

In terms of numbers, ambitiousness, and complexity, the world space effort continued to gain momentum. A total of 160 payloads was successfully orbited in a total of 102 launches. Thus more than one third of the payloads was in multiple-payload launches. The U.S. program was up to 94 payloads in 61 launches in 1965 from 76 payloads in 60 launches in 1964 (DOD: 67 payloads in 39 launches, compared with 53/38; NASA: 27 payloads—counting 2 rendezvous pods ejected from manned Gemini spacecraft—in 22 launches, compared with 23/22). The U.S.S.R. once again almost doubled its program from a numerical standpoint, with 66 payloads in 50 launches, compared with 35/30 in 1964. Multiple payload launches increased: the U.S. had 46 payloads on 15 multiple-payload launches; DOD had 40 payloads on its 12 multi-payload launchings and NASA had 6 on 3; U.S.S.R. had 23 on 7. Areas of concentration were also obvious: the U.S. took the absolute lead in manned spaceflight, with five 2-manned Gemini flights to the one Soviet 2-manned flight. But the Soviets more than doubled the U.S. deep-space effort, with 5 lunar flights and two planetary flights to the U.S. 2 and 1.

As we have cautioned in previous years, the "Remarks" column of these appendixes is never complete because of the inescapable lag behind each flight of the analysis and interpretation of scientific results.

576　　ASTRONAUTICS AND AERONAUTICS, 1965

Launch Date	Name	International Designation	Vehicle	Payload Data	Apogee (st. mi.)	Perigee (st. mi.)	Period (minutes)	Inclination	Remarks
Jan. 11	COSMOS LII (U.S.S.R.)	1965-1A	Not available	Total weight: Not available. Objective: Continuation of Cosmos scientific satellite series. Payload: Not available.	189	127	89.5	65°	Reentered 1/19/65.
Jan. 15	DOD Spacecraft (United States)	1965-2A	Thor-Agena D	Total weight: Not available. Objective: Develop spaceflight techniques and technology. Payload: Not available.	265	117	90.5	74.94°	Reentered 2/9/65.
Jan. 19	DOD Spacecraft (United States)	1965-3A	Thor-Altair	Total weight: Not available. Objective: Develop spaceflight techniques and technology. Payload: Not available.	516	286	97.6	98.77°	Still in orbit.
Jan. 22	TIROS IX (United States)	1965-4A	Thor-Delta	Total weight: 305 lbs. Objective: Test new cartwheel configuration and operating mode to be used in future operational weather satellites. Payload: 22″ × 42″ cylindrical 18-sided polygon, containing 2 TV cameras with 104° lens, photo storage, and 2-sec. scan transmission system; control system and horizon scanners; 9,100 n-on-p solar cells; 63 nickel-cadmium batteries.	1,602	426	81.6	119°	Complicated 3-dogleg maneuver put TIROS IX in near-polar orbit to enable global weather coverage; 11-sec. extra burn of Delta 2nd stage caused elliptical orbit instead of 460-mi. circular one. Cartwheel operational mode was successful, as were new components. One camera ceased operation 4/65.
Jan. 23	DOD Spacecraft (United States)	1965-5A	Atlas-Agena D	Total weight: Not available. Objective: Develop spaceflight techniques and technology. Payload: Not available.	152	89	89	102.5°	Reentered 1/28-29/65.
Jan. 30	COSMOS LIII (U.S.S.R.)	1965-6A	Not available	Total weight: Not available. Objective: Continuation of Cosmos scientific satellite series. Payload: Not available.	741	141	98.7	48.8°	Still in orbit.
Feb. 3	OSO II (United States)	1965-7A	Thor-Delta	Total weight: 545 lbs. Objective: Continue OSO I studies of solar x-ray, gamma ray, and ultraviolet emission, with added capability to scan entire solar disc and part of corona. Payload: Top part of spacecraft a 22″-radius semicircular sail continu-	393	343	97	33°	6 of 8 sun-study experiments functioned well, but OSO II did not achieve all its primary objectives. Satellite was turned off in 11/65, after exceeding its lifetime expectancy by 50% and returning some 9,020,000,000 bits of data. Still in orbit.

Date	Name (Country)	Designation	Vehicle	Payload				Remarks
Feb. 11	Titan IIIA (United States)	1965-8A	Titan IIIA	Total weight: 11,500 lbs., including 7,000-lb. transtage and 1,070 lbs. of truss ballast. Objective: Develop spaceflight techniques and technology. Payload: Transtage, ballast, and engineering telemetry.	1,737	1,721	145.6	32.15°	Transtage performed well, demonstrating by triple ignition of motors the ability to change orbit several times; also demonstrated multiple satellite ejection by orbiting LES I and the chunk of truss ballast. Still in orbit.
	LES I and	1965-8B		Total weight: 69 lbs. Objective: Develop spaceflight techniques and technology. Payload: Experimental communications equipment.	1,740	1,722	145.7	32.15°	LES I was to have fired solid-propellant motor to achieve 11,500-mi.-apogee elliptical orbit, but motor did not fire. Still in orbit.
Feb. 16	PEGASUS I (United States)	1965-9A	Saturn I	Total weight: 3,200 lbs. (plus 19,800-lb. S-IV 2nd stage, plus 10,000-lb. Apollo boilerplate, command and service modules; total in orbit, 33,200 lbs.) Objective: Study distribution, size, and velocity of meteoroids in near-earth orbit; continue development of Saturn I launch vehicle. Payload: 96′ (when unfolded in orbit) × 14′ unfolding wings composed of electrically charged panels sensitive to meteoroid hits; center section contains motor for unfolding wings, telemetry, solar cells, batteries; all this affixed to 41½′-long S-IV 2d stage of the Saturn launch vehicle.	462	308	97	31.7°	Boilerplate Apollo separated from S-IV stage on schedule; PEGASUS I then deployed its wings; by 9/65 it had recorded meteoroid hit rates of 44 per sq. meter per year. SA-9 vehicle performed excellently, marking 8th successful Saturn I flight in 8 attempts. PEGASUS I still in orbit.
	Apollo and	1965-9B		Total weight: 10,000 lbs. Objective: Test separation techniques and total vehicle balance. Payload: 12′10″ (dia. at base) × 11′2″ conical boilerplate Apollo command module, attached to 12′10″ (dia.) × 13′2″ boilerplate service module.	453	309	97.1	31.76°	Separation was on schedule. Still in orbit.

Launch Date	Name	International Designation	Vehicle	Payload Data	Apogee (st. mi.)	Perigee (st. mi.)	Period (minutes)	Inclination	Remarks
Feb. 17	RANGER VIII (United States)	1965-10A	Atlas-Agena B	Total weight: 806.8 lbs. Objective: By means of close-in photography, contribute to scientific understanding of lunar surface and support Surveyor and Apollo soft-lunar-landing programs. Payload: 15'-wide and 10'4"-tall (cruise position, with solar panels extended) structure. Hexagonal base contains conical midcourse motor, retro-rocket; other elements are command system, 1 radio receiver and 3 transmitters, telemetry system, 4 batteries, 6 TV cameras, 9,793 solar cells, 2 antennas, attitude control system.	Impacted on moon.				RANGER VIII transmitted to earth some 7,000 close-in photos of the moon's maria before impacting on the moon's Sea of Tranquillity 2/20/65, after excellent flight. Photo time was last 23 min. of flight (compared with last 17 min. of RANGER VII) to provide photos from far enough above the moon's surface to be comparable to ones obtainable from earth. Results indicated the Sea of Tranquillity was similar in structure to Sea of Clouds photographed by RANGER VII, suggested surface had consistency of crunchy snow.
Feb. 21	COSMOS LIV (U.S.S.R.)	1965-11A	Not available	Total weight: Not available. Objective: Continuation of Cosmos scientific satellite series. Payload: Not available.	1,151	173	106.2	56.07°	Still in orbit. Three satellites launched with single vehicle.
	and COSMOS LV	1965-11B		Total weight: Not available. Objective: Continuation of Cosmos scientific satellite series. Payload: Not available.	1,151	173	106.2	56.07°	Still in orbit.
	and COSMOS LVI	1965-11C		Total weight: Not available. Objective: Continuation of Cosmos scientific satellite series. Payload: Not available.	1,151	173	106.2	56.07°	Still in orbit.
Feb. 22	COSMOS LVII (U.S.S.R.)	1965-12A	Not available	Total weight: Not available. Objective: Continuation of Cosmos scientific satellite series. Payload: Not available.	317	109	91.1	64.77°	Reentered 2/22/65. Accompanying vehicle disintegrated into more than 160 fragments, the last of which had reentered by 3/28/65.
Feb. 25	DOD Spacecraft (United States)	1965-13A	Thor-Agena D	Total weight: Not available. Objective: Develop spaceflight techniques and technology. Payload: Not available.	228	111	89.9	75.07°	Reentered 3/18/65.
Feb. 26	COSMOS LVIII (U.S.S.R.)	1965-14A	Not available	Total weight: Not available. Objective: Continuation of Cos-	409	360	96.8	65°	Still in orbit.

ASTRONAUTICS AND AERONAUTICS, 1965

Date	Name	Designation	Launch vehicle	Description					
Mar. 7	COSMOS LIX (U.S.S.R.)	1965-15A	Not available	mos scientific satellite series. Payload: Not available. Total weight: Not available. Objective: Continuation of Cosmos scientific satellite series. Payload: Not available.	211	130	89.7	65°	Reentered 3/15/65.
Mar. 9	GREB VI (United States)	1965-16A	Thor-Agena D	Total weight: Not available. Objective: Develop spaceflight techniques and technology. Payload: Instruments for measurement of solar radiation; telemetry.	585	562	103.5	70.06°	Still in orbit. Eight satellites launched with single vehicle.
	GGSE II and	1965-16B		Total weight: Not available. Objective: Develop spaceflight techniques and technology. Payload: Gravity gradient stabilization experiment.	583	564	103.5	70.08°	Still in orbit.
	GGSE III and	1965-16C		Total weight: Not available. Objective: Develop spaceflight techniques and technology. Payload: Gravity gradient stabilization experiment.	582	564	103.5	70.08°	Still in orbit.
	SOLRAD and	1965-16D		Total weight: Not available. Objective: Develop spaceflight techniques and technology. Payload: Instruments for measurement of solar radiation.	583	564	103.5	70.08°	Still in orbit.
	SECOR III and	1965-16E		Total weight: 40 lbs. Objective: Develop spaceflight techniques and technology. Payload: Geodetic satellite.	583	563	103.5	70.11°	Still in orbit.
	OSCAR III and	1965-16F		Total weight: 33 lbs. Objective: Develop spaceflight techniques and technology. Payload: Amateur radio communications satellite.	583	564	103.5	70.08°	Still in orbit.
	SURCAL and	1965-16G		Total weight: Not available. Objective: Develop spaceflight techniques and technology. Payload: Calibration satellite for Spasur tracking system.	583	564	103.5	70.08°	Still in orbit.
	SURCAL and	1965-16H		Total weight: Not available. Objective: Develop spaceflight techniques and technology. Payload: Calibration satellite for Spasur tracking system (dodecahedron shape).	585	562	103.5	70.08°	Still in orbit.

Launch Date	Name	International Designation	Vehicle	Payload Data	Apogee (st. mi.)	Perigee (st. mi.)	Period (minutes)	Inclination	Remarks
Mar. 11	DOD Spacecraft (United States) and SECOR II	1965-17A 1965-17B	Thor-Able-Star	Total weight: Not available. Objective: Develop spaceflight techniques and technology. Payload: Not available. Total weight: 40 lbs. Objective: Develop spaceflight techniques and technology. Payload: Geodetic satellite.	568 642	127 177	96.4 97.9	89.97° 89.99°	Reentered 6/14/65. Two spacecraft launched with single vehicle. Still in orbit.
Mar. 12	COSMOS LX (U.S.S.R.)	1965-18A	Not available	Total weight: Not available. Objective: Continuation of Cosmos scientific satellite series. Payload: Not available.	178	125	89.1	69.7°	Reentered 3/17/65.
Mar. 12	DOD Spacecraft (United States)	1965-19A	Atlas-Agena D	Total weight: Not available. Objective: Develop spaceflight techniques and technology. Payload: Not available.	151	97	88.5	107.6°	Reentered 3/17/65.
Mar. 15	COSMOS LXI (U.S.S.R.) and COSMOS LXII and COSMOS LXIII	1965-20A 1965-20B 1965-20C	Not available	Total weight: Not available. Objective: Continuation of Cosmos scientific satellite series. Payload: Not available. Total weight: Not available. Objective: Continuation of Cosmos scientific satellite series. Payload: Not available. Total weight: Not available. Objective: Continuation of Cosmos scientific satellite series. Payload: Not available.	1,139 1,139 1,139	169 169 169	106 106 106	56° 56° 56°	Still in orbit. Three spacecraft launched with single vehicle. Accompanying vehicle disintegrated into more than 120 fragments, of which most were still in orbit. Still in orbit. Still in orbit.
Mar. 18	DOD Spacecraft (United States)	1965-21A	Thor-Altair	Total weight: Not available. Objective: Develop spaceflight techniques and technology. Payload: Not available.	469	328	97.5	99.03°	Still in orbit.
Mar. 18	VOSKHOD II (U.S.S.R.)	1965-22A	Not available	Total weight: 12,529 lbs. Objective: Continuation of study of manned spaceflight; conduct extravehicular activity.	309	108	91	65°	First manned extravehicular activity was performed by co-pilot L/Col. Aleksey Leonov. VOSKHOD II was launched from

Date	Name	Designation	Launch Vehicle	Payload/Objective	Weight (lbs)	Perigee (mi)	Apogee (mi)	Inclination	Remarks
									Baikonur, Kazakhstan; during the 2nd orbit Col. Leonov moved out of the airlock on a 16-ft. tether, spent 12 min. free-floating in space. VOSKHOD II reentered 3/19, after 17 orbits and 26 hrs. 2 min. in flight; landed near Perm, U.S.S.R. Pilot was Col. Pavel Belyayev.
Mar. 21	RANGER IX (United States)	1965-23A	Atlas-Agena B	Payload: 2-man spacecraft in 2 modules; command module contains crew compartment; life-support equipment; airlock; 2 UHF transmitters and 2 HF receivers; 2 UHF transmitters and 2 UHF receivers; broadcast receiver; tape recorder; 2 TV cameras outside spacecraft and 2 TV cameras and video check device inside spacecraft; instrument module containing control rockets and fuel; liquid-fuel retro-rocket and reserve solid-fuel retro-rocket. Total weight: 808.8 lbs. Objective: By means of close-in photography, contribute to scientific understanding of lunar surface and support Surveyor and Apollo soft-lunar-landing programs. Payload: 15'-wide and 10'4"-tall (cruise position, with solar panels extended) structure. Hexagonal base contains conical midcourse motor, retrorocket; other elements are command system, 1 radio receiver and 3 transmitters, telemetry system, 4 batteries, 6 TV cameras, 9,793 solar cells, 2 antennas, attitude-control system.		Impacted on moon			RANGER IX transmitted to earth some 5,814 close-in photos of the moon's mountains and craters before impacting the moon's crater Alphonsus 3/24/65, only 4 mi. off target. Network TV broadcast "live" photos from last 10 min. of flight. Results suggested parts of lunar highlands might be harder and smoother than the maria but that crater floors might be dangerously soft. This was final Ranger flight.
Mar. 23	GEMINI III (United States)	1965-24A	Titan II	Total weight: 7,111 lbs. (includes reentry and adapter modules). Objective: Demonstrate orbital manned flight capability of Gemini spacecraft, including maneuver in orbit and controlled reentry and landing. Payload: 18'5" × 10'2-module bell-shaped spacecraft, containing 2 astronauts; guidance and control equipment; 2 cameras; 1 HF and 1 UHF transceiver, high and low frequency telemetry transmitters, tracking and recovery communications; batteries; environmental control system; reentry and recovery systems.	140	100	88.2	32.6°	GEMINI III, 1st U.S. two-man spaceflight and 1st manned flight in the Gemini program, successfully flew the planned 3 orbits of the earth. Astronauts Virgil I. Grissom and John W. Young made world's first piloted orbital changes (in 2 in-plane maneuvers, dropped apogee from 140 to 105 mi., later dropped perigee from 99 to 52 mi., in out-of-plane maneuver, changed plane by .02°). Made controlled reentry and landing after flight of 4 hrs. 53 min., impacting in the Atlantic 58 mi. short when spacecraft developed less lift on reentry than was expected; crew was picked up by helicopter.

Launch Date	Name	International Designation	Vehicle	Payload Data	Apogee (st. mi.)	Perigee (st. mi.)	Period (minutes)	Inclination	Remarks
Mar. 25	COSMOS LXIV (U.S.S.R.)	1965-25A	Not available	Total weight: Not available. Objective: Continuation of Cosmos scientific satellite series. Payload: Not available.	167	127	89.2	65°	Reentered 4/2/65.
Mar. 25	DOD Spacecraft (United States)	1965-26A	Thor-Agena D	Total weight: Not available. Objective: Develop spaceflight techniques and technology. Payload: Not available.	147	112	88.9	96.02°	Reentered 4/5/65.
Mar. 30	DOD Probe (United States)		Blue Scout, Jr.	Total weight: 47 lbs. Objective: Measure variations in radiation trapped in earth's magnetic field. Payload: Sensors, batteries, telemetry.	Orbit not intended.				Flight was successful; rose to altitude of 9,700 mi., impacted in Indian Ocean.
Apr. 3	SNAPSHOT I (United States)	1965-27A	Atlas-Agena D	Total weight: 970 lbs. Objective: Test operation and life of a remotely controlled nuclear reactor in space; also test ion propulsion. Payload: Snap 10A nuclear reactor for power source; ion propulsion unit; telemetry.	820	788	111.5	90.17°	12 hrs. after launch, Snap 10A was producing 600 watts of power. Operated until 5/20/65, when reactor shut down automatically for reasons external to its own operation. Ion propulsion unit did not perform adequately. Still in orbit.
	and SECOR IV (United States)	1965-27B		Total weight: 40 lbs. Objective: Develop spaceflight techniques and technology. Payload: Geodetic satellite.	817	788	111.4	90.21°	Still in orbit.
Apr. 6	EARLY BIRD I (United States)	1965-28A	Thrust-Augmented Delta	Total weight: 87 lbs. Objective: Launch a communications satellite into synchronous orbit; cooperate in its operation as a commercial communications relay point. Payload: 28½" X 23¼" cylindrical satellite, with 2 traveling-wave-tube transmitters, 2 VHF transmitters, telemetry; 6 antennas; 6,000 n-on-p solar cells; 2 nickel-cadmium batteries; apogee motor; attitude control system.	22,765	21,774	1,436.4	0.13°	Preliminary orbit was excellent; 5 days before schedule, on 4/9, Communications Satellite Corporation, for whom NASA launched the satellite, could fire the apogee motor to go into synchronous orbit. Within a week of launch EARLY BIRD I began communications relay between Europe and North America; has offered continuous voice, TV, and data transmission over its 240 duplex channels. Still in orbit.

Date	Name (Country)	ID	Vehicle	Payload/Objective	Orbit intended			Incl.	Remarks
Apr. 9	DOD Probe (United States)		Blue Scout, Jr.	Total weight: Not available. Objective: Measure space environment effects on biological samples. Payload: Biological equivalent ionization chamber; magnetic electron spectrometer; telemetry.	Orbit not intended.				Ascended to 18,000 mi. altitude, reentered over South Atlantic; telemetry received for 15 min. of flight only.
Apr. 17	COSMOS LXV (U.S.S.R.)	1965-29A	Not available	Total weight: Not available. Objective: Continuation of Cosmos scientific satellite series. Payload: Not available.	212	130	89.8	65°	Reentered 4/25/65.
Apr. 23	MOLNIYA I (U.S.S.R.)	1965-30A	Not available	Total weight: Not available. Objective: Transmit TV, 2-way multichannel telephone, phototelegraphic, and telegraphic communication. Payload: Hermetically sealed cylindrical satellite with conical ends, one containing sun and earth sensors, the other the apogee motor; 6 solar paddles extend like wheel spokes; 2 parabolic antennas; transmitters; command system; attitude control system; solar cells; batteries.	24,470	309	728	65°	Experimental communications satellite MOLNIYA I functioned well; established regular TV and telephone relay periods between Moscow and Vladivostok; experimented with color TV. On 5/2 an apogee rocket firing raised the orbit slightly (apogee by 567 km, perigee by 51 km.). Joint experiments were conducted with France on possible future international comsat system. Still in orbit.
Apr. 28	DOD Spacecraft (United States) and DOD Spacecraft	1965-31A	Atlas-Agena D	Total weight: Not available. Objective: Develop spaceflight techniques and technology. Payload: Not available.	171	95	88.1	95°	Two spacecraft launched with single vehicle. Reentered 5/3/65.
		1965-31B		Total weight: Not available. Objective: Develop spaceflight techniques and technology. Payload: Not available.	348	303	95	95.2°	Still in orbit.
Apr. 29	EXPLORER XXVII (United States)	1965-32A	Scout	Total weight: 132 lbs. Objective: Map irregularities in the earth's gravitational field; continue collection of global electron counts for cross sections of the ionosphere; continue experiments with laser for tracking, geodetic, and communications purposes. Payload: 18" X 12" octagonal satellite, with 4 solar panels extending like windmill blades; 4 radio transmitters, 4 antennas; magnetometer; nickel-cadmium batteries; 360 1-in. glass-prism reflectors and laser signal detector; 2 bar magnets.	819	584	108	41°	EXPLORER XXVII was 2nd in series of 5 satellites in geodetic research. Its results would supplement those of EXPLORER XXII (making same measurements at 80° inclination). Electron-count experiment would involve 86 ground stations in 36 countries, largest international effort to date. Still in orbit.

Launch Date	Name	International Designation	Vehicle	Payload Data	Apogee (st. mi.)	Perigee (st. mi.)	Period (minutes)	Inclination	Remarks
Apr. 29	DOD Spacecraft (United States)	1965-33A	Thor-Agena D	Total weight: Not available. Objective: Develop spaceflight techniques and technology. Payload: Not available.	290	112	90.9	85°	Reentered 5/26/65. A part of the payload separated, was designated 33B, reentered 6/8/65.
May 6	DOD Spacecraft (United States)	1965-34A	Titan IIIA	Total weight: 7,000 lbs. Objective: Develop spaceflight techniques and technology. Payload: Not available.	2,319	1,725	157	32.07°	Two spacecraft launched with 1 vehicle. Still in orbit.
	LES II and	1965-34B		Total weight: 82 lbs. Objective: Develop spaceflight techniques and technology. Payload: Experimental communications satellite.	9,364	1,753	315.3	31.35°	Still in orbit.
	LCS I	1965-34C		Total weight: 75 lbs. Objective: Develop spaceflight techniques and technology. Payload: Radar calibration sphere.	1,729	1,721	145.6	32.11°	Still in orbit.
May 7	COSMOS LXVI (U.S.S.R.)	1965-35A	Not available	Total weight: Not available. Objective: Continuation of Cosmos seientific satellite series. Payload: Not available.	181	122	89.3	65°	Reentered 5/15/65.
May 9	LUNA V (U.S.S.R.)	1965-36A	Not available	Total weight: 3,254 lbs. Objective: Attempt soft landing on lunar surface; take scientific observations of lunar environment. Payload: Telemetry; scientific instruments; retrorocket.	Impacted on moon				LUNA V impacted the moon in the Sea of Clouds on 5/12/65; attempted soft landing failed.
May 12	DOD Probe (United States)		Blue Scout, Jr.	Total weight: 47 lbs. Objective: Measure pitch angle and magnetic field intensity to 3 earth radii. Payload: Sensors, batteries, telemetry.	Orbit not intended.				Ascended to 8,536-mi. altitude in 3 hr. 50 min. flight, impacted in Indian Ocean after returning good data.
May 18	DOD Spacecraft (United States)	1965-37A	Thor-Agena D	Total weight: Not available. Objective: Develop spaceflight techniques and technology. Payload: Not available.	191	123	98.6	75°	Reentered 6/15/65.
May 20	DOD Spacecraft (United States)	1965-38A	Thor FW4S	Total weight: Not available. Objective: Develop spaceflight	597	345	100.0	98.62°	Still in orbit.

Date	Satellite	1965-	Launch vehicle	Remarks				
May 25	PEGASUS II (United States)	1965-39A	Saturn I	techniques and technology. Payload: Not available. Total weight: 3,200 lbs. (plus 19,900-lb. S-IV 2nd stage, plus 9,700-lb. Apollo boilerplate command and service modules; total in orbit, 33,800 lbs.). Objective: Study distribution, size, and velocity of meteoroids in near-earth orbit; continue development of Saturn I launch vehicle. Payload: 96' (when unfolded in orbit) × 14' unfolding wings composed of electrically charged panels sensitive to meteoroid hits; center section contains motor for unfolding wings, telemetry, solar cells, batteries; all this affixed to 41½'-long S-IV 2nd stage of the Saturn launch vehicle.	466	314	97.29	31.76°	By 9/65 PEGASUS II had recorded meteoroid hits penetrating its 1.5-mil, 8-mil, and 16-mil-thickness panels at the respective rates of 70, 5.5, and 1.45 per sq. meter per year. SA-8 vehicle performed excellently, the 9th successful Saturn I flight in 9 attempts. Still in orbit.
	Apollo and	1965-39B		Total weight: 9,700 lbs. Objective: Test separation techniques and total vehicle dynamics. Payload: 12'10" (dia. at base) × 1'2" conical boilerplate Apollo command module, attached to 12'10" (dia.) × 13'2" boilerplate service module.	467	315	97.29	31.77°	Boilerplate Apollo separated from PEGASUS II on schedule. Still in orbit.
May 25	COSMOS LXVII (U.S.S.R.)	1965-40A	Not available	Total weight: Not available. Objective: Continuation of Cosmos scientific satellite series. Payload: Not available.	217	128	90.1	51.8°	Reentered 6/2/65.
May 27	DOD Spacecraft (United States)	1965-41A	Atlas-Agena D	Total weight: Not available. Objective: Develop spaceflight techniques and technology. Payload: Not available.	164	94	88.6	95.77°	Reentered 6/1/65.
May 29	EXPLORER XXVIII (United States)	1965-42A	Thor-Delta	Total weight: 130 lbs. Objective: In very eccentric orbit, study magnetosphere, cosmic radiation, and solar wind. Payload: 28" (dia.) × 8" octagonal spacecraft, with 6' boom deploying rubidium-vapor magnetometer; 2 7' booms,deploying flux-gate magnetometers; 4 cosmic ray detectors; 3 solar-wind analyzers; 4 solar paddles mounting 6,144 n-on-p solar cells; 13 silver-cadmium batteries; transmitter.	164,000	120	8,520	34°	Slightly long burn of the booster engines put EXPLORER XXVIII in even more elliptical orbit (164,000-120 mi.) than planned (130,000-120 mi.). This was 3rd in Imp series (others were EXPLORERS XVIII and XXI). Still in orbit.

Launch Date	Name	International Designation	Vehicle	Payload Data	Apogee (st. mi.)	Perigee (st. mi.)	Period (minutes)	Inclination	Remarks
Jun. 3	GEMINI IV (United States)	1965-43A	Titan II	Total weight: 7,879 lbs. (includes reentry and adapter modules). Objective: Demonstrate performance of spacecraft and crew in spaceflight exceeding 4 days' duration. Payload: 18'5" X 10' 2-module bell-shaped spacecraft, containing 2 astronauts; guidance and control equipment; 3 cameras; 1 HF and 1 UHF transceiver, high and low frequency telemetry transmitters, tracking and recovery communications; batteries; environmental control system; reentry and recovery systems.	184	100	88.9	32.5°	GEMINI IV completed the planned 62 revolutions (97 hrs. 56 min.), longest U.S. manned spaceflight to date. Astronauts James A. McDivitt and Edward H. White tried to rendezvous with their Titan booster on 1st orbit, but gave up when it began consuming too much maneuver fuel. On the 3rd orbit White made 1st U.S. walk in space, world's 1st extravehicular activity in which man could control his movements; planned 10-min. EVA lasted 22 min. GEMINI IV landed 6/7/65 in the Atlantic 48 mi. short; crew was picked up by helicopter.
	and Rendezvous Stage	1965-43B		Total weight: 5,200 lbs. Objective: Serve as passive rendezvous object after performing as 2nd stage of Titan II launch vehicle. Payload: Signal beacon.	152	87	88.9	32.5°	Reentered 6/5/65.
Jun. 8	LUNA VI (U.S.S.R.)	1965-44A	Not available	Total weight: 3,179 lbs. Objective: Investigate the moon; develop techniques and technology for lunar investigation. Payload: Radio; telemetry; scientific instrumentation.	In heliocentric orbit.				LUNA VI passed 100,000 mi. from the moon on 6/11; Soviet sources blame the lunar miss on a midcourse motor which failed to turn off after making the midcourse correction. No Soviet statement of an intended soft landing on the moon was made about LUNA VI as it had been about LUNA V. In heliocentric orbit.
Jun. 9	DOD Probe (United States)		Blue Scout, Jr.	Total weight: 31 lbs. Objective: Measure effects of space radiation on human tissue equivalents. Payload: Tissue-equivalent ion chamber; magnetic electron spectrometer; batteries; telemetry.	Orbit not intended.				Probe ascended to 10,897 mi. altitude, impacted in Indian Ocean after 4 hr. 32 min. flight; returned good data.

Date	Spacecraft	Cospar ID	Launch Vehicle	Description					Remarks
Jun. 9	DOD Spacecraft (United States)	1965-45A	Thor-Agena D	Total weight: Not available. Objective: Develop spaceflight techniques and technology. Payload: Not available.	206	109	89.7	75.07°	Reentered 6/22/65.
Jun. 15	COSMOS LXVIII (U.S.S.R.)	1965-46A	Not available	Total weight: Not available. Objective: Continuation of Cosmos scientific satellite series. Payload: Not available.	207	127	89.7	65°	Reentered 6/23/65.
Jun. 18	DOD Spacecraft (United States)	1965-47A	Titan IIIC	Total weight: 29,300 (including dummy payload of 21,400 lbs, of which 21,000 lbs. was lead ballast). Objective: Develop spaceflight techniques and technology. Payload: Dummy payload of ballast.	118	104	88.1	32.14°	After 6 hrs. in flight, dummy payload was separated from transtage. Reentered 6/29/65.
Jun. 24	DOD Spacecraft (United States)	1965-48A	Thor-Able-Star	Total weight: Not available. Objective: Develop spaceflight techniques and technology. Payload: Not available.	704	640	106.9	90°	Still in orbit.
Jun. 25	COSMOS LXIX (U.S.S.R.)	1965-49A	Not available	Total weight: Not available. Objective: Continuation of Cosmos scientific satellite series. Payload: Not available.	201	131	89.7	65°	Reentered 7/3/65.
Jun. 25	DOD Spacecraft (United States)	1965-50A	Atlas-Agena D	Total weight: Not available. Objective: Develop spaceflight techniques and technology. Payload: Not available.	315	309	94.7	107.65°	Still in orbit.
Jun. 25	and DOD Spacecraft	1965-50B		Total weight: Not available. Objective: Develop spaceflight techniques and technology. Payload: Not available.	158	93	88.5	107.60°	Reentered 6/30/65.
Jul. 2	TIROS X (United States)	1965-51A	Thrust-Augmented Delta	Total weight: 290 lbs. Objective: Continue development of a meteorological satellite system; provide continuity in weather observation; provide maximum coverage of hurricane season. Payload: 22" × 42" cylindrical 18-sided polygon, containing 2 TV cameras with 104° lens, photo storage; transmission system; magnetic attitude control system; 9,100 p-on-n solar cells; 63 nickel-cadmium batteries.	517	458	100.6	81.4°	TIROS X was dogleg-launched into near-polar orbit; 1st Weather Bureau-funded satellite; photo from TIROS X was 1st identification of Hurricane Betsy. Still in orbit.

Launch Date	Name	International Designation	Vehicle	Payload Data	Apogee (st. mi.)	Perigee (st. mi.)	Period (minutes)	Inclination	Remarks
Jul. 2	COSMOS LXX (U.S.S.R.)	1965-52A	Not available	Total weight: Not available. Objective: Continuation of Cosmos scientific satellite series. Payload: Not available.	717	142	98.3	48.8°	Still in orbit.
Jul. 16	COSMOS LXXI (U.S.S.R.)	1965-53A	Not available	Total weight: Not available. Objective: Continuation of Cosmos scientific satellite series. Payload: Not available.	337	324	95	56°	Five satellites launched with single launch vehicle. Still in orbit.
	and COSMOS LXXII	1965-53B		Total weight: Not available. Objective: Continuation of Cosmos scientific satellite series. Payload: Not available.	365	334	95	56°	Still in orbit.
	and COSMOS LXXIII	1965-53C		Total weight: Not available. Objective: Continuation of Cosmos scientific satellite series. Payload: Not available.	350	330	95	56°	Still in orbit.
	and COSMOS LXXIV	1965-53D		Total weight: Not available. Objective: Continuation of Cosmos scientific satellite series. Payload: Not available.	384	334	95	56°	Still in orbit.
	and COSMOS LXXV	1965-53E		Total weight: Not available. Objective: Continuation of Cosmos scientific satellite series. Payload: Not available.	400	335	95	56°	Still in orbit.
Jul. 16	PROTON I (U.S.S.R.)	1965-54A	Not available	Total weight: 26,880 lbs. Objective: Study space radiation. Payload: Cylindrical satellite, containing ionization calorimeter, gamma ray telescope, and other scientific instruments; 4 solar paddles; batteries; telemetry.	375	114	92	63.5°	Termed the heaviest payload ever orbited, PROTON I was launched by a new and more powerful booster said to generate more than 600,000 hp. It was called a scientific space station. Reentered 10/11/65.
Jul. 17	DoD Spacecraft (United States)	1965-55A	Thor-Agena D	Total weight: Not available. Objective: Develop spaceflight techniques and technology. Payload: Not available.	318	292	94.4	70.17°	Still in orbit.

Date	Name	Designation	Launch Vehicle	Description					Remarks
Jul. 18	ZOND III (U.S.S.R.)	1965-56A	Not available	Total weight: Not available. Objective: Photograph back side of the moon. Payload: Cameras; telemetry.	In heliocentric orbit.				ZOND III photographed the back side of the moon as it passed some 6,200 mi. away on 7/20; transmitted 8 photos to earth on 7/29. In heliocentric orbit.
Jul. 19	DOD Spacecraft (United States)	1965-57A	Thor-Agena D	Total weight: Not available. Objective: Develop spaceflight techniques and technology. Payload: Not available.	275	112	90.7	85.05°	Reentered 8/18/65.
Jul. 20	VELA HOTEL V (United States)	1965-58A	Atlas-Agena D	Total weight: 524 lbs. (including 334 lbs. of spacecraft, 190 lbs. of kick motor). Objective: Develop techniques for monitoring nuclear explosions detectable from space. Payload: 6 x-ray, 6 gamma-ray, 1 neutron, and 2 "$\bar{z}$" particle detectors; 1 x-ray analyzer; 1 electron-proton spectrometer; 2 Geiger-Mueller tubes; 1 magnetometer; extreme ultraviolet detectors; solar cells; batteries; telemetry.	59,644	54,913	5,148.2	35.26°	Three satellites launched with single launch vehicle. Still in orbit.
	and VELA HOTEL VI	1965-58B		Total weight: 524 lbs. (including 334 lbs. of spacecraft, 190 lbs. of kick motor). Objective: Develop techniques for monitoring nuclear explosions detectable from space. Payload: 6 x-ray, 6 gamma-ray, 1 neutron, and 2 "$\bar{z}$" particle detectors; 1 x-ray analyzer; 1 electron-proton spectrometer; 2 Geiger-Mueller tubes; 1 magnetometer; extreme ultraviolet detectors; solar cells; batteries; telemetry.	75,483	63,138	6,726.1	34.98°	Still in orbit.
	and ORS III-1	1965-58C		Total weight: 12 lbs. Objective: Measure background radiation in the Van Allen belts. Payload: Octahedron satellite, with radiation detectors; solar cells; batteries; telemetry.	69,870	95	2,610.5	34.39°	Still in orbit.
Jul. 23	COSMOS LXXVI (U.S.S.R.)	1965-59A	Not available	Total weight: Not available. Objective: Continuation of Cosmos scientific satellite series. Payload: Not available.	261	162	92.2	49°	Still in orbit.

Launch Date	Name	International Designation	Vehicle	Payload Data	Apogee (st. mi.)	Perigee (st. mi.)	Period (minutes)	Inclination	Remarks
Jul. 30	PEGASUS III (United States)	1965-60A	Saturn I	Total weight: 3,200 lbs. (plus 19,900-lb. S-IV 2nd stage, plus 9,700-lb. Apollo boilerplate command and service modules; total in orbit: 33,800 lbs.). Objective: Continue study of distribution, size, and velocity of meteoroids in near-earth orbit; continue development of Saturn I launch vehicle. Payload: 96' (when unfolded in orbit) × 14' unfolding wings, composed of electrically charged removable panels sensitive to meteoroid hits; center section contains motor for unfolding the wings, telemetry, solar cells, batteries; all this affixed to 41½'-long S-IV 2nd stage of the Saturn I launch vehicle.	331	328	95.3	28.9°	PEGASUS III was orbited with removable detection panels, for possible retrieval by a future astronaut flight; results to 10/8/65 had hits recorded on the 1.5-mil, 8-mil, and 16-mil-thickness panels at the respective rates of 58, 5.1, and 1.45 per sq. meter per year. This launch completed the Pegasus series, with 3 successes out of 3 attempts. SA-10 launch vehicle performed excellently, completing the launch development program for Saturn I with a perfect record of 10 successes out of 10 attempts.
	Apollo and	1965-60B		Total weight: 9,700 lbs. Objective: Test separation techniques and total vehicle dynamics. Payload: 12'10" (dia. at base) × 11'2" conical boilerplate Apollo command module attached to 12'10" (dia.) × 13'2" boilerplate service module.	331	327	95.3	28.86°	Boilerplate Apollo separated from PEGASUS III on schedule. Still in orbit.
Aug. 3	COSMOS LXXVII (U.S.S.R.)	1965-61A	Not available	Total weight: Not available. Objective: Continuation of Cosmos scientific satellite series. Payload: Not available.	187	114	89.3	51.8°	Reentered 8/11/65.
Aug. 3	DOD Spacecraft (United States) and DOD Spacecraft	1965-62A	Atlas-Agena D	Total weight: Not available. Objective: Develop spaceflight techniques and technology. Payload: Not available.	171	94	88.7	107.4°	Two satellites launched with single launch vehicle. Reentered 8/7/65.
	DOD Spacecraft	1965-62B		Total weight: Not available. Objective: Develop spaceflight techniques and technology. Payload: Not available.	316	312	94.7	107.3°	Still in orbit.
Aug. 10	Scout-SECOR V (United States)	1965-63A	Scout FW4S	Total weight: 44 lbs. Objective: Orbit a Secor satellite; flight-quality and demonstrate major new components of the Scout launch vehicle. Payload: Spherical satellite, containing transponder, solar cells and batteries; telemetry; antennas.	1,504	702	122.18	69.23°	Scout (SEV-A) performed excellently in its test flight, including the new motors in 2nd and 4th stages and new spacecraft separation system. The U.S. Army's SECOR V geodetic satellite operated satisfactorily. Still in orbit.

Date	Name	Designation	Launch Vehicle	Description	Weight/Apogee	Perigee	Period	Inclination	Remarks
Aug. 11	Centaur-Surveyor (United States)	1965-64A	Atlas-Centaur	Total weight: 6,230 lbs. (including 2,084 lb. simulated Surveyor spacecraft). Objective: In full-scale simulation, determine capability of Atlas-Centaur vehicle to launch a Surveyor spacecraft on a lunar transfer trajectory; test new systems on launch vehicle. Payload: Surveyor Dynamic Model (SD-2), consisting of Surveyor spaceframe, simulated retromotor assembly, S-band transponder, spacecraft separation assembly.	510,861	104	31 days	28.58°	Simulated Surveyor spacecraft was launched into a precise, deliberately offset lunar transfer orbit; guidance was so accurate that an actual spacecraft would have hit the moon without midcourse correction. First Centaur flight with the uprated engines in the Atlas (total of 389,000 lbs. thrust); 4th success in 6 Atlas-Centaur flights. Still in orbit.
Aug. 13	SURCAL (United States)	1965-65B	Thor-Able-Star	Total weight: Not available. Objective: Develop spaceflight techniques and technology. Payload: Calibration satellite for Spasur system; Long Rod (200 ft. in length) intended to separate from carrier rocket and deploy.	719	675	107.8	90.02°	Six satellites launched with 1 vehicle. The Long Rod failed to separate from the last stage of Thor-Able-Star and to deploy to full length. Still in orbit.
	and SURCAL	1965-65C		Total weight: 9 lbs. Objective: Develop spaceflight techniques and technology. Payload: Calibration satellite for Spasur system; dodecapod shape.	733	679	108.1	90.02°	Still in orbit.
	and SURCAL	1965-65E		Total weight: 20 lbs. Objective: Develop spaceflight techniques and technology. Payload: Calibration satellite for Spasur system; black sphere.	738	680	108.2	90°	Still in orbit.
	and DoD Spacecraft	1965-65F		Total weight: Not available. Objective: Develop spaceflight techniques and technology. Payload: Not available.	740	676	108.1	90.02°	Still in orbit.
	and SURCAL	1965-65H		Total weight: 5 lbs. Objective: Develop spaceflight techniques and technology. Payload: Calibration satellite for Spasur system; white sphere.	737	678	108.1	90.04°	Still in orbit.
	and SURCAL	1965-65L		Total weight: 13 lbs. Objective: Develop spaceflight techniques and technology. Payload: Calibration satellite for Spasur system, with transponder.	741	674	108.1	90.03°	Still in orbit

Launch Date	Name	International Designation	Vehicle	Payload Data	Apogee (st. mi.)	Perigee (st. mi.)	Period (minutes)	Inclination	Remarks
Aug. 14	COSMOS LXXVIII (U.S.S.R.)	1965-66A	Not available	Total weight: Not available. Objective: Continuation of Cosmos scientific satellite series. Payload: Not available.	204	130	89.8	60°	Reentered 8/22/65.
Aug. 17	DOD Spacecraft (United States)	1965-67A	Thor-Agena D	Total weight: Not available. Objective: Develop spaceflight techniques and technology. Payload: Not available.	262	131	90.9	70.01°	Reentered 10/11/65.
Aug. 21	GEMINI V (United States)	1965-68A	Titan II	Total weight: 7,947 lbs. (including reentry and adapter modules). Objective: Demonstrate performance of spacecraft and crew in 8-day spaceflight; evaluate performance of rendezvous guidance and navigation system using the radar evaluation pod. Payload: 18'5" × 10' 2-module bell-shaped spacecraft, containing 2 astronauts; guidance and control equipment; rendezvous radar and navigation system; cameras; 1 HF and 1 UHF transceiver, high and low frequency telemetry transmitters, tracking and recovery communications; fuel cell; environmental control system; recovery and reentry systems.	217 later: 193	101 124	89.58	32.6°	GEMINI V set new world record for longest manned spaceflight (190 hrs. 55 min., 120 revolutions), confirmed that astronauts were physically capable of lunar flight duration and, in spite of minor problems, confirmed operation of the fuel cell and rendezvous radar. Astronauts L. Gordon Cooper, Jr., and Charles Conrad, Jr., performed 16 of 17 planned experiments. Human error by ground crew caused GEMINI V to land 90 mi. short on 8/29; helicopter picked up crew.
	and Rendezvous evaluation pod	1965-68C		Total weight: 76 lbs. Objective: Serve as rendezvous target for Gemini 5. Payload: Radar transponder; flashing xenon lights.	217	101	89.68	32.6°	Still in orbit.
Aug. 25	COSMOS LXXIX (U.S.S.R.)	1965-69A	Not available	Total weight: Not available. Objective: Continuation of Cosmos scientific satellite series. Payload: Not available.	223	131	89.7	64.9°	Reentered 9/7/65.
Sep. 3	COSMOS LXXX (U.S.S.R.)	1965-70A	Not available	Total weight: Not available. Objective: Continuation of Cosmos scientific satellite series. Payload: Not available.	965	842	115	56°	Five satellites launched with single launch vehicle. One of the satellites was powered by a radioisotope device. Still in orbit.

ASTRONAUTICS AND AERONAUTICS, 1965

Date	Name	Designation	Launch vehicle	Description				Remarks
	COSMOS LXXXI and	1965-70B		Total weight: Not available. Objective: Continuation of Cosmos scientific satellite series. Payload: Not available.	907	861	115	56°	Still in orbit.
	COSMOS LXXXII and	1965-70C		Total weight: Not available. Objective: Continuation of Cosmos scientific satellite series. Payload: Not available.	971	876	115	56°	Still in orbit.
	COSMOS LXXXIII and	1965-70D		Total weight: Not available. Objective: Continuation of Cosmos scientific satellite series. Payload: Not available.	975	874	116	56°	Still in orbit.
	COSMOS LXXXIV	1965-70E		Total weight: Not available. Objective: Continuation of Cosmos scientific satellite series. Payload: Not available.	978	912	116	56°	Still in orbit.
Sep. 9	COSMOS LXXXV (U.S.S.R.)	1965-71A	Not available	Total weight: Not available. Objective: Continuation of Cosmos scientific satellite series. Payload: Not available.	199	131	89.6	65°	Reentered 9/17/65.
Sep. 10	DOD Spacecraft (United States)	1965-72A	Thor FW4S	Total weight: Not available. Objective: Develop spaceflight techniques and technology. Payload: Not available.	655	401	101.9	98.65°	Still in orbit.
Sep. 18	COSMOS LXXXVI (U.S.S.R.)	1965-73A	Not available	Total weight: Not available. Objective: Continuation of Cosmos scientific satellite series. Payload: Not available.	1,050	857	116.7	57°	Five satellites launched with single launch vehicle. One of the satellites was powered by a radioisotope device. Still in orbit.
	COSMOS LXXXVII and	1965-73B		Total weight: Not available. Objective: Continuation of Cosmos scientific satellite series. Payload: Not available.	1,050	857	116.7	57°	Still in orbit.
	COSMOS LXXXVIII and	1965-73C		Total weight: Not available. Objective: Continuation of Cosmos scientific satellite series. Payload: Not available.	1,050	857	116.7	57°	Still in orbit.
	COSMOS LXXXIX and	1965-73D		Total weight: Not available. Objective: Continuation of Cosmos scientific satellite series. Payload: Not available.	1,050	857	116.7	57°	Still in orbit.
	COSMOS XC	1965-73E		Total weight: Not available. Objective: Continuation of Cosmos scientific satellite series. Payload: Not available.	1,050	857	116.7	57°	Still in orbit.

Launch Date	Name	International Designation	Vehicle	Payload Data	Apogee (st. mi.)	Perigee (st. mi.)	Period (minutes)	Inclination	Remarks
Sep. 22	DOD Spacecraft (United States)	1965-74A	Thor-Agena D	Total weight: Not available. Objective: Develop spaceflight techniques and technology. Payload: Not available.	222	109	89.9	80.05°	Reentered 10/11/65.
Sep. 23	COSMOS XCI (U.S.S.R.)	1965-75A	Not available	Total weight: Not available. Objective: Continuation of Cosmos scientific satellite series. Payload: Not available.	212	131	89.8	65°	Reentered 10/1/65.
Sep. 30	DOD Spacecraft (United States)	1965-76A	Atlas-Agena D	Total weight: Not available. Objective: Develop spaceflight techniques and technology. Payload: Not available.	188	88	88.6	95.55°	Reentered 10/5/65.
Oct. 4	LUNA VII (U.S.S.R.)	1965-77A	Not available	Total weight: 3,313 lts. Objective: Soft-land on the moon; take measurements of lunar environment. Payload: Radio; telemetry; scientific instrumentation.	Impacted on moon				LUNA VII impacted the moon and was destroyed 10/7/65; apparently the retrorockets which were to have fired and slowed the spacecraft to a soft landing did fire but not soon enough.
Oct. 5	AR8 (United States)	1965-78A	Atlas	Total weight: Not available. Objective: Develop spaceflight techniques and technology. Payload: Instruments to map and monitor energetic particles; mass spectrometers; detectors imbedded in radiation shield; instruments measuring dose rates in tissue-equivalent medium; telemetry.	2,141	255	125.7	144.3°	Aerospace Research Satellite was carried piggyback on regular missile launch; it entered the first highly retrograde orbit. Returned data successfully. Still in orbit.
Oct. 5	DOD Spacecraft (United States)	1965-79A	Thor-Agena D	Total weight: Not available. Objective: Develop spaceflight techniques and technology. Payload: Not available.	195	125	89.6	75.02°	Reentered 10/29/65.
Oct. 14*	MOLNIYA I-2 (U.S.S.R.)	1965-80A	Not available	Total weight: Not available. Objective: Transmit TV, 2-way multichannel telephone, phototelegraphic, and telegraphic communication.	24,800	310	719	65°	Highly elliptical orbit was intended to keep MOLNIYA I-2 in range of Soviet ground stations for longer periods of time. In conjunction with MOLNIYA I

Date	Name	Launch Vehicle	Cat. No.	Payload / Objective	Weight (lbs)	Apogee/Perigee	Period (min)	Inclination	Remarks
Oct. 14	OGO II (United States)	Thrust-Augmented Thor-Agena D	1965-81A	Payload: Hermetically sealed cylindrical satellite with conical ends; one containing sun and earth sensors, the other the apogee motor; 6 solar paddles extend like wheel-spokes; 2 parabolic antennas; transmitters; command system; attitude control system; solar cells; batteries. Total weight: 1,150 lbs. Objective: Take geophysical measurements of the near-earth environment during a period of low solar activity in study of correlative aspects of the relationship between the sun and the earth environment. Payload: 67″ × 32″ × 31″ rectangular parallelepiped spacecraft, containing many of the experiments and the subsystems such as telemetry, attitude control, temperature control, nickel-cadmium batteries; deployed from the spacecraft are 2 22′ booms and 4 4′ booms on the ends of which experiments are mounted; extended from the spacecraft are 2 large rotatable solar paddles faced with 33,000 solar cells; 20 experiments.	1,197	253	104.3	87.4°	(launched 4/23/65), the comsat would be used both for routine communications and to experiment with working of a comsat net to provide 24-hr. communications service to all points in U.S.S.R. Still in orbit. OGO II was launched into higher orbit than planned because of failure of primary launch vehicle guidance; 19 of the 20 experiments operated and returned data, but trouble with the horizon scanners depleted the stabilization gas supply, which in turn caused electrical power depletion. OGO II ceased transmitting 10/24/65, was ruled a failure by NASA. Still in orbit.
Oct. 15	Titan IIIC (United States)	Titan IIIC	1965-82A	Total weight: Not available (OV2-1 portion of payload weighed 375 lbs.). Objective: Develop spaceflight techniques and technology. Payload: Three satellites to be deployed by the transtage. First satellite was unmanned DOD spacecraft; 2nd was LCS II, a radar calibration sphere. Third was OV2-1, a radiation measurement satellite containing 3 energetic particle detectors, 5 spectrometers, 2 flux-gate magnetometers, Faraday cup, Cerenkov counter, 2 plasma probes, proton detector, tissue-equivalent ion chamber, 3 proton spectrometers, low energy magnetic electron spectrometer, DE/DX telescope; VLF receiver; telemetry.	615	324	99.9	31.85°	Transtage with satellites got into orbit, but about the time of the 2nd burn of transtage motor the whole payload exploded, into hundreds of fragments. Satellites were statistically counted as 3 separate payloads but never functioned. Primary mission, launch vehicle testing, was achieved. Fragments still in orbit.

Launch Date	Name	International Designation	Vehicle	Payload Data	Apogee (st. mi.)	Perigee (st. mi.)	Period (minutes)	Inclination	Remarks
Oct. 16	COSMOS XCII (U.S.S.R.)	1965-83A	Not available	Total weight: Not available. Objective: Continuation of Cosmos scientific satellite series. Payload: Not available.	204	123	89.8	65.02°	Reentered 10/24/65.
Oct. 19	COSMOS XCIII (U.S.S.R.)	1965-84A	Not available	Total weight: Not available. Objective: Continuation of Cosmos scientific satellite series. Payload: Not available.	324	137	91.6	48.24°	Reentered 1/3/66.
Oct. 28	COSMOS XCIV (U.S.S.R.)	1965-85A	Not available	Total weight: Not available. Objective: Continuation of Cosmos scientific satellite series. Payload: Not available.	182	131	89.3	65°	Reentered 11/5/65.
Oct. 28	DOD Spacecraft (United States)	1965-86A	Thor-Agena D	Total weight: Not available. Objective: Develop spaceflight techniques and technology. Payload: Not available.	268	107	90.6	74.97°	Reentered 11/17/65.
Nov. 2	PROTON II (U.S.S.R.)	1965-87A	Not available	Total weight: 26,900 lbs. Objective: Orbit unmanned space station to study cosmic radiation. Payload: Cylindrical satellite, containing ionization calorimeter, gamma ray telescope, and other scientific instruments; 4 solar paddles; batteries; telemetry.	396	119	92.6	63.5°	PROTON II was to study energetic particles of super-high energies, cosmic rays and their radiation danger, and nuclear interaction of cosmic particles with super-high energies of up to 1,000 billion electron volts. The size and weight of PROTON I (launched 7/16/65) and PROTON II, plus Soviet statements, argued that these were not simply scientific satellites but prototypes of a new generation of Soviet spacecraft—possibly a manned space station. Still in orbit.
Nov. 4	COSMOS XCV (U.S.S.R.)	1965-88A	Not available	Total weight: Not available. Objective: Continuation of Cosmos scientific satellite series. Payload: Not available.	323	128	91.7	48°	Still in orbit.

Date	Name	Designation	Launch Vehicle	Description	Weight (lbs)	Perigee	Apogee	Incl.	Remarks
Nov. 6	EXPLORER XXIX (United States)	1965-89A	Improved Thrust-Augmented Delta	Total weight: 385 lbs. Objective: Take geodetic measurements relating continental and local geodetic datums into a single global datum related to the earth's common center of mass; map the structure of the earth's gravitational field. Payload: 48" (dia.) octagon topped by octagonal cone, for a total spacecraft height of 32"; below the spacecraft protrudes a 24"-dia. hemisphere antenna; inside the spacecraft are a clock, memory computer system, telemetry, 3 power systems, and geodetic instruments; the face of the satellite is surfaced with 4,992 solar cells, 4 flashing-light beacons, 4 silvered quartz corner reflectors for laser experiments; 60' boom deploys for gravity-gradient stabilization; antennas.	1,410	695	120.31	59.38°	EXPLORER XXIX orbited with higher apogee than planned when 2nd stage motor did not cut off on ground command. All experiments functioned. First use of Improved Thrust-Augmented Delta launch vehicle. On 11/28/65 large cameras at KSC photographed flashing lights of EXPLORER XXIX, part of experiments to provide better geodetic positioning of U.S. launching and tracking sites. Still in orbit.
Nov. 8	DOD Spacecraft (United States) and DOD Spacecraft	1965-90A	Atlas-Agena D	Total weight: Not available. Objective: Develop spaceflight techniques and technology. Payload: Not available.	174	88	88.7	93.9°	Reentered 11/11/65.
		1965-90B		Total weight: Not available. Objective: Develop spaceflight techniques and technology. Payload: Not available.	176	96	88.9	93.9°	Reentered 11/9/65.
Nov. 12	VENUS II (U.S.S.R.)	1965-91A	Not available	Total weight: 2,123 lbs. Objective: Make flyby of planet Venus, take TV photos of planet and relay them back to earth; take scientific measurements in interplanetary space. Payload: 2-compartment spacecraft, one containing TV cameras, the other the scientific instruments for measuring interplanetary cosmic rays, solar wind, magnetic fields, and micrometeoroids.	In heliocentric orbit.				VENUS II passed 14,912 mi. from the planet Venus on 2/27/66, but telemetry failed prior to flyby; TV photos that were to have been taken of the planet were not received. Data on scientific measurements of interplanetary space were transmitted throughout the flight. No midcourse correction was necessary.

Launch Date	Name	International Designation	Vehicle	Payload Data	Apogee (st. mi.)	Perigee (st. mi.)	Period (minutes)	Inclination	Remarks
Nov. 16	VENUS III (U.S.S.R.)	1965-92A	Not available	Total weight: 2,116 lbs. Objective: Eject and parachute-land a capsule on the surface of Venus to take scientific measurements; take scientific measurements in interplanetary space. Payload: 2-compartment spacecraft, one containing 23.6"-dia. spherical capsule and parachute apparatus, the capsule containing instruments to measure Venutian surface temperature and pressure; the other compartment containing the scientific instruments for measuring interplanetary cosmic rays, solar wind, magnetic fields, and micrometeoroids.	Impacted on Venus.				VENUS III impacted on the planet Venus on 3/1/66 within 500 mi. of its aiming point, but communications failure prior to the spacecraft's arrival in the vicinity of Venus prevented receipt of data on the Venutian atmosphere; also no confirmation that VENUS III ejected the instrument capsule which was to have parachuted to the planet's surface. Data on scientific measurements of interplanetary space, taken of the same phenomena as those measured by VENUS II and in the same areas of space and only a few days later but with slightly different instruments, were transmitted throughout the flight. Midcourse correction had been made 12/27/65. The capsule was sterilized of earth microorganisms but apparently not the spacecraft.
Nov. 18*	EXPLORER XXX (United States)	1965-93A	Scout	Total weight: 125 lbs. Objective: Monitor the sun's x-ray emission; measure the time history of x-ray intensity in relation to solar flares; provide real-time solar data to participants in IQSY. Payload: 2 24"-dia. hemispheres separated by 3½" band in which are mounted 12 photometers and 4 telemetry antennas; hemispheres contain data storage system, 2 transmitters and 2 receivers; solar cells, nickel-cadmium batteries.	548	440	100.8	59.7°	EXPLORER XXX, 9th Solrad satellite developed by the Naval Research Laboratory, was launched by NASA for the U.S. Navy as part of U.S. participation in IQSY. All experiments functioned well. Still in orbit.
Nov. 23	COSMOS XCVI (U.S.S.R.)	1965-94A	Not available	Total weight: Not available. Objective: Continuation of Cosmos scientific satellite series. Payload: Not available.	193	141	89.6	51.54°	Reentered 12/9/65.

Date	Name	Designation	Launch Vehicle	Payload/Objective	Weight (lbs)	Perigee/Apogee	Period (min)	Inclination	Remarks
Nov. 26	COSMOS XCVII (U.S.S.R.)	1965-95A	Not available	Total weight: Not available. Objective: Continuation of Cosmos scientific satellite series. Payload: Not available.	1,304	137	108.3	49°	Still in orbit.
Nov. 26	A-1 (France)	1965-96A	Diamant	Total weight: 88 lbs. Objective: Test the Diamant launch vehicle system; orbit satellite. Payload: Cylindrical spacecraft, coned on both ends, with 4 extendable antennas spaced around the center of the cylinder; transmitter; batteries.	1,008	326	108	53°	First satellite launched by France; 1st to be completely built and launched by country other than U.S.S.R. and U.S. Satellite carried no scientific equipment, was to send signal for 2-week lifetime. Signal was very weak, probably caused by damage to antennas during launch from Hammaguir Range, Algeria. Still in orbit.
Nov. 27	COSMOS XCVIII (U.S.S.R.)	1965-97A	Not available	Total weight: Not available. Objective: Continuation of Cosmos scientific satellite series. Payload: Not available.	354	134	92	65°	Reentered 12/5/65.
Nov. 29	ALOUETTE II (United States-Canada)	1965-98A	Thor-Agena B	Total weight: 323 lbs. Objective: As part of ISIS-X program, simultaneously with a companion spacecraft take measurements of ionospheric characteristics, both as a topside sounder and in the vicinity of the spacecraft. Payload: 42" (dia.) × 34" near-oval spacecraft, containing 5 experiments, transmitter, receiver, nickel-cadmium batteries; surface covered with 6,480 n-on-p solar cells; extending from the spacecraft are 2 dipole antennas, one 75', the other 240'; 5 whip antennas.	1,850	310	121.4	79.7°	ALOUETTE II, Canadian-built satellite, was launched with EXPLORER XXXI by a single Thor-Agena booster as part of the joint U.S.-Canadian ionospheric research program. Satellite performance was excellent. Still in orbit.
	and EXPLORER XXXI (United States)	1965-98B		Total weight: 218 lbs. Objective: As part of ISIS-X program, simultaneously with a companion spacecraft take measurements of ionospheric density and temperature of ions and electrons, composition of ions, and corpuscular radiation. Payload: 30" (dia.) × 25" octagonal spacecraft with spherical ion mass spectrometer protruding 21" above it; inside are most of 8 experiments, transmitter, telemetry, battery; outside are solar cells, antenna.	1,850	310	121.4	79.7°	EXPLORER XXXI, U.S. companion satellite to Canadian ALOUETTE II, was launched along with ALOUETTE by a single Thor-Agena booster as part of the joint U.S.-Canadian ionospheric research program. This was 1st of 4 projected launches. EXPLORER XXXI was performing well. Still in orbit.

Launch Date	Name	International Designation	Vehicle	Payload Data	Apogee (st. mi.)	Perigee (st. mi.)	Period (minutes)	Inclination	Remarks
Dec. 3	LUNA VIII (U.S.S.R.)	1965-99A	Not available	Total weight: 3,422 lbs. Objective: Test a lunar soft-landing system; take scientific measurements. Payload: Capsule for soft landing, containing instruments, telemetry, retrorocket; parent spacecraft containing telemetry, attitude control system, command system, batteries.	Impacted on the moon.				LUNA VIII impacted on the moon on 12/7, when apparently the retrorockets used in slowing the capsule down for the soft landing attempt did not have full effect.
Dec. 4	GEMINI VII (United States)	1965-100A	Titan II	Total weight: 8,076 lbs. (includes reentry and adapter modules). Objective: Demonstrate performance of Gemini spacecraft and crew in 14-day flight; serve as target vehicle for rendezvous attempt. Payload: 18'5" × 10' 2-module bell-shaped spacecraft, containing 2 astronauts; guidance and control equipment; rendezvous radar; cameras; 1 HF and 1 UHF transceiver; high and low frequency telemetry transmitters, tracking and recovery communications; fuel cell; environmental control system; recovery and reentry systems.	204	100	89.39	28.89°	GEMINI VII broke GEMINI V's world record for longest manned spaceflight (330 hrs. 35 min, 206 revolutions); returned Astronauts Frank Borman and James A. Lovell in excellent condition confirming human adaptability to conditions of spaceflight; participated as target vehicle for GEMINI VI's successful space rendezvous; collected valuable data on all 20 planned experiments; landed 12/18 9 mi. from U.S.S. *Wasp*; astronauts were picked up by helicopter.
Dec. 6	FR-1 (United States-France)	1965-101A	Scout	Total weight: 135 lbs. Objective: Study the very-low-frequency wave field in the magnetosphere and irregularities in distribution of ionization in the magnetosphere. Payload: 27" (dia. across corners) × 52" spacecraft consisting of octagonal central section faced with 2 octagonal prisms covered with solar cells; 19" boom with electron-density probe deploys below spacecraft; above the spacecraft extends magnetic-field antenna; from the sides extend 4 6½' dipole antennas for electrical field measurement; 4 telemetry antennas; magnetometer; 2 transmitters; 2 nickel-cadmium batteries.	480	462	100	75.9°	FR-1 was launched for France by NASA under joint U.S.-French space research program. Experiments with FR-1 would be conducted at both French and U.S. ground stations using radio equipment furnished by France. Spacecraft was functioning well. Still in orbit.

Dec. 9	DOD Spacecraft (United States)	1965-102A	Thor-Agena D	Total weight: Not available. Objective: Develop spaceflight techniques and technology. Payload: Not available.	260	112	90.5	90.02°	Reentered 12/26/65.
Dec. 10	COSMOS XCIX (U.S.S.R.)	1965-103A	Not available	Total weight: Not available. Objective: Continuation of Cosmos scientific satellite series. Payload: Not available.	199	124	89.6	65°	Reentered 12/18/65.
Dec. 15	GEMINI VI (United States)	1965-104A	Titan II	Total weight: 7,817 lbs. (includes reentry and adapter modules). Objective: Perform space rendezvous to within a few feet of target vehicle, GEMINI VII. Payload: 13'5" × 10' 2-module bell-shaped spacecraft, containing 2 astronauts; guidance and control equipment; rendezvous radar and navigation equipment; cameras; 1 HF and 1 UHF transceiver, high and low frequency telemetry transmitters, tracking and recovery communications; batteries; environmental control system; recovery and reentry systems.	161	100	88.7	28.97°	GEMINI VI performed world's first piloted rendezvous in space; beginning in 2nd revolution, Astronauts Walter M. Schirra, Jr., and Thomas P. Stafford started a series of spacecraft maneuvers to catch up with GEMINI VII; 5 maneuvers produced circular orbit (172/171) after 3 hrs. 47 min. of flight; radar lockon with GEMINI VII was confirmed at 4 hrs. 16 min., at a distance of 235 mi.; 4 more maneuvers effected rendezvous; for 5 hrs. 16 min. rendezvous was maintained, the two spacecraft coming as close to each other as 1 ft. GEMINI VI reentered and landed 12/16 after 25 hrs. 51 min. (16 revolutions), impacting in the Atlantic 7.6 mi. from aiming point; astronauts were hoisted aboard U.S.S. *Wasp* in their spacecraft.
Dec. 16	PIONEER VI (United States)	1965-105A	Improved Thrust-Augmented Delta	Total weight: 140 lbs. Objective: Obtain scientific data on interplanetary phenomena at points closer to the sun than is the earth's orbit. Payload: 37" (dia.) × 35" cylindrical spacecraft, with a boom protruding from the top containing communications antennas; the sides covered with solar cells except for a narrow band in which are placed the experiments and 3 booms, 2 for orientation jets and 1 for the magnetometer; data storage; transmitter; batteries; 6 experiments.	Aphelion, .98	Perihelion, .81	311 days	.2° (to ecliptic)	PIONEER VI was successfully placed in heliocentric orbit; 1st of projected 4 Pioneer interplanetary spacecraft studying the solar wind, solar physics, magnetic fields of the sun, and interactions of high-energy particles and magnetic fields. Spacecraft was performing well. Still in orbit.

Launch Date	Name	International Designation	Vehicle	Payload Data	Apogee (st. mi.)	Perigee (st. mi.)	Period (minutes)	Inclination	Remarks
Dec. 17	COSMOS C (U.S.S.R.)	1965-106A	Not available	Total weight: Not available. Objective: Continuation of Cosmos scientific satellite series. Payload: Not available.	404	404	97.7	65°	Still in orbit.
Dec. 21	COSMOS CI (U.S.S.R.)	1965-107A	Not available	Total weight: Not available. Objective: Continuation of Cosmos scientific satellite series. Payload: Not available.	341	165	92.4	49°	Still in orbit.
Dec. 21	Titan IIIC (United States)	1965-108A	Titan IIIC	Total weight: Not available. Objective: Develop spaceflight techniques and technology. Payload: Transtage with 4 satellite payloads.	20,857	110	589.7	26.38°	In 3d vehicle development flight for Titan IIIC, the transtage was to have attained low circular orbit (which it did, at 105 mi.), then refire to eccentric orbit (which it did, as listed). After 5 hours it was to refire again to circularize into a near-synchronous one and release 4 small satellites. A defective valve caused tumbling, prevented restart. 3 of 4 satellites did achieve release. Still in orbit.
	ov2-3 and			Total weight: 427 lbs. Objective: Develop spaceflight techniques and technology. Payload: Instruments to measure solar and geomagnetic activity, cosmic rays, trapped particle fluxes; magnetometers, spectrometers; telemetry.	20,857	110	589.7	26.38°	Was never released from Titan IIIC transtage. Still in orbit.
	LES IV and	1965-108B		Total weight: 115 lbs. Objective: Develop spaceflight techniques and technology. Payload: Experimental communications satellite, with solid-state transponder in 8,000 mc region.	20,844	124	589.6	26.6°	LES IV was released from the transtage, telemetry indicated. After a few days in orbit, internal power built up to the point where the satellite could be used. Still in orbit.
	OSCAR IV and	1965-108C		Total weight: 41 lbs. Objective: Develop spaceflight techniques and technology. Payload: Amateur radio communications satellite, with transponder; power supply.	20,723	101	587.5	26.8°	OSCAR IV was released from the transtage, telemetry indicated. It was functioning well. Orbital parameters took some time to determine; those given are as of 1/15/66. Still in orbit.

	LES III and	1965-108D		Total weight: 35 lbs. Objective: Develop spaceflight techniques and technology. Payload: Experimental communications satellite, with signal generator in 300 mc UHF band.	20,477	121	581	26.46°	LES III was released from the transtage, telemetry indicated. Signals were received from the satellite, but its identity and orbital parameters were difficult to determine; those given are as of 1/31/66. Still in orbit.
Dec. 22	DOD Spacecraft (United States)	1965-109A	Scout	Total weight: Not available. Objective: Develop spaceflight techniques and technology. Payload: Not available.	674	563	105	89.11°	Still in orbit.
Dec. 24	DOD Spacecraft (United States)	1965-110A	Thor-Agena D	Total weight: Not available. Objective: Develop spaceflight techniques and technology. Payload: Not available.	268	112	90.7	80.01°	Still in orbit.
Dec. 28*	COSMOS CII (U.S.S.R.)	1965-111A	Not available	Total weight: Not available. Objective: Continuation of Cosmos scientific satellite series. Payload: Not available.	172	135	89.24	65°	Reentered 1/13/66.
Dec. 28	COSMOS CIII (U.S.S.R.)	1965-112A	Not available	Total weight: Not available. Objective: Continuation of Cosmos scientific satellite series. Payload: Not available.	372	372	97	56.04°	Still in orbit.

Appendix B

CHRONOLOGY OF MAJOR NASA LAUNCHINGS

JANUARY 1, 1965, THROUGH DECEMBER 31, 1965

This chronology of major NASA launchings in 1965 is intended to provide an accurate and ready historical reference, one compiling and verifying information previously scattered over several sources. It includes launchings of all rocket vehicles larger than sounding rockets launched either by NASA or under "NASA direction" (e.g., NASA provided vehicle, launch facilities, tracking facilities, and performed the launch for the French satellite, FR-1).

An attempt has been made to classify the performance of both the launch vehicle and the payload and to summarize total results in terms of primary mission. Three categories have been used for vehicle performance and mission results—successful (S), partially successful (P), and unsuccessful (U). A fourth category, unknown (Unk), has been provided for payloads where vehicle malfunctions did not give the payload a chance to exercise its main experiments. These divisions are necessarily arbitrary, since many of the results cannot be neatly categorized. Also they ignore the fact that a great deal was learned from shots that may have been classified as unsuccessful.

A few unique items require separate treatment. Their dates have been kept in sequence but their history has been relegated to footnotes. Date of launch is referenced to local time at the launch site.

Sources used were all open ones, verified where in doubt from the project offices in NASA Hq. and from the NASA Centers. For further information on each item, see Appendix A of this volume and the entries in the main chronology as referenced in the index. Prepared January 1966 by Dr. Frank W. Anderson, Jr., Deputy NASA Historian (EPH).

Date	Name (NASA Code)	General Mission	Launch Vehicle (Site)	Vehicle	Payload	Mission	Remarks
1965							
Jan 19	GEMINI (GT-2)	Suborbital Gemini spacecraft test.	Titan II (ETR)	S	S	S	Suborbital flight of Gemini flight-model spacecraft confirmed systems performance and launch and reentry sequences, provided equipment okay for 1st manned flight; reached altitude of 98.9 mi.; impacted 2,127 mi. downrange, 16 mi. short.
Jan 22	TIROS IX (Tiros I (eye)) [A-54]	Meteorological earth satellite.	Thor-Delta (ETR)	S	S	S	First flight-test of "rolling wheel" design plus 3-dogleg maneuver to achieve polar orbit. R&D version of operational weather satellite. 2nd stage Delta burned 11 sec. too long, creating elliptical orbit (1,602 mi.) instead of planned circular one (460 mi.). Photos excellent.
Feb 3	OSO II (Oso-B2) [S-17]	Scientific satellite, solar.	Thor-Delta (ETR)	S	S	S	6 of 8 sun-study experiments functioned well, representing university-industry-Government effort seeking data on solar x-rays, gamma rays, and ultraviolet radiation; OSO II contained parts from Oso-B, damaged 4/64 in testing accident.
Feb 16	PEGASUS I (SA-9)	Scientific satellite, meteoroid. Launch vehicle development test.	Saturn I (ETR)	S	S	S	8th successful Saturn I test flight. S-IV 2nd stage separated Apollo bollerplate command and service modules (10,000 lbs.) into separate orbit, then itself orbited with 3,200-lb. 96-ft.-wingspread PEGASUS I meteoroid detection satellite (subtotal 23,000 lbs.; total in orbit (33,200 lbs.)
Feb 17	RANGER VIII (Ranger C) [RA-8]	Scientific lunar probe, photographic.	Atlas-Agena B (ETR)	S	S	S	Transmitted 7,000 photos of moon's maria to earth before impacting the moon in Sea of Tranquillity, 2/20/65.
Mar 21	RANGER IX (Ranger D) [RA-9]	Scientific lunar probe, photographic.	Atlas-Agena B (ETR)	S	S	S	Transmitted 5,814 photos of moon's mountains and craters to earth before impacting the moon in crater Alphonsus, 3/24/65. Last Ranger flight.
Mar 23	GEMINI III (GT-3)	Orbital manned Gemini flight.	Titan II (ETR)	S	S	S	First U.S. 2-man space flight; Astronauts Virgil I. Grissom and John W. Young flew spacecraft "Molly Brown" 3 orbits, made world's first piloted orbital changes (dropped apogee from 139 mi. to 105 mi., shifted orbital plane by .02°, then dropped perigee from 100 mi. to 52 mi.); reentered and landed 58 mi. short of U.S.S. *Intrepid*, were picked up by helicopter.
Apr 6 Apr 29	(?) EXPLORER XXVII (BE-C)	Scientific satellite, geodetic.	Scout (WS)	S	S	S	Primary mission was to map irregularities in earth's magnetic field for geodetic purposes; also to make ionospheric measurements and laser tracking and communications experiments.
May 19	APOLLO (BP-22)	Suborbital Apollo capsule test.	Little Joe II (WSMR)	U	S	P	Little Joe II launch vehicle developed abnormal roll rate shortly after launch, building up centrifugal forces that destroyed vehicle at 14,000 ft. (112,000 ft. planned altitude). Launch escape system operated automatically, pulled Apollo capsule free to 19,000 ft.; capsule landed safely 3 mi. from launch site (110 mi. planned). Did not achieve primary flight objective of testing launch escape system stabilization at maximum alti-

See footnotes at end of table.

Date	Name	Purpose	Vehicle (site)				Results
May 22	FIRE II	37,000 fps reentry test	Atlas D (ETR)	S	S	S	tude; did achieve other primary objective of testing launch escape system flight with heat shield forward; did achieve bonus in successful launch escape system operation in unscheduled emergency. Payload rose 500 mi. in ballistic trajectory, achieved 25,400 mph reentry speed, landed 5,130 mi. from launch site in 32 min. Flight results excellent, telemetry received all the way.
May 25	PEGASUS II (SA-8)	Scientific satellite, meteoroid. Launch vehicle development test.	Saturn I (ETR)	S	S	S	9th successful Saturn I test flight. S-IV 2nd stage separated Apollo boilerplate command (BP-26) and service modules (9,700 lbs.) into separate orbit, then itself orbited with 3,200-lb., 96-ft.-wingspread PEGASUS II meteoroid detection satellite (subtotal 23,100 lbs.; total in orbit 33,800 lbs.).
May 29	EXPLORER XXVIII (Imp-C)	Scientific satellite, radiation monitor.	Thor-Delta (ETR)	S	S	S	Slightly long burn by booster engines put satellite in advantageous higher apogee than planned (164,000 mi. vs. 130,000).
Jun 3	GEMINI IV (GT-4)	Orbital manned Gemini flight.	Titan II (ETR)	S	S	S	Astronauts James A. McDivitt and Edward H. White flew 62 revolutions in 97 hrs. 56 min., landed 4/7 in Atlantic 53 mi. from U.S.S. *Wasp*, were picked up by helicopter. On 3d orbit, Astronaut White "walked" in space for 22 min. Attempted near-rendezvous with Titan 2nd stage was abandoned when too much fuel was consumed in maneuvering. Flight was longer than all other U.S. manned flights combined.
Jun 29 Jul 2	(?) TIROS X (OT-1)	Meteorological earth satellite.	Thrust-Augmented Delta (ETR)	S	S	S	Launched into sun-synchronous orbit, as was TIROS IX. First of interim operational Tiros satellites funded by Weather Bureau to provide continuity of photo coverage pending TOS in 1966. Provided first identification of Hurricane Betsy.
Jul 30	PEGASUS III (SA-10)	Scientific satellite, meteoroid.	Saturn I (ETR)	S	S	S	PEGASUS III orbited with removable "coupons" for possible astronaut recovery; ended Saturn I launches, with record of 10 successes out of 10 launches, 1st perfect launch record in U.S. space program.
Aug 10	Scout (Sev-A)	Launch vehicle development test.	Scout (WS)	S	S	S	Provided evaluation test of vehicle modifications including uprated 2nd and FW4S 4th stage engines and dogleg-launch capability; also launched Army SECOR V geodetic satellite.
Aug 11	Atlas-Centaur (AC-6)	Launch vehicle development test.	Atlas-Centaur (ETR)	S	S	S	4th success of 6 Centaur launches to date; placed dynamic test model Surveyor spacecraft in simulated lunar transfer orbit, confirmed accuracy of deep-space guidance system.
Aug 21	GEMINI V (Gemini 5)	Orbital manned Gemini flight.	Titan II (ETR)	S	S	S	8-day mission by Astronauts L. Gordon Cooper and Charles Conrad, Jr., was 1st U.S. manned flight to make no-hold launch; confirmed time-duration feasibility of lunar landing mission; set 8 world space records, including longest manned space flight (120 revolutions, 190 hrs., 55 min.). Problems with fuel-cell-related systems caused abandonment of pod-rendezvous experiment; 16 out of 17 scientific experiments were performed. Landed 90 mi. off target in Atlantic, 8/29; airlifted by helicopter to U.S.S. *Lake Champlain*.

See footnotes at end of table.

Date	Name (NASA Code)	General Mission	Launch Vehicle (Site)	Performance			Remarks
				Vehicle	Payload	Mission	
1965							
Aug 25	Oso-C	Scientific satellite, solar.	Thor-Delta (ETR)	U	Unk	U	Did not achieve orbit because premature firing of 3rd stage prevented attainment of orbital speed; fell into the Atlantic.
Oct 14	OGO II (Ogo-C)	Scientific earth satellite, geophysical.	Thrust-Augmented Thor-Agena D (WTR)	P	P	P	Nineteen out of twenty experiments operated but by 10/24/65 malfunction of horizon scanners had depleted stabilization gas and electrical power; spacecraft ceased transmission.
Oct 25	AGENA (GATV)	Orbital manned Gemini flight, rendezvous and docking.	Atlas-GATV (ETR)	U		U	Agena stage target vehicle for Gemini 6 rendezvous, disintegrated at time of ignition of main Agena engine; did not orbit. Caused postponement of launch of Gemini 6, which was later rescheduled to rendezvous with Gemini 7.
Nov 6	EXPLORER XXIX (Geos A)	Scientific earth satellite, geodetic.	Improved Thrust-Augmented Delta (ETR)	S	S	S	First launch of Improved Thrust-Augmented Delta; first gravity-gradient-stabilized satellite launched by NASA; all experiments functioned.
Nov 19 Nov 29	(²) ALOUETTE II (Alouette-B) and EXPLORER XXXI (Direct Measurement Explorer-A)	Scientific earth satellites, ionospheric.	Thor-Agena B (WTR)	S S	S S	S S	Project known as Isis-X. The two satellites orbited close to each other, would complement each other's data. All experiments on both satellites functioned well.
Dec 4	GEMINI VII (Gemini VII)	Orbital manned Gemini flight, endurance.	Titan II (ETR)	S	S	S	Longest spaceflight to date (330 hrs. 35 min, 206 revolutions); with GEMINI VI, first space rendezvous; first U.S. spaceflight in which part of flight was made without spacesuit. Astronauts Frank Borman and James Lovell landed 12/18, some 9 miles from U.S.S. Wasp, were picked up by helicopter.
Dec 6 Dec 15	(⁴) GEMINI VI (Gemini VI)	Orbital manned Gemini flight, rendezvous.	Titan II (ETR)	S	S	S	Astronauts Walter M. Schirra and Thomas P. Stafford conducted first piloted space rendezvous, maintained close rendezvous with GEMINI VII for some 6 hrs., approaching as near as 1 ft.; reentered 12/16 after 25 hrs. 51 min., landed within 12 miles of U.S.S. Wasp and were hoisted aboard in GEMINI VI.
Dec 16	PIONEER VI (Pioneer A)	Interplanetary probe, scientific.	Improved Thrust-Augmented Delta (ETR)	S	S	S	In heliocentric orbit between orbits of earth and Venus; all experiments functioned.

¹ Apr. 6, 1965. ComSatCorp's EARLY BIRD I communications satellite was launched with a Thrust-Augmented Delta (Tad) booster from ETR, with NASA performing launch for ComSatCorp on reimbursable basis. EARLY BIRD was put into synchronous orbit over the Atlantic Ocean, spent several months in extensive testing and sample TV, telephone, and data transmission. On June 28, 1965, the satellite went on a regular transmission schedule, becoming the world's first operational commercial satellite.
² June 29, 1965. Apollo pad abort test (BP-23A) was successful at WSMR. Launch vehicle escape rocket boosted Apollo boilerplate capsule about 5,000 ft. off pad, from whence it parachuted to safe landing. Capsule was first in Apollo tests to be reused.
³ EXPLORER XXX (Se-A), launched by NASA for U.S. Navy, was successfully launched from WS on a Scout booster; part of the IQSY program, this was the 9th Solrad satellite in NRL series. All experiments functioned.
⁴ FR-1 (FR1A), French ionospheric satellite, was successfully launched by NASA from WTR on a Scout booster; experiments performed well.

Appendix C

CHRONOLOGY OF MANNED SPACE FLIGHT 1961-1965

This chronology contains basic information on manned spaceflights by the United States and the Soviet Union through 1965. The information was compiled by William D. Putnam, Assistant NASA Historian for Manned Space Flight. The scope and pace of the manned spaceflight effort has increased to the extent that a table devoted exclusively to manned flights seems a useful reference tool.

By the end of 1965, the United States had conducted 11 manned spaceflights, nine of these orbital, with a total of 13 different crewmen. Three of the 13 American astronauts had participated in two flights. The Soviet Union had conducted eight manned flights, all orbital, with 11 different crewmen. No Soviet cosmonaut had yet experienced two space flights. Cumulative totals for manned spacecraft hours in flight had reached 704 hours, 4 minutes for the United States and 431 hours, 53 minutes for the Soviet Union. Cumulative total manhours in space were 1,354 hours, 14 minutes, and 506 hours, 29 minutes, respectively. The American lead in this area came from two-man Gemini flights; these increased in length from almost five hours in March 1965 to over 330 hours for the two-week mission of GEMINI VII in December.

Data on United States flights are the latest available to date within NASA. Soviet data are unofficial and derived from open sources, relying heavily on the excellent work done on this subject by Dr. Charles S. Sheldon II, professional staff member of the National Aeronautics and Space Council. Details are always subject to revision as information is refined, but the major aspects of United States flights have been subject to direct observation by the interested citizens of the world and scientific findings have been presented to professional audiences at symposia and in the professional literature.

Since specific astronaut-cosmonaut ranks and titles are subject to change over time, they are omitted from the chronology in the interest of simplicity. All crew members listed, except Feoktistov and Yegorov of VOSKHOD I, held military rank. Weight given is the weight of the total spacecraft as placed in orbit. Maximum altitude is chosen from many possible measurements because it represents a world record category as recognized by the Fédération Aéronautique Internationale.

610 ASTRONAUTICS AND AERONAUTICS, 1965

Date		Designation	Crew	Weight (lbs.)	Revolutions	Max. Apogee (st. ml.)	Duration	Remarks
Launched	Recovered							
1961								
Apr. 12	Apr. 12	VOSTOK I	Yuri A. Gagarin	10,419	1	203	1 hr. 48 min.	*First man in space, conducted radio and television communication with earth. Spacecraft contained life support systems, telemetry, ejection seat, and recovery system. Cosmonaut and spacecraft landed at "pre-selected" area. Automatic control with cosmonaut available as backup. Two-gas air supply at sea-level pressure.*
May 5	May 5	FREEDOM 7	Alan B. Shepard, Jr.	2,855	—	116	15 min.	Mercury spacecraft launched in ballistic suborbital trajectory by Redstone booster (Mercury-Redstone 3). Downrange distance 302 st. ml. First U.S. man in space; weightless for 5 min. with no ill effects. Astronaut exercised manual control of spacecraft.
Jul. 21	Jul. 21	LIBERTY BELL 7	Virgil I. Grissom	2,836	—	118	15 min.	Mercury spacecraft launched into ballistic suborbital trajectory by Redstone booster (Mercury-Redstone 4). Downrange distance 303 st. ml. Five minutes of weightlessness with no ill effects. Hatch opened prematurely during recovery, spacecraft filled with water and sank in 2,500 fathoms of water in Atlantic Ocean. Astronaut was recovered safely.
Aug. 6	Aug. 7	VOSTOK II	Gherman S. Titov	10,432	16	152	25 hrs. 11 min.	*First test of prolonged weightlessness; cosmonaut ate, worked, and slept in space. Monitored by TV and radio. Vestibular disturbances produced motion sickness but apparently no significant after-effects.*

1962						
Feb. 20	FRIENDSHIP 7	John H. Glenn, Jr.	2,987	3	162	4 hrs. 55 min.	Mercury spacecraft launched into orbit by Atlas booster (Mercury-Atlas 6). First U.S. manned orbital flight, achieving original Project Mercury objectives of placing man in orbit, observing his reactions to space environment, and recovering him safely. No adverse physiological effects from 4½ hrs. of weightlessness. Manual control systems used for orbital attitude; retrorocket pack retained through reentry as safety measure when faulty signal light seemed to indicate that heat shield was loose.
May 24	AURORA 7	M. Scott Carpenter	2,975	3	167	4 hrs. 56 min.	Mercury spacecraft launched into orbit by Atlas booster (Mercury-Atlas 7). Flight plan similar to FRIENDSHIP 7, but 77 min. was reprogramed to drifting flight when a fuel shortage arose from accidental use of two control systems simultaneously. Manual attitude control exercised during retrofire. Landed 250 mi. beyond predicted impact point, caused by yaw error of 25 degrees and retrofire 3 sec. late. Of two experiments, balloon to measure drag and provide visibility data failed to deploy properly; experiment on behavior of liquid in weightless state went as anticipated.
Aug. 11	VOSTOK III	Andrian G. Nikolayev	10,412	60	146	94 hrs. 22 min.	*Launched as first half of "tandem" mission with VOSTOK IV. Radiation measurements taken and effects on biological specimens studied. Extensive physiological measurements recorded during prolonged exposure to space environment with no adverse effects noted. TV monitoring, radio and television contact with VOSTOK IV. Cosmonaut floated free of restraint for a total of 3.5 hrs., was ejected to land separately from spacecraft.*

Date Launched	Date Recovered	Designation	Crew	Weight (lbs.)	Revolutions	Max. Apogee (st. mi.)	Duration	Remarks
1962								
Aug. 12	*Aug. 15*	*VOSTOK IV*	*Pavel R. Popovich*	*10,425*	*45*	*147*	*70 hrs. 57 min.*	*Launched from same facility as VOSTOK III into "tandem" flight. Passed to within 4 mi. of VOSTOK III at closest point. Extensive radiation and physiological experiments conducted. Television monitoring, radio and television contact with VOSTOK III. Cosmonaut floated free of restraint for a total of about 3 hrs. Ejected and landed separately from spacecraft.*
Oct. 3	Oct. 3	SIGMA 7	Walter M. Schirra, Jr.	3,029	6	176	9 hrs. 13 min.	Mercury spacecraft launched into orbit by Atlas booster (Mercury-Atlas 8) with modifications to reaction control system to conserve fuel and added HF antennas to improve communication. Minor problem in adjusting suit temperature; drifting flight for 136 min. to study fuel conservation techniques. Automatic reentry mode utilized, landed 9,000 yds. from prime recovery ship for first recovery in Pacific Ocean. Experiments included camera, radiation sensors, and ablative samples on the neck of spacecraft.
1963								
May 15	May 16	FAITH 7	L. Gordon Cooper, Jr.	3,033	21	166	34 hrs. 20 min.	Mercury spacecraft launched into orbit by Atlas booster (Mercury-Atlas 9). Long-endurance mission with no adverse physiological effects as astronaut ate, slept, and worked in the space environment. An indicator malfunction in 19th orbit precipitated decision to use the complete manual reentry sequence with astronaut orienting spacecraft and firing retro-rockets manually. A very accurate landing was made in the Pacific

Date	Spacecraft	Crew			Duration	Remarks
Jun. 14	VOSTOK V	Valery F. Bykovsky	10,408	76	119 hrs. 6 min.	Ocean, 7,000 yds. from prime recovery ship; astronaut remained in spacecraft until it was hoisted to carrier deck. Experiments included infrared and standard photography, radiation measurement, ejecting flashing light in 3rd orbit (which was sighted in 5th and 6th orbits), direct television transmission, and observation of lights on the ground.
Jun. 16	VOSTOK VI	Valentina V. Tereshkova	10,392	45	70 hrs. 50 min.	New time and distance spaceflight records set, last 45 revolutions in tandem with VOSTOK VI. Radio contact with VOSTOK VI. Direct television broadcast to earth in real time. Extensive medical and biological experiments conducted.
						Orbited first woman and first person not trained as a pilot. Passed within proximity of VOSTOK V on first orbit, in different orbital plane. Radio contact with VOSTOK V. Direct TV broadcast to earth in real time. Parachuted separately from spacecraft on landing.
1964						
Oct. 12	VOSKHOD I	Vladimir M. Komarov Konstantin P. Feoktistov Boris B. Yegorov	11,731	16	24 hrs. 17 min.	First multi-manned spaceflight; first "shirt sleeve" environment in spaceflight, no pressure suits; record altitude for manned spaceflight. Feoktistov "scientist" and Yegorov "physician" as experimenters and observers in addition to Komarov as pilot. Both Feoktistov and Yegorov experienced motion sickness and disorientation illusions. Crew worked, ate, drank, and slept unrestrained. Tests included attitude control by ion engine and navigation by cosmonauts. Crew remained in spacecraft through landing.

Date		Designation	Crew	Weight (lbs.)	Revolutions	Max. Apogee (st. mi.)	Duration	Remarks
Launched	Recovered							
1965								
Mar. 18	Mar. 19	VOSKHOD II	*Pavel I. Belyayev Aleksey A. Leonov*	*12,529*	*16*	*308*	*26 hrs. 2 min.*	*New record altitude for manned spaceflight. During 2nd orbit Leonov took first "spacewalk" with autonomous life support system, moved 17 ft. from spacecraft on tether, 23 min. in space environment, including 11 min. in airlock through which he exited from spacecraft. While outside Leonov was observed through the spacecraft TV camera. First Soviet landing sequence using manual system in lieu of automatic. Crew remained in spacecraft through landing. Planned landing point was overshot; landed in deep snow.*
Mar. 23	Mar. 23	GEMINI III	Virgil I. Grissom John W. Young	7,111	3	140	4 hrs. 53 min.	Gemini spacecraft launched into orbit by modified Titan II booster (Gemini-Titan 3). First U.S. two-man spaceflight, Grissom first man in space for 2nd time. One orbital maneuver was conducted in each of the three orbits. During 1st orbit the apogee was lowered from 139 mi. to 105 mi. and the perigee from 100 mi. to 98 mi.; in 2nd orbit a series of maneuvers produced a translational movement changing inclination by 1/50 of a degree; in 3rd orbit the perigee was lowered to 52 mi. Manual control was exercised throughout the reentry phase, using the limited lifting characteristics of the spacecraft to steer toward touchdown. Two of three experiments were performed; experiment on effect of weightlessness on sea-urchin eggs during fertilization and cell division failed; successful experiments were effect of weightlessness in interaction with

Jun. 3	GEMINI IV	James A. McDivitt Edward H. White II	7,879	62	184	97 hrs. 56 min.	Gemini spacecraft launched into orbit by modified Titan II booster (Gemini-Titan 4). Four-day endurance mission to test crew and spacecraft in buildup to longer missions. Eleven scientific and engineering experiments, in addition to basic medical checks on four days of exposure to the space environment, measured radiation, electrostatic charge, proton-electron flow, and geomagnetic fields. Synoptic weather and terrain photography was accomplished, as well as two-color photographs of the earth's limb. Navigation measurements were conducted, an inflight exerciser was evaluated, inflight phonocardiograms taken, and bone demineralization measurements made. Attempt to rendezvous with spent 2nd stage was abandoned when spacecraft fuel consumption became excessive. White spent 22 min. outside the spacecraft in extravehicular activity, propelling himself with an oxygen-jet gun. During spacewalk he was attached to the spacecraft with a 25-ft. tether-and-oxygen line; he carried emergency oxygen and camera. Spacecraft computer malfunction necessitated a ballistic reentry sequence. First mission controlled from new Mission Control Center at Manned Spacecraft Center, Houston, Texas.
Aug. 21	GEMINI V	L. Gordon Cooper, Jr. Charles Conrad, Jr.	7,947	120	217	190 hrs. 55 min.	Gemini spacecraft launched into orbit by modified Titan II booster (Gemini-Titan 5). Eight-day endurance mission confirmed the physiological feasibility of Apollo lunar landing mission. Several major records set by U.S.: longest manned spaceflight in time and distance, total man hours in space, and most manned flights. First flight

(continued from previous entry:) radiation on white blood cells and injection of fluid into the reentry plasma sheath in attempt to attenuate communication blackout.

Date		Designation	Crew	Weight (lbs.)	Revolutions	Max. Apogee (st. ml.)	Duration	Remarks
Launched	Recovered							
1965								of fuel cell electrical power system. During first orbit, perigee was raised to 106 ml. from initial 100 ml. At beginning of 2nd orbit, a Radar Evaluation Pod was ejected; the GEMINI V rendezvous radar furnished range and range rate data on this target for 40 minutes. On 8/23 a simulated rendezvous was conducted with a series of 4 maneuvers through 2 orbits which raised the perigee to 124 ml. and lowered the apogee to 192.6 ml. Sixteen of 17 experiments were successfully conducted: 5 medical experiments measured physiological effects; extensive weather and terrain photography was conducted; visual observations of missile launches and ground patterns were made; and zodiacal light was photographed. Only experiment canceled was photography of the Radar Evaluation Pod. Voice communication was conducted with Sea Lab II under the Pacific Ocean.
Dec. 4	Dec. 18	GEMINI VII	Frank Borman James A. Lovell, Jr.	8,076	206	204	330 hrs. 35 min.	Gemini spacecraft boosted into orbit by modified Titan II booster (Gemini-Titan 7). Fourteen-day endurance mission, longest scheduled flight in Gemini program. Station-keeping conducted with booster 2nd stage after separation; spacecraft maneuvered to raise apogee from original 100 ml. to 138 ml. to serve as rendezvous target for Gemini 6. On 12/7 orbit was adjusted to 197 ml. apogee, 145 ml. perigee, then on 12/9 perigee was raised to 185.8 ml. and apogee lowered to 188.3 ml. Missile reentry was observed over

Dec. 15					Pacific on 12/14. Accurate controlled reentry was made, landing 7.6 mi. from target. One astronaut at a time doffed his pressure suit to enhance comfort. No adverse physiological effects from 2 weeks in the space environment. Twenty experiments were incorporated into the mission, ranging from detailed medical measurements and sleep analysis to earth photography and radiometry measurements.
Dec. 16	GEMINI VI	Walter M. Schirra, Jr. Thomas P. Stafford	7,817	16	193	25 hrs. 51 min.	Gemini spacecraft launched into orbit by modified Titan II booster (Gemini-Titan 6). This spacecraft-booster combination was originally ready for launch on 10/25 but held when Agena target vehicle failed to orbit. Rescheduled to rendezvous with GEMINI VII as target, launch attempted on 12/12 but aborted when the 1st stage engine shut down immediately after ignition. Initial orbit was: apogee, 161 mi.; perigee 100 mi.; a series of posigrade maneuvers began the rendezvous chase and shortened the initial 1,200-mi. separation distance to 730 mi. During 2nd revolution the distance was reduced to 300 mi. with another thruster firing, followed by a plane change to match GEMINI VII's orbit. Orbit was adjusted to near circular at beginning of 3rd revolution; as 4th revolution began, the terminal phase of rendezvous was initiated, resulting in approach to within 1 ft. of GEMINI VII. Station keeping and fly-around maneuvers were conducted for 5 hrs. 19 min. Controlled reentry to landing within 13 mi. of prime recovery ship.

Appendix D

ABBREVIATIONS OF REFERENCES

Listed here are abbreviations for sources cited in the text. This list does not include all sources provided in the chronology, for some of the references cited are not abbreviated. Only those references which appear in abbreviated form are listed below. Abbreviations used in the chronology entries themselves are cross-referenced in the Index.

A&A	AIAA magazine, *Astronautics & Aeronautics*
A&A 65	NASA *Astronautics and Aeronautics, 1965* [this publication]
ABC	American Broadcasting Company
AEC Release	Atomic Energy Commission News Release
AF Info. Pol. Ltr.	Air Force Information Policy Letter for Commanders
AF/SD	*Air Force and Space Digest* magazine
AFOSR Release	Air Force Office of Scientific Research News Release
AFSC Release	Air Force Systems Command News Release
AIAA Release	American Institute of Aeronautics and Astronautics News Release
AIP News	*American Institute of Physics News*
AP	Associated Press
ARC Release	NASA Ames Research Center News Release
Atlanta J/Const.	*Atlanta Journal and Constitution* newspaper
Atomic Energy Programs, 1965	AEC *Major Activities in the Atomic Energy Programs, 1965*
ATSS-T Trans.	Translation by NASA Scientific and Technical Information Div., Translators
Av. Daily	*Aviation Daily* newsletter
Av. Wk.	*Aviation Week and Space Technology* magazine
Balt. *Sun*	Baltimore *Sun* newspaper
CBS	Columbia Broadcasting System
Chic. Trib.	*Chicago Tribune* newspaper
CNS	Copley News Service
Commerce Dept. Release WB-	Dept. of Commerce, Weather Bureau, News Release
ComSatCorp Release	Communications Satellite Corporation News Release
CR	*Congressional Record*
CSM	*Christian Science Monitor* newspaper
CTNS	Chicago Tribune News Service
DAC Release	Douglas Aircraft Co. News Release
DJNS	Dow Jones News Service
DMSSD *Apogee*	Douglas Missile and Space Systems Div. *Apogee*
DOD Release	Dept. of Defense News Release
FAA Release	Federal Aviation Agency News Release
FonF	*Facts on File*
FR	*Federal Register*
FRC Release	NASA Flight Research Center News Release
FRC X-Press	NASA Flight Research Center *FRC X-Press*
GE Forum	*General Electric Forum* magazine
Goddard News	NASA Goddard Space Flight Center *Goddard News*

GSFC Release	NASA Goddard Space Flight Center News Release
GSFC SSR	NASA Goddard Space Flight Center *Satellite Situation Report*
HHR-8	NASA (EPH) *Chronology of Major NASA Launches* (HHR-8)
Houston Chron.	*Houston Chronicle* newspaper
HTNS	New York Herald Tribune News Service
Ind. Off. Approp. Hearings	U.S. Congress, *Hearings on Independent Offices Appropriations* [FY 1966]
Int. Sci. & Tech.	*International Science and Technology* magazine
JAMA	*Journal of the American Medical Association*
J/Armed Forces	*Journal of the Armed Forces*
J/Spacecraft and Rockets	*Journal of Spacecraft and Rockets*
KSC Release	John F. Kennedy Space Center, NASA, News Release
Langley Researcher	NASA Langley Research Center *Langley Researcher*
LARC Release	NASA Langley Research Center News Release
L.A. Times	*Los Angeles Times* newspaper
Lewis News	NASA Lewis Research Center *Lewis News*
LRC Release	NASA Lewis Research Center News Release
M&R	*Missiles and Rockets* magazine
Marshall Star	NASA George C. Marshall Space Flight Center *Marshall Star*
Miami Her.	*Miami Herald* newspaper
Milwaukee J.	*Milwaukee Journal* newspaper
Minn. Trib.	*Minneapolis Tribune* newspaper
MSC Release	NASA Manned Spacecraft Center News Release
MSC Roundup	NASA Manned Spacecraft Center *Space News Roundup*
M/S Daily	*Missile Space Daily* newsletter
NAA Release	North American Aviation, Inc., News Release
NAA *S&ID Skywriter*	North American Aviation, Inc., Space and Information Systems Div. *S&ID Skywriter*
NANA	North American Newspaper Alliance
NASA Ann.	NASA Announcement
NASA Auth. Hearings	U.S. Congress, *Hearings on NASA Authorization* [FY 1966]
NASA Hq. *Bull.*	NASA Headquarters *Weekly Bulletin*
NASA LAR IV/50	NASA Legislative Activities Report, Vol. IV, No. 50
NASA Off. Int. Aff.	NASA Office of International Affairs
NASA Release	NASA Headquarters News Release
NASA Rpt. SRL	NASA Report of Sounding Rocket Launching
NASA SP-5018	NASA Special Publication #5018
NASA Proj. Off.	NASA Project Office
NASA X-15 Proj. Off.	NASA (Headquarters) X-15 Project Office
NASC Release	National Aeronautics and Space Council News Release
NAS-NRC-NAE *News Report*	National Academy of Sciences-National Research Council-National Academy of Engineering *News Report*
NBC	National Broadcasting Company
NSC Newsletter	National Space Club Newsletter
N.Y. Her. Trib.	*New York Herald Tribune* newspaper
N.Y. J. Amer.	New York *Journal American* newspaper
NYT	*New York Times* newspaper
NYTNS	New York Times News Service
OAR Release	Office of Aerospace Research (USAF) News Release
Orl. Sent.	*Orlando Sentinel* newspaper
OSSA, OART S. Rkt. Proj. Off.	NASA Office of Space Science and Applications, Office of Advanced Research and Technology Sounding Rocket Project Offices
Phil. Eve. Bull.	Philadelphia *Evening Bulletin* newspaper
Phil. Inq.	*Philadelphia Inquirer* newspaper
Pres. Doc.	National Archives and Records Service *Weekly Compilation of President Documents*
Sat. Eve. Post.	*Saturday Evening Post* magazine

SBD	*Space Business Daily* newsletter
Sci. Amer.	*Scientific American* magazine
Sci. Serv.	Science Service
S.F. Chron.	*San Francisco Chronicle* newspaper
SR	*Saturday Review* magazine
Testimony	Congressional testimony, prepared statements
Text	Prepared report or speech text
Transcript	Official transcript of news conference or Congressional hearing
UPI	United Press International
U.S. Naval Inst. Proc.	*U.S. Naval Institute Proceedings*
U.S. News	*U.S. News and World Report* magazine
U.S. Aeron. & Space Act., 1965	President's Report to Congress, *United States Aeronautics and Space Activities, 1965*
Wallops Release	NASA Wallops Station News Release
Wash. *Daily News*	Washington *Daily News* newspaper
Wash. *Eve. Star*/Wash. *Sun. Star*	Washington *Evening/Sunday Star* newspaper
Wash. Post	*Washington Post* newspaper
WBE Sci. Serv.	World Book Encyclopedia Science Service
WSJ	*Wall Street Journal* newspaper

Index

A-1 (French satellite), 456, 526, 527, 570
A-7 (attack aircraft), 20
AAAS. See American Association for the Advancement of Science.
Aafss. See Advanced Aerial Fire Support System.
AAS. See American Astronautical Society.
Abbot, Dr. Charles G., 335
ABC. See American Broadcasting Co.
Abelson, Dr. Philip H., 97, 337
ABL. See Automated Biological Laboratory.
Able, Charles R., 446
Abraham, Karl, 183
Abres. See Advanced Ballistic Reentry Systems.
Acceleration, 20, 23, 123, 181, 191
Accident, 123, 204, 302, 345, 373, 377, 398, 414, 492
Achilles, Theodore C., 94
Acs. See Attitude Control System.
AC Spark Plug Div. (General Motors Corp.), 139
Actuator, 14, 156
Adams, Rep. Brock, 46
Adams, Dr. Mac C., 386, 460, 564
Adams, Maj. Michael J. (USAF), 514
Adenosine triphosphate (Atp), 371
Adler, Joel, 566
ADN (E. German press agency), 232
Adp. See Automatic data processing.
Advanced Aerial Fire Support System (Aafss), 504
Advanced Ballistic Reentry Systems (Abres), 144
Advanced Flight Simulation Laboratory, 311
Advanced Manned Strategic Aircraft (Amsa), 84, 333, 485
Advanced Orbiting Solar Observatory (Aoso), 332, 554
Advanced Research Projects Agency (ARPA), 157, 345, 373, 525
Advanced Vidicon Camera System (Avcs), 408
"Advances in Aircraft Technology," 518
AEC. See Atomic Energy Commission.
AEDC. See Arnold Engineering Development Center.
Aero Club of Washington, 364, 491
Aero Spacelines, Inc., 234, 437, 562

Aerobee (sounding rocket), 215, 321
150—108, 127, 132, 170, 264, 455, 521, 534
150A—15, 261, 439, 449, 483
300A—139
350—289, 450
Aerojet-General Corp., 49, 55, 192, 270, 286, 347
Aetron Div., 115
contract, 274, 282, 572
strike, 248, 260
Aeronautics, 33, 77, 90, 327
award, 32, 186, 292
research, 18–19, 35–37, 40–42, 49–50, 57, 77–78, 84, 89, 90, 112, 170, 176, 178, 231, 294, 463–465, 482, 496, 497, 518, 519, 524, 543, 568
Aeronutronic Div., Philco Corp., 485
Aerospace (subject). See entries under Space.
Aerospace, 407, 569
Aerospace Corp., 216, 221, 419–420
Aerospace Industries Association of America, Inc., 407, 415, 562
Aerospace industry, 13, 52, 54, 70, 110, 111, 113, 114, 141, 152, 175, 569
contracts, 123, 572
employment, 418
exports, 7, 164
market forecast, 110
Aerospace Instrumentation Symposium, 461
Aerospace Research Laboratory, 237
Aerospace Research Satellite (Ars), 461
Aerospace Research Vehicle (Arv) (satellite), 24
Aerospace Transporter (spacecraft), 206
Aerozine-50, 547
AES. See Apollo Extension System.
Aetron Div., Aerojet-General Corp., 115
AFA. See Air Force Association.
AFCMD. See USAF Contract Management Division.
AFCRL. See Air Force Cambridge Research Laboratories.
Africa, 554
AFSC. See Air Force Systems Command.
AFSSD. See Air Force Space Systems Div.
Agena (booster) (see also Atlas-Agena, Thor-Agena), 23, 165, 166, 190
B, 142
D, 5, 269, 289

623

Agena Review Board, 491
Agena Target Vehicle, 23, 347, 487, 489, 491, 502, 509, 511, 524, 571
Agriculture, Dept. of (USDA), 174
AGU. See American Geophysical Union.
AIAA. See American Institute of Aeronautics and Astronautics.
Aiken, Sen. George D., 46
AIP. See American Institute of Physics.
"Air bus" (transport aircraft), 314
Air cushion vehicle, 34, 374
Air Density Explorer (spacecraft), 486
Air Force and Space Digest, 361, 498
Air Force Association (AFA), 304, 440, 442
Air Force Cambridge Research Laboratories (AFCRL), 24, 117, 124, 149, 206, 456, 566
  balloon launch, 47, 78
  Space Physics Laboratory, 47
Air Force/Industry Planning Seminar, 147
Air Force Scientific Advisory Board, 451
Air Force Space Systems Div. (AFSSD), 490
Air Force Systems Command (AFSC), 18, 66, 114, 128, 147, 170, 264, 354, 384, 406, 408
  Aeronautical Systems Div., 84
  Aerospace Test Wing, 6595th, 144
  Agena (booster), 23, 166, 190
    Review Board, 491
    Target Vehicle, 23, 524
  agreement, 484
  airspace defense control system, 241, 411
  Apollo (program), 484
  astronaut training, 246, 399
  contract, 8, 84, 121, 166, 190, 537
  Contract Management Div., 5
  Manned Orbiting Laboratory, 231, 483
  research
    aeronautical, 84, 121, 204, 381, 537
    space systems, 58, 105–107, 204, 537
  rocket engine, 23, 157, 165
  test, 84, 157, 182
Air France, 237
Air India, 237
Air pollution, 271
Air Products and Chemicals, Inc., 445
Air Rescue Service, 390
Air Research Manufacturing Co., 438
Air traffic control, 152, 174, 353, 384, 500
Air Transport Association of America (ATA), 114, 118
Air transportation, 175, 205, 264, 335, 345, 481, 518
Aircraft, 2, 106, 128, 165, 222, 322, 350, 361, 480, 555
  bomber, 6, 18, 34, 73, 84, 87, 168, 189, 278, 290, 291, 329, 333, 381, 484, 544
  cargo, 18, 20, 41, 50, 121, 283–285, 406, 481, 562
  fighter, 6, 19, 52, 63, 109, 178, 288, 393, 433, 510, 512
  supersonic, 6, 14, 18, 19, 20, 52, 90, 109, 121, 122, 172, 178, 180, 183, 241, 307, 325, 474, 500, 537, 549
  foreign, 18, 34, 52, 63, 87, 189, 196, 234, 241, 275, 283–285, 292, 314, 315, 316, 318, 329, 337, 360–361, 372, 383, 393, 528
  helicopter, 32, 39, 114, 122, 231, 254, 297, 315, 371, 372, 382, 504, 521
  hovercraft, 33, 374, 528
  hypersonic 35, 90, 203, 349, 464, 543
  interceptor, 19, 144, 213, 221
  reconnaissance, 18, 52, 87, 124, 504, 537
  record, 122, 144, 150, 189, 213, 221, 233, 245, 371, 453
  research, 18–19, 40–41, 77, 84, 170, 191, 204–206, 231, 242, 352, 518
  Stol, 50, 52, 63, 197
  training, 234, 245, 275
  transport (see also Aircraft, Stol, V/Stol, and Vtol; Aircraft, supersonic transport), 52, 63, 115, 121, 124, 157, 275, 294, 314, 315, 318, 337
    jet, 14, 73, 84, 94, 131, 189, 222, 241, 258, 292, 302, 340, 393, 469, 516, 555
    military, 52, 335, 454, 482
  variable-sweep-wing, 14, 59, 121, 234, 241, 307, 334
  V/Stol, 15, 41, 50, 54, 90, 204, 248, 322, 349, 387, 399
  Vtol, 15, 36, 41, 50, 52, 63, 69, 178, 393
  X-15 (rocket research). See X-15.
Aircraft Noise Research Program, 242
Aircraft Operating Problems Committee, 228
Aircraft, supersonic transport (Sst) (see also Concorde (U.K.-France) supersonic transport), 25, 90, 228, 287, 330, 417, 481, 490, 521
  airworthiness, 48, 73
  benefits, 58, 185
  contract, 8, 257, 312
  criticism, 545
  development, 33–34, 37, 40, 50, 59, 297, 309, 316, 491, 516
  Fausst group, 48, 73
  foreign, 73, 87–88, 204, 205–206, 237, 279, 315–316, 318–319, 336, 417, 479, 490
  hazards, 555
  sonic boom, 15, 48, 73, 199, 237, 571
L'Aire Liquide, 202
Airglow, 157
Airlines, 25, 49, 83, 112, 114, 185–186, 237, 469, 501, 514, 517
  safety, 233
Airlock, 153, 157, 231, 430
The Airman, 534
Airport, 198, 275, 335, 415, 420, 495, 549
Airport surveillance radar (Asr), 513
Akron, Ohio, 131
Akulinichev, Dr. I., 3

INDEX

Alabama, 173, 255
Alabama, Univ. of, 232–233
Alamogordo, N. Mex., 332
Alaska, 58, 214, 233, 276
Alaska, Univ. of, 525
Albert, Rep. Carl, 325
Albuquerque, N. Mex., 237, 241, 514, 549
Alcock (comet), 452
Alcock, G. E. D., 452
Aldebaran (star), 57, 366
Alden Electronics Corp., 282
Aldrin, Edwin E., Jr., 10, 321
Alessandrini, Federico, 152
Alexander, W. M., 330
Alexandria, Va., 294
Alfven, Hannes, 198, 573
Algatron (life support system), 232
Algeria, 76, 151
Alitalia, 25
Alkaid (star), 215, 261
All-Union Conference on Planetary Cosmogony (Moscow), 343
Alldredge, Dr. Leroy R., 523
Allen, H. Julian, 444, 462
Allen, J. Denton, 355
Allen, John G., Jr., 360
Allen, William M., 37
Aller, Dr. L. H., 221
Aller, Robert, 563
Allis-Chalmers Research Div., 238
Alloy, 535
ALOUETTE (Canadian satellite)
  I, 239, 448, 452
  II, 523, 527, 546, 570
Alphonsus (moon crater), 140, 143, 146, 148, 149, 515
Alsep. See Apollo Lunar Surface Experiments Packages.
Alter, Dinsmore, 357
Amazonis (Martian desert), 68
Amchitka Island, 495, 525
American Academy of Arts and Sciences, 119
American Airlines, 59, 237
American Association for the Advancement of Science (AAAS), 3, 11, 564, 566
American Association of Clinical Chemists, 410
American Astronautical Society (AAS), 59, 62, 64, 133, 154, 216, 383, 488
American Bar Association, 371
American Broadcasting Co. (ABC), 229, 249, 254, 271, 445
American Car and Foundry Co., Inc., 282
American Chemical Society, 168, 229, 425, 427, 436
American College of Preventive Medicine, 494
American Congress on Surveying and Mapping, 159
American Electric Power Co., 407
American Federation of Musicians, 561
American Geophysical Union (AGU), 189, 190, 191, 193, 194, 196, 414
American Heart Association, 9
American Helicopter Society, 231
American Institute of Aeronautics and Astronautics (AIAA), 35, 154
  awards, 471, 507
  International Aerospace Abstracts, 471
  meeting, 5, 22, 101–102, 119, 173, 194–195, 347, 492, 505, 519
  Research Award, 32
  Wyld, James H., Propulsion Award, 286
American Institute of Chemical Engineers, 63
American Institute of Industrial Engineers, 138
American Institute of Physics (AIP), 310
American Legion, 101, 371
American Machine and Foundry Co., 92, 191, 341
American Management Association, 106
American Meteorological Society, 38
American Museum of Natural History, 78
American Newspaper Publishers Association, 495, 532
American Nuclear Society-Atomic Industrial Forum, 516, 517
American Ordnance Association, 494
American Physical Society, 38, 202, 208, 209, 495
American Society for Testing and Materials (ASTM), 283
American Society of Mechanical Engineers (ASME), 128, 131, 463, 512
American Society of Newspaper Editors, 423
American Society of Photogrammetry, 159
American Society of Tool and Manufacturing Engineers, 156
American Telephone and Telegraph Co. (AT&T), 81, 201, 265, 278, 334, 532
American Trial Lawyers Association, 351
American Univ., 514
Ames, Milton B., Jr., 107, 233, 234, 310
Ames Research Center (ARC), 36, 49, 50, 64, 77, 78, 93–94, 119, 138, 169, 295, 443, 518, 556
  award, 356, 462
  Biosatellite, 59, 138
  contract, 191, 311
  experiments, 36, 207
  facilities, 311, 565
  Mission Analysis Division, 69
  research, 69, 323, 352, 425, 449, 468, 555
  test, 41, 516
AMR. See Atlantic Missile Range and Eastern Test Range.
Amsa. See Advanced Manned Strategic Aircraft.
AMSE. See American Society of Mechanical Engineers.
An-22 (U.S.S.R. transport aircraft), 283, 337, 361
An-24B (U.S.S.R. transport aircraft), 361

626 INDEX

Anders, Edward, 448
Anders, William, 321
Anderson, C. A., 78
Anderson, Sen. Clinton P., 62, 155
Anderson, Omer, 32
Anderson, Orvil A., 513
Anderson, Rep. William R., 46
Andes, 416
Andover, Me., 172, 180, 197, 228, 555
Andre, l/c Daniel (usaf), 213
Andrews afb, Md., 532
Ann Arbor, Mich., 283
anna ib (geodetic satellite), 274
Anniversary, 1, 7, 42, 107, 121, 127, 138, 147, 162–163, 165, 222, 243, 297, 298, 325, 335, 360, 371, 381–382, 512, 527, 546
Anniversary of the Bolshevik Revolution, 512, 527
Antarctic Treaty, 531
Antarctica, 56, 115, 174, 181, 523, 531, 532, 557
  dust "core sampling," 191
Antares (rocket), 244
Antenna, 16, 30, 67, 92, 108, 115, 126, 139, 149, 197, 228, 348, 364, 386, 467, 492, 510, 537, 556, 573
Anticosmos, 107
Antideuteron, 107, 279, 573
Antimatter, 107, 258, 279, 573
Antimissile missile, 39, 86, 156, 161, 310, 447, 452, 522, 541, 553
Antirock, 258
Antisatellite defense, 242
Antiworld, 279
Antonov, Oleg, 283, 361
Aonius Sinus (Mars), 68
Aoso. See Advanced Orbiting Solar Observatory.
ap. See Associated Press.
apl. See Applied Physics Laboratory.
Apollo (program), 29–30, 109, 115, 174, 177, 482, 506
  astronaut, 10, 56, 133
  criticism, 11, 97, 181
  experiment, 45, 524
  facilities, 11, 490, 502
  funds for, 27, 75, 109, 171, 204, 219, 482, 557
  impact, 394
  lunar landing program, 5, 7, 45, 75, 97, 109, 165, 246, 351, 467
  management, 1, 20, 80, 479
  plans for, 7, 27, 29, 43, 75, 171, 174, 295, 335, 341, 346, 358–359, 367, 418, 429, 447, 469–470
  progress, 23–24, 29–30, 43, 96, 100, 116, 123, 129, 162, 188, 198, 209, 301, 325, 348, 478, 490, 517
  space hazards, 160, 196
  test, 7, 27, 43, 54, 55, 72, 73, 80, 112, 162
  tracking, 92, 107, 120, 174, 466
Apollo (spacecraft), 77, 92, 113, 174, 227
Apollo-X, 27, 29–30, 77, 174

boilerplate model, 55, 72, 193, 365, 485
  contract, 33, 91, 94, 109, 117, 139, 174, 177, 182, 185, 549
  equipment, 115 160–161, 165
  escape system, 112, 238, 540
  facilities, 11, 48, 52, 54, 92, 115, 119, 178–179, 184
  launch vehicle, 29–30, 39, 54, 56, 72, 73, 80, 82, 91, 93, 96, 97, 100, 123, 162, 177, 182, 188, 191, 197, 198, 209
  module, 55, 185
    command, 72, 113, 139, 174, 177, 193, 244, 247, 301, 490, 492, 540
    lunar excursion (lem), 7, 33, 48, 75, 112, 113, 119, 139, 152, 160, 177, 186, 315, 345, 352
    test, 43, 54, 55, 56, 79–80, 110, 112, 162, 177, 186, 188, 197, 198, 228, 238, 540, 570
Apollo Extension System (aes), 27, 29, 77, 174, 217, 302, 341, 343, 434, 444, 449
Apollo Lunar Surface Experiments Packages (Alsep), 366, 476
Appel, Frederick, 222, 289
Applications Technology Satellite (Ats), 275, 289, 519, 550
Applied Physics Laboratory (apl) (Johns Hopkins Univ.), 508, 515
Apt. See Automatic Picture Transmission system.
Aquanaut, 402, 417, 421, 425, 434, 443, 450, 467, 471
Aral Sea, 369
arc. See Ames Research Center.
Arcas (meteorological rocket), 121, 236, 420
Arctic and Antarctic Institute (U.S.S.R.), 387
Arctic Circle, 35, 58
Arecibo, Puerto Rico, 194
Arecibo Ionospheric Observatory, 109, 194, 229
Argentina, 58, 225, 236, 307, 370, 416, 511
Argo d-4 (Javelin) (sounding rocket), 15, 239, 249, 447, 468
Argo d-8 (sounding rocket), 184, 302
ariel ii (UK satellite), 328
Aristarchus (moon crater), 515
Arizona, Univ. of, 4, 57, 84, 85
Arkansas, 255
Armed Forces Communications and Electronics Association, 249
Armed Forces News Service, 473
Armed Forces Week, 233
Arms control, 32, 105, 416, 427
Armstrong, Neil A., 58, 334, 339, 382, 444
Army Air Corps, 513
Army Corps of Engineers, 88
  contract, 14, 37, 105, 175, 453
  secor v, 374
Arnold Engineering Development Center (aedc), 80, 535

INDEX 627

Arnold, James R., 448
ARPA. See Advanced Research Projects Agency.
Ars. See Aerospace Research Satellite.
Artemyev, Yevgeny, 129
Artificial life, 427, 436, 466
Artron (artificial neuron), 204
Artsimovich, Lev A., 183
Arv. See Aerospace Research Vehicle.
Asahi Shimbum, 236
Ascension Island, 39
ASME. See American Society of Mechanical Engineers.
Aspen, Colo., 549, 553
Asr. See Airport surveillance radar.
Asset, Project, 88, 92, 106
Associated Press (AP), 96, 186, 452, 527, 553, 555
Association of Military Surgeons of the U.S., 517
Asteroid, 472, 506, 507
ASTM. See American Society for Testing and Materials.
Astrogeologist, 496
Astronaut (see also Cosmonaut), 363, 402, 442, 461, 535, 562
  extravehicular activity, 190, 208, 247, 248, 255, 256, 265, 269, 273, 510, 570
  former, 5, 345, 460, 474, 476
  goodwill tour, 424, 438, 440, 441, 446, 448, 452, 460, 463, 474
  GT-3 flight, 15, 152
  GEMINI VI flight, 483, 493, 511, 547, 551-552, 554, 556, 558
  GEMINI VII flight, 493, 511, 538, 554, 558, 560
  Gemini VIII flight, 444
  Gemini IX flight, 510
  honored, 155, 156, 158, 159, 259, 278, 283, 286, 404, 433, 437, 448-449, 472, 474, 476, 518-519, 567
  message to, 147, 271, 403, 548
  performance, 147, 512, 566, 567
  physical examination, 118, 377, 382, 403, 533, 567
  pilot/astronaut, 423
  press conference, 470, 500
  promotion, 294, 331, 548
  scientist-astronaut, 5, 299, 300, 343, 383, 398, 423
  scientists, 5, 63, 102, 123, 132-133
  selection, 19-20, 118, 123, 269, 299, 495, 514, 535
  training, 10, 97, 246, 321, 334, 399, 530
  women as, 63, 123
Astronautics and Aeronautics, 44, 98, 305
Astronautics Award, 471
Astronomical Journal, 401
Astronomical Unit (AU), 258
Astronomy, 4, 43, 67, 108, 126, 183, 223, 261, 264, 281, 311, 319, 321, 342, 343, 344
Astrophysical Observatory, Cambridge, Mass., 67, 82, 376

ATA. See Air Transport Association of America.
AT&T. See American Telephone and Telegraph Co.
Athena (missile), 49, 67, 426, 473
Athena (reentry program), 290
Athens, Greece, 390
Atlantic City, N.J., 238, 353, 488
Atlantic Council, 94
Atlantic Missile Range (AMR) (see also Eastern Test Range), 87
Atlantic Ocean, 22, 39, 116, 132, 146
Atlantic Research Corp., 97
Atlas (booster), 106, 132, 144, 275, 289
  D, 244, 461
  SLV-3, 183
  SLV-3X, 123, 275
Atlas (missile), 57, 66, 73, 161, 254, 257, 522
Atlas-Agena (booster) (see also Agena), 165, 206, 332, 339, 490, 509
  B, 140
  D, 27, 120, 206, 341, 454
Atlas-Centaur (booster) (see also Centaur), 103, 120, 197, 348, 374, 421, 570
Atluri, C. R., 258
Atmosphere
  artificial, 36, 128
  contamination of, 305
  density, 6, 15, 55, 78
  oxygen-nitrogen, 145
  pressure, 78
  temperature, 55, 78, 124, 374
  upper measurement, 35, 58, 110, 121, 132, 262
Atom, 517
"Atomic age," 335
Atomic bomb, 232, 236, 332, 527
Atomic clock, 284
Atomic Energy Commission (AEC), 3, 18, 39, 44, 45, 63, 107, 122, 135, 180, 325, 332, 536, 546
  Brookhaven National Laboratory, 107, 279
  cooperation, 61, 240, 243, 292
  Division of Space Nuclear Systems, 292
  funds, 3, 28, 62, 98, 279
  launch, 166
  Mound Laboratory, 99
  Oak Ridge National Laboratory, 167, 547
  reactor
    Kiwi, 13, 25, 117, 306, 519
    NRX, 240, 255, 306
    Phoebus, 49, 282, 296, 519
    Snap, 7, 28, 30, 59, 62, 100, 141, 166, 172, 243, 292, 306, 510, 569
  satellite, 39-40, 242, 341
  spacecraft, 167, 172
Atomics International, 18, 30, 272
Atp. See Adenosine triphosphate.
Ats. See Applications Technology Satellite.

AT&T. See American Telephone and Telegraph Co.
Attitude Control System (Acs), 15, 261
Atwood, J. L., 336, 514
Auburn, Mass., 121
Ault, G. Mervin, 283, 519
Aurora, 525
AURORA 7 (spacecraft), 173
Austin, Tex., 546
Australia, 76, 85, 353, 520, 540, 545
Australian National University, 85
Auter, Henry F., 505
Autoflare (automatic landing system), 275
Automated Biological Laboratory (Abl), 60, 201
Automatic data processing (Adp), 561
Automatic Picture Transmission System (Apt), 408
Avco Corp., 460, 536
Avcs. See Advanced Vidicon Camera System.
Aviaexport (U.S.S.R.), 314
Aviation, 49, 77
 commercial, 49, 50, 84, 88, 114, 118, 185
Aviation and Space Conference, ASME, 129
Aviation Daily, 447
Aviation Hall of Fame, 555
Aviation/Space Writers' Association, 50
 meeting, 183, 455
Aviation/Space Writers' Association Conference, 237, 241, 244, 245, 545, 562
Aviation Week and Space Technology, 3, 20, 86, 124, 142, 170, 190, 215, 235, 300, 364, 380, 405, 418, 427, 444, 451, 481, 482, 498, 500, 501, 541
Aviatsiya i Kosmonavtika, 372, 532
Awards, 156, 199, 472
 civic, 156, 199, 472
 government, 15, 17, 48, 57, 58, 71, 134, 194, 195, 239, 256, 290, 295, 301, 343, 406, 431, 453–454, 478
 industry, 101, 331, 411
 institutions, 25, 30, 73, 129, 156, 283, 485
 society, 9, 17, 101
  aeronautics, 292, 486, 506–507
  astronautics, 131, 390, 431, 486, 506–507
  astronomy, 43, 67, 202–203
  aviation, 13, 32, 186, 292, 526
  engineering, 156, 209
  foreign, 73, 448–449, 474
  space, 32, 98, 471
Azores, 327
Azores Weather Bureau, 324
Azusa, Calif., 260

B-17 (Fortress), 284
B-52 (Stratofortress), 84, 278, 290, 329, 333, 544
B-58 (Hustler, bomber), 6, 189, 284, 316, 544
B-70 (bomber), 232
 XB-70, 19, 77, 84, 206, 284, 547, 568
 XB-70A, 72, 150, 191, 284, 308, 336, 351, 453, 478, 482, 562, 563
BAC-111 (jet airliner), 318
Backup Interceptor Control (Buic), 411
Bader, Dr. Michel, 490
Baffles, 107
Bahamas, 260
Baikonur Cosmodrome, 131, 139, 140, 155, 189
Bain, Gordon, 386
Baker-Nunn Camera Station, 528
Bakinskiy rabochiy, 372
Balakshin, Leonid, 386
Balboa, Panama Canal Zone, 121
Balch, Jackson M., 221, 505
Baldwin, Hanson, 45
Ball Brothers Research Corp., 188, 405, 447
Ballard, William F. R., 495
Balloon
 flight, 78, 326, 513, 520
 launch facilities, 350
 record, 243
 recovery system, 456
 steerable, 556
 weather, 47, 175, 220, 271
Baltimore, Md., 317
Baltimore Sun, 106, 441, 527
Bandung, West Java, 375
Barbados, 501–502
Barbour, Laura Taber, Air Safety Award, 186
Barking Sands missile tracking facility, 49
Barling, Walter Henry, 168
Barnes Engineering Co., 414
Bartlett, Sen. E. L., 276
"Basic Research and National Goals" (report), 199
Bassett, Capt. Charles A. (USAF), 321, 510
Bates, Dr. Thomas F., 124
Battelle Memorial Institute, 187, 283, 408
Baumann, Robert C., 544
Baxter Construction Co., 453
Bazooka, 127
BBC. See British Broadcasting Co.
BEA. See British European Airways.
Bealmear, Sister M. Margaret, 124
Bean, Allen, 10
Bedford, Mass., 220
Behavioral studies, 174
"Belching spider" (lunar landing simulator), 360
Belgium, 259
Belgrade, Yugoslavia, 401
Bell Aerospace Corp., 157, 161, 375
Bell Aerosystems Co., 165, 186, 210, 342, 360, 374, 399, 465
Bell, Alexander Graham, 555
Bell Telephone Laboratories, Inc., 2, 228, 261
Bell Telephone System, 236

Beller, William S., 380, 450
Belyayev, Col. Pavel (U.S.S.R.), 131, 132, 134, 136, 139, 140, 141, 147, 153, 154, 156, 222, 323, 395, 430, 438–439, 457, 460, 473
Ben Bella, President Ahmed (Algeria), 151
Bendix Corp., 131, 240, 342
Bendix Systems Div., 367
Bendix Field Engineering Corp., 70, 448, 461
Benioff, Hugo, 193
Bennett, Dr. Geoffrey, 204–205
Berg, B/G Russell A. (USAF), 406
Bergman, Jules, 271, 438
Berkeley, Calif., 77, 281, 565, 566
Berkner, Dr. Lloyd V., 397
Berkone (Sealab support ship), 450
Berl, Dr. Walter G., 99
Berlin, Germany, 188
Berliner Zeitung, 346, 379
Bermuda, 12, 175, 559
Berry, Dr. Charles A., 201, 267, 405, 424, 431, 488, 533, 558, 560, 566
Beta (U.S.S.R. nuclear power system), 123
Betelgeuse (star), 57, 366
Bethesda, Md., 368
Bhabha, Dr. Homi J., 527
Bikle, Paul F., 278, 464
Bilhorn, Thomas, 349
Billings, Mont., 473
Bioastronautics, 59, 115, 132, 146, 181, 188, 373, 410, 483
Bioletti, Carlton, 138
Biological clock, 116, 281, 470
Biology, 281, 297, 470
Bioluminescence, 119
Biomedicine, 174, 196, 201, 266, 267, 283, 313, 443
Bionics, 204
Biophysical Society, 481
Biosatellite (program), 59, 133, 138, 150, 207
Bioscience, 4, 54, 115, 132, 344
BIS. See British Interplanetary Society.
Bisplinghoff, Dr. Raymond L., 103, 206, 231, 252, 253, 295, 349, 350, 379, 386, 498
Black Brant (rocket), 277, 290
Black Knight (research rocket), 526
Blagonravov, Dr. Anatoli A., 111, 227, 262
Bleymaier, B/G Joseph S. (USAF), 12, 107, 183, 290, 291, 354, 406
Blind, aids for, 90
Blizard, Dr. Jane, 202
Blount Brothers Corp., 26, 312
Blue Angels (Aerial Team), 316
Blue Book Project, 38, 327
Blue Scout (rocket), 157
Blue Scout, Jr. (rocket), 39, 176, 227, 272
Blue Streak (British rocket), 143, 365, 545
Blumenthal, Fred, 259

Blumrich, Josef F., 165
BMwF. See Federal German Ministry for Scientific Research.
BOAC. See British Overseas Airways Corp.
Boat, solar, 299
BOB. See Budget, Bureau of.
Bochum Institute for Satellite and Space Research, 562–563
Bochum Observatory (W. Germany), 292, 311, 336
Boeing 377 (Stratocruiser), 234
Boeing 707 (Stratoliner), 109, 131, 228, 302, 334
Boeing 707–349C (jet aircraft), 516
Boeing 720 (jetliner), 131
Boeing 727–22 (jetliner), 221
Boeing 727–QC (jet aircraft), 346
Boeing 737 (jetliner), 84
Boeing Co., 41, 55, 84, 100, 120, 231, 314, 428, 430
  contract, 315, 393, 435, 447, 572
  landing system, 131
  Molab, 218
  Saturn V, 191, 447, 525, 546
  sonic boom study, 65
  strike, 437, 445, 462–463
  supersonic transport, 37, 59, 163, 214, 257, 309, 515
Boeing Scientific Research Laboratories, 8
Bogard, L/G Frank A. (USAF, Ret.), 368
Bolender, Col. C. H. (USAF), 297
Bollerud, Col. Jack (USAF), 297
Bomber aircraft, 6, 34, 84, 168, 189, 190, 290, 291, 329, 333, 381, 484, 544
  foreign, 17, 34, 87
  noise, 6, 72
  supersonic, 19, 72, 84, 150, 191, 199, 232, 264, 308, 336, 478, 482, 562, 563
Bombsight, 284
Bond, Capt. George (USN), 471
Bone, demineralization, 489, 540
Bonn, W. Germany, 178
Bonney, Walter T., 350
Boosted-Dart (sounding rocket), 236
Borisov, T., 230
Borman, L/C Frank (USAF), 208, 308, 317, 493, 500, 511, 530, 533, 557, 558, 560, 563 567
Boron compounds, 19, 170
Boston, Mass., 292, 505
Boston College, 198
Boston Sunday Globe, 2
Bostrom, Dr. C. O., 118
Bothmer, Clyde, 68
Bourdeau, Robert E., 375
Bowie, William, Medal, 193
Bowman, B/G Julian H. (USAF, Ret.), 324, 368
Boy Scouts, 109
Boyden Station, South Africa, 516
Brademas, Rep. John, 399
Bradley, Gen. Mark E. (USAF, Ret.), 477
Bradley, Gen. Omar N. (USA, Ret.), 352

Brandeis Univ., 509
Brandt, Mayor Willie, 188
Branscomb, Dr. Lewis, 169
Bravo (buoy station), 564
Brazil, 58, 307, 325, 365, 370, 513, 515, 553, 557
Brazilian Space Commission (CNAE), 220, 307, 515, 553
Breaux, John J., 179
Breguet 121 (French aircraft), 234
Brennan, Dr. John J., Jr., 12
Brevard Engineering College, 519
Brewster Flat, Wash., 320, 326, 385, 462, 463, 500, 527, 537
Brezhnev, Leonid I., 140, 313
Bridge, Dr. H. S., 330
Brigham City, Utah, 32
British Aircraft Corp., 87, 121, 215, 234, 237, 413
British Association for the Advancement of Science, 413, 492
British Broadcasting Co. (BBC), 246, 277
British European Airways (BEA), 275
British Interplanetary Society (BIS), 154
British Overseas Airways Corp. (BOAC), 237, 314, 357
Broglio, Prof. Luigi, 340
Brookhaven National Laboratory, 107, 279
Brooklyn College, 182
Brooks AFB, Tex., 305, 494, 532
Brooks, Harvey, 125, 310
Brouwer, Prof. Dirk, 416
Brown Engineering Corp., 109, 319
Brown, Gov. Edmund G., 9, 113, 406, 457
Brown, Dr. Frank A., Jr., 115–116, 495
Brown, Rep. George E., 46
Brown, Dr. Harold, 46, 158, 322, 424, 458, 498
Brown, Samuel P., 102
Brown Univ., 269
Brunn, Robert R., 312, 475
Brussels, Belgium, 76
Bryson Construction Co., Inc., 533
Buchanan, Don, 517
Buckley, Edmond C., 88, 105, 561
Bucyrus-Erie, 422
Budapest, Hungary, 272
Budget, Bureau of (BOB), 45, 77, 192
Budiardjo, Air V/A (Indonesia), 6
Bueche, Dr. Arthur M., 410
Buell, David N., 244
Buffalo, N.Y., 240
Buic. See Backup Interceptor Control.
Building Products Executives Conference, 463
Bull, Dr. Gerald V., 502
Bull Pup (missile), 315
Bulova Watch Co., 352
Bunker-Ramo Corp., 484
Buoy, 564
Buoy network, 38, 220–221
Burch, John L., 455
Burcham, Dr. Donald P., 533
Bureau of Naval Weapons (USN), 453
Burgers, Dr. Johannes M., 466

Burgess, Eric, 378
Burke, Walter F., 493
"Burner II" (launch vehicle stage), 435
Burns, Gov. Haydon, 394, 404
Burton, Dr. Russell R., 181
Buryy, V. V., 345
Bush, Robert A., 505
Butler, Ralph, 326
Bykovsky, Valentina F., 66, 182
Bykovsky, Valery F., 66, 182, 397
Byrd Station, Antarctica, 174
Byram, E. T., 212

C-1 (rocket engine), 65
C-5A (cargo transport), 20, 120, 335, 454, 481, 484, 491
C-119 (Flying Boxcar) (cargo aircraft), 116, 358
C-130 (Hercules) (transport aircraft), 52, 63, 112, 114, 124
C-130E (transport aircraft), 63
C-133B (Cargomaster) (transport aircraft), 490
C-141 (Starlifter) (cargo jet), 2, 198
C-141A (cargo jet), 41
CAB. See Civil Aeronautics Board.
Cabell, Gen. Charles P. (USAF, Ret.), 312
Cabot, Godfrey L., Award, 292
Cadiz, Ohio, 243
Cairo, U.A.R., 8, 187
Caldecote, Lord, 215
Calhoun, Dr. John C., Jr., 124
California, 9, 106, 135, 216, 405, 457
California Institute of Technology (Cal Tech), 61, 67, 206, 371, 481, 495, 507, 520
California, Univ. of, 85, 96, 103, 232, 258, 439, 448, 502, 566
California, Univ. of, at Los Angeles (UCLA), 161, 221, 431
Callahan, Fred, 179
Calle, Paul, 202
Cal Tech. See California Institute of Technology.
Calverton, N.Y., 307, 530
Calvin, Melvin, 223
Cambodia, 381
Cambridge, Mass., 12, 49, 67, 376, 382
Cambridge Redevelopment Authority, 186
Cambridge Research Laboratories. See Air Force Cambridge Research Laboratories.
Camera, 233
  Cosmos, 86
  GEMINI IV, 255
  MARINER IV, 68
  RANGER VIII, 73
  RANGER IX, 140, 149
  television, 68, 73, 132, 149
  TIROS IX, 168
  ZOND III, 392
Campbell, Joseph, 95, 216, 221
Canada, 14, 58, 75, 81, 325, 328
  aircraft, 197
  ALOUETTE, 239, 523, 527

Black Brant, 277, 290
EXAMETNET, 370
launch, 277, 527
Mid-Canada Warning Line, 165
National Defence Research Council, 91
Canadian Defence Research Board, 528
Capacitor, 311
Canary Islands, 76, 130
Canaveral Council of Technical Societies, Space Congress of, 171
Canberra, Australia, 108, 337
Cannon, Sen. Howard W., 453
Canoga Park, Calif., 30, 101, 132
Canopus (star), 164
Canopus sensor, 65, 102
Canton Island, 48, 552
Cape Kennedy, Fla. (see also Eastern Test Range and Kennedy Space Center), 57, 59, 89, 120, 162, 164, 168, 177, 186, 189, 193, 208, 242, 266, 460, 549
  construction 37, 52, 54, 80, 92, 119
  labor relations, 49
  organization, 1, 177
  strike, 39, 49, 133, 157, 447
  test, 136
  visit 86, 146-147, 188
Capella (star), 366
Caravelle (jet transport), 318
Carbon, 10, 258
Carbon dioxide, 134-135
Career Service Award, 194, 240
Cargo aircraft, 2, 17, 20, 41, 120, 198, 283, 346, 406, 481, 562
Carlson, Harry W., 15, 241
Carnarvon, Australia, 89, 269
Carnegie Institution of Washington, 97, 177, 559
Carpenter, Cdr. M Scott (USN), 128, 299, 339, 389, 402, 425, 450, 518, 540
Carroll, Thomas, 545
Carronade (rocket launching ship), 459
Carswell AFB, N. Mex., 6
Case Institute of Technology, 376, 379-380, 503
Case Laboratory for Space Engineering Research, 376
Casey, Rep. Robert, 108
Cassani, John R., 462
Castro, Raul, 134
Cat. See Clear Air Turbulence.
Catholic Univ., 134, 258
CBS. See Columbia Broadcasting System.
Ccms. See Central Cardiac Monitoring System.
Celescope, 82
Celestial mechanics, 68
Centaur (booster) (see also Atlas-Centaur), 60, 77, 103-104, 130, 197, 204, 227, 237, 374
Centaure (French sounding rocket), 352, 543
Center for European Nuclear Research (CERN), 57
Center for Sensory Aids Evaluation and Development, MIT, 90
Center for the Study of Democratic Institutions, 103, 558
Central Cardiac Monitoring System (Ccms), 532
Central Computer Facility, Slidell, La., 105
Central Connecticut State College, 536
Central Inertial Guidance Test Facility (CIGTF), 31
Central Intelligence Agency (CIA), 45, 463
Central Radio Propagation Laboratory, 229, 322-323, 326, 410
Centre National d'Etudes Spatiales (CNES), 519, 540
Centrifuge, 63, 311
CERN. See Center for European Nuclear Research.
Cernan, LCdr. Eugene M. (USN), 321, 510
Ceylon, 59
CGA Corp., 553
Chaffee, Roger, 10, 321
Chamical, Argentina, 236, 243, 370, 511
Chang, Dr. C. C., 134
Chanute, Octave, Award, 471
Chao, Dr. E. C. T., 507
Chapman, Dr. Sydney, 382
Charlotte, N.C., 37
Charyk, Dr. Joseph V., 42, 249, 254
Chebotarev, Gleb, 372
Cheryomuckhin, Aleksey, 372
Chicago, Ill., 6, 83, 159, 216, 222, 236, 278
Chicago Bridge and Iron Co., 26, 159
Chicago, Univ. of, 32, 245, 473
Chieo, Calif., 47, 78
Childe, V. Gordon, 50
China-Burma-India World War II Service Group, 372
China, Communist, 398, 405, 553
  meteorite, 550
  nuclear bomb, 206-207, 232, 236, 243, 248, 567
  nuclear test, 178
  rocket program, 260-261
  U.S. aircraft shot down, 504
China Lake, Calif., 66
Christensen, E. E., 382
Christensen, Charles H., 414
Christiakov, Dr. N. I., 545
Christian Science Monitor, 4, 39, 105, 177, 312, 475, 533
Chrysler Corp., 94, 96, 115, 123, 129, 244, 250, 319, 378, 434, 490, 543
Chubb, T. A., 212
Churchill, Manitoba, 91
Churchill Research Range, Canada, 321
Churchill, Sir Winston, 41
Cibola National Forest, 514
CIGTF. See Central Inertial Guidance Test Facility.
Cincinnati, Univ. of, Observatory, 202-203

Civil Aeronautics Board (CAB), 118, 131, 222, 297, 374, 501
Civilian-Military Liaison Committee (CMLC), 351
Clarion River (rocket launching ship), 459
Clark AFB, Philippines, 139
Clark, Dr. Barry, 371
Clark, David, Co., 407–408
Clark, Evert, 314, 316, 335, 424
Clark, Dr. John F., 188, 343, 375
Clark, Ramsey, 416
Clark Univ., 419, 478
Clarke, Arthur C., 22, 244
Clarson, R. E., Inc., 37, 52
Clear Air Turbulence (Cat), 161
Clearinghouse for Federal Scientific and Technical Information, 425
Clemence, Gerald M., 82
Clermont (NASA tugboat), 483
Cleveland Natural Science Museum, 406
Cleveland, Ohio, 545
Cleveland *Plain Dealer*, 291
Clock, atomic, 284
Cloud photographs, 25–26, 81, 124, 163
Cloudcroft, N. Mex., 280
"Cluge" (photographic device), 330
CMLC. See Civilian-Military Liaison Committee.
CNAE. See Brazilian Space Commission.
CNES. See Centre National d'Etudes Spatiales.
CNIE. See Comisión Nacional de Investigaciones Espaciales.
CNRS. See National Center for Radioactivity Research.
Coating, 98, 319
Cochran, Jacqueline, 134, 246
Cocoa Beach, Fla., 405
Coesite, 96
Cohen, William, 98
Colburn, Robert, 456–457
Coleman, Sidney, 262
College Park, Md., 368
Colleges. See Universities.
Collier Trophy, 486,
Collier, W. A., 364
Collins, Maj. Michael (USAF), 308, 511
Collins Radio Co., 92
Colorado Springs, Colo., 286
Colorado, Univ. of, 483, 488
Columbia Broadcasting System (CBS), 240, 282, 444
Columbia Univ., 223,
Nevis Cyclotron Laboratory, 279. 572
Columbus Association, Genoa, 448–449
Columbus Prize, 448–449
Combustion Engineering, 18
Comet, 67, 452, 502, 509, 513, 516, 518
Comisión Nacional de Investigaciones Espaciales (CNIE), 236, 511
Comissão Nacional de Atividades Espaciais (CNAE). See Brazilian Space Commission.

Commerce, Dept. of, 174, 425, 517, 565
Awards Program, 71
metric system study, 328, 395
patent system, 299
Committee on Space Research (COSPAR), 225, 226, 227
Commoner, Dr. Barry, 3
Commonwealth Scientific and Industrial Research Organization (Australia), 321
Communication (see also Communication satellite systems), 4, 121, 201, 207, 440
blackout, 126, 143, 146
deep space, 1, 377, 434
global, 30, 58, 87, 172, 246, 314–315, 361, 381, 393, 428, 440, 555
international, 51, 76, 186, 204
laser use in, 73, 434, 538
NASA facilities, 75, 87, 107, 136, 175, 496, 502, 520
satellites, 1, 22, 57, 65, 81, 95, 101, 105, 114, 136, 147, 172, 176, 197, 229, 232, 236, 385, 440–441
system, 2, 249, 492
balloon-satellite, 41-42
military, 27, 102, 441
transmissions via, 1, 22, 41, 180, 197, 214, 246, 250, 268, 273, 277, 282, 449–450, 478, 484, 530, 555
S-band method, 92, 363
tests, 110, 136, 172, 180, 197, 214, 436
Communications Satellite Act, 186, 249
Communications Satellite Corp. (ComSatCorp), 69, 128, 272
contract, 31, 51, 81, 283, 381, 412, 527, 549, 555, 568
criticism, 51, 146
EARLY BIRD, 42, 95, 101, 122, 128, 172, 176, 180, 184, 193, 197, 214, 240, 246, 260, 267, 271, 282, 289, 300, 303, 334, 344, 451, 571
tariff, 245, 256-257, 277, 282, 343
ground station, 197, 201, 228, 232, 278, 314, 320, 326, 343, 344, 359, 360, 385, 393, 412, 462, 463, 503, 527
international applications, 58, 250, 268, 278, 300, 322
management, 66, 169, 204, 318, 479
military applications, 31, 51, 80, 93
rates, 245, 256–257, 277, 282, 343
satellite program (see also EARLY BIRD), 7, 30, 80, 95, 101, 122, 229, 249, 254, 303, 358-359, 455, 555, 568
services, 186, 289, 334, 451, 495, 500, 532
test, 172, 180, 197
Communications satellite system, international (see also Communications Satellite Corp.), 2, 7, 22, 58, 81, 93, 121–122, 128, 147–148, 412, 428, 462, 464, 465, 480, 495
military use, 31, 51, 93, 102
Composites, structural, 274
Compressed Gas Association, 21

Computers, 131, 175, 228, 275, 555
  foreign use, 110, 119
  NASA, 105, 177, 295, 337, 485, 542, 561
  spacecraft, 266, 275, 293, 295, 522, 542
  universities, 9, 100, 108, 193
  use of, in Federal Government, 13, 14, 280
ComSatCorp. See Communications Satellite Corp.
Conable, Rep. Barber B., Jr., 46
Concord, N.C., 520
Concorde (U.K.-France) supersonic transport, 73, 204, 206, 237, 316, 319, 337, 417, 479, 490
Confer, Harold E., 189
Conference on Aerospace Engineering, 129
Conference on Aircraft Operating Problems, 223
Conference on Civilian and Military Uses of Aerospace, 16
Conference on Selected Technology for the Petroleum Industry, 545
Conference on the Peaceful Uses of Space, 251, 252, 253
Congo Brazzaville, 76
Congress, 218, 351, 557
  conference committee, 366
  Joint Committee on Atomic Energy, 61, 63, 107
  Joint Economic Committee, 161
  NASA budget, 28, 29, 61, 74, 75, 76, 81, 100, 108, 109-110, 114, 139, 141, 192
  report to, 39-40, 69, 246, 487
Congress, House of Representatives, 72, 134, 397, 446
  bills introduced, 357, 422
  bills passed, 218
  Committee on Appropriations, 232, 275, 276, 438, 481
    Subcommittee on Dept. of Defense Appropriations, 158, 160, 169, 173
    Subcommittee on Independent Offices, 169, 172, 173, 276
  Committee on Armed Services, 86, 377, 419, 481
    Special Investigations Subcommittee, 216, 221
    Subcommittee on Real Estate, 257
  Committee on Education and Labor, 530
  Committee on Foreign Affairs, 34
  Committee on Government Operations,
    Subcommittee on Foreign Operations and Government Information, 144
    Subcommittee on Military Operations, 267
  Committee on Interstate and Foreign Commerce, 292
  Committee on Rules, 422
  Committee on Science and Astronautics, 40, 46, 89, 123, 164, 171, 198, 199, 204, 242, 255, 271, 290, 419, 465, 478, 509
    hearings 33, 74, 75, 76, 79, 81, 89, 90, 119
    Panel on Science and Technology, 175
    seminar, 33
    Subcommittee on Advanced Research and Tracking, 105
    Subcommittee on Manned Space Flight, 404, 567
    Subcommittee on NASA Oversight, 501, 502
    Subcommittee on Science, Research, and Development, 205
    Subcommittee on Space Sciences and Applications, 107, 108, 125, 130
  Committee on Ways and Means, 434
Congress, Senate, 40, 84, 249, 270, 296, 303, 324, 327, 453, 478, 487
  bills introduced, 7, 51, 117, 202, 216, 276, 287, 487
  bills passed, 107, 264, 287, 325
  Committee on Aeronautical and Space Sciences, 24, 36, 40, 46, 62, 110, 112, 116, 164, 242, 397, 398
  Committee on Armed Services, 159, 331
  Committee on Commerce, 129, 204, 328, 438
    Subcommittee on Aviation, 114, 118, 231
  Committee on the Judiciary, 7
    Subcommittee on Patents, Trademarks and Copyrights, 264
Congressional Record, 32, 89, 315, 320, 399, 422, 487, 488
Connor, John T., 439
Conrad, LCdr. Charles, Jr. (USN), 415, 433
  GEMINI V flight, 58, 317, 339, 382, 385, 387, 391, 402, 412, 415, 420, 422, 500
  Gemini VIII flight, 444
  goodwill tour, 435, 437, 438-439, 441, 442, 446, 448, 452
  honors, 403, 431, 449
  news conference, 339, 421
  physical examination, 404, 405, 410, 411
  training, 10, 334
  Washington, D.C., visit, 433, 437
"Conrad" (radio navigation aid), 448
Considine, Bob, 300
Consolidated Systems Corp., 68, 476
Constantine, King of Greece, 440
Consultants and Designers, Inc., 309
Conte, Rep. Silvio, 32
Continental Airlines, 237
Contract (see also under agencies, such as NASA, USAF, etc.), 141, 572
  cost-plus-award-fee, 70, 94, 195, 308, 368, 447, 474, 561
  cost-plus-fixed-fee, 80, 334, 447, 484
  cost-plus-incentive-fee, 5, 66, 91, 109, 383

Contract—Continued
  fixed price, 73, 228, 240, 272, 274, 283, 297, 312, 454, 502
  geographical distribution of, 106, 115, 134
  incentive, 5, 40, 49, 77, 91, 139
  letter, 91, 121
  military, 5, 121, 240, 264, 265, 274, 283, 297, 302, 447, 448, 450, 453, 454, 485, 492, 536, 565
  space, 141, 234, 239, 297, 319, 450, 494, 562, 564
  university, 4, 56, 149, 182, 274, 311, 370, 562
Control Data Corp., 334, 375
Convair 990 A (jet transport), 258
Cooby Creek, Australia, 543
Cooke, Richard P., 241
Cooney, Capt. James (USAF), 213
Cooper, Maj. Gordon L. (USAF), 415, 433
  GEMINI V flight, 58, 317, 334, 339, 382, 385, 387, 391, 402, 412, 420, 422, 500
  goodwill tour, 435, 437, 438-439, 441, 442, 446, 448, 452
  honors, 403, 431, 449
  news conference, 339, 421
  physical examination, 404, 405, 410, 411, 415
Cooper's Island, Bermuda, 175
Coordinating Research Council (CRC), 18
Coral Gables, Fla., 30
Coralie (rocket engine), 495-496
Corliss, William R., 47
Corman, Rep. James C., 223
Cornell Aeronautical Laboratory, 524
Cornell Univ., 1, 187, 194, 206, 229
Corona, 10
Coronograph, 107
Corps of Engineers (USA). See Army Corps of Engineers.
Corpus Christi, Tex., 363, 544
Correale, James V., 470
Corrosion detector, 274, 381
Cortright, Edgar M., 102, 356
Cosmic dust, 342, 502
Cosmic radiation, 56, 211, 372, 431, 554, 555
Cosmonaut, 19, 31, 45, 98, 139, 140, 141, 151, 156, 323, 413, 512, 514, 528, 548, 559
  death, 173, 505
  extravehicular activity, 132, 135, 138, 140-141, 154, 571
  interview, 161, 270, 419, 480, 491, 568
  visits, 473, 491
Cosmos (U.S.S.R. satellite program), 86
COSMOS III, 217
COSMOS IV, 217
COSMOS XLI, 217
COSMOS L, 125
COSMOS LII, 12
COSMOS LIII, 41
COSMOS LIV, 86
COSMOS LV, 86
COSMOS LVI, 86
COSMOS LVII, 86, 124-125, 159
COSMOS LVIII, 94
COSMOS LIX, 109
COSMOS LX, 120, 571
COSMOS LXI, 123, 238
COSMOS LXII, 123, 238
COSMOS LXIII, 123, 238
COSMOS LXIV, 152
COSMOS LXV, 189
COSMOS LXVI, 221, 238
COSMOS LXVII, 248
COSMOS LXVIII, 282
COSMOS LXIX, 296
COSMOS LXX, 311
COSMOS LXXI, 333
COSMOS LXXII, 333
COSMOS LXXIII, 333
COSMOS LXXIV, 333
COSMOS LXXV, 333
COSMOS LXXVI, 345
COSMOS LXXVII, 364
COSMOS LXXVIII, 377
COSMOS LXXIX, 397
COSMOS LXXX, 413
COSMOS LXXXI, 413
COSMOS LXXXII, 413
COSMOS LXXXIII, 413
COSMOS LXXXIV, 413
COSMOS LXXXV, 421
COSMOS LXXXVI, 441
COSMOS LXXXVII, 441
COSMOS LXXXVIII, 441
COSMOS LXXXIX, 441
COSMOS XC, 441
COSMOS XCI, 448
COSMOS XCII, 479
COSMOS XCIII, 483
COSMOS XCIV, 492
COSMOS XCV, 504
COSMOS XCVI, 525, 571
COSMOS XCVII, 526
COSMOS XCVIII, 527
COSMOS XCIX, 546
COSMOS C, 556
COSMOS CI, 560
COSMOS CII, 565
COSMOS CIII, 565
COSPAR. See Committee on Space Research.
Cost Reduction and Management Improvement Seminar, 248
Cotton, Col. Joseph (USAF), 336
Cotton, Paul E., 368
Coughlin, William J., 101, 144, 180, 203, 215, 235, 501, 524
Council of the National Economy, U.S.S.R., 119
Courter, Robert F., Jr., 485
Cousteau, Capt. Jacques-Yves, 280, 475
Couve de Murville, Maurice, 502
Covell, Charles, 425
Cowan, Dr. Clyde, 258
Cowen, Robert C., 4

INDEX

Crabhill, Donald E., 192
Crane, Les, Show (TV), 27
Crawler-transporter, 381, 422, 481, 517
CRC. See Coordinating Research Council.
Cressman, Dr. George P., 420, 471
Crews, Maj. Albert H. (USAF), 514
Crimean Astrophysical Observatory (U.S.S.R.), 462
"Criteria for Federal Support of Research and Development," 10
Cromley, Ray, 541
Cronyn, Willard M., 137
Crossfield, A. Scott, 134
Cryomagnet, 256
CTA-102 (radio source), 180, 182, 236
Cuba, 1, 134, 398, 405, 552
Cubic Corp., 457
Culbertson, Philip E., 426
Culver City, Calif., 39
Cumberland, U.K., 365
Cunningham, Alfred Austell, 555
Cunningham, R. Walter, 125, 321, 341
Curtiss-Wright Corp., 275
Cushman, Ralph E., 553
CV-7A (Stol transport), 197
Cybernation, 103
Czechoslovak Academy of Science, 157

D-1 (French satellite), 456
D-1B, 456
D-558 (research aircraft), 350
Daddario, Rep. Emilio Q., 205
Daily Express (London), 8
Dallas, Tex., 132, 186
Dampier (survey vessel), 525
Dana, William H., 347, 504
Daniel, Orville H., 211
Daniel, Maj. Walter F. (USAF), 213
Data, 305
Data acquisition system, 174, 177
Data processing, 105
D'Auitolo, Charles T., 376
Davis, L/G W. A. (USAF), 66
Davenport Times-Democrat, 89
David, Heather M., 450
Davies, David Arthur, 162
Davis, L/G Leighton I. (USAF), 563
Davis-Monthan AFB, Ariz., 14, 84
Day, LeRoy E., 479
Dayton, Ohio, 147
DC-8-61 (jet liner), 189
DC-9 (jet liner), 14, 94
Dearborn, Mich., 243
Dearing, Waldo H., 505
Debus, Dr. Kurt H., 183, 331, 333
Deep Space Net communications system, 364, 377
Deep Space Planetary Probe System, 197
Deer Valley, Ariz., 176
Defelice, J., 191
Defender, Project, (ARPA), 531
Defense Communications Agency, 66, 283, 441

Defense, Dept. of (DOD) (see also USA, USAF, USN), 5, 14, 45, 131, 223, 301, 360, 398, 473
  aircraft, 63, 232, 544
  budget, 27, 109, 158, 159
  communications satellite system, 1, 28, 51, 80, 82, 283, 319, 571
  contract, 5, 14, 51, 120, 131, 157, 190, 197, 228, 283, 474
  cooperation, 174, 486
  NASA, 1, 29, 45, 68, 74, 77, 88, 102, 111, 123, 182, 203, 207, 268, 298
  criticism, 63, 153, 168
  expenditures, 125, 218, 232
  Manned Orbiting Laboratory, 267, 398, 401, 514
  missiles, 16, 86
  personnel, 458
  R&D, 99, 153
  space projects, 106, 123, 217, 268, 384, 531, 541
Defense Electronic Products, 283
Defense Satellite Communications System, 66
Defense Supply Agency (DSA), 207, 294
DeFrance, Dr. Smith J., 443, 462
DeGaulle, President Charles (France), 205, 291
Delormé, Jean, 202
Delta (booster), 25, 28, 74, 380, 397
Delta, Thrust Augmented (TAD) (booster), 172
Delta, Thrust Augmented Improved (booster), 507, 553
Denmark, 170, 466
Denver, Colo., 59, 62
Denver Post, 210
Denver, Univ. of, 157, 424
De Orsey, C. Leo, 210
Dept. of Scientific and Industrial Research, U.K., 110, 164
Derring, Eldridge H., 27
Detroit, Mich., 25
Deuterium, 269
Diamant (French booster), 98, 113, 229, 272, 436, 456, 526
Dicke, Dr. Robert H., 261, 401
"Dictionary of Scientific Biography," 110
Dietz, David, 244
Dimona (nuclear reactor), 122
Disher, John H., 370
Distinguished Civilian Service Award (DOD), 223
Distinguished Lecture Series (Metropolitan Washington Board of Trade), 486
Distinguished Service Medal (NASA), 155, 567
Dixon, Dr. Franklin P., 174
DKfW. See German Space Research Commission.
DMS, Inc., 109
Dna (nucleic acid), 323
Dobrynin, Ambassador Anatoli F., 544

Docking, 444, 510
Documentation, Inc., 334, 368
DOD. See Defense, Dept. of,
Dominick, Sen. Peter H., 488
Donn, Dr. Bertram, 182
Donn, Dr. William I., 182
Donner, Frederick G., 169, 204
Doolittle, Gen. James H. (USAFR), 134, 540
Dornberger, Walter R., 210, 499
Douglas Aircraft Co., Inc., 8, 446
  air bus, 314
  contract, 80, 120, 296, 397, 572
  DC-8, 189
  DC-9, 14, 94
  F-5D, 41
  Manned Orbiting Laboratory, 396
  Missile and Space Systems Div., 191
  S-IVB stage, 209, 296, 426, 533
  Saturn V test, 182
  studies, 63
Douglas, Col. William K. (USAF), 471
Douglas, Justice William O., 186
Dragon (French sounding rocket), 413
Drake, Frank, 1
Draper, Dr. Charles S., 58, 94, 390, 429, 431
Draper, Henry, Medal, 203
Drop test, 110
Drummond, Dr. A. J., 31
Dryden, Dr. Hugh L., 122, 127, 146, 158, 409, 532, 561
  awards to, 73, 283, 547
  death, 534, 539, 540, 565, 569
  international cooperation, 111
  space program, 75, 119, 133, 162, 251, 346, 426, 458, 512, 546
DSA. See Defense Supply Agency.
Dubinin, Nikolay P., 372
DuBridge, Dr. Lee A., 263
Duke Univ., 198
Dulk, George A., 1
Dulles International Airport, 164, 292
Dupree, Prof. A. Hunter, 281
Durham, Franklin P., 306
Dutton, Richard E., 178
Dwarf star, 70
Dwight, Capt. Edward J., Jr. (USAF), 269
Dyce, Dr. Rolf H., 194
Dyson, Freeman J., 321

E-22 (U.S.S.R. jet trainer), 245-246
E-166 (U.S.S.R. aircraft), 213
E-266 (U.S.S.R. aircraft), 189
Ea-1 (U.S.S.R. helicopter), 372
Eaker, L/G Ira C. (USAF, Ret.), 122, 309, 329, 491
Earhart, Amelia, 134
EARLY BIRD I (ComSatCorp communications satellite), 42, 122, 176, 225, 303, 434, 502, 568, 571
  channels, lease of, 271, 334
  launch, 101, 172
  license, 272, 334
  orbit, 176
  performance, 180, 184, 197, 214
  rates, 245, 256-257, 277, 289, 343
  use of, 95, 128, 240, 246, 260, 267, 282, 289, 300-301, 340, 360, 436, 484, 555, 562
Earth
  age, 182-183
  gravity, 124
  magnetosphere, 81
  Mohole, Project, 37-38
  origin, 37-38
  photograph, 101, 142
  shape, 174
  structure, 37-38
Earthquakes, 37-38, 523, 525, 529
"East Meets West" (British TV program), 147-148
Eastern Air Lines, 189, 551
Eastern Test Range (ETR) (see also Cape Kennedy and Kennedy Space Center), 12, 290, 458, 490
  contract, 66, 340
  launch, 87, 242
    Asset, 88
    Blue Scout, Jr., 272
    booster, 64-65, 219-220, 260, 288, 477, 559
    Saturn I, 71-72, 247, 357
  Fire II, 244, 254
  Gemini spacecraft, 103
    GT-2, 21
    GT-3, 145
    GT-4, 265
    GEMINI V, 387
    GEMINI VI, 489, 547, 551
    GEMINI VII, 537
  Polaris, 513
  PIONEER VI, 553
  satellite, 25, 26, 53, 64-65, 71, 72, 172, 247, 310, 341, 357, 396
    Explorer, 257, 507, 521, 528
  test, 88, 110, 136
Eaton, William W., 299
Ebony magazine, 269
Eccles, Sir John, 85
Echo (communications satellite), 233
ECHO I, 376
ECHO II, 41, 76
Eclipse
  lunar, 8
  solar, 126, 258, 259
Economic Club of Detroit, 70-71
Economichesky Gazetta, 591
The Economist, 121-122
Ecumenical Council, 448
Eddington Medal (Royal Astronomical Society), 43
Edelson, Dr. Burton I., 22
Edmonson, Dr. Frank K., 95
Education, 103
  computer, use of, in, 9

INDEX 637

space, impact of, on, 17
space science courses, 134
television, use of, in, 57
U.S.S.R., 147
Edwards AFB, Calif., 41, 94, 210, 300
  Aerospace Research Pilot School, 246, 399
  flights, 72-73, 191, 206, 284, 287, 293, 302, 308, 336, 351, 453, 478, 562, 563
    record, 213
  test, 322, 414, 449
Eggers, Dr. Alfred J., Jr., 50, 455, 470–471
Eglin AFB, Fla., 240, 511
Egypt. See United Arab Republic.
Ehricke, Dr. Kraft A., 498
Eisenhower, President Dwight D., 163, 232, 445
Eisele, Donn F., 10
Eklund, Sigvard, 548
El Centro, Calif., 112
El Diablo (meteorite crater), 290
El Segundo, Calif., 64
ELDO. See European Launcher Development Organization.
Electric propulsion, 16, 44, 116–117
Electro-Mechanical Research, Inc., 309
Electro-Optical Systems, Inc. (EOS), 5, 255, 348, 375
Electron, 118, 139, 167, 169, 337, 438
Electron (U.S.S.R. satellite), 217
ELECTRON I, 360
ELECTRON II, 360
Electronic clock, 352
Electronic Industries Association, 124
Electronics, 503, 569
  research, 2, 24, 49
Electronics, 87
Electronics Research Center (ERC) (NASA), 2, 11, 49, 186, 207, 303, 349, 382, 468, 593
Electrophylic gas, 127
Ellingson, Col. Harold V. (USAF), 494
Ellington AFB, Tex., 97, 275, 411
Ellis, Lt. Frank K. (USN), 495, 535
Elms, James C., 368
Ely, Eugene Burton, 555
Emeraude (French rocket stage), 98, 113, 229
Emme, Dr. Eugene M., 176.
Empire State Building, 124
Engine
  aircraft, 32, 121, 283, 519
  cryogenic, 398
  electric, 30, 55, 116–117, 519
  gas turbine, 32, 482
  hypergolic, 65, 186, 474
  hypersonic, 169, 203, 349, 464, 593
  ion, 5, 116, 167, 172, 243, 255, 349, 472
  jet, 18, 41, 84, 254, 486
  liquid fuel, 33, 65, 92, 127, 219, 315
    fluorine-oxygen, 132
    hydrogen, 31, 141, 358, 365, 526
    hydrogen-oxygen, 81, 104, 322, 375
  ramjet, 169, 203, 464, 543
  rocket, 2, 13, 24, 27, 32, 40, 42, 49, 52, 60, 64, 81, 95, 107, 108, 112, 116, 157, 165, 179, 180, 192, 219, 247, 269–270, 291, 319, 320, 321, 339, 373, 496
  supersonic transport, 214, 257, 311
  vernier, 119
Engineer of the Year, 98
Engineers
  awards, 58, 98, 156
  education, 9, 57
  women, 92
England. See United Kingdom.
Engle, Capt. Joseph H. (USAF), 13, 49, 197, 255, 284, 302, 331, 374, 417, 476
Engstrom, Elmer W., 209
Environment (see also Space environment; Weightlessness), 59–61
  atmospheric pressures, 115–116, 134–135, 313
  heart research, 9
  oxygen tests, 128
  temperature, 116, 313
Environment, hazards of, 59, 104, 132–133, 372
  acceleration, 19, 23, 181, 191
  isolation, 66, 121, 170
  motion sickness, 70
  oxygen, 63, 135, 158, 190, 494
  radiation, 7, 59, 167, 202
  space cabin, simulation, 129, 173, 190, 550
  space, simulation, 36, 159, 181
Environmental Science Services Administration (ESSA), 229, 326, 386, 565
Eniwetok, 48
Eniwetok Lagoon, 511
EOS. See Electro-Optical Systems.
Eppley Laboratory, Newport, R.I., 31
Epstein, Julius, 300
ERC. See Electronics Research Center.
Escape system, 66, 301, 513
Eshkol, Premier Levi (Israel), 6
ESRO. See European Space Research Organization.
Esro 1 (ESRO satellite), 182
Esp. See Extravehicular Support Pack.
ESSA. See Environmental Science Services Administration.
Esso Research and Engineering Co., 87
ETR. See Eastern Test Range.
Europa I (ELDO booster), 365, 496
Europe, 214, 215, 224, 267, 354, 361, 534, 563
European Broadcasting Union, 245
European Launcher Development Organization (ELDO), 68, 76, 143, 365, 418, 496
European Organization for Nuclear Research, 95
European Post and Telecommunications Congress, 277
European Space Research Organization (ESRO), 68, 76, 80, 182, 276, 353, 354, 375, 434, 543, 550

Eurospace, 202, 205, 209, 215
Eurovision, 148
Eva. See Extravehicular activity.
Evans, B/G Harry L. (USAF), 406
Evans, Dr. John W., 223
Evanston College, 498
Evensen, Jean, 434
Everett, W. L., 204
Evvard, Dr. John C., 16, 519
Exactel Instrument Co., 93
EXAMETNET. See Inter-American Experimental Meteorological Sounding Rocket Network.
Exceptional Scientific Achievement Medal (NASA), 155
Exceptional Service Medal (NASA), 155, 431
Exhibit, 156, 173, 202, 292
Exobiology, 344
Explorer (U.S. satellite), 104, 154, 370
EXPLORER I, 42
Explorer II (balloon), 513
EXPLORER XVI, 311, 376
EXPLORER XVIII, 193
EXPLORER XX, 76
EXPLORER XXII, 24, 27, 36, 67, 276
EXPLORER XXIII, 233, 311, 376
EXPLORER XXVII, 207, 276
EXPLORER XXVIII, 257
EXPLORER XXIX, 507, 521, 528
EXPLORER XXX, 521, 571
EXPLORER XXXI, 527, 546
Explosion
  meteorite, 2
  nuclear, 2, 18, 39–40, 81, 232, 236, 242, 243, 248, 258
Extraterrestrial life, 4, 17, 36, 47, 61, 168, 180, 182, 214, 246, 571
Extravehicular activity (Eva), 141, 152, 172, 190, 208, 317, 339, 500, 510, 531
  Leonov, Aleksey, 132, 135, 138, 140, 153, 155, 171, 216, 278, 571
  White, Edward H., 248, 255, 256, 266, 269, 273, 286
Extravehicular Support Pack (Esp), 522
Eyeball Mark One (navigational instrument), 318
"Eyewitness to Space" (exhibit), 202
Eyring, Henry, 488

F-1 (engine), 100, 112, 178, 188, 219, 275, 291, 319
F-4 (Phantom II) (fighter aircraft), 52, 63, 510
F-4C (Phantom), 287
F-5A (supersonic aircraft), 438
F-5D (aircraft), 41
F-104 (Starfighter) (aircraft), 178
F-111 (supersonic fighter), 14, 17, 19, 20, 52, 90, 109, 121, 172, 234, 474, 500, 537, 544, 549
F-111A, 6, 121, 179, 180, 183, 537

F-111B, 183, 241, 307, 500, 537
FAA. See Federal Aviation Agency.
FAI. See Fédération Aéronautique Internationale.
Fairbanks, Alaska, 76, 276
Fairchild Hiller Corp., 309
Fairhall, Minister of Supply Allen, 520, 543
Falcon (fanjet aircraft), 131
Falmouth, Mass., 341
Fang, Dr. P. H., 484
Fausst. See French-Anglo-United States Supersonic Transport.
FCC. See Federal Communications Commission.
Federal Aviation Agency (FAA), 6, 41, 49, 131, 164, 188, 194, 259, 345, 386, 488
  Administrator, 205, 237, 303, 309
  air traffic control, 152, 353, 500, 513
  aircraft certification, 41, 188
  airports, 198, 415
  award, 186, 302
  contract, 8, 164, 176, 214, 345, 513, 550
  cooperation, 40, 131, 231
  regulations, 5, 144, 296, 455
  statistics, 49, 152, 198, 475
  test, 176
    sonic boom, 48, 65, 274
  transport, supersonic, 25, 48, 58, 73, 210, 214, 297, 309, 311
    design and development, 8, 34, 59, 163, 214, 297, 492, 515
Federal Civil Service Employee of the Year, 14
Federal Communications Commission (FCC)
  applications to, 186, 271, 303
  approvals, 228, 232, 289, 463
  briefs filed with, 495, 500, 532
  ComSatCorp, 249, 344
  contract, 31, 51, 80, 81
  ground stations, 201
  requests to, 268, 271, 277, 303, 445, 451
Federal Electric Co., 308
"Federal Funds for Research, Development, and other Scientific Activities" (report), 99
Federal German Ministry for Scientific Research (BMWF), 339
Federal Housing and Home Financing Agency, 495
Federal Mediation and Conciliation Service, 523, 525–26
Federal Urban Renewal Administration, 186
Fédération Aéronautique Internationale (FAI), 18, 213, 561
Fedotov, A. N., 189
Fedynskiy, V. V., 345
Fellowship of Reconciliation, 103
Fels Planetarium, 19

Feoktistov, Konstantin P., 548
Ferguson, Gen. James (USAF), 160, 369
Fermi, Enrico, Institute of Nuclear Studies, 32
Fermi, Enrico, Medal, 17
Feynman, Richard, 487
Filshner Ice Shelf, 532
Findlay, John W., 109
Fine, Dr. Samuel, 168
Finger, Dr. Harold B., 3, 116, 305, 520
Finley, Lt. John L. (USN), 514
Finney, John, 559
Fire II (reentry test), 244, 570
Firefly (life detection instrument), 107
Fireman, E. L., 191
Fish Bowl, Project, 337
Fisher Construction Co., 175
Flagstaff, Ariz., 509
Flare, solar, 126, 163, 224
Flax, Dr. Alexander, 36
Fleming, John Adam, Award, 193
Flemming, Arthur S., Award, 57
Flight Research Center (FRC) (NASA), 48, 163, 360, 518
  aircraft research, 73, 169, 464, 524, 569
  award, 82
  contract, 105, 209
Flight Safety Foundation, 9, 176, 186, 557
Florida, 460
Florida Research and Development Center, 137
Florida, Univ. of, 353, 460
Flox (liquid flourine and liquid oxygen), 132
Flying belt, 485
Flying saucers, 25, 38, 52, 324, 326, 366, 374, 438
Fog dispersal, 387
Fokker (aircraft), 131
Foothill Jr. College, Calif., 114
Ford Foundation, 416
Ford, Henry, 331
Foreign Affairs Journal, 45
Forrestal, James, Memorial Award, 153
Ft. Benning, Ga., 179
Ft. Campbell, Ky., 473
Ft. Churchill, Canada, 167, 439, 510, 511, 525
Ft. Monmouth, N.J., 73
Ft. Wingate, N.Mex., 514
Fortune magazine, 361
Fortune, W. C., 275
Foss, Dr. Ted, 10
Foster, Dr. John S., Jr., 424, 458
Foster, William C., 134
Fowler, William A., 206
FR-1 (French satellite), 437, 439, 449, 540, 544, 571
France, 38, 381, 519, 552
  agreement, 205, 254
  aircraft, 234
  supersonic transport "Concorde" (France-U.K.), 48, 73, 205, 417, 479
  Atomic Energy Commission, 329
  booster, 98, 113, 229, 272, 436, 456, 467, 526
  isolation test, 66, 121
  launch, 113, 98, 229, 413, 467, 478, 519, 525, 526, 540
  Ministry of Scientific Research, 27
  satellite, 437, 525, 526, 527, 540, 544, 570, 571
  sounding rocket, 413, 478
  space program, 229, 272, 320, 486, 564
  tracking, 14, 76
Frangible Arcas (sounding rocket), 97
Frank, Paul, 12
Franklin Institute, 23, 399
FRC. See Flight Research Center.
Freche, John C., 56
FREEDOM 7, 461
Freeman, Faith L., 25
Freeman, Capt. Theodore C., Memorial Library of Astronautics, 25
Freitag, Capt. Robert F. (USN, Ret.), 228
French-Anglo-United States Supersonic Transport (Fausst) meeting, 48
French, L/C Frederic C. (USA), 505
French Guiana, 38, 320, 559
French National Weather Center, 282
Friedl, George Jr., 90, 91, 264
Friedman, Dr. Herbert, 196, 206
Friendly, Fred W., 387
Friendship (Dutch aircraft), 318
FRIENDSHIP 7, 82
Fritz, Dr. Sigmund, 71
Frutkin, Arnold W., 226, 544
Fryklund, Richard, 221, 250
Fubini, Dr. Eugene G., 301, 535
Fuel (see also Propellant; Propulsion), 112, 127, 132, 135–136
  hydrogen, 81, 171
  hypergolic, 65, 186, 474
  liquid, 27, 31, 65, 127, 132, 141, 315, 365, 474, 526
  nuclear, 269
  solid, 24, 27, 81, 107, 210, 219, 223, 269, 288, 297, 435, 449, 550, 562
  thorium, 135, 565
Fuel cell, 87, 113, 207, 365, 391, 410, 521, 523, 538,
Fuji (icebreaker), 523
Fulbright, Sen. J. W., 167
Fulgham, Maj. Dan (USAF), 112
Fulton, Rep. James G., 218
Fulton, Langdon H., 156
Funk, M/G Ben I. (USAF), 290, 490
Future Space Programs panel, 503

Gagarin, Col. Yuri A., 273, 275, 290, 292, 314, 568
Gainesville, Fla., 159
Galaxy, 223, 277, 281
"Galileo" (jet aircraft), 480
Galitskaya, E. B., 318
Gallup Poll, 343
Galveston Bay, Tex., 97, 394
Gamma Cassiopeia (star), 319
GAO. See Government Accounting Office.
Gape, Project (General Aviation Pilot Education), 345
Garrett Corp., 169
Garriott, Owen K., 299, 300
Garrison, Lindley M., 165
Gatland, Kenneth, 154
Gatv. See Gemini Agena Target Vehicle.
Gault, Donald E., 555
Gazenko, Dr. O. Z., 225
GCA. See Geophysics Corp. of America.
GEAV. See Guidance Error Analysis Vehicles.
Geiger-Mueller tube, 102
Gemini (program), 29, 79, 89, 141, 171, 173, 177, 214, 289, 325, 567
  criticism, 142, 265
  development of, 116, 557, 570
  experiments, 21, 146
  management, 177, 297, 314, 479
  plans for, 68, 79, 109, 146, 234, 248, 291, 339, 346
  tests, 22, 23, 65, 66, 79, 110, 112, 114, 116
  tracking stations, 107, 130
Gemini Agena Target Vehicle (Gatv), 487, 509, 524, 571
Gemini GT-3 (flight), 15, 68, 110, 130, 142, 148, 151, 152, 172, 177
  astronaut honors, 155, 156, 159
  experiment, 144, 145, 146, 483
  landing, 145, 193, 214
  launch, 145
  mission simulation, 110, 116, 136
  spacecraft, 3, 116, 130, 145, 177, 186, 193, 289
Gemini GT-4 (flight), 177, 190, 196, 208, 229, 245, 254, 267, 268, 270, 276, 277, 288, 308
  astronaut honors, 283, 286
  experiment, 483
  extravehicular activity, 265, 269, 273, 570
  landing, 267, 275, 293
  launch, 265-67
  news coverage, 196, 276, 288
  plans for, 168, 229, 245, 248, 256, 260, 288
  rendezvous, 265, 322
  spacecraft, 190, 248, 254, 255, 260, 275, 308, 322, 331, 366
GEMINI V (flight), 412, 420, 421
  achievement, 395, 402, 452
  criticism, 394, 427
  launch, 387-390
    postponed, 385
  medical aspects, 377, 410, 412, 421, 422, 426, 433, 488-489
  plans for, 58, 152, 208, 297, 317, 365, 382, 383
  record, 391, 404
  spacecraft, 290, 317, 340, 344, 360, 387-389, 391
GEMINI VI (flight), 552, 570
  launch, 551
    failure, 547, 548
    postponed, 489, 491, 501, 509
  plans for, 49, 170, 317, 436, 449, 500, 505, 511, 546, 550
  spacecraft, 317, 318, 546, 571
  U.S.S.R. invitation to observe, 396, 397, 412, 420
GEMINI VII (flight), 500, 533, 544, 552
  achievement, 539, 570
  launch, 538
  plans for, 308, 317, 456, 493, 500, 505, 511, 521
  spacecraft, 308, 317, 467, 523
Gemini VIII (flight), 444, 522
Gemini IX (flight), 510
Gemini (spacecraft), 40, 48, 65, 76, 97, 128, 130, 156, 173
  astronaut training, 334, 530
  escape system, 238
  extravehicular equipment, 522, 531
  GT-2, 21, 239
  parachute landing system, 65, 112, 114, 116
  photographic equipment, 254, 483
  rendezvous, 23, 49, 170, 255, 265, 322, 340, 444, 493, 500, 501, 509, 510, 537, 551, 552, 557, 558, 563, 567, 570
  test, 21, 23, 65, 66, 112, 114, 116
General Aviation Pilot Education. See Gape, Project.
General Dynamics/Convair, 454, 551, 572
General Dynamics Corp., 2, 6, 498
  Atlas SLV x3, 123
  Bravo (buoy), 564
  contract, 121, 417, 567
  F-111, 14, 17, 19, 20, 52, 90, 109, 121, 172, 500
  F-111A, 179, 180, 183, 121
  life support system, 359
General Electric Co., 98, 120, 163, 169, 214, 231, 257, 283, 308, 309, 312, 396, 410, 454, 531, 549
General Motors Corp., 190, 139, 342, 448
General Precision, Inc., 513
General Purpose Airborne Simulator (Gpas), 524
General Services Administration (GSA), 415

Generator, nuclear, 7, 87
Geneva Agreement, 25
Geneva, Switzerland, 85
Geodesy, 208, 374, 523
Geodetic satellite (Geos), 36, 503, 507, 521
Geological Society of America Bulletin, 183
Geological Society of America Symposium, 507
Geomagnetism, 76, 267, 523
Geophysics, 344
Geophysics Corp. of America (GCA), 220, 287
George Washington Univ., 517
  Program of Policy Studies in Science and Technology, 351
  School of Medicine, 45
Geos. See Geodetic satellite.
German-Soviet Friendship Society, 473
German Space Research Commission (DKfW), 417
Germany, East, 379, 460, 550
Germany, West, 6, 33, 318, 321, 325, 337, 354, 369, 393, 417, 523, 550, 560
  Bochum Observatory, 292, 311, 336
  Defense Ministry, 393
  Ministry for Scientific Research, 339
Gerstenkorn, H., 198
Gessow, Alfred, 21
Getlein, Frank, 202
Getting, Dr. Ivan A., 221
Getze, George, 221
GGSE II (Gravity Gradient Stabilization Experiment), 114
GGSE III, 114
Gibson, Edward G., 299, 300
Gillespie, Dr. Charles C., 110
Gillett, Horace W., 283
Gilruth, Dr. Robert R., 54, 108, 129, 208, 211, 372, 409, 421, 491, 493, 558
GIMRADA. See U.S. Army Engineer Geodesy, Intelligence and Mapping Research and Development Agency.
Ginzburg, Vitaly, 125
Giovinetto, M. B., 191
Glassey, Eugene A., 93
Glenn, Col. John H. Jr., (USMC, Ret), 5, 82, 95, 134, 345, 441, 452, 460, 474, 533, 540
Glennan, Dr. T. Keith, 163, 231, 462, 540
Glew, Dr. Donald H., 517
Glider, 88, 92, 106, 414, 506
Gluhareff, Michael, 32
Gobi Desert, 550
Goddard Institute for Space Studies, 23
Goddard Memorial Dinner, 137
Goddard, Dr. Robert Hutchings, 7, 121, 127, 131, 144, 239, 332, 419
Goddard, Mrs. Robert H., 127, 131, 137
Goddard, Robert Hutchings, Day, 107, 121, 127

Goddard, Robert Hutchings, Memorial Library, 240, 335
Goddard, Robert H., Memorial Trophy, 137
Goddard Space Flight Center (GSFC) (NASA), 30, 44, 57, 62, 83, 85, 95, 96, 100, 127, 182, 187, 190, 193, 210, 220, 229, 328, 343, 371, 450, 484, 537
  award, 193, 462
  contract, 45, 48, 174, 177, 272, 308, 332, 374, 375, 474, 550, 564
  experiment, 62, 82, 107, 110, 120, 121, 132, 136, 180, 215, 276, 289
  management, 44, 53, 375
  Magnetic Field Components Test Facility, 318
  National Space Science Data Center, 125
  satellite monitoring, 106, 136, 168
  strike, 308, 353
  test, 85, 100, 334, 437
  tracking, 174
Goelet, Robert G., 297
Goett, Dr. Harry J., 343
Gold, Prof. Thomas, 104, 187, 194, 281
Goldberg, Arthur J., 559
Goldberger, Marvin L., 169
Goldhaber, Dr. Maurice, 107
Goldstein, Charles M., 401
Goldstone Tracking Station, 149, 337, 364, 404, 467
Goldwater, Barry, 367
Goleta, Calif., 78
Golueke, Dr. Clarence G., 232
Goodrich, B. F., Corp., 125
Goonhilly Downs, England, 180, 197
Gordon, LCDR Richard F., Jr. (USN), 10, 444
Gordon, William E., 109
Gorki Univ., U.S.S.R., 43
Gould, Jack, 224
Gould, R. Gordon, 545
Government Accounting Office (GAO), 44, 98
Governor's Conference on Oceanography and Astronautics, 458
Gpas. See General Purpose Airborne Simulator.
Graduate Research Center of the Southwest, 369
Grand Prairie, Tex., 54
Grand Turk Island, 146
Grants
  facilities, 353, 376, 420, 424
  Federal, 185, 327, 426
  multidisciplinary, 103
  Predoctoral Traineeship, 103
  summer space science program, 223
  Technical Utilization Program, 96
  universities, 229, 232, 274, 371
Graphite, 117, 274

Grasse, France, 121
Graveline, Dr. Duane E., 299, 300, 383
Gravimeter, 124
Gravity, 100, 181, 274
Gravity Gradient Stabilization Experiment. See GGSE II.
Gray, Dr. Edward Z., 172, 351
Gray, Harold E., 292
Graybiel, Capt. Ashton (USN), 70, 120
Great Britain. See United Kingdom.
Great Falls, Mont., 535
Great Lakes, 129
Greater Akron Safety Conference, 131
GREB VI (solar x-ray monitor satellite), 114
Green Bank, W. Va., 109, 371, 475
Green River, Utah, 18, 49, 67, 386, 504
Greenbelt, Md., 187
Greenland, 233, 466
Greenstein, Jesse L., 206
Grenade experiments, 35, 54, 58, 372
Grey, Robert, 458
Grigorov, Nikolai I., 342, 390
Grissom, Maj. Virgil I. (USAF), 68, 116, 130, 142, 145, 146, 148, 150, 153, 155, 156, 158, 159, 170, 171, 172, 193, 259, 268, 298, 449
Gromyko, Andrei A., 502
Groom, Nelson J., 143
Ground effects machine, 34
Ground station, 197, 201, 228, 232, 278, 314, 320, 326, 343, 344, 359, 360, 385, 393, 412, 462, 463, 503, 527
Grumman Aircraft Engineering Corp., 177, 224, 228, 474, 500, 530, 572
GSA. See General Services Administration.
GSFC. See Goddard Space Flight Center.
Guimarro, Constance, 261
GT-2–Gemini IX. See Gemini.
Guam, 318
Guaymas, Mex., 107
Guggenheim, Daniel, Medal, 131
Guggenheim, Daniel and Florence, Foundation, 532
Guggenheim, Daniel and Florence, International Astronautics Award, 431
Guggenheim, Harry F., 32
Guidance and control, 103, 176–177, 177
Guidance Error Analysis Vehicles (GEAV), 31
Gulf Stream, 564
Gun, light-gas, 207
Gursel, President Cemal (Turkey), 442

H–1 (rocket engine), 71, 92, 197, 291, 490
Hafnium, 535–536
Hafstad, Dr. L. R., 190
Hagerty, Dr. William W., 435
Haggerty, James J., Jr., 42, 168, 449
Haifa, Israel, 189
Haile Selassie, Emperor of Ethiopia, 446

Halaby, Najeeb E., 6, 34, 41, 58, 73, 131, 164, 194, 199, 204, 237, 274, 303, 435
Hall, Dr. Albert C., 343, 350, 411
Hall, Lawrence B., 61
Hamilton Standard Div. (United Aircraft Corp.), 125, 363, 506, 540
Hammaguir Range (Algeria), 98, 229, 467, 478, 526, 527
Hampton County (S.C.) Watermelon Festival, 297
Hampton, Va., 512
Hancock County, Miss., 483
Handley Page Herald (British aircraft), 318
"Handling Hazardous Materials," 498
Hanes, Maj. Gen. Horace A. (USAF), 233
Haney, Paul P., 196, 317, 371, 421, 444, 510, 520
Hanscom Field, Mass., 67
Hardy, Rep. Porter, Jr., 221
Harald, Crown Prince (Norway), 467
Harper, Charles W., 50
Harr, Dr. Karl G., 70, 407, 419, 562
Harriman, W. Averell, 339
Harris, Sen. Fred., 325
Harris, Rep. Oren, 292
Harris poll, 501
Harris, R. J., 306
Harrison, Gov. Albertis S., Jr., 553
Harrison, Lewis P., 71
Hart, Sen. Phillip A., 487–88
Hartford Courant, 234
Hartford Rotary Club, 460
Harvard Business School Club, New York, 184
Harvard College Observatory, 4, 106, 505
Harvard Divinity School, 168
Harvard Engineers Club, 49
Harvard Univ., 125, 262, 310, 373, 481, 487, 562
Harvey, Dr. Mose L., 100
Haseltine, William A., 96
Haskins, Dr. Caryl P., 559
Hasler Research Center, 78
Hasp (meteorological rocket), 420
Hatcher, Norman M., 143
Hatheway, E. A., Co., 311
Haughton, Daniel J., 347
Havana, Cuba, 1
Hawaii, 42, 319, 320
Hawk (missile), 189
Hawker Siddeley Aviation Co., 275
Hawker Siddeley group, 52
Haworth, Leland, 70
Hayes, Al J., 186, 248
Hayes, Dr. Wallace D., 32
Hayes International Corp., 109
Hays, Edward L., 190
Hazleton Laboratories, 60
Healey, Denis, 17, 178, 549
Health, Education and Welfare, Dept. of (HEW), 131

Heart, 9, 63
Heart, artificial, 547
Heat probe, 244
Heat shield, 222, 231, 239
Heidelberg, Germany, 321
Helicopter, 254, 271, 521
  Advanced Aerial Fire Support System, 504
  commercial lines, 83, 113, 118, 231, 297, 354
  crane, 275
  multipurpose, 32
  pressure jet, 39
  record, 35, 275, 371, 372
  Sperry Award, 32
Heliostation, 318
Heller, Dr. John H., 514
Hellyer, Paul, 165
Helwan Observatory, Cairo, 8
Henderson, C. Williams, 430
Henry, Thomas, 453
Heos. See Highly Eccentric Orbit Satellite.
Herget, Dr. Paul, 202-203
Hess, Dr. Harry H., 397
HEW. See Health, Education, and Welfare, Dept. of.
Hibex. See High Acceleration Experimental Booster.
Hickam AFB, Hawaii, 480
Hicks, Jim, 438
High Acceleration Experimental Booster (Hibex), 94
High Resolution Infrared Radiometer (Hrir), 408
High-Speed Ground Transportation Act of 1965, 454
High Temperature Instruments Corp., 334
Highly Eccentric Orbit Satellite (Heos), 550
Hilburn, Earl D., 462
Hildred, Sir William, 490
Himmel, Seymour C., 462
Hindsight, Project, 482
Hindustani Times, 273
Hines, William, 52, 191, 268, 288, 332, 342, 410, 422, 431, 448, 477, 486, 505, 536, 547, 563, 565
Hiroshima, Japan, 371
Hixon, S. Walter, 14
Hjornevik, Wesley L., 57
Hodgkins Medal, 381
Hoffman, David H., 143, 244
Hoffmeister, Cuno, 550
Holex, Inc., 315
Holland, Sen. Spessard, 394
Holloman AFB, N. Mex., 339
Hollomon, J. Herbert, 366
Holloway, Dr. John T., 533
Hollywood, Fla., 564
Holmdell, N. J., 261
Holmes and Narver, Inc., 393
Holter, Edward F., 202

Honest John-Nike Hydac (research rocket), 511
Honeysuckle Creek, Australia, 353, 543
Honeywell, Inc., 421, 535
Hoover Institute of War, Revolution, and Peace, 300
Hopkins, H. Z., Jr., 453
Hopkins, R. U. F., 510
"Hopper" (Lunar Flying Vehicle), 342
Horizon scanner, 126, 255, 302, 370, 399, 421, 451, 476, 492, 505
Horner, Richard E., 35
Hornig, Dr. Donald F., 13, 160, 452, 536
Horowitz, Dr. Norman H., 474
Hosenball, S. Neil, 563
Hot Springs, Ark., 517
Hotz, Robert, 20, 181, 215, 235, 292, 380, 418, 427, 444, 501
Hound Dog (missile), 504, 514
House of Commons, U.K., 117
Housing and Urban Redevelopment Act, 495
Houston, Tex., 17, 25, 63, 159, 208, 506, 510, 563
Houston Baptist College, 25
Houston Chronicle, 17, 201, 228
Houston Junior Chamber of Commerce, 18
Houston Post, 436, 440
Hovercraft, 34, 374, 528
Howard, Larry Dean, 101
Howard Univ., 530
Howe, Dr. Everett D., 77
Hoyle, Prof. Fred, 110, 206, 269, 467
Hrir. See High Resolution Infrared Radiometer.
HS-681 (British military transport), 52
Hsinhau (press agency), 165, 232
Hughes Aircraft Co., 51, 54, 64, 80, 121, 232, 283, 302, 466, 483, 500
Hughes Tool Co., 39, 521
Human Events, 390
Humphrey, Vice President Hubert H., 40, 105, 127, 198, 295, 394
  astronauts, visits with, 278, 431
  Dryden, Dr. Hugh L., tribute, 535
  International Air Show, 290, 291, 292, 314, 315
  NASA visit, 92, 122, 145, 163, 325
  research institutions, 541
  space program, 85-86, 137, 147, 186, 243-4, 432, 517, 529
Hungary, 272
Hunsaker, Dr. Jerome C., 249
Huntington Beach, Calif., 182
Huntsville, Ala., 173
Huntsville Times, 23, 365
Hurricane Betsy, 420, 422, 425, 428, 478
Hurricane Cleo, 142
Hurricane Dora, 142
Hurricane Ethel, 142
Hurricane Florence, 142
Hurricane research, 307, 308

Huss, Pierre H., 320
Huston, M/G Vincent (USAF), 490
Hutchins, Dr. Robert Maynard, 558
Hyatt, Abraham, 129
Hydra-Iris (sounding rocket), 31
Hydrogen, 125, 179
  fuel, 81, 104, 171
  gun, 207
  liquid, 49, 141, 385
Hydrogen bomb, 337, 495
Hydroskimmer, 34
Hydrotest, 179
Hypersonic aircraft, 35, 90, 203, 349, 464, 543
Hypersonic flow theory, 32
Hypersonic Ramjet Experiment Project, 169

IAEA. See International Atomic Energy Agency.
IAM. See International Association of Machinists.
IAU. See International Astronomical Union.
IBM. See International Business Machines Corp.
Icarus (asteroid), 507
Icarus, 261
ICBM. See Missile, ballistic, intercontinental.
Iceland, 321, 413
ICSC. See Interim Communications Satellite Committee.
IEEE. See Institute of Electrical and Electronics Engineers.
IGY. See International Geophysical Year.
IIT Research Institute (Illinois Institute of Technology), 216, 536
Ikeya-Seki (comet), 447, 480, 484, 485, 490, 502, 509, 513, 516, 572
IL-62 (jet liner), 241, 275, 318
Iliff, Robert, 24
Illinois, 134
Illinois Institute of Technology. See IIT Research Institute.
Illinois, Univ. of, 435, 553
Imp (Interplanetary Monitoring Platform), 257, 258
"The Impact of Space Exploration on Society," 383
Incentive contract, 5, 40, 49, 77, 80, 91, 139
India, 307, 325, 527
Indian Aviation, 161
Indian National Commission for Space Research, 307
Indian Ocean, 106, 136
Indiana, 95, 134
Indiana, Univ. of, 95
Indianapolis, 323
Indianapolis Star, 45, 277, 323
Indonesia, 6, 21, 59, 312, 373, 375

Information
  distribution and dissemination of, 107
  exchange of, 111, 131
  retrieval, 484
Infrared detector, 97
Infrared Interferometer Spectrometer (Iris), 374
Infrared sensing instrument, 143
Initial Defense Communications Satellite Project, 102
Injun Explorer (satellite), 485
Institute for Advanced Study (Princeton, N.J.), 321
Institute for Earth Sciences, 523
Institute for Research into Problems of the Future, 558
Institute for Satellite and Space Research, Bochum, W. Germany, 311
Institute of Air and Space Law (McGill Univ.), 536
Institute of Earth Physics (U.S.S.R.), 187
Institute of Electrical and Electronics Engineers (IEEE), 143, 519
Institute of Navigation, 208
Institute of Oceanography, 565
Institute of Space and Aeronautical Science, Japan, 200
Institute of Strategic Studies, 522
Institute of Theoretical Astronomy, U.K., 110, 447
Institute on Man and Science, 346
Instrument, 461
  coronograph, 107
  gravimeter, 124
  infrared, 26, 97, 143, 374
  life detector, 107
  penetrometer, 82
  seismometer, 149, 473
  spectrograph, 120, 132
  spectroheliograph, 179
  spectrometer, 62, 106, 215, 485
  spectrophotometer, 62
Intelsat. See International Telecommunications Satellite Consortium.
Interact Conference (Rotary), 178
Inter-American Experimental Meteorological Sounding Rocket Network (EXAMETNET), 236, 307, 365
Inter-American Skyway, 529
Interavia Air Letter, 447
Interim Communications Satellite Committee (ICSC), 502
Interior, Dept. of, 174
"Interlopers" (stellar objects), 38
Internal Zero Defects Program, AFSC, 114
International Academy of Astronautics, 390, 431
International Aerospace and Science Exposition, 164
International Aerospace Hall of Fame, San Diego, 134

International Air Show (Paris), 275, 279, 283, 284, 290, 291, 292, 314, 315, 318, 337
International Air Transport Association (IATA), 490
International Alliance of Theatrical and Stage Employees, 295, 301
International Association of Machinists (IAM), 186, 248, 260, 295, 447, 520
International Association of Machinists and Aerospace Workers, 437
International Astronautical Federation Congress, 390, 429, 437, 440, 448, 470
International Astronomical Union (IAU), 187
International Atomic Energy Agency (IAEA), 69, 122, 548
International Business Machines Corp. (IBM), 173, 275
International Christian Leadership World Conference, 320
International Civil Aviation Organization (ICAO), 233
International Commission for Space Research, 418
International Committee on Space Research, 68
International Conference on Aircraft Design and Technology, 519
International Conference on Phenomena in Ionized Gases, Seventh, 401
International Congress on Air Technology, 517
International cooperation, 129
  aircraft, 314
  astronomy, 3, 183, 187, 338
  civil aviation, 8, 48, 73
  communications, 22, 31, 58, 69, 95, 121, 249, 278, 322, 428, 434, 503, 541, 545, 568
  IQSY, 10, 163, 226, 511, 553
  meteorology, 24–5, 147, 162, 220, 282, 307, 365, 453
  military, 96, 178, 347
  nuclear, 57, 69, 95, 122
  science and technology, 314, 339, 568
  space (see also International space programs), 27, 35, 55, 58, 75, 77, 151, 286, 321, 370, 375, 431, 528, 531, 545
    U.S.-Argentina, 243, 511
    Australia, 353
    Brazil, 220, 515, 553
    Canada, 546
    Europe, 75, 143, 182, 361, 563
    France, 437, 540
    Germany, West, 339
    India, 307
    U.K., 549
    U.S.S.R., 9, 24, 76, 111, 153, 272, 279, 446
    U.S.S.R.-France, 486, 502, 519
  tracking, 418
  tracking station, 89, 136, 543
International Cooperation Year, 220
International Federation for Documentation, 471
International Geophysical Year (IGY), 67
International Latex Corp., 125, 506
International law, 559
International Lunar Geological Conference, 401
International Powder Metallurgy Conference, 285
International Radio and Television Society, 466
International Satellites for Ionospheric Studies (Isis-X), 523, 528
International Science and Technology, 47, 211, 456
International Scientific Radio Union, 196
International Space Electronics Symposium, 503
International Space Patrol, 105
International space programs (see also International cooperation, space), 68, 75, 182, 202, 210, 321, 352–54, 370, 375, 418, 466, 528, 543, 549, 560, 562
  U.K.-U.S.S.R., 76
  U.S.-Argentina, 243, 365, 421
  Australia, 353, 543, 545
  Brazil, 220, 365, 421, 515, 553
  Canada, 81, 91, 528, 546
  France, 282, 437, 540
  Germany, West, 339
  India, 325, 527
  Mexico, 107
  Netherlands, 23
  South Africa, 300
  U.K., 75, 486
  U.S.S.R., 9, 76, 111, 272, 279, 429, 465, 470
  U.S.S.R.-France, 486, 502, 519, 564
International Symposium on Advances in Gas Chromatography, Third, 482
International Symposium on Basic Environmental Problems of Man in Space, Second, 279
International Telecommunications Satellite Consortium (Intelsat), 502
International Telecommunications Union (ITU), 545
International Telephone and Telegraph Corp. (ITT), 182, 217, 308, 334, 340, 344
International Transport Fair, 333, 411
International Union of United Plant Guard Workers of America, 308
International Year of the Quiet Sun (IQSY), 10, 161, 163, 195, 196, 226, 521, 523, 553, 556
Interplanetary Monitoring Probe (Imp), 257
Inventions, 8, 156
Ion, 132
Ion propulsion, 167, 472
Ion thrustor, 5, 116
Iona College, 129

Ionization, 126
Ionized gas, 401
Ionosphere, 30
  measurement, 53, 110, 169, 249, 289, 523, 546
  research, 121, 163, 207, 220
Iowa, 85, 134
Iowa, Univ. of, 485
IQSY. See International Year of the Quiet Sun.
Iraq, 59
Iris. See Infrared Interferometer Spectrometer.
Irls. See Interrogation Recording and Location System.
Irradiation, 146
Isayeva, L., 371
Iselin, Dr. Columbus, 348
Isis. See International Satellites for Ionospheric Studies.
Isolation test, 66, 237, 314
Isotope, radioactive, 197
Israel, 6, 189, 321, 325
  atomic energy, 122
Italian National Committee on Space Research, 340
Italy, 25, 75, 259, 552
  Air Force, 340, 375
  Defense Ministry, 375
  launch, 543
  SAN MARCO I, 2
Itokawa, Dr. Hideo, 200
ITT. See International Telephone and Telegraph Corp.
ITT World Communications, Inc., 201
ITU. See International Telecommunications Union.
Ivanchenko, Prof. V., 467
Ives, Whitehead & Co., Inc., 297
Izvestia, 138, 269, 467, 492, 552, 559

J-2 (rocket engine), 31, 198, 209, 322, 527, 535, 543
Jackass Flats, Nev., 49, 240, 255
Jackson, Sen. Henry M., 232
Jackson, Nelson P., Aerospace Award, 137
Jacksonville, Fla., 353, 389
Jacquet, Mark, 73
Jaffe, Leonard, 57, 384
James, Jack N., 245, 356
Jane's All the World's Aircraft, 528
Japan, 477
  launch, 42
  Ministry of Telecommunications, 2
  nuclear capability, 325
  satellite, 21, 227
  sounding rocket, 42, 133–4, 312, 375
  space program, 200, 227, 260
  U.S. missile practice firing, 463
Japan Society for Aeronautical and Space Sciences, 519
Japanese Trade Ministry, 312
Jarry Hydraulics, Ltd., 14

Jastrow, Dr. Robert, 23, 57, 531
Javelin (sounding rocket). See Argo D-4.
Jeffries, John, Award, 471
Jenkins, Dale W., 59
Jenkins, Roy W., 73, 117, 314, 486
Jenks, Arthur E., 186
Jeppson, John, 336
Jet engine analyzing system, 84
Jet Propulsion Laboratory (JPL) (Cal Tech), 61, 303, 382, 461, 511, 556
  award, 137, 462
  Deep Space Network, 364, 377
  Goldstone Tracking Station, 149, 337, 364, 467
  Mariner project, 1, 64, 67, 81, 92, 102, 184, 189, 190, 326, 330, 336, 337, 363, 364, 497
  Ranger project, 4, 73, 78, 80, 84, 89, 92, 140, 148, 175, 245
  Surveyor, 466, 524
  Voyager, 341, 533
Jetport, 292
JMSPO. See Joint Meteorological Satellite Program Office.
Jodrell Bank Experimental Station (U.K.), 78, 180, 222, 464, 542
Johannesburg, South Africa, 55, 327, 330
Johannesburg *Sunday Times*, 297
John XXIII, Pope, 103
Johns Hopkins Univ., 99, 118, 508, 515
Johnson, Katherine, 428
Johnson, David, 71
Johnson City, Tex., 534
Johnson, Gifford K., 369
Johnson, John A., 434
Johnson, President Lyndon B., 22, 138, 161, 164, 246, 280, 286, 296, 370, 377, 413, 436, 446, 450, 509
  appointments, 169, 204, 206, 287, 294, 359
  astronaut, 423, 431, 460
  awards by, 48, 58, 286, 355, 431, 547
  budget request, 27, 100, 301
  Civilian-Military Liaison Committee, 351
  communications satellite system, 69, 92, 343
  defense, 20, 39, 242
  EARLY BIRD I message, 225, 282
  Environmental Science Services Administration (proposed), 229
  Gemini space flight
    GT-3, 147, 152, 155
    GT-4, 254, 289
    GEMINI V, 390, 402, 404, 405
    GEMINI VI, 395, 397
    GEMINI VII, 493, 538, 552
  Manned Orbiting Laboratory, 325, 395, 400, 410, 424, 427
  NASA visit, 92, 122, 276
  proclamations, 120, 501
  science, 101, 298, 473
  Science Advisory Committee, 169

space program, 136, 139, 150, 160, 171, 177, 286, 366, 396, 487
  international cooperation, 650
  supersonic transport, 237, 287, 309, 312, 316, 336
  transportation, 198, 237, 454
  tributes, 423, 533, 547
  VOSKHOD II, 138
Johnson, Richard L., 6
Johnson, Roy William, 345
Johnson, Vern W., & Sons, Inc., 527
Johnston Island, 49
Johnston, Richard S., 325, 470
Joint Meteorological Satellite Programs Office (JMSPO), 182
Jonas, Rep. Charles R., 173
Jones, David M., 370
Jones, J. A., Construction Co., 478
Jordan, 59
Jordan, Sen. Len B., 46
Jordan, Sen. B. Everett, 449
Journal of Spacecraft and Rockets, 456
Journal of the Armed Forces, 168, 449
Journeyman (sounding rocket), 184, 302
JP-4 jet fuel, 18
JPL. See Jet Propulsion Laboratory.
Jungk, Dr. Robert, 558
Junkers Flugzeug und Motorenwerke AG, 550
Jupiter (planet), 181, 197, 261, 348, 366, 374, 382, 399
Jupiter Calibration Experiment, 528
Justice, Dept. of, 359

Kaanapali, Hawaii, 42
Kaiser Aluminum and Chemical Sales, 92
Kaminin, L/G Nikolai (U.S.S.R.), 173, 395, 457, 489
Kaminski, Heinz, 311
Kanowski, CWO Mitch (USN), 112
Kansas City, Mo., 106
Kansas City Times, 109
Kaplan, Dr. Joseph, 382
Kaplanov, Prof. M., 480
Kappa 8L (Japanese sounding rocket), 375
Kardashev, Dr. Nikolai, 180
Karst formation, 95
Karth, Rep. Joseph E., 115, 422, 466, 501
Kazakhstan, U.S.S.R., 131
KC-135 (Stratolifter), 340
Keil, Dr. Klaus, 112
Keldysh, Prof. Mstislav V., 113, 156, 390, 392, 405, 420, 431
Kelley, Dr. Albert J., 304
Kelly, Lloyd L., 357
Kennedy, President John F., 76, 509
Kennedy, Dr. Joseph W., 305
Kennedy, Sen. Robert F., 118, 474
Kennedy Space Center, NASA (KSC), 3, 115, 193, 210, 245, 269, 331, 353, 366, 381, 390, 513
  astronaut debriefing, 410, 411
  award, 331, 568

  contract, 12, 91, 118, 158, 163, 311, 340, 450
  facilities, 11, 37, 178, 345, 454
  Launch Operations Div., GSFC, 458
  Pacific Launch Operations Office, 458
  Saturn, 96, 162, 224, 250, 298, 319, 341, 406, 426, 437, 443, 533
  spacecraft, 3, 168, 467, 490, 492
  spaceport, 178, 422
  strike, 39, 49, 118, 133, 159, 248, 277, 295, 302
Kenton, Frank, 527
Kenyatta, President Jomo (Kenya), 448
Kepler (lunar crater), 467
Kerch Strait, 141
Kerr, Breene M., 223
Kerwin, LCDR Joseph P. (USN), 299, 300
Keshishian, Dr. John M., 45
Khabarovsk, U.S.S.R., 88
Kharchenko, Boris I., 24, 382
Khodarev, Yuri, 392
Khrushchev, Premier Nikita (U.S.S.R.), 98, 480
Kiev Institute of Civil Aviation Engineers (U.S.S.R.), 345
Kiewit, Peter, Sons Co., 14
Kiilsgaard, Thor H., 97
Kiladze, Rolan, 329
Kilauea, Hawaii, 10
Kilgour, Frederick C., 108
Kimes, Capt. Charles H., 302
Kimmons, William L., 455
Kimzey, John H., 63
Kinard, William H., 376
Kincheloe, Iven C., Award, 454
King, Dr. Jean I. F., 220
King Salmon AFB, Alaska, 298
Kirchner, Dr. Werner R., 192, 286
Kirkland AFB, N. Mex., 549
Kiruna, Sweden, 352
Kiselev, M. I., 318
Kistiakowsky, Dr. George B., 200, 206, 310
Kiwi (nuclear reactor), 13, 112, 306, 519
Klein, Dr. Edmund, 168
Kleinknecht, Kenneth S., 346, 347
Klemperer, Dr. Wolfgang B., 154
Kliore, Dr. A. J., 414
Klokov, Ivan V., 478, 492
Klystra, Dr. Johannes H., 240
Knabenshue, A. Roy, 556
Knight, Capt. William J. (USAF), 346, 454
Knoblock, L/C Edward C. (USA), 410
Knoxville News-Sentinel, 244
Kock, Dr. Winston E., 49, 208, 349, 382
Kolcum, Edward H., 86, 517
Komarov, Col. Vladimir (U.S.S.R.), 1, 436
Kondratief, Prof. K. I., 85
Konecci, Dr. Eugene B., 70, 216, 471
Konrad, John, 54
Konstantinov, B., 3
Korea, 245

Korff, Dr. Serge A., 516
Kosberg, Semyon A., 3
Kosygin, Premier Alexi (U.S.S.R.), 542
Kozlovskaya, Sofia, 342
Kozyrev, N., 515
Kraft, Christopher C., 68, 177, 265, 397, 411, 512, 520
Krakow, Poland, 257
Krasnaya Zvezda, (See also *Red Star*) 57, 197, 367, 398, 568
Krichagin, Vladimir, 134
Krumb, Henry, School of Mines (Columbia Univ.), 124
Krylov, Marshal Nikolai I. (U.S.S.R.), 57
Kryl'ya rodiny, 360
KSC. See Kennedy Space Center, NASA.
Kuczma, Julius E., 8
Kuettner, Dr. Joachim P., 386
Kuhn, Richard E., 463
Kuiper, Dr. Gerard P., 84, 95, 149, 187
Kuo, Dr. John T. F., 124
Kurzweg, Dr. Herman N. H., 126
Kuwait, 58
Kuzmin, Dr. Arkady, 371
Kybal, Dalimil, 361
Kyushu, Japan, 42

Labor, Dept. of, 131
Labor relations, 8, 39, 49, 133, 157, 186
Laboratoire Central de Télécommunications (LCT), 182
LaChance, Dr. Paul, 455
LaGow, Herman E., 375
Laika (dog, U.S.S.R.), 445
LaJolla, Calif., 95, 299, 389
Lalli, Vincent R., 17
Lally, Eugene, 348
Lamda (Japanese sounding rocket), 133
Lamda III-2 (Japanese sounding rocket), 42
Lamont Geological Observatory, 182
Lance (missile), 125
Landing pads, 165
Landing system, automatic, 131
Lange, Dr. Oswald H., 450
Langley Research Center (LaRC) (NASA), 12, 283, 314, 518
  awards, 14, 494
  contract, 35, 80, 82
  experiment, 169, 559
  Lunar Landing Research Facility, 303
  management, 80, 82
  meeting, 223, 271
  research, 127, 143, 468, 553
  Research Staff Office, 27
  sonic boom study, 15, 65
  supersonic transport experiments, 334, 516
Langway, C. C., Jr., 191
Lansbergh, Dr. M. P., 23
Lapp, Dr. Ralph E., 536
LaRC. See Langley Research Center.
Laredo, Tex., 389, 538
Largos. See Laser Activated Reflecting Geodetic Optical Satellite.

Larsen, Dr. Finn J., 535
Las Cruces, N. Mex., 22
Las Vegas, Nev., 541
Laser, 168
  deep space, 207, 503
  tracking techniques, 24, 27, 276
  patent, 545
  use of, 67, 73, 168, 209, 227, 228, 276, 317, 497, 517
Laser Activated Reflecting Geodetic Optical Satellite (Largos), 24
Lasker, Albert, Medical Journalism Awards, 243
Launch Complex 16, 11, 54
Launch Complex 19, 523, 539
Launch Complex 34, 37, 52, 92, 94, 118, 345
Launch Complex 36B, 103
Launch Complex 37, 92, 94, 312, 385
Launch Complex 39, 48, 94, 164, 184, 319, 346, 454, 544
Launch vehicle, 178, 501
Launch sites, deactivation, 57
  operations and capability, 19, 54, 72, 75, 80, 82
  reliability 36, 60, 71-72
  U.S.S.R., 73, 137, 222, 345, 348, 532
  military, 29, 66, 74, 287, 288, 291
  cost, 33, 77, 384
  development, 60, 74, 77, 88, 130, 132, 163, 191, 210, 275, 301, 352, 365
  contract, 5, 80, 182, 274, 275, 289, 296, 340, 526
Laures, Josiane, 66, 121
Lawrence, David, 438
Lawyer, Capt. Richard E. (USAF), 514
LCS 11 (radar calibration sphere), 477
LCT. See Laboratoire Central de Télécommunications.
Lear, John, 104
Lear, William, 361
Lebanon, 59, 76
Lebedev, Victor, 119
Lebedev Institute of Physics, Moscow, 371
Lebedinskiy, Prof. Aleksandr, 479
LeBourget, France, 275
Lederberg, Dr. Joshua, 372
Lederer, Jerome, 526, 557
Ledford, Col. Otto C., (USAF), 290
Lee, John G., 209
Lee, Dr. William A., 7
Legion of Merit, 519
Leighton, Dr. Robert B., 354, 355, 356
Lem. See Lunar Excursion Module.
LeMay, Gen. Curtis E. (USAF, Ret.), 42, 48, 84, 153, 250, 486
Leningrad, 31, 372, 387
Leningrad, Univ. of, 85
Leonid meteor shower, 518, 520, 521
Leonov, L/c Aleksey (U.S.S.R.), 131, 132, 134, 135, 138, 139, 140, 147, 150, 153, 155, 156, 216, 217, 225, 270, 278, 395, 419, 430, 438, 442, 457, 460, 473, 514, 571

LES I (Lincoln Laboratory Experimental Satellite), 65
LES III, 560
LES IV, 560, 565
Lesher, Dr. Richard L., 223
Levin, Boris J., 187
Levin, Gilbert V., 60
Levin, Kenneth L., 360
Levitt, Dr. I. M., 19
Lewis, David S., 363
Lewis Research Center (LRC) (NASA), 17, 49, 63, 123, 275, 519
  Centaur, 103, 377, 417
  contract, 26
  experiment, 158, 237
  Plum Brook Station, 26, 180
  research, 49, 56, 96, 169, 256, 258, 285, 334, 438
  test, 128, 132, 296, 357, 377
Lewis, Richard, 23
Lfv. See Lunar Flying Vehicle.
Libby, Dr. Willard F., 103, 210, 258, 461
Libraries, 25, 108
Library of Congress, 487
  Legislative Reference Service, 271
  Science Policy Research Div., 259
Libya, 58
Lick Observatory, 502
Life magazine, 178, 244, 288, 438
Life science (see also Bioscience), 70, 118, 120, 246, 334, 538, 564
  acceleration forces, 123, 181, 191
  artificial life, 427, 466
  atmospheric pressures, 115-116, 134-135, 173, 475
  blood pressure, 190, 566
  ear, 23, 70, 345, 421, 443
  environment, effects of, 66, 115, 129, 132-133, 190, 452, 470
  extraterrestrial life, contamination by, 246
  heart, 9, 63, 267, 566
    artificial, 546
  isolation experiment, 237, 314
  life support system, 359, 363, 430, 476, 494, 506, 540
  nutrition, 56, 128, 153, 172, 237, 421, 422
  radiation, 59, 167, 202, 217, 248, 266, 313, 450, 566
  temperature, 116, 313
  weightlessness, 19, 45, 55, 59, 63, 132, 138, 146, 267, 279, 412, 527, 531, 540
Life Sciences Research Laboratory (NASA) 564
Life support system, 132, 232, 363, 506, 540
Lifting body vehicle, 209
Lil. See Lunar International Laboratory.
Lilly, William, 462
Lima, Peru, 121
Limonite, 330
Lincoln Laboratory, MIT
  LES I, 65

LES III, 560
LES IV, 560, 565
Lindberg, Charles A., 131, 134, 331
Linde Co., 445
Lindsay, Mayor John V., 562
Ling-Temco-Vought, 15, 54, 80, 119, 195
  contract, 447, 502, 561
  Lance, 125
  Scout, 502, 541
  XC-142A, 15, 54, 322
Liquid hydrogen, 445
Lisbon, Portugal, 277
Little, Arthur D., Inc., 13, 52
Little Joe II (booster), 238, 513, 571
Little John (missile), 463
Little Rock AFB, Ark., 14
Livingston Electronic Co., 128
Llrv. See Lunar Landing Research Vehicle.
Local Scientific Survey Module (Lssm), 315
Lockheed Aircraft Corp., 264, 309, 414, 550
  C-5A, 454, 491
  C-130, 52, 63, 112, 114, 124
  C-141, 2, 41, 198
  contract, 120, 214, 257, 393, 454
  helicopter, 315, 504
  Orion, 2
  707-349C, 516
  supersonic transport, 41, 163
  U-2, 124
Lockheed-California Co., 525
Lockheed Electronics Co., 195, 309, 474
Lockheed Jet Star, 524
Lockheed Missiles and Space Co., 5, 23, 165, 190, 269, 289, 361, 491, 524, 550
Lockheed Propulsion Co., 24, 32
Loewe, Dr. Erhard, 205
Loewy, Robert G., 451, 457
Logandale, Nev., 78
London, U.K., 17, 67, 73, 95, 275, 314, 492
Long, Sen. Russell B., 216
Longshot, Project, 495, 525
Los Alamos, N. Mex., 335
Los Alamos Scientific Laboratory, 306, 450
Los Angeles, Calif., 54, 68, 78, 83, 170, 222, 228, 272, 454, 514, 564
Los Angeles Air Force Station, Calif., 5
Los Angeles Times, 221
Losey, Robert M., Award, 471
"Loss of a Leader" (editorial), 539
Louisiana, 255
Louisville, Univ. of, 9
Lovelace Clinic, 553
Lovelace Foundation, 435
Lovelace, Dr. W. Randolph, II, 118, 296, 549, 553, 554
Lovell, Sir Bernard, 222, 464, 542
Lovell, Cdr. James A. (USN), 208, 308, 317, 493, 500, 511, 530, 533, 557, 558, 560, 563

Lovelock, Dr. J. E., 482
Low, Charles A., Jr., 258
Low, Dr. Frank J., 57
Lowell Observatory, 96
LRC. See Lewis Research Center.
LSSM. See Local Scientific Survey Module.
Lubell, Samuel, 143
Lucas, Dr. William R., 506
Lucet, Charles, 544
Luftfahrttechnik, Raumfahrttechnik, 456
Lufthansa German Airlines, 84
LUNA V (U.S.S.R. lunar probe), 222, 223, 227, 230, 232, 234, 235, 571
LUNA VI, 272, 274, 284, 571
LUNA VII, 460, 464, 467, 571
LUNA VIII, 536, 539, 542, 571
Lunar (see also Moon)
  base, 174, 239
  crater, 143, 149, 154, 187, 194, 323, 379, 392, 507, 515, 555
  dust, 85, 187, 194
  eclipse, 8, 525
  exploration, 143, 152, 174, 179, 180, 252, 253, 342, 358, 380
  laboratory, 327, 430, 431
  landing research facility, 303
  landing research vehicle, 360
  law, 530
  photographs, 8, 68, 80, 84, 89, 95, 97, 100, 104, 108, 137, 143, 148, 153, 175, 232, 325, 379, 382, 392, 570
  probe, 4, 74, 79, 84, 140, 146, 148, 222, 223, 227, 232, 272, 274, 285, 379, 382, 460, 463, 464, 477, 536, 539, 540, 542, 570, 571
  surface, 10, 27, 43, 80, 81, 84, 85, 96, 104, 143, 153, 160, 187, 227, 234, 298, 392, 474, 487, 496, 555, 565
Lunar and Planetary Laboratory, Univ. of Arizona, 57
Lunar Excursion Module (Lem), 33, 48, 75, 152, 160, 187, 196, 352
  contract, 139, 177, 315, 345
  test, 113, 186
Lunar exploration vehicle, 530
Lunar Flying Vehicle (Lfv), 342
Lunar International Laboratory (Lil), 431
Lunar laboratory, 3, 27, 430
Lunar Landing Research Facility (LaRC), 303
Lunar Landing Research Vehicle (Llrv), 82, 360, 465
Lunar materials, 27, 209
Lunar Mission and Space Exploration Facility, 175
Lunar Mobile Laboratory (Molab), 315
Lunar Orbiter program, 5, 181, 204, 275, 487
Lunar program (U.S.) (see also Moon and Lunar), 23, 27, 29, 40, 43, 75, 80, 151, 160, 173, 181
  appropriations, 171
  cost of, 175, 530
  criticism, 97, 164, 178, 186, 530
  importance of, 29, 97
  manned flight and landing 75, 77, 152
Lundin, Bruce T., 462
LUNK III (U.S.S.R. lunar probe), 379
Lupenko, Trofim D., 297
Luster, Project, 518, 521
Luyten, Dr. Willem J., 70

M-1 (rocket engine), 27, 100, 141
M1-4 (Russian helicopter), 122
M2-F2 (manned lifting body research vehicle), 278
M-110 (helicopter), 275
McAdams, Alfred, 202
McCall, G. J. H., 401
McCall, Dr. J. C., 359
McClellan, Sen. John L., 359
McCollum, John S., 464
McCollum-Pratt Institute, 119
McConnell AFB, Kan., 14
McConnell, Gen. John Paul (USAF), 48, 217, 331, 440
McCormack, M/G James (USAF, Ret.), 479
McCormack, Rep. John W., 33, 155
McCullough, Warren, 281
McCusker, Donald F., 414
McDivitt, L/C James A. (USAF)
  GT-4 spaceflight, 268, 269, 371, 375
  launch, 266, 267, 268
  plans, 168, 229, 248, 254, 255, 256
  honors, 270, 278, 283, 286, 331
  International Air Show visit, 290, 292, 314, 316
  interview, 196, 208
  messages to
    Gagarin, Yuri, 273
    Johnson, President Lyndon B., 271
  promotion, 276
  United Nations visit, 463
McDonald, Adm. David L. (USN), 535
McDonald, Dr. Gordon J. F., 169, 193, 397
McDonnell Aircraft Corp., 40, 363, 493, 567
  contract, 209, 393, 524, 572
  F4C, 2, 287
  GT-2, 239
  GT-4, 168
  Phantom II, 52
  strike, 510, 520, 523, 525
Macelwane, James B., Award, 193
McElroy, Dr. William D., 119
McGill Univ., 502, 536
McKay, John B., 94, 206, 236, 293, 319, 411, 433, 451, 492
MacKay Trophy, 340
McKee, Gen. William F. (USAF, Ret.), 287, 294, 303, 309, 322, 481, 491
McKinney, Marion O., Jr., 463
McLean, Francis E., 15, 241
McLean, George, 161
Macleay, Capt. Lachlan (USAF), 598

MacLeod, Norman E., 167
McLeod, Norman J., 228
McMillan, Dr. Brockway, 322
McMurdo Station, Antarctica, 174
McNamara, Robert S., 14, 27, 39, 86, 144, 168, 172, 221, 235, 316, 346, 347, 367, 438, 451, 454, 458, 474, 486, 537, 566
Macomber, Frank, 241, 346
Macy, John W., Jr., 455
Madagascar (Malagasy Republic), 551
Madison, Stuart, 15
Madrid, Spain, 330
Maeda, K., 227
Magnesium-lithium alloys, 408
Magnetic field, 36, 110, 157, 207, 223, 256, 257, 267, 318, 553, 557
Magnetic Field Components Test Facility (GSFC), 318
Magnetometer, 36
Magnitogorsk, U.S.S.R., 369
Magnuson, Sen. Warren G., 51, 309
Malvern, U. K., 41
Management, 1, 14, 77, 91, 185, 190
Management Services, Inc., 109
La Mañana, 25
Manchester Guardian, 273, 480
Maneuvering Ballistic Re-entry Vehicle, 383
Manned Flying System (Mfs), 342
Manned maneuvering unit (Mmu), 510
Manned Orbiting Laboratory (Mol), 29, 115, 267, 290, 295, 299, 325, 346, 441, 484, 514, 567
  cooperation, 484, 494
  criticism, 269, 423, 427, 544
  design studies, 27, 28, 66, 231
  Gemini, 171, 183, 239
  launch plans, 549
  news censorship, 410, 488
  purpose, 86, 101, 102, 330, 416, 424, 488, 494
  test, 183, 531, 549, 560, 571
Manned space flight, 1, 11, 222
  achievements, 146, 150, 151, 153, 156, 265, 268, 403, 418, 538, 550–553, 570
  appropriations, 28, 75
  capability, 29, 77, 177, 410
  cost of, 177, 457
  criticism, 142, 172, 271
  extravehicular activity, 132, 135, 138, 152, 155, 171, 190, 208, 570
  hazards, 19, 63, 133, 139, 372, 375, 536
  lunar landing (see also Lunar program), 3, 45, 143, 181, 383, 429, 436
    cost of, 7, 62, 457
    criticism, 97, 164, 167, 177, 185
    goal, 7, 28, 29, 75, 79, 108, 160
    hazards, 19, 59, 63, 70, 104, 196
    man's role in, 196, 396, 414, 416
    objectives, 25, 79, 177, 211, 429, 456, 461, 528
    program, 7, 15, 23, 27, 40, 68, 108, 116, 139, 151, 168, 170, 171, 174, 177, 188, 251, 317, 365, 367, 382, 461, 549
    R&D, 317
    support of, 62, 97, 509
    training, 10, 133
Manned Spacecraft Center (MSC) (NASA), 10, 54, 115, 128, 168, 196, 208, 276, 297, 347, 367, 371, 394, 440, 493
  astronaut, 68, 129, 159, 170, 196, 298, 510
    debriefing, 412, 415
    training, 10, 56, 97, 321
  awards, 57, 301, 372
  contract, 308, 474
  facilities, 1, 54, 175, 177, 308
  manned space missions, 7, 80, 116, 177, 195
    Apollo, 55, 56, 109, 113, 185
    Gemini, 58, 65, 67, 109, 116, 170, 314, 347
      GT-2, 21, 239
      GT-3, 68
      GT-4, 168, 190, 196, 208, 265, 268, 276, 322
      GEMINI V, 58, 152, 208, 339
      GEMINI VI, 49, 170, 523
      GEMINI VII, 523
  research, 63, 160, 358, 456, 476
  scientist-astronaut program, 563
  spacecraft, 160, 201, 485, 522
Mantz, Paul, 321
Mao Tze-tung, 140
Mars (planet)
  atmosphere, 36, 244, 313, 328, 338, 354, 414, 482, 507, 572
  canals, 109, 336, 378
  contamination of, 36, 60, 97, 246, 262
  craters, 336, 354, 448, 507
  exploration, 62, 77, 201, 213 238, 252, 338, 382, 399, 429
  flight to
    cost, 356, 394, 429
    manned, 160, 194, 342, 351, 394, 429, 469, 472
    unmanned, 1, 2, 7, 17, 18, 24, 39, 55, 61, 68, 73, 75, 81, 112, 160, 184, 189, 190, 201, 228, 235, 245, 258, 277, 293, 300, 311, 317, 323, 326, 328, 329, 333, 336, 338, 339, 359, 380, 382, 384, 398, 440, 519
  life on, 4, 17, 36, 64, 96, 97, 158, 202, 281, 324, 330, 336, 355, 372, 443, 474
  magnetic field, 190, 330, 355, 572
  photographs, 293, 300, 328, 330, 336, 338, 342, 346, 354, 355, 363, 378
  study of, 32, 120, 200, 201, 262
  surface, 355, 481
  water on, 4, 244, 324, 354, 372, 448, 507
Mardel Plata, Argentina, 225
Marconi Company, Ltd., 549
Mare Sirenum (Mars), 68
Margaret, Princess (U.K.), 11
Mariana Islands, 42, 134
Marine Contracting, Inc., 407

Mariner (program), 102
MARINER II (Venus probe), 184, 241, 258, 323
MARINER IV (Mars probe), 1, 66, 67, 73, 81, 92, 113, 156, 326, 337, 355, 565
  experiment, 129, 444
  performance, 2, 7, 17, 24, 39, 55, 64, 82, 83, 102, 108, 118, 152, 165, 174, 184, 190, 207, 218, 228, 242, 244, 258, 316–317, 323, 333, 435, 443, 497
  photographs, 1, 68, 245, 293, 300, 325–326, 329, 330, 336, 338, 346, 348, 354, 360, 404, 474, 570
  progress report, 190
  results, 354–355, 359, 497, 572
Markov, Prof. Alexander, 153, 323
Marion Power Shovel Co., 39, 381, 422, 517
Markow, Edward G., 530
Marmain, J., 361
Marquardt Corp., 169
MARS I (U.S.S.R. probe), 207
Mars spacecraft, 16
Mars station, 467
Marshall Space Flight Center (MSFC) (NASA), 8, 94, 96, 128, 165, 239, 253, 255, 275, 319, 373, 411, 429, 573
  award, 411, 455, 506
  Computer Operations Office, 179
  Computation Laboratory, 312
  contract, 92, 526
    study, 302, 342, 343, 358
    support services, 115, 173, 179, 191, 204, 234, 312, 315, 319, 447, 533
  J–2 (rocket engine) facilities, 105, 319
  launch vehicle, 8
    Saturn, 39, 54, 100, 123, 162, 178, 188, 197, 198, 219, 241, 275, 292, 296, 319, 326, 348, 363, 373, 458, 459, 490, 525, 543, 554, 557
  Lunar Flying Vehicle, 342
  management, 248, 446
  Manpower Utilization and Administration Office, 340
  meeting, 194, 248, 255
  Mississippi Test Facility, 220, 482, 505
  Office of Technology Utilization, 503
  personnel, 173, 273, 340, 359, 569
  personnel honored, 129, 156, 326, 506
  Technical Systems Office, 359
  West Test Area, 54, 319
Marsography, 423
Martin Co., 58, 87, 106, 157, 246, 344, 347, 387, 524, 536, 548, 567
Martin, Minta, Lecture, 129
Martz, Emmett L., 455
Maryland, Univ. of, 129, 202, 486
Masevich, Alla, 392
Mason, Dr. B. H., 78
Mason, John F., 87
Mason-Rust Co., 179, 550
Mass spectrometer, 68, 76
Massachusetts, 106
Massachusetts Department of Commerce and Development, 468

Massachusetts Institute of Technology (MIT), 73, 545
  Center for Sensory Aids Evaluation and Development, 90
  Dept. of Aeronautics and Astronautics, 94
  Lincoln Laboratory, 65
Masursky, Harold, 507
Materials, 170, 274, 570
  composite, 157, 170
  heat-resistant, 56, 92
  high temperature, 19, 519, 535
  lightweight, 19
"Materials and Tomorrow's Air Force," 274
Materials testing, 65, 533
Mathews, Charles W., 286, 325, 558
Mathias, Rep. Charles McC., 357
MATS. See Military Air Transport Service.
Mauna Kea, Hawaii, 10
Mauna Loa, Hawaii, 10
Max Planck Institute (Germany), 321
Maxwell, B/G Jewell C. (USAF), 387, 515
Mechling Barge Lines, Inc., 239
Medical electronics, 488–489
Medical Society of the State of New York, 70
Melbourne, Australia, 269
Melbourne, Fla., 45
Melpar, Inc., 310
Memorandum of understanding
  international, 236, 339, 568
  national, 74, 135
Mendel, Gregor, 297
"Mendel Rivers Day," 298
Mengel, John T., 567
Menzies, Sir Robert, 136
Mercury (planet), 194, 229, 382
Mercury, Project, 12, 68, 82, 89, 133, 146, 501, 546
Mercury (spacecraft), 21, 173, 183, 394, 462
Meredith, Leslie H., 462
Merritt Island Launch Area (MILA) (KSC), 57, 179, 188, 245
  facilities, 199
    construction, 157
    support services, 48
  launch complex, 92, 164, 184
  strike, 8, 39, 49, 133, 157
Mesoscale structures, 124
Mesosphere, 226
Meteor, 8, 221
Meteorite, 2, 85, 100, 258, 415, 421, 550
Meteoroid, 117, 216, 247, 248, 249, 376, 507
Meteoroid detection, 71, 90, 94, 97, 100, 106, 505, 511
  micrometeoroid, 193
Meteorological Institute (Sweden), 321
Meteorological Satellite Laboratory (U.S. Weather Bureau), 71
Meteorology (see also Weather), 38, 65, 71, 126, 220, 332, 516

cooperation, 77, 112, 182, 243, 307, 365
forecasting, 228–229, 267
probe, 216, 257
satellite, 28, 81, 147, 175, 220, 226, 301, 570
   achievement, 81, 208
   communications, use of, 282
   manned, 85
sounding rocket experiments, 214, 227, 239, 249, 289, 294
Methanol, 87
Metric system, 246, 329, 365, 393
Metropolitan Washington Board of Trade, 486, 514
Mettler, Dr. Ruben F., 492
Mexico, 107, 370
Mfs. See Manned Flying System.
Miami, Fla., 366
Miami Beach, Fla., 274
Miami Herald, 71
Miami, Univ. of, 30
   Center for Advanced International Studies, 100
Miami Weather Bureau, 366
Michel, Dr. Frank C., 299, 300, 341
Michelson, Albert A., Award, 485
Michigan, 134
Michigan, Univ. of, 121, 136, 283, 372, 510, 511
Michigan Education Association Meeting,
Michoud Assembly Facility, 308, 319, 340, 504
   363, 428, 478, 512, 543, 550, 561
Michoud Operations (NASA), 105, 123, 130, 162, 179, 250, 308
Mickelsen, William R., 258
Micrometeorite, 421
Micrometeoroid, 507
Micrometeoroid detection, 24, 193
Microwave research, 543
Mid-Canada Warning Line, 165
Middle East Airlines, 237
Midway Island, 49
Midwest, 399
MiG–17 (U.S.S.R., jet fighter), 287, 291
Mikhailov, Alexander A., 342, 392
Mikoyan, Anastas, 138, 153, 405
MILA. See Merritt Island Launch Area.
Military Air Transport Service (MATS), 573
Military Electronics Convention, 54
"The Military Implications of Space," 459
Military Review, 260
Military technology, 151
Miller, Col. David V. (USAF), 290
Miller, Rep. George P., 72, 89, 108, 155, 198, 255, 422, 509, 564
Mils. See Missile Impact Location Station.
Milwaukee Public Museum, 391
Mines, Bureau of, 209
Mining, extraterrestrial, 312
"Minilab," 157
Minneapolis, Minn., 119, 174, 312, 353

Minners, Dr. Howard, 555
Minnesota, 134, 223
Minnesota, Univ. of, 129, 184
Minuteman (missile), 16, 312, 313
   contract, 197
   Guidance Error Analysis Vehicle use, 31
   launch, 24, 54, 63, 101, 152, 511
   II, 19, 541
   launch vehicle use, 203, 541
Minuteman Mark IIA reentry vehicle, 536
Mirage IV (French aircraft), 329
Mirny Observatory, 387
Missile, 16, 20, 31, 49, 57, 67, 125, 203, 257, 315, 479, 527, 531
   air-to-surface, 504, 514
   antimissile, 39, 86, 156, 224, 235, 310, 447, 452, 522, 541, 553
   ballistic, 16, 37, 224
      intercontinental (Icbm), 16, 24, 31, 54, 63, 101, 109, 144, 152, 229, 250, 310, 312, 515, 522, 527, 528
   military application, foreign, 189, 312, 347
   nuclear, 463, 512
   orbital, 512, 521, 543–544
   reliability, 59, 125
   tracking, 49, 233, 414
   U.S.S.R., 222, 224, 235, 250, 312, 313, 321, 515, 521, 528, 544
Missile Impact Location Station (Mils), 49
Missile launch sites, 88, 257
Missile launching submarine, 59
Missile Sites Labor Commission, 49, 302
Missile Space Daily, 351, 352, 354, 463
Missiles and Rockets, 31, 60, 113, 144, 203, 210, 215, 217, 227, 235, 373, 380, 450, 456, 461, 517, 524
Mississippi, 255
Mississippi Test Facility (MTF), 220, 275, 308, 425, 482, 505, 512
Mississippi Test Facility Task Force, 220
Mississippi Test Operations, 105, 115, 131, 308
Missouri, 255
MIT. See Massachusetts Institute of Technology.
Mitcham, Grady L., 430
Mitchell, Ind., 259
Mmu. See Manned maneuvering unit.
Mobile Range Facility, 110
Molab. See Lunar Mobile Laboratory.
Mock, Jerrie, 453
Mod 1 (computer guidance system), 445
Moeller, Rep. Walter H., 46
Mohole, Project, 38, 346, 452
Mojave Desert Ground Station, Calif., 450
Mol. See Manned Orbiting Laboratory.
"Molly Brown" (GEMINI III), 130, 145
MOLNIYA I (communications satellite, U.S.S.R.), 197, 221, 228, 254, 259, 260, 313, 456, 478, 492, 530, 571
Monaco, 58

Mondale, Sen. Walter F., 46
Le Monde, 456
Monroney, Sen. A. S. Mike, 34, 83, 114, 185, 231, 302
Monrovia, Calif., 68
Monsanto Chemical Co., 237
Monsanto Research Corp., 157
Monsoon, 226
Montevideo, Uruguay, 31
Montgomery, Ala., 26
Montreal, Canada, 3
Monument, space, 178
Moon (see also Lunar program)
  atmosphere, 507
  composition, 100, 221
  crater, 8, 143, 148, 149, 194, 379, 392, 507, 515, 555
  eclipse, 8, 525
  exploration of, 142, 174, 180, 181, 253, 262, 342, 358, 530
  landing, 487
    manned, 3, 29, 75, 97, 113, 160, 227, 232, 246, 252, 324, 398, 491, 509
    soft, 227, 230, 232, 465, 524, 536, 542
    unmanned, 228, 275, 303
  mapping of, 495, 507
  meteoroid effect on, 117, 507
  origin, 85, 187, 474
  photographs, 9, 67, 74, 80, 84, 89, 95, 97, 100, 108, 137, 153, 175, 232, 325, 378, 379, 382, 392, 570, 571
  probe, 222, 223, 272
  surface of, 8, 10, 43, 44, 80, 81, 84, 85, 96, 104, 143, 187, 227, 234, 290, 298, 378, 392, 434, 474, 487, 507, 555, 566
  water on, 507
Moonport, 178
Moore, Wendell F., 399
Moorhead, Rep. William S., 446
Moritz, Bernard, 563
Moro, Prime Minister Aldo (Italy), 476
Morocco, 59
Morris, Thomas D., 458
Morse, Thomas M., 113
Morse, Sen. Wayne, 287
Moscovsky Komsomolets, 66
Moscow, 78, 98, 134, 135, 140, 147, 222, 241, 371, 379, 502, 509, 512, 521
Moscow News, 486
Mossbauer radiation, 208
"Mother-daughter" experiment, 15
Motion sickness, 70
Motor, brushless, 48
Mound Laboratory, 99
Moyers, William D., 420, 493
Mt. Fuji, Japan, 463
Mt. Hamilton, Calif., 502
Mt. Palomar Observatory, 38, 206, 236, 277, 516
Mount Siple, 532
Mt. Wilson Observatory, 38, 206, 277, 495
MSC. See Manned Spacecraft Center.
MSFC. See Marshall Space Flight Center.
MTF. See Mississippi Test Facility.
Mu (rocket), 227

Mu Cephei (star), 57
Mueller, Dr. George E.
  Apollo, 79, 80, 351
  extravehicular activity, 172, 208
  Gemini, 79, 255, 267, 283
  Saturn V, 188
  space flight, manned, 116, 177, 195, 196, 251, 282, 483
  space program, 460, 461
Mullard, Ltd., 549
Mundt, Sen. Karl E., 487
Munich, 57, 331, 333
Murphy, Dr. Charles H., 501
Murphy, Daniel, 400
Murray, Prof. Bruce C., 61, 355, 507
Museum of Natural History, 483
Myers, Myron L., 505

NACA See National Advisory Committee for Aeronautics.
Nadge (NATO Air Defense Ground Environment system), 96
NAS. See National Academy of Sciences.
NASA. See National Aeronautics and Space Administration.
NASA Advanced Technology Program, 438
NASA-AEC Space Nuclear Propulsion Office, 541
NASA-AIAA Manned Space Flight Meeting, 469
NASA Apollo Applications Office, 469
NASA Applications Technology Satellite Program, 102
NASA Communication and Navigation Programs, 384
NASA Conference on Aircraft Operating Problems, 223
NASA Deep Space Network, 458
NASA Design Certification Review Board, 510
NASA Exceptional Service Award, 286
NASA Facilities Management Office, 553
NASA Future Programs Task Group, 164
NASA Manned Space Flight Field Center Development, 228
NASA Mission Analysis Div., 69, 77
NASA Office of Advanced Research and Technology, 69, 306, 376, 472
NASA Office of Grants and Research Contracts, 103
NASA Office of Industry Affairs, 68
NASA Office of Lunar and Planetary Programs, 108, 356
NASA Office of Manned Space Flight, 27, 297, 370
NASA Office of Space Science and Applications, 59, 341, 344, 356
NASA Office of Technology Utilization, 161, 408, 498
NASA Office of Tracking and Data Acquisition, 105, 561
NASA Pacific Launch Operations Office, 458
NASA Predoctoral Traineeship Grant, 103
NASA Science Advisory Committee, 348

NASA Scientific and Technical Information Facility, 368
NASA Space Nuclear Propulsion Office, 3, 292, 305
NASA Space Radiation Effects Laboratory, 553
NASA Technology Utilization Div., 457
NASA Technology Utilization Program, 408, 545
NASA-Univ. of Alabama Educational Symposium, 255
NASA-Univ. of Virginia Bio-Space Technology Training Program, 373
NASA University Explorers Program, 56, 94
NASA-University Program Review Conference, 103, 106
NASA-Western University Conference, 510
NASC. See National Aeronautics and Space Council.
Nascom (NASA Communications Network), 136
Natal, Brazil, 220, 370
Natal Range, Brazil, 552, 557
Nathan, Dr. Robert, 100
National Academy of Engineering, 206, 209, 487
National Academy of Sciences (NAS), 129, 193, 203, 205, 226, 339, 432, 474, 569
  Committee on Science and Public Policy, 310
  Mars exploration, 201, 213
  report, 200, 397
  scientist-astronaut recommendations, 63, 299
  Space Science Board, 17, 200, 246, 293, 397
National Aeronautics and Space Act, 291
National Aeronautics and Space Administration (NASA) (see also under NASA centers, programs, and satellites such as Ames Research Center, Apollo, MARINER IV, etc.)
  agreement, 29, 74, 294, 558
    international, 91, 106, 307, 365, 466, 515, 520
  anniversary, 127, 162, 164, 506
  appropriations, 204, 218, 219, 251, 264, 324, 325, 366-7, 487
  astronaut, 5, 10, 15, 25, 49, 58, 59, 117, 124, 152, 155, 158, 168, 172, 196, 208, 209, 294, 308, 321, 322, 500, 509, 510, 514, 519, 529, 533, 535, 537, 539, 547, 550, 552, 554, 555, 557, 559-562, 566-7, 569
  award, 8, 14, 57-8, 73-4, 119, 156, 193-4, 239, 256, 282, 285, 287, 291, 295, 301, 335, 356, 372, 406, 462-3, 494, 506, 509, 513, 529, 539, 543, 547, 567
  budget, 17, 27-29, 61, 74-76, 100, 103, 108-9, 114, 133, 139, 141, 168, 178, 192, 218-19, 231, 251, 404, 481, 535, 545, 571
  conference, 102, 105, 223, 251, 255, 520
  contract, 9, 570, 571
    administration, 91, 114, 130, 141
    engine, 49, 65, 92, 163, 169, 173, 182, 269-70, 275, 289, 296, 474, 526, 535
    facilities, 52, 105, 115, 175, 191, 465, 533
    incentive, 5, 40, 49, 76, 91, 133
    nuclear power, 282, 476
    spacecraft, 40, 138, 176, 188, 195, 228, 494, 525, 549, 563
    space equipment, 63, 82, 498, 524, 549-50
    study, 9, 18, 48, 158, 209, 301-2, 562
    support services, 48, 67, 72, 80, 91-2, 109, 117, 138, 164, 179, 234, 239, 308, 311, 447, 512, 561
    tracking, 69, 174, 460, 466
  cooperation, 53, 110, 174, 515
    AEC, 240, 291, 296
    DOD, 1, 29-30, 45, 68, 74, 88, 102, 110, 174, 182, 203, 207, 267, 298, 329
    FAA, 555
    USA, 68, 295
    USAF, 177, 218, 300, 305, 444, 464
    USN, 177
    Weather Bureau, 25, 162
  cooperation, international, 76, 226, 370, 529, 544, 556
    Argentina, 236, 243, 511
    Australia, 545
    Brazil, 220, 307, 365, 553, 557
    Canada, 81, 91, 527
    Europe, 214, 563
    France, 75, 437, 540
    Germany, West, 339
    India, 307
    Italy, 2
    Mexico, 107
    Netherlands, 441
    U.K., 76, 525
    U.S.S.R., 10, 76, 111, 152, 466, 470
  criticism of, 1, 3, 5, 11, 44-5, 61, 64, 95, 140, 141, 167, 172, 180, 211, 231, 293, 302, 324, 397, 423, 465, 495
  economy, 11, 54, 76, 98, 455
  education, 14, 17, 57
  exhibit, 156, 173, 502, 557
  expenditures, 3, 29, 105, 192, 276, 324
  experiments, 1, 15, 36, 45, 58, 60, 62, 68, 75, 76, 103, 112, 114, 143, 146, 190, 215, 249, 289, 294, 302, 321, 325, 437, 439, 443, 449, 475, 483, 515, 530, 540, 546, 552, 559, 564
  facilities, 123, 175, 179, 533, 553, 564
    construction, 11, 25, 49, 105
  funds, 17, 40, 75, 100, 103, 123, 130, 141, 192
  grants, 96, 102, 222, 232, 275, 300, 466, 529, 537, 546

656 INDEX

National Aeronautics and Space Administration (Continued)
  information dissemination, 96, 98, 102, 114, 140, 141, 170, 194, 203, 209, 401
  information exchange, 112
  information retrieval, 471, 484, 560
  labor relations, 39, 49, 133, 156
  launch, 7, 510–11, 569
    balloon, 6
    booster, 188, 373–4
    failure, 39, 102–3, 120, 141, 322, 468, 489, 491–2, 509, 547–8
    manned, 145, 265, 387, 537, 551
    postponed, 154, 301, 383, 489, 503, 522, 549
    probe, lunar, 73, 140
    satellite, 25, 54, 65, 73, 140, 172, 207, 257, 307, 357, 475, 489, 503, 507, 515, 519, 520, 546, 553, 569, 570
    sounding rocket, 6, 15, 35, 53, 58, 108, 112, 132, 140, 167, 169, 179, 184, 195, 213, 215, 238, 239, 249, 287, 289, 302, 396, 437, 439, 443, 447, 473, 484, 511, 520, 525, 534, 552, 570, 571
    test
      Apollo, 27, 55, 56, 72, 73, 185, 521, 522
      Gemini, 22, 23, 64, 66, 71, 521
      reentry, 244
      Saturn, 162, 197, 207, 247, 368, 491
  management, 1, 14, 78, 80, 112, 123, 184, 294, 504, 505, 532, 553, 556
  organization, 1, 68, 69, 76, 176, 458, 566
  patents, 8, 92, 164, 264, 288
  personnel, 1, 22, 23, 27, 30, 57, 58, 75, 76, 77, 79, 80, 81, 85, 88, 89, 90, 217, 220, 231, 239, 255, 282, 296, 335, 367, 373, 380, 443, 511, 512, 533, 534, 546, 547, 552, 554, 559
    appointments, 7, 68, 69, 70, 94, 95, 100, 138, 209, 222, 264, 296, 297, 312, 324, 340, 368, 371, 385, 458, 478, 505, 553, 563
  procurement, 68, 110, 115, 185, 302
  programs, 121, 164, 173, 200, 275, 294, 455, 458
    Apollo, 23, 27, 29, 30, 45, 75, 79, 80, 97, 112, 116, 164, 167, 173, 351, 363, 468, 478, 502, 506, 517, 524, 526, 557, 570
    astronomy, 27, 29, 30, 36, 53, 54, 62, 68, 82, 119, 177, 188, 228, 275, 332, 529, 554
    bioscience program, 3–4, 115, 132, 181, 196
    Centaur, 130, 375
    Gemini, 68, 79, 97, 110, 116, 143, 145, 147, 151, 168, 193, 196, 208, 234, 265–268, 288, 289, 325, 387, 397, 493, 500–1, 505, 509, 521, 522, 533, 537, 540, 544, 546, 548, 551, 552, 554, 557, 559, 563, 570
    geodetic satellite, 35, 125
    Lunar Orbiter, 5, 177, 181, 204, 487
    Mars, 213, 355, 398
    meteorology, 26, 44–5, 72, 156, 162, 182
    Mol, 28, 29, 66, 86, 101, 115, 171
    nuclear propulsion, 3, 13, 44, 49, 87, 104, 116, 296, 305
    Pegasus, 72, 89, 94, 97, 106, 193
    Ranger, 4, 5, 9, 84, 100, 108, 148, 149, 175, 181, 188
    rocket motor, solid propellant, 40, 44, 95, 100
    Saturn, 7, 27, 30, 71, 72, 75, 112, 357, 365, 472
    sounding rocket, 169, 195
    space, 7, 24, 27, 28, 29, 35, 43, 54, 55, 60, 75, 77, 79, 86, 101, 112, 137, 138, 151, 173, 185, 227, 461, 463, 471, 518, 538, 570
    Surveyor, 135, 181, 204, 374, 466, 477, 501
    Syncom, 1, 2, 88, 102, 106, 136, 147, 173
    tracking and data acquisition, 89, 174, 448, 461
    Voyager, 4, 18, 27, 60, 61, 75, 130, 200, 444, 476, 533, 562
  research 74, 219, 229, 555, 570
    aeronautical, 36, 49, 56, 68, 78, 90, 92, 112, 176, 177, 231, 295, 303, 463, 482, 519, 524, 543, 569
    electronics, 2, 24, 49, 126, 543
    fuel, 112, 132
    nuclear, 13, 112
    propulsion, 16, 169
    sonic boom, 15, 572
  Science Advisory Committee, 348
  scientist astronaut, 5, 63, 300
  studies, 8, 18, 29, 48, 103
  supersonic transport, 73, 90, 515, 571
  test, 27, 41, 209
    booster, 23, 27, 104, 112, 155, 186, 188, 197, 219, 241, 248, 386, 449, 490, 522, 526, 533, 546, 554
    communications, 42, 76, 127, 136
    nuclear, 13
    spacecraft, 17, 21, 27, 65, 104, 116, 126, 227, 238, 244, 293, 328, 522, 523
  tracking station, 70, 76, 89, 107, 174, 497, 520
  universities, 56, 103, 106, 111, 130, 185, 213, 255, 468, 510, 521, 541, 555, 561, 572
    grants, 96, 103, 223, 233, 303, 311
  x–15, 13, 15, 49, 56, 74, 94, 112, 169, 173, 191, 197, 206, 464, 482, 503, 504
National Aeronautics and Space Council (NASC), 11, 27, 52, 70, 73, 85, 105, 113, 127, 216, 243, 252, 295, 366, 444, 463, 471, 494, 497, 503, 516, 517, 535
National Aeronautics Association, 526, 561
National Air Museum, 43, 55, 128, 156, 162
National Aircraft Noise Symposium, 274

National Association of Broadcasters, 151
National Association of Retail Druggists, 471
National Association of Science Writers, 205
National Aviation Facilities Experimental Center, 488
National Broadcasting Co. (NBC), 268
National Bureau of Standards, 169, 326, 365, 534
National Center for Atmospheric Research (NCAR), 1, 350, 520
National Center for Radioactivity Research (CNRS), France, 321
National Citizens' Commission on International Cooperation, 528
National Civil Service League, 194, 239, 240
National Commander's Award for Distinguished Service (American Legion), 101
National Commission on Technology, Automation and Economic Progress, 40
National Communications System, 428
National Conference on Spacecraft Sterilization Technology, 520
National Conference on the Peaceful Uses of Space, Fifth, 251
National Defense Transportation Association, 37
National Engineers' Week, 98
National Foundation on the Arts and Humanities (proposed), 117, 202
National Gallery of Art, 121, 202
National Geodetic Satellite Program, 528
National Geographic Society, 414, 513
National Goals in Space, 1971-1985, 397
National Governor's Conference, 353
National Humanities Foundation, 185
National Labor Relations Board (NLRB), 157
National Medal of Science, 58, 547
National Merit Scholarship Award, 531
National Oceanographic Council, 51, 129
National Operational Meteorological Satellite System (Nomss), 71
National Park Service, 269
National Press Club, 217, 518, 563
National Radio Astronomy Observatory, 109, 371, 475
National Research Council, 125
National Research Council (Canada), 91
National Science Fair, 179
National Science Foundation (NSF), 56, 69, 115, 191, 205, 362
 annual report, 70, 106, 246
 Antarctic station, 557
 budget, 27
 grant, 67, 108, 110, 174, 212, 274, 426
 memorandum of understanding, Yugoslavia, 568
 Mohole Project, 37
 R&D funds report, 99
 role of, 191, 271

National Science Teachers Association, 136
National security, 18, 19, 38, 55, 105, 137, 144, 147, 153, 236, 270, 298
National Security Industrial Association, 66, 153, 186
National Space Club, 31, 82, 137, 192, 239, 240, 282, 343, 382, 446, 484
National Space Science Data Center, 125
National Symposium on Reliability and Quality Control, 17
National Weather Satellite Center (NWSC), 28, 38, 386
National Youth Science Camp, 320
National Youth Science Congress, 136
National Zeitung, 318
NATO. See North Atlantic Treaty Organization.
Nature, 467
Naval Air Engineering Center, 128, 190
Naval Air Facility, 112
Naval Aviation Center, 230
Naval Ordnance Laboratory, 421
Naval Ordnance Test Station, 66
Naval Research Laboratory (NRL), 10, 106, 196, 206, 212, 484, 521
Naval School of Aviation Medicine, 23
Navigation satellite system, 13
Navy Oceanographic and Meteorological Automatic Device (Nomad), 59
NBC. See National Broadcasting Co.
NCAR. See National Center for Atmospheric Research.
NCNA. See New China News Agency.
Nebraska Wesleyan Univ., 55
Nedelin, Marshal Mitrofan I. (U.S.S.R.), 480
Nedelya, 383
Nellis AFB, Nev., 84
Neon, 435
Nerva. See Nuclear Engine for Rocket Vehicle Application.
Ness, Norman F., 193
Netherlands, 23, 259, 541
Netherlands Organization for the Advancement of Pure Research (ZWO), 541
Neubeck, Capt. F. Gregory (USAF), 514
Neumann, Robert J., 376
Neumann, Temple W., 60
Neutron, 53
Nevada, 536
Nevis Cyclotron Laboratory, 279
New China News Agency (NCNA), 21
New England, 124
New England Aero Club, 292
New Guinea, 525
New Hampshire, Univ. of, 121
New Mexico, 161, 538
New Mexico State Univ., 174
New Orleans, La., 58
New Orleans Times-Picayune, 377
The New Republic, 424
The New Scientist, 41, 154
New York, 106

New York, N.Y., 7, 9, 83, 92, 184, 185, 222, 541
New York Academy of Sciences, 12, 502, 533
New York Airways, 297
New York City Planning Commission, 495
New York City Youth Board, 531
New York Herald Tribune, 244
New York Journal American, 320
New York State Atomic and Space Development Authority, 341
New York, State Univ. of, 240
New York Times, 21, 121, 213, 235, 236, 279, 290, 379, 382, 383, 395, 398, 414, 417, 427, 436, 479, 488, 489, 523, 530, 542, 547, 548, 559
New York Univ., 516
New York World Telegram, 541, 567
New York World's Fair, 173, 327, 366, 402
  Space Park, 173, 327, 366
New Zealand, 58, 85
Newcomb, Arthur L., Jr., 143
Newell, Dr. Homer E., 44, 194, 202, 211, 239, 521
  budget, 81, 108, 114, 130, 142
  moon, 181
  space programs, scientific, 3, 4, 60, 81, 108, 125–126, 132–133, 177
  space results, 81, 208, 344, 503
Newell, N. Dak., 101
Newport, R.I., 31
Newport News Shipbuilding & Drydock Co., 179, 269
Newsweek, 144
Newton, Dr. Robert R., 274
Niagara Falls, N.Y., 248
Nickel, 170
Nicks, Oran W., 356
Nicolaides, Dr. John D., 31
Nicolet, Prof. Marcel, 382
Nightglow, 455
Nike-Apache (sounding rocket)
  launch, 110, 132, 158
    Argentina, 511
    Brazil, 557
    Canada, 167
    Netherlands, 441
    U.S., 6, 53, 169, 195, 278, 287, 289, 294, 393, 410, 413, 439, 510, 511, 521, 525, 553
Nike-Cajun (sounding rocket)
  launch
    foreign, 372
    U.S., 35, 54, 58, 195, 214, 372, 511
Nike-Tomahawk (sounding rocket), 136
Nike-X (antimissile missile system), 5, 39, 86, 452
Nike-Zeus (antimissile missile system), 144, 310, 452
Nikolayev, L/C Andrian (U.S.S.R.), 139, 491
Nikolayeva-Tereshkova, Valentina, 151, 491

Nimbus (meteorological satellite program), 142
NIMBUS I, 142, 408, 532
Nimbus B, 44, 45, 87
Nitric acid, 547
Nitrogen tetroxide, 547
Nitze, Paul H., 214
NLRB. See National Labor Relations Board.
Nobel Prize, 487
Noise level studies, 228, 242, 243
Nomad. See Navy Oceanographic and Meteorological Automatic Device.
Nomss. See National Operational Meteorological Satellite System.
NORAD. See North American Air Defense Command.
North American Air Defense Command (NORAD), 13, 366, 543
North American Aviation, Inc., 56, 84, 110, 272, 336, 342, 453, 457, 471
  contract, 84, 121, 264, 572
  H–1 rocket engine, 31, 92, 197
  labor, 435, 436, 468
  program, Apollo, 56, 325, 485, 490, 492
  Rocketdyne Div., 31, 92, 132, 197, 291, 490, 526
  test laboratory, 198, 373
  XB–70, 72, 84, 150, 191, 206, 478, 547
North Atlantic Search and Rescue Seminar, 230
North Atlantic Treaty Organization (NATO), 34, 96, 347, 523
Nordberg, Dr. William, 220, 462
Norden, Carl L., 284
Norman, Okla., 366
Normyle, William J., 170
North Star Research and Development Institute, 96
North Vietnam, 291, 398, 573
Northeast Electronics Research and Engineering Meeting, 505
Northeastern Univ., 168
Northrop Corp., 5, 109, 438
Northrop-Norair, 209, 278
Northrop Space Labs., 358, 524
Northwestern Univ., 497
Norton AFB, Calif., 16
Norway, 80, 81, 466, 467
Notre Dame, Univ. of, 31
Novosti Press, 161
Nozzle, 49, 96, 112
NRDS. See Nuclear Rocket Development Station.
NRL. See Naval Research Laboratory.
NRX A–2 (nuclear reactor), 306
NRX A–3 (nuclear reactor), 197, 240, 255
NSF. See National Science Foundation.
Nuclear bomb, 206–207, 232, 236, 243
Nuclear Detection Program, 459
Nuclear energy, 13, 17
Nuclear Engine for Rocket Vehicle Application (Nerva), 116–117, 197, 240, 255, 519, 541
Nuclear explosion, 2, 3, 18, 39, 40, 81

Nuclear generator, 87
Nuclear propulsion, 3, 16, 17, 28, 63, 64, 104, 243, 296, 305, 306
Nuclear reactor, 13, 122, 135–136, 180, 227, 240, 243, 255, 272, 282, 305, 548, 565
Nuclear research center, 57
Nuclear Rocket Development Station (NRDS), 49, 282, 296
Nuclear submarine, 59, 78
Nuclear test, 178, 536–537
Nuclear test ban treaty, 3, 34, 39, 40, 81, 341
Nucleotides, 323, 324
Nutrition, 56, 128, 153, 172, 237, 421, 422, 449, 456
NWSC. See National Weather Satellite Center.

Oahu, Hawaii, 326
Oak Park, Ill., 26
Oak Ridge National Laboratory, 167, 547
Oakland, Calif., 374
Oao. See Orbiting Astronomical Observatory.
OAR. See USAF Office of Aerospace Research
Oba, King of Benin, 452
Obata, Gyo, 43
Oberth, Hermann, 332
Oberth, Hermann, Award, 506–507
O'Brien, Dr. Brian J., 56
Ocean Science and Ocean Engineering Conference, 280
Oceanography, 37–38, 51–52, 129, 175, 299, 332–333, 346, 525, 471, 487, 518, 519, 529
O'Connor, B/G Edmund F. (USAF), 347–48
OECD. See Organization of Economic Cooperation and Development.
Office of Aerospace Research (OAR), 525
Office of Science and Technology (President's), 13, 46, 271
Ogden, Dr. Eric, 9
Ogo. See Orbiting Geophysical Observatory.
Ogonek, 313
O'Keefe, Dr. John A., 537
O'Keefe, Dr. Walter, 85
Oklahoma City, Okla., 65, 190
Oliver, Bernard M., 348
Olson, Dr. Walter T., 334
Operation Firefly, 371
"Operation Moon Harvest," 100
Operation Zero Defect, 363
"Opportunities for Participation in Space Flight Investigations," 455
Optical Technology Satellite (Ots), 302
Orbit, 12, 24
  geostationary, 81
  libration, 113
  lunar, 113
  sun-synchronous, 25

Orbiting Astronomical Observatory (Oao), 82, 204, 228, 275
  Oao A–2, 1, 177
Orbiting Geophysical Observatory (Ogo), 60, 126, 204, 275
  OGO I, 107, 184, 440
  OGO II, 475–476, 489, 491, 571
Orbiting Solar Observatory (Oso), 53, 62, 405
  OSO I, 396
  OSO II, 106–107, 396, 529, 571
  Oso C, 126, 383, 396, 571
  Oso D, 188
  Oso E, 188
Order of Merit, 476
Organization for Economic Cooperation and Development (OECD), 534
Orion (barge), 209
Orion (constellation), 108, 120, 132, 474
Orion, Project, 3, 321, 322, 390, 391, 461
Orlando, Fla., 178, 199
Orlando *Evening Star*, 367
Orroral Valley, Australia, 353
ORS (Octahedron Research Satellite), 341
Oscar (Orbiting Satellite Carrying Amateur Radio), 238
OSCAR III (communications satellite), 114
OSCAR IV, 560, 562
Osmundeen, John U., 398
Oso. See Orbiting Solar Observatory.
L'Osservatore della Domenica, 152, 273
Ostrander, M/G Don R. (USAF), 62, 171, 203, 281, 478
Oswald, Dr. William J., 232
Ots. See Optical Technology Satellite.
Otto, E. W., 63
Ousley, Gilbert W., 9
Outstanding Leadership Medal (NASA), 256
OV2–1 (radiation sensor satellite), 477
Overseas Writers Club, 206
Owen, Kenneth, 161
Owings Mills, Md., 70
Owl (satellite), 56, 370
Oxford Univ., England, 562
Oxidation reduction, 305
Oxygen, 63
  deficiency, 158
  environmental test, 129
  FAA regulation, 455
  physiological effect, 190, 494
Oyster Point, Va., 553
Ozernoi, Leonid, 125
Ozone, 121

P–1154 (vertical-take off fighter), 17, 52
Pacem in Terris (encyclical), 103
Pacific Crane and Rigging Co., 48
Pacific Missile Range (PMR) (see also Western Test Range), 49
Pacific Ocean, 9, 42, 109, 110, 134, 473, 495, 526, 550
Pads, landing, 165

Page Communications Engineers, Inc., 490
Page, M/G Jerry D. (USAF), 504
Pageos (passive geodetic satellite), 35
Pais, Dr. Abraham, 38
Pake, Dr. George, 169
Pakistan, 78, 322, 337
Palace of Pioneers, Moscow, 514
Palaemon (NASA barge), 239, 363, 437, 459, 543
Palestine, Texas, 326, 520
Palmdale, Calif., 72
Palo Alto, Calif., 203
Pan American World Airways, 131, 222, 237, 273, 292, 302, 335, 346, 434
Parachute
  jumping, 300
  landing system, 22, 358
  test, 21, 22, 65, 112
  use of, 456
Paraglider, 414, 506
Paris, France, 121, 279, 541, 545
Paris Convention for the Protection of Industrial Property, 129
Park, President Chung Hee (Korea), 245
Parke-Bernet Galleries, 246
Parker, Dr. Eugene N., 32
Parkinson's disease, 334
Parsons, John F., 444
Particle accelerator, 57
Passive Geodetic Satellite. See Pageos.
Patents, 359
  application, 8, 545
  award, 2, 123, 156, 304, 305, 360, 414, 485, 506
  convention, 129
  international, 299
  legislation, 216, 249, 264, 487
  license, 93, 94
Patrick AFB, Fla., 57, 490
Patterson, Richard C., 156
Patterson, W. A., 545
Paul VI, Pope, 9, 138, 271, 362, 451, 474
Paul, Norman S., 458
Paumalu, Hawaii, 385, 503, 534
Payne, James, 456
PDP-5 (computer), 280
Peace, 103, 362, 559
Peconic, L.I., 224
Pegasus, Project, 502, 505
PEGASUS I (meteoroid detection satellite), 72, 89, 94, 97, 106, 193, 234, 247, 311, 319, 376
PEGASUS II, 247, 249, 311, 319, 358, 376
PEGASUS III, 357-58, 376
Pegasus B (micrometeoroid detection satellite), 193
Pegasus C, 294, 319
Peking Aeronautical Engineering College, 261
Peking, China, 165, 504
Pell, Sen. Claiborne, 117, 202, 246-247, 328-329
Pelligra, Dr. Ralph, 449
Pendray, G. Edward, 127

Penetrometer, 82
Penkovsky, Col. Oleg V., 480, 505
Pennsylvania State Univ., 124, 239
Pensacola Air Station, Fla., 23
Pensacola, Fla., 70
Pentagon, 458
"The Pentagon, the 'Madmen,' and the Moon," 10
Penzias, Dr. Arno A., 261
Pepin, Dr. I., 429
Perkin-Elmer Corp., 434, 476
Perkinson, William J., 512
Perm, Russia, 136, 140, 155
Perry, Robert, 350
Pershing (missile), 224
Peru, 370
Petrovich, Prof. Georgi V., 479, 532
Petrushkin, I. P., 313
Pettingill, Dr. Gordon H., 194
Pfaffe, H., 379
Phaethontis (Martian desert), 68
Phantom II (fighter aircraft), 52, 63
Phased Project Planning, 91
Philadelphia, Pa., 128, 186, 190, 205, 229
Philadelphia Aquarama, 531-532
Philadelphia Evening Bulletin, 284
Philadelphia Inquirer, 19, 442
Philadelphia Rotary Club, 186
Philadelphia Society for Paint Technology, 229
Philco Corp., 51, 54, 60, 80, 82, 93, 194, 283, 375, 485, 494
Philippines, 552
Phillips, M/G Samuel C. (USAF), 27, 295
Phoebus (nuclear reactor), 49, 282, 296, 306, 519
Phoenix, Ariz., 9
Phoenix (missile), 500
Photographs, cloud, 163, 168-169, 175
Photography
  computer use in, 100-101
  of Earth, 101, 142
  Ranger, 100, 101, 104, 108, 137, 148-149, 152, 153, 156
Photovoltaic Specialists Conference, 484
Physical Review Letters, 279
Physics, 183, 344, 573
  fluid, 127
  history, 310
  Nobel prize, 487
  plasma, 125
Piccard, Dr. Jeanette, 243
Pickering, Dr. William H., 43, 137, 146, 325, 328, 330, 336, 347, 355, 356, 382, 409, 431, 440, 461
Pieper, George F., Jr., 375
Pillsbury Co., 422
Pilot-astronaut, 423
Pilz, Prof. Wolfgang, 6, 321
Pimentel, George C., 96
PIONEER VI (interplanetary probe), 553-554, 565, 570
Pioneer Chain of the Compass Card with Diamonds (aviation decoration), 411
Pioneer of the Wind-Rose Award, 333

Pittendrigh, Dr. Colin, 200, 281, 443
Pittsburgh, Univ. of, 521
Pitzer, Kenneth, 169
Planetoid, 550
Planets, life on, 3, 4, 17, 36, 47, 61, 97–98, 133, 168, 180, 182, 214, 246, 261–262, 572
Planning Research Corp., 564
Plasma sheath, 127, 146
Platner, John L., 238
Plessey-UK, Ltd., 549
Pleumer-Bodou, France, 180, 197
Plotkin, Dr. Henry, 276
Plss. See Portable Life Support System.
Plum Brook Reactor Facility (LRC), 26–27, 180
Pluto (nuclear reactor program), 28
Pluto (planet), 197, 217, 372
Plutonium, 304–305
PMR. See Pacific Missile Range and Western Test Range.
Pocomoke City, Md., 565
Pogo. See Polar Orbiting Geophysical Observatory.
Pohn, Howard, 509, 516
Point Arguello, Calif., 48
Point Barrow, Alaska, 35, 54, 299
Point Mugu, Calif., 13
Point Pillar tracking station, Calif., 48
Poland, 257, 284
Polar Orbiting Geophysical Observatory (Pogo), 174
Polaris (missile) (USN), 57, 339, 513–514
  A–1, 184
  A–3, 59, 210, 538
  Poseidon (B–3), 20
Polaris (star), 184, 513
Polish Astronautical Society, 257
Ponnamperuma, Dr. Cyril S., 36, 323, 425
Ponomarev, Leonid, 391, 394
Poodle (rocket engine), 197
Pope AFB, N.C., 340
Popovich, L/C Pavel R., 135, 559
Port Tobacco, Md., 515
Portable Life Support System (Plss), 363
Porter, Dr. Richard W., 226
Portugal, 324, 327
Poseidon (NASA barge), 544
Poseidon (Polaris B–3 missile), 20
Potassium, 547
"Potential Hazards of Back Contamination from the Planets," 246
Potomac Electric Power Co., 95
Potrero, Calif., 550
Potter, V. R., 51
Pound, Prof. Robert V., 43
Powell, Rep. Adam Clayton, 530
Power Source Conference, 238
Prague, Czechoslovakia, 228
Prahl, Val E., 6
Pratt and Whitney Div. (United Aircraft), 137
  contract, 120, 163, 214, 257, 309, 312
Pravda, 3, 390, 412, 467, 544

Pravda Ukrainy, 346
Pre-Continent III (underwater sphere), 475
Pregnant Guppy (aircraft), 224, 234, 294, 347, 492, 562
Presidential Citation, 301
President's Advisory Committee on Supersonic Transport, 309
President's Science Advisory Committee (PSAC), 119, 169
Press conference, 36, 182,
  extravehicular activity, 135, 208
  Gemini flights
    GT–3, 147, 152
    GT–4, 248, 255, 276
    GEMINI V, 385, 397
    GEMINI VIII, 444
    GEMINI IX, 510
  lunar landing, 62, 227, 389, 436, 444
  nuclear test, 522
  supersonic transport, 287
  U.S.S.R. military strength, 78, 489, 521
  U.S.S.R. space activities, 140, 154, 345
Press Wireless, Inc., 282
Preston, G. Merritt, 490
Pretoria, South Africa, 14
Price, Dr. Charles C., 427, 436
Price, Don K., 564
Price, Capt. F. H., Jr. (USN), 14
Price, Rep. Melvin, 63–64, 481
Princess Martha Coast, 532
Princeton Univ., 32, 112, 169, 202, 211, 261, 281, 283, 326, 443, 573
Princeton Univ. Observatory, 108, 264
Probe, 136, 139, 334
  lunar, 4, 73, 78, 84, 142, 143, 222, 223, 227, 232, 272, 274, 460, 463, 464, 477
  Lunar Orbiter, 5, 181, 204, 275, 487
  Mars, 4, 17, 24, 36, 55, 61, 64, 67, 68, 82, 83, 108, 112, 118, 129, 152, 174, 184, 201, 207, 216, 228, 235, 241, 244, 258, 277, 293, 386, 398, 435, 533
  planetary, 73, 81, 113, 197, 216, 570, 571
  solar, 102, 244
  space, 67, 81, 177, 179
  Venus, 4, 61, 81, 513, 515, 518, 535, 568, 571
Problems of Space Biology, 419
Project Fire II (spacecraft), 244, 254
Prokhanova, Natasha, 245
Promise (NASA barge), 96, 239, 250, 376, 378, 425
Propellant, 65, 95, 96, 474
Propulsion, 16, 112, 128–129, 137, 156, 352
  chemical, 16, 464, 519
  electric, 16, 30, 44, 55, 116, 521, 547, 571, 572
  ion, 5, 116, 166, 172, 348–349, 472, 571, 572
  jet, 18, 78, 84, 465, 486
  nuclear, 3, 17, 28, 44, 49, 63–64, 116, 122, 165–66, 282, 296, 305, 321, 322, 461, 464, 472, 517, 519, 520, 571

Propulsion (Continued)
  ramjet, 169, 203, 349, 571
  rocket, 2, 13, 24, 28, 32, 40, 42, 49, 52, 60, 63, 64, 81, 95–96, 107, 108, 112, 116, 117, 157, 165, 178, 179, 180, 192, 288, 291, 292, 321, 322, 490–91
Propulsion, Fourth Symposium on Advanced Concepts, 203
Propulsion Joint Specialists' Conference, 286
Proton, 196, 279
PROTON I (U.S.S.R. space station), 333, 336, 339, 342, 343, 345, 348, 356, 369, 371, 431, 571
PROTON II, 502, 571
Proxmire amendment, 324
Proxmire, Sen. William F., 118, 324, 327
PSAC. See President's Science Advisory Committee.
Psurtsev, Nikolai D., 216
Publishers' Lunch Club, 7
Pucinski, Rep. Roman C., 397
Puerto Rico, 109
Pulkovo Observatory (U.S.S.R.), 153, 343, 392
Pulse motor, 24, 321
Purdue Univ., 185, 195
Purdy, William G., 348

Qantas Airlines, 109, 237
Quaide, William L., 555
Quantum electrodynamics, 487
"Quarks" (matter theory), 38
Quarles, Donald A., 445
Quasar (quasi-stellar object), 38, 67, 206, 236–237, 467, 569, 572
  CTA–102, 236
  3C–9, 236
Quatinetz, Max, 285
Queens Chamber of Commerce, 292
Quinn Construction Co., 105

Radar, 41–42, 50, 175, 240, 241, 317, 322, 420, 513
  Pinetree system, 165
Radar calibration sphere. See LCS II.
Radiation, 8, 118, 146, 160–161
  cosmic, 30, 56, 555, 572
  effects, 59, 167, 226, 202, 215, 217, 337, 450, 553
  Mars, 4, 572
  measurement, 56, 102, 129, 157, 169, 184, 266
  nuclear, 7, 13, 167, 206
  solar, 10, 53, 56, 67, 80, 106, 126, 169, 196
  shielding from, 160–161, 202
  space, 217, 553
  Van Allen belt, 31, 56, 81, 157, 184
Radiation, Inc., 45, 272, 492
Radiation spectrometer, 330
Radio astronomy, 1, 13, 14, 42, 43, 370–71
Radio Astronomy Explorer (radio astronomy satellite), 564
  RAE-A, 30

RAE-B, 30
Radio attenuation, 127
Radio Corporation of America (RCA), 156, 174, 209, 249, 268, 283, 440, 572
  Communications Systems Div., 201, 268, 334
  Service Co., 109
Radio Prague, 548
Radio signal, 13, 41–42, 65, 67, 81, 180, 182, 210, 277, 278, 494, 510, 515, 562–563
Radio waves, 261
Radioactive fallout, 34, 248, 522
Radioactive isotope, 99, 197, 258
Radiodiffusion-Television Française, 260
Radiotelescope, 475
Radnofsky, Matthew I., 470
Rads. See Ryukyu Air Defense System.
RAE. See Radio Astronomy Explorer.
RAF. See Royal Air Force.
Raisting, W. Germany, 180, 197
Ram, Project, 143–144
Ramjet, 169, 203, 349, 571
RAND Corp., 350
Ranger (program), 4–5, 100, 108, 148, 149, 150, 175, 180–181, 188, 300, 497
Ranger I (lunar probe), 148–149
Ranger II, 148–149
RANGER III, 149
RANGER IV, 149
RANGER V, 149
RANGER VI, 149, 160
RANGER VII, 4, 137, 160, 187, 360
  photographs, 9, 68, 74, 80, 84, 143, 153, 360
RANGER VIII, 4, 92, 156
  launch, 73–74, 78–79
  impact, 84, 188
  photographs, 84, 89, 95, 97, 104–105, 143, 570
RANGER IX, 4, 101, 156, 181
  launch, 140
  impact, 148, 149, 188
  performance, 142, 146
  photographs, 149–150, 152, 175, 187–188, 570
Rao, M. S. V., 226
Rapid City, S. Dak., 513
Raymond, Arthur E., 32
Raymondville, Tex., 228
Raytheon Co., 117, 368
RCA. See Radio Corporation of America.
RCA Communications Systems Div., 201, 268, 334
Reactor
  atomic, 122
  graphite, 117
  nuclear, 13, 87, 122, 135–136, 180, 227, 240, 243, 255, 272, 282, 304–305, 571
  organic-cooled, 18
  space, 27–28, 30, 44–45, 61–62, 63–64, 112, 117, 166–167, 197
Read, Albert Cushing, 556
Rebka, Dr. Glen A., Jr., 43
Rechtin, Eberhardt, 462

INDEX

Record
   altitude, 213, 245
   flight of visible object, 207
   helicopter, 122
   space communication, 184, 207
   spacecraft, 18, 131, 557, 561
   speed, 144, 150, 189, 213, 453
   weight, 150
   women's, 245
Recoverable spacecraft booster, 8
Recovery technique, 12
Red Star (See also *Krasnaya Zvezda*), 147, 390, 427
Redlands, Calif., 24
Redstone (missile), 373, 445, 531, 553
Redstone Arsenal, Ala., 569
Reed College, 245
Reed, Sylvanus Albert, Award, 32
Reeder, John P., 463
Reentry
   control, 139–140, 141, 145–146, 147, 552
   Gemini, 539, 552
   glider, 88, 92
   heating, 127, 244, 254
   research, 18, 570
   Scout Reentry Heating Project, 559
   Start program, 106
   vehicle, 18, 20, 48, 106, 278, 414–415
   Venus gravity field, use of, 19
Reese, David E., Jr., 64
Refractory metals, 283
Reiger, Siegfried H., 172
Reinartz, Stanley R., 237
Relay (communications satellite), 173
RELAY I, 2
Relay, Md., 344
Reliability, 17, 37, 60, 113, 123, 385, 503, 533, 553, 564, 570
Religion and space, 168
Rendezvous, 23, 141–142, 289, 444, 501, 557
   GT-4, 255, 265, 322
   GEMINI V, 339
   GEMINI VI, 49, 170, 493, 500, 509–510, 538, 551–552, 554, 557, 563, 567, 568, 570
   GEMINI VII, 493, 500, 509–510, 538, 551–552, 554, 557, 563, 567, 568, 570
   GEMINI VIII, 444
   GEMINI IX, 510
Rendezvous, 210
Rendezvous Evaluation Pod (Rep), 317–318, 340
Rennels, Col. Fred L., Jr. (USAF), 5, 408
Rensselaerville, N.Y., 346
Rep. See Rendezvous Evaluation Pod.
Reporter, 288
Republic Aviation Corp., 332, 393
Rescue device, 230
Research and development, 2, 9, 49–50, 69, 446, 473, 534
   aeronautical, 36, 41, 50, 77, 90, 112, 177
   benefits, 198, 446
   Federal support, 10, 125, 426
   funds for, 12, 39, 74, 106, 134, 185, 200, 262
   geographic distribution, 115, 134
   NASA grants, 3, 90, 103, 182, 193, 206
   industry, 130, 190, 262, 305
   information, distribution of, 128
   management, 77, 190
   weapon systems, 5, 39, 130, 170
Research Institute for Advanced Studies, 344
Research Triangle, 217
Reston, James, 542
Retrorocket, 47, 68, 464, 467
Reusable Orbital Module-Booster and Utility Shuttle (Rombus), 8
Re-Usable Orbital Transport (rocket "plane"), 253
Revolución, 405
Reynolds, James, 159
Reynolds, Robert V., 322
RF-111A (supersonic reconnaissance aircraft), 537
Ribonucleic acid (RNA), 323, 425
Rice Univ., 56, 370
Richard, L. G., 359
Richardson, Charles E., 450
Rickenbacker, Edward V., 555
Rickover, V/A Hyman G. (USN), 7, 135, 423, 492
Riehlman, R. Walter, 7, 264, 506
Rieke, William B., 564
Riesel, Victor, 133, 277
Rigel (star), 366
Ritchey, Harold W., 107
Ritland, M/G O. J. (USAF), 101, 530
Rivers, Rep. L. Mendel, 298
RNA. See Ribonucleic acid.
Roanoke, Va., 407
Roberts, Chalmers M., 189
Robertson, Sen. A. Willis, 309
Robinson, Maj. Hugh (USA), 424
Robinson, John H., 301
Robinson, Thomas, 520
Rochelle, Dr. Robert W., 544
Rochester, Univ. of, 223, 376
Rock, Dr. Vincent P., 351
Rockefeller, Gov. Nelson, 341
Rockefeller Institute, 38
Rocket, 2, 9, 313
   braking, 230, 232, 389, 467
   chemical, 16
   design, multi-stage, 52, 127
   foreign, 9, 42, 372, 375, 478, 557
   nuclear, 13, 28, 44, 112, 272, 282, 426, 510, 517, 520, 541, 548
   solid, 2, 24, 81, 107, 179, 210, 219, 223, 244, 270, 288, 297, 435, 449, 550, 562
Rocket engine, 19, 157, 165, 179, 291
   clustering, 365, 491
   electric, 30, 519
   hypergolic, 65, 186, 474
   ion, 116, 166, 172, 255, 472
   liquid fluorine-oxygen, 132
   liquid hydrogen, 27, 31, 141, 365, 526
   liquid propellant, 33, 65, 127, 315, 474

Rocket engine (Continued)
  nuclear, 13, 28, 49, 63, 112, 117, 141, 243, 255, 272, 510, 519, 520, 541, 548
  radioisotope, 197
  solid propellant, 24, 27, 32, 40, 81, 95, 98, 107, 141, 192, 210, 219, 223, 269, 288, 297, 435, 449, 550, 562
  U.S.S.R. "new type," 42, 52, 479
Rocket Engine Test Site (MSFC), 291
Rocketdyne Div. See North American Aviation, Inc., 31, 526
Rockets in Defense of Peace (Soviet TV film), 226
Rockwell Standard Corp., 516
Rodewisch, E. Germany, 232
Rolls Royce, Ltd., 234
Roma, Tex., 228
Romashka (Daisy) nuclear reactor, 548
Rombus. See Reusable Orbital Module-Booster and Utility Shuttle.
Rose Polytechnic Institute, 213
Rosen, Dr. Harold, 232
Roswell Park Memorial Institute for Cancer Research, 168
Rotational stress, 120
Roth, Dr. Emmanuel, 435
Roush, Rep. J. Edward, 106
Rover (nuclear rocket engine), 204, 541
Royal Aeronautical Society, 519
Royal Air Force (RAF), 172
Royal Astronomical Society, 43, 67
Royal Institute of Technology, Stockholm, 198
Royal Radar Establishment, 41
Rozenberg, Prof. Georgiy, 318
RS-70 (bomber), 34
Rudnev, Konstantin N., 339
Rumsfeld, Rep. Donald, 327
Rumford Prize, 119
Rushworth, L/C Robert (USAF), 74, 341, 365
Rusk, Secretary of State Dean, 369
Russell, Sen. Richard B., 40
Russiyan, T., 371
Rust Engineering Co., 109
Ryan Aeronautical Co., 204
Ryle, Martin, 203
Ryukyu Air Defense System (Rads), 241
Ryukyu Islands, 241

SAC. See Strategic Air Command.
Sacramento, Calif., 533
Sacramento Test Center, 409
SAE. See Society of Automotive Engineers.
Safety, 18, 68, 84, 131
Safety Steel Services, Inc., 544
Sagan, Dr. Carl, 262, 372, 481
Sage (Semi-Automatic Ground Environment), 411
Sahara, 113, 229
St. Francis River (rocket launching ship), 459
St. Joseph's Day, 138

St. Louis, Mo., 3, 157, 179, 363
St. Louis Bicentennial Space Symposium, 251, 252, 253, 254
St. Nicholas Island, 1
St. Petersburg, Fla., 52
Salisbury, Australia, 106
Salisbury, Dr. John W., 566
Salpeter, Edwin E., 206
Salto di Quirra, Sardinia, 375, 543
Saltonstall, Sen. Leverett, 7
Samos (satellite), 416, 451, 486, 524
Sampson, M/G George P. (USA), 66, 344
San Antonio Express, 309
San Clemente Island, 49
San Diego, Calif., 133
San Diego Evening Tribune, 133
San Diego Express, 492
San Diego Union, 122, 183
San Francisco, Calif., 83, 109, 255, 374, 383, 535
San Francisco Sunday Chronicle, 179
SAN MARCO I (Italian satellite), 2
San Salvador, 261
Sandage, Dr. Allan R., 38, 237, 277
Sanders, Myrl E., 505
Sands, B/G H. J. (USAF), 479
Santa Clara County, Calif., 50
Santa Fe Engineers, Inc., 105
Santa Susana, Calif., 198, 373
Saphir (French rocket), 467
Sardinia, Italy, 375, 543
Sarnoff, David, 101, 249, 250, 440
Satellite (see also under names of individual satellites and satellite projects), 42, 82, 417, 461, 567
  armed, interception and destruction, 210, 512
  cooperation, international, 2, 38, 58–59, 77, 111–112, 246
  destruction of, 124, 183
  foreign, 12, 21, 41, 76, 86, 94, 109, 120, 123, 124, 152, 189, 197, 216, 227, 260, 282, 311, 337, 339, 364, 379, 421, 483, 504, 525, 526, 527–528, 546, 556, 560, 565
  multiple launching, 238, 341
  surveillance system, 86, 179, 271, 390, 416, 451, 486, 524
  tracking, 12, 14, 89, 105, 114, 119, 174, 202, 228, 236, 276, 292, 363, 418, 497
  use of, 10, 13, 24, 25, 72, 80, 82, 85, 89, 94, 101, 105, 106, 112, 118, 142, 147, 162, 168–169, 193, 197, 207, 226–227, 232, 234, 273, 278, 282, 313, 384, 424
Satellite, communications, 2, 7, 65, 224, 240, 244, 250
  contract, 31, 51, 81, 172, 283, 381, 393, 483, 490, 534, 537
  cooperation, international, 7, 58–59, 95, 250, 254, 322, 502, 545, 555, 570–571
  design, 254, 260, 283, 506

Satellite, communications (Continued)
  foreign, 204, 216, 229, 249, 260, 313, 361, 476, 480, 492
  ground stations, 201, 232, 315, 320, 385, 393, 412, 462, 463, 490, 497, 503, 527, 534, 537
  military use of, 1, 31, 51, 81, 93, 101, 241–242, 283, 319, 441–442
  performance, 172, 176, 180, 197, 214, 221, 228, 236, 313
  rates, 245, 250, 257, 277, 282, 289
  television channels, lease of, 268, 271, 334, 364
  use of, 2, 22, 31, 81, 186, 193, 214, 226, 229, 240, 246, 254, 268, 277, 289, 300–301, 313, 334, 344, 440, 446, 449–450, 495, 532, 541, 562, 569
Satellite, geodetic, 35, 158, 274
Satellite, ionosphere, 24, 207
Satellite, meteoroid detection. See Pegasus.
Satellite, meteorological, 72, 156, 173
  cooperation,
    international, 76, 162, 215
    NASA-Weather Bureau, 26, 163
  funds for, 44
  Nimbus, 44, 45, 87, 144, 532
  plans for, 45, 104, 162, 220
  TIROS IX, 26, 162, 168, 169
  TIROS X, 307, 570
  use of, 81, 162, 169, 226, 245, 424
  U.S.S.R., 76, 85, 245
Satellite, military, 179, 210, 238
Satellite, navigational, 14, 215, 226, 384, 453
Satellite, orbiting observatory (see also Oao, Ogo, Oso), 12, 28, 82, 126
Satellite, polar orbit, 35
Satellite, radio astronomy, 28, 30
Satellite, reconnaissance, 86, 157, 175, 179, 189, 271, 353
Satellite Situation Report, 144, 159
Satellite, solar-powered, 140
Satellite, solar x-ray monitor, 10
Satellite Telemetry Automatic Reduction System (Stars), 105
Satellite, unidentified
  U.S.,
    launch vehicle,
      Atlas-Agena D, 27, 206, 254, 341, 454, 509
      Scout, 562
      Thor-Able-Star, 119, 295, 377
      Thor-Agena D, 18, 93, 114, 152, 207, 236, 238, 336, 447–448, 462, 493, 546, 564
      Thor F4ws, 240, 423
Satellite, weather. See Satellite, meteorological.
Saturday Evening Post, 10, 211
Saturday Review, 1
Saturn (planet), 197, 216
Saturn (program),
  achievements, 72, 97

  contracts, 37, 48, 52, 91, 92, 94, 109, 115, 118, 173, 182, 204
  facilities, construction of, 37, 39, 48, 115
  plans for, 7, 43, 54, 60, 75, 79, 130
  progress, 31, 39, 54, 56, 96, 100, 112, 123, 129, 162, 178, 188, 191, 197, 209
  propulsion, 30, 82
Saturn I (booster), 112, 239, 288, 365, 373, 502, 570
  contract, 94
  launch, 72, 227, 294, 570
    SA-7, 261
    SA-8, 247
    SA-9, 72, 97
    SA-10, 227
  stage, 96, 250
Saturn IB (booster), 31, 82, 237, 377
  contract, 37, 52, 92, 94, 118, 173, 204, 292, 296, 527, 562
  development, 30, 60, 112, 115, 123, 347–348
  engine,
    H-1, 92, 197, 291–292, 490
    J-2, 31, 209, 322, 527
  equipment, 37, 173, 204, 239, 459
  facilities, 37, 311–312, 345
  plans for, 7, 43, 54, 60, 75, 79, 130, 298, 472, 533, 556
  stage, 556, 570
    S-IB-1, 123, 129–130, 162, 341, 376, 377, 543
    S-IB-2, 363
    S-IB-3, 437, 491
    S-IVB, 31, 182, 191, 209, 296, 298, 373, 406, 562
  test, 319, 543
Saturn V (booster), 54, 56, 75, 79, 80, 82, 106, 123, 178, 326, 348, 467, 473, 517
  contract, 48, 92, 109, 115, 173, 182, 204, 447, 527
  development, 30, 348
  engine,
    F-1, 112, 178, 188, 219, 291, 368
    J-2, 198, 322, 527, 535
  equipment, 39, 173, 182, 204, 319, 373
  facilities, 39, 48, 115, 239, 381, 422, 447, 544, 562
  plans for, 7, 43, 54, 295, 429, 476, 533, 556
  stage, 112, 570
    S-IC, 54, 188, 219, 291, 450, 544
    S-IC-T, 54, 100, 241, 275–276, 368, 525, 554
    S-II, 56, 198, 298, 322, 326, 373, 421, 451, 492
    S-IVB, 191, 224, 296, 298, 386, 406, 535, 562
Savage, Melvyn, 370
Savenko, Prof., 431
Saxon, D. R., 306
Scads. See Scanning Celestial Attitude Determination System.
Scanner, horizon, 126, 255, 302, 370, 374, 399, 421, 476, 505

Scanning Celestial Attitude Determination System (Scads), 334
Schawlow, Dr. Arthur L., 545
Scheer, Julian, 356, 438–439, 514
Schirra, Capt. Walter M., Jr., (USN), 49, 170, 294, 325, 449, 489, 493, 511, 547, 548, 550–553, 556, 557, 567
Schisler, Rep. Gale, 46
Schjedahl, G. T., Co., 35
Schmidhauser, Rep. John R., 89
Schmidt, Dr. Maarten, 237
Schmidt, R. A., 191
Schmitt, Dr. Harrison H., 299, 300, 341
Schneider, William C., 479, 567
Schneiderman, Dan, 102, 356, 448, 462
Schriever, Gen. Bernard A. (USAF), 5, 18–19, 106, 128, 147, 151, 170, 203, 241–242, 264, 274, 320, 382, 406, 408, 452, 459, 484
Scriven, B/G George P. (USA), 165
Schult, Eugene, 494
Schurmeier, Harris M., 155, 341
Schutt, Dr. John B., 229
Schwartz, Harry, 280, 339, 427, 548
Schwartz, Leonard, 105
Schweickart, Russell, 321
Schwinger, Julian, 487
Schwinghammer, Robert J., 156
Science, 10, 21, 33, 69, 101, 185, 213, 262, 271, 280, 360, 426, 461, 473, 493, 516, 529, 531, 560, 562
Science (magazine), 97, 198, 322, 337, 415, 441, 448, 488, 496, 564
Science and Technology Advisory Council (New York City), 562
Science Fair, Greater St. Louis, 179
"Science in the Sixties" (seminar), 280–281
Science Museum, London, 461
Science teachers, Soviet, 123
Scientific American, 360
Scientific and Technical Aerospace Reports (STAR), 472
Scientific Engineering Institute, Boston, 280
Scientific Research Radiophysics Institute, Gorki, U.S.S.R., 434
Scientists, 531
 and engineers, 190
 as astronauts, 5, 63, 132, 299, 300, 383
 awards, 58, 67, 73
 interdisciplinarian, 514
 President's Science Advisory Committee, 119, 169
 space program, 17, 30
 training, 88, 133
 training, lunar mission, 10
Scorpio (rocket engine), 157
Scott, Maj. David R. (USAF), 10, 321, 444
Scout (booster), 113, 207, 370, 373, 486, 521, 540, 541, 544, 562
 system management, 80
Scout Evaluation Vehicle (Sev), 373

Scout Reentry Heating Project, 559
Scr. See Silicone-controlled rectifier.
Scramjet. See Supersonic combustion ramjet.
Scribner, 110
Sea of Clouds (moon), 85, 227
Sea of Storms (moon), 467
Sea of Tranquillity (moon), 74, 84–85, 95
Sea Test Range, 49
Sea Vixen (British aircraft), 52
Sea Water Conversion Laboratory, Univ. of Calif., 77
Seaborg, Dr. Glenn T., 135, 304–305, 325, 546–547
Sealab II, 299, 340, 389, 402, 417, 421, 425, 434, 450, 467, 471, 519
Seals, 274
Seamans, Dr. Robert C., Jr., 9, 77, 112, 146, 155, 156, 292, 402, 421, 460, 520, 546, 561, 566
Searcy, Ark., 373
Seattle, Wash., 55
Secor (Sequential Collation Of Range) (geodetic satellite) (USA), 457
SECOR II, 118
SECOR III, 114
SECOR V, 374, 571
Sedov, Prof. Leonid I., 431
See, Elliot M., Jr., 334, 339, 382
Sehlstedt, Albert, Jr., 106
Seife, Alvin, 64
Seismology, 18, 57, 523, 525
Seismometer, 149, 473
Seitz, Dr. Frederick, 185
Seleznez, Vasily, 135
Selfridge AFB, Mich., 233
Selfridge, Thomas Etholen, 555
Seliakov, Leonid, 87–88
Semantics, 423
Semipalatinsk region, U.S.S.R., 18
Senni, Antoine, 66, 170
Senter, Raymond, 424
Sequential Collation Of Range. See Secor.
Serling, Robert J., 316
Sert (See Space Electric Rocket Test), 17
SERT I (spacecraft), 17, 116
Servo-system, 93
Ses. See Space Environment Simulator.
SETP. See Society of Experimental Test Pilots.
Sev. See Scout Evaluation Vehicle.
Seville, Spain, 563
Seychelles Islands (West Indian Ocean), 319
Shai, Charles M., 229
Shank, Robert J., 231
Shapley, Willis H., 343–344
Sharp, Prof. Robert P., 355, 507
Sharpless, Dr. Stewart, 223
Shea, Dr. Joseph F., 43, 57, 113, 185, 238, 325

Sheldon, Dr. Charles S., II, 27, 73, 113, 497, 505
Shepard, Capt. Alan B., Jr. (USN), 443, 461
Shepard, Van, 150, 206, 453, 547, 563
Sherman, Windsor L., 12
Shinn, R/A Allan M. (USN), 275
Shirk, James S., 96
Shklovsky, Dr. I. S., 199
Shneour, Élie A., 61
Shoemaker, Dr. Eugene, 175, 187, 496, 507
Shorthill, Dr. Richard, 8
Short Range Attack Missile (Sram), 20, 24
Siegel, Dr. Peter, 387
Siegel, Dr. Sanford M., 158
Sight Lecture, Second Annual, 249
Sigma Delta Chi National Convention, 514
Sikorsky (aircraft), 131
Sikorsky, Igor I., 32, 372
Silicon-controlled rectifier (Scr), 308
Silliman, Benjamin, Jr., 331
Simkin, William E., 523
Simons, Howard, 11, 189, 246, 401, 412, 434, 528, 563
Simpson, Dr. George L., Jr., 217, 296
Simpson, Dr. John A., 189, 245
Simpson, Ron, 322
Sims, Theo E., 143
Simulator, flight, 191
Singer, Dr. S. Fred, 71, 288
Sinkiang, China, 550
Sinoite, 78
Sinton, Dr. William M., 96
Sissakian, N. M., 313
Sjogren, W. L., 160
Skogasandur, Iceland, 414
Skua (U.K. sounding rocket), 161
Skuridin, Gennadii, 216
Skylark (U.K. sounding rocket), 226, 352
Skyraider (bomber aircraft), 291, 381
Slayton, Donald K., 567
Slidell, La., 105
Slish, Vyacheslev, 494
Sloan, Richard K., 190, 355
Sloop, John L., 131
"Slowdown in the Pentagon," 45
Small Unified Reactor Facility with Systems for Isotopes (Surfside), 341
Smathers, Sen. George, 394
Smith and Sapp Construction Co., 228
Smith and Sons, Ltd., 275
Smith, C. R., 59
Smith, Francis B., 164
Smith, Gerald L., 119
Smith, Joan Merriam, 78
Smith, Sen. Margaret Chase, 24, 238
Smith Mountain Dam, Va., 407
Smith, William D., 489
Smithson, James, 296
Smithsonian Astrophysical Observatory, 67, 82, 262, 372, 376, 447, 505, 528
Smithsonian Institution, 255, 296, 502
  Bicentennial Celebration, 436
  Museum of History and Technology, 445
  National Air Museum, 43, 55, 128, 156, 162, 329
Smoluchowski, Dr. Roman, 202
Smull, Dr. Thomas L. K., 103, 533
Snap (System for Nuclear Auxiliary Power), 122, 292
  funds for, 28
Snap-7D, 59
Snap-8, 27, 61, 100, 141, 219, 272
Snap-9A, 7
Snap-10A, 30, 166, 172, 243, 511, 569
Snap-19, 197
Snap-50, 63
SNAPSHOT (satellite), 167, 243, 569
Snowden, Earl of (Antony Armstrong-Jones), 511
SNPO. See Space Nuclear Propulsion Office.
"Social Implications of Space Exploration," 384
Société des Ateliers d'Aviation Louis Brequet, 234
Société Générale Aéronautique Marcelle Dassault, 234
Society of Automotive Engineers (SAE), 185, 368, 463
Society of Experimental Test Pilots (SETP), 453–454
Society of Women Engineers, 92
Sodium, liquid, 296
Sodium-vapor experiment, 6
Soesterberg, Netherlands, 23
Sofar. See Sound Fixing and Ranging Device.
Soffen, Gerald A., 61
Sofia, Bulgaria, 228
Sohl, Gordon, 348
Sohn, Robert L., 19, 61, 429
Sokolovsky, Marshal Vasily (U.S.S.R.), 78
Solar boat, 299
Solar cell, 67, 74, 484
Solar eclipse, 126, 154
Solar energy, 3
Solar flare, 55, 80, 163, 224
Solar mill, 318
Solar plasma probe, 102, 217
Solar Radiation satellite. See Solrad.
Solar still, 77–78
Solar system, 60, 62, 312, 372
Solar wind, 32, 36, 257, 330
Solid propellant, 2, 32, 40, 81, 95, 98, 100, 104, 141, 192, 288, 321, 335
Solomon Islands, 525
Solrad (Solar Radiation satellite), 10
SOLRAD IX, 521, 571
Sonic boom (see also Noise), 65, 199, 379
  supersonic bomber, 6, 73, 159
  supersonic transport, 15, 48, 73, 199, 241, 572

Sound Fixing and Ranging Device (Sofar), 230
Sounding rocket
  experiments, 6, 15, 31, 35, 53, 58, 108, 120, 132, 167, 169, 180, 184, 195, 214, 215, 239, 249, 264, 278, 287, 289, 294, 302, 397, 439, 447–448, 534
  international programs, 163, 195, 236, 243, 439, 443, 447–448
  launch, 112, 163, 571
    Brazil, 557
    France, 478
    Indonesia, 21, 373, 375
    Italy, 375
    NASA, 6, 15, 35, 53, 58, 108, 120, 132, 139, 167, 169, 180, 184, 195, 196, 214, 215, 239, 249, 264, 278, 287, 294, 302, 397, 439, 447–448, 473, 484, 510, 511, 525, 534, 571
    Netherlands, 443
    USAF, 39
    USN, 31, 132, 163, 169
  use of, 6, 15, 31, 35, 53, 58, 120, 132, 161, 167, 169, 180, 184, 195, 211, 450
Soundovac, 179
South Africa, 14, 55, 58, 76, 297, 300, 347, 438, 497
South Atlantic anomaly, 266
South Pole, 57, 387, 557
South Vietnam, 290, 314, 573
Southern Governor's Conference, 164
Southern Interstate Nuclear Board, 164
Sotheby's, 246
Soudan Formation, 223
Southern Methodist Univ., 414
Southwest Center for Advanced Studies, 410
Sovetskaya Rossiya, 382
Soviet Academy of Sciences, 45, 76, 78, 183, 187, 225, 278, 313, 335, 342, 379, 390, 392, 396, 405, 412, 415, 420, 431, 462
Soviet Central Scientific Research Institute of Communications, 497
Soviet Life, 262
Soviet Space Law Commission, 367
Soviet Union. See U.S.S.R.
Sovietsky Patriot, 323
Space Business Daily, 519
Space Conference, Fifth Annual, 393
Space Congress of the Canaveral Council of Technical Societies, Second, 171
Space Corp., 164
Space Defense Center, 210
Space Detection and Tracking System (SPADATS) (NORAD), 210, 228
Space Electric Rocket Test. See Sert.
Space environment, 62, 81, 129, 137, 385
  simulated, 36, 129, 134–135, 181
  testing, 68
Space Environment Simulator (Ses), 328
Space, exploration of, 177, 178, 384, 461, 471, 488
Space Fair, San Diego, 134
Space garden, 188

Space-General Corp., 289, 348, 367, 405
Space, impact of, 17, 27, 29, 30, 119, 137–138, 142, 151, 153, 179, 185, 208
Space junk, 13, 57, 456, 543, 563
Space laboratory, 102
Space law, 367, 531–532, 536, 559
Space medicine, 45, 55, 70, 412, 415, 433, 489, 494, 533, 538, 548, 566
Space, military use of (see also Manned Orbiting Laboratory), 10, 46, 102, 137, 168, 238, 365, 369, 459, 475, 534
  anti-satellite defense, 241–242
  manned space flight, 42, 171, 241–242, 544
  missile detection and warning, 241–242
  nuclear detonation detection, 241–242
  objectives, 62, 239, 327, 459
  space station, 151, 290
  U.S.S.R., 78, 153, 171
Space Monument, Moscow, 178
Space Nuclear Propulsion Office (SNPO) (NASA-AEC), 3, 44
Space, peaceful use of, 35, 55, 69, 75, 105, 138, 245, 251–254, 413
Space program, national
  accomplishments, 4, 35, 36, 43, 54, 69, 72, 76–77, 82, 83, 89, 108, 117, 118, 147, 150, 163, 226–227, 245, 251, 252, 267, 323, 336, 358, 371, 380, 432, 459, 503, 549–550, 552, 558, 559
  budget, 28–29, 40, 62, 75, 100, 141, 192, 219, 535
  cost of, 4, 54, 91, 96–97, 106, 177, 191–192, 210, 217, 250, 276, 356, 380, 501, 530
  criticism, 1, 3, 11, 44, 46, 63–64, 66, 97, 121–122, 139, 167–168, 210, 254, 273, 420, 438, 487, 495, 499, 530, 534, 545, 557
  economic impact, 460, 569
  future developments, 228, 251, 253, 279, 331, 348, 382, 393, 460, 464, 498
  military, 20, 27, 88, 102, 108, 137, 153, 168, 182, 218, 242, 290, 347, 350, 366, 367, 369, 371, 400, 401, 438, 504
  need, 119–129, 133, 138, 446, 514
  objectives, 27, 60, 69, 77, 86, 97, 101, 102, 125–126, 137, 139, 174, 201, 210, 214, 240, 255, 293, 350, 351–352, 371, 396, 431, 520, 529, 539, 560–561
  policy, 13, 55, 123, 182, 217–218, 366–367, 401, 444, 487
  support of, 62, 86, 89, 150, 228, 433
  universities, 103, 213, 504
"Space Programs and the Federal Budget," 192
Space race, 11, 243, 283, 471, 473
  booster, 73, 347
  manned space flight, 143, 278
  moon, 181, 228, 280, 327, 457, 466, 545
  military, 154, 159, 168, 177, 278, 313, 327, 488
  payload, 252, 348
  probe, 32

U.S.-U.S.S.R., 32, 56, 73, 142, 144, 151, 178, 204, 214, 235–236, 252, 278, 486, 548
Space Radiation Effects Laboratory, 553
Space research, 90, 102, 104, 105, 115, 136, 146, 147, 151, 156, 161, 179, 534
Space Research Coordination Center, 521
Space Research Summer Study, 293, 397
Space results, 54, 81, 101, 112, 114, 150, 270, 344, 383, 384, 405, 407, 459, 483, 488, 492, 499, 518–519, 534, 570
   astronomical data, 15, 55, 81, 108, 129, 226, 344, 359, 431, 442, 483, 571
   bioscience, 115, 132–133, 146, 167, 267, 483, 503
   communication, 226, 273, 313, 483, 503
   education, 213, 270, 503
   geology, 344, 359, 432, 532
   laser signals, 24, 67
   lunar topography, 4, 43, 80, 84, 104, 108, 149, 153, 187, 379, 380, 392
   Mars, 36, 293, 326, 327, 330, 333, 335, 338, 355
   meteorology, 110, 216, 226, 245, 267, 318, 424, 432, 455, 483, 503.
   micrometeoroid density, 193, 310, 376,
   navigation, 226, 384, 504
   radiation, 80, 81, 94, 106–107, 118, 157, 483
   science, 179, 262, 503
Space science, 27, 30–31, 78, 82, 214, 344, 384, 412, 461
   curriculum, 134
Space Science Award, 32
Space Science Board (NAS), 17, 201, 246, 293, 397
Space station, manned, 151, 174, 318, 383, 464, 499, 536, 544, 554
   foreign, 346, 442
   inspection of, 105
   MoI, 28, 29, 30, 66, 86, 101–102, 115, 171, 183, 549, 560, 567
Space suit, 125, 128, 132, 134, 138, 152, 190, 208, 216, 237, 248, 421, 430, 470, 500, 506, 512, 540
Space Surveillance Calibration. See SURCAL.
Space Technology Laboratories, 19, 33, 61, 65, 229, 283, 429
Space Tracking and Data Acquisition Network (STADAN), 448, 461
Space World, 161
Spacecraft
   braking, 19, 68, 232, 467, 542
   design, 18, 94, 104, 123, 126, 142, 153, 157, 222, 231, 247, 387, 517, 528
   development testing, 21–22, 25, 55, 79, 88, 97, 110, 128, 182, 227–228, 239, 244, 293–294, 301, 383
   electrical equipment, 503, 547
   environment, simulated, 173, 432, 501, 550
   escape system, 66, 301, 513
   exhibit, 156, 173, 366, 461
   experiments pallet, 524
   extravehicular equipment, 522, 531
   heating, 21, 22, 36, 222, 230, 239, 244, 254, 323
   instrumentation, 143, 494–495, 509, 515
   landing system, 65, 112, 228, 323, 358, 506, 550–551
   maneuverability, 68, 79, 145, 151, 155, 216, 318, 388, 464
   military missions, 20, 29, 62, 66, 369, 416, 451, 459, 504
   propulsion systems, 16, 40, 42, 44, 83, 296, 347, 358, 391, 417, 435, 438, 464, 472, 476, 519, 523
   nuclear, 390, 461, 510, 517
   record, 131, 403, 427, 447, 557
   recovery systems, 12, 87, 563
   reliability, 17, 60, 503
   rendezvous, 388, 449, 493, 500, 501, 563, 567
   reusable, 268, 398, 464
Spacecraft Technology and Advanced Reentry Tests program (Start), 106
Spaceport, 178, 422, 464
Spaco, Inc., 92, 109, 503
SPADATS. See Space Detection and Tracking System (NORAD).
Spadeadam Rocket Establishment, England, 365
Spain, 325, 563
Spangler, Eugene, 470
Sparrow-Arcas (sounding rocket), 511
Spectrogram, 264, 299
Spectrograph, 120, 132
Spectroheliograph, 107, 179
Spectrometer, 62, 107, 215, 485
Spectrophotometer, 62
Spectroscopy, 215
Speed Scientific School, 9
Spence, Roderick W., 306
Spero, Donald, 279
Sperry, Elmer A., Award, 32
Sperry Farragut, 48
Sperry Rand Corp., 109
   contract, 121
   Univac Div., 73, 555
Sphere experiment, 372
Spica (star), 215, 261
Der Spiegel, 318–319
Spilhaus, Dr. Athelstan, 129
Spirit of St. Louis Medal, 129
Spitsbergen, 76, 80–81
Spray, Norm, 201
Sprint (antimissile missile), 156, 522
Sputnik (U.S.S.R. satellite), 52, 56
SPUTNIK I, 139, 445, 460
SPUTNIK II, 139
Sram. See Short-range attack missile.
S.S. France, 278
SSt. See Supersonic transport.
SST Development Corp., 297
Stability and control, 127
Stabilization, attitude, 227, 236
Stack, John, 350

STADAN. See Space Tracking and Data Acquisition Network.
Staff. See Stellar Acquisition Flight Feasibility Test.
Stafford, Maj. Thomas P. (USAF), 170, 449, 489, 510, 511, 547-548, 551-552, 556, 557, 567
Stalony-Dobrzanski, J., 5
Stamford Museum and Nature Center, 233
Standard Telephone Labs., 549
Stanford, Neal, 39
Stanford Research Institute, 514
Stanford Univ., 61, 174, 300, 372, 420, 545
Stanton, Dr. Frank, 240
Stanyukovich, K., 507
Star, 30, 38, 206
  dwarf, 70
  halved, 57
  intensity, 82
  radiation, 215, 261
  study of, 223, 512
  variable, 223
STAR. See *Scientific and Technical Aerospace Reports.*
Star collision, 281-282
"Starfish" electrons, 337-338
Stars. See Satellite Telemetry Automatic Reduction System.
Start. See Spacecraft Technology and Advanced Reentry Tests program.
Stassinopoulos, E. G., 338
State Committee of Inventions, U.S.S.R., 129
State, Dept. of, 121, 435, 517, 522
  Policy Planning Council, 100
Staten Island Public Health Service, 527
Statistical standards, 354
Stecker, E. J., 315
Steel Executive (cargo vessel), 406, 443
Steg, Dr. Leo, 98
Steinmetz, Charles Proteus, Centennial Medal, 209
Stellar Acquisition Feasibility Flight (Staff) program, 184, 513
Stellar Inertial Guidance System (Stings), 184, 513
Stendahl, Dr. Krister, 168
Stennis, Sen. John, 84
Stephens, Col. Robert L. (USAF), 213
Sternberg Astronomical Institute, Moscow, 180, 182, 199
Stevens, Capt. Albert W. (USA), 513
Stewart, Dr. Harris B., Jr., 565-566
Still, solar, 77-78
Stings. See Stellar Inertial Guidance System.
Stockholm, Sweden, 198
Stoiko, Michael, 429-430, 536
Stol aircraft, 50, 52, 63, 197
Stormfury, Project, 308
Stormy Spring, Project, 124
Strategic Air Command, (SAC), 63, 304, 391

"Strategies for Survival in the Aerospace Industry," 13, 52
Stratoscope II (balloon), 326
Stratosphere, 318
Strebig, James J., Memorial Award, 244
Stress analysis, 17
Stress, Rotational, 120
Strikes
  Boeing Co., 444-445, 447, 462
  Cape Kennedy, 8, 39, 49, 133, 157, 159, 295, 301, 302
  Goddard Space Flight Center, 308
  Kennedy Space Center, 277, 437, 444-445, 447, 527
  McDonnell Aircraft Corp., 510, 512-513, 520, 523, 524
  Mississippi Test Operations, 131
  North American Aviation, Inc., 435-436
Strong, Dr. John, 393
Strother, Fred P. 109
Strughold, Dr. Hubertus, 324-325
Stuhlinger, Dr. Ernst, 505
SU-6 (behavior of matter theory), 38
Submarine
  missile bearing, 541
  nuclear, 59, 78, 538
Sud-Aviation (France), 237, 479
Sudan, 59
Suitland, Md., 24
Sudets, Soviet Air Marshal Vladimir Aleksandrovich, 383
Sullivan, Francis J., 225
Sullivan, Walter, 213, 237, 326, 338
Sulphuric oxide, 318
Summerfield, Martin, 211
Sun (see also Solar cell, etc.), 31, 47, 182, 188, 259
  exploration, 217, 243, 563
  IQSY, 10, 163, 194, 196, 197, 226, 511, 521, 523, 553, 555, 556
  magnetic field, 223-224, 565
  radiation, 10, 53, 56, 67, 80, 107, 126, 169, 196
  satellite data, 10, 521
"The Sunday Show" (TV program), 85
Sunspots, 163, 196
Super Guppy (aircraft), 406, 449, 562
Supernovae, 573
Supersonic combustion ramjet (Scramjet), 203-204, 543
Supersonic flow theory, 32
Supersonic transport (Sst) (see also aircraft, supersonic transport), 8-9, 15, 16, 28, 34, 36, 41, 228, 309, 310, 311, 316, 417, 481, 490, 491, 515, 522, 545, 555, 571
SURCAL (Space Surveillance Calibration satellite), 114, 238, 377
Surfside. See Small Unified Facility with Systems for Isotopes.
Surveillance, airspace, 241
Surveyor (program), 118, 181, 204, 342, 466, 477, 501

Surveyor (spacecraft), 64, 103, 188, 374, 570.
Sutro, Louis, 281
sv-5 (wingless aircraft), 106
Sweden, 325, 352, 466
Swiss Federal Institute of Technology, 73
Switzerland, 325, 552
Sydney, Australia, 109
Sylvania Electric Products, Inc., 537
Sylvester, Arthur, 474
Symington, Sen. Stuart, 103, 127, 513
Symmetry group theory, 38
Symposium on Advanced Propulsion Concepts, Fourth, 203
Symposium on Fluid Dynamics and Plasma, 466
Symposium on Meteor Orbits and Dust, 376
Symposium on Post-Apollo Space Exploration, 216, 235
Symposium on Space and Ballistic Missile Technology, 366
Symposium on Unmanned Exploration of the Solar System, 59, 60, 61, 62, 64
Synchroton, 279
Syncom (communications satellite), 106, 147, 173
SYNCOM II, 1, 88, 102, 106, 136, 319, 570
SYNCOM III, 1, 2, 82, 88, 319, 570
Syria, 38
Syverton, Clarence A., 69

T-33 (jet trainer), 222
T-38 Talon (jet trainer), 246
Tabanera, Teofilo, 236
Tactical Air Command (TAC), 340
Tactical Fighter Experimental (Tfx). See F-111.
Tad. See Delta, Thrust Augmented.
Talcott, Rep. Burt L., 315
Tanner, Carol S., 228
Tantalum, 535–536
Tascher, John, 137
Tass, 42, 43, 125, 221, 222, 223, 254, 274, 311, 313, 318, 329, 337, 342, 343, 356, 364, 371, 377, 378, 387, 391, 394, 413, 418, 460, 462, 463, 473, 479, 483, 494, 509, 512, 513, 526, 527, 530, 535, 539, 540, 548, 550, 560, 568
Tate, Mayor James H. J., 472
Taylor, Charles Edward, 555–556
Taylor, Capt. James M. (USAF), 514
Taylor, Rep. Royal A., 46
Taylor, William B., 469
Taz-8 (alloy), 56
Teague, Rep. Olin E., 108, 171, 219, 404, 567
Tech/Courier Corp., 334
Technology, 7, 10, 16, 18, 21, 33, 41, 185, 213, 250, 260–261, 281, 473, 492, 498, 516, 529, 534, 558, 559, 562
Technology and Culture, 127
Technology Status and Trends Symposium, 194

Technology utilization, space, 14, 27, 30–31, 35, 90, 101, 114, 118, 125–126, 162, 171, 174, 175–176, 185, 194, 503
Technology Utilization Program, 96, 161, 369, 456, 571
Tektite, 85, 96, 507
Tel Aviv, Univ. of, 321
Telecommunication, 108
Telecomputing Services, Inc., 474, 512
Telefunken, 205
Telemetry, 1, 15, 92, 105, 136, 142, 243
Telescope, 4–5, 311
  orbiting, 12, 82
Television, 41, 57, 128, 132, 214, 245, 246, 250, 254, 360, 383, 387, 456, 484, 503, 530, 539, 555, 562
  color, 254, 383
Television Infrared Observation Satellite. See Tiros.
Telstar (communications satellite), 173, 221
TELSTAR II, 41, 236
Temm, Peter, 200
Tennessee, 255
Tennis, Richard, 186
Teplinskiy, M/G G. (U.S.S.R.), 10
Tepper, Dr. Morris, 163, 225
Test facility, 26, 100, 105, 115, 123, 160, 533
Texas Instruments, Inc., 374, 513
Texas Women's Univ., 540
Textron Corp., 165
Tfx (Tactical Fighter Experimental). See F-111.
Thayer, Sylvanus, 88
Theoretical Astronomy Institute, Leningrad, 372
Thiokol Chemical Corp., 65, 95, 107, 179, 492
  contract, 197, 270, 297, 315, 453, 474, 567
  Reaction Motors Div., 315
  Wasatch Div., 32
Thomas, Rep. Albert D., 172, 276
Thomas, David D., 303
Thomas, John E., 159
Thomas, Lowell, 408
Thompson, Dr. Floyd L., 16, 518
Thompson, Milton O., 15, 247, 287, 370, 397
Thompson-Ramo-Wooldrige, Inc. (see also TRW, Inc.), 61, 65, 197
  Space Technology Laboratories Div., 19, 33, 61, 195
Thompson, Robert F., 297, 314
Thompson, Thomas W., 193
Thor (missile), 57
Thor-Able (booster), 163
Thor-Able-Star (booster), 13, 74, 118. 295, 377
Thor-Agena (booster), 332
  launch, 272, 546
  B, 527
  D, 18, 93, 114, 152, 207, 236, 238, 336, 339, 382, 447, 462, 493, 546, 564

Thor-Altair (booster), 20, 132
Thor-Delta (booster), 257
Thor FW4S (booster), 240, 423
Thorium, 135
Thrust-Augmented Improved Delta, 507
Thumba, India, experiments, 226
Thunderbird Aerial Team, 316
Thurmond, Sen. Strom, 320
Thurston, Robert, Lecture, 511
Tidbinbilla, Australia, 136, 337, 353
Tides, 505, 564
Tillich, Paul, 362
Tilton, Thomas D., 156
Time magazine, 232
Tipton, Stuart G., 114, 118, 469
Tiros (meteorological satellite), 72, 156, 162, 173, 307
TIROS I, 162
TIROS VII, 290, 420
TIROS VIII, 420
TIROS IX, 26, 162, 168, 169, 282, 307, 420, 570
TIROS X, 301–302, 307, 413, 420
Tiros Operational Satellite system (Tos), 26, 162
Tischler, Adelbert O., 472
Tison, R/A James C. (USN), 420
Titan I (missile), 16, 57, 109, 161
Titan II (booster), 21, 57, 66, 128, 145, 288, 347, 373, 493, 538, 547, 548, 551
 launch facility, 11, 14
Titan II (missile), 304, 377
Titan III (booster), 12, 28, 66, 89
 cost, 385
Titan III-A (booster), 12, 19, 65, 220, 227
Titan III-C, 287, 288, 290, 291, 295, 329, 347, 348, 396, 459, 477, 498–499, 559–560, 562–563, 564, 571
Titan III-X, 274
Titanium, 58
Tito, Pres. Josip Broz, 413
Titov, L/C Gherman S., 272
Tnt. See Transient Nuclear Test.
Tokyo Japan, 523
Tokyo, Univ., 260
Tokyo Univ. Aeronautical Institute, 133
Tolansky, Samuel, 290
Tolubko, Col. Gen. Vladimir, 423, 515
Tombaugh, Dr. Clyde W., 109
Tomonaga, Shinichero, 487
Topaze (rocket), 229
Torrejon AFB, Spain, 284
Tos. See Tiros Operational Satellite.
Townes, Dr. Charles H., 209, 545
Townsend, Dr. John W., 344, 375
Towson, Md., 240
Tracking, 13, 89, 114, 119, 202, 363, 450
 deep space, 174, 364, 377
 laser beam, 276
 missile, 49, 233
 stations, 114, 117, 160
  Alaska, 276
  Australia, 89, 108, 136, 337, 520, 543
  Denmark, 170

France, 14, 75
Germany, West, 292, 377
Mexico, 107
Norway, 81
South Africa, 14, 55, 297, 497
Spitsbergen, 80–81
U.K., 78
U.S., 48–49, 87, 89, 174, 228, 467
 training, 418
Trampoline bed, 55
Transient Nuclear Test (Tnt), 13
Transit (satellite), 14
Transponder, 456
Transportation, 198, 237, 454, 457
Trans-World Airlines, 119, 297
Trask, D. W., 160
Travis AFB, Calif., 198, 302
Treaty, nuclear test ban, 3, 34, 40–41, 81
Treib, Albert J., 366
Tri-State Roofing Co., 478
Trident (aircraft), 275
Troitsky, Vsevolod, 43, 434
Trud, 497
Truly, Lt. Richard H. (USN), 514
Truszynski, Gerald M., 88
TRW, Inc. (see also Thompson-Ramo-Wooldridge, Inc.), 229, 367, 492, 549, 555
 Space Technology Laboratories Div., 229
Trybura Ludu, 339
Tschudi, President Hans Peter (Switzerland), 301
Tsinghua Univ., 261
TSR-2 (tactical strike-reconnaissance aircraft), 17, 52, 87, 549
TU-134 (jetliner), 17, 52, 87, 549
TU-144 (supersonic transport), 279, 318, 337
Tula, Russia, 300
Tullahoma, Tenn., 535
Tulsa, Okla., 185
Tunguska (meteorite), 258
Tunisia, 59
Tupolev, Andrei N., 87, 372
Turbocompressor, 438
Turbomeca, 234
Turbulence, atmospheric, 551
Turkey, 441, 442
Twiss, Peter M., 456
Tycho (lunar crater), 8
Tydings, Sen. Joseph, 46
Typhoon Ruby, 142
Typhoon Sally, 142

U-2 (photographic airplane), 124
U-235, 243
U.A.R. See United Arab Republic.
UAW. See United Auto Workers.
Uchinoura, Japan, 133–134
Uchitel'skaya Gazeta, 345
UCLA. See California, Univ. of, at Los Angeles.
Udall, Interior Secretary Stewart L., 124
Ufo. See Unidentified flying objects.

INDEX 673

U.K. See United Kingdom.
Ulrich, Dr. Bruce T., 495
U.N. See United Nations.
Underground nuclear test, 18, 34, 81, 466, 524
UNESCO (United Nations Educational, Scientific and Cultural Organization), 541, 545
Unidentified flying objects (Ufo), 38, 52, 324, 327, 366, 374, 438
Union Carbide Corp., 157, 445
Union Carbide Research Institute, 157, 261
Union of Soviet Socialist Republics. See U.S.S.R.
United Aerospace Workers Union, 435, 468
United Aircraft Corp., 209
   Hamilton Standard Div., 125, 506, 540
   Pratt and Whitney Div., 120, 137, 163, 309, 312
   Vtol study, 393
United Airlines, 189, 221, 387, 545, 555
United Arab Republic (U.A.R.), 6, 32, 59, 187, 325, 393
United Association of Plumbers and Pipefitters, 159
United Auto Workers (UAW), 115, 186–187, 295
United Kingdom (U.K.), 2, 75, 117, 164, 234, 246, 267, 522, 525
   aircraft, 2, 17, 34, 52, 63, 87, 188, 274, 314, 318, 357, 486
   Defense Ministry, 17, 34
   Dept. of Education and Science, 164
   Dept. of Scientific and Industrial Research, 110, 164
   General Post Office, 549
   House of Commons, 117
   launch, 143, 526
   Meteorological Office, 161
   military, 63, 178, 234, 426, 486
   Ministry of Aviation, 117, 486
   Ministry of Technology, 164
   rocket, 2, 143, 365, 526, 545
   Science Research Council, 164
   space program, 164, 354
   supersonic transport, 48, 413
      Concorde, 73, 204, 237, 316, 319, 337, 417, 479
United Nations (U.N.), 68, 147, 298, 344, 418, 424, 463, 474
   Committee on Peaceful Uses of Outer Space, 254
   General Assembly, 163, 367, 559
   resolution against weapons in space, 521–522, 544
United Press International (UPI), 246, 401
United States (U.S.) (see also appropriate government agencies),
   budget, 12, 27, 54, 60, 61, 74, 75, 86, 90, 192, 313
   defense, 13, 102
   economy, 86, 98
   expenditures, 9, 536
   goal, 17, 69, 200, 201, 246, 366, 371, 427
   Government
      criticism of, 142, 152, 167, 177, 423, 427, 558
      R&D, 10, 92, 97, 124, 134, 185, 190, 203, 360, 426, 473, 536
      science and technology, 20, 69, 92, 101, 103, 119, 169, 185, 271, 298, 357, 426, 434, 497, 515
   information, exchange of, 121, 131, 273, 343, 494
   information, freedom of, 128, 203
   international relations, 121, 259, 299, 517
   manpower, 131, 147
   military, 20, 21, 37, 39, 42, 45, 69, 74, 84, 105, 170, 210, 369
   patents, 264, 287, 299, 359, 487
   peace, 55, 86, 103, 366, 432
   policy, 92, 121-2, 128, 181, 216, 237, 264, 432, 446
   security, national, 19, 20–21, 38, 55, 105, 137, 144, 147, 153, 241, 270, 298
   space
      accomplishments, 4, 36, 43, 51, 54, 72, 81, 105, 112, 146, 149, 152, 155, 227, 252, 279, 380, 383, 412, 414, 418, 458–9
      activities, 35, 75, 79, 94, 116
      capability, 29, 42, 54, 80, 105, 138, 175, 252, 335, 390, 416
      censorship, 410
      power, 79, 153, 171
      race, 11, 31, 73, 108, 110, 113, 137, 138, 142, 144, 150, 153, 167, 182, 211, 235, 243, 251, 278, 292, 312, 321, 327, 351, 403–4, 427, 441, 445, 473
United Technology Center (UTC), 288, 297, 541
Universe, 416, 569, 572–573
Universities, 57, 111, 213, 223, 226, 239, 240, 261, 264, 269, 273, 279, 281, 299, 311, 321, 326, 435, 473, 486, 509, 554, 572
   computer use by, 9, 108
   graduate training program, 130, 555
   grants, 96, 103, 134, 185, 223, 232–233, 274, 303, 327, 353, 370, 376, 420, 424, 530, 538
   NASA-Western University Conference, 510
   space exploration, impact on, 17, 255, 260, 270, 373, 468
University Explorers Program, 56, 94
University Program Review Conference, 103, 106
Unmanned Spacecraft Meeting, 101–102
UPI. See United Press International.
Upper Volta, 76
Uranium, 305
Urey, Dr. Harold C., 58, 85, 95, 149, 187, 206, 474

Uruguay, 25
Urumchi, Sinkiang, China, 550
U.S. See United States.
USA. See U.S. Army.
USA Ballistic Research Laboratories, 501
USAF. See U.S. Air Force.
USAF Contract Management Div. (AFCMD), 408
USAF Flight Safety Review Board, 509–510
USAF Office of Aerospace Research (USAF OAR), 62, 478
USAF San Bernardino Air Materiel Area, 161
USAF School of Aerospace Medicine, 167, 173, 305, 494, 532
U.S. Air Force (USAF) (see also Defense, Dept. of, and individual commands and laboratories), 48, 153, 440, 442, 453, 458, 550, 563, 573
   Aerospace Corporation, 377, 419
   Agena Target Vehicle, 55, 166, 509, 524
   aircraft, 15, 19, 54, 144, 178, 191, 198, 213, 221, 248, 284, 290, 322, 329, 333, 454, 481, 484, 491, 515, 537
     F–111, 14, 17, 19, 20, 52, 90, 109, 121, 172, 325, 333, 429, 537
     F–111A, 6, 121, 179, 180, 183, 537
     XB–70, 19, 73, 77, 84, 111, 191, 206, 568
     XB–70A, 77, 150, 284, 308, 351, 482, 562
   aircraft defense, 411
   astronaut, 302, 331, 399, 406, 514
   Atlas SLV–3 (booster), 123, 183
   awards, 13, 223, 354, 471
   contract, 5, 84, 144, 160, 240, 302, 377, 387, 408, 451, 492
   booster, 8, 32, 66, 165, 183, 265, 274, 297, 435, 498, 559, 567
   cooperation,
     AEC, 243
     NASA, 178, 218, 300, 305, 444, 452, 464, 482, 483, 493
     NASA-USN, 41, 178
     USA, 473
   laser program, 24, 67
   launch,
     failure, 39, 184, 254, 287
     missile, 49, 101, 144, 310, 426, 514
     operational, 24, 54, 63, 67, 152
     R&D, 49, 109
   nuclear reactor, 243
   probe, 176, 227, 272
   reentry vehicle, 18, 88, 92
   rocket, 64, 114, 220, 511
   satellite, 18, 20, 24, 27, 93, 114, 118, 120, 132, 152, 207, 237, 238, 240, 254, 272, 295, 296, 336, 341, 423, 447, 448, 454, 462, 493, 510, 546, 562, 564
   management, 4, 114, 456
   meeting, 147, 274
   missile program, 48, 57, 101, 161, 184, 304, 377, 473, 479, 514, 541
   Mol, 28, 29, 66, 86, 171, 183, 290, 295, 329–330, 346, 367, 396, 400, 401, 406, 416, 424, 441, 463, 484, 494, 498, 549
   rocket motor, solid propellant, 321, 157, 165, 297
   Scramjet, 203, 543
   sonic boom tests, 6, 73, 159
   space program, 12, 58, 97, 106, 144, 145, 147, 152, 160, 167, 217, 242, 353, 368, 459, 475, 498, 549
   test, 63, 65, 84, 88, 106, 183, 184, 219, 513
   Titan II (booster), 21, 66, 128
   Titan III, 12, 66, 89, 274, 385, 471
   Titan III-A, 12, 19, 65, 229
   Titan III-C, 287, 288, 290, 291, 295, 329, 347, 348, 396, 459, 498, 560, 564, 565, 571
   Ufo, 38, 52, 327, 366, 374
U.S. Air Force, Aerospace Research Pilot School, 246
U.S. Air Force Scientific Advisory Board, 451
U.S. Arms Control and Disarmament Agency, 34
U.S. Army (USA), 256, 450, 506
   Advanced Aerial Fire Support System, 504
   antimissile missile, 156, 309, 452, 522
   contract, 4, 48, 452, 454, 521
   helicopter, 39, 254, 521
   missile, 125, 373, 444, 463, 531, 553
   research, 73, 295
   satellite, 118, 373, 571
   satellite communications, 319, 441–442
   V/Stol aircraft, 41, 90, 204, 248
U.S. Army Ballistic Research Laboratories, 501
U.S. Army Electronics Command, 73
U.S. Army Materiel Command, 69, 295
U.S. Army Engineer Geodesy, Intelligence and Mapping Research and Development Agency (GIMRADA), 457
U.S. Army Nike-X Project Office, 447, 450
U.S. Army Satellite Communications Agency, 102
U.S. Army Strategic Communications Command, 319
U.S. Army Transportation Research Command, 39
U.S. Bureau of Mines, 312
U.S. Chamber of Commerce, 10
U.S. Civil Service Commission, 455
U.S. Coast and Geodetic Survey, 229, 326, 565–66
U.S. Coast Guard (USCG), 230
U.S. Court of Customs and Patent Appeals, 545
USDA. See Agriculture, Dept. of.
U.S. Dept. of Defense. See Defense, Dept. of.
U.S. Dept. of Labor. See Labor, Dept. of.
U.S. Geological Survey, 97, 149, 187, 496, 507

U.S. Junior Chamber of Commerce, 13
U.S. Labor-Management Government Commission, 8
U.S. Marine Corps (USMC), 190, 555
  astronaut, 406
U.S. Military Academy, 88
USN. See U.S. Navy.
U.S. National Museum, 506
U.S. Naval Academy, 175, 556
U.S. Naval Air Engineering Center, 128, 190
U.S. Naval Air Facility (El Centro, Calif.), 112
U.S. Naval School of Aviation Medicine, 70, 120
U.S. Navy (USN), 59, 190, 284, 346, 453, 549, 564
  aircraft, 20, 34, 177, 248, 308, 500, 515, 537
  astronaut, 214, 411, 535
  launch,
    missile, 58, 513, 537
    sounding rocket, 31, 120, 169, 177
  Pacific Missile Range, 48, 210
  Polaris, 340, 512
    A-1, 513
    A-3, 210, 537
    B-3 (Poseidon), 20
  rocket launching ships, 121, 163, 170, 176, 196
  satellite, 7, 10, 13, 66, 452, 570
  Sealab II, 402, 425, 467, 475
  tracking station, 48, 228
U.S. Navy Electronics Laboratory, 510
U.S. Navy Field Office for Manned Orbiting Laboratory, 115
U.S. Navy Pacific Missile Facility, 210
U.S. News and World Report, 171, 278, 325, 438, 530, 545
USNS *Coastal Sentry*, 136, 147
USNS *Croatan*, 110, 121, 163, 169, 195
USNS *Intrepid*, 146
USNS *Kingsport*, 136
USNS *Point Barrow*, 298, 458
U.S. Public Health Service, Div. of Radiological Health, 248
U.S.S. *Benjamin Franklin*, 538
U.S.S. *Lake Champlain*, 22, 340, 389, 403
U.S.S. *Long Beach*, 14
U.S.S.R. (Union of Soviet Socialist Republics) (see also Soviet Academy of Sciences), 78, 81, 171, 178, 360, 380, 387, 405, 413, 421, 426, 489
  aircraft, 241, 283, 314, 315, 337, 360, 382, 392, 528
  antiaircraft defense, 78, 382-3
  astronomy, 43, 123, 153-4, 180, 182, 183, 199, 343, 372, 392, 495
  bioscience, 19, 45, 278, 313, 372, 413, 419
  capability, 167, 288, 400, 544
  communications system, 204, 249, 478, 480, 490
  computers, 119, 346
  cooperation, 3, 9, 76, 339
    agreements, 25, 76, 204
    communications, 76, 545
    meteorology, 24, 76
    space, 9, 153, 156, 273, 279, 429, 465, 564
  cosmonaut, 1, 19, 31, 45, 66, 98, 134, 135, 139, 141, 147, 151, 154, 173, 217, 225, 314, 323, 419, 460, 473, 491, 505, 548, 559, 568
  Council of the National Economy, 119
  environmental effects tests, 19, 252
  launch (see also COSMOS L-CIII; LUNA V-VIII; MOLNIYA I; PROTON I; VENUS II and III; VOSKHOD II; ZOND II and III)
    COSMOS, 12, 41, 86, 94, 109, 123, 125, 189, 221, 248, 282, 296, 311, 333, 345, 364, 377, 397, 421, 483, 504, 525, 526, 527, 546, 556, 560, 565
    LUNA V, 222
    LUNA VI, 272
    LUNA VII, 460
    LUNA VIII, 536
    MOLNIYA I, 197, 476
    "new type" booster, 42
    PROTON I, 333, 532
    VENUS II, 513
    VENUS III, 518
    VOSKHOD II, 131
    ZOND II, 112
    ZOND III, 337
  launch vehicles, 230, 338, 348
  lunar flight, manned, 1, 3, 181, 201, 227, 252, 392, 455, 491, 498
  military space program, 86, 158, 179, 424, 427, 428, 541, 544
  missile and rocket program, 78, 222, 224, 236, 312, 313, 320, 509, 512, 515, 521, 522, 541
  Ministry of Communications, 313
  Navy, 78
  nuclear power, 58, 78, 313
  nuclear testing, 18, 31, 81
  probe, 73, 113, 216, 236, 329, 337, 339, 379, 442, 492, 535, 540, 571
  record,
    air, 122, 189, 245, 300
    spacecraft, 442
  satellite program, 57, 85, 87, 179, 183, 197, 200, 216, 252, 338, 476, 505
  science, 123, 140, 297, 336
  space activities, 147, 151, 216, 351, 444, 445, 520, 534, 551
  space boosters, 108, 147, 371, 479, 526, 550
  space effort, 31, 139, 486, 501, 542, 570, 571
  space failure, 31, 157, 173, 216, 274, 285, 300, 338, 464, 467
  space flight, manned, 1, 18, 98, 112, 157, 252, 418, 570, 571

U.S.S.R. (Union of Soviet Socialist Republics) (Continued)
  VOSKHOD II, 124, 131, 132, 134, 136, 138, 139, 140, 142, 147, 153, 156, 157, 254
  space plans, 200, 210, 419, 429, 497, 507, 542
  space station, 346
  spacecraft, 56, 209, 230, 323, 338, 419, 528
  State Committee of Inventions, 129
  supersonic transport, 279, 315, 316
  weapons, 57, 78, 153, 252, 288, 320, 321
U.S. Space Park, 327
U.S.S. *Wasp*, 267, 273, 484, 552, 555, 559
U.S. Weather Bureau, 26, 27, 38, 71, 229, 259, 308, 326, 370, 412, 420, 471, 516
  Tiros, 26, 72, 156, 163, 169, 307, 570
Utah, Univ. of, 488, 562
UTC. See United Technology Center.
U Thant, U.N. Secretary Gen., 156–157, 463

V-2 (missile), 50
VAB. See Vehicle Assembly Building.
Vacuum test chamber, 160
Vaeth, J. Gordon, 38
Valentine, Dr. Wilbur G., 182
Valiant (U.K. bomber), 34
Vallat, Sir Francis, 536
Valparaiso, Chile, 195
Valve technology, 161
Van Allen, Dr. James A., 190, 330, 486
Van Allen radiation belt, 31, 56, 81, 157
Van Winkle, C. W. G., 98
Vandenberg AFB, Calif., 189, 332, 401, 549
  launch, 145
    AEC spacecraft, 167
    missile,
      Atlas, 24
      Atlas D, 310
      Minuteman, 24, 63, 389, 510, 538, 542
      Titan II, 304
    satellite,
      ARV, 24
      Atlas-Agena, 206, 296, 510
      Thor-Able, 295
      Thor-Agena D, 18, 114, 152, 206, 207, 236, 272, 339, 382, 448, 475, 546
      Thor-Altair, 20
Vanguard (satellite), 233, 380, 445, 501
VANGUARD I, 67
Vanguard Computing Center, 203
Variable-sweep wing, 59, 121, 234, 307, 325
Varian Associates, 284
Vatican City, 9, 58
VC-10 (U.K. transport), 314, 318
Vechernaya Moskva, 87–88
Vega program, 497
Vehicle Assembly Building (VAB), MILA, 184

Vela (satellite), 40, 158, 242, 341
Vela Hotel (Sentry). See Vela (satellite).
Velikovsky, Immanuel, 416
Venus (planet),
  atmosphere, 269, 370–371, 393
  flights to, 81, 238, 325, 382, 384, 399, 497, 518, 562
  gravity field, 19
  life on, 370–371
  study of, 32, 262, 342
  surface, 242, 370–371
VENUS II (U.S.S.R. interplanetary probe), 513, 515, 535, 568, 571
VENUS III, 518, 535, 568, 571
"Venus Flytrap," 321
Vernon, France, 496
Vernov, Prof. S. N., 342
Verwoerd, Prime Minister Henrik S. (South Africa), 297, 300, 497
Vesta (sounding rocket), 478
Vienna, Austria, 558
Vierling, Bernard J., 522
Viet Nam, 314, 398, 438, 563, 573
Virginia Polytechnic Institute, 167, 393
Virginia Technological Institute Space Conference, 399
Vitro Corp. of America, 109, 309
Vivian, Rep. Weston E., 46, 115
Vladivostok, U.S.S.R., 228, 313, 513
Vlasic, Prof. Ivan A., 435
Vnukovo Airport (Moscow), 241
Volcanoes, 318
Von Braun, Dr. Wernher, 129, 134, 239, 240, 253, 320, 357, 373, 411, 440, 506
Von Eshleman, R., 414
Van Hassel, Kai Uwe, 178
Von Kármán, Theodore, Lecture, 349
Von Kármán, Theodore, Memorial Seminar, 228
Vortex flow, 41
Vose, Dr. George P., 540
Voskhod (program), 56
VOSKHOD I (U.S.S.R. spacecraft), 1, 18, 98, 230, 288, 548
VOSKHOD II, 124, 131, 132, 134, 136, 137, 138, 139, 140, 141, 142, 147, 153, 154, 156, 157, 178, 216, 226, 230, 231, 254, 278, 288, 323, 447, 571
Vostok (U.S.S.R. spacecraft), 56, 157, 209, 223, 230, 315
VOSTOK I, 230, 380, 528
VOSTOK III, 568
VOSTOK IV, 559, 568
VOSTOK VI, 230, 318
Voyager (spacecraft), 3–4, 27, 75, 200, 341, 468, 482
  contract, 18, 195
  experiment, 60, 325, 440
  launch vehicle, 60, 130, 417, 476, 533
V/Stol aircraft, 15, 41, 50, 54, 90, 204, 248, 349, 387, 399
Vtol aircraft, 15, 36, 41, 50, 52, 69, 178, 393

INDEX

Wackenhut Services, Inc., 308, 353
Waging Peace, 1956–1961, 445
Wagner, Mayor Robert (N.Y.), 156
Wahl, Dr. Arthur C., 304–305
Wake Island, 49
Wakelin, Dr. James H., 280
Waldheim, Dr. Kurt, 254
Walker, Joseph A., 518
Wall Street Journal, 241, 335, 422–423, 545
Wallis, Dr. Barnes, 234, 413
Wallops Island, Va., 565
Wallops Station (NASA), 123, 243, 307, 365, 458, 477, 559
   Bio-Space Technology Training Program, 373
   Inter-American Experimental Meteorological Rocket Network meeting, 370
   launch,
      satellite,
         EXPLORER XXIII, 233
         EXPLORER XXVII, 154
         EXPLORER XXX, 521
         SECOR V, 373
      sounding rocket, 220, 226, 239
         Aerobee, 139
         Aerobee 150A, 15, 261, 439
         Aerobee 300A, 139
         Aerobee 350, 289
         Argo D–4 (Javelin), 15, 235, 302, 447
         Argo D–8 (Journeyman), 184, 302
         Nike-Apache, 6, 53, 132, 158, 195, 278, 287, 289, 294, 410, 413, 552
         Nike-Cajun, 54, 58, 195
         Nike-Tomahawk, 136
   Owl satellite, 56, 370
   Scout Evaluation Vehicle, 373
Walter Reed Army Institute of Research, 410
Ward, Bob, 365
Walter, Louis, 96
Wark, Dr. D. Q., 220
Warren AFB, Wyo., 304
Warren, Chief Justice Earl, 529
Warsaw Convention, 517
Warsaw, Poland, 228
"Was the Ranger Worth the Cost?" 4
Washington Board of Trade, 227
Washington, D.C., 73, 278, 540, 565
   exposition, 164
   meetings, 31, 34, 128, 133, 136, 186, 189, 193, 196, 206, 217, 232, 239, 249, 264, 324, 382, 505, 520, 537, 563
   museum, 43, 55, 255
   news conference, 208
Washington Daily News, 543, 550
Washington *Evening Star*, 52, 191–192, 221, 250, 288, 291, 369–370, 394, 410, 422, 425, 431, 437, 453, 477, 486, 491, 512, 536, 563, 564
Washington National Airport, 545
Washington Post, 11, 189, 246, 279, 360, 374, 395, 401, 412, 416–417, 434, 486, 488, 528, 534, 554, 563
Washington *Sunday Star*, 200, 539–540, 558
Washington Univ., 157, 169
Water desalinization, 77–78, 370
Water pollution, 280
Waters, Capt. John M. (USCG), 230
Waters, William J., 56
Watson, James Craig, Medal, 203
Wayne State Univ., 25
WDMV (radio station), 565
Weapon system, 20, 512
   laser, 517
   nuclear, 57–58, 78, 125
   space, 153, 288, 320–321
Weather (see also Meteorology; Satellite, meteorological), 65–66, 101, 111, 122, 124, 142, 143, 175, 213, 220, 243, 246, 259, 271, 282
   modification, 312, 357, 434
Weather Bureau. See U.S. Weather Bureau.
Weather satellite. See Satellite, meteorological.
Weather station, 162
   nuclear powered, 59, 123
Weaver, Robert C., 495
Weaver, Dr. Warren, 530, 545
Webb, Jack G., 488
Webb, James E., 88, 95, 122, 146, 155, 159, 172, 175, 190, 200, 203, 209, 231, 264, 303, 312, 335, 343, 369, 443, 477, 535
   appropriations, 74, 100, 139, 169, 173, 284
   award by, 462, 530, 567
   award to, 25, 30, 509
   Gemini program, 289, 297, 402
   GT 3 flight, 15, 152
   GEMINI VI flight, 396, 493, 552
   GEMINI VII flight, 410, 552
   International Air Show, 290, 316
   launch vehicles, 25, 227
   lunar landing, manned, 40, 152, 389
   Marshall Space Flight Center, 255, 173
   nuclear propulsion, 61, 209–210
   space cooperation, 88, 218, 298
   space flights, 244
   space program, 54, 56, 105, 111–112, 137–138, 184–185, 294, 356, 546
   space program, military, 168
   university program, 213, 232–233, 353, 420, 561
Weber Aircraft Co., 65–66
Webster, Grove, 264
Weddell Sea, 532
Wehr und Wirtschaft, 361
Weightlessness, effects of,
   animals, 59
   humans, 19, 45, 55, 63, 132, 138, 146, 216, 267, 279, 412, 527, 531

Weightlessness, effects of, (continued)
  plants, 59
  primate, 59
  space flight, 107, 132, 138, 267, 279, 480, 540
Weizman Institute of Science, 541
Welsh, Leo D., 318, 479
Welsh, Dr. Edward C., 11, 46, 252, 270, 333, 335, 366, 444, 463, 494, 503, 517, 535, 552
Wendover Air Force Auxiliary Field, Utah, 531
West, Charles F., 462
West Germany. See Germany, West.
West Java, 21, 375
West Point, N.Y., 88
West Test Area (MSFC), 54, 319
Western Australia, Univ. of, 401
Western Electric Co., 5, 447
Western Reserve Univ., 231
Western Test Range (WTR),
  Vandenberg AFB, 66, 189, 458
  launch, 74, 144
    AEC spacecraft, 166
    missile, 24, 63
    satellite, 18, 20, 24, 27, 93, 114, 118, 120, 132, 152, 158, 206, 240, 254, 336, 365, 377, 423, 447, 454, 462, 493, 528, 540, 545, 564
    test, 97
Western Union International, Inc., 201, 334, 436
Westinghouse Electric Corp., 121, 317, 358
Westinghouse scholarship, 101
Westinghouse science talent search, 101
Westminster Abbey, 546
Wetherill, John Price, Medal, 399
Wharton, U.K., 87
Wheaton, Elmer P., 215, 361
Whipple, Dr. Fred L., 188, 505
Whitaker, Dr. Ewan A., 85, 149, 187
White, Alvin S., 150, 206, 454, 471, 547
White, L/C Edward H., II (USAF), 168, 255, 363
  GT-4 flight, 265–268, 269, 375, 402, 570
    plans, 168, 229, 248, 256
  GEMINI VII flight, 309, 511
  honors, 270, 278, 283, 286, 331
  International Air Show visit, 290, 293, 314, 316
  interview, 196, 208
  messages to
    Gagarin, Yuri, 273
    Johnson, President Lyndon B., 271
  promotion, 276
White House, 48, 155, 169, 287, 309, 325, 390, 424, 431, 434, 454, 493
White House Conference on International Cooperation, 528–529, 530
White, M/G M. S. (USAF), 387
White, Marvin L., 47
White River (rocket launching ship), 459
White, Dr. Robert M., 163, 517

White Sands Missile Range (WSMR), N. Mex., 18, 49, 67, 161, 290, 447, 514, 538,
  launch,
    Aerobee, 108, 120, 132, 179, 264, 299, 474, 518
    Nike-Apache, 132, 521, 534
  test,
    Apollo, 55, 238, 301
    Athena, 49, 426
    Biosatellite, 149, 207
    Hibex, 94
    Lance, 125
    Lem engine, 185
    sonic boom, 188
    Sprint, 156, 522
White Sands Operations, 308
White Sands Test Facility, 308, 345
Whithaus, Douglas A., 179
Whitney, John Hay, 297
Whittle, Sir Frank, 32
Wible, Keith, 340
Wiesner, Dr. Jerome B., 509
Wilcox Electric Co., 500
Williams, Clifton C., 10, 321
Williams, Dr. D. J., 118
Williams, Franklin, 326
Williams, Walter C., 350
Wilson, Charles, 445
Wilson, Charles A., 138
Wilson, George C., 500
Wilson, Prime Minister Harold, 52, 301
Wilson, Richard, 428
Wilson, Dr. Robert W., 261
Wilson, President T. Woodrow, 165
Wilson, L/G Walter K. (USA), 256
Wind tunnel, 112, 121
Windsor, Duke of, 477
Wing, design,
  "s"-shaped, 41
  supersonic transport, 36, 41, 59
  variable geometry, 170
  variable-sweep, 59, 121, 234, 307
  V/Stol, 15, 54
Wingrove, Rodney C., 356
Wings Club, 249
Winston, Jay S., 71
Wirtz, W. Willard, 524
Wisconsin, 134
Wisconsin, Univ. of, 82, 176, 260
Wiswell, George C., Jr., 407
Witkin, Richard, 121
Wittanen, Theodore, 24
Witze, Claude, 499
Wolff, Rep. Lester L., 46
Wolfie, Dael, 360
Women as scientist-astronauts, 63
Women as scientists, 92
Wood, L/C James W. (USAF), 224
Woods Hole, Mass., 293, 382
Woollard, Dr. George P., 293
Woomera Rocket Range, Australia, 353, 526, 545
Woomera Weapons Research Range, Australia, 553

INDEX

Worcester, Mass., 128
Worcester Polytechnic Institute, 128
World Affairs Council, 170
World Book Encyclopedia Science Service, Inc., 104
World Exhibition of Transport and Communications, First, 331
"World Geophysical Intervals," 161
World Meteorological Day, 147
World Meteorological Organization, 147, 163
World Peace through Law Conference, 435, 440
"World Was There" (film), 121
World's Fair, 173
Wright Brothers Day, 501
Wright Brothers Medal, 463
Wright Brothers Memorial Dinner, 557
Wright Brothers Memorial Trophy, 527, 557
Wright, Orville, 134
Wright-Patterson AFB, Ohio, 160, 237, 455–466
Wright, Wilbur, 134
WTOP (radio station), 278
WTR. See Western Test Range.
Wuhan, China, 261
Wydler, Rep. John W., 290–291, 502
Wyman-Gordon Co., 345
Wyman, Louis C., 438

X-1 (rocket aircraft), 350
X-2 (rocket airplane), 350
XV-5A (V/Stol aircraft), 41, 90, 204
XV-9A (pressure jet helicopter), 39
X-15 (rocket research aircraft), 13, 169, 173, 191–192, 253, 319, 331, 347, 349, 464, 570
  flight, 112, 177
    No. 1, 247–248, 287, 370, 421, 445, 447, 454, 476, 504–505
    No. 2, 74, 94, 206, 236, 293, 319, 365, 411
    No. 3, 15, 49, 255, 284, 302, 374, 433, 451, 471, 492, 503
  pilots, 331, 347, 417
  test, 15, 197, 255, 374, 421, 433, 451, 503
    atmospheric pressure, 476, 504–505
    heat-resisting materials, 49, 56
    inertial guidance system, 247–248, 287
    infrared horizon scanner, 255, 302, 370, 374, 399, 421, 451, 476, 492, 504–505
    landing gear, 206, 236, 293, 411
    noise, 255, 284, 374, 399, 451, 492

  reaction augmentation, 365, 411
  skin friction, 15, 374
  stability, 206, 236, 293, 370, 492
X-15 Technical Conference, 464
X-19 (V/Stol aircraft), 399
X-22A (V/Stol aircraft), 248
XB-70 (supersonic aircraft), 19, 73, 77, 84, 111, 206, 228, 547, 568
XB-70A (supersonic aircraft), 77, 150, 191, 284, 308, 336, 351, 453, 478, 482, 562, 563
XC-142A (V/Stol aircraft), 15, 54, 322
XH-51A (helicopter), 254
XS-1 (rocket aircraft), 350

Yaffee, Michael L., 482
Yale-Columbia Observatory, 416
Yale Univ., 67–68
Yarborough, Sen. Ralph, 128, 249
Yardley, John F., 493
Yegorov, Dr. Boris B., 161, 480, 548
Yemen, 322
YF-12A (A-11 aircraft), 19, 144, 213, 221
Young, Andrew T., 4
Young, Dr. Donald, 449
Young, Dr. John, 36
Young, Cdr. John W. (USN), 68, 116, 130, 142, 145, 147–148, 151, 153, 155, 156, 158, 159, 170, 171, 199, 214, 268, 294, 298, 449
Young, John, Award, 199
Youth Opportunity Campaign, 255
Youth Science Congress, 136
Yugoslav Academy of Sciences, 568
Yugoslavia, 325, 401, 568
Yuma, Ariz., 321

Za Rubezhom, 423
Zehnder, Dr. Alfred, 73
Zeitschrift für Astrophysik, 198
Zeus 1 (amateur rocket), 179
Zeus 2 (amateur rocket), 179
Zeus (antimissile missile), 452, 522
Zharikov, Vyacheslav, 300
Zhukov, Genadii, 367
Zhurkov, Serafim Nikolayevich, 336
Zimmer, Harold, 369
Zoeckler, B/G John L. (USAF), 224
Zolnierz Wolnosci, 419
ZOND II (U.S.S.R. space probe), 73, 112–113, 216, 235, 329, 494
ZOND III, 337, 339, 379, 442, 492, 571
Zuckert, Eugene M., 6, 329, 377, 406, 442, 453
ZWO. See Netherlands Organization for the Advancement of Pure Research.

NASA HISTORICAL PUBLICATIONS

HISTORIES

An Administrative History of NASA, 1958–1963, by Robert L. Rosholt, with Foreword by James E. Webb, NASA SP–4101, 1966; for sale by Supt. of Documents ($4.00).

HISTORICAL STUDIES

History of Rocket Technology, special issue of *Technology and Culture* edited by the NASA Historian, Fall, 1963, republished as book by Society for the History of Technology (Detroit: Wayne State Univ., 1964).

Space Medicine in Project Mercury, by Mae Mills Link, with Foreword by Hugh L. Dryden and Introduction by W. Randolph Lovelace, II, NASA SP–4003, 1965; for sale by Supt. of Documents ($1.00).

CHRONOLOGIES AND SPECIAL STUDIES

Project Mercury: A Chronology, by James Grimwood, with Foreword by Hugh L. Dryden, NASA SP–4001, 1963, for sale by Supt. of Documents ($1.50).

Aeronautics and Astronautics: An American Chronology of Science and Technology in the Exploration of Space, 1915–1960, with Foreword by Hugh L. Dryden, Washington: GPO, 1961; for sale by Supt. of Documents ($1.75).

Aeronautical and Astronautical Events of 1961, with Foreword by James E. Webb, published by the House Committee on Science and Astronautics, 1962 (out of print).

Astronautical and Aeronautical Events of 1962, with Foreword by George L. Simpson, Jr., published by the House Committee on Science and Astronautics, 1963; for sale by Supt. of Documents ($1.00).

Astronautics and Aeronautics, 1963, with Foreword by Hugh L. Dryden, NASA SP–4004, 1964; for sale by Supt. of Documents ($1.75).

Astronautics and Aeronautics, 1964, with Foreword by Robert C. Seamans, Jr., NASA SP–4005, 1965; for sale by Supt. of Documents ($1.75).

A Historical Sketch of NASA, prepared by the NASA Historical Staff, NASA EP–29, 1965; for sale by Supt. of Documents ($.25).

www.ingramcontent.com/pod-product-compliance
Lightning Source LLC
Chambersburg PA
CBHW081713170526
45167CB00009B/3561